HELL & CONFUSION

GALLIPOLI DAY *by* DAY

VOL. 1

'Alive with Death'

AUGUST 1914 — APRIL 1915

by

JIM GRUNDY

Copyright © 2024 Jim Grundy.

Images copyright expired or no known copyright restrictions.

All rights reserved. No part of this book may be reproduced in any form by electronic or mechanical means, including information storage and retrieval systems, without permission in writing from the publisher, except by a reviewer who may quote brief passages in a review.

Published August 2024.

A catalogue record for this book is available from the National Library of Australia.

ISBN 978-1-7636268-1-2 (paperback)
ISBN 978-1-7636268-0-5 (ebook)

Little Gully Publishing
littlegully.com

This book is dedicated to

my father
JOHN GRUNDY
(1933–2003)

my father-in-law
DONALD REYNOLDS
(1925–2022)

And everyone caught up in that one word:

GALLIPOLI.

ACKNOWLEDGEMENTS

Firstly, my thanks are due to Peter Hart, not only for contributing his very generous foreword, but for introducing me to the battlefields of Gallipoli in 2011. I had brought with me some accounts by men, local to me, who served during the campaign. At X Beach, he asked me to read one man's description of the landing there. After I had finished, Peter said that was great but, just one little thing… it was rubbish (though that wasn't the word used). Though a tale was told of landing under heavy fire, of men killed and wounded in the boats as they were rowed ashore, by the time that man got there the 12 Turks that had guarded the area were long gone. It was a lesson in not believing everything you read.

I have been lucky to meet and been helped by many people at Gallipoli. My particular thanks go to Haluk Oral for his friendship, generosity and hospitality both on the peninsula and at Istanbul. Bill and Serpil Sellars have introduced me to some of the less frequently visited parts of the peninsula. Their collection of rare photographs is one of the best there is and I am extremely grateful that they have allowed me to make use of some of their images in this book.

There are special memories of a shared visit with my friend Wayne Osborne. But no visit to the battlefields would have been possible without Ramazan Altuntaş, Bulent Yilmaz Korkmaz and everyone at Crowded House Tours. The irrepressible Burak Yetimoğlu is a true friend to all visitors there — and how can I repay him for introducing me to Valhalla? Mehmet Kibil has been a source of constant support, while Halil Çakar, Hasan Günduğar and Çağrı Tok are always happy to help. And Abdurrahim Boz, who makes the best coffee in town. The late Yücel Özkorucu is much missed and I will always remember that gentle man.

I would like to extend my sincere appreciation to Bernard de Broglio at Little Gully Publishing. It is only through his dedication and patience that this book has been brought together.

But, and I have never written a truer word, this book would not exist without the support of my partner, Liz Watson. If anyone takes anything from this book, know that it is down to her.

— *Jim Grundy*

CONTENTS

Foreword i

Preface 1

The Taba Crisis, 1906 3

August 1914 14

September 1914 57

October 1914 99

November 1914 140

December 1914 180

January 1915 217

February 1915 263

March 1915 321

April 1915 393

Afterword 655

Maps 658

Abbreviations & Acronyms 661

Select Bibliography 664

Turkish coastal fort at Kilid Bahr on the Dardanelles.

FOREWORD

This brilliant new book on the Gallipoli campaign expands our horizons as it explores the origins of the campaign reaching right back to Duckworth's abortive mission through the Dardanelles in 1807. A hundred years later a report of the Committee of Imperial Defence considering a repeat performance found the accumulated difficulties made success an unlikely prospect. But my favourite quote comes from a Grenadier Guards officer, Henry Woods, who unofficially explored the Peninsula on foot. He warned of the consequences of commencing a campaign without the proper military resources, summing it up pithily as 'If you mean to bark be first assured that you can effectively bite!' It could act as an epitaph for the campaigns of 1915.

When it comes to the outbreak of war, the day-by-day format allows us to get right inside the dramatic events, ranging from the 'escape' of the *Goeben* and *Breslau*, the manoeuvrings between the 'Great Powers' and politicians on all sides, the recruitment and build-up of military forces, then the Anglo-French naval operations culminating in the doomed assault on the Dardanelles on 18 March 1915. Throughout the tension builds up to a nerve-jangling climax with the Gallipoli landings on 25 April 1915 and the immediate aftermath. All the drama, the action, the courage, the tragedies, the deaths and the pathos are here. You thought that story was done to death? Well, think again! New voices galore add layers to our picture of events, helping our understanding of what happened and why.

Using the words of the 'men who were there' we can vicariously experience 'Hell and Confusion' at Gallipoli. A wonderful kaleidoscopic picture in which different perspectives abound — you cannot help but be drawn in. Grundy's diligent research scouring local newspapers and archives has uncovered many accounts that flesh out the story, reaching into nooks and crannies omitted in previous books. It is a tremendous tribute to the men on all sides who fought so hard at Gallipoli.

— *Peter Hart*

* * *

Le Vice-Amiral Anglais Duckworth à la Porte, le 3 mars 1807.

The Royal Navy's failure to overawe the Turks was mocked in a contemporary French cartoon.

PREFACE

On 19th February 1807, Vice-Admiral Sir John Duckworth succeeded in taking his squadron through the Narrows of the Dardanelles. Sailing through the Sea of Marmora, he anchored off the Princes' Islands to await the Sultan's expected agreement to British demands. It never came and he was forced to run the gauntlet a second time.

In his report to the Admiralty on 6th March 1807, Duckworth told of his narrow escape three days previously.

> "The fire of the two inner castles had, on our going up, been severe; but, I am sorry to say, the effects they have had on our ships returning, has proved them to be doubly formidable; in short, had they been allowed another week to complete their defences throughout the channel, it would have been a very doubtful point whether a return lay open to us at all. The manner in which they employed the interval of our absence has proved their assiduity. I transmit your lordship an account of the damages sustained by the respective ships; as also their loss in killed and wounded, which your lordship will perceive is far from trifling. The mainmast of the *Windsor Castle* being more than three-quarters cut through by a granite shot of eight hundred weight, we have found great difficulty in saving it."[1]

Captain John Fane, 3rd Dragoon Guards, had been aboard the *Windsor Castle*.

> "We are just returned from Constantinople, after two severe actions in passing the Dardanelles, and after having been frightened at the fleet, etc., of the Porte. I fear we cannot boast of more than of having fought bravely ... But let whatever have been in our power, nothing can excuse the English Minister having sent seven ships of line to impose the most disgraceful terms upon an empire of such resources as the Turks. How a Government could have asked an empire to "deliver up their fleet," "to renounce all connection with France," and to make peace, and a disgraceful one, with Russia, with only seven ships of the line and not a single soldier to enforce their terms, is to me incomprehensible... The natural inference would be that those who conceived such a plan must be mad; but in England they are Ministers, and in this part of the world we are only to put their plans into execution if we can."[2]

Madness, indeed.

---

[1] Duckworth quoted in Howard, Edward, *Memoirs of Admiral Sir Sidney Smith, K.C.B.*, vol. 2, pp. 60–61, Richard Bentley (London) 1839.

[2] John Fane, Lord Burghersh, quoted in Weigall, Rachel (Ed.), *Correspondence of Lord Burghersh, Afterwards Eleventh Earl of Westmorland, 1808–1840*, pp. 7–8, John Murray (London) 1912.

THE TABA CRISIS, 1906

"A stupid misunderstanding."
Captain Robert Phipps Hornby, RN, HMS *Diana*

In January 1906, a dispute developed over Egypt's eastern border around the Bay of Aqaba on the Red Sea. Some Turkish troops had established themselves in the area of Taba, a place that sent everyone to their maps to locate and almost sent the British and Turks to war.

Taba came under Cairo's jurisdiction according to the British. As Egypt itself was still, officially at least, part of the Ottoman Empire, that was not how the Turks saw it. Its status had not troubled anyone much prior to the arrival of Wilfred Jennings-Bramly, the senior British official in Sinai, and four Egyptian policemen sent on 10th January to establish guard posts in the area.

After an uncomfortable initial encounter with Ottoman forces there and a fruitless exchange of diplomatic messages, in time-honoured fashion, on 25th January a member of the British Embassy staff in Cairo, Mansfeldt de Cardonnel Findlay, suggested that the Navy be sent in. His advice came with a caveat, a consideration that coloured all British thinking in the region at the time.

> "In the event of its appearing that the attitude of the Turks is not merely "bluff," instructions might perhaps be given to His Majesty's ship "Diana," now at Port Soudan, to proceed to Akaba. I am telegraphing to Lord Cromer a summary of the situation.
>
> "On the other hand, considering the strong feeling which would be excited among Moslems in Egypt by this step, it would, if possible, be well to avoid anything in the nature of a naval demonstration against Turkey."[1]

Captain Robert Phipps Hornby, RN, aboard the cruiser HMS *Diana*, arrived off shore on 19th February. He discovered the Turks had not expected he had come for a chat.

> "I left Faroun. at 9 a.m ... and passed along the coast. A considerable number of troops (about fifty) were seen at Taba, another party of about eighty half a mile further on, and then several other small bodies of fifteen to thirty men in each. It looked as though there were a chain of posts all along the west side of the Gulf.

---

[1] Affairs of Egypt and Sudan. Further Correspondence Part LXIV 1906 January-March. TNA FO 407/165.

> "On approaching Akaba there was something like a panic. About three battalions streamed off into the hills, and other troops lined the ridges to the west of the town and the walls of the town itself. They evidently expected an attack."[2]

The local Turkish commander, Rüştü Pasha, looked on.

> "The cruiser anchored close to the port of Aqaba. A lifeboat approached the harbour. An Egyptian officer wearing a fez came out of the lifeboat. He said that he was one of the Egyptian Military Officers and his name was Naum Efendi. He brought the card of *Monsieur* Hornby, Commander of the Cruiser *Diana*. I asked what he came for. He talked a little about the issue. I told him to convey my greetings to the commander of the cruiser and to inform him that since this issue is an Ottoman–Egyptian issue, I do not have the authority to meet with British officials about the issue, but since a ship of a friendly state has arrived in our port and has shown a desire to visit me, I will show him the necessary hospitality."[3]

Hornby was impressed by his Turkish counterpart.

> "On meeting Rushdi Pasha he was decidedly nervous and not cordial. After conversation had proceeded for some time he said he would like to ask why I had come, and if as an enemy. I assured him I had not. He then said that he had understood from Mr. Bramly that if ships came up they would come as enemies, and drive him out of the place; and in my opinion a great part of the present difficulty is due to the use, unintentionally or through faulty interpretation, of threats on the part of Mr. Bramly and Saad Bey Rifaat. Directly the Pasha was reassured on this point he became much more friendly, and said he wished to talk about the matter."[4]

The respect seems to have been mutual, as Rüştü Pasha recorded.

> "Later, I went to the ship for a return visit. The ship's commander is a well-mannered and competent person… He said that the British have always been friends of the Turks, showing a picture of his father participating in the Crimean war as an admiral.[5] I said that my father was one of the pashas who participated in the Crimean war, that our fathers once fought side by side in Crimea, that while England was an ally of the Ottomans at that time, now England is… always opposite of us."[6]

Both men understood that the matter could not be settled between them. To the British, Sinai was the buffer protecting the Suez Canal and Egypt. Nevertheless, Lord Cromer, the British Consul-General in Cairo, could not understand why such an insignificant place could have attained such importance for the Turks. Unless, that was, there were other factors in play. He wrote to Sir Edward Grey, the Foreign Secretary, on 17th March.

---

2   Affairs of Egypt and Sudan, TNA FO 407/165.
3   Öztürk, Mustafa, *Rüştü Pasha and the Aqaba Affair*, pp. 25–26, Fırat University Publications (Elazığ) 1997. Translated by Dr. Mehmet Kibil.
4   Affairs of Egypt and Sudan, TNA FO 407/165.
5   Admiral of the Fleet Sir Geoffrey Phipps Hornby, GCB, served in the Crimea and the Dardanelles during the Russo-Turkish War of 1877/78. He died in 1895.
6   Öztürk, *Rüştü Pasha and the Aqaba Affair*, p. 26.

> "I do not want to be unduly alarmist, but I cannot say that I feel at all comfortable about this Akaba affair. It is possible that we are in for a bigger business than the mere local settlement of the frontier. The language of the Sultan and the Grand Vizier to O'Conor [British Ambassador in Istanbul] appears to me to be most unsatisfactory. That their attitude in this and other matters is generally due to the fact that they have the Germans behind them, can scarcely be doubted."[7]

Cromer admitted he had little evidence of any direct German involvement in the affair but he definitely considered they were attempting to undermine the British position in Egypt.

> "German agents are extremely active. They employ a certain Baron Oppenheim, who is in constant communication with Moukhtar Pasha[8] and with the leaders of the Pan-Islamic party in this country."[9]

Captain Hornby had shared Cromer's suspicions about German involvement but, writing on 25th March 1906, thought he knew why things had deteriorated so quickly. It was less of an international conspiracy, more a local personality clash.

> "The longer I am here, and the more I hear locally, the more sure I am that the present impasse arose solely through a stupid misunderstanding between Rushdi Pasha and Mr. Bramly. Rushdi had somehow got it into his head that England and Egypt meant to seize the whole of the head of the Gulf of Akaba, though what started this bee in his bonnet I cannot say, and then when he met Mr. Bramly each thought the other was threatening him and meant to attack. Rushdi had the force under his hand, and, being a soldier, seized the strategic points and held them, following this up by an alarmist report to Constantinople and a request for reinforcements. From what I have seen of Rushdi I believe that he is a soldier and a gentleman who would keep any promise he made, and he is just as anxious as I am to see this matter amicably settled and get away."[10]

Irrespective of any likely trivial origins of the controversy, it had proved to be the catalyst for much bigger issues to come to the fore. Hornby thought a resolution could only be found elsewhere.

> "Wherefore I venture the opinion that any display here of additional force will do more harm than good, as it will not intimidate Rushdi, but merely cause him to make fresh preparations, and it will not frighten the Sultan as would a little pressure nearer the Dardanelles."[11]

---

7   Foreign Office, Private Offices: Various Ministers' and Officials' Papers. Grey, Sir Edward (Viscount). Egypt. TNA FO 800/46.
8   Moukhtar Pasha, the Turkish High Commissioner in Cairo.
9   Foreign Office papers, Grey, Sir Edward, TNA FO 800/46.
10 Affairs of Egypt and Sudan, TNA FO 407/165.
11 *ibid.*

In Cairo, Cromer agreed, writing to Grey on 4th April.

> "I would also take the opinion of the competent authorities as to the manner in which pressure could be brought to bear on the Sultan by naval action in the Mediterranean; and if these authorities could devise a plan of action, a marked effect would probably be produced by a move of the British fleet towards the point where that action would be taken."[12]

While Hornby and Rüştü Pasha had negotiated a partial Turkish withdrawal, there was no progress in the exchanges between the governments. And when he wrote to Cromer on 9th April it was clear that matters had taken a turn for the worse.

> "From Rushdi's whole manner he is evidently expecting some unpleasant news to come from us and the Turks are I am sure expecting to be attacked at any time. They watch our every move and even noticed that we manned and armed boats the other morning because our steamboat on her way back from Akaba on Monday stopped off a point N of Taba to ascertain its position, some soldiers at once turned out and came rushing along the beach loading their rifles, and on Wednesday Rushdi said apologetically they had reported that they thought that the boat was about to land a party and his orders were that no one was to be allowed to land…
>
> "I most thoroughly and entirely agree with your view of the futility of a display of force here by trying to seize Taba with small Naval Force, in fact I think it would be most unwise as I am quite sure that Rushdi would resist our landing and would I expect immediately advance on Nekhl as he let fall a remark the other day to the effect that "should occasion arise he could be in Nekhl in 2 days" and I expect his plans for this are all cut and dried. I think I told you that the first time I went to see him, Rushdi told Naum Bey as he was landing after returning my call that when he saw us approaching in the morning "he hardly thought he would have been alive at that time in the afternoon."[13] I only tell you these trifling incidents in order to try and show you the kind of man we have to deal with here. Much may be done with him by fair and honest dealing, but threats would convert him into a mule. Taba is a nasty place to hold, as it is surrounded by hills, but none of them is of a form to hold with a small force. A force at or near the wells could and would be incessantly sniped at from the adjacent hills. It is noteworthy that nearly all the hills about here, whilst precipitous and difficult from the South, are easy of approach from the N and W."[14]

He thought a proposal to send another ship, HMS *Minerva*, could provoke rather than placate the Turks and precipitate not prevent an attack on the Suez Canal.

> "If she [*Minerva*] came here now, Rushdi would at once get back the troops he sent away, and many more too. I do not think that Sir N. O'Conor can realize that the Turks had all the strategic points here, and are fully aware of it, and could be on the

---

12   Affairs of Egypt and Sudan, TNA FO 407/165.
13   Rüştü Pasha fought throughout the Balkan wars, the Great War and Turkish War of Independence. He was executed on 13th July 1926 for his part in a plot to assassinate Mustafa Kemal.
14   Foreign Office papers, Grey, Sir Edward, TNA FO 800/46.

banks of the Canal long before a sufficiently large English or Anglo-Indian force could be assembled to stop them, as I do not suppose the few white troops in Egypt could be moved out of the towns for this purpose."[15]

Cromer, ever-sensitive to the threat of radical Islam, cabled Grey on 22nd April. His bête-noir, Moukhtar Pasha, had been busy.

"The day before yesterday several Mullahs of influence preached sermons in the mosques calculated to inflame Moslem sentiment, and then proceeded to visit Moukhtar Pasha and the report what they had done. The Khedive is the only person who can put a stop to these proceedings. Failing the adoption of very strong repressive measures, which would not be justified at present, I will mention the matter to the Khedive when he returns from Alexandria, although I have not much hope that he will be of any real assistance. Two of the most violent Mullahs are connected with religious establishments which he directly controls."[16]

He added his thoughts on the outcome he looked for in a further note to Grey on 28th April.

"It is really almost a necessity of our situation here that, in coming out of this business, the Egyptians should be fully convinced that we have got the best of it, and that the Turks have had to climb down. Unless this is made quite clear, I anticipate great difficulties in the future… As you know, I do not, on its own merits, attach much importance to the possession of this post. On the other hand, it has been so much talked about, that it has become, in public opinion here, rather the outward and visible sign by which the result of our present quarrel will be judged. Indeed, the papers inspired by Moukhtar are already beginning to dwell on this point… I regard it, therefore, as very essential that the Turks should be made to turn out of Taba, even if, eventually, we allow them to come back in again."[17]

Cromer asked for reinforcements to be sent to Egypt as a sign of British determination not to back down. Garrisons around the Mediterranean were closest and on 29th April amongst those sent was 1st Battalion Lancashire Fusiliers, including Captain Richard Willis and Lt Thomas Maunsell.[18]

That day, HMS *Minerva*, under the command of Captain Raymond Waymouth, RN, approached the north-eastern Egyptian coast to investigate reports of Turkish activity at Rafeh. Waymouth was unimpressed with his reception or, rather, the absence of one.

"We have the honour to state that, having waited for five hours in the telegraph house opposite your camp for an interview without a reply from you or your having come

---

15 Foreign Office papers, Grey, Sir Edward, TNA FO 800/46.
16 Affairs of Egypt and Sudan, TNA FO 407/165.
17 Foreign Office papers, Grey, Sir Edward, TNA FO 800/46.
18 Captain Willis was awarded the VC for his bravery on W Beach on 25th April 1915. Maunsell, by then promoted Captain, was killed in action the same day. He is buried in Lancashire Landing Cemetery, Cape Helles.

yourself, we returned to the ship. We observe that the boundary-pillars on both sides of the tree near which you camp have been removed. We also observe that the Egyptian telegraph poles from the boundary line to the Rufieh Road have been replaced by others.

"On behalf of the Egyptian and British Governments we protest most strongly against this procedure, and request that you will restore the boundary-pillars and telegraph poles to their places, and keep to the recognized boundary limits. A copy of this letter will be forwarded to the Egyptian and British authorities in Cairo."[19]

Patience exhausted, Sir Nicholas O'Conor, the British Ambassador in Istanbul, delivered an ultimatum to the Sultan on 3rd May.

"Both the substance and tone of the Grand Vizier's communication to the Khedive have made further negotiations at Cairo impossible. The contentions as to the frontier put forward in the Grand Vizier's reply are quite inadmissible; if admitted, they would seriously prejudice the position as regards the Suez Canal and Egypt. Negotiations have now been prolonged over several weeks not only without progress, but with increasing claims on the part of the Porte, to the prejudice of the administrative frontier of Egypt…

"I have consequently the honour to inform your Excellency that I have received from His Majesty's Principal Secretary of State for Foreign Affairs instructions to request that the Ottoman Government will agree to the demarcation of the line from Rafeh to the head of the Gulf of Akaba on the basis of the aforesaid telegram of the 8th April, 1892, and that, pending such demarcation, Taba shall be evacuated. Further delay must increase the difficulties of the situation, and I am therefore to add that if this request should not have been complied with within a period of ten days, the position will become grave."[20]

The next day, Reuter's reported the concentration and planned departure that night of the British Mediterranean Fleet from Malta.

"H.M. battleships *Formidable* and *Irresistible*, and the cruiser *Leviathan* have arrived from a cruise, Vice-Admiral Lord Charles Beresford is on board the *Formidable*, and Rear-Admiral Sir Hedworth Lambton is on the *Leviathan*, and Rear-Admiral Bridgeman is on the *Irresistible*. The cruisers *Carnarvon*, *Barham*, *Suffolk*, and the *Venus*, and the special torpedo vessel *Vulcan*, with destroyers, are proceeding to the Piraeus to-night."[21]

As British forces concentrated in the Eastern Mediterranean, O'Conor was visited on 5th May.

---

19   Affairs of Egypt and Sudan, TNA FO 407/165.
20   *Correspondence Respecting the Turco-Egyptian Frontier in the Sinai Peninsula*, p. 11, HMSO (London) 1906.
21   *Sheffield Evening Telegraph*, 5th May 1906.

> "Nedjib Pasha called on me bearing a message from the Sultan. It was to the effect that His Imperial Majesty considered that the action of His Majesty's Government was directed against his prestige as Caliph and the integrity of the sacred Vilayet of the Hedjaz, and that he was greatly perturbed thereat.
>
> "I replied that I was simply amazed at the Sultan taking this view having regard to the representations I had personally made to His Majesty on the 2nd March, and to those which had repeatedly been made at the Porte."[22]

The British secured diplomatic support, initially from Russia and France, and then German diplomats confirmed their backing. Simultaneously, further military reinforcements were dispatched; 4th Battalion Worcestershire Regiment left Malta on 8th May. But even as they began their journey aboard SS *Dilwara*, Cromer thought he needed more. There were other potential enemies to contend with than just the Turks.

> "It has to be borne in mind that, in addition to the question of disturbances in Egypt, there is a great probability of disaffection extending to the Egyptian army if there is a real outburst of Mahommedan fanaticism, although the accounts which have so far reached me of the state of feeling among the troops are satisfactory."[23]

Sir Edward Grey wrote to the British Ambassador in Paris on 9th May about the steps to be taken in the event of no agreement being reached.

> "I told M. Cambon to-day that we had not yet received any answer to our note at Constantinople. It, was possible, therefore, that the ten days might elapse without result, in which case we did not propose, at first, to take any step more violent than a naval demonstration."[24]

Grey expanded upon British intentions in a note to Cromer.

> "If no satisfactory reply is received, it is proposed, according to present arrangements, that Mitylene and Lemnos should be occupied on Monday [14th May 1906], and, if that is not effective, that action should be taken in the Mediterranean with a view to stopping Turkish transports."[25]

In addition to those targets, Admiral Lord Charles Beresford, the commander of the Mediterranean Fleet, queried whether he should tackle another.

> "I am informed that the Germans are actively engaged with concessions which have been granted to them in Thasos, and that it is their intention to establish a wireless telegraph station on that island. I am sending to ascertain exactly how matters stand. In view of the fact that Thasos belongs to Egypt, and is situated opposite to Kavala, which lies within the sphere of influence of the British Macedonian gendarmerie,

---

22 Affairs of Egypt and Sudan, TNA FO 407/165.
23 *ibid.*
24 *ibid.*
25 *ibid.*

would it be advisable for me to send a ship to hoist the British and Egyptian flags on the island, if hostile eventualities really occur, and I am obliged to take action?"[26]

Cromer told Beresford on 10th May that Thasos was not Egyptian but Turkish and not to pick a fight with the Germans and the Turks at the same time.

Politically isolated and facing a growing naval and military threat, the Turks finally acceded to the British demands. Taba was evacuated on 13th May and a joint commission appointed to define Egypt's eastern borders. The Crescent might still have flown over Egypt but its token status was never more apparent.

Though war had been averted, there had been several episodes where an 'incident' could have developed into something much more serious. It did not take long for questions to be asked about what might have happened; how effective would British military and naval intervention have been had the shooting started?

In London, the Committee of Imperial Defence (CID) discussed the matter during its meeting on 11th May. It was decided to commission a study of the feasibility of attacking, not around the periphery of the Ottoman Empire, but at its heart: the Dardanelles.

Had the CID been in touch with Hornby aboard the *Diana*, he could have told them what his father, Admiral Sir Geoffrey Phipps Hornby, told Lord Derby in 1877 when asked a similar question. At that time it was feared that the Russians might occupy the Gallipoli peninsula. In those circumstances, what could the Navy do to expel them?

> "If you will send for the chart of the Dardanelles, No. 2429, you will see that from three and a half miles below Kilid Bahar to Ak Bashi Imian, six and a half miles above it, an almost continuous cliff overhangs the shore-line, while the Straits close to half a mile in one part, and are never more than two miles wide. An enemy in possession of the peninsula would be sure to put guns on commanding points of those cliffs. All the more if the present batteries, which are a *fleur d'eau*, were destroyed. Such guns could not fail to stop transports and colliers, and would be most difficult for men-of-war to silence. We should have to fire at them with considerable elevation. Shots which were a trifle low would lodge harmlessly in the sandstone cliffs; those a trifle high would fly into the country, without the slightest effect on the gunners except amusement."[27]

After arriving in the Dardanelles, Admiral Hornby wrote from there on 19th February 1878.

> "There seems to be an idea that this fleet can keep the Dardanelles and Bosphorus open. Nothing can be more visionary. Not all the fleets in the world can keep them open for unarmoured ships. Small earthworks on the cliffs would always prevent their passage."[28]

But that verdict was around 40 years old. Much had changed in the interim, so were Hornby's views still valid?

---

26 Affairs of Egypt and Sudan, TNA FO 407/165.
27 Vice-Admiral Sir Geoffrey Phipps Hornby, RN, quoted in Egerton, Mary Augusta Phipps, *Admiral of the Fleet: Sir Geoffrey Phipps Hornby, a Biography*, p. 217, William Blackwood & Sons (London) 1896.
28 Admiral Hornby quoted in Egerton, *Admiral of the Fleet: Sir Geoffrey Phipps Hornby*, p. 253.

The report produced for the CID was completed in December 1906. It outlined what might happen to a fleet attempting to force the Dardanelles — a repeat of Duckworth's experiences in 1807 or, potentially, worse.

> "Even if... it were feasible to rush a number of His Majesty's least valuable ships past the batteries lining the Dardanelles and over the minefields which are believed to exist in the channel, their arrival off Constantinople would be no guarantee that the Sultan would be thereby brought to reason...

> "A defenceless and undefended town, however otherwise important, offers no objective to ships of war. Without a large military force ready to land at a moment's notice, there would be no means of controlling the populace... The Turkish communications furthermore between Europe and Asia would remain virtually intact, and our squadron might find itself face to face with the necessity of again running the gauntlet of the Dardanelles under circumstances which might lead to its destruction."[29]

The report had been drafted by Major-General Charles Callwell. He held out few hopes for the khaki component of any possible operation.

> "When the question of dispatching a military expedition force to the Gallipoli Peninsula comes to be passed in review, the first point to be considered is the general one of whether a landing is possible at all in face of active opposition under modern conditions. In regard to this, history affords no guide. The whole conditions of war have been revolutionised since such an operation was last attempted. Military opinion, however, will certainly lean strongly to the view that no landing could nowadays be effected in presence of an enemy, unless the co-operating naval squadron was in a position to guarantee with its guns that the men, horses, and vehicles of the landing force should reach the shore unmolested, and that they should find after disembarkation a sufficiently extended area free from hostile fire to enable them to form up for battle on suitable ground...

> "... the successful conclusion of a military enterprise directed against the Gallipoli Peninsula must hinge... upon the ability of the fleet not only to dominate the Turkish defences with gun fire, and to crush their field troops during that period of helplessness which exists whilst an army in actual process of disembarkation, but also to cover the advance of the troops once ashore they could gain a foothold, and establish themselves upon the high ground in rear of the coast defences of the Dardanelles."[30]

The report's conclusion was unequivocal.

> "However brilliant as a combination of war, and however fruitful in its consequences, such an operation would be, were it crowned with success, the General Staff, in view of the risks involved, are not prepared to recommend its being attempted."[31]

---

29   The Possibility of a Joint Naval and Military Attack Upon the Dardanelles, December 1906, TNA CAB 38/12/60.
30   *ibid*.
31   *ibid*.

The paper was considered by the CID at its meeting held in London on 28th February 1907. Chaired by the Prime Minister, Henry Campbell-Bannerman (who died on 22nd April 1908), the attendees included his successor, Herbert Henry Asquith, then Chancellor of the Exchequer; Richard Haldane, the Secretary of State for War; Sir Edward Grey, the Foreign Secretary; Admiral Sir John Fisher, RN, the First Sea Lord; and General Sir John French.

Haldane said he "understood that the Admiralty agreed with the General Staff as to the gravity of the risks involved, but that they considered that circumstances might arise in which it might be desirable to attempt the operation."[32]

Grey thought "it essential that these conclusions should be kept secret… that the Department, in dealing with Turkey, should recognize the limitations of our powers of offence against that country. [Though] fear of the possibility of our landing on the Gallipoli Peninsula might be a powerful lever in our hands."[33] He added that "the forcing of the passage of the Dardanelles should be ruled out as an impracticable operation… that other means of bringing pressure to bear on Turkey would be considered."[34]

The joint conclusion reached by the CID was "that the operation of landing an expeditionary force on or near the Gallipoli Peninsula would involve great risk, and should not be undertaken if other means of bringing pressure to bear on Turkey were available."[35]

Haldane had alluded to opinion within the Admiralty not being wholly against an attempt on the Dardanelles. Captain Charles Langdale Ottley, RN, the Director of Naval Intelligence, who had been present at the meeting, was one of them. He later remarked that there was "no reason to despair of success."[36] (Later, in 1912, Ottley took up a directorship of Armstrong Whitworth and its Turkish subsidiary, the Imperial Ottoman Docks Company.)

A sub-committee was established to consider the options available to apply pressure to the Turks in any future confrontation. In the meantime, intelligence gathering about the Dardanelles continued, as it had for decades. In 1905 and 1906, a Grenadier Guards officer, Henry Charles Woods, explored the peninsula on foot and reported his findings to the British authorities.

> "The hills rise precipitously in many places sheer up from the water's edge, reaching an altitude of some hundreds of feet, and present a rocky, barren, uninhabited appearance. The roads are bad, and scarcely any communication by land is possible, nearly everybody travelling from place to place by sea.

---

32   Haldane quoted by Clarke, George, Committee of Imperial Defence. Minutes of 96th Meeting, February 28, 1907. CAB 38/13.
33   Grey quoted by Clarke, Committee of Imperial Defence, CAB 38/13.
34   *ibid.*
35   Clarke, Committee of Imperial Defence, CAB 38/13.
36   Ottley quoted in Gilbert, Martin, *The Challenge of War. Winston S. Churchill 1914–1916*, p. 294, Minerva (London) 1990.

> "There are but few villages worthy of the name, all being small and scattered. The hills are intersected by many small streams which drain these heights towards the sea in both directions."[37]

Woods then considered the locations of the forts and possible approaches for a force to outflank them from the land.

> "These forts defend the narrows and sweep them with fire in all directions. One of the most important of these forts lies south-west of Kilid Bahr, and partly owes its great strength to its height above the sea and its field of fire, and the consequent difficulty there is of damaging it from the water. [He echoed Admiral Hornby here.]
>
> "These forts are nearly all dominated from the higher hills behind them; many are open-backed, and could be menaced by a force occupying this vantage ground.
>
> "It would be of great assistance to a fleet attempting to force a passage if a small force were landed on the north-western shore near Gaba Tepe, and seized these hills in rear of the forts. This landing would be a matter of the utmost difficulty, and unless it came as a complete surprise it would probably be strongly opposed, which would render a landing and any advance to the hills quite impossible. A force once gaining these hills could cut off the water supply of many of the forts, which they obtain through pipes from the hills in rear, and would also greatly lessen the effect of the fire from their guns towards the Straits. A landing would, however, be rendered specially difficult by the small fort or look-out station situated on Gaba Tepe."[38]

Doubtful of the chances of a landing force achieving the surprise required to attack the forts, Woods, noting the limitations of naval power, held out few hopes for an attack by sea either. And concluded with the most important question of all that had been raised by the whole Taba affair.

> "The politicians of England should remember that though her fleet may occupy the islands of Mitylene, Chios, Samos, and Rhodes, her ships cannot ascend the passes of the Rhodope Balkans, or penetrate the mountains of Macedonia, — or probably even force the passage of the Dardanelles...
>
> "The strength of the Dardanelles is one of the most important factors to be considered when discussing the Near Eastern Question... and when advocating that this pressure should be carried to the length of armed coercion, — a policy so persistently pressed upon the British Government by certain sections of the community. If you mean to bark be first assured that you can effectively bite. Are we assured of this?"[39]

---

37 Woods, Henry Charles, *Washed by Four Seas. An English Officer's Travels in the Near East*, p. 58, T. Fisher Unwin (London) 1908.
38 *ibid.*, pp. 61–62.
39 *ibid.*, pp. 67–68.

AUGUST 1914

"Unto us a son is born!"
Enver Pasha

The German battlecruiser, SMS *Goeben*.

At the end of July 1914, with the European crisis worsening, the Admiralty issued its instructions to the commander of the Mediterranean Fleet, Admiral Sir Archibald Berkeley Milne.

> "Should war break out… it now seems possible that Italy will remain neutral… The attitude of Italy, however, is uncertain, and it is especially important that your squadron should not be seriously engaged with Austrian ships before we know what Italy will do. Your first task should be to aid the French in the transportation of their African Army by covering, and if possible, bringing to action individual fast German ships, particularly *Goeben,* which may interfere with that transportation… Do not at this stage be brought to action against superior forces except in combination with the French as part of a general battle… We shall hope later to reinforce the Mediterranean, and you must husband your force at the outset."[1]
>
> THE ADMIRALTY.

So, Berkeley Milne had to protect French troop convoys, attack enemy ships but take care not to risk his own. What constituted a superior force was not defined.

1 AUGUST 1914

ENGLAND

As war threatened to spread far beyond Serbia's borders, Herbert Asquith believed the fate of Belgium, rather than events in the Balkans, would determine what happened next…

> "The main controversy now pivots upon Belgium and its neutrality."[2]
>
> HERBERT ASQUITH, MP, BRITISH PRIME MINISTER.

Winston Churchill felt differently. For him, the ship, the Fleet, had already sailed before receiving the news of Germany's declaration of war on Russia.

> "I found the Prime Minister upstairs… with him were Sir Edward Grey, Lord Haldane and Lord Crewe… I said that I intended instantly to mobilize the Fleet notwithstanding the Cabinet decision, and that I would take full personal responsibility to the Cabinet the next morning. The Prime Minister, who felt himself bound to the Cabinet, said not a single word, but I was clear from his look that he was quite content… I went back to the Admiralty and gave forthwith the order to mobilize. We had no legal authority for calling up the Naval Reserves… but we were quite sure that the Fleet men would unquestioningly obey the summons."[3]
>
> WINSTON CHURCHILL, MP, FIRST LORD OF THE ADMIRALTY.

---

1   Admiralty to Berkeley Milne, 30th July 1914, in Lumby, E.W.R. (Ed.), *Policy and Operations in the Mediterranean 1912-14*, p. 146, Navy Records Society (London) 1970.

2   Asquith quoted in Spender, John Alfred & Asquith, Cyril, *Life of Herbert Henry Asquith, Lord Oxford and Asquith*, vol. 2, pp. 98-99, Hutchinson (London) 1932.

3   Churchill, Winston, *The World Crisis 1911-1914*, p. 217, Thornton Butterworth Ltd (London) 1923.

2 AUGUST 1914

TURKEY

Negotiations between the Ottomans and Germany were concluded by the signing of an agreement this day. Without a trace of irony, Russians felt that Turkish fears concerning their territorial integrity had been exploited.

> "The German Government has offered us an alliance, and as the proposal seems to us in the interests of the country we have signed the compact with Ambassador von Wangenheim to-day... It is an agreement which had due regard for the interests of both parties, and secures their rights in a manner which no other Government has yet done."[4]

SAID HALIM PASHA, OTTOMAN GRAND VIZIER.

> "There is no doubt that, fearing us, and suspecting, by reason of the calumnies of our enemies, that we will attack her, in her heart she desires the success of Germany. This feeling is strongly sustained by the efforts of the officials of the German military commission remaining in Turkey. This element is a highly undesirable one, as they are constantly inciting the Turks against us, but I suppose the Porte will not decide to send them away until the result of our struggle with Germany is made known."[5]

MIKHAIL NIKOLAYEVICH DE GIERS, RUSSIAN AMBASSADOR, ISTANBUL.

MALTA

With the mobilisation of the Fleet underway, so began the operation to find, follow and, if need be, engage the *Goeben* and *Breslau*.

> "I received information that the *Goeben* had been coaling at Brindisi. At 5.12 p.m. the *Chatham*... had sailed from Malta with instructions to search for the *Goeben* in the Strait of Messina, and subsequently to join the Rear-Admiral's squadron. Four destroyers were patrolling the Malta Channel. My force at Malta was thus reduced to the battle cruiser *Inflexible* (flag), two light cruisers and small craft."[6]

ADMIRAL SIR BERKELEY MILNE, RN, C-IN-C MEDITERRANEAN FLEET, HMS *INFLEXIBLE*.

> "I was ashore on the Sunday afternoon with some chums; we were going for a bike ride, and we had got about two miles from the harbour when we heard the booming of guns. Of course, we had been warned before going ashore that if we heard the guns fired from the ships we were to return aboard at once, so we turned about and

---

[4] Said Halim Pasha quoted in Djemal Pasha, *Memories of a Turkish Statesman*, p. 109.
[5] Scott, James Brown (Ed.), *Diplomatic Documents Relating to the Outbreak of the European War*, pp. 1386–1387, Oxford University Press, American Branch, (New York) 1916.
[6] Berkeley Milne, Admiral Sir Archibald, *The Flight of the 'Goeben' and the 'Breslau.' An Episode in Naval History*, pp. 53–54, Eveleigh Nash Company Ltd. (London) 1921.

got aboard the ship in nearly no time, and when we did so we found everything in a state of bustle, and left Malta that night, we and the *Indomitable*. There were various buzzes as to where we were going, but the captain had the ship's company "fall in" on deck and told us we had to find and shadow the two German cruisers, so that in case war did break out with Germany we should know where to lay our hands on them. The last place they were heard of was at Messina, so we sent the *Chatham* in there to find out, and she came back to say they had left some hours before."[7]

L/STO. HARVEY TWEEDALE, RN, HMS *INDEFATIGABLE*.

Berkeley Milne circulated his orders. Rear-Admiral Ernest Troubridge was to take note.

"The Admiralty have informed me, that should we become engaged in war, it will be important at first to husband the naval force in the Mediterranean and, in the earlier stages, I am to avoid being brought to action against superior forces. You are to be guided by this should war be declared."[8]

ADMIRAL SIR BERKELEY MILNE, RN, C-IN-C MEDITERRANEAN FLEET, HMS *INFLEXIBLE*.

3 AUGUST 1914

ENGLAND

The British Cabinet was divided about entering the war. Some felt the potential violation of Belgian neutrality was a pretext, rather than a genuine justification for intervention.

"The precipitate and peremptory blaze about Belgium was due less to indignation at the violation of a Treaty than to natural perception of the plea that it would furnish for intervention on behalf of France, for [an] expeditionary force and all the rest. Belgium was to take the place that had been taken before, as pleas for war, by Morocco and Agadir."[9]

JOHN MORLEY, 1ST VISCOUNT MORLEY OF BLACKBURN, LORD PRESIDENT OF THE COUNCIL.

Following their defeats in the Balkans and against Italy, the Turks decided to strengthen their navy. Two battleships were being built for them in British yards, the *Reshadieh* and *Sultan Osman I*. But, with war imminent, the temptation for Churchill to add them to the Royal Navy proved too much and on 31st July he notified the shipyard owners of his intention to commandeer the vessels.[10] British diplomats were now told to inform the Turks of this development.

---

7   *Rochdale Observer*, 17th October 1914.
8   Lumby, *Policy and Operations in the Mediterranean 1912–14*, p. 274.
9   Morley, John Viscount, *Memorandum on Resignation*, p. 14, MacMillan (New York) 1928.
10 Taking over ships under construction in British yards for other nations in the event of war was not a new idea. Churchill raised the possibility during a discussion of the German naval building programme at the Committee of Imperial Defence on 4th July 1912.

"Arrangements are being made with the firm of Armstrong, Whitworth, and Co. for His Majesty's Government to take over the Turkish battleship "Osman I" now building with that firm.

"Please inform Turkish Government that His Majesty's Government are anxious to take over the contract."[11]

SIR EDWARD GREY, MP, BRITISH FOREIGN SECRETARY.

TURKEY

"Grand Vizier and Minister of the Interior spoke to me with some vexation of the detention of Turkish ship, which they seemed to consider an unfriendly act as Turkey is not at war. Minister of the Interior referred to the very heavy financial sacrifices by which this ship had been paid for with money borrowed at a rate amounting to interest at 20 per cent."[12]

HENRY BEAUMONT, BRITISH EMBASSY, ISTANBUL.

Badge of Turkish battleship *Sultan Osman I*, kept as a souvenir when the ship was taken over by the Royal Navy and commissioned as HMS *Agincourt*.

11 *Correspondence Respecting Events Leading to the Rupture of Relations with Turkey*, p. 1, HMSO (London) 1914.
12 *ibid*.

The diplomatic language employed could not hide how badly the take-over of the ships had been received. Given the circumstances, with some of the money having been raised through public subscription, this could not have been surprising. At a stroke, these ships became 'Belgium' for the Ottomans.

MEDITERRANEAN

The British assessment was that the German ships would attempt to break out into the Atlantic or seek to join up with their expected soon to be Austrian allies. The latest intelligence was that the German ships, having left the Adriatic, were sailing west.

> "At 7 a.m., the *Chatham* reported that neither the *Goeben* nor the *Breslau* was in the Strait of Messina. At the same time I received information that *Goeben* and *Breslau* had been sighted early on the previous morning off Cape Trion, the southern horn of the Gulf of Taranto, heading south-west... In order both to maintain the watch on the Adriatic and to find *Goeben* and *Breslau*, at about 8.30 a.m. I ordered Rear-Admiral Troubridge... to send the light cruiser *Gloucester* and the eight destroyers to the mouth of the Adriatic, while the rest of his squadron was to pass south of Sicily and to the westward. The light cruiser *Chatham* was ordered to pass westward along the north coast of Sicily. The light cruisers *Dublin* and *Weymouth* were set to watch the Malta Channel... My own impression was that the Germans would turn westward was confirmed by a report that a German collier was waiting at Majorca."[13]
>
> ADMIRAL SIR BERKELEY MILNE, RN, C-IN-C MEDITERRANEAN FLEET, HMS *INFLEXIBLE*.

It appeared to Berkeley Milne that the Admiralty agreed with his appreciation.

> "At 8.30 p.m. I received instructions from the Admiralty to send two battle cruisers to Gibraltar at high speed to prevent the *Goeben* from leaving the Mediterranean. *Indomitable* and *Indefatigable* were already on their way westward, and they were ordered to proceed at 22 knots to Gibraltar. The *Chatham*, which was then rounding Sicily, and which had nothing to report, was ordered to Malta to coal."[14]
>
> ADMIRAL SIR BERKELEY MILNE, RN, C-IN-C MEDITERRANEAN FLEET, HMS *INFLEXIBLE*.

4 AUGUST 1914

ENGLAND

Confirmation that German troops had crossed into Belgium helped resolve the doubts of most (but not all) waverers within the Cabinet. Some could not wait for it all to begin.

> "We... got the news that the Germans had entered Belgium and had announced that if necessary they would push their way through by force of arms. This simplifies matters.

---

13   Berkeley Milne, *The Flight of the 'Goeben' and the 'Breslau'*, pp. 56–59.
14   *ibid.*, p. 59.

So we sent the Germans an ultimatum to expire at midnight requesting them to give a like assurance with the French that they would respect Belgian neutrality... Winston has got on all his war-paint, is longing for sea-fight in the early hours of the morning to result in the sinking of the *Goeben*. The whole thing fills me with sadness."[15]

HERBERT ASQUITH, MP, BRITISH PRIME MINISTER.

The Foreign Office was trying to contain Turkish anger over the acquisition of their ships.

"I am sure Turkish Government will understand necessity for His Majesty's Government to keep all warships available in England for their own needs in this crisis.

"Financial and other loss to Turkey will receive all due consideration, and is subject of sincere regret to His Majesty's Government. You should inform Grand Vizier."[16]

SIR EDWARD GREY, MP, BRITISH FOREIGN SECRETARY.

GERMANY

After the signing of the agreement with the Ottomans, the Germans saw an opportunity to not only prevent the loss of their ships in the Mediterranean but also to bring to the Turks closer to their camp. The signal was sent at 1.35 a.m.

"Alliance with Turkey. *Goeben* and *Breslau* proceed immediately to Constantinople."[17]

GERMAN ADMIRALTY STAFF.

ALGERIA

Before complying with the order, the *Goeben* and *Breslau* had business in French north Africa.

"The *Breslau* received orders to bombard Bone; we laid a course for Philippeville... Flying the Russian flag, the *Goeben* slowly approached the harbour... The fortifications stood picturesquely on the heights above the town. The guns were still covered with tarpaulins as protection against the weather... Now a number of boats laden with bananas, pineapples and coconuts put out towards us...

"The Russian flag ran down, and in its place rose the German war-flag; and already the first salvoes of our 15cm guns were crashing out against the barracks, harbour works, warehouses, and mole. The secondary armament was sufficient for our purpose... Never had we seen oarsmen pull with such desperate vigour as those fruit-sellers...

"On the shore beyond flames were rising; our shells were causing heavy damage...

---

15   Asquith, Herbert, *Memories and Reflections 1852–1927*, vol. 2, p. 21, Cassell (London) 1928.
16   *Correspondence Respecting Events Leading to the Rupture of Relations with Turkey*, p. 2.
17   James, William Milbourne, *The Eyes of the Navy. A Biographical Study of Admiral Sir Reginald Hall, K.C.M.G., C.B., LL.D., D.C.L.*, p. 60, Methuen (London) 1955.

After about ten minutes the admiral ordered the cease-fire. The bombardment had done its work."[18]

WIRELESS OPERATOR GEORGE KOPP, SMS *GOEBEN*.

"At five o'clock in the morning, we were awakened by the noise of shell fire. Before I could get on deck I heard a shell burst on board my ship amidships. It was the sixth shell they fired, and it pierced our funnel at the base on the starboard side.

"Exploding inside the funnel, the shell ripped it open for ten feet on the port side, splinters of the projectile smashing the combing of the main hatch and perforating the wooden screen round the bridge. A boat aft the funnel was also damaged. Fortunately, we had all got below, and nobody was hurt. We watched the remainder of the firing from the deck. The Germans confined their shots to the quay, which they swept from end to end, doing as much damage as they could.

"The only casualties were caused when a shell hit a forage shed containing a quantity of ammunition and about thirty sleeping reservists. Thirteen of these were killed and about the same number wounded. The shell hit the ammunition and the shed was soon in flames."[19]

CAPTAIN ERNEST GRANDIN, MASTER, SS *ISLE OF HASTINGS*, PHILIPPEVILLE.

Those on the *Isle of Hastings* could be forgiven for thinking that the shelling was aimed at them.

"I think the Germans must have made our funnel their target, for a shell which meant to drop amidships came hurtling across, carrying away large part of the funnel, and also part of the fiddle deck and several of the boiler pipes. That was the only shot which struck us, but the bombardment lasted half-an-hour, doing great damage among the houses on the foreshore.

"The cook [James Codrington] on board our ship had just left his galley when the shell smashed part of the erection. He had a very narrow escape, because had he remained inside he must surely have been crushed by the debris.

"There was fort Philippeville, but it was not manned. When the roar of the shells was heard, however, the soldiers hastened with all speed and commenced to reply to the German fire. When they had fired three shots the *Goeben* seemed to think she was in dangerous company, and she turned tail and vanished out of sight into the Mediterranean Sea. I do not know whether any of the shells from the fort struck her or not.

18   Kopp, Georg, *Two Lone Ships. 'Goeben' & 'Breslau'*, pp. 23–24, Hutchinson & Co. Ltd. (London) 1931.
19   *Ottawa Free Press* (Canada), 19th September 1914.

> "The bombardment came as a great shock to us, for most of us were in our bunks. We had four Germans among us, and when they saw the craft which was shelling the town they dressed themselves in their best clothes and came on deck to see what was going on."[20]
>
> THIRD ENGINEER ALEXANDER MORRISON, SS *ISLE OF HASTINGS*, PHILIPPEVILLE.[21]

Berkeley Milne was informed of the attack later that morning. But neither he nor the Admiralty knew a British ship had been fired upon (nor the Germans, for that matter).

> "Where were the *Goeben* and *Breslau*? No one knew. At 7 a.m… the *Chatham* had reported they were not in the Strait of Messina… At 8.30 a.m… I received information that Bona had been bombarded by the German ships.
>
> "At 9.32 a.m. *Indomitable* and *Indefatigable* off Bona, on the Algerian coast, sighted the *Goeben* and *Breslau*, which were steering to the eastward… Had a state of war then existed, it is probable there would have been a very different end to that meeting."[22]
>
> ADMIRAL SIR BERKELEY MILNE, RN, C-IN-C MEDITERRANEAN FLEET, HMS *INFLEXIBLE*.

One British sailor recorded the encounter.

> "We kept them in view… we were within two miles of them, but we could not fire because war was not then declared. Everything was excitement aboard, guns were loaded and trained on them, and torpedoes ready to fire, but we had orders not fire unless they fired first."[23]
>
> L/STO. HARVEY TWEEDALE, RN, HMS *INDEFATIGABLE*.

> "During the afternoon, *Dublin* joined *Indomitable* and *Indefatigable* at a point north of Bizerta. In the meantime, *Goeben* and *Breslau*, steaming at their utmost speed, were drawing away from the British battle cruisers, which presently lost sight of the German ships. The *Dublin* picked them up about 5 p.m., and kept them in sight until nearly 10 p.m., when she lost them off the Cape San Vito, on the north coast of Sicily, and turned back to rejoin the battle cruisers."[24]
>
> ADMIRAL SIR BERKELEY MILNE, RN, C-IN-C MEDITERRANEAN FLEET, HMS *INFLEXIBLE*.

Within two hours of the *Dublin* losing contact the British ultimatum to Germany expired and a state of war now existed. In the meantime, with the fugitive ships nearing Italian shores, Berkeley Milne received a warning.

---

20 *Dundee Evening Telegraph*, 25th November 1914.
21 Morrison had served for five years with 4th Bn Royal Highlanders (Black Watch) pre-war. After returning to Britain, he rejoined his old regiment. He landed in France on 24th February 1915; promoted Sergeant on 11th March 1915; on 2nd October 1915 reported to have been wounded; commissioned on 29th May 1917.
22 Berkeley Milne, *The Flight of the 'Goeben' and the 'Breslau'*, pp. 60–61.
23 *Rochdale Observer*, 17th October 1914.
24 Berkeley Milne, *The Flight of the 'Goeben' and the 'Breslau'*, p. 62.

> "Italian Government have declared neutrality. You are to respect this neutrality rigidly and should not allow any of His Majesty's ships to come within six miles of Italian coast."[25]

THE ADMIRALTY.

> "The effect of the order was to bar the Strait of Messina… certainly to British ships. If the *Goeben* and *Breslau* entered the Strait, they could not be followed. They might break back westward or they might turn south through the Strait, and then either turn eastward to the Adriatic, or west through the channel between Africa and Sicily."[26]

ADMIRAL SIR BERKELEY MILNE, RN, C-IN-C MEDITERRANEAN FLEET, HMS *INFLEXIBLE*.

Berkeley Milne was still focused on blocking any German exit into the Atlantic.

> "I considered that the German ships would avoid both the north of Corsica and the Strait of Bonifacio, for fear of French cruisers, destroyers and submarines. In all probability they would, therefore, try to pass south of Sardinia, and thence to Majorca, where a German collier was waiting at Palma.

> "In these circumstances, Rear-Admiral Troubridge was ordered… to detach *Gloucester* to watch the southern end of the Strait of Messina… [and] with four armoured cruisers and eight destroyers… to continue to watch the mouth of the Adriatic. The two battle cruisers… and three destroyers were ordered to join my flag off Pantellaria Island at 11 a.m. on the following day… These dispositions were communicated to the Admiralty at 8.30 p.m.[27]

ADMIRAL SIR BERKELEY MILNE, RN, C-IN-C MEDITERRANEAN FLEET, HMS *INFLEXIBLE*.

5 AUGUST 1914

GERMANY

> "Revolution in India and Egypt, and also the Caucasus is of the highest importance. The Treaty with Turkey will make it possible for the [German] Foreign Office to realise the idea and to awaken the fanaticism of Islam."[28]

GERMAN GENERAL STAFF.

---

25   Churchill, *The World Crisis 1911–1914*, p. 226.
26   Berkeley Milne, *The Flight of the 'Goeben' and the 'Breslau'*, p. 67.
27   ibid., pp. 70–71.
28   Macfield, A.L., *The End of the Ottoman Empire: 1908–1923*, p. 141, Longman (London) 1998.

ENGLAND

The post of Secretary of State for War had been vacant since the Curragh Mutiny. With war declared, Asquith called on one of Britain's most celebrated soldiers.

> "I have taken an important decision to-day: to give up the War Office and install Kitchener there as an emergency man until the war comes to an end. It was quite impossible for me to go on now that war is actually in being. It requires the undivided time and thought of any man to do the job properly, and I hate scamped work. K. was, to do him justice, not at all anxious to come in, but when it was presented to him as a duty he agreed. It is clearly understood that he has no politics and that his place at Cairo is kept open so that he can return to it when peace comes back. It is a hazardous experiment, but the best in the circumstances, I think…

> "Oddly enough, there is no authentic war news either by land or sea. All that appears in the papers is invention. Winston's mouth waters for the *Goeben,* but so far she is still at large."[29]

HERBERT ASQUITH, MP, BRITISH PRIME MINISTER.

> "Official notice received mines being laid in Bosphorus and Dardanelles and lights extinguished."[30]

THE ADMIRALTY.

TURKEY

The report that Turkish waters had been mined was confirmed when a British ship carrying Russian grain to Germany entered the Bosphorus.

> "The ship lifted bodily forward, and a column of water shot up over the top of the foremast and completely buried the ship. I was on the bridge at the time with the captain, and the volume of water lifted us right off our feet. Luckily, none of the crew happened to be round the fore-deck at the time, but the hands in the forecastle were thrown out of their bunks, and a lot of the wooden bulkheads in the forecastle were smashed up. Pieces of lead and of copper cable were smashed on to the deck, supposedly from the mine."[31]

ALFRED STABLEFORD, CHIEF MATE, SS *CRAIGFORTH,* ISTANBUL.

The *Craigforth* had to be beached but no-one onboard was harmed.

---

29 Asquith, *Memories and Reflections,* vol. 2, pp. 24–25.
30 Lumby, *Policy and Operations in the Mediterranean 1912–14,* p. 167.
31 *Newcastle Daily Chronicle,* 1st December 1914.

ITALY

The location of the *Goeben* and *Breslau* was revealed by their need to take on more coal. They had gone where Berkeley Milne had just been ordered to avoid.

> "Following sent to Consul at Messina: — Urgent. Have learnt privately that German ship *Breslau* is attempting to procure coal from British collier at Messina. Captain should be warned to supply no coal to belligerent. If force being used protest to local authorities."[32]

SIR JAMES RODD, BRITISH AMBASSADOR, ROME.

MEDITERRANEAN

British ships had to coal too. They remained focussed on blocking any German attempt to breakout to the west, while keeping an eye on the Adriatic.

> "At about 11 a.m… *Inflexible* (flag), *Indomitable*, *Indefatigable*, *Dublin*, *Weymouth*, *Chatham* and three destroyers were assembled off Pantellaria Island, midway in the channel between the African coast and Sicily. *Dublin* was sent back to Malta, there to coal and thence to proceed with two destroyers to join Rear-Admiral Troubridge at the mouth of the Adriatic. *Indomitable* and three destroyers went into Bizerta to coal. *Inflexible* (flag) with *Indefatigable*, *Chatham* and *Weymouth*, patrolled on a line northward from Bizerta, being thus disposed to intercept the German ships should they attempt to escape westwards."[33]

ADMIRAL SIR BERKELEY MILNE, RN, C-IN-C MEDITERRANEAN FLEET, HMS *INFLEXIBLE*.

After more evidence that *Goeben* was at Messina, the Admiralty informed Berkeley Milne that the Dardanelles had been mined.

> "At 5 p.m… I received a report from *Gloucester* that, judging by wireless signals intercepted, the *Goeben* appeared to be at Messina… At 7 p.m. I received information from the Admiralty that mines had been laid in the Dardanelles… and that the Dardanelles lights had been extinguished."[34]

ADMIRAL SIR BERKELEY MILNE, RN, C-IN-C MEDITERRANEAN FLEET, HMS *INFLEXIBLE*.

Having taken steps to cover the Adriatic, the mining of the Dardanelles may well have convinced Berkeley Milne that an escape to the west was now the only option open to the German ships. He was not privy to cables from the German Admiralty.

---

[32]   Sir James Rennell Rodd, quoted in Lumby, *Policy and Operations in the Mediterranean 1912–14*, p. 167.
[33]   Berkeley Milne, *The Flight of the 'Goeben' and the 'Breslau'*, pp. 71–72.
[34]   *ibid.*, pp. 82–83.

6 AUGUST 1914

ENGLAND

Politicians in London formed their plans and received offers to help implement them. After agreeing to send the British Expeditionary Force to France, dealing with Germany's Pacific colonies was high on the list.

> "We had our usual Cabinet this morning, and decided with much less demur than I expected to sanction the dispatch of the Expeditionary Force of four divisions. We also discussed a number of smaller schemes for attacking German ports and wireless stations in East and West Africa and the China Seas. Indeed, I had to remark that we looked more like a gang of Elizabethan buccaneers than a meek collection of black-coated Liberal Ministers."[35]

HERBERT ASQUITH, MP, BRITISH PRIME MINISTER.

> "His Majesty's Government gratefully accept offer of your [Australian] Ministers to send a force of twenty thousand men to this country, and would be glad if it could be despatched as soon as possible. Secret. If your Ministers at the same time desire and feel themselves able to seize German Wireless Stations at Yprop in Marshall Islands, Nauru or Pleasant Island and New Guinea we should feel this a great and urgent Imperial service. You will, however, realize that any territory now occupied must be at the disposal of the Imperial Government for purposes of ultimate settlement at the conclusion of the war. Other Dominions are acting in similar way on the same understanding, in particular suggestion is being made to New Zealand in regard to Samoa."[36]

LEWIS HARCOURT, SECRETARY OF STATE FOR THE COLONIES.

Germany's South Pacific colonies posed no threat to British interests and could have been taken at leisure. They were quite literally going nowhere. By contrast, the German East Asia Squadron had the potential to inflict serious harm across a wide area. The buccaneers had picked the wrong target.

NORTH SEA

Rapid technological developments meant that navies went to war with weapons they had neither used nor faced. An Australian witnessed the loss of the first Royal Navy warship.

> "At 6.30 [a.m.] our leader, the *Amphion*, struck a mine. We heard and felt a heavy explosion, and observed yellow smoke coming up from under the *Amphion*'s bows. All destroyers closed on her to give assistance; lifeboats were lowered and took everyone off. The last man had to jump for it, but everyone was as cool as if nothing

---

35   *The Sun* (Sydney, New South Wales), 21st July 1928.
36   Typescripts; comprise questions to be put to Sir Brudenell White, AWM38 3DRL 6673/153.

was wrong. No sooner had the boats shoved off than a terrific explosion occurred, and all the fore part of the ship went into the air. Men that were put out in the first explosion, including German prisoners, guns, bridge, &c., were blown to atoms. The last explosion reached about 1,000 feet, and then all was still. It is wonderful how all the destroyers escaped, as we were steaming over a bed of mines. After that we carried on as usual."[37]

AB ERNEST ABFALTER, RN, HMS *LOUIS*.

ITALY

The *Goeben* had been coaling all night but was still at Messina when dawn broke. It would have to leave before the sun set.

> "Up to now 1,200 tons of coal had been taken aboard. But it was not enough… Coaling from these totally ill-adapted steamers, devoid of any kind of suitable gear, was one of the most difficult and exhausting tasks we had ever been faced with…
>
> "It was now midday. Only few hands were now capable of further work. The majority were lying like dead about the deck, overcome with exhaustion, the shovels still gripped in their black and blistered hands. Then the admiral gave the order to break off coaling. There was no need for this order, as no more was being done. We had shipped 1,500 tons."[38]

WIRELESS OPERATOR GEORGE KOPP, SMS *GOEBEN*.

MEDITERRANEAN

After taking his strongest units north of Sicily, Berkeley Milne received the news that the German ships were not bound by similarly rigid orders regarding Italian neutrality.

> "These dispositions had scarcely been made when, half an hour later, the *Gloucester*, which was watching the southern entrance to the Strait, reported that the *Goeben* was coming out of the Strait of Messina, the *Breslau* following her one mile astern, steering eastward. The position was then as follows: If *Goeben* and *Breslau* attempted to enter the Adriatic, Rear-Admiral Troubridge, with the First Cruiser Squadron and ten destroyers, would prevent them; if the German ships, followed by *Gloucester*, escaped her in the night and turned westwards, my squadron of battle cruisers must be so placed as to intercept them. As my instructions strictly forbade me to enter the Strait of Messina, I was obliged, in order to take up the requisite position, to come down the west coast of Sicily… Further reports from *Gloucester*, which was pursuing the German ships, stated that they were steering eastward, then north-eastward. I therefore continued on my course to Malta, in order to coal there and to continue the chase… In the meantime, at 11 p.m… I had received a telegram from the Admiralty countermanding previous instructions and ordering me, if the *Goeben* went south,

---

37 *Northern Argus* (Clare, South Australia), 23rd April 1915.
38 Kopp, *Two Lone Ships*, pp. 38–39.

to follow her through the Strait of Messina. Unfortunately, by the time the new instructions reached me, it was too late to fulfil them."[39]

ADMIRAL SIR BERKELEY MILNE, RN, C-IN-C MEDITERRANEAN FLEET, HMS *INFLEXIBLE*.

7 AUGUST 1914

MEDITERRANEAN

With the *Goeben* emerging from the Straits of Messina, the 1st Cruiser Squadron was best placed to intercept. Before doing so, Troubridge was approached by his Flag Captain.

> "I feel, Sir, it is my duty from a gunnery point of view purely, to say that if you engage the *Goeben* under these circumstances you are committing absolute suicide of the whole Squadron. I have myself seen the *Goeben*'s trial results of her guns: they make an absolutely perfect pattern at 15,000 yards; at 14,000 yards is her most efficient and effective range."[40]
>
> CAPTAIN FAWCETT WRAY, RN, HMS *DEFENCE*, 1ST CRUISER SQUADRON.

Troubridge realised what was at stake. More than his reputation hung on the assessment of whether the *Goeben* constituted a "superior force." In battle practice the previous month, the *Defence* had failed to hit any target beyond 8,000 yards. He signalled Berkeley Milne.

> "Being only able to meet *Goeben* outside the range of our guns and inside his I have abandoned the chase with my squadron… *Goeben* evidently going to Eastern Mediterranean. I had hoped to have met her before daylight."[41]
>
> REAR-ADMIRAL SIR ERNEST TROUBRIDGE, RN, HMS *DEFENCE*, 1ST CRUISER SQUADRON.

Berkeley Milne challenged Troubridge to justify his decision. They both understood what refusing action could entail.

> "With visibility at the time I could have been sighted from twenty to twenty-five miles away and could never have got nearer unless *Goeben* wished to bring me to action which she could have done under circumstances most advantageous to her. I could never have brought her to action. I had hoped to engage her at three thirty in the morning in dim light but had gone north first with the object of engaging her in the entrance to the Adriatic… I would consider it a great imprudence to place squadron in such a position as to be picked off at leisure and sunk while unable to effectively reply."[42]
>
> REAR-ADMIRAL SIR ERNEST TROUBRIDGE, RN, HMS *DEFENCE*, 1ST CRUISER SQUADRON.

---

39   Berkeley Milne, *The Flight of the 'Goeben' and the 'Breslau'*, pp. 86–88.
40   Wray quoted in Lumby, *Policy and Operations in the Mediterranean 1912–14*, p. 259.
41   Troubridge quoted in Lumby, *Policy and Operations in the Mediterranean 1912–14*, p. 181.
42   *ibid.*, p. 183.

These were far from the last words on the matter. And now only the light cruiser *Gloucester* was left to keep tabs on the German ships.

> "About eight o'clock the morning stillness was broken by the whirring of the alarm bells! To the eastward a column of smoke had been sighted, and soon the hull of a warship could be distinguished, bearing down upon us... it turned out to be the *Breslau*! She was coming from a direction where we had never expected to see her."[43]
>
> WIRELESS OPERATOR GEORGE KOPP, SMS *GOEBEN*.

Gloucester did catch up with them around 1.00 p.m.

> "The smoke column grew larger and larger. Soon we knew who it was we had astern of us — it was again the *Gloucester*, following at full steam... We reduced speed and let her come nearer. The *Breslau* was steaming behind us, slightly on the beam.
>
> "Suddenly there were signs of life in the enemy cruiser — a flash — a couple of darting tongues of fire, and a moment later the air was rent by the thunder of guns...
>
> "The splashes rose like fountains far astern of the *Breslau*. Then came a line of flashes from her own guns. Near the English ship, thin water columns rose aloft — an action between the two cruisers was in progress... The *Goeben* now began to turn about, but the enemy, at once seeing her intention, broke off the action immediately and steamed away.
>
> "The *Breslau* reported in Morse that she had suffered a hit in her water-line armour, but otherwise had taken no hurt."[44]
>
> WIRELESS OPERATOR GEORGE KOPP, SMS *GOEBEN*.

Damage to both *Breslau* and *Gloucester* was slight during this brief exchange, despite a belief in the efficacy of unofficial additions to the charges used.

> "The *Gloucester* opened fire upon the *Breslau*, but her first shot fell short. The *Breslau* replied with 30 shots, of which two only took effect, two of the *Gloucester*'s boats being smashed on the davits.
>
> "After the first shot our lads were quite happy, and kept firing as quickly as possible. One chap nearly swallowed his chew of baccy when the first shot fell short. The next one he spat on for luck, and it took half the *Breslau*'s funnel away. He repeated the operation on the next shot, which cleared the *Breslau*'s quarter-deck, and put her aft gun out of action."[45]
>
> L/STO. TOM MARSDEN, RN, HMS *GLOUCESTER*.

The action had been witnessed by the passengers and crew of an Italian steamer.

> "Between 10.30 and 11 in the morning attention was called to three vessels, all hull down, on our starboard beam. We were then, I suppose, off Zante, and as I remember

---

43   Kopp, *Two Lone Ships*, pp. 49–50.
44   *ibid.*, p. 51.
45   *Nottingham Evening Post*, 26th August 1914.

the three boats seemed to be steering approximately S.E. We watched them with some interest for some time. Our courses seemed to be converging, and, as they grew nearer it appeared to us that they were three warships, probably some British patrol vessels. They were, of course, too far to distinguish, but I rather thought that the second, if British, might belong to the *Devonshire* class and the last to the *Weymouth* class, a vessel with which I have become very well acquainted during her six months' stay in Constantinople during the Balkan War…

"Interest in them naturally decreased as time went on, until suddenly the sound of a gun again attracted the attention of the whole ship's company to the vessels, which in the morning had caused considerable excitement and wonder. The last of the three vessels and the second exchanged about half a dozen shots. We then, of course, knew immediately that the two leading ships were the *Goeben* and the *Breslau* respectively. It was not until 4 o'clock that we knew that the last of the three was the *Gloucester*. We watched the shell of the *Breslau* fall short. They did not secure a single hit, but we were unable to tell whether the *Gloucester*'s replies had hit the mark."[46]

MIDDLETON EDWARD, PASSENGER, SS *SICILIA*.

The Germans turned away and put on speed.

"Once more began a fierce struggle, an exciting tussle of boilers and engines. Once more every man who could in any way be spared was sent to the bunkers and stokeholds. Once more we had to rely for help on the superior power of our turbines…

"Smarting and irritating, the fine coal-dust in the bunkers penetrated the nose and clogged the throat. Lungs laboured heavily as the men struggled at their work. A crust of coal-dust would form in the throat and cause a racking cough. Coffee and lemonade were constantly taken below and greedily gulped down. But the relief did not last. The coal-dust would attack the nose and throat worse than before, form a fresh coating, and find its way into the eyes, bringing tears and inflammation."[47]

WIRELESS OPERATOR GEORGE KOPP, SMS *GOEBEN*.

Gloucester approached the spectators of the fight.

"At about 4 o'clock the *Gloucester* came alongside to ask who we were and where we were going to, and having satisfied herself that everything was in order we passed on, and I secured a photograph of her as a recollection of the circumstance. The smoke of the German vessels finally disappeared below the horizon just about sunset."[48]

MIDDLETON EDWARD, PASSENGER, SS *SICILIA*.

---

46   *Gloucester Journal*, 19th February 1915.
47   Kopp, *Two Lone Ships*, p. 52.
48   *Gloucester Journal*, 19th February 1915.

"I... ordered Captain Kelly, who was, I knew, getting short of coal, and who ran great risk of capture, to stop pursuit at Cape Matapan and to rejoin the Rear-Admiral. At 4.40 p.m., then, Captain Kelly turned, while the German ships held on through the Cervi Channel, between the southern extremity of Greece and the island of Kithera."[49]

ADMIRAL SIR BERKELEY MILNE, RN, C-IN-C MEDITERRANEAN FLEET, HMS *INFLEXIBLE*.

"Quite unexpectedly, the *Gloucester* now abandoned the chase. She seemed to think it too dangerous to follow us into the Archipelago, fearing she might be drawn into a trap in these cramped waters."[50]

WIRELESS OPERATOR GEORGE KOPP, SMS *GOEBEN*.

8 AUGUST 1914

TURKEY

The Turks confirmed their agreement to allow the fugitive German ships to enter the Dardanelles.

"Captain Hamann, Naval Attaché to the German Embassy, came to the Ministry. He informed me that the German Mediterranean squadron, pursued by the English, was withdrawing in the direction of the Dardanelles, and as, to judge by his reports, the *Goeben* had practically no more coal, they were compelled to send some from Constantinople. But as there was no English coal available he asked me to lend him five or six thousand tons from our naval depots. I immediately telephoned to the Grand Vizier, Enver Pasha, and Talaat Bey to ascertain their opinion.

"They replied that I should agree. I ordered that the required quantity of coal should be supplied from the Derindji depots, and also sent a naval labour section to assist with the loading of the vessel. It was loaded within a few hours, and then set out for the Aegean Sea."[51]

AHMED DJEMAL PASHA, OTTOMAN NAVAL MINISTER.

RUSSIA

If the British remained uncertain about the German ships' destination, Russian diplomats in Istanbul were warned to expect the *Goeben* and *Breslau* to head their way.

"It is reported that the cruisers "Goeben" and "Breslau," having rounded Matapan, are apparently headed for the Dardanelles. You will please, in concert with the French and British Ambassadors, make the most earnest representations to the Porte relative to the responsibility it would assume in permitting the passage of these vessels

---

49   Berkeley Milne, *The Flight of the 'Goeben' and the 'Breslau'*, pp. 103–105.
50   Kopp, *Two Lone Ships*, p. 53.
51   Djemal Pasha, *Memories of a Turkish Statesman*, p. 118.

through the Dardanelles, and insist upon their leaving the Straits, or that they shall be disarmed, without carrying the matter to an open rupture."[52]

SERGEI DMITRYEVICH SAZONOV, RUSSIAN MINISTER OF FOREIGN AFFAIRS.

MEDITERRANEAN

As Berkeley Milne considered his next steps, he received a signal at 1.58 p.m. which decided them.

"Commence hostilities at once against Austria."[53]

THE ADMIRALTY.

"Acting instantly upon the instructions provided for that contingency, I proceeded to a position in which I could support Rear-Admiral Troubridge's squadron, then watching the mouth of the Adriatic, and issued orders concentrating the Fleet.

"In order to execute these dispositions, it was necessary to turn north-westwards, and to steer for a rendezvous 100 miles south-west of Cephalonia, at which the Rear-Admiral was ordered to join my flag… those instructions involved the concentration of the Fleet and consequently the entire abandonment of the pursuit of the German vessels."[54]

ADMIRAL SIR BERKELEY MILNE, RN, C-IN-C MEDITERRANEAN FLEET, HMS *INFLEXIBLE*.

The Admiralty signal was a mistake but by the time the error was corrected it was too late.

9 AUGUST 1914

RUSSIA

"If the "Goeben" proceeds through the Dardanelles under the German flag, it will be left to Admiral Eberhardt to use all means within his power to bar its exit into the Black Sea, and to destroy the "Goeben." Still, the Admiral will be instructed to avoid, except in case of extreme necessity, taking any action directly against Turkey."[55]

SERGEI DMITRYEVICH SAZONOV, RUSSIAN MINISTER OF FOREIGN AFFAIRS.

TURKEY

The non-arrival of the ships taken over by the British was proving to be a source of growing discontent in Istanbul.

"An official communiqué was recently published here which showed a distinctly hostile tone towards Great Britain. This communiqué dealt with the requisition

---

52 Scott, *Diplomatic Documents Relating to the Outbreak of the European War*, p. 1389.
53 Lumby, *Policy and Operations in the Mediterranean 1912–14*, p. 189.
54 Berkeley Milne, *The Flight of the 'Goeben' and the 'Breslau'*, pp. 119–124.
55 Scott, *Diplomatic Documents Relating to the Outbreak of the European War*, p. 1391.

of the Turkish warships by His Majesty's Government. The Grand Vizier has told me that Turkish Government had to pretend to the Turkish public, as the latter had subscribed towards the purchase money for the vessels, that they were taking a stronger line than really was the case. He said, however, that we should not attach too much importance to publications of this kind.

"Public opinion is daily growing more excited, and I think that if His Majesty's Government were able to give an assurance that Turkey would have the ships, if possible, on the conclusion of hostilities, such an assurance would have a soothing effect."[56]

HENRY BEAUMONT, BRITISH EMBASSY, ISTANBUL.

MEDITERRANEAN

As Berkeley Milne considered where the German ships might be, the Dardanelles did not feature in his speculations.

"At noon... I received a telegram from the Admiralty stating definitely that Great Britain was not at war with Austria, and instructing me to resume the pursuit of the *Goeben*. Twenty-four hours had thus elapsed since the arrival of the order from the Admiralty to begin hostilities against Austria, compelling me to alter the whole of my dispositions and thus to relinquish the search for the German vessels, which were therefore twenty-four hours steaming further away...

"As the movements and intentions of the German vessels were utterly unknown to me it was necessary to take measures in accordance with probable contingencies. These were: (1) That the *Goeben* might attempt to take refuge in a Greek port, where Admiral Souchon could rely on the good offices of Greece. (The German Admiral was accustomed to use Phalerum in time of peace.) (2) That the *Goeben* might proceed to Salonika to attack that port and thus destroy the Serbian supplies. (3) That she might turn south to attack the south-eastern trade and to destroy British shipping at Alexandria and Port Said. (4) That she might attempt to return westward and to leave the Mediterranean."[57]

ADMIRAL SIR BERKELEY MILNE, RN, C-IN-C MEDITERRANEAN FLEET, HMS *INFLEXIBLE*.

"We are at sea between Greece and Malta coaling from a collier, and she is taking this letter back with her. As yet no harm has befallen me or any of us here. We are waiting to see how Austria and Italy are going to place themselves. Only one ship here, a German ship called the *Goeben*, has more speed than our fleet. When she is put out of the way we will have little to fear from any."[58]

PTE FREDERICK MORGAN, RMLI, HMS *WARRIOR*.[59]

---

56 *Correspondence Respecting Events Leading to the Rupture of Relations with Turkey*, p. 2.
57 Berkeley Milne, *The Flight of the 'Goeben' and the 'Breslau'*, pp. 124–133.
58 *Belfast Weekly Telegraph*, 30th January 1915.
59 Morgan survived the loss of his ship at Jutland.

The *Goeben* and *Breslau* were coaling too, having rendezvoused with the collier *General* at the Greek island of Donousa in the Cyclades.

> "It was not large, but had high mountains and a deep, deserted bay wonderfully protected by the sheer, towering cliffs…
>
> "Slowly and cautiously the *Goeben* turned and steered into the bight, the long massive hull pushing its way little by little through the entrance and then turning broadside towards it, to be ready for any unexpected and unwelcome attack. The collier made fast to her side and the *Breslau* took station behind her. The ship was cleared for coaling."[60]
>
> WIRELESS OPERATOR GEORGE KOPP, SMS *GOEBEN*.

10 AUGUST 1914

RUSSIA

> "The German warships, *Goeben* and *Breslau*, according to intelligence that has reached here, have passed Greece, apparently directing their course to the Dardanelles."[61]
>
> REUTER'S, ST. PETERSBURG.

MEDITERRANEAN

The Admiralty passed on no information about the *Goeben*'s likely destination, not even the Reuter's cable. The searchers picked up German wireless transmissions but could not trace their origin.

> "At 4 a.m. we rounded Cape Malea and spread fifteen miles, on a N.E. search line. At 10 a.m. we were startled by intercepting a wireless signal, force 12, i.e. maximum intensity, and evidently originating from a high-power installation near to us (? *Goeben*). Raised steam for full-speed and hoped for the best. Sent the *Weymouth* on to search the islands to the north and east of the Gulf of Athens."[62]
>
> LT-CDR RUDOLF VERNER, RN, HMS *INFLEXIBLE*.

The *Goeben* left its Greek refuge and headed for the Dardanelles.

> "In the bright morning sunshine our secret coaling-place was soon left far astern.
>
> "Our eyes swept the wide surface of the Aegean Sea stretched out under the scorching sky… The look-outs at their posts scanned the horizon, a particularly sharp watch being kept astern for the smoke of enemy fighting ships developing out of the shimmering haze.
>
> "Thus hour after hour passed. We held our course at high speed northward, It was an

---

60   Kopp, *Two Lone Ships*, p. 57.
61   *Daily News* (London), 10th August 1914.
62   Lt.-Cdr. Verner quoted in Chatterton, E. Keble, *Dardanelles Dilemma*, p. 36, Rich & Cowan Ltd. (London) 1935.

anxious cruise... With even, monotonous roar, the two ships moved swiftly forward. Tenedos lay already lay astern...

"We were nearing the Dardanelles. It was five o'clock in the afternoon... Already the coast of Asia Minor divided itself from the narrow tongue of Gallipoli, leaving open a gleaming passage... The entrance to the Dardanelles Straits lay ahead."[63]

WIRELESS OPERATOR GEORGE KOPP, SMS *GOEBEN*.

TURKEY

Sir Edwin Pears was a long-term resident of Istanbul (though he never learned Turkish) and, doubtless, was a fine lawyer but he was fooling himself if he felt public feeling over the take-over of the Turkish ships could be disposed of as he would a point of law.

"The impression amongst the Turks that the British Government, in taking possession of the two ships the "Sultan Osman" and "Rechadié," has not only acted illegally, but has played a trick.

"I am not merely convinced, as an Englishman, that this is not the case, but I have recollections... the seizure of two ships by the Germans at the opening of the Franco-German War in 1870 at, I think, Calais. I recollect also the taking possession of two battle ships which were being built in England, which had been built for some South American Government some ten or twelve years ago, the latter occurring when both Great Britain and the country to which they belonged were at peace...

"It may be that England is acting solely on this provision, but it might be useful if I could know whether the cases suggested are relied on... and by so doing the really ugly feeling which has arisen from the the suggestions of bad faith on the part of "perfide Albon" might be allayed, because, after all, there are sensible men amongst the Turks who at present are honestly under a delusion."[64]

SIR EDWIN PEARS, ISTANBUL.

11 AUGUST 1914

ENGLAND

"I learn that at 8-30 p.m. last night "Goeben" and "Breslau" reached the Dardanelles. These ships should not be allowed to pass through the Straits, and they should either leave within twenty-four hours, or be disarmed and laid up. You should point out to the Turkish Government that these are the dates entailed upon them by their neutrality, and that His Majesty's Government expect that they will act up to their obligations."[65]

SIR EDWARD GREY, MP, BRITISH FOREIGN SECRETARY.

63 Kopp, *Two Lone Ships*, pp. 64–66.
64 Ryan, Sir Andrew: Miscellaneous Correspondence, TNA FO 800/240.
65 *Correspondence Respecting Events Leading to the Rupture of Relations with Turkey*, p. 3.

Enver Bey c.1910, later Enver Pasha, Ottoman Minister for War.

TURKEY

The mood in Istanbul contrasted with that in London.

> "Enver Pasha... remarked with that quiet smile which was peculiar to him: "Unto us a son is born!... The *Goeben* and *Breslau* appeared off the Dardanelles this morning, and as they were being followed by the English fleet, they asked that they should be allowed to pass through the Narrows. I granted the permission, as I did not wish to condemn the ships of an allied State to certain destruction, and by now the ships are in the Dardanelles under the protection of the forts of the Narrows. The sequel is that we are faced with a political problem...
>
> "After an hour of lively discussion between the Grand Vizier, Talaat Bey, and the [German] Ambassador, the latter promised to get into communication with Berlin the same night and get a favourable answer before morning.... It came about four o'clock in the morning. It empowered us, on condition that we accepted Admiral Souchon in the Ottoman service, to say that the ships had been sold to Turkey. It was not a real, but merely fictitious sale."[66]
>
> AHMED DJEMAL PASHA, OTTOMAN NAVAL MINISTER.

The message to the British was that the two German vessels had been acquired to further Turkish claims to islands seized by the Greeks during the Balkan wars.

> "The Ottoman Government have bought "Goeben" and "Breslau." Officers and men will be allowed to return to Germany. Grand Vizier told me that purchase was due to our detention of "Sultan Osman." They must have ships to bargain with regard to question of the islands on equal terms with Greece, and it was in no way directed against Russia, the idea of which he scouted.
>
> "He formally asked that the British naval mission might be allowed to remain."[67]
>
> HENRY BEAUMONT, BRITISH EMBASSY, ISTANBUL.

DARDANELLES

> "Just after sunset two large ships of war were seen approaching very slowly, preceded by a pilot boat. Immediately the greatest excitement was aroused, and the rumour spread that they were the *Goeben* and the *Breslau*. When they arrived alongside us, however, it was already too dark to see. Next morning, 11th August, our doubts, were dispelled. The two cruisers were the *Goeben* and the *Breslau*. They were flying the German colours, and anchored near us. I examined them very closely through my field glasses, and they showed no signs of having suffered in any way. About 11 a.m. the *Goeben* sent out a steam pinnace with three officers and a body of sailors carrying rifles. The boat went alongside several ships, including the *Saghalien* and *Henri Traissinet*.

---

66 Djemal Pasha, *Memories of a Turkish Statesman*, pp. 119–120.
67 *Correspondence Respecting Events Leading to the Rupture of Relations with Turkey*, p. 3.

"The German officers also went on board a small English salvage boat. We subsequently learned that they had destroyed the wireless installation of the *Saghalien*, and had warned all ships they passed not to move. Turkish officers visited the *Goeben* and *Breslau* during the day. I saw them shaking hands most cordially with officers of the cruisers and offering them cigarettes. Everyone was astonished, and believed that Turkey must have concluded an alliance with Germany."[68]

REUTER'S.

MEDITERRANEAN

"Between 4 a.m. and 5 a.m.... the wireless signals of the German armed auxiliary *General* increased in strength; whereupon *Weymouth* was ordered to proceed to examine the Gulf of Smyrna; further instructions to search among the islands were given to *Chatham;* and the battle cruisers, keeping within searchlight signalling distance, were spread to search. At 6 a.m., *Dublin*, sent to join my flag by the Rear-Admiral, rounded Cape Malea and was sent to Milo Island to coal…

"It was at 10.30 a.m… that I received the first definite information of the position of the German vessels since the *Gloucester*, upon turning back, had seen *Goeben* and *Breslau* enter the Cervi Channel at 4.40 p.m. on Friday, 7th August. At 10.30 on the morning of the 11th August, then, I received a telegram from Malta informing me that *Goeben* and *Breslau* had arrived at the Dardanelles at nine o'clock on the previous evening."[69]

ADMIRAL SIR BERKELEY MILNE, RN, C-IN-C MEDITERRANEAN FLEET, HMS *INFLEXIBLE*.

12 AUGUST 1914

ENGLAND

Asquith and Grey took a relaxed view about the German escape from the Royal Navy. It could be taken as a practised display of insouciance, an indication the Turks were not to be taken seriously or a bit of both.

"The only interesting thing is the arrival of the *Goeben* in the Dardanelles, and her sale to Turkey. The Turks are very angry at Winston's seizure of their battleships here. As we shall insist that the *Goeben* shall be manned by a Turkish instead of a German crew it does not much matter, as the Turkish sailors cannot navigate her except on to rocks or mines."[70]

HERBERT ASQUITH, MP, BRITISH PRIME MINISTER.

"Turkey will probably purchase *Goeben* and *Breslau* and I do not consider that this

---

68   *The Telegraph* (Brisbane), 7th November 1914.
69   Berkeley Milne, *The Flight of the 'Goeben' and the 'Breslau'*, pp. 134–137.
70   *The Sun* (Sydney, New South Wales), 21st July 1928.

means any immediate departure from neutrality; nor does it follow that she will attack Egypt.

"You should therefore prepare quietly for contingencies, there is in my opinion no need for alarm at present respecting Egypt."[71]

SIR EDWARD GREY, MP, BRITISH FOREIGN SECRETARY.

TURKEY

The claim that the German crews had been allowed to remain because they were very tired was met with a mixture of diplomatic scepticism, anger and frustration.

"I sent to the papers an official communiqué referring to the purchase of the *Goeben* and *Breslau*... I asked the Press to speak enthusiastically of the circumstance that we had obtained possession of the ships as compensation for the *Sultan Osman* and the *Reschadieh*, of which the English had robbed us...

"I had a report from [Admiral Limpus, Head of the British Naval Mission in Istanbul]... He congratulated the Ottoman Government on having secured possession of two such vessels, and assured me that as the two ships came under his direct command, he would have the selected officers and men ready within a month to manoeuvre with these most modern units. I asked the Admiral to call on me, and began to discuss the matter with him. I asked him that, in view of the fact that the German Admiral and ships' companies were very exhausted, so that the date on which they would leave the ship was still uncertain, he would occupy himself in making out the list of officers and men who were to be employed on them."[72]

AHMED DJEMAL PASHA, OTTOMAN NAVAL MINISTER.

"I saw the Grand Vizier this morning and made strong representations to him against restrictions of free passage of the Straits, which the military authorities are now imposing under various pretexts. I said they had been holding up passenger and grain ships in the Dardanelles, refusing to deliver papers to ships wishing to leave Constantinople, and ordering grain ships to return to Constantinople at their caprice...

"It seems that the Minister of War has now got entirely out of hand, and I gather that he alone is responsible for the present situation. Matters are undoubtedly becoming serious, but a Cabinet Council is being held this afternoon, and I hope I may be able afterwards to report some improvement."[73]

HENRY BEAUMONT, BRITISH EMBASSY, ISTANBUL.

"The "Goeben" and the "Breslau" are still at Nogara [sic]. The cruisers, which were greeted by the Turks with enthusiasm, are taking coal here from the Turkish authorities. Turkish torpedo boats are passing out to sea from the Dardanelles, and are reporting to the German vessels the result of their reconnaissance. The German

71 Foreign Office. Private Offices. Grey, Sir Edward (Viscount), TNA FO 800/48.
72 Djemal Pasha, *Memories of a Turkish Statesman*, pp. 120–121.
73 *Correspondence Respecting Events Leading to the Rupture of Relations with Turkey*, p. 4.

sailors made a strict search of the French, British, and Greek merchant vessels lying in the Dardanelles, and took a wireless telegraph outfit by force from the French steamer "Saghalien," under threat of blowing up the ship."[74]

MIKHAIL NIKOLAYEVICH DE GIERS, RUSSIAN AMBASSADOR, ISTANBUL.

AEGEAN

One naval officer recorded how the news of Germans' escape was received.

"The Ward Room was a rather comic sight this morning. We had just finished our 'Action Stations' when the news arrived, and the sight of the 'Control' and 'Quarters' Officers sitting dejectedly about the room made one think of the Lions' House at feeding time and a butchers' strike just announced."[75]

LT-CDR RUDOLF VERNER, RN, HMS *INFLEXIBLE*.[76]

13 AUGUST 1914

DARDANELLES

"On the 13th we were still lying at Chanak, the governor of the Dardanelles refusing to give the captain of the *Roumania* a written order to discharge his cargo. The captain wired to Constantinople, but got no reply, and later he sent a letter to the Italian Embassy by an Italian steamer that was going to Constantinople. A small steamer — the *Virginia* — with a Greek captain and crew, but flying the American flag, came in in the afternoon, and as I learned it was going out, shortly, I decided to go on board and got away. I went on board with a British officer, who was returning to Egypt to rejoin his regiment, and an American. At 4 o'clock we decided to start, but immediately a pilot boat set off from the shore and told us to anchor again. At 5 o'clock we signalled "May we go" and this time we were allowed to start. Outside the Dardanelles we saw three British cruisers, one of which signalled us to show our flag. When we had proceeded some distance the cruiser… came alongside. The captain shouted through a megaphone. 'Have you any news of German ships in the Dardanelles?' I shouted back to tell him to send a man abroad. The order was at once given to get the cutter ready, and in a few minutes the cutter came along side.

"An officer and lieutenant, who were loudly cheered by the numerous Greeks on the steamer, came on board, and we gave them all the information they required."[77]

REUTER'S.

---

74   Scott, *Diplomatic Documents Relating to the Outbreak of the European War*, p. 1393.
75   Lt.-Cdr. Verner quoted in Chatterton, E. Keble, *Dardanelles Dilemma*, p. 37, Rich & Cowan Ltd. (London) 1935.
76   Verner, then a Commander, was killed aboard the *Inflexible* during the attempt for force the Dardanelles on 18th March 1915. Commemorated on the Chatham Naval Memorial, he was the 32 year-old son of Colonel Willoughby Verner and the Hon. Elizabeth Verner, of Old Yard Farm, Ditchling, Sussex.
77   *The Telegraph* (Brisbane), 7th November 1914.

14 AUGUST 1914

TURKEY

"Admiral Limpus has received a promise from Minister of Marine that his Excellency will make crews for the "Goeben" and "Breslau." This will take time, but nevertheless it will be done; and his Excellency has undertaken to hand over the two ships bodily to the British admiral.

"Admiral Limpus informs me that a month will probably elapse before "Sultan Selim" (late "Goeben") can be even moved by the Turkish crew; but the formalities of transfer may be complete technically in a day or two. Further delay in taking delivery from the Germans is unavoidable.

"Minister of Marine declared there was no intention of sending the ships outside Sea of Marmora until the end of the war."[78]

HENRY BEAUMONT, BRITISH EMBASSY, ISTANBUL.

"The *Goeben* and the *Breslau*, renamed, respectively, *Sultan Selim the Grim* [*Yavuz Sultan Selim*] and *Mytilene* [*Midilli*], will arrive here shortly from the Dardanelles, after having coaled.

"I am informed that the Turkish Government have given assurances to the Russian Ambassador that the German officers and crew will immediately disembark, that their places will be filled by Turkish seamen, and that the vessels will be put under the command of Rear-Admiral Limpus for the Porte, and that there is no intention to employ them against Russia."[79]

PHILIP GRAVES, BRITISH JOURNALIST, ISTANBUL.

15 AUGUST 1914

TURKEY

Despite pledging to remove the German crews, it was the British Naval Mission in Istanbul that was dismissed.

"Admiral Limpus and all officers of British Naval Mission have suddenly been replaced in their executive command by Turkish officers, and have been ordered to continue work at Ministry of Marine if they remain. Although I have been given to understand by a member of the Government that they are still anxious to get officers and crew of the "Goeben" and "Breslau" out of Turkey, this will probably

---

78 *Correspondence Respecting Events Leading to the Rupture of Relations with Turkey*, p. 5.
79 *Evening Irish Times*, 15th August 1914.

mean retention of mechanics and technical experts at least, which will create most dangerous situation here."[80]

HENRY BEAUMONT, BRITISH EMBASSY, ISTANBUL.

CORAL SEA

The location of other German ships remained unknown. Where was Admiral von Spee's German East Asia Squadron, principally the *Scharnhorst* and *Gneisenau*?

> "We… are looking out for German cruisers. They consist, as far as we know, of two 11,000 ton cruisers, and a good few smaller ones, but we are prepared and ready in case of an attack. We go about at night in pitch darkness, all dead lights on the mess decks, etc., are closed, and no one is allowed to strike a match on deck. There are always five guns manned, and you get cells if found asleep on watch."[81]

BOY 1ST CLASS GORDON MITCHELL CAMPBELL, RAN,
HMAS *ENCOUNTER*, RUSSELL ISLAND.

16 AUGUST 1914

TURKEY

As the Grand Vizier made reassuring noises about remaining neutral, reports came in to show that the minefields guarding the Dardanelles had been strengthened.

> "The "Goeben" and "Breslau" are at present lying off Constantinople. The Grand Vizier has assured me that there is no intention of moving them from Marmora. They are now flying the Ottoman flag under nominal command of Turkish officer, and have been transferred. This at least is a good sign."[82] …

> "I have received the following telegram, dated the 15th August, from His Majesty's vice-consul at Dardanelles: —

> "A new field of mines has been laid in the zone formerly sown with mines of observation type. It may be assumed that these latter had previously been removed.

> "The new contact mines, to the number of forty-one, were laid by the 'Mtibah' from Kephez to Suandere in a double line. Seven were kept on the ship, and the twenty-four from the 'Selanik,' which is proceeding to Constantinople, were also taken on board."[83]

HENRY BEAUMONT, BRITISH EMBASSY, ISTANBUL.

---

80   *Correspondence Respecting Events Leading to the Rupture of Relations with Turkey*, p. 5.
81   *Casino and Kyogle Courier and North Coast Advertiser* (New South Wales), 9th September 1914.
82   *Correspondence Respecting Events Leading to the Rupture of Relations with Turkey*, pp. 5–6.
83   *ibid.*, p. 6.

17 AUGUST 1914

ENGLAND

The unresolved problem of the *Goeben* and *Breslau* encouraged speculation about Turkish intentions. Churchill was prepared to attack regardless; others preached caution.

> "Turkey has come into the foreground, threatens vaguely enterprises against Egypt, and seems disposed to play a double game about the *Goeben* and the *Breslau*. Winston, in his most bellicose mood, is all for sending a torpedo flotilla through the Dardanelles to threaten and, if necessary, to sink the *Goeben* and her consort. Crewe and Kitchener very much against it. In the interests of the Mussulmans in India and Egypt they are against our doing anything at all which could be interpreted as meaning that we are taking the initiative against Turkey. She ought to be compelled to strike the first blow. I agreed to this, but the Turks must be obliged to come out and tell us whether they are going at once to dismiss the German crews."[84]
>
> HERBERT ASQUITH, MP, BRITISH PRIME MINISTER.

TURKEY

> "The "Goeben" and "Breslau" yesterday left Ismid to repair damages received: two shot holes on the starboard side of the "Goeben" and on the prow of the "Breslau." Repairs will require about ten days. Thirty-eight wounded men were put ashore from the "Goeben." Yesterday, 200 men of the German crews were removed from both vessels. The remainder, according to assurances given by the Minister of the Marine, will be removed upon the arrival from England of the Turkish crew returning on the Turkish cruiser "Reshid."[85]
>
> MIKHAIL NIKOLAYEVICH DE GIERS, RUSSIAN AMBASSADOR, ISTANBUL.

18 AUGUST 1914

TURKEY

The British Ambassador returned to Istanbul facing a full in-tray. He trusted the Grand Vizier's good faith, not tempered by Russian fears of his weakening influence.

> "I have been accorded most cordial reception upon my return to my post by the Grand Vizier, of whom I enquired whether the German crews would be removed soon and what guarantee he would give that the "Goeben" and "Breslau" would be used neither against Great Britain nor against Russia. I also expressed my surprise that the Turkish Government should be apparently entirely under German influence, and that they

---

84   *The Sun* (Sydney, New South Wales), 21st July 1928.
85   Scott, *Diplomatic Documents Relating to the Outbreak of the European War*, p. 1396.

should have committed such a serious breach of neutrality as was involved by their action in the matter of the German ships.

"His Highness said that he deeply deplored this breach of neutrality, which he could not deny. He begged me to give him time to get rid of German crews, which he promised he would do gradually, but, until arrival of Turkish transport with crews from London, Turkish Government had no crew to replace Germans… and he assured me that I need not be anxious lest Turkey should be drawn into war with Great Britain or with Russia. The present crisis would pass.

"I am convinced of the absolute personal sincerity of Grand Vizier in these utterances."[86]

SIR LOUIS MALLET, BRITISH AMBASSADOR, ISTANBUL.

19 AUGUST 1914

ENGLAND

Churchill wrote to Enver Pasha spelling out a package of compensation for the ships taken over from the Turks.

"I deeply regretted necessity for detaining Turkish ships because I knew the patriotism with which the money had been raised all over Turkey. As a soldier you know what military necessity compels in war. I am willing to propose to His Majesty's Government the following arrangement: —

"(1) both ships to be delivered to Turkey at the end of the war after being thoroughly repaired at our expense in British Dockyards; (2) if either is sunk we will pay the full value to Turkey immediately on the declaration of peace; (3) we will also pay at once the actual extra expense caused to Turkey by sending out crews and other incidentals as determined by an arbitrator; (4) as a compensation to Turkey for the delay in getting the ships we will pay £1,000 a day in weekly instalments for every day we keep them, dating retrospectively from when we took them over.

"This arrangement will come into force on the day when the last German officer and man belonging to the *Goeben* and *Breslau* shall have left Turkish territory definitely and finally, and will continue binding so long as Turkey maintains a loyal and impartial neutrality in this war and favours neither one side nor the other.

"Do you agree?"[87]

WINSTON CHURCHILL, MP, FIRST LORD OF THE ADMIRALTY.

---

86   *Correspondence Respecting Events Leading to the Rupture of Relations with Turkey*, pp. 6–7.
87   Churchill's note to Enver Pasha, via Grey (Foreign Secretary) to Mallet (British Ambassador, Istanbul), TNA ADM 137/800.

20 AUGUST 1914

ENGLAND

If some in the Admiralty harboured the belief that Churchill had been unwise to take over the Turkish battleships, at least one of his Staff regarded him less charitably than that.

> "I really believe Churchill is not sane. His entire energies have since last Monday been devoted to forming a naval battalion for shore service, consisting of the RNVR not embarked, the stokers, etc of the RNR & some odds & ends of marines & marine artillerymen...
>
> "These men who are thus to be employed in soldiering know nothing of the business. They are all amateurs of the most marked kind... Such use as they could be in the stokeholds of the ships is left out of account... The whole thing is so wicked that Churchill ought to be hanged before he should be allowed to do such a thing."[88]

CAPTAIN HERBERT RICHMOND, RN, ASSISTANT DIRECTOR OF OPERATIONS, ADMIRALTY.

TURKEY

A call for Muslims to prepare themselves for holy war was not the sole cause for allied pessimism. The voices of moderation were not being heard.

> "O Moslem people, I gathered you here at the beginning of the Holy Ramazan to call you to an external holy war. A war against Atheist Europe. Perhaps some of you think it is not proper for me to preach about politics under this sacred dome. But I tell you Mohammedan religion is a political religion and this the most important problem about which one can ever preach in this sacred place is politics.
>
> "Real politics is to defend the nation against internal & external enemies. So I come to declare to you, O ye Moslem people, to get ready for the holy war. We are at the beginning of the holy Bairam. I bring you good tidings, this is also the dawn of victory."[89]

OUBEID ULLAH EFFENDI, HAGIA SOPHIA, ISTANBUL.

> "I must say that the situation here is of the greatest gravity for the reason that all affairs are in the hands of the military, who allow themselves license in everything and are openly exerting themselves, under German pressure, to draw Turkey into a war with us. Djavid Bey, the Grand Vizier, and, to some extent, Djemal Pasha, oppose this, but I am by no means convinced that the last word rests with them."[90]

MIKHAIL NIKOLAYEVICH DE GIERS, RUSSIAN AMBASSADOR, ISTANBUL.

---

88   Marder, Arthur Jacob (Ed.), *Portrait of an Admiral: The Life and Papers of Sir Herbert Richmond*, p. 100, Harvard University Press (Cambridge, Massachusetts) 1952.
89   Ryan, Sir Andrew: Miscellaneous Correspondence, TNA FO 800/240.
90   Scott, *Diplomatic Documents Relating to the Outbreak of the European War*, p. 1397.

21 AUGUST 1914

ENGLAND

The call from Hagia Sophia for holy war would not have been known in London but the potential implications were never far from the minds of Britain's leaders.

> "We had a long Cabinet this morning, mostly about details connected with the War. The real centre of interest, political, not military, at the moment is Turkey, and the two darkest horses in the European stable, Italy and Rumania. The different points of view of different people are rather amusing, Kitchener strong that Rumania is the real pivot of the situation; Masterman eagerly pro-Bulgarian, but very much against any aggressive action vis-à-vis Turkey which would excite our Mussulmans in India and Egypt; Lloyd George keen for Balkan Confederation, Grey judicious and critical all round; Haldane instructive; and the "Beagles" and "Bobtails" silent and bewildered."[91]
>
> HERBERT ASQUITH, MP, BRITISH PRIME MINISTER.

TURKEY

> "I was informed by the Grand Vizier to-night that he wanted all the support that the Triple Entente could give him, and that the sooner they could give a written declaration respecting the independence and integrity of Turkey the better.
>
> "A sharp struggle, which may come to a head at any moment, is in progress between the Moderates and the German party, headed by the Minister for War, and is meanwhile creating anarchy here."[92]
>
> SIR LOUIS MALLET, BRITISH AMBASSADOR, ISTANBUL.

AUSTRALIA

Meanwhile, new recruits were getting used to their new routine.

> "This life is a change, I am sure it will do me good. I am with fellows that I travelled with, and we are getting to know each other, and have some good fun together…
>
> "I will give you an outline of a day's programme. Up at 6 a.m. and get coffee and dog biscuits if smart enough to get in before the crowd, but it is an awful rush, and have to go without sometimes. We fall in soon after for roll call, then have physical drill for an hour or so out on the parade ground. A person feels the benefit of it by breakfast time about eight a.m. Two orderlies are appointed, we take it turn about to go to the cookhouse with [a] bucket for tea or coffee, and dishes for whatever happens to be on. We all breakfast together then clear up the tent, fold blanket and make the place ship shape. By the time we have done this it is about time to fall in for parade. We

---

91   Herbert Asquith quoted in Spender, J.A. & Asquith, Cyril, *Life of Herbert Henry Asquith, Lord Oxford and Asquith*, vol. 2, p. 125, Hutchinson & Co. (London) 1932.

92   *Correspondence Respecting Events Leading to the Rupture of Relations with Turkey*, p. 9.

drill then till about 12 noon; then there is another raid on the cook house and dinner proceeds… The afternoon is taken up with drill, and everyone seems to be coming on really well. By the time we have had a few days of it we will be about broken in. About five p.m. we are dismissed, and then have a little time to ourselves to wash and brush up. Then tea time and the rest of the evening till nine is our own. Roll call then, and after, bed time and lights out, and peaceful slumber if you can sleep under the circumstances, rolled up in a single blanket with all clothes on, only boots off on water proof sheets with a little straw under it, and the hard ground under all."[93]

PTE JOHN MALCOLM, 7TH BATTALION AIF, BROADMEADOWS CAMP.[94]

22 AUGUST 1914

ENGLAND

"The demands made by the Turkish Government are excessive; we do not, however, wish to refuse all discussion, and you may therefore… address the following communication to the Porte: —

"If the Turkish Government will repatriate immediately the German officers and crews of the "Goeben" and "Breslau," will give a written assurance that all facilities shall be furnished for the peaceful and uninterrupted passage of merchant vessels, and that all the obligations, of neutrality shall be observed by Turkey during the present war, the three allied Powers will in return agree, with regard to the Capitulations, to withdraw their extra-territorial jurisdiction, as soon as a scheme of judicial administration, which will satisfy modern conditions, is set up.

"They will further give a joint guarantee in writing that they will respect the independence and integrity of Turkey, and will engage that no conditions in the terms of peace at the end of the war shall prejudice this independence and integrity."[95]

SIR EDWARD GREY, MP, BRITISH FOREIGN SECRETARY.

AUSTRALIA

One group from Dimboola visited Broadmeadows Camp to see how their local volunteers were doing.

"The camp is prettily situated on rising ground, and the numbers of white tents extending over acres of green fields made a fine picture. After passing the tents of the Army Medical Corps, etc., we came to the infantry lines… the first one to greet us

---

93   *Kerang New Times* (Victoria), 28th August 1914.
94   Malcolm, a former grocer, was killed in action on 25th April 1915. Commemorated on the Lone Pine Memorial, he was the 23 year-old son of James William and Jane Malcolm, of Scoresby Street, Kerang, Victoria.
95   *Correspondence Respecting Events Leading to the Rupture of Relations with Turkey*, p. 9.

was our dear friend Henry D'Alton.[96] We had informed him on the Saturday of our proposed visit, so of course he was waiting for us, and, as were all the others, was very pleased to see someone from Dimboola and hear all the news…

"I had letters to Lieut.-Colonel Bolton[97] and Captain Coulter,[98] the latter of whom was adjutant for that day. After informing us as to the best means of seeing everything, he very kindly invited my friend and I to join him for afternoon tea in the officers' tent, which we greatly appreciated after walking all through the camp lines for an hour or so…

"All the boys wished to be remembered to all their friends and relatives. They expect to sail in about three or four weeks time, and have been informed that they will have three days leave granted them to come back and say 'good-bye' to all. We… hope to see them all before they leave, and if anyone from up this way is visiting town it is well worth while visiting the camp at Broadmeadows, as it is a sight ever to be remembered."[99]

CEDRIC BENNETT, DIMBOOLA, VICTORIA.

23 AUGUST 1914

TURKEY

"I saw Minister of Marine, as the Turkish transport [*Neshid Pasha*][100] has now arrived, and asked him when the crews of the "Goeben" and "Breslau" would be repatriated.

"He said that it depended upon the Grand Vizier. He was himself in favour of their repatriation.

"I shall press the matter strongly, but do not know whether the Moderates are sufficiently strong to insist upon such a step."[101]

SIR LOUIS MALLET, BRITISH AMBASSADOR, ISTANBUL.

---

96 Pte Henry St. Eloy D'Alton, 8th Bn, AIF, died of wounds aboard the transport 'Seam Choow' between 27th and 29th April 1915. He is commemorated on the Lone Pine Memorial. His brother, Pte Charles Edward (Eddie) D'Alton, 8th Bn, AIF, was killed in action on 6th August 1915 and is buried in Shrapnel Valley Cemetery. They were the sons of St. Eloy D'Alton and Ann D'Alton, of Dimboola, Victoria.
97 Lt-Col. William Kinsey Bolton, 8th Bn, AIF, asked to be relieved of his command at Gallipoli on 18th May 1915.
98 Capt., later Lt-Col. Graham Coulter, 8th Bn, AIF, was evacuated from the peninsula on 16th November 1915 suffering from jaundice.
99 *Dimboola Banner and Wimmera and Mallee Advertiser* (Victoria), 28th August 1914.
100 This was the ship that took the Turkish crew that was intended to bring the *Sultan Osman I* to Turkey. The *Neshid Pasha* left the Tyne on 7th August 1914 missing one man. The body of Kadri, son of Mehmet, aged around 20, was found in the river on 11th August. At the inquest, held two days later, no explanation was found for how he got there.
101 *Correspondence Respecting Events Leading to the Rupture of Relations with Turkey*, p. 10.

24 AUGUST 1914

ENGLAND

Australian and New Zealand troops could not be brought to Europe while they were still in the process of taking over Germany's Pacific colonies and the German East Asiatic Squadron remained at large.

> "Admiralty consider it to be most inadvisable that any portion of Expeditionary Force should be sent at once. There are on the route of transports German warships which have not been yet definitely located and until they are destroyed or hunted off route transports should not leave without a convoy. Convoy is not at present practicable as the greater part of the Australian and New Zealand squadrons are engaged in offensive operations in the Pacific. When the Force does start it should go preferably in one convoy and probably by Suez Canal."[102]
>
> LEWIS HARCOURT, SECRETARY OF STATE FOR THE COLONIES.

TURKEY

> "I hear that a further contingent of German officers has recently arrived via Sophia for service here."[103]
>
> SIR LOUIS MALLET, BRITISH AMBASSADOR, ISTANBUL.

25 AUGUST 1914

ENGLAND

It was decided to try a direct message from King George V to Sultan Mehmed V in an effort to placate local opinion regarding the seizure of the Turkish battleships.

> "His Majesty the King desires that your Excellency should convey to His Imperial Majesty the Sultan of Turkey a personal message from His Majesty, expressing his deep regret at the sorrow caused to the Turkish people by the detention of the two warships which His Imperial Majesty's subjects had made such sacrifices to acquire. His Majesty the King wishes the Sultan to understand that the exigencies of the defence of his dominions are the only cause of the detention of these ships, which His Majesty hopes will not be for long, it being the intention of His Majesty's Government to restore them to the Ottoman Government at the end of the war, in the event of the

---

102   Typescripts; comprise questions to be put to Sir Brudenell White, AWM38 3DRL 6673/153.
103   *Correspondence Respecting Events Leading to the Rupture of Relations with Turkey*, p. 10.

maintenance of a strict neutrality by Turkey without favour to the King's enemies, as at present shown by the Ottoman Government."[104]

SIR EDWARD GREY, MP, BRITISH FOREIGN SECRETARY.

TURKEY

The latest meeting between the British Ambassador and the Grand Vizier saw Enver Pasha absent through alleged illness. Would the Dardanelles be open to British shipping?

"I said that His Majesty's Government would not tolerate that the Turkish fleet, as well as the Turkish army, should be in the hands of Germany, warning his Highness that the British fleet would not leave the Dardanelles until His Majesty's Government were satisfied that the Turkish Government had loyally carried out the condition laid down, and until British merchantmen could navigate Turkish waters without either delay or molestation. It was therefore obvious that if there was any idea of manning the Turkish fleet with German officers and men it must be given up… I [said] that I should not be satisfied with less than the actual departure of the German crews."[105]

SIR LOUIS MALLET, BRITISH AMBASSADOR, ISTANBUL.

AUSTRALIA

"The camp is very rough, but we shall soon get used to it, and then we shall be prepared for what we may have to put up with later. The G Company is composed of a fine set of fellows. I am proud to be in it…. When we were passing the Victoria markets yesterday on the march out, a butcher was standing out in front holding aloft a spear stuck through two German sausages. As we passed he called out, 'This is the way to do it, boys — two at a time. That's the Kaiser on top.' A little further on, the Governor, Sir Munro Ferguson, passed in a car. A wag in our section said, 'The Gov.'s eyesight seems to he fading. He didn't recognise me.' All the nap we had last night was on a waterproof sheet with one small blanket on the bare ground — no straw. Very few of us slept last night, but we'll soon come to it."[106]

PTE ALEXANDER KENNEDY, 5TH BATTALION, AIF, BROADMEADOWS CAMP.

26 AUGUST 1914

ENGLAND

The British way of war had traditionally involved utilising sea power to intervene on the margins of the main theatre of war. Royal Marines had been despatched to Ostend.

---

104   *Correspondence Respecting Events Leading to the Rupture of Relations with Turkey*, p. 11.
105   *ibid.*, pp. 10–11.
106   *Bendigo Advertiser* (Victoria), 25th August 1914.

> "I had a long visit from Winston and Kitchener and we summoned Edward Grey into our councils. They were bitten by an idea of Hankey's to dispatch a brigade of Marines, about 3,000, conveyed and escorted in battleships to Ostend, to land there and take possession of the town and scout about in the neighbourhood. This would please the Belgians and annoy and harass the Germans, who would certainly take it to be the pioneer of a larger force, and it would further be quite a safe operation as the Marines could at any moment re-embark. Grey and I commented, and the little force is probably at this moment disembarking at Ostend. Winston, I need not say, was full of ardour about his Marines and takes the whole adventure, of which the Cabinet only heard for the first time an hour ago, very seriously."[107]

HERBERT ASQUITH, MP, BRITISH PRIME MINISTER.

TURKEY

British diplomats were holding on to the hope that matters could improve.

> "Weber Pasha, who is in command at the Dardanelles, is said to be urging closing of the Straits. I have brought this to the notice of the Grand Vizier. His Highness most positively repudiated any such idea, and begged me to have patience, as this situation would not last, and he was gaining authority... Straits would be entirely closed, and, according to the German Ambassador, quite impossible to force, since Germans have taken special measures to make them impregnable.
>
> "To sum up, the situation is most unsatisfactory, though not actually desperate."[108]

SIR LOUIS MALLET, BRITISH AMBASSADOR, ISTANBUL.

27 AUGUST 1914

TURKEY

> "According to persistent rumours, the exit of the "Goeben" and "Breslau" into the Black Sea will take place before long... I had an interview to-day with the Grand Vizier... He gave his word that the "Goeben" would not leave for any point, and promised to again insist upon the prompt removal of all German crews from vessels. I believe the Grand Vizier is sincere, but his influence is greatly waning, and his final fall may come at any time. For that reason I freely admit the possibility of the departure of the "Goeben" into the Black Sea under German pressure, with crew part German, and flying the Turkish flag."[109]

MIKHAIL NIKOLAYEVICH DE GIERS, RUSSIAN AMBASSADOR, ISTANBUL.

---

107   Asquith, *Memories and Reflections*, vol. 2, pp. 28–29.
108   *Correspondence Respecting Events Leading to the Rupture of Relations with Turkey*, pp. 12–13.
109   Scott, *Diplomatic Documents Relating to the Outbreak of the European War*, p. 1401.

Dardanelles fort at the Narrows.

The continued failure to remove the German crews led some to consider the options for military and naval intervention.

"[T]o command situation properly at Dardanelles, requires also use of military force and point arises whether substantial enterprises should be attempted in quite a subsidiary theatre of war. Moreover military operations against Turks would be far easier in Persian Gulf or Syria where Turkish forces are almost negligible. Should decision be eventually taken for a fleet movement I need hardly impress that for local reasons there should be no mistake as to rapidity of execution and minimum risk of failure."[110]

LIT-COL. FREDERICK CUNLIFFE-OWEN, BRITISH MILITARY ATTACHÉ, ISTANBUL.

28 AUGUST 1914

TURKEY

"From information that has reached me, there is no doubt that in course of time the whole area of the Dardanelles, Constantinople, and the Bosphorus will become nothing more nor less than a sort of German enclave. Sailors recently arrived from Sophia will be sent to Straits forts and more will follow. This is over and above German military reservists already allotted to garrison those forts.

"I hear that, although Turks have not yet any ordnance of the more modern type for mounting in Straits defences,[111] it is very probable that consignment of guns will arrive in the near future from Germany and Austria through Constanza."[112]

SIR LOUIS MALLET, BRITISH AMBASSADOR, ISTANBUL.

29 AUGUST 1914

TURKEY

"During the past half of August, the efforts of Germany to draw Turkey into war and to excite Turkish public opinion against the Powers of the Entente have become more and more plainly apparent. The Wolff Agency is industriously circulating in Turkey

---

110 Foreign Office: Political Departments: General Correspondence from 1906–66, Turkey (War). TNA FO 371/2138.
111 The British were in possession of detailed information regarding the number, location and armament of the forts, not only from members of their own Naval Mission, but from a 1912 Russian report secured by Lieutenant John Fitzwilliams, Royal Horse Artillery. Later promoted Major, Fitzwilliams was killed in action in France on 30th August 1918.
112 *Correspondence Respecting Events Leading to the Rupture of Relations with Turkey*, p. 15.

a manifesto of the Sultan and a warlike order from the army of Enver Pasha, calling upon the army to wash away the shame of the Balkan war."[113]

MIKHAIL NIKOLAYEVICH DE GIERS, RUSSIAN AMBASSADOR, ISTANBUL.

SAMOA

After an ultimatum to the German governor at Apia was delivered by Rear-Admiral George Patey, RN, no resistance was offered to the allied forces.

> "As our troopship proceeded up the harbor we were greeted with a salute from the guns of the French flagship *Montcalm*, as our ship passed their band played "God Save the King" while ours replied with "The Marseillaise." But the climax was reached when the troopship came alongside the breastwork. The populace went mad with delight, and all that afternoon and evening troops could be seen with their hats on a line hauling up postcards, coins, cigars, fruit, confectionery, all sorts of keepsakes, and exchanging autographs...

> "I might state that it came as a great surprise to us, as also to the officers, that no resistance was offered, for we fully expected fighting. HMS *Australia* had its 13-inch guns trained on the town while we were landing."[114]

PTE THOMAS ALDERTON, SAMOAN EXPEDITIONARY FORCE.

30 AUGUST 1914

TURKEY

After weeks of pledges to remove the German crews, nothing had happened. But the British Ambassador was struggling to accept that the Turks would fight.

> "I and my colleagues still do not regard situation as hopeless, and are of opinion that we should go on as long as possible without provoking a rupture. I find it hard to believe that, when it comes to the point, Turks would declare war on Russia or on ourselves."[115]

SIR LOUIS MALLET, BRITISH AMBASSADOR, ISTANBUL.

AUSTRALIA

> "There have been quite a number of desertions from the various companies since our arrival in camp. Perhaps they found the work too hard; perhaps the daily reports from the front of bayonet charges, etc., may have had a dampening effect on their warlike

---

113 Scott, *Diplomatic Documents Relating to the Outbreak of the European War*, pp. 1402–1403.
114 *Marlborough Express* (New Zealand), 16th October 1914.
115 *Correspondence Respecting Events Leading to the Rupture of Relations with Turkey*, p. 16.

spirit. You see we have not yet been sworn in, so the authorities have actually no hold over us, but after this week those who remain will have to go right through with it. I'm remaining… Although no one is supposed to know where we are proceeding to, the 'heads' let drop remarks which make us certain that we are bound for England for further training before being drafted to the front."[116]

WALTER MORLEY, BELLEVUE CAMP.[117]

31 AUGUST 1914

ENGLAND

One Australian was looking forward to the arrival of his fellow countrymen.

"I am… as happy as a sand boy. I have good reason to be. You see, I am on a good ship. I really believe it is the best ship in the flotilla. Our officers are made of the real stuff, and the men have proved themselves to be A1. There is no doubt about it. It is the same throughout the Royal Navy. I am proud to be with them. I am anxiously awaiting the arrival of the Australian troops. I may have the opportunity of meeting some I know after this lot is over."[118]

AB ERNEST ABFALTER, RN, HMS *LOUIS*, HARWICH.[119]

NEW ZEALAND

The Press was thanked for its co-operation in not publishing sensitive military and naval information. The public was asked for its understanding.

"I am very grateful for the loyal way in which they have carried out our desires. I am quite aware that our journalists might have published things of great interest to their readers, but they refrained in the interests of the men and the mission. Our only object in keeping the matter secret was to avoid danger or loss or life. If it had been published to the world that the vessels had gone to Samoa, there is a chance that they would not have got there without resistance. The public should back the Press in helping Ministers in silence and secrecy when it is necessary in the interests of the lives of the men and the interests of the grave Empire mission we have had placed in our hands. We will not be secret unless there is need."[120]

HON. JAMES ALLEN, NEW ZEALAND DEFENCE MINISTER.

116   *Kalgoorlie Miner* (Western Australia), 6th May 1915.
117   The letter was dated "August 1914." Later posted to 11th Bn, AIF, Morley, a former tram conductor from Kalgoorlie, Western Australia, died of wounds aboard HMT *Lutzow* on 28th April 1915. He is commemorated on the Lone Pine Memorial.
118   *Blyth Agriculturalist* (South Australia), 30th October 1914.
119   HMS *Louis* was wrecked at Suvla, driven ashore in a storm on 1st November 1915. Its Captain, Lt-Commander Harold Dallas Adair-Hall was tried by court martial 3rd/4th December 1915 and found negligent and 'dismissed ship.'
120   *Oamaru Mail* (New Zealand), 1st September 1914.

The German Embassy, Pera, Istanbul.

SEPTEMBER 1914

"I hear that Germans are now dominant."
Sir Louis Mallet, British Ambassador, Istanbul

Constantinople – Ambassade allemande à Péra.

1 SEPTEMBER 1914

ENGLAND

Kaiser Wilhelm II once described the Suez Canal as the "jugular vein of the British Empire." Sensitive to this, care was also taken to reassure the Turks that defensive preparations there were not a prelude to more aggressive measures in future.

> "In order that there may be no room for misconception, you should inform Turkish Government that the Egyptian Government are taking measures to patrol Suez Canal on both banks and that this step is necessary to protect the safe and proper working of the Canal. You should add that no advance into Sinai, nor military operations in that region, are under contemplation."[1]
>
> SIR EDWARD GREY, MP, BRITISH FOREIGN SECRETARY.

TURKEY

Though the international situation was unclear to them, British residents in Istanbul detected growing German influence and fell back on some caricatured imagery of what could be expected of the 'Turk.'

> "We are practically cut off from the world here, and get tons of false news… Turkey has not yet declared war, but is acting as if she intended doing so, as the place (army particularly) is absolutely under the thumb of Germany. It would be an awful sin to throw this country into war, as the poor fellows are taken off the fields, and a rifle is put into their hands, and they are sent to be shot at. Moreover, they have to provide their own food; and I believe at the present moment that half the army is in a state of starvation…
>
> "You will have heard about the two German warships, the *Goeben* and *Breslau*, which escaped from the Mediterranean. Well, they dodged behind a French passenger boat so as to avoid the British gun, and then slipped into the Dardanelles and escaped, and were immediately sold to Turkey. Another of Germany's fly moves. The women here are feeling very nervous, as the Turks are overfond of a massacre: but there's no fear of that to-day, as they would have to deal with a people who hit back, and hit hard."[2]
>
> H. H. BELL, TELEPHONE ENGINEER, ISTANBUL.

NEW ZEALAND

As men came forward to volunteer across the British Commonwealth and Empire, the issue of Maori recruitment was raised in the Parliament in Wellington.

---

1   *Correspondence Respecting Events Leading to the Rupture of Relations with Turkey*, p. 17.
2   *Taranaki Daily News* (New Zealand), 17th November 1914.

> "I regret to say that although I have a good deal of sympathy with the suggestion, there has always been a sort of understanding that coloured races were not to be employed in any European war, but now we have the precedent of Indians being taken to Europe for the purpose of taking part in the present war. The Maoris are free citizens of the Empire, sharing all the privileges and benefits of British citizenship, and I do not think they should be denied the opportunity of fighting for the Empire. I shall be very glad to represent to his Excellency the Governor the suggestion made by [G. M. Thomas] the member for Dunedin North."[3]

WILLIAM MASSEY, NEW ZEALAND PRIME MINISTER.

2 SEPTEMBER 1914

ENGLAND

> "We had our regular Cabinet in his [Kitchener's] absence this morning, and the kind of thing that comes up day by day can easily be judged by this short synopsis of topics: naval air reconnoitring at Dunkirk; protection of London against bomb-throwing from Zeppelins; Greece and Turkey on the verge of war; proposed offer of financial help to Rumania and Servia; Japan: Can she help with her fleet or army, or both? Scheme for pledging State credit to deal with the discounting of post-moratorium bills, etc., etc., etc."[4]

HERBERT ASQUITH, MP, BRITISH PRIME MINISTER.

TURKEY

Advice was sought from London regarding the attitude of the British Government towards the German ships should they venture outside Turkish waters. The expected reply was received the following day.

> "I should be glad to learn whether British Admiral has instructions in case "Goeben" went into Mediterranean under Turkish flag. Should I tell Turkish Government that, so long as she has Germans on board, we shall regard her as a German ship and treat her as such, and that, before she goes out into Mediterranean, Admiral Limpus must be allowed to assure himself that there are no Germans on board?
>
> "I do not anticipate her going out, but should like to make it clear beforehand what our attitude would be in case she does so."[5]

SIR LOUIS MALLET, BRITISH AMBASSADOR, ISTANBUL.

---

3 *Evening Post* (New Zealand), 2nd September 1914.
4 Asquith, *Memories and Reflections*, vol. 2, p. 30.
5 *Correspondence Respecting Events Leading to the Rupture of Relations with Turkey*, p. 17.

3 SEPTEMBER 1914

ENGLAND

> "So long as German crews have not been sent away, "Goeben" will certainly be treated as a German ship if she comes out of the Straits. It was only on express condition that German crews would be sent away that we waived the demand, to which we were strictly entitled, that ship should be interned until the end of the war."[6]
>
> SIR EDWARD GREY, MP, BRITISH FOREIGN SECRETARY.

Churchill cast some bread upon the waters, espousing the merits of using Greek troops and British ships to take control of the Dardanelles. Not everyone was convinced.

> "W.S.C. [Churchill] elaborated a plan for combining with the Greeks to seize Gallipoli, and thus dominate Constantinople before the Germans have armed the forts."[7]
>
> CHARLES HOBHOUSE, MP, POSTMASTER-GENERAL.

> "[I]t ought to be clearly understood that an attack upon the Gallipoli Peninsula from the sea side (outside the Straits) is likely to prove an extremely difficult operation of war."[8]
>
> GENERAL CHARLES CALLWELL, DIRECTOR, MILITARY OPERATIONS, WAR OFFICE.

New volunteers' horizons were more limited; their concerns more immediate.

> "The whole business had been done in a rush of exaltation that didn't allow me to think. But when I stepped out into the crowded streets with that two shillings rattling in my pocket I felt a very sober man. I knew nothing whatever of soldiering. I hardly even knew a corporal from a private or a rifle from a ramrod, and here I was Trooper A. H. Gibbs, 9th Lancers, with the sullen, rumble of heavy guns just across the Channel — growing louder…

> "Bad smells, bad beer, bad women, bad language!… We lay at night side by side on adjoining bunks, fifty of us in a room. They had spent their two days' pay on beer, bad beer. The weather was hot. Most of them were stark naked. I'd had a bath that morning. They hadn't."[9]
>
> PTE ARTHUR GIBBS, 9TH LANCERS, WOOLWICH.

---

[6] *Correspondence Respecting Events Leading to the Rupture of Relations with Turkey*, p. 17.
[7] David, Edward (Ed.), *Inside Asquith's Cabinet. From the Diaries of Charles Hobhouse*, p. 187, St. Martin's Press (New York) 1978.
[8] *Cambridge Daily News*, 20th March 1917.
[9] Gibbs, Arthur Hamilton, *Gun Fodder. The Diary of Four Years of War*, p. 7, Little, Brown and Co. (Boston) 1919.

4 SEPTEMBER 1914

ENGLAND

After airing the possibility with Cabinet colleagues, Churchill contacted Vice-Admiral Mark Kerr, commanding Royal Hellenic Navy, for his views on the scope for Greek participation in an attack on the Dardanelles.

> "If you are addressed by the Greek Government, you are authorised to take part in this discussion on behalf of the Admiralty, whose views, on general lines, are as follows: —
>
> "In principle the Admiralty would proposed to reinforce the Greek Fleet by a squadron and a flotilla of strength sufficient to give a decisive and unquestionable superiority over the Turkish and German vessels. They would proposed that you should hoist your flag in the British battle-cruiser *Indomitable*, and that the whole command of the combined fleets should be vested in you. They would reinforce you to any extent and with any vessels that circumstances may render necessary. The method of attacking Turkey, which is both self-evident and right, is by an immediate blow at the heart. To strike such a blow it would be necessary for the Greek Army, under superior sea predominance, to seize the Gallipoli Peninsula and thus to open the Dardanelles, so that the Anglo-Greek Fleet should be admitted to the Sea of Marmora, where the Turko-German ships can be brought to action and sunk, and whence the whole situation can be dominated in combination with the Russian Black Sea Fleet and the Russian military forces."[10]

WINSTON CHURCHILL, MP, FIRST LORD OF THE ADMIRALTY.

5 SEPTEMBER 1914

NORTH SEA

As the *Amphion*'s fate demonstrated the danger posed by mines, so the German *U-21* showed what submarines were capable of.

> "I had just gone up into the steam cutter with a mate when I caught sight of a torpedo coming straight for the ship. The torpedo was then about fifty yards away. The order was given for the engines to be stopped and reversed, but before this could be done the torpedo struck the bows of the ship under the bridge. There was a terrific explosion, which caused the magazine to blow up, and immediately all the fore part of the vessel appeared in flames."[11]

STO. 1 WILLIAM HUGHES, RN, HMS *PATHFINDER*.

---

10 Winston Churchill quoted in Kerr, Admiral Mark, *The Navy in My Time*, pp. 181–182, Rich & Cowan Ltd. (London) 1933.
11 *Western Mail*, 24th September 1914.

"I saw a flash, and the ship seemed to lift right out of the water. Down went the mast and forward funnel and forward part of the ship… I scrambled to the quarter deck which was littered with mangled corpses, and looked about for something to cling to… It was now every man for himself, and I at once pulled off my boots, coat, and trousers, and over I went. I think I broke all swimming records trying to put as much space between myself and the ship, being afraid of suction. Turning round, the last I saw of the ship was about 50 yards away. The after-end was sticking upright in the air about a hundred feet. Then she gradually reeled over towards me and sank… It was all over in about five minutes from the start."[12]

ERA3 JAMES EDWARD HEATH, RN, HMS *PATHFINDER*.

TURKEY

Mallet could be accused of hearing what he wished to this day and reporting it the next.

"I have to-day gone over the whole ground with the Minister of the Interior [Talaat Pasha], who seems more inclined to be reasonable. I think there is an improvement in the situation.

"Minister quite understands that "Goeben" will be treated as a German ship if she goes out. They assure me that Turkish fleet will not leave the Dardanelles on any account."[13]

SIR LOUIS MALLET, BRITISH AMBASSADOR, ISTANBUL.

6 SEPTEMBER 1914

TURKEY

"I had a long conversation with Minister of Interior yesterday. He assured me that there was no question of Turkey going to war, but when I pressed him about Greece, he admitted that, unless Turkish Government could get some real satisfaction about islands, by which I gathered he would be satisfied with regime at Samos, they would go to war on land… It was not likely that a war with Greece would long remain localized, and it seemed extremely probable that Turkey in the long run would find herself up against the Triple Entente. I told him His Majesty's Government regarded Turkish Fleet as annex of German Fleet, and that if it went out into the Aegean we should sink it. He quite realized this, and said that it had no intention of leaving Dardanelles."[14]

SIR LOUIS MALLET, BRITISH AMBASSADOR, ISTANBUL.

---

12 *Hull Daily Mail*, 19th September 1914.
13 *Correspondence Respecting Events Leading to the Rupture of Relations with Turkey*, p. 18.
14 Sir Louis Mallet quoted in Viscount Grey of Fallodon, K.G., *Twenty Five Years 1892–1916*, Volume 2, p. 182, Frederick A. Stokes Company (New York) 1925.

ENGLAND

Churchill received the news that some German naval codes had been captured by the Russians. These were to be sent to London.

> "On September 6 the Russian Naval Attaché came to see me. He had received a message from Petrograd telling him what had happened, and that the Russian Admiralty with the aid of the cypher and signal books had been able to decode portions at least of the German naval messages. The Russians felt that as the leading naval Power, the British Admiralty ought to have these books and charts. If we would send a vessel to Alexandrov, the Russian officers in charge of the books would bring them to England."[15]
>
> WINSTON CHURCHILL, MP, FIRST LORD OF THE ADMIRALTY.

7 SEPTEMBER 1914

ENGLAND

Although the Admiralty had originally accepted the explanations given for the failure to intercept *Goeben* and *Breslau*, that changed.

> "The escape of the *Goeben* must ever remain a shameful episode in the war. The flag officer who is responsible for this failure cannot be entrusted with any further command afloat and his continuance in such command constitutes a danger to the State.
>
> "I therefore propose that Rear-Admiral Troubridge be directed to return to England forthwith… Admiral Sir Berkeley Milne should be directed to attend as a witness."[16]
>
> ADMIRAL PRINCE LOUIS VON BATTENBERG, RN, FIRST SEA LORD.

GREECE

Plans had been drawn up by the Greeks for an assault upon the Dardanelles and the Gallipoli peninsula during the Balkan wars. Vice-Admiral Kerr was given the go ahead to share these with the British but King Constantine wanted to keep Greece out of the war.

> "I had a long conversation with His Majesty this morning on the subject of a telegram from the First Lord of the Admiralty, dated London, September 4th.
>
> "His Majesty stated that unless attacked by Turkey, he did not wish to go to war. He said that Germany wished him to remain neutral in any case that might arise, but although he wished to avoid quarrelling with the German Emperor, he had replied

---

15   Churchill, *The World Crisis 1911–14*, p. 462.
16   Battenberg quoted in Lumby, *Policy and Operations in the Mediterranean 1912–14*, p. 237.

that in the event of Bulgaria's attacking Servia, he was bound by a Treaty to assist, and this he would do…

"His Majesty promised that the General Staff would supply me with the results of their deliberations as to the point of attack in the event of Turkey becoming aggressive, and also of the quantity of transport available, which information I hope to get in the course of two or three days."[17]

VICE-ADMIRAL MARK KERR, RN, COMMANDING ROYAL HELLENIC NAVY.

8 SEPTEMBER 1914

ENGLAND

Others less senior were more anxious to get involved, even if they began to wonder what they had gotten themselves into.

"I left the office of *The Scout*, 28 Maiden Lane, W.C., on September 8th, 1914, took leave of the editor and the staff, said farewell to my little camp in the beech-woods of Buckinghamshire and to my woodcraft scouts, bade good-bye to my father, and went off to enlist in the Royal Army Medical Corps.

"I made my way to the Marylebone recruiting office, and after waiting about two hours, I went at last upstairs and "stripped out" with a lot of other men for the medical examination.

"The smell of human sweat was overpowering in the little ante-room. Some of the men had hearts and anchors and ships and dancing-girls tattooed in blue on their chests and arms. Some were skinny and others too fat. Very few looked fit. I remarked upon the shyness they suffered in walking about naked."[18]

JOHN HARGRAVE.

By this time, the Admiralty considered the position of the British Naval Mission in Istanbul to be untenable. The Foreign Office asked how that course of action would be interpreted there.

"Before any decision respecting the recall of the mission is taken by His Majesty's Government, I wish to have your views on the subject. I am reluctant to take any step, however justified it may be, that would precipitate unfavourable developments, as long as there is a reasonable chance of avoiding them. What effect do you consider that withdrawal of mission would have upon the political situation?

"The Admiralty are of opinion that the position of the mission may become unsafe and that it is already undignified. They therefore wish it to be recalled and attached to the

---

17 Kerr, Admiral Mark, *The Navy in My Time*, pp. 184–185, Rich & Cowan Ltd. (London) 1933.
18 Hargrave, John, *At Suvla Bay. Being the Notes and Sketches of Scenes, Characters and Adventures of the Dardanelles Campaign*, p. 1, Constable & Co. Ltd. (London) 1916. Hargrave joined "A" Section, 32nd Field Ambulance, RAMC, 10th (Irish) Division.

embassy until you can arrange a safe passage home for Admiral Limpus and the other officers. There is clearly ample justification for the view taken by the Admiralty."[19]

SIR EDWARD GREY, MP, BRITISH FOREIGN SECRETARY.

TURKEY

"In many respects the situation seems to show improvement, but unless His Majesty's Government wish the mission to remain indefinitely it seems to me that the present would be a suitable moment to withdraw it. The Turks could not regard this step as a grievance as it is obviously justified by their conduct. The mission are at present treated as non-existent, and their position is consequently both false and invidious. German hold on the navy is becoming stronger daily, and there is no sign of German crews leaving. As a matter of fact, far from being disadvantageous to us, this is becoming embarrassing to the Turkish Government, who are at last beginning to realise that the Germans are not an unmixed blessing. Great discontent reigns among Turkish naval officers, so Admiral Limpus tells me, as they dislike German officers, and they even hint that they would rather mutiny than serve under them."[20]

SIR LOUIS MALLET, BRITISH AMBASSADOR, ISTANBUL.

AUSTRALIA

New recruits told their families how they were getting on in their new surroundings.

"I thought you would perhaps like to hear how the Tat. [Tatura] boys are getting on. Well, we all passed the doctor, and are in camp together, and it is really grand. You cannot realise what it is like without seeing it. There are about 7000 men in camp, more or less advanced, and they are coming in every day. Still, when we landed, I was offered a berth in the crack regiment. Only two men were required to fill it up, but I wanted to be with the boys, so we all got in together. It is funny getting into camp ways. Eleven of us occupy one tent; we have good straw mattresses, one for each man, one blanket, and a waterproof cover, so we do not fare too badly. I would not miss it for the world. We get up at six in the morning (they give us five minutes to dress), and then roll call. After that hot tomato soup, then we fall in and go on parade (all the infantry). We are lined up at 9, and go out for three hours' drill, returning at 12 for dinner and two hours' rest. We then go out for three more hours, returning at 5. We are free till 8 o'clock, at which time we must be in bed; lights out at 10. But you can't sleep too much on account of the noise from the other tents, some of them going strong till morning. You can tell the Tat. folk they are missing the chance of a lifetime."[21]

PTE JOHN WADESON, 7TH BATTALION, AIF, BROADMEADOWS CAMP.[22]

19   *Correspondence Respecting Events Leading to the Rupture of Relations with Turkey*, pp. 21–22.
20   ibid., p. 22.
21   *Kyabram Free Press and Rodney and Deakin Shire Advocate* (Victoria), 8th September 1914.
22   Wadeson, a former grocer from Tatura, Victoria, was wounded on 8th/9th August 1915.

9 SEPTEMBER 1914

ENGLAND

As the recently enlisted were just beginning their military service, older Territorials were about to leave for Egypt.

> "I reported at Fulham. More hours of waiting. I discovered an old postman who had also enlisted in the R.A.M.C., and as he "knew the ropes" I stuck to him like a leech. In the afternoon an old recruiting sergeant with a husky voice fell us in, and we marched, a mob of civilians, through the London streets to the railway station. Although this was quite a short distance, the sergeant fell us out near a public-house, and he and a lot more disappeared inside.
>
> "What a motley crowd we were: clerks in bowler hats; "knuts" in brown suits, brown ties, brown shoes, and a horse-shoe tie-pin; tramp-like looking men in rags and tatters and smelling of dirt and beer and rank twist."[23]
>
> JOHN HARGRAVE, RAMC.

> "We left Turton, and had a 16 hours' railway journey, nothing to eat, but only seven in a compartment. We passed through Todmorden, and eventually landed at Southampton. There we were shipped on board T.S.S. 'Saturnia,' for Egypt. There were 2,000 troops on board, besides officers and crew, about 200 men, so you can see we were packed like herrings."[24]
>
> PTE HAROLD SOUTHWELL, 1/6TH BN LANCASHIRE FUSILIERS, HMT *SATURNIA*.

GREECE

While the Greek King wanted to keep out of the war completely, the Greek General Staff had conditions of their own before considering taking part in any Dardanelles expedition. Kerr was quite confident a joint military and naval attack could succeed.

> "I have had a consultation with the Greek General Staff on the subject of the telegram received from the First Lord of the Admiralty. They are of opinion that if Bulgaria does not attack Greece, the force at the disposal of Greece is sufficient to take Gallipoli. I agree with this. They will not, however, trust Bulgaria unless she also attacks Turkey at the same time with her whole force. An undertaking on her part to remain neutral would not be regarded as sufficient guarantee.
>
> "If the above conditions are fulfilled, the plan for taking the Straits is ready. Greece has sufficient transports to convey the troops. The assistance of a British squadron of two battle cruisers, one armoured cruiser, three light cruisers, and a flotilla of destroyers and mine-sweepers will be needed. This was the plan originally devised

---

23   Hargrave, *At Suvla Bay*, pp. 4–5.
24   *Todmorden Advertiser and Hebden Bridge Newsletter*, 16th October 1914.

by the General Staff and myself without any assistance from outside. The operation has now, however, become a greater one, since Turkey has obtained German ships and mobilised her forces."[25]

VICE-ADMIRAL MARK KERR, RN, COMMANDING ROYAL HELLENIC NAVY.

TURKEY

The British were told that the Capitulations which provided a degree of autonomy to British subjects and other foreign nationals within the Ottoman Empire were to be abolished. The Turks were drifting away from the allied camp.

"Grand Vizier admitted this morning that the Turkish Government were going to abolish Capitulations... The Capitulations and conventions were not a unilateral agreement; we had on a former occasion informed the Turkish Government that we were willing to consider any request they might put forward in a generous spirit, but I did not imagine that my Government would acquiesce in their total abolition by a stroke of the pen. We were now under martial law. Did he expect us to allow British subjects to be judged by court-martial, especially so long as army was in hands of Germans?"[26]

SIR LOUIS MALLET, BRITISH AMBASSADOR, ISTANBUL.

10 SEPTEMBER 1914

TURKEY

Russian sources continued to give a clearer appreciation of what was going on than their British counterparts.

"A person who is in close touch with the Ottoman Ministers expresses the opinion that Turkey is bound to Germany by an agreement, which is supposed to have been effected mainly at the instance of Enver Pasha. This agreement does not, however, bind Turkey to an immediate declaration of war against us, which explains the position they are now taking, notwithstanding all the efforts made by the Germans to hasten matters to a conclusion...

"According to a communication from the Dardanelles, active strengthening of the forts there is going on under the direction of Weber Pasha and a newly arrived German high officer."[27]

MIKHAIL NIKOLAYEVICH DE GIERS, RUSSIAN AMBASSADOR, ISTANBUL.

---

25   Kerr, Admiral Mark, *The Navy in My Time*, pp. 183–184, Rich & Cowan Ltd. (London) 1933.
26   *Correspondence Respecting Events Leading to the Rupture of Relations with Turkey*, p. 22.
27   Scott, *Diplomatic Documents Relating to the Outbreak of the European War*, p. 1409.

Meanwhile, frustration was growing within the British Embassy at the Turks' unilateral actions regarding the Capitulations. Or was it an awareness of their own impotence?

> "Note abolishing all the Capitulations was received last night. All my colleagues, including German and Austrian Ambassadors, have to-day addressed identic[al] notes to the Sublime Porte stating that, while communicating to our respective Governments note respecting abolition of Capitulations, we must point out that capitulatory régime is not an autonomous institution of the Empire, but the resultant of international treaties, diplomatic agreements, and contractual acts of different kinds. It cannot be abolished in any part, a fortiori wholly, without consent of contracting parties. Therefore, in, the absence of understanding arrived at before 1st October between Ottoman Government and our respective Governments, we cannot recognise executory force after that date of a unilateral decision of Sublime Porte."[28]

SIR LOUIS MALLET, BRITISH AMBASSADOR, ISTANBUL.

INDIAN OCEAN

The *Emden*, detached from the German East Asia Squadron, stopped a Greek ship, the SS *Pontoporos*, in the Bay of Bengal. Being neutral, the captain might have expected to be safe. But what was he carrying?

> "He tried to stall me, and explained that he was bound from Calcutta to Karachi with a cargo of coal... Reluctantly he brought [the charter paper] forth. There I found, as I had suspected from the first, that the cargo was destined for the British Government...
>
> "Coal was the commodity we wanted above all... The coal was of the Indian variety — much too soft; it clogged our boilers, and we could not make our accustomed speed with it."[29]

OBERLEUTNANT ZUR SEE DER RESERVE JULIUS LAUTERBACH, SMS *EMDEN*.

Taken as a prize, the *Pontoporos*' chief engineer was questioned about its workings.

> "I was then escorted down to the engine-room and stokeholes, and, for the space of two hours, explained the engines. A German engineer and a guard were left in the engine-room and a German officer took charge of the boat. I was then taken to the *Emden*, and, as I stepped on deck, the German chief engineer came forward and shook hands with me, saying 'Mr. Chief, you will be treated like a gentleman. We can never tell, but we may be prisoners next.' All the crew raised their caps to me, and the skipper came down from the bridge and shook hands. He also assured me that I should be well treated. And we were treated like gentlemen; rank gave cabin room to rank. My own brothers could not have treated us better."[30]

BENJAMIN FORBISTER, CHIEF ENGINEER, SS *PONTOPOROS*.

---

28   *Correspondence Respecting Events Leading to the Rupture of Relations with Turkey*, p. 23.
29   Julius Lauterbach quoted in Thomas, Lowell, *Lauterbach of the China Sea: The Escapes and Adventures of a Seagoing Falstaff*, p. 45, Doubleday, Doran & Company (New York) 1930.
30   *Liverpool Journal of Commerce*, 19th October 1914.

11 SEPTEMBER 1914

ENGLAND

An American observer recorded some of the temporary arrangements made to accommodate units mobilising in Britain. There were more men than uniforms.

> "There appeared to be four battalions of the Essex Regt. (Territorials) and one battalion at least of the Bedford Regt. (Territorials) in addition to the 3, Cos. of R.G.A., and 2 Cos. Engineers above referred to, in Harwich and vicinity. All positions that were reserved for military purposes were marked "closed" and strongly guarded. There was also a strong guard round the Parkeston Quay reservation. The troops were mostly quartered in buildings but guard detachments, etc. were in tents, and all kinds of old white canvas in use as tentage. There was a large hospital at Shotley. The large Great Eastern Hotel facing Harwich Pier had also been converted into a hospital. The troops at Harwich were not busy (Saturday afternoon) and many were on liberty in Harwich and Dovercourt, and playing football and other games. There appeared to be nearly 5,000 recruits in Harwich and Dovercourt, and they were all in civilian clothes."[31]

MAJOR THOMAS C. TREADWELL, USMC, AMERICAN EMBASSY, LONDON.

TURKEY

It was reported that the abolition of capitulations was greeted with public celebrations in Istanbul. Many Turks regarded the Capitulations as colonial impositions by foreign powers.

> "Street processions and public rejoicings on the occasion of the abolition of the capitulations were continued last night. The principal arteries of the city were decorated with flags and illuminated. A mass meeting was held in Stamboul, and in the evening a banquet of 200 covers, at which Cabinet Ministers were present, was given at the City Prefecture to celebrate the proclamation of national independence."[32]

REUTER'S.

AUSTRALIA

The growing numbers of Australians being recruited was reported.

> "Some little misapprehension appears to exist as to the number of additional troops it is desired to enlist for service in Europe. The number required for the first expeditionary force under the command of General Bridges, viz., 20,000, has been completed. Some little time since, however, the Ministry decided that enrolment should go on continuously, with a view to supplying wastage in the above force and other

---

31 Naval Attache's Reports, Office of Naval Intelligence, Unpublished Manuscript, US Naval War College, September 1914.
32 *Coventry Evening Telegraph*, 12th September 1914.

purposes — to form new units from time to time. The men now being enrolled will go to form — (a) The first, reinforcements for the main force; (b) a line of communication unit; and (c) a brigade of infantry and brigade of light horse, with necessary attached units. These will involve about 10,000 men, and it is thought, acting on the advice of military officers, that it would be better to despatch contingents of about this number from time to time, rather than wait until a larger force has been organised. It is anticipated that enrolment for this will be completed and embarkation arrangements made by the end of October. This does not mean, however, that Australia's effort will close with the despatch of the second contingent now being organised. As stated, enlistment will go on continually, and subject to transport arrangements, units will be despatched as they are completed. This process can, of course, be repeated as long as the necessity exists, and the Commonwealth Government approves."[33]

EDWARD MILLEN, AUSTRALIAN MINISTER FOR DEFENCE.

12 SEPTEMBER 1914

INDIAN OCEAN

The *Emden* continued its operations in the Bay of Bengal. It had recently taken control of a neutral ship found to be transporting an allied cargo. How would it deal with an enemy vessel but carrying merchandise bound for a neutral nation?

"We were all startled at 11.30 p.m. by a terrific shot, which was placed across the bows of our ship by the German cruiser. Naturally, we were all out on deck like lightning, and thought it would be a Britisher or a Frenchman. Anyhow she flashed out with her Morse lamp, "Do not work your wireless," and then "Stop your engines." But our curiosity was soon satisfied when the German boat got alongside and about 30 German sailors, including about ten officers, stepped aboard. I was at the top of the gangway, when the lieutenant [asked]: "Where's the wireless cabin?" Of course I then guessed what would happen. They soon put the apparatus out of order, and demanded all my books and searched and ransacked the place. They hauled down the aerial, cut it in various pieces, and pitched it overboard…

"The Germans gave us orders to pack our boxes, or rather a small bag… But suddenly the commander of the *Emden* changed his mind. After the ships had been inspected, for reasons which we can only surmise he said he would not sink her, but put all the crews of the ships he had already sunk on one ship. One thing, we had the captain's wife and kiddy aboard, and then again we had an American cargo, and also plenty of space to accommodate the crews."[34]

WIRELESS OPERATOR WILLIAM WESELBY, SS *KABINGA*.[35]

33   *Bendigo Advertiser* (Victoria), 11th September 1914.
34   *Leicester Daily Mercury*, 2nd November 1914.
35   The *Emden*'s men did not find all the spare components on the *Kabinga*. Weselby was later praised by the Marconi International Marine Communication Company for reconstructing a wireless set and re-establishing contact with a coastal station.

German commerce raider *Emden* with captured Greek merchantman *Pontoporos*, employed as a collier, photographed from *Kabinga*.

"Late that night we came upon the S.S. [*Kabinga*]… She was laden with a cargo of jute and bound for New York under charter to an American firm. The captain and I were in his chart room looking over the ship's papers. Suddenly a woman in a night dress with a little boy by her side appeared in the doorway. It was the captain's wife and little son. She seemed terribly frightened…

"The British with their atrocity propaganda led many of our enforced guests to suspect us of everything that is unholy… I communicated to Captain Mueller the fact that the boat was under charter to an American firm, and he ordered us not to sink her."[36]

OBERLEUTNANT ZUR SEE DER RESERVE JULIUS LAUTERBACH, SMS EMDEN.

AUSTRALIA

There were casualties long before any shot was fired in anger even thousands of miles away from the battlefields.

"Bill Edgar, Ben Hayes, and Bill Plunkett were coming along the road, back to camp, when a 'speed maniac' came along in a motor at a terrific pace. The three men dodged on to the side of the road to get out of the way, but the maniac in the motor went on the same side with the result that he sent Bill Edgar[37] and Ben Hayes sprawling. Plunkett just missed it. Bill and Ben had to be taken to the hospital. Edgar got his shoulder dislocated and a lot bruises on his head and face. Ben Hayes had his leg hurt and received other bruises. Plunkett[38] reckoned the reason that he missed the car, was because he was too narrow for it to run into him."[39]

PTE HENRY BUSWELL, 11TH BATTALION, AIF, BELLEVUE CAMP.

36   Julius Lauterbach quoted in Thomas, *Lauterbach of the China Sea*, p. 47.
37   Possibly Pte Wolverton Edgar, 11th Bn, AIF, a lumper from Bunbury, Western Australia, originally, Preston, Lancashire, enlisted on 10th September 1914. A time-expired veteran of the East Surrey Regt, he was killed in action on 25th April 1915. Buried in Plugge's Plateau Cemetery, Anzac, he was the 36 year-old brother of John Edgar, of Great Billing, Hampshire.
38   Pte William Plunkett, MG Section, 11th Bn, AIF. The labourer, originally from Liverpool, England, enlisted on 10th September 1914; wounded on at Leane's Trench, 31st July 1915; promoted Sergeant on 20th January 1917; awarded the DCM (*London Gazette*, 13th February 1917); and was killed in action on 16th April 1917. He is commemorated on the Villers-Bretonneux Memorial, France.
39   *Southern Times* (Bunbury, Western Australia), 15th September 1914.

13 SEPTEMBER 1914

TURKEY

As tales of German influence grew, the British decided to withdraw their naval mission. The Russians knew where power lay and where it did not.

> "I hear that Germans are now dominant at Alexandretta, and secretly suggest and control everything. From 7th September to morning of 12th September, 24 mountain guns, 400 horses and mules, 500 artillery troops belonging to service of 6th Army Corps, and large quantity of ammunition passed through Alexandretta, proceeding by railway to Constantinople."[40]
>
> SIR LOUIS MALLET, BRITISH AMBASSADOR, ISTANBUL.

> "I warned the Grand Vizier that the appearance in the Black Sea of the "Goeben" and "Breslau" might lead to complications, the more so because the German officers on board those vessels would try to bring about such complications in order to draw Turkey into a war with us. The Grand Vizier answered me to the effect that he had no information concerning the departure of the vessels, and he did not see any reason for sending them into the Black Sea. I believe the Grand Vizier will oppose the departure of the "Goeben" and "Breslau" into the Black Sea, but, unfortunately, his voice has no longer any decisive significance…
>
> "To-day the British Ambassador informed the Grand Vizier that England had recalled Admiral Limpus and all British officers in view of the impossible situation created for the British naval commission in Turkey."[41]
>
> MIKHAIL NIKOLAYEVICH DE GIERS, RUSSIAN AMBASSADOR, ISTANBUL.

14 SEPTEMBER 1914

TURKEY

The ending of Capitulations continued to be celebrated locally and protested internationally.

> "[A]ll the Powers are protesting against Turkey's abolition of capitulations, as expressed in the sheltering of the warships *Goeben* and *Breslau* and various German forces. The official German version declares that the Triple Entente offered to consent to the abolition if the Porte would remain neutral. The Porte replied that neutrality could not be bought, and thereupon decreed abolition."[42]
>
> REUTER'S.

---

40   *Correspondence Respecting Events Leading to the Rupture of Relations with Turkey*, p. 24.
41   Scott, *Diplomatic Documents Relating to the Outbreak of the European War*, pp. 1411–1412.
42   *Hamilton Spectator* (Victoria), 14th September 1914.

"An agreement in which any group of States is concerned cannot be terminated at a fortnight's notice on the decision of one of those States alone. We are confronted once again with the problem of the "scrap of paper"; are treaties binding, or are they to be torn up at will? I note that on this occasion Germany and Austria are prepared to uphold the permanence of their character, though it may be doubted whether their protest will carry as much conviction as that of Great Britain, who has shown herself ready to sacrifice her men and her millions in recognition of her treaty obligations."[43]

GERTRUDE BELL.

SAMOA

The appearance of the German East Asia Squadron caused alarm at Samoa. In London, Wellington, Sydney and elsewhere too, no doubt.

"[W]e had a visit from two German warships, and were fully expecting some fighting. After having the guns trained on the town for about an hour they left without firing a shot."[44]

PTE THOMAS ALDERTON, SAMOAN EXPEDITIONARY FORCE.

15 SEPTEMBER 1914

ENGLAND

There was yet more evidence of the shortage of everything required to transform Kitchener's volunteers into a 'New Army'. It would be many months before these men would be ready to go to war.

"On the afternoon of the 15th they were drilling everywhere. At least 30,000 men must have been drilling at one time. On Queen's Parade, the drill ground in the town [Aldershot], more than 5,000 men were drilling in detachments. Drills were seen in extended order, company drill, manual of arms, school of the recruit, entrenching, signals, etc., but very few of the recruits were drilling with arms. There seemed to be shortage of non-commissioned officers to conduct those drills, for many of the drill musters appeared to have larger detachments than they could well handle, while others did not seem to be competent."[45]

MAJOR THOMAS C. TREADWELL, USMC, AMERICAN EMBASSY, LONDON.

---

43   *Yorkshire Post and Leeds Intelligencer*, 21st September 1914.
44   *Marlborough Express* (New Zealand), 16th October 1914.
45   Naval Attache's Reports, Office of Naval Intelligence, Unpublished Manuscript, US Naval War College, September 1914.

TURKEY

As British diplomats held onto the hope that the movement towards war was not inevitable, it was clear that they understood how little influence they had.

> "Fleet is now entirely in German hands, and Minister of Marine is powerless. Germans consider that Dardanelles are now impassable, and they are impressing this upon military authorities. It is said that, if the Turkish fleet moved into the Black Sea, Straits would be entirely closed by additional mines, which have just been sent there on the "Nilufer."
>
> "Though I do not say that this coup will actually come off, danger is undoubtedly greater since news has been received of the recent successes of the allies, as the Germans are all the more anxious to create a diversion. My impression is that majority of the Cabinet and the Grand Vizier himself are entirely opposed to any such adventure, and that they are doing their utmost to prevent it; but they are finding out, though they will not admit it, that they are powerless to stop matters."[46]
>
> SIR LOUIS MALLET, BRITISH AMBASSADOR, ISTANBUL.

AUSTRALIA

But some new recruits, even those who had been to war previously, just wanted to get started.

> "Well, at present our chief worry is that the Germans will be belted out before we get a chance to show our mettle, for the Allies are (according to all reports) giving them a rough time. Every man in the camp is anxious to have a smack at the Germans. We may not be as well drilled at the enemy, but taking the shooting we have done, as a guide, we can promise the party of Germans we bump up against a pretty hot time for our shooting in the majority of cases was deadly. The South West members of the Battalion are deadly in their shooting. Some of the lads surprised the old hands; Bullocky Blythe,[47] and Roy Earl[48] in particular Tommy Stokes[49] is a fine shot too. We have a lot of Rifle Club, and old kangaroo shots in the Battalion, and they are all experts with the rifle. All the boys from around Bunbury can hold up their end of the stick…

---

46   *Correspondence Respecting Events Leading to the Rupture of Relations with Turkey*, p. 24.
47   Brothers, Privates Drummond James Blythe and Francis Albert Blythe enlisted together on 7th September 1914 and posted to 11th Bn, AIF. Drummond died of wounds aboard HMHS *Sicilia* on 6th June 1915. The former telephone lineman is commemorated on the Lone Pine Memorial. Francis, a former salesman, died in England on 30th July 1916 from wounds received at Pozieres. He is buried Chicester Cemetery, Sussex. They were the son of James and Eliza Blythe, of Spencer Street, Bunbury, Western Australia.
48   A farmer from Bunbury, Western Australia, Pte Roy Earl, 11th Bn, AIF, was wounded on 25th April 1915. He later rose to the rank of Captain with 51st Bn and decorated, receiving the Military Cross and Bar and the Military Medal.
49   Pte Thomas Stokes, an engine cleaner, was posted to 11th Bn, AIF. He was killed in action on 2nd May 1915. Buried in Shell Green Cemetery, he was the 21 year-old son of Thomas and Jane Stokes, of Bunbury, Western Australia.

"We had a sham fight on Friday, and the men showed they had a good grasp of war tactics, thanks to previous training and the great amount of energy that all officers have put into present training, as they are a fine lot…

"After the sham fight, we scouts went up to the position of the enemy was supposed to have held to examine the effects of the shell fire, and found that the gunners had done some fine shooting. They have fine guns in the eighteen pounders, they are a wonderful improvement to the old fifteen pounder."[50]

PTE HENRY BUSWELL, 11TH BATTALION, AIF, BELLEVUE CAMP.[51]

16 SEPTEMBER 1914

ENGLAND

The continuing activities of the German East Asia Squadron, particularly the detached *Emden*, was having an impact on British plans.

"Situation changed by appearance of "Scharnhorst" and "Gneisenau" at Samoa on 14th September and "Emden" in Bay of Bengal. "Australia" and "Montcalm" to cover "Encounter" and Expeditionary Force from attack and then search for the two cruisers; "Melbourne" to be used at Rear-Admiral's discretion; "Sydney" to return for convoy of Australian troops to Aden."

"Hampshire" and "Yarmouth" to sink "Emden". "Minotaur" to arrive at Fremantle by 4th October for Australian convoy, one Japanese cruiser to accompany "Minotaur"."[52]

THE ADMIRALTY.

ATLANTIC OCEAN

On their way to Egypt, the experience of life aboard a troopship was new to most but brought back memories to others who had seen it all before.

"I expect you will welcome a line or two from any of your local Territorials. As you are well aware, we left Littleboro' at 5.30 p.m. on Sept. 9th en route for Southampton. We were received with the greatest enthusiasm all through England, especially in Yorkshire, particularly at Brighouse, Elland and Hebden Bridge. Our first stop was at midnight at Kettering, our next in the early hours of the morning at Sittingbourne. We got to Southampton at 7-30 p.m., and the embarkation officer had us on our troopship 15 minutes later. I sailed from here 13 years ago to South Africa, but the sight at this port now will live long in the memory of every man belonging the East

---

50   *Southern Times* (Bunbury, Western Australia), 15th September 1914.
51   A veteran of the Boer War, Pte, later Sgt Henry Buswell, MM, was killed in action at Pozieres on 23rd July 1916. The former lumper, also from Bunbury, is commemorated on the Villers-Bretonneux Memorial, France.
52   Typescripts; comprise questions to be put to Sir Brudenell White, AWM38 3DRL 6673/153.

Lancashire Division. We are on one of the Anchor liners, the "Caledonia," and I may say that she is an excellent boat. We have on board 1,000 the 5th Manchesters and 1,000 of the Middlesex Regiment, all Terriers. We sailed out of Southampton at 5.30 p.m. on Thursday, the 10th September, but only got as far as Eddystone Lighthouse, where we stopped until 10-30 on Friday, the 11th. We sailed then with 15 transport ships, all conveying Terriers for foreign lands. Altogether there are 30,000 Terriers sailing together…

"The picture now as I am writing is splendid, it is more than I can describe. The division is moving very slowly. We have been on board now almost a week, and have only seen land once, that was yesterday, when we got a view of the coast of Spain. I must say a word or two about our passage through the Bay of Biscay. Practically all my Company were placed *hors de combat* for a day or two. Just fancy having two full days tossed about mid-ocean. One of my sergeants (George Beazant)[53] wanted the ship to go down. Of course, he wasn't the only one who said this. I myself personally was not affected, but I know well enough what seasickness is, for I have had some. But to see the lads now, they are made over again. Every man is in the pink of condition. We are already in a very hot climate. It is just like midsummer at home. All are now quite settled to the ship, but anxious to be seeing land again. We are expecting to be in Gibraltar to-morrow, where we shall land the Middlesex Regiment, who are going to relieve a regular regiment. As you know, we are going out to Egypt, but I am not in a position to disclose our place of disembarkation. I may tell you that one of the Manchester Regiments is for Alexandria, one for Cairo, one for Khartoum, and one for Cyprus. Anyhow, I will let you know when we get to our quarters. The clothing that we are wearing is too warm for us, and when we get to Egypt we shall wear fine khaki uniform and helmet, for we can hardly stand this sun now, and are not yet at Gibraltar… Who would have thought a couple of months ago that we should be on our way to Egypt?"[54]

CSM WALTER SPENCER, 1/5TH BN MANCHESTER REGT, HMT *CALEDONIA*.[55]

17 SEPTEMBER 1914

TURKEY

The Russians in Istanbul sought advice about an offer they had received to guarantee future Turkish neutrality.

"Citing the alleged successful contest with the war party in the Cabinet, Djavid Bey yesterday expressed his assurance that the Powers of the Triple Entente could bring about a demobilisation of Turkey if they would make it conditional upon their

---

53   Sgt George Beazant, 1/5th Bn Manchester Regt. He was discharged to a commission with 1/6th Bn on 8th October 1918.
54   *Leigh Chronicle and Weekly District Advertiser*, 2nd October 1914.
55   CSM Walter Spencer, 1/5th Bn Manchester Regt, was killed in action on 28th May 1915. Buried in Redoubt Cemetery, he was the husband of Johannah Spencer.

consent to the suppression of economic, as well as juridical, capitulations. In view of the endeavour of the Turks after the retaking of Adrianople to place before Europe an accomplished fact from which they afterward refused to recede; and in view of the fact that, without declaring war against them, the Powers at the present time have no means of exerting pressure upon them, I ask instructions, with the least possible delay, concerning the attitude the Imperial Government would maintain toward the proposal of Djavid Bey, if it were renewed in the name of the entire Cabinet, and with a secret clause providing that the regime without capitulations might be applied to foreigners only after the formulation of new rules effectually guaranteeing the inviolability of persons and dwellings of foreigners."[56]

MIKHAIL NIKOLAYEVICH DE GIERS, RUSSIAN AMBASSADOR, ISTANBUL.

18 SEPTEMBER 1914

NEW POMERANIA

"We are busy moving about from island to island, and at present are quartered at Herbertsshoe, New Pomerania. Our troops are a most likely lot, and comprise infantry and naval reserves. Some are stationed at Rabal [Rabaul], 15 miles back from here, and the balance are operating here. We have had some very long marches through the country, but, with the exception of last Friday — when six men were killed and a few wounded on our side, and eighteen natives and two Germans killed of the enemy — we have had little opposition. The enemy have retreated on each occasion. We had a few shots fired on our flank on Friday afternoon, but no one was injured. On Monday we put up a record in regard to marches in tropical climates. We started out from camp at 5.30 a.m. and were continually on the march till 10.15 p.m., with the exception of two short halts for food. We covered about 21 miles, and considering that the country is thick with scrub and entanglements of all kinds the march was wonderful, as no men fell out, although everybody was fit to drop. In the afternoon rain fell in torrents, and we were in heavy marching order. The officer in charge (Colonel Watson), who mapped out the march, was proud of the achievement. The country around is beautiful — cocoanuts, pineapples, bananas, and paw-paws in abundance."[57]

SGT ARTHUR GRAY, AUSTRALIAN NAVAL & MILITARY EXPEDITIONARY FORCE.[58]

---

56   Scott, *Diplomatic Documents Relating to the Outbreak of the European War*, pp. 1413–1414.
57   *The Albury Banner and Wodonga Express* (New South Wales), 9th October 1914.
58   The Boer War veteran was promoted RSM at Gallipoli on 9th October 1915 with 19th Bn, AIF; awarded the DCM for his bravery at Pozieres; and killed in action on 10th November 1916. Commemorated on the Villers Bretonneux Memorial, he was the 33 year-old son of the late George and Mary Ann Eliza Gray, originally of Yass, New South Wales.

19 SEPTEMBER 1914

TURKEY

The British Ambassador continued to believe that war was not inevitable but noted the continued build up in the Dardanelles.

> "In conversation with the President of the Chamber to-day, I said that if it was really Turkey's intention to go to war with Russia, I considered such a policy absolute madness.

> "President said that, even if Turkish fleet went into Black Sea, it would not be with any hostile intention towards Russia, with whom they were not going to war. I pointed out to him that Germany was pressing Turkey to send their fleet into the Black Sea with one object only, namely, that war might be provoked by some incident. I therefore urged him most strongly against any such action. He said that he was against it, and that he saw the force of my argument, to which I replied that as the Minister of War was supreme it was unfortunately no guarantee that it would not be done. President told me that the Cabinet had their own policy, which was to remain neutral, and that they were all alive to the aims of Germany. I pressed him hard as to what was the policy of the Minister of War.

> "I do not regard situation as hopeless. Party in favour of neutrality is growing, but it would be unsafe to rely on their power to restrain war party.

> "I hear that 156 more mines and the minelayer "Ghairet" have been sent to Roumeli Kanak, on the Bosphorus. Turkish fleet went to Halki yesterday for review, and will probably remain there till next week, when the "Hamidieh" and "Messudiyeh" will be ready. German officers and men continue to arrive by train. It is probable that there are German reservists resident in Turkey who have been incorporated in Turkish army. 200 Germans arrived at the Dardanelles on September 17th."[59]

SIR LOUIS MALLET, BRITISH AMBASSADOR, ISTANBUL.

ENGLAND

> "The Ottoman Government informed his Majesty's Ambassador in August that the reason for the detention of certain ships in the Dardanelles was that some mines had become detached from their moorings, and that vessels had been prevented from continuing their voyage in order to avoid accident until the mines had been dragged up."[60]

FRANCIS DYKE ACLAND, MP, UNDER-SECRETARY OF STATE FOR FOREIGN AFFAIRS.

---

59 *Correspondence Respecting Events Leading to the Rupture of Relations with Turkey*, p. 27.
60 *London Evening Standard*, 19th September 1914.

AUSTRALIA

"Last night a terrific storm passed over Queen's Park, resulting in tents being washed out and many a sleepless night being passed. It was real funny to see practically naked figures hovering about with a spade and lamp digging trenches to leave the water off. Our tent escaped, more by good luck than good management. However, everyone appeared to take the whole affair in a good spirit. After all, I suppose, it is only one of the many experiences we will have before our return to dear old Orange. Claude West[61] and Jack Earls,[62] of Orange, who are in the medical corps, wish to be remembered to Orange friends. Both, like myself, are looking well and quite used to the hard conditions — at any rate, they appear hard to us fellows, after leading a life of ease."[63]

DVR THOMAS NICHOLSON, 1ST FIELD AMBULANCE, AAMC, AIF, QUEEN'S PARK CAMP.

20 SEPTEMBER 1914

TURKEY

Mallet reported back to London his belief that the group in favour of peace was growing but that Enver Pasha was simply able to ignore them all.

"I have just had an animated interview with the Grand Vizier, and I am convinced that he is sincere. Other Ministers are all peaceably inclined, with the exception of the Minister of War. So long as the latter remains supreme an incident may occur at any moment. I tackled the Grand Vizier on the subject of the "Breslau" entering the Black Sea. He vehemently disclaimed any intention of attacking Russia, and said that Turkish Government had a right to send their fleet into the Black Sea if they wished to. I reminded him that neither the "Goeben" nor the "Breslau" were Turkish ships according to international law, and said that if they left the Dardanelles we would most certainly treat them as enemy ships. He replied that I had told him this often before, and there was no question of the ships leaving the Dardanelles. I then said that information had reached me that Council of Ministers, in order to avoid risk of an incident, had come to the wise decision that the "Goeben" and the "Breslau" should not go into the Black Sea; and yet, on the very day on which this decision had been reached by the Cabinet, it was totally disregarded by the Minister of War, as his Highness was doubtless aware. This showed how much control his Highness now exercised…

---

61 Driver Claude West, 1st Field Ambulance. Granted a commission, the former clerk died of wounds while serving with 1st Bn, AIF, 16th May 1917. Buried in Etaples Military Cemetery, he was the son of Arthur West, of 44 Hill Street, Orange, New South Wales.
62 Pte John Earls, 1st Field Ambulance, was killed in action in Belgium with 4th Bn, AIF, on 12th October 1916. Buried in Bedford House Cemetery, Zillebeke, he was the 23 year-old son of John and Edth Earls, of "Albacutya," Nile Street, Orange, New South Wales.
63 *Leader* (Orange, New South Wales), 21st September 1914.

> "I said that doubtless peace party was growing, but, nevertheless, Minister of War was pushing forward warlike preparations uninterruptedly. I was receiving constant information respecting British official war news being stopped, cases of requisitions, &c., and I knew as a fact that intrigues against Egypt were being carried on. If his Highness could stop these things, why did he not do so, and when would he be able to do so? His Highness gave me to understand that if a crisis did come there would be a means of stopping Minister of War."[64]
>
> SIR LOUIS MALLET, BRITISH AMBASSADOR, ISTANBUL.

Charles Lister saw no positive outcome for the Turks. The British had made themselves unpopular but what would a German victory mean for them?

> "Our holding up these Turkish ships and the arrival of the *Goeben* has created a strong feeling in her favour, and the military mission have been consequently able to get the whole thing into their hands. The Turks are en mauvaise voie. If the Germans win, Turkey will become a sort of German Egypt. If the Germans lose, Turkey will have to face the victorious Triple Entente Powers in no mood for trifling."[65]
>
> CHARLES LISTER, BRITISH EMBASSY, ISTANBUL.

21 SEPTEMBER 1914

TURKEY

It was the Russians' turn to repeat what had been said many times about the status of the *Goeben* and *Breslau*. They, like the British, were speaking to the wrong man and they knew it.

> "To-day I pointed out to the Grand Vizier that the voyage to-day of the "Goeben," even for a short time, into the Black Sea, did not coincide with his declaration that the Turkish fleet would not go there. I reminded him of my previous statement, that such a move might lead to incidents and results, the responsibility for which would fall upon Turkey. I did not fail to call his attention anew to the fact that the international position of the "Goeben" and "Breslau" under the Turkish flag could not be deemed correct. The Grand Vizier strove to justify himself on the score that the departure of one large vessel for target practice, alleged to be dangerous in the Sea of Marmora, was not the departure of the fleet, and that he could only repeat the assurance that a general exit of the whole Turkish fleet would not take place.
>
> "His assertions, even if approved by the Council of Ministers, do not, however, constitute a serious guaranty, as the fleet, now in the hands of the Germans and

---

64 *Correspondence Respecting Events Leading to the Rupture of Relations with Turkey*, pp. 27–28.
65 Charles Lister quoted by Ribblesdale, Thomas Lister, 4th Baron, *Charles Lister. Letters and Recollections, With a Memoir by His Father, Lord Ribblesdale*, pp. 125–126, T. Fisher Unwin Ltd. (London) 1917.

under the immediate command of the Vice-Generalissimo Enver Pasha, can sail out even without the consent of the other Ministers."[66]

MIKHAIL NIKOLAYEVICH DE GIERS, RUSSIAN AMBASSADOR, ISTANBUL.

AUSTRALIA

As the political situation in the Eastern Mediterranean worsened, Australians prepared to leave for war.

> "It is a red-letter day to-day in South Australia. This is the day the South Australian first lot of troops are leaving for England. I have just seen three trainloads of lads go through Alberton on their way to the troopships. They are leaving here for Freemantle (West Australia), and from there to England. They are going to be escorted part of the way by a Japanese man-of-war. Our first contribution is 20,000 men, and we are getting up a second and a third contingent. J. Garland[67] leaves to-day with the first contingent, and expects to arrive in England about Christmas. He expects to see you all before he goes to the front."[68]

HENRY CLARK, ALBERTON, SOUTH AUSTRALIA.

22 SEPTEMBER 1914

NORTH SEA

This was to be a very bad day for the Admiralty and its leadership. It was a whole lot worse for the men on three obsolescent cruisers, *Aboukir*, *Hogue* and *Cressy*, patrolling off the Dutch coast. Referred to as the 'Live Bait Squadron,' their crews were made up largely by recalled reservists and young sailors; some of the latter at sea for the very first time. One German submarine found them that morning.

> "When I first sighted them they were near enough for torpedo work, but I wanted to make my aim sure, so I went down and in on them. I had taken the position of the three ships before submerging and I succeeded in getting another flash through my periscope before I began action. I soon reached what I regarded as a good shooting point.

> "Then I loosed one of my torpedoes at the middle ship. I was then about 12 feet under water and got the shot off in good shape, my men handling the boat as if she had been a skiff. I climbed to the surface to get a sight through my tube of the effect, and discovered that the shot had gone straight and true, striking the ship... under one of her magazines...

---

66   Scott, *Diplomatic Documents Relating to the Outbreak of the European War*, pp. 1414–1415.
67   Twenty-five year-old tram conductor, James Garland, like Henry Clark, was born in Hull, Yorkshire. He enlisted on 20th August 1914 and was posted to 3rd Field Ambulance, Australian Field Ambulance Corps.
68   *Hull Daily Mail*, 7th December 1914.

"There was a fountain of water, a burst of smoke, a flash of fire and part of the cruiser rose in the air. Then I heard a roar and felt reverberations sent through the water by the detonation. She had been broken apart and sank in a few minutes."[69]

KAPITÄNLEUTNANT OTTO WEDDINGEN, U-9.

"I was lying in my hammock, awake, before being called at seven o'clock. There was an explosion like a terrific clap of thunder, accompanied with a flash like lightning as the *Aboukir* was torpedoed. She had been struck upon the port beam. There was no visible damage around me, but we quickly begun dressing. There was no confusion or panic, and orders were given to close all dead lights and the watertight doors. I was lying just outside the wardroom, some distance from the place where the torpedo struck the ship, and after this I proceeded to the boat deck, where men were gathered on the starboard side. It was a cool, grey morning, and the sun was just rising. The *Aboukir* had a list to the port of from 50 to 60 degrees. Captain Drummond gave the order that our four sick men (on stretchers) should be transferred to the only boat that could be lowered, and this was carried out clamly [sic]. Other sailors followed, and the boat having been lowered into the sea, it floated away with about 87 in it…

"The *Hogue* was standing about 400 yards away from the *Aboukir*, and the *Cressy* some distance beyond that. I climbed over the starboard rail and lowered myself on to a six-inch gun, and stood upon it. Remaining there until about 6.45, I plunged into the water. Making straight for the *Hogue*, I reached her quite safely, held on to one of their ropes, and was hauled up on to the deck."[70]

OS ALFRED BROWN, RN, HMS *ABOUKIR*.

"The [*Hogue* and *Cressy*] came on a mission of inquiry and rescue, for many of the *Aboukir*'s crew were now in the water… l sent a second charge at the nearest of the oncoming vessels, which was the *Hogue*. The English were playing my game, for I had scarcely to move out of my position, which was a great aid, since it helped to keep me from detection…

"The attack on the *Hogue* went true. But this time I did not have the advantageous aid of having the torpedo detonate under the magazine, so for 20 minutes the *Hogue* lay wounded and helpless on the surface before she heaved, half turned over and sank."[71]

KAPITÄNLEUTNANT OTTO WEDDINGEN, U-9.

"Immediately I reached the deck two tremendous explosions occurred, one about five seconds after the other. The ship began to sink very rapidly, and I really had no time to get clear again. She listed to starboard, and I made an ineffectual attempt to leap over the port side, but the movement of the ship was so quick that I was thrown off the deck into the water on the starboard side. We were carried underneath with the suction, and seemed to be a tremendous time before reaching the surface again….

---

69 *The Birmingham Age Herald* (Birmingham, Alabama), 14th October 1914.
70 *Bucks Advertiser & Aylesbury News*, 3rd October 1914.
71 *The Birmingham Age Herald* (Birmingham, Alabama), 14th October 1914.

The *Hogue* floated keel upwards, and as I looked round I saw the *Cressy* standing by... I turned round and swam to her. It took a tremendous struggle to reach the *Cressy*, and I was pretty well exhausted by the time I was taken on the quarter deck."[72]

OS ALFRED BROWN, RN, HMS *ABOUKIR*.

"By this time the third cruiser knew of course that the enemy was upon her and she sought as best she could to defend herself. She loosed her torpedo defense batteries on boats, starboard and port, and stood her ground as if more anxious to help the many sailors who were in the water than to save herself. In common with the method of defending herself against a submarine attack, she steamed in a zig-zag course, and this made it necessary... for me to get nearer to the *Cressy*...

"When I got within suitable range I sent away my third attack. This time I sent a second torpedo after the first to make the strike doubly certain. My crew were aiming like sharpshooters and both torpedoes went to their bull's eye. My luck was with me again, for the enemy was made useless and at once began sinking by her head. Then she careened far over but all the while her men stayed at the guns looking for their invisible foe. They were brave and true to their country's sea traditions. Then she eventually suffered a boiler explosion and completely turned turtle."[73]

KAPITÄNLEUTNANT OTTO WEDDINGEN, U-9.[74]

"I was lying down on the quarter deck, having swallowed a lot of water, when the *Cressy* was first struck. The explosion threw a shower of water into the air... Shortly after this a second explosion occurred, which caused a great deal more damage than the first, and the boat began slowly to list to port. The crew then threw every piece of loose woodwork and all floatable gear overboard for the purpose of saving life. I remained on the *Cressy* until the last minute, when I was obliged to dive into the water again."[75]

OS ALFRED BROWN, RN, HMS *ABOUKIR*.

In a single hour before breakfast 62 officers and 1,397 men had been lost. The Admiralty issued instructions that never again should British ships stop to pick up survivors, presenting themselves as such perfect targets. But what happened this day was not so easily dealt with. What were the ships doing there in the first place and why had so many young, inexperienced men crewed them? Who was responsible for their deployment?

Stories of survival and loss fed the shock that many felt at this time. Herbert Penn was on the *Cressy* with his two brothers but only he lived to tell the tale.

"The sea was literally alive with men struggling for anything to save themselves with, while to add to the horror of the scene, the Germans kept firing their torpedoes at us.

---

72 *Bucks Advertiser & Aylesbury News*, 3rd October 1914.
73 *The Birmingham Age Herald* (Birmingham, Alabama), 14th October 1914.
74 Weddigen, commanding *U-29*, was killed in the Pentland Firth when it was rammed by HMS *Dreadnought* on 29th March 1915. The submarine was cut in two. There were no survivors.
75 *Bucks Advertiser & Aylesbury News*, 3rd October 1914.

"I was just going to jump into the sea when I saw dear brother Alfred coming along the deck, which was then awash. Together we lingered for moment, shook hands, and told each other whoever was saved to tell dear mother that our last thoughts were of her.

"We then kissed, wished each other goodbye, and plunged into the sea together. We never saw each other again,[76] nor did we see any sign of brother Louis."[77]

SEAMAN HERBERT PENN, RNR, HMS *CRESSY*.

TURKEY

After the ending of Capitulations, the British learned that the independent national postal services were to be closed. The message being delivered to them was clear.

"A letter was yesterday received by British postmaster from a subordinate official in the Turkish postal administration. In this letter postmaster was informed that foreign post offices in Turkey would be abolished as from 1st October next. I instructed British postmaster to return the letter, and to say that matter had been referred to his Ambassador.

"This discourteous manner of communication was my first official information of any intention to abolish foreign post offices in Turkey. I accordingly saw Grand Vizier at once, and said that I resented the manner of communication, and had instructed British postmaster to return the letter. Post offices did not depend upon the Capitulations, and if Turkish Government wished to see the system modified, they should approach His Majesty's Government through the usual diplomatic channel. I warned him that His Majesty's Government would not allow themselves to be ignored in this manner, and I would not, unless by your instructions, consent to summary closing of British post offices on 1st October unless Turkish Government had given guarantees for safeguarding British interests."[78]

SIR LOUIS MALLET, BRITISH AMBASSADOR, ISTANBUL.

INDIA

As the day drew to a close, the *Emden* appeared off the coast of Madras. It had a message that lit up the night sky.

"Along about nine in the evening our searchlight swept in an arc over the great metropolis of South India. We could see the domes of the British high court, and the tall Hindu temples. To our right were the tanks belonging to the Standard Oil. To our

---

76 The brothers, Seaman Alfred and Seaman Louis Penn, RNR, HMS *Cressy*, died on 22nd September 1914. Commemorated on the Chatham Naval Memorial, they were the sons of Elizabeth Penn, of 22 Farrier Street, Deal, Kent.
77 *Dundee Evening Telegraph*, 25th September 1914.
78 *Correspondence Respecting Events Leading to the Rupture of Relations with Turkey*, pp. 28–29.

left were the huge storage containers of the British. Almost as soon as our searchlight beams picked them up we cut loose at a range of three thousand yards. The first shell fell short and landed on a steamer inside the breakwater.

"Then pandemonium broke loose. Our next burst of fire landed on one of the big oil reservoirs with an effect that was magnificent.

"A shattering blast tore asunder the black curtain of the night, and our searchlight, by comparison, seemed like the faint flicker of a candle. The whole city and harbor were lighted as by a gigantic torch; and the a fast-spreading mass of smoke — inky, heavy, ominous — hung like a pall of gloom."[79]

OBERLEUTNANT ZUR SEE DER RESERVE JULIUS LAUTERBACH, SMS *EMDEN*.

"It was about nine o'clock… and I had just gone to bed, when I heard a noise like thunder, and shortly afterwards saw a bright light in the sky. I at once got up, and informed my neighbours who live in the flat below me. One of them telephoned to a friend who has a car, and the car quickly came round. Having got to the harbour, we found two immense oil tanks ablaze, and two buildings wrecked. About 150 yards from the burning oil tanks was the petrol store. We succeeded in getting the petrol to a place of safety, though the heat was intense, and several of the petrol tins had begun to leak. Luckily, we were to windward of the blazing tanks, and the wind was rising.

"The hut belonging to the watchman of the Burmah Oil Company was a mass of wreckage… The manager's bungalow of the same company was damaged, but the manager was unhurt. There was a hole in one of the walls of the National Bank of India, and I obtained fragments of shell…

"Madras was obviously taken unprepared, but after the damage had been done the fort replied, and the first shot fell short. The second also fell short, and the *Emden* then extinguished her searchlight and steamed away under cover of darkness and of the smoke of the burning oil, which was being blown out to sea."[80]

GUY RANSOM WARWICK, BRITISH MERCHANT, MADRAS.

Three cruisers lost before breakfast and Madras shelled after dinner. Questions were being asked: just what was going on at the Admiralty?

AUSTRALIA

Men continued to volunteer for service; men from every background.

"Along with a score or more of others, I lined up in front of the drill hall to await my turn. My form being filled in and passed as satisfactory by one officer, another drafted me into the Infantry and as ready to proceed. Some time later we lined up, and after many pauses, gaps, stops, and false starts, we were rightly 'dressed' and numbered, and scurrying to form fours we 'righted' and set forth for Kensington, a

---

79 Julius Lauterbach quoted in Thomas, *Lauterbach of the China Sea*, p. 53.
80 *Nottingham Evening Post*, 22nd October 1914.

squad of about 150 all told. A motley crowd we were — a heterogeneous collection here, the out of work clerk, decked out in silk bow tie, with admirably pressed suit... the man who could lay his hand on a private income of a few hundreds, going out to see the fun. The other side, however, was more in evidence — the unwashed larrikin from Wooloomooloo, the loafer from the streets whose language was strong enough... and dirty enough to go down the sewer, enticed by the prospect of easy work, a smart uniform and 5/ per diem, not to mention the 1/ deferred...

"The men are well bedded, albeit like horses, with a bale of straw to three men; two good woollen blankets, one waterproof rubber rug together with a mug, plate, knife and fork are given each man. We have a tin of jam and a double loaf of bread in the morning for breakfast; a bully beef, vegetable and other mixture for dinner, with the said double loaf... again; one tin of jam, said loaf again, and one candle (evidently not for mastication) at night for tea. Coffee is served at 6.15 a.m., tea at 8 o'clock breakfast, soup at dinner, and tea at 5 o'clock. The routine of the work is as follows:

"6 o'clock reveille, 6.15 coffee, 6.45 parade and drill, 7.45 tent orderlies, 8 a.m. breakfast. 9 a.m. sick parade, 9.15 battalion and company drill, 12.45 dismiss, 12.45 tent orderlies. 1 p.m. dinner, 2 o'clock fall in, 2.15 battalion and company drill, 4.30 dismiss, 4.45 tent orderlies, 5 p.m. tea, 7.30 parade, 8.30 dismiss, 9 p.m. first post, 10 o'clock tattoo, 10.15 lights out."

L/CPL GEORGE TIDEX, 13TH BATTALION, AIF, ROSEBERY PARK CAMP.[81]

23 SEPTEMBER 1914

SCOTLAND

David Beatty had once commanded HMS *Aboukir*. He was sure where the blame lay for its loss.

"Three fine ships and the greater part of 2,200 men that can be ill spared. It was bound to happen. Our cruisers had no conceivable right to be where they were. It is not being wise after the event, but I had frequently discussed with others that sooner or later they would surely be caught by submarines or battle cruisers, if they continued to occupy that position. It was inevitable and faulty strategy on the part of the Admiralty."[82]

REAR-ADMIRAL SIR DAVID BEATTY, RN, 1ST BATTLECRUISER SQUADRON.

---

81   Tidex, a former cinematograph operator, was wounded at Gallipoli on 7th May 1915. He returned to Australia to be discharged, leaving Suez on 5th July 1915.

82   David Beatty, quoted in Chalmers, W. S., *The Life and Letters of David, Early Beatty. Admiral of the Fleet*, p. 157, Hodder and Stoughton (London) 1951.

ENGLAND

The tone of the language to be used with the Turks hardened into one of undisguised threat. If the Grand Vizier, so long and loud in his protestations of good faith, wished the Ottomans to be regarded as neutral he would have to control Enver Pasha. The Foreign Office sent this advice to the British Ambassador.

> "His Majesty's Government regard state of things at Constantinople as most unsatisfactory. On behalf of His Majesty's Government you should speak in the following sense to the Grand Vizier: —

> "British Government contemplate no hostile act towards Turkey by British fleet, and they have no desire to precipitate a conflict with her. But the fact that Great Britain has not taken any hostile action against her must not mislead Turkish Government into supposing that His Majesty's Government consider Turkey's attitude is consistent with the obligations imposed upon her by the neutrality which she has officially declared. German officers and men are participating increasingly in Turkish fleet and Dardanelles defences, and not only has Turkey failed to send away the German officers and crews, as she promised, but she has admitted more overland, and they are now in active control of the "Goeben" and "Breslau." The capital is undoubtedly now under the control of the Germans. If His Majesty's Government so desired, present state of things affords ample justification for protesting against violation of neutrality. Great Britain has not, however, so far taken action, as she cherishes the hope that the peace party will win the day. It should, however, be realised by the Grand Vizier and his supporters that unless they soon succeed in getting the situation in hand and bringing it within the limits of neutrality, it will become clear that Constantinople is no longer under Turkish but German control, and that open hostility will be forced on them by Germany."[83]

> SIR EDWARD GREY, MP, BRITISH FOREIGN SECRETARY.

Threats had to be backed up by force. An American report recorded the formation of the Royal Naval Division in London. Major Treadwell appeared impressed with what he saw.

> "The Admiralty took over the Crystal Palace about Sept. 8th, to be used in connection with the organization of a Royal Naval Division to be composed of three brigades, consisting of a Marine Brigade and two Naval Brigades, the nucleus of which was found in existing reserves for the Naval Service which could not be used afloat at present…

> "The Royal Naval Division will be completely equipped by the Admiralty, with field hospitals, transport, ammunition columns, signal companies, cyclists, motor-cars, and machine guns. An aeroplane squadron from the naval wing will be available when required.

---

83   *Correspondence Respecting Events Leading to the Rupture of Relations with Turkey*, p. 29.

> "If at any time the Naval situation becomes sufficiently favorable to enable this force to be definitely released by the Admiralty for military duty, it will be handed over intact to the army for general service. It is thought that the training, discipline, experience and quality which a large part of the personnel already posses, should after five or six months' special instruction in field duties, produce a highly efficient division, and that the prospects of the Royal Naval Division winning distinction on the Continent are therefore good…
>
> "The men appeared to be of good quality and were in blue naval uniforms and caps, with "Royal Naval Division" on capband, It is understood that before being put in the field the Naval Brigade are to be provided with a khaki uniform similar in cut to the Naval uniform. Few of the men were provided with arms. There were detachments of about 200 recruits that had just come in."[84]

MAJOR THOMAS C. TREADWELL, USMC, AMERICAN EMBASSY, LONDON.

24 SEPTEMBER 1914

ENGLAND

There were fewer more sensitive areas than Suez for the British. Australian and New Zealand sensitivities about the presence of German cruisers, the danger they represented to their troopships, had to be recognised too.

> "I hear that Egyptian frontier has been violated by armed mounted Arabs said to be encouraged by Turkish troops, and also that Hedjaz line is being reserved for troops. British military authorities consider that breach of the peace on Egyptian frontier is imminent, whether with or without sanction of Turkish Government. You should bring these facts to the knowledge of the Grand Vizier and of the Khedive, who is at present at Constantinople."[85]

SIR EDWARD GREY, MP, BRITISH FOREIGN SECRETARY.

> "Admiralty adhere to opinion despatch of transports from New Zealand [&] Australian Ports to point of concentration at Fremantle is an operation free from undue risk but in view of anxiety felt by your Ministers and Government of Australian Commonwealth they propose to send "Minotaur" and "Ibuki" to Wellington to fetch New Zealand convoy and escort it westward along the Australian coast picking up Australian transports on way and bringing the whole to their destination. This will involve about three weeks' delay."[86]

LEWIS HARCOURT, MP, SECRETARY OF STATE FOR THE COLONIES.

---

84 Naval Attache's Reports, Office of Naval Intelligence, Unpublished Manuscript, US Naval War College, September 1914.
85 *Correspondence Respecting Events Leading to the Rupture of Relations with Turkey*, pp. 29–30.
86 Typescripts; comprise questions to be put to Sir Brudenell White, AWM38 3DRL 6673/153.

TURKEY

Mallet had passed on the Foreign Secretary's warning to the Grand Vizier. He was struggling to understand why the Turks felt able to ignore the British Empire.

> "I warned him that the information respecting Turkish preparations against Egypt would infallibly produce a most serious impression upon His Majesty's Government…
>
> "I cannot believe that they are not alive to the disastrous consequences of going to war with us, or that they seriously can contemplate an expedition against Egypt. They have undoubtedly been strongly urged to send such an expedition by the Germans, and I think that they have allowed preparations to be made, partly to profit as much as possible by German connection and by allowing the Germans to think that they will act, and partly in order to be ready, if Great Britain sustains a serious defeat by land or sea."[87]

SIR LOUIS MALLET, BRITISH AMBASSADOR, ISTANBUL.

CYPRUS

British Territorials were on their way to garrison Cyprus. A local police inspector did not think they would have much to occupy them on the island.

> "Germans are very scarce here — only about a dozen all told; and, in spite of the trouble across in Asia Minor, comparative calm reigns here. Our navy is at present lent to convey transports. The Manchester Terriers are arriving to-day and to-morrow. They've come to help us to eat the grapes we can't export. There is little else for them to do, so long as the "Goeben" is bottled up…
>
> "The cost of living is exceedingly low, and has fallen lately, owing to the surplus food produced, of which export is prohibited."[88]

INSPECTOR GILBERT MOODY, DEPUTY INSTRUCTOR, CYPRUS POLICE.

25 SEPTEMBER 1914

TURKEY

British frustrations were on show during the latest meeting between the Ambassador and Grand Vizier.

> "I have again seen Grand Vizier, and pointed out to him as earnestly as is within my power the fatal result to the Turkish Empire of persisting in a course of veiled hostility and petty intrigue against the British Empire. I recalled to him that time and again he had undertaken that the German crews of the "Goeben" and the "Breslau" should

---

87 *Correspondence Respecting Events Leading to the Rupture of Relations with Turkey*, p. 30.
88 *The Fleetwood Chronicle and Fylde Advertiser*, 23rd October 1914.

be sent out of Turkey, and that not only had these promises been broken, but further German officers and men had actually arrived. This proved conclusively that he was either insincere in his assurances or that he was powerless. His Highness begged that I would credit him with the fact that for eight weeks he had kept the peace. He assured me that he had every intention of seeing to it that peace was maintained. I replied that it was not his good intentions that I doubted, but I did distinctly doubt his ability to control the situation."[89]

SIR LOUIS MALLET, BRITISH AMBASSADOR, ISTANBUL.

EGYPT

The men of the East Lancashire Division took in the scene as they arrived in Alexandria.

"We landed at Alexandria on Friday, the 25th September. We had all been looking forward eagerly for a sight of Egypt, but did not expect to see anything like the scene that lay before us: bright blue sea, with the sunlight rippling on it, and in the distance the huge stone mosques with their domes and minarets. We had a long time to take in the scene before us, as other ships had to be moored before ours. Eventually our turn came, and then the fun commenced. Natives swarmed all over the ship selling fruit, which looked good, but which we were forbidden to buy. The boys, however, persisted in buying until the officers threw buckets of water over when the natives would retaliate by rowing away with the lads' money. Some of us were throwing money on the quay, and watching the natives scramble for it. The native police tried to drive them away, and their attempts to do so drove us into roars of laughter. We thought their treatment of them rather cruel, as they used sticks rather freely, and threw stones when they got out of their reach."[90]

PRIVATES JAMES CAVANAGH, SYDNEY JAMES COOK, MICHAEL GAVAGHAN, WILLIE EDMONDSON, L/CPL SAMUEL SHAW, 1/6TH BN LANCASHIRE FUSILIERS, EAST LANCASHIRE DIVISION.

AUSTRALIA

The Australian Prime Minister confirmed his insistence that no troop convoy should leave without escort while German ships remained active along its potential route. Preparations by the men themselves continued, meanwhile, in anticipation of their departure.

"Am strongly of opinion that no transport ships with troops on board should proceed to sea unprotected by effective Convoy if enemy ships are within striking distance."[91]

ANDREW FISHER, PRIME MINISTER OF AUSTRALIA.

---

89   *Correspondence Respecting Events Leading to the Rupture of Relations with Turkey*, p. 31.
90   *Todmorden Advertiser and Hebden Bridge Newsletter*, 11th December 1914.
91   Typescripts; comprise questions to be put to Sir Brudenell White, AWM38 3DRL 6673/153.

"We have been working very hard for the last few days, getting our kits ready for the road. Lismore boys were delighted to see Mr. Franklin on Sunday, as they had no idea he was coming. I was weighed and am 11 stone 10 lbs, so camp life is agreeing with me."[92]

TPR DONALD GRAHAM, MG SECTION, 2ND AUSTRALIAN LIGHT HORSE REGIMENT, AIF, ENOGGERA CAMP.[93]

26 SEPTEMBER 1914

TURKEY

The British Ambassador told the Grand Vizier that war would ensue if there was any incursion into Egypt. This time Enver Pasha was in the room. His response was unrecorded.

"I warned his Highness that if... preparations against Egypt were allowed to continue, serious consequences would ensue. Minister of War was with Grand Vizier when I made these representations, and his Highness informed me that he fully realised the importance of the question, with which he was occupying himself. I have taken steps to enlighten influential people with what is being done as regards Egypt, and I have seen Minister of Interior and left a memorandum with him on the subject; I have also put the facts before other prominent members of the Cabinet."[94]

SIR LOUIS MALLET, BRITISH AMBASSADOR, ISTANBUL.

AUSTRALIA

Though the Australian Prime Minister was not prepared to sanction to despatch of a troop convoy at this time, the commander of the Australian Imperial Force considered the threat posed by the likes of the *Emden* to be low.

"At present we can make some prediction as to the position of the enemy's ships and that position does not impose on our transports any appreciable risk. Each day of delay brings the enemy nearer, enables him to seek further information, and lay his plans...

"The supposition that the enemy will sacrifice his ships in an attempt to destroy our transports is based on the theory that it is only a question of time before he will be captured or destroyed, and that therefore he must seek to inflict the utmost damage

---

92   *Northern Star* (New South Wales), 25th September 1914.
93   Graham was evacuated from Gallipoli on 16th September 1915, first to Mudros and then to England, suffering from influenza. Transferred to 5th Medium Trench Mortar Battery, Australian Field Artillery, he was killed by a misfire on 8th November 1916. Buried in Cite Bonjean Military Cemetery, Armentieres, France, he was the 21 year-old son of Robinson and Mary Eleanor Graham, of High Street, Lismore, New South Wales.
94   *Correspondence Respecting Events Leading to the Rupture of Relations with Turkey*, p. 32.

within the time at his disposal. Upon such premises, isolated transports do not present sufficiently tempting objectives; if they did those transports now at sea on the east coast are exposed to the greatest danger."[95]

MAJOR-GENERAL WILLIAM THROSBY BRIDGES, COMMANDING AUSTRALIAN IMPERIAL FORCE.

27 SEPTEMBER 1914

TURKEY

The British, French and Russian Ambassadors visited the Grand Vizier together to protest the closure of the Dardanelles.

"An incident has occurred outside the Dardanelles. At 6 o'clock this evening I heard that a Turkish destroyer was stopped last night outside the Dardanelles and turned back by one of our destroyers. Upon this, Commandant of the Dardanelles closed the Straits. When the news arrived, the Russian and French Ambassadors were with me, and we at once went to see the Grand Vizier. When I arrived the Grand Vizier was in a state of some perturbation. He said sudden action of British fleet had given rise to the belief that an immediate attack was contemplated. Having reassured his Highness that any such belief was unfounded, I said that it seemed to me highly desirable that the Dardanelles should be opened at once, for should the incident become known, it would certainly create the impression that some desperate step was intended by Turkish Government. I explained to his Highness that we were naturally apprehensive lest Germans on Turkish destroyers might endeavour to torpedo or mine our ships, and that it was for that reason that British fleet had been instructed to prevent any Turkish ships from leaving the Dardanelles, so long as any German officers or crews remained."[96]

SIR LOUIS MALLET, BRITISH AMBASSADOR, ISTANBUL.

The worsening situation led British residents to look to their safety in the event of hostilities breaking out.

"On the 27th of September, the Dardanelles were definitely closed to any traffic and the Ambassador became anxious as to the actual safety of the British in Constantinople. It was thought highly probable that in consequence of the closure of the Straits after the entrance of the *Goeben* and *Breslau*, the Allied fleets might take some offensive action, in which case the situation would probably develop in a somewhat dangerous manner, more especially should the mob get out of hand and attack the Europeans. All the Entene stationnaires had, of course, left, and we were more or less now at the mercy of the Turks, with no means of getting away should the need arise.

---

95   Typescripts; comprise questions to be put to Sir Brudenell White, AWM38 3DRL 6673/153.
96   *Correspondence Respecting Events Leading to the Rupture of Relations with Turkey*, p. 32.

"In case of emergency a stock of rifles was always kept at our Embassy in Pera, and F.[97] was instructed to remove these secretly to Therapia. In conjunction with this, Captain Macauley, the Commandant of the American stationnaire (*Scorpion*), and F. together devised a scheme whereby everyone could take refuge in the Robert College at Rumili Hissar, and here the *Scorpion* decided likewise to move so as to afford us protection. A supply of stores was surreptitiously put into the college and each of us warned to be in readiness to leave at a moment's notice, while at the same time we were shown the quickest and the most secluded route to the college should danger arise."[98]

BETTY CUNLIFFE-OWEN, ISTANBUL.

EGYPT

The atmosphere was more relaxed here, as the newly-arrived Lancashire Fusiliers settled into their new surroundings.

"[B]eing Sunday, we had no drill. Some of us spent the morning wandering round the barracks, while others watched the departure of the 3rd Dragoons for Aldershot. The afternoon was spent in getting into our new quarters, which had been occupied by the 3rd Dragoon Guards. Our barracks is called Abbass Hilmi. It is composed of a number of fine stone buildings, with large stone balconies. We are in the married quarters, and each quarter is called a mess. A mess contains three large rooms, with scullery, cupboard, and w.c. About 20 men occupy a mess, in which they eat, sleep and live. The barracks are built on the soft desert sand, which is inches deep, and raises a cloud of dust at every step, so that you can imagine the amount of dust raised by a battalion on the march. There is a large canteen in barracks, run by Dickeson, of London. There are also fine officers' messes. It is winter here, but the weather is like an ideal English summer, excepting that we have seen no rain since leaving England. To see the sun rising and setting in Egypt is a sight one rarely forgets, and is quite beyond description. When the sun has set the sky retains for a time a vivid red hue, whilst from the east the sky gradually darkens, and half-an-hour after the sun has gone down all is darkness."[99]

PRIVATES JAMES CAVANAGH, SYDNEY JAMES COOK, MICHAEL GAVAGHAN, WILLIE EDMONDSON, L/CPL SAMUEL SHAW, 1/6TH BN LANCASHIRE FUSILIERS, EAST LANCASHIRE DIVISION.

---

97 Her husband, Frederick Cunliffe-Owen.
98 Cunliffe-Owen, Betty, *Thro' the Gates of Memory (From the Bosphorus to Baghdad)*, pp. 52–53, Hutchinson & Co. (London) 1923.
99 *Todmorden Advertiser and Hebden Bridge Newsletter*, 11th December 1914.

INDIAN OCEAN

Just before dawn, the *Emden* spotted the lights of another potential prize.

> "Along came the S.S. *Buresk*, on her maiden trip, carrying five thousand tons of the best Cardiff coal for the British Asiatic Squadron. We ran right up to within fifty yards and signaled with the siren for her to stop. I was the boarding officer, and I hardly expected to find as our prize a supply of coal that would last us a month.
>
> "We transferred her crew to the *Ghyfwale* [sic], and told the captain of the prison ship to sail for Cochin on a certain course. He was informed that if he was caught off that course he would be torpedoed."[100]
>
> OBERLEUTNANT ZUR SEE DER RESERVE JULIUS LAUTERBACH, SMS *EMDEN*.

> "They gave us half an hour to clear off the ship. They put us on board an English steamer, the *Gryfvale* [sic], together with the crews of four other steamers, which they had already sunk. They did not sink our ship. I guess they wanted the coal, but I suppose they eventually sank her."[101]
>
> ARCHIBALD EWART ADAMS, SECOND MATE, SS *BURESK*.

The *Buresk* was not sunk, its coal being far too valuable, and retained to serve as a collier for the *Emden*. Just after midday, the British steamer *Ribera* was intercepted and, after it was found to be empty, sunk; its crew were transferred to the *Gryfevale*. A Dutch ship, *Djoca*, was stopped next but released the same afternoon. At sunset the fourth ship that day to be sighted was found to be a British one, the *Foyle*.

> "Seeing smoke on the horizon, the captain thought a ship was on fire, and he altered his course to investigate. It was dark before we got near. The next thing we knew was that a vessel having no lights loomed into view from the darkness. There was a peremptory signal to stop, which we obeyed: a small pinnace drew alongside, but until the men boarded us we had no idea they were Germans.
>
> "We were told that the ship would be blown up in 20 minutes, and we were allowed to pack as much as we could in that time. We were only a day's run from Colombo at the time."[102]
>
> BRACEWELL JOHN LOMAX, ENGINEER, SS *FOYLE*.

Lomax and the others from the *Foyle* were taken off and looked on from the *Gryfevale* as their ship disappeared. They spent only a few hours as prisoners of war, as at 10.00 p.m. the *Gryfevale* was released with instructions to make for Colombo. The former captives cheered von Müller and the *Emden*.

---

100 Julius Lauterbach quoted in Thomas, *Lauterbach of the China Sea*, p. 59.
101 *South Wales Weekly Argus and Monmouthshire Advertiser*, 7th November 1914.
102 *Northern Daily Telegraph*, 4th November 1914.

28 SEPTEMBER 1914

TURKEY

The Turks promised to reopen the Dardanelles if the British ships guarding its entrance were withdrawn. The British, convinced that the Ottoman Navy was under German control, were not about to comply.

> "Yesterday the Grand Vizier requested the British Ambassador to withdraw the British squadron to some distance from its alleged position at the very entrance of the Dardanelles, promising, in the event of compliance, that he would at once open the Straits. Sir Louis Mallet transmitted this request to London. It is of the highest importance to us that if a withdrawal of the British squadron to some distance is deemed admissible, the latter shall take place only upon the absolute condition of the admission of all decisive measures necessary to preclude the possibility of the entrance of any ship of an enemy into the Dardanelles."[103]
>
> MIKHAIL NIKOLAYEVICH DE GIERS, RUSSIAN AMBASSADOR, ISTANBUL.

> "The Dardanelles were closed yesterday until further notice, both for entrance and exit. The reason given by the Ottoman Government is that the fleets of England and France, belligerent governments, are cruising at the entrance of the Dardanelles and visiting vessels of commerce to the prejudice of the advantages resulting from the straits being left open. It is announced that this closure will be maintained until the above mentioned fleets withdraw from the entrance to the strait and bring to an end this unnatural state of affairs.
>
> "The real reason is that the British Government had notified the Ottoman Government that the fact that the Turkish Naval vessels were being partially manned by German officers and men, was looked on with disfavor and suspicion, and that the Turkish men of war so manned could not be allowed outside the Dardanelles. I am informed, on good authority, that a Turkish Torpedo Boat proceeded out of the Dardanelles on the 26th, and was turned back by the British Fleet. Thereupon the General commanding at the Dardanelles, who is German, after conference with the Admiral on the *Sultan Selim* (*Goeben*), Admiral Souchon, who is also German, and in command of the Turkish Fleet, closed the strait and planted the additional mines necessary to close the channel."[104]
>
> LT-CDR EDWARD MCCAULEY, USN, USS *SCORPION*, ISTANBUL.

---

103 Scott, *Diplomatic Documents Relating to the Outbreak of the European War*, pp. 1420–1421.
104 Naval Attache's Reports, Office of Naval Intelligence, Unpublished Manuscript, US Naval War College, September 1914.

If a news agency expected the closure to last only two or three days, others did not share that optimism.

> "The port authorities of the Dardanelles notified to-day the closing of the Dardanelles for navigation. The duration of the closure was not stated, but there is good reason to believe the Dardanelles will be reopened to navigation in two or three days."[105]
>
> REUTER'S.

> "Well, not much this week; things seem shaping for war, and it looks as if the Turks had decided to commit suicide. The excitements here leave me quite cold now, as is natural after what has been a two-months-long crisis at crisis pressure of work. It is not over yet."[106]
>
> CHARLES LISTER, BRITISH EMBASSY, ISTANBUL.

CHINA

One German aviator witnessed the opening of the naval bombardment at Tsing Tao. In the one hour bombardment beginning at 9.00 a.m, 148 shells were fired. The sight, he claimed, was spectacular; the results less so.

> "The first bombardment of Kiao-Chow… was particularly impressive. The crashing and bursting of the shells, with the accompanying roar, was accentuated by the echo from the surrounding mountains. Crash followed upon crash, and we had the impression that the whole of Kiao-Chow was being turned into a heap of ruins, It was a weird feeling, but we soon got used to it. One is completely helpless in the face of exploding shells, and can but wait until it is all over…

> "The enemy ships stood so far out that our guns could not reach them. Therefore, they were quite safe. In the van steamed three Japanese battleships, and under Japanese command, at the rear, the English battleship *Triumph*…

> "Thank God, the damage cause by the bombardment was not of much consequence, and from then on we awaited their cannonading with the greatest calm."[107]
>
> KAPITÄNLEUTNANT GUNTHER PLÜSCHOW, *DEUTSCHE LUFTSTREITKRÄFTE.*

29 SEPTEMBER 1914

TURKEY

The Dardanelles were closed. The same could have been said of any negotiations that could have led to their reopening.

---

105   *Yorkshire Post and Leeds Intelligencer*, 30th September 1914.
106   Charles Lister quoted by Ribblesdale, *Charles Lister*, p. 128.
107   Plüschow, Gunther, *My Escape from Donnington Hall, Preceded by an Account of the Siege of Kiao-Chow in 1915*, pp. 53–54, John Lane The Bodley Head Ltd. (London) 1922.

> "Germans are making capital out of closure of the Straits, and I hear on good authority that great pressure is being exerted by them to induce Turkey to attack Russia in the Black Sea. Turks have however, refused so far to fall in with this scheme.
>
> "Great umbrage has been caused to the Turks by fact that it was upon the German Ambassador's order that the "Breslau" went into the Black Sea the other day.
>
> "Grand Vizier is most anxious to reopen the Straits, and has again, begged me this morning to let him know whether His Majesty's Government would not consent to. move British fleet a little further off."[108]
>
> SIR LOUIS MALLET, BRITISH AMBASSADOR, ISTANBUL.

30 SEPTEMBER 1914

ENGLAND

> "Dardanelles were closed unnecessarily by Turkish authorities, and there is no reason why they should not be reopened. Turkish Government are well aware that we have no intention of initiating any aggressive action against Turkey.
>
> "The watch maintained by British fleet outside Dardanelles cannot be withdrawn so long as German officers and men remain in Turkish waters and are in control of Turkish fleet. Until, therefore, the German officers and crews are repatriated, the request that the fleet should be moved cannot be entertained."[109]
>
> SIR EDWARD GREY, MP, BRITISH FOREIGN SECRETARY.

A U.S. Marine Major continued to observe the progress of the Royal Naval Division.

> "There was no training or drilling going on at the time of my visit [to Shorncliffe], Saturday afternoon. According to Army orders the training of the men of the New Army should be for six days of the week, eight hours per day. However, battalions and companies were being marched to the beach, each man with his towel over his shoulder, where all the men stripped and were sent into the water.
>
> "I saw several thousand men stripped, affording an opportunity of judging the physical characteristics of the troops. The physical average was higher than that of American recruits, in chest development these men seemed to be superior.
>
> "Judging from the civilian clothing worn, from the bearing and appearance of those men, I would say that in intelligence, adaptability and social habits, these men are inferior to an equal number of American recruits."[110]
>
> MAJOR THOMAS C. TREADWELL, USMC, AMERICAN EMBASSY, LONDON.

---

108 *Correspondence Respecting Events Leading to the Rupture of Relations with Turkey*, p. 33.
109 *ibid.*, p. 34.
110 Naval Attache's Reports, Office of Naval Intelligence, Unpublished Manuscript, US Naval War College, September 1914.

Burning oil tanks at Novorossiysk.

OCTOBER 1914

"Flames rose... a dense, dark smoke hung over the fortress."
Wireless Operator George Kopp, SMS *Goeben*

1 OCTOBER 1914

ENGLAND

The still-forming Royal Naval Division was beginning its initial training while an early, and very premature, deployment came under consideration.

> "At Deal, which is the Depot of the RMLI, I was admitted to the parade ground of the South Barracks, and there saw seven hundred men paraded. These men were mostly recruits under training, but all were in uniform, in contrast with conditions at army posts I have visited, where a majority of the men have been without uniforms. These Marines were very well set up, clean, smart and military, but averaged very young and in physique were inferior to the army recruits. Especial efforts are being made to recruit for the light infantry, which have not been observed in the case of the artillery…
>
> "The brigade of the Royal Naval Division at the Crystal Palace has reached over 5000 in strength. It is very popular, there being competition to enlist. The organization has been completed and according to report no more men will be accepted. The men are quartered in some of the exposition buildings and sleep in hammocks suspended from frames. They were being trained in infantry tactics and in signalling."[1]

LT-COL. RUFUS H. LANE, USMC, LONDON.

> "This morning I was drilling like Hell. This afternoon I had to inspect two hundred rifles. Now I'm lying on my camp bed snatching a rest, before dressing for a 'night-attack'. We start out at 4.30, and have to 'fight' and march through the night, returning at dawn. There'll be a full moon, and it'll be damned cold. In front, in the sunlight, a sentry with fixed bayonet is marching up and down. All these men are in naval uniform still. I — *I* — am in control of some fifty of them; awful ruffians, nearly all from remote parts of Scotland or — oh! — Ireland. When I try to get their names, they say 'Mgchngchchch'. What a tongue, the Celtic! But they're very nice, and I get fond of them."[2]

SUB-LT RUPERT BROOKE, RNVR, ANSON BATTALION, ROYAL NAVAL DIVISION, BETTESHANGER CAMP.

> "The Germans are pounding away with their big guns at Antwerp, and though the Belgians are in a large numerical superiority they seem for the moment to have lost morale. One cannot be surprised at this, or blame them, for the Germans have been unusually active the last few days in burning their villages and shooting the inhabitants… The commander telegraphs that he does not think they can hold out

---

[1] Naval Attache's Reports, Office of Naval Intelligence, Unpublished Manuscript, US Naval War College, October 1914.

[2] Rupert Brooke quoted in Keynes, Geoffrey (Ed.), *The Letters of Rupert Brooke*, p. 620, Harcourt, Brace & World (New York) 1968.

for more than another three days. Of course it would be idle butchery to send a force like Winston's little army there. If anything is to be done it must be by Regulars in sufficient numbers."³

HERBERT ASQUITH, MP, BRITISH PRIME MINISTER.

CHINA

One British pre-dreadnought battleship, HMS *Triumph*, had supported the Japanese assault on the German enclave at Tsing Tao. The view offshore often hid more than it revealed. It was a lesson others would have to learn.

"We made a start... on Monday, joining with the Japanese Admiral in the attack. The place in strongly fortified, there being quite 6,000 German soldiers there in addition to garrison fort units. We bombarded them under a very heavy artillery fire from their forts, which concentrated fire on us. Shot and shell were bursting around the ships and across them, but we stuck to it and we silenced a few of their forts, and did a deal of damage...

"We are in Wei-hai-wei now, coaling and taking ammunition, and we leave again to-morrow for Tsing-tao. We have the worst to go through yet... I hope to be able to tell you in the next letter that we have done Tsing-tao in."⁴

PTE WILLIAM SHAW, RMLI, HMS *TRIUMPH*.

NEW ZEALAND

"I am so sorry I have not written to you before; but never mind dear sister I have not forgotten you. I got your letter the other night, and I was so glad to hear from you. I am having such a fine time in Wellington... I proposed to Della in a letter, and she has accepted to be my future wife, if I returned. She is such a darling kid in her Photo and she is always talking about you. If I do not come back from the front you can claim my money. I have fixed it all up. I shall send you a Photo of me in a few days. We [are] staying here at least three weeks... I am a mess orderly on my ship, and we get to feed seven hundred men."⁵

PTE JOSEPH REMNANT, WELLINGTON BN, NZEF, HMNZT *MAUNGANUI*.⁶

---

3   Asquith, *Memories and Reflections*, vol. 2, p. 40.
4   *Leicester Evening Mail*, 5th November 1914.
5   Military personnel files for REMNANT, Joseph Stanley, Archives New Zealand R20805199.
6   Remnant, a former farmer, left the ship at Albany, Western Australia. He died in hospital of pneumonia on 19th March 1915. Buried in Albany Cemetery (Old), he was the brother of Emma Remnant, of 'Te Mata,' Whakaronga, Palmerston, New Zealand.

2 OCTOBER 1914

TURKEY

Russian sources informed them that they should expect to be attacked, mostly likely by ships in the Black Sea.

> "From absolutely reliable sources I learn that the Austro-Hungarian Ambassador declared to the Grand Vizier that Turkey ought now to proceed against Russia. The Grand Vizier replied that Turkey was ready to proceed, but did not know in what direction action should begin; whereupon, the Ambassador pointed out that the fleet should be used."[7]
>
> MIKHAIL NIKOLAYEVICH DE GIERS, RUSSIAN AMBASSADOR, ISTANBUL.

RABAUL

German warships still at large in the Pacific and Indian Oceans were preventing the transfer of men to Europe. The enforced inaction was causing problems.

> "There is some talk of our being relieved from here as soon as the German warships in the vicinity have been captured or destroyed. As it is now we are always prepared for attack, for it is very probable that the enemy's warships will make some determined attempt to wrest these possessions from us. Our fleet is laying out in the bay ever watchful. A big French cruiser is with us. I am sorry to say that the Australian submarine *AE1*[8] has been missing about ten days, and no hope is held out for her. Whether she was sunk by the enemy or went down to the depths never to rise again, as has so often happened to her kind, we do not know. One of my best naval friends was torpedo instructor aboard her, so you can understand how her loss affects me."[9]
>
> PTE EDWARD SCOTT-HOLLAND, AUSTRALIAN NAVAL & MILITARY EXPEDITIONARY FORCE.

AUSTRALIA

> "In view of the unrest which the present delay in the despatch of the Australian Imperial Force is occasioning, and the ill effect upon training of an indeterminate period of detention, I think it very desirable that the Admiralty should be asked if existing conditions admit of the departure of transports."[10]
>
> MAJOR-GENERAL WILLIAM BRIDGES, COMMANDING AUSTRALIAN IMPERIAL FORCE.

---

7   Scott, *Diplomatic Documents Relating to the Outbreak of the European War*, pp. 1420–1421.
8   The submarine, under the command of Lt-Cdr Thomas Fleming Besant, RN, attached RAN, was reported missing on 14th September 1914 off Rabaul and later presumed lost with all hands. Its fate was not confirmed until the wreck was discovered in December 2017 near the Duke of York Islands. It is believed a ventilation valve had been left open after diving causing catastrophic flooding, sending the *AE1* to the bottom.
9   *Newcastle Morning Herald and Miners' Advocate* (New South Wales), 16th October 1914.
10 Typescripts; comprise questions to be put to Sir Brudenell White, AWM38 3DRL 6673/153.

3 OCTOBER 1914

TURKEY

After discovering the existence of a treaty between the Ottomans and Germany, with little confidence in any organised opposition to Enver Pasha, the Russian Ambassador concluded that war was practically inevitable.

> "The general situation in Constantinople for the past few weeks has developed itself in one direction — increased preparation of Turkey for war. The appearance of the "Goeben" and "Breslau" completely turned the heads of the Turks, a fact of which the Germans and Austrians were not slow to take advantage, finally to win Turkey over to their side. As you already know through my telegram of August 27, even a treaty was concluded between them. Since that time, the Minister of War, appointed generalissimo of the army and navy, completely turned them over, the one as well as the other, to German hands...
>
> "A struggle is on in the Council of Ministers between the conservative party and Enver, sometimes supported by Talaat Bey — a struggle wavering continually according to the tenor of the news received from the seat of war. In the country, which is without doubt being plundered for war purposes, great dissatisfaction is arising, and in the army discontent is growing against the German hegemony. But there is no one with energy enough to head the movement. This alarming uncertainty of the situation may continue until we shall have achieved complete success in the war, when the present Ministers will have the hardihood to liberate themselves from Enver and from the Germans. But the most probable outcome is that the Germans themselves will create an incident to precipitate Turkey into war."[11]
>
> MIKHAIL NIKOLAYEVICH DE GIERS, RUSSIAN AMBASSADOR, ISTANBUL.

ENGLAND

Despite considering it "idle butchery" to send the untrained Royal Naval Division to Antwerp two days previously, the worsening situation there changed Asquith's mind to the dismay of some naval officers.

> "The Belgian Government, notwithstanding that we are sending them heavy guns and trying hard to get troops to raise the siege of Antwerp, resolved yesterday to throw up the sponge and leave to-day for Ostend.... So we at once replied urging them to hold out and promising Winston's Marines to-morrow, with the hope of help from the main army and reinforcements from there...

---

11   Scott, *Diplomatic Documents Relating to the Outbreak of the European War*, p. 1424.

> "[T]he intrepid Winston set off at midnight and ought to have reached Antwerp at about 9 this morning.... we are... rather anxiously waiting Winston's report. I don't know how fluent he is in French, but, if he was able to do himself justice in a foreign tongue, the Belges will have listened to a discourse the like of which they have never heard before. I cannot but think that he will stiffen them up."[12]

HERBERT ASQUITH, MP, BRITISH PRIME MINISTER.

4 OCTOBER 1914

ENGLAND

> "The siege of Antwerp looks ugly. I hope it may hold out. The 1st Sea Lord is sending his army there; I don't mind *his* tuppenny untrained rabble but I do strongly object to 2,000 invaluable marines being sent to be locked up in the fortress. They are our last reserve. No Board of Admiralty of two-pennyworth of knowledge & backbone would have allowed marines to be used in such a way… It is a tragedy that the Navy should be in such lunatic hands at this time."[13]

CAPT. HERBERT RICHMOND, RN, ASSISTANT DIRECTOR OF OPERATIONS, ADMIRALTY.

BELGIUM

Churchill viewed the developing crisis at Antwerp as his opportunity to take a leading role in the action. He asked to be allowed to resign from the Admiralty and be given command of all allied forces there. Such was his fantasy.

> "If it is thought by H.M. Government that I can be of service here, I am willing to resign my office and undertake command of relieving and defensive forces assigned to Antwerp in conjunction with Belgian Army, provided that I am given necessary military rank and authority, and full powers of a commander of a detached force in the field. I feel it my duty to offer my services, because I am sure this arrangement will afford the best prospects of a victorious result to an enterprise in which I am deeply involved. I should require complete staff proportionate to the force employed, as I have had to use all the officers now here in positions of urgency. I wait your reply. Runciman would do Admiralty well."[14]

WINSTON CHURCHILL, MP, FIRST LORD OF THE ADMIRALTY.

---

12   Asquith, *Memories and Reflections*, vol. 2, p. 41.
13   Marder, *Portrait of an Admiral: The Life and Papers of Sir Herbert Richmond*, pp. 111–112.
14   Churchill quoted in Beaverbrook, Lord, *Politicians and the War 1914–1916*, p. 48, Oldbourne Book Co. Ltd. (London) 1960.

5 OCTOBER 1914

ENGLAND

Asquith's response to Churchill's suggestion that he be appointed the commander of the forces defending Antwerp was diplomatic. It is doubtful if Kitchener, for one, would have felt the need to be so polite.

> "I found when I arrived here this morning a telegram from Winston, who proposes to resign his office in order to take the command in the field of this great military force. Of course, without consulting anybody, I at once telegraphed to him warm appreciation of his mission and his offer, with a most decided negative, saying we could not spare him at the Admiralty. I had not meant to read it at the Cabinet, but as everybody, including K., began to ask how soon he was going to return, I was at last obliged to do so. Winston is an ex-lieutenant of Hussars, and would, if his proposal had been accepted, have been in command of two distinguished major-generals, not to mention brigadiers, colonels, etc., while the Navy were only contributing their little brigades."[15]

HERBERT ASQUITH, MP, BRITISH PRIME MINISTER.

BELGIUM

Nevertheless, the First Lord's sense of his own importance was undented by Asquith's rebuff.

> "It is my duty to remain here and continue my direction of affairs unless relieved by some person of consequence, in view of the situation and developing German attack. Prospects will not be unfavourable if we can hold out for next three days."[16]

WINSTON CHURCHILL, MP, FIRST LORD OF THE ADMIRALTY.

TURKEY

The diplomatic dance continued in Istanbul.

> "A month ago the British Ambassador, Sir Louis Mallet, returned to Constantinople. Shortly after his arrival, Sir Louis solicited an audience with His Majesty the Sultan. His Majesty gave the Ambassador an extraordinarily amiable reception and told him Turkey desired to observe strict neutrality and did not cherish the slightest unfriendly design against any of the Foreign Powers, and that the German crews brought with the war vessels obtained from Germany might be sent away within a few days.
>
> "In response to this conciliatory declaration. Sir Louis Mallet in his turn, informed His Majesty that he was authorised to state that at the close of the war the Government

---

15   Asquith, *Memories and Reflections*, vol. 2, pp. 41–42.
16   Churchill, *The World Crisis 1911–1914*, pp. 351–352.

of Great Britain would return to Turkey the two dreadnoughts which it has under detention."[17]

MIKHAIL NIKOLAYEVICH DE GIERS, RUSSIAN AMBASSADOR, ISTANBUL.

NEW ZEALAND

The threat posed by the German East Asia Squadron continued to disrupt the planned departure of the troopships to Europe.

"Some of our infantry transports departed from here September 24th but were recalled on receipt of the message to the effect that very serious risk was being incurred. Another message was received from England to the effect that adequate escort was being arranged to escort our transports from Wellington to junction with Australian Expeditionary Forces. Great public anxiety felt as regards what had taken place and I gave assurance that the ships should not go till the escort should arrive. Communications are passing between the New Zealand Government and the Imperial Government but my Government expects good faith to be kept with regard to escort and as soon as word comes we can be ready to leave at probably 24 hours' notice."[18]

WILLIAM MASSEY, PRIME MINISTER OF NEW ZEALAND.

6 OCTOBER 1914

ENGLAND

Churchill's adventures in Antwerp meant that Admiralty responsibilities had to be taken up by the Prime Minister. Asquith did not order him home.

"Winston persists in remaining there, which leaves the Admiralty here without a head, and I have had to tell them to submit all decisions to me. I think that Winston ought to return now that a capable general is arriving. He has done good service."[19]

HERBERT ASQUITH, MP, BRITISH PRIME MINISTER.

BELGIUM

"Colonel Maxwell was quite close to me when he was knocked over. We were all in our trenches digging ourselves in. About 9 a.m… I was standing on the parapet in front with a spade, making head cover, when a shell came whistling by. Colonel Maxwell had just been along my Company's trenches, and we were looking at a Taube about 3000 feet above us. He said — "He's got our position; so look out for some shells." He

---

17   Scott, *Diplomatic Documents Relating to the Outbreak of the European War*, pp. 1425–1426.
18   Typescripts; comprise questions to be put to Sir Brudenell White, AWM38 3DRL 6673/153.
19   Asquith, *Memories and Reflections*, vol. 2, p. 42.

went along to the next trenches, and the first shell pitched just behind them. He was then going across to his dug-out to get some breakfast, when a shell burst close to him (common shell, I think, for it did not burst like shrapnel). Colonel Maxwell was knocked over, and four men near him were killed at once. They shelled us heavily, but Lieut. Carlisle[20]… and Petty Officer Mutram (who was himself wounded) ran out and carried him in. As soon as I got all my men stowed away in the dug-outs, I crawled along the connecting trench to Colonel Maxwell's dug-out with some bandages; but I could see he was badly hit and was unconscious.[21] We got him off to the Antwerp Hospital as soon as possible, and tried to keep it from the men as long as we could."[22]

LT GEORGE HAMMICK, RNVR, COLLINGWOOD BATTALION,
1ST (NAVAL) BDE, ROYAL NAVAL DIVISION.

TURKEY

"I learn through reliable sources that yesterday the Austro-Hungarian Ambassador advised the Grand Vizier that the allied German and Austrian Governments deem that the time has arrived for hostile action against us, and that the Turkish fleet should now be attacking the Black Sea coast. The Grand Vizier is alleged to have answered in an evasive manner, expressing the opinion that action by the fleet cannot have decisive results in the present state of affairs."[23]

MIKHAIL NIKOLAYEVICH DE GIERS, RUSSIAN AMBASSADOR, ISTANBUL.

7 OCTOBER 1914

ENGLAND

Despite the evolving disaster at Antwerp, Churchill's appetite for military command had not diminished. With the Navy's work "practically over," he considered his about to begin.

"I have had a long call from Winston, who, after dilating in great detail on the actual situation, became suddenly very confidential and implored me not to take a conventional view of his future. Having, as he says, tasted blood these last few days, he is beginning, like a tiger, to raven for more, and begs that sooner or later — and the sooner the better — he may be relieved of his present office and put in some kind of military command. I told him that he could not be spared from the Admiralty, but he scoffs at that, alleging that the naval part of the business is practically over, as our superiority will grow greater and greater every month. His mouth waters at the sight

---

20 Sub-Lt William Carlisle, RNVR, Collingwood Bn, later a prisoner of war.
21 Lt-Col. Aymer Maxwell, Royal Marines, Commanding Collingwood Bn, died of wounds on 9th October 1914. Buried Schoonselhof Cemetery, Belgium, he was the 36 year-old son of the Rt. Hon. Sir Herbert Maxwell, 7th Bart., of Monreith, Wigtownshire, and Lady Maxwell; husband of Lady Mary Maxwell, of House of Elrig, Portwilliam, Wigtownshire.
22 George Hammick quoted in Maxwell, Herbert, *Evening Memories*, pp. 344–345, Alexander Maclehose & Co. (London) 1932.
23 Scott, *Diplomatic Documents Relating to the Outbreak of the European War*, p. 1428.

and thought of K.'s new armies. Are these "glittering commands" to be entrusted to "dug-out trash" bred on the obsolete tactics of twenty-five years ago, "mediocrities who have led a sheltered life mouldering in military routine," etc., etc.? For about a quarter of an hour he poured forth a ceaseless cataract of invective and appeal, and I much regretted that there was no shorthand writer within hearing, as some of his unpremeditated phrases were quite priceless. He was, however, three parts serious and declared that a political career was nothing to him in comparison with military glory."[24]

HERBERT ASQUITH, MP, BRITISH PRIME MINISTER.

BELGIUM

One British woman would have been able to describe the sounds glory made, or of shells at Antwerp at least.

"Some people have described the noise as being a scream, and others have called it a yell, and we get such expressions as "whizzing" and "whistling," but I do not think any of these words quite describe it. It is a curious sound of rending, increasing in violence as the missile comes towards one, and giving one plenty of time to wonder, if one feels so disposed, whether it intends to hit one or not. This has its useful side if one is inclined to take cover, but it certainly adds a little to the mental discomfort which being under a prolonged bombardment involves…

"As soon as the shells began to come over, the helpless wounded all began to scream, while some of those who, we imagined, would not walk again leapt out of bed. The nurses quieted everybody, and an assurance that we did not mean to desert them seemed to bring a curious sense of safety to the men — as if a handful of women could protect them from bursting shells!"[25]

SARAH MACNAUGHTON, WOMEN'S SICK AND WOUNDED CONVOY CORPS, ANTWERP.

8 OCTOBER 1914

ENGLAND

The fighting around Antwerp was nearing its end. An unmitigated disaster, it demonstrated the folly of sending barely trained men, without artillery, into action against an enemy that knew its business.

"The news this morning from Antwerp was distinctly bad. The Germans had been bombarding away all night, and General Paris, who commands Winston's Naval Division, talked of evacuating the trenches….

"Kitchener has just been with us and is coming again to confer with me and Winston. I have still some hope that things may come out right…

24   Asquith, *Memories and Reflections*, vol. 2, pp. 45–46.
25   Macnaughton, Sarah, *A Woman's Diary of the War*, pp. 41–42, Thomas Nelson and Sons (London) 1916.

"Later. Just had a conference with K. and Winston about Belgium. There is, I fear, nothing to be done but to order our naval men to evacuate the trenches to-night... Poor Winston is very depressed, as he feels that his mission has been in vain."[26]

HERBERT ASQUITH, MP, BRITISH PRIME MINISTER.

BELGIUM

Belgians had more reason to be depressed than Antwerp's erstwhile saviour.

"The city is in a panic. All [tram] cars have stopped running, and the business houses have closed. After a lot of trouble we have managed to get a ticket for Rotterdam. This afternoon we left. The scenes in the city and noises of the guns were awful."[27]

LILLIE MATHILDA DEMAEGHT, ANTWERP.

9 OCTOBER 1914

TURKEY

American intelligence identified five lines of mines blocking the Dardanelles. The German crews remained aboard their ships.

"The Dardanelles is now actually closed, not only by decree but by mines.

"The Dardanelles is mined as follows:

1st line,	From Kephez Light across	— 20 mines
2nd line,	Across from between Kephez Light and Kephez Point	— 20 mines
3rd line,	Across a little above Kephez Point	— 29 mines
4th line,	Across from Sari Siglar Buoy	— 22 mines
5th line,	On the 1st of October, just after the Dardanelles had been closed, I received a report from a good, and reliable, authority that a 5th line of 29 mines had been placed between the lighthouse on Namazieh Point across in the direction of the Fort Hamidieh, Asia, leaving passage by which the battleship *Messoudieh*, now anchored in Sari Siglar Bay, might come in as needed. The 2nd, 3rd, and 4th lines of mines have been prolonged by the addition of 18 mines, 6 for each line, towards the coast of Asia. Total mines placed up to October 1st — 192...	

"Last Sunday I saw about 200 German sailors and 2 German officers from the *Sultan Selim (Goeben)*, on shore on the Asiatic side. They wore fezzes but still the German uniform.

---

26   Asquith, *Memories and Reflections*, vol. 2, pp. 43–44. Dated by Sir Martin Gilbert to 7th October 1914.
27   *The Register* (Adelaide, South Australia), 8th December 1914.

"The *Sultan Selim (Goeben)* when getting underway last week, played the German National Air, in spite of the fact that they were flying the Turkish flag."[28]

LT-CDR EDWARD MCCAULEY, USN, USS *SCORPION*, ISTANBUL.

10 OCTOBER 1914

EGYPT

As the training of British Territorials continued, news of the progress made by the Turks since the Balkan wars was noted.

"There is no doubt that very considerable progress is being made in [the Ottoman army's] efficiency, and that it will be far superior to that in existence before the Balkan war. The continuous training… and the time which has elapsed for the deliberate organisation of mobilisation and administrative arrangements must cause the Turkish forces to be now regarded as a factor… to be taken seriously into account."[29]

LT-COL. FREDERICK CUNLIFFE-OWEN, ROYAL FIELD ARTILLERY,
FORMER BRITISH MILITARY ATTACHÉ, ISTANBUL.

"We have a very good instructor over us. He takes us out on a part of the desert and then we act as though there were a few thousand Germans in front of us… I don't care how soon they send us to the front; we are all dying to have a pop at the Germans… All the trained men sent a paper in to the officer last week, volunteering for the front, but he told them they would be there soon enough. I don't think we shall see the front all, but we might go to France to guard the lines of communication. We are all in the best of health so far."[30]

PTE FRANCIS WHARAM, 1/6TH BN LANCASHIRE FUSILIERS, EAST LANCASHIRE DIVISION.

11 OCTOBER 1914

TURKEY

The price for Turkish participation in the war having been agreed, the attitude shown towards the allied nations took another turn for the worse.

"I received from von Wangenheim an invitation to an intimate lunch in the Embassy at Therapia. When I arrived I found the Grand Vizier present, with Talaat Bey, Halil Bey, and Enver Pasha. Von Kuhlmann, recently appointed Councillor of Embassy, was also there. After lunch we all went to the Ambassador's private room. Wangenheim, with a

---

28 Naval Attache's Reports, Office of Naval Intelligence, Unpublished Manuscript, US Naval War College, October 1914.
29 Frederick Cunliffe-Owen quoted in Strachan, Hew, *The First World War, Volume One: To Arms*, p. 693, Oxford University Press, (Oxford) 2001.
30 *Rochdale Observer*, 7th November 1914.

very doleful face, told us that Germany had accepted all our financial conditions, and looked at us as much as to say: "Now don't start thinking of any more objections!"[31]

AHMED DJEMAL PASHA, OTTOMAN NAVAL MINISTER.

"The Porte has decided to promulgate at an early date a law for the subjection of all foreign schools, whether secular or clerical, to governmental control. I will protest conjointly with the French and British Ambassadors."[32]

MIKHAIL NIKOLAYEVICH DE GIERS, RUSSIAN AMBASSADOR, ISTANBUL.

EGYPT

With the increasing likelihood of war with the Ottomans, troops were being prepared to provide a garrison for Cyprus.

"When you receive this we shall probably have left Egypt, but as yet I cannot say definitely. A day or two ago we were informed that the B and C Companies must hold themselves in readiness to move, and we expect before many days are gone to be on our way to [Cyprus]. We continue to drill, and drill, but know nothing. I am beginning to think it will be quite two years before I see Southport again. From such news as we can gather, I calculate the war to last a considerable time, and the longer it lasts the better chance we have of seeing some firing… We have an awful lot of work — six hours a day for the men, and about twelve for the officers. There are no pyramids about here, or any ancient buildings at all. I hope to get a chance of going to Cairo before I return. Everything I say about the future is merely surmise, but there is quite a chance we shall see Jerusalem. I look like seeing a bit of the world on this trip, especially if we are moved out to India."[33]

LT EDWARD HORSFALL, 1/8TH BN MANCHESTER REGT, EAST LANCASHIRE DIVISION.

12 OCTOBER 1914

TURKEY

The leaders of the Ottoman Empire met to consider their options. They had two choices. Peace was not one of them:

"1. Immediate intervention in the World War.

"2. To send Halil Bey, accompanied by Hakki Bey and the Deputy Chief of the General Staff, to convince the Germans of the necessity of maintaining neutrality for another six months.

---

31 Djemal Pasha, *Memories of a Turkish Statesman*, p. 129.
32 Scott, *Diplomatic Documents Relating to the Outbreak of the European War*, p. 1428.
33 *Todmorden Advertiser and Hebden Bridge Newsletter*, 6th November 1914.

"The second alternative was advocated by Djavid Bey, but the other Ministers stood by the first. For the first time the Grand Vizier showed himself undecided.

"At that moment Enver Pasha told us that in consequence of the numerous and very justified protests of the Admiral, on military grounds he could no longer oppose the cruise of the *Goeben* and *Breslau* into the Black Sea. Yet the excursion of these two warships, accompanied by other Ottoman vessels, would inevitably involve our participation in the war."[34]

AHMED DJEMAL PASHA, OTTOMAN NAVAL MINISTER.

ENGLAND

Asquith learned that untrained and ill-equipped men had been sent to Antwerp. Someone had lied to him. He may have been guilty of lying to himself.

"Oc[35] came to lunch yesterday and I had a long talk with him after midnight, in the course of which he gave me a full and vivid account of the expedition to Antwerp and the retirement. Marines of course are splendid troops and can go anywhere and can do anything, but Winston ought never to have sent the two Naval Brigades. I was assured that all the recruits were being left behind, and that the main body at any rate consisted of seasoned Naval Reserve men. As a matter of fact only about one quarter were Reservists and the rest were a callow crowd of the rawest recruits, most of whom had never fired off a rifle, while none of them had even handled an entrenching tool."[36]

HERBERT ASQUITH, MP, BRITISH PRIME MINISTER.

AUSTRALIA

The order to move was received at Broadmeadows.

"The bomb has fallen. We have got news at last. We break camp on Thursday to leave for Melbourne, and from there to proceed (after a period of training) to the old land, thence to Germany...

"Yesterday, our last Sunday in camp we had a big crowd — loving mothers, sisters and brothers, coming to bid their loved ones so-long, and incidentally bringing certain stocks of goodly provisions."[37]

A/CPL GEORGE TIDEX, 13TH BATTALION, AIF, BROADMEADOWS CAMP.

---

34   Djemal Pasha, *Memories of a Turkish Statesman*, pp. 129–130.
35   Asquith's son, Sub-Lt Arthur Melland Asquith, RNVR, Hood Bn. He was wounded at Helles on 6th May 1915 and again, in France, on 19th October 1916.
36   Asquith, *Memories and Reflections*, vol. 2, p. 45.
37   *The Leader and Stock and Station News, Morning Daily* (Orange, New South Wales), 16th October 1914.

13 OCTOBER 1914

ENGLAND

Churchill's popularity was far from universal. Recent events had not improved his standing in some quarters.

> "And the time has come for more plain and definite speaking. What has to be said, then, is that the attempt to relieve Antwerp by a small force of marines and naval volunteers was a costly blunder for which Mr Winston Spencer Churchill must be held responsible on the present evidence…
>
> "The other day Mr. Churchill used the expression that the German navy were to be 'dug out of their hole like rats' if they could not be got out otherwise. Now it is obvious that either this was an idle boast or that it foreshadowed a combined military and naval operation. If the former, then boasting is not appropriate to a British minister's position; if the latter, why betray a secret of such importance?"[38]

HOWELL GWYNNE, EDITORIAL, MORNING POST.

TURKEY

Residents of Istanbul awoke to the sound of gunfire.

> "At about 5.30 [a.m.], just before daylight heavy gun firing was heard in the direction of the Black Sea. It lasted for about twenty minutes and was continuous, I sent up the Bosphorus, as soon as daylight permitted, to find out particulars of the firing, but nothing was known of its purport. It had, however, been heard by everyone in the vicinity and in the city. However, the batteries in the upper Bosphorus had not fired, and the Turkish Fleet were all inside having come in the evening before, or that morning. A Russian merchantman, which came in from the Black Sea that morning, had also heard the firing, but knew nothing of its cause. I myself believe it was one or more of the Turkish Coast Batteries on the Black Sea firing on some native, or innocent, vessel which they took to be a Russian torpedo boat or other Russian war vessel which they thought was attempting to enter the Bosphorus. Many rumors were circulated for none of which I find any foundation, and the Government authorities said it was target practice and exercises. This I do not believe owing to the unusual and unreasonable hour, and the extent of the firing. However, I feel sure that firing was not hostile."[39]

LT-CDR EDWARD MCCAULEY, USN, USS *SCORPION*, ISTANBUL.

---

38   *The Morning Post*, 13th October 1914.
39   Naval Attache's Reports, Office of Naval Intelligence, Unpublished Manuscript, US Naval War College, October 1914.

14 OCTOBER 1914

TURKEY

News agencies reported the deteriorating situation from the Bosphorus to the Dardanelles.

> "Sir Louis Mallet, the British Ambassador, in a circular telegram to the British Consuls throughout the Ottoman Empire, pointed out that the Dardanelles had been closed by the commander because a Turkish destroyer, with officers on board, had been turned back by the British Fleet, the British Government having previously informed the Ottoman Government that as long as the Ottoman Fleet was officered and manned by Germans it must regard it as part of the German Fleet...

> "The circular added that it was estimated that there were three thousand German officers and men in the Turkish Fleet, and that the closing of the Dardanelles would not affect British interests to a large extent, as British commerce had ceased for some weeks past owing to the military requisitions. The circular has attracted much attention, and has produced a disagreeable impression in Turkish official circles."[40]

REUTER'S.

NEW ZEALAND

As orders for embarkation were issued, families said their goodbyes.

> "We go out into the stream to-morrow morning early, so I am writing you now, and will post it before we leave the wharf. Pater came on board with mater to say good-bye last night, and I took them over our quarters. Mater was vastly impressed with the fact that we have the first-class accommodation of the vessel. We have our mess in the huge saloon, but all the marble panelling is covered over with rough deal boards and all the regular tables have been taken out, and we now sit at boards that are as rough as our manners. You know we have a mixed company, and bank clerks and big run holders sit cheek by jowl with shearers and roadmen. All of them are jolly good fellows though, and all as keen as mustard to get away...

> "The pater is a funny old bird, isn't he? When he had talked a bit to me about keeping my nut down when it wasn't wanted up, he said he had a lot of writing to do for to-morrow's English mail. Then he shook hands rather hurriedly and went down the gangway and along the wharf without even once looking back. His figure faded into a mist as he got near the end, and I had to take a pull on myself and talk hard to mater, who had not gone ashore... Neither of us felt too cheery, but mater is the bravest little woman in the world, and she kissed me and went down on to the wharf with the cheeriest of smiles on her face. She waited for a while at the barrier and waved."[41]

L/CPL NOËL ROSS, CANTERBURY BN, NZEF, HMNZT *ARAWA*.

---

40   *Birmingham Daily Post*, 19th October 1914.
41   Ross, Malcolm and Ross, Noël, *Light and Shade in War*, pp. 28–29, Edward Arnold (London) 1916.

15 OCTOBER 1914

ENGLAND

Asquith's peacetime laissez-faire attitude allowing ministers to run their departments with minimal intervention was not working in wartime. No-one could be in doubt who was being referenced here.

> "I believe that the public do ask for and require adequate assurances that unnecessary risks will not be run as the result of efforts, however well meant, of single individuals, uncontrolled by the Cabinet as a whole, and, even worse, without the assent of the leaders in the field."[42]
>
> WALTER LONG, MP.

RABAUL

New soldiers had much to learn. Some did not get the opportunity to learn much.

> "A chap was fooling around with his rifle, which exploded, the bullet entering his thigh, carrying away portion of the bone. He died that night from shock.[43] He was buried with full military honors. The coffin was carried on a gun carriage, drawn by blue jackets, to the cemetery, where the troops lined up. After the burial service a firing party then fired a volley and then the 'Last Post' was sounded. It was very impressive."[44]
>
> L/CPL HAROLD ATKINS, AUSTRALIAN NAVAL & MILITARY EXPEDITIONARY FORCE.[45]

16 OCTOBER 1914

TURKEY

As the British Ambassador struggled to accept the Ottomans would be so foolish as to opt for war, his Russian counterpart knew better.

> "I cannot give up hope that if we still continue to exercise patience and if we still have successes as I do not doubt, we may pull it off, and that although we are at the mercy of an incident, it is not I but Wangenheim who will have to leave first. I confess that I should hate to be beaten now by Wangenheim, who is a typically unscrupulous and

---

42 *Leicester Evening Mail*, 15th October 1914.
43 Pte Albert Wates, Australian Naval & Military Expeditionary Force, died on 14th October 1914. He is buried in Rabaul (Bita Paka) War Cemetery. The former butcher was the brother of Elizabeth Ellen Harvey, of 78 Wellington Street, Collingwood, Victoria.
44 *National Advocate* (Bathurst, New South Wales), 3rd November 1914.
45 Commissioned, 2/Lt Atkins, 1st Bn, AIF, was killed in action at Pozieres on 23rd July 1916. Commemorated on the Villers-Bretonneux Memorial, France, he was the 27 year-old son of John and Mary Atkins, of Bathurst, New South Wales.

contemptible form of Teuton. It must not be thought that we could have prevented the abolition of the capitulations or the Post Offices or anything else by taking a different line."[46]

SIR LOUIS MALLET, BRITISH AMBASSADOR, ISTANBUL.

"I learned from an authentic source that on September 28 [October 11[47]], a meeting took place at the German Ambassador's, in which Enver Pasha and Talaat Bey took part. A special document even was signed, by virtue of which Turkey obligated herself to open hostilities against us upon receipt of a financial subsidy from Germany. The first instalment of the latter has been received."[48]

MIKHAIL NIKOLAYEVICH DE GIERS, RUSSIAN AMBASSADOR, ISTANBUL.

AUSTRALIA

The departure of the expeditionary forces could not come quickly enough for some.

"I hope from time to time to be able to report to the Minister for Defence the conduct of the Australian troops both in camp and in the field is all that it should be, and worthy of the trust imposed by the people of the Commonwealth. The men are a fine lot, soldierly and patriotic. I am grateful to the soldiers and citizens for the help given me in organizing and preparing the force which is now about to do its part for the good of the Empire. In saying good-bye I would like to express my hope that no matter how great the demand upon their patience the Australian people will see to it that there is no diminution in their determination to face their responsibilities. This spirit cannot fail then to pervade the troops."[49]

MAJOR-GENERAL SIR WILLIAM THROSBY BRIDGES,
COMMANDING AUSTRALIAN IMPERIAL FORCE.

"So far we have not left the shores of this sunny land, although the day of sailing was fixed for over a fortnight ago, but according to the latest we can hear we will be on the war path about the end of the coming week. The boys were somewhat disappointed at the postponement, and the camp is generally losing its attractiveness… It is getting close on eight weeks since joining camp, and I'm afraid another similar stay would find a good number of the lads asking for their discharges. As far as can be learnt our first move is to Aldershot, and after a period of training there we will see the actual thing, but it is expected that it will be close on April before that comes about. The health of our boys has been pretty good, but Arthur Clarkson[50] has been laid up with

---

46 Mallet quoted in Heller, Joseph, *British Policy Towards the Ottoman Empire*, p. 150, Frank Cass and Company Ltd. (London) 1983.
47 Russia used the Julian calendar until 1918.
48 Scott, *Diplomatic Documents Relating to the Outbreak of the European War*, p. 1429.
49 *The Narracoorte Herald* (South Australia), 24th November 1914.
50 Tpr Arthur Clarkson, 4th Australian Light Horse Regt, was wounded at Gallipoli on 30th August 1915. He was evacuated from the peninsula suffering from enteritis, being treated first at Malta and then England.

influenza the last three days. As a matter of fact they have all, with the exception of myself, been under the doctor with colds, but only for a day or so."[51]

SGT GEORGE CAMPIGLI, 4TH AUSTRALIAN LIGHT HORSE REGIMENT, AIF, BROADMEADOWS CAMP.[52]

NEW ZEALAND

"At about half-past eight the first of our escort, the flagship, steamed slowly out of the harbour, and one by one our big grey transports followed, until we brought up the end of the line. It was a wonderful sight, and the few people who had the luck to be about saw something that they will remember all their lives… On the other side of the harbour faint cheers came from shore-wards, and a flag dipped to us as we passed the Forts. On through the rocky headlands and out away westwards, past the long white beaches, and then the iron-bound coast of Terawhiti. The strait was calm as glass, and as the long line of sixteen vessels swung out, the smoke from their funnels towered in black columns above them. The six transports on the starboard side slowed down (we could hear the bells of their engine-room telegraphs) until the last six of the line came up abreast of them. Then in double file, with the black signal cones showing that they were making "required speed," they forged slowly ahead.

"Away to starboard, a big grey cruiser [*Ibuki*] belched black smoke (she burns oil), and the flag at her stern was the Rising Sun of Japan."[53]

L/CPL NOËL ROSS, CANTERBURY BN, NZEF, HMNZT *ARAWA*.

17 OCTOBER 1914

ENGLAND

Churchill's defence of the Antwerp expedition included the claim that training could be described as 'incomplete' when this meant a man had never previously picked up a rifle.

"Naval brigades were chosen because the need for them was urgent and bitter; because mobile troops could not be spared for fortress duties; because they were the nearest to the scene and could be embarked the quickest, and because their training, though incomplete, was as far advanced as that of a large portion not only of the forces defending Antwerp, but of the enemy's forces attacking that place."[54]

WINSTON CHURCHILL, MP, FIRST LORD OF THE ADMIRALTY.

---

51   *Seymour Express and Goulburn Valley, Avenel, Graytown, Nagambie, Tallarook and Yea Advertiser* (Victoria), 16th October 1914.
52   The former railway clerk was invalided from Gallipoli with gastritis on 17th June 1915.
      Granted a commission in the British Army, Campigli ended the war as a Lt-Col., Deputy Assistant Director Railway Traffic, Palestine.
53   Ross, Malcolm and Ross, Noël, *Light and Shade in War*, pp. 30–31, Edward Arnold (London) 1916.
54   *The Sun* (New York), 18th October 1914.

AEGEAN SEA

As things began to move on the other side of the world, the Royal Navy maintained its watch for any sign of the *Goeben* and *Breslau* coming out. A change of headgear had not disguised the nationality of the crews.

> "We are still outside the Dardanelles, waiting for something to happen, and let us hope it is soon, as it is getting a bit monotonous waiting out here and doing nothing. We are living in hopes of those two German ships coming out again; but it don't seem as if they like to risk the experiment, and I don't think they will get very far.

> "They were supposed to have sold them to Turkey, but they have still got the German crews on board, only they are wearing Turkish fezs.

> "I received the papers; they are very acceptable, I can tell you, and get passed around a lot. They generally finish up by going aboard a merchant ship, some of whom have been up the Dardanelles for two months, and have heard nothing of the war. We have to board everyone that comes out to see if she is German or not."[55]

AB HORACE WOODCOCK, RN, HMS *RENARD*.

AUSTRALIA

> "We have been disappointed as to our dates of sailing, but hope now that it will not be long. We have all settled down to our work, and the boys are doing very well. This is no idle boast, as No 3 section of G Coy won the battalion section competition. There are 32 sections in the battalion, so I consider it a very good performance, and I feel very proud of my men. In our field work we have come in for our share of favorable criticism. We are all a happy family. Some, of course, come in for punishment, but very few, and those who do take it smiling. We have no bad offenders in the company, and the offences are only minor ones. We all feel proud to be able to represent Bendigo, and I feel sure the country boys will not be far behind. However, time will tell its tale, and I am sure we all hope to worthily represent the city we are so proud of."[56]

CAPT. HERBERT HUNTER, 7TH BATTALION, AIF, BROADMEADOWS CAMP.[57]

---

55 *Herne Bay Press*, 17th October 1914.
56 *Bendigo Advertiser* (Victoria), 17th October 1914.
57 Capt. Hunter was killed at Helles between 8th and 12th May 1915. Commemorated on the Helles Memorial, he was the 33 year-old son of the late George Frederick and Elizabeth Hunter, of Bendigo, Victoria.

18 OCTOBER 1914

TURKEY

"According to reliable information which has reached me, another remittance of the money promised to Turkey for the attack she has obligated herself to make upon Russia will arrive in Constantinople from Germany October 8 [21st October]. Thereupon, Enver Pasha and Talaat Bey will demand to know of the Grand Vizier whether he approves of immediate action, and, if not, they will demand his removal."[58]

MIKHAIL NIKOLAYEVICH DE GIERS, RUSSIAN AMBASSADOR, ISTANBUL.

AUSTRALIA

Men from the 2nd Battalion boarded their transport after a public farewell in the pouring rain.

"[T]he troops were ready to leave Kensington by 7 o'clock and at 7.30, headed by their own band, they marched out to the accompaniment of cheers from their comrades of the 1st. The day was not at all auspicious. Heavy rain clouds overhung the sky and it looked bad for a long march, but out in the street a surprise awaited them for trams were drawn up ready to take them to Darlinghurst. Sergeant Grayston and two other color-sergeants were escort for the colors. The troops got aboard the trams and the cheering that went on was deafening.

"A great crowd awaited them at Darlinghurst, and rain began to fall. The men got off the trams in quick time, and were soon marching independently in companies to the boat, being followed by crowds of people, despite the rain. The embarkation was quickly carried out, the whole battalion being seated at the mess tables within 20 minutes after arrival on the wharf. At 2 o'clock all the boys came up and lined the rigging, every available point of vantage being occupied, even to the staging round the top of the funnel. A great crowd of people lined the hill facing the wharf, and as the T.S.S. *Suffolk* began to move away the cheering and cock-a-doodle doos were repeated with renewed vigor. Handkerchiefs waved on both sides in galore, and the bugles blow call after call. Little boats, big boats, boats of medium size, motor boats, steam boats and pulling boats all swarmed round to bid the troops farewell.

"The vessel did not go straight out, but anchored for a short time in the stream. At 4 o'clock it began to move out and shortly passed through the heads. Then the fun began. A high sea was running at the time and the boat rocked a good deal, consequently mal-de-mer was very prevalent."[59]

QMS WILLIAM GRAYSTON, 2ND BATTALION, AIF, HMAT *SUFFOLK*.[60]

---

58 Scott, *Diplomatic Documents Relating to the Outbreak of the European War*, pp. 1429–1430.
59 *The Tamworth Daily Observer* (New South Wales), 10th November 1914.
60 Grayston was wounded at Walker's Ridge on 25th April 1915. His wounds led to his discharge as medically unfit.

19 OCTOBER 1914

ENGLAND

The sheer volume of volunteers far exceeded the capacity to equip and accommodate them adequately, let alone to train them quickly.

> "There were many complaints of inadequate provision of shelter, food, and clothing for recruits at training centres. That there was and is an inadequate supply of clothing and arms appears unquestionable. Many battalions are yet only partially uniformed, and some are wearing the old blue and red uniforms. That there were amy [sic] cases of inadequate shelter and food seem probable, as was inevitable under the circumstances. The public was called upon to contribute blankets, and many thousands were so furnished. However, the emergency was well met, considering the conditions, and it is doubtful if the complaints have had much effect on recruiting."[61]
>
> LT-COL. RUFUS H. LANE, USMC, LONDON.

As the men who had avoided either capture by the Germans or internment in Holland returned to England from Antwerp, stories of what had gone on there began to circulate.

> "I saw one whole section on my right completely destroyed, while one piece of a burst shrapnel killed three men quite near me. The exasperating part was that we could not retaliate. All we could do was use our rifles upon the entrenched German infantry... There was no need for the enemy's infantry to attack; their big guns were doing the work for them, and in a deadly fashion...
>
> "Towards Antwerp the scene was horrible. The shells bursting in the darkness, the oil tanks one after another sending up their huge sheets of flame as one after another they caught fire, the river aflame with the burning oil, it was as though the heavens had burst. What a different picture from that which a few days before I had looked upon as we passed through the city. Then Antwerp had been as tranquil as Crediton is to-day! Now it was a seething furnace."[62]
>
> L/CPL FRANK RADFORD, PLYMOUTH BATTALION, RMLI, RM BDE, ROYAL NAVAL DIVISION.

INDIAN OCEAN

Fresh from camp and onto transport ships, the men had to adjust to a new routine.

> "At two o'clock we drew anchor and sailed, passing through the rip at 5.45 with the *Loongana*, which had caught up to us alongside; but we very soon lost sight of her, as we both changed our course. The following day we had to try and get ourselves into

---

61 Naval Attache's Reports, Office of Naval Intelligence, Unpublished Manuscript, US Naval War College, October 1914.
62 *Western Times*, 19th October 1914.

the routine work of the boat, which was not too easy, considering that the men for the most part had not been accustomed to anything like it before. The daily routine times were as follows: — Reveille, 6 a.m.; breakfast, 7.15 a.m.; morning parade, 9 a.m.; dismissal, 11 a.m.; dinner, noon; afternoon parade, 2 p.m.; dismissal, 4 p.m.; tea, 5 p.m.; hammock swinging, 8 p.m.; lights out, 9 p.m. This hammock-swinging and sleeping business is not much good for the back till you get used to it; one got in something of a U shape."[63]

PTE RAY YOULDEN, 8TH BATTALION, AIF, HMAT BENALLA.

RABAUL

Others had to get used to a very different set of new surroundings.

"This is a very funny place; it is surrounded with mountains and two volcanoes, one of which is still active and smouldering, and a hot sulphur spring comes from the bottom. In some places it is too hot to bear the hand in, but where it enters the sea it is bosker to swim in…

"The soil seems to be very rich, although it is mostly used for cocoa-nut growing. We are here just in the right time, as all the fruit is now ripe.

"We had a funeral here last Sunday. One of our chaps shot himself accidentally while cleaning his rifle…

"Our only trouble is water, as we have to condense the sea-water, and, having only small condensers, we have to spare it."[64]

PTE ERNEST TAYLOR, AUSTRALIAN NAVAL & MILITARY EXPEDITIONARY FORCE.[65]

AUSTRALIA

But not everyone was prepared to accept a change of their surroundings, at least not without a fight.

"On 19th October 1914 I was escort to 888 Private J. Stonnell.[66] While at a public house in Port Melbourne Sergeant Munro gave the accused into my custody. I told Stonnell to come aboard with me. He said "I will not" at the same time he made a hit at me. I caught hold of him. He hit me with his head on my mouth. He struggled all the way

---

63   *Maryborough and Dunolly Advertiser* (Victoria), 25th December 1914.
64   *Molong Express and Western District Advertiser* (New South Wales), 21st November 1914.
65   Taylor arrived on the peninsula as a reinforcement to 2nd Bn, AIF, on 26th May 1915. He was evacuated to Egypt aboard HMHS *Grantully Castle* on 27th July suffering from diarrhoea and the malaria he may well have contracted at Rabaul.
66   Pte John Stonnell, originally from Cardiff, was sent back to Australia from Egypt and discharged due to misconduct: "This man seems to have no personal sense of responsibility, is a chronic leave breaker, and when not absent reports sick for trivial causes in order to avoid his work." NAA: B2455, STONNELL, JOHN.

to the boat [HMAT *Omrah*]. Another man was helping me to bring the accused along. The other man was also a Military Policeman named John McLeod[67] of E. Coy."[68]

L/CPL ALEXANDER MCLEOD, 9TH BATTALION, AIF, HMAT *OMRAH*.[69]

20 OCTOBER 1914

ENGLAND

Patience with the Ottoman Turks ended, the decision was taken by the British Government to abandon any pledge to respect the empire's territorial integrity. But had any consideration been given as to what would replace it?

"The Cabinet are of opinion that we ought to take a vigorous offensive against Turkey and to make every effort to bring in Bulgaria, Greece, and above all Rumania. Henceforward Great Britain must finally abandon the formula of 'Ottoman Integrity,' whether in Europe or in Asia."[70]

HERBERT ASQUITH, MP, BRITISH PRIME MINISTER.

INDIAN OCEAN

"After the strenuous eight or ten weeks at Randwick and Kensington, the 1st Battalion are at last afloat and steaming slowly south…

"The first day out a good number suffered with seasickness, and their ailment was not improved by those who were well counting them out. One or two of the more daring when they felt the symptoms, endeavoured to count themselves out, but it generally ended in disaster to those sitting near them, and with a cry of "Chuck him out," they would be hustled to the deck. After the first day we had beautiful calm weather, and everyone commenced to look more pleased…

"We have a daily newspaper printed on board and edited by Walter Wade (A.J.A.),[71] and sub-editor by Sergeant Buckleton.[72] It is bright and breezy and a great acquisition to everyone, and is eagerly snapped up at the Sydney price of 1d. It seems strange to hear the cry: "Paper, Sir," in the good old regular style."[73]

SGT OSWALD MOORE, 1ST BATTALION, AIF, HMT *AFRIC*.

---

67   L/Cpl John Alexander McLeod, 9th Bn, AIF, was wounded on 25th April 1915.

68   John Stonnell's service record, NAA: B2455, STONNELL JOHN.

69   L/Cpl Alexander McLeod, a former policeman, was wounded on 25th April 1915 or shortly afterwards; admitted to No. 2 Australian General Hospital, Mena, Egypt, on 30th April; arrived back in Australia on 10th September; and discharged as medically unfit 25th November 1915.

70   Herbert Asquith quoted in Spender, J.A. & Asquith, Cyril, *Life of Herbert Henry Asquith, Lord Oxford and Asquith*, vol. 2, p. 129, Hutchinson & Co. (London) 1932.

71   Pte Walter Wade, 1st Bn, AIF. The former journalist deserted from Mena Camp on 12th December 1914.

72   Sgt Sydney Buckleton, 1st Bn, AIF, a former printer returned to Australia, leaving Suez on 22nd March 1915 due to illness.

73   *Evening News* (Sydney, New South Wales), 19th November 1914.

The novelty of life on ship wore off quickly. Fortunately, a blind eye was taken to some forms of 'illegal' entertainment.

> "An order has come out to prevent gambling, and only a game called house and a sweep stake with a limit of one shilling on the ship's daily run is permitted, but the officers are very tactful, and in a quiet corner they overlook a small party having a quiet game, for it is the heritage of the Australian soldier to have a small "bit on," and regulations will be of a very stringent kind to completely check it."[74]
>
> SGT OSWALD MOORE, 1ST BATTALION, AIF, HMT *AFRIC*.

Meanwhile, the times were changing. Literally.

> "During our sleep the ship's time is put back 20 minutes each night, and it is blamed for that tired feeling at reveille; during the next few weeks we have to lose 10 hours, to drop back into the home time, so that although growing older each day, we are gradually getting younger than our cousins in Sydney."[75]
>
> SGT OSWALD MOORE, 1ST BATTALION, AIF, HMT *AFRIC*.[76]

21 OCTOBER 1914

AUSTRALIA

> "The *Arawa* dropped anchor in Hobart harbour at 12 noon. Took Hobart by surprise. The weather was very fine. We steamed to the wharf at 6.30 p.m. During the afternoon a boat race took place, the Artillery winning, Mounted Rifles second, and Infantry third, after which bathing was allowed and the men really enjoyed it."[77]
>
> LT WILLIAM JANSON, WELLINGTON MOUNTED RIFLES, NZEF, HMNZT *ARAWA*.

EGYPT

> "Every day we have sham fights. We march about five miles into the desert, and then one company acts defensive and the other as attacking party, and we have a right good fight. Lord Rochdale said we did very well, but did not drop down quick enough

---

74 *Evening News* (Sydney, New South Wales), 19th November 1914.
75 ibid.
76 Sgt Oswald Moore, 1st Bn, AIF, was wounded 25th–29th April 1915. Rejoining the battalion on 18th May, he was wounded again on 24th May, dying later that day. Commemorated on the Lone Pine Memorial, he was the 34 year-old son of Sarah Lydia Moore, of "East Anglia," Maroubra Bay Road, South Randwick, New South Wales, and the late Richard James John Moore, originally of Dunedin, Otago, New Zealand.
77 *Hawera Star* (New Zealand), 24th April 1928.

after each rush. He said we should soon learn that when the bullets were flying over our heads. Skirmishing is all right, but it is hard work running over the loose sand."[78]

PTE FRANCIS WHARAM, 1/6TH BN LANCASHIRE FUSILIERS, EAST LANCASHIRE DIVISION.[79]

22 OCTOBER 1914

AUSTRALIA

"The route march through Hobart and suburbs took place, a distance of about 10 miles, and we received a good send-off by the Hobart people. The march took three hours, from 8.30 to 11.30. At 12.30 the *Arawa* steamed from the wharf out into the stream, ready to sail at 4 p.m. for Albany. We left at 4 p.m. with a head wind and a good swell; we sailed south-west of Tasmania."[80]

LT WILLIAM JANSON, WELLINGTON MOUNTED RIFLES, NZEF, HMNZT *ARAWA*.

23 OCTOBER 1914

ENGLAND

"[David Lloyd George] is rather disgusted with Winston still about Antwerp, and think that the P.M. [Prime Minister] is too. Having taken untrained men over there, he left them in the lurch. He behaved in rather a swaggering way when over there, standing for photographers & cinematographers with shells bursting near him, & actually promoting his pals on the field of action."[81]

FRANCES STEVENSON, LLOYD GEORGE'S SECRETARY.

EGYPT

"Major E. Fletcher[82] arrived here yesterday, and looks very well after the voyage out. We are kept very busy here, and are now having route marches daily. We are also doing lot of field work, musketry, physical drill, bayonet fighting, &c., and shall soon be able to tickle the Germans with the latter weapon…

"We are greatly troubled with flies and other inserts; bugs are here in great abundance, and we have also a few mosquitoes, and one of our boys killed a scorpion the other

---

78 *Rochdale Observer*, 7th November 1914.
79 Surviving Gallipoli, Wharam was transferred to 2nd Bn Border Regt and while serving with that unit he died of wounds on 24th December 1916. Buried in Etaples Military Cemetery, he was the 24 year-old son of Godfrey and Emily Wharam, of 4 Feathershall Road, Littleborough, Lancashire.
80 *Hawera Star* (New Zealand), 24th April 1928.
81 Frances Stevenson quoted in Taylor, A.J.P. (Ed.), *Lloyd George. A Diary*, p. 6, Hutchinson (London) 1972.
82 Major Ernest Fletcher, 1/5th Bn Manchester Regt.

day. Camels are for a considerable number of jobs, but mules and donkeys are used in great numbers. As you are no doubt aware, we are 'hanging out' about six miles out of town (on the place where the battle of Alexandria was fought), but there is a splendid service of cars on the electric railway, and we can get down to town in a very short time when we wish, or rather when our C.O. wishes."[83]

PTE FRED SIMKIN PRESCOTT, 1/5TH BATTALION MANCHESTER REGIMENT, EAST LANCASHIRE DIVISION.[84]

TASMAN SEA

"The morning broke dull and grey, with a very heavy swell. The position of boats is similar, ecept [sic] that the Japanese cruiser *Ibuki* is steaming on our port side, the *Sydney* being on our starboard side. Our first horse went down today and there is very little hope of saving him. He is a chestnut belonging to the signal corps. Am feeling very much better to-day, and feel that I have got my sea legs. To-day I first took part in physical drill which is arranged for the officers."[85]

LT WILLIAM JANSON, WELLINGTON MOUNTED RIFLES, NZEF, HMNZT *ARAWA*.

24 OCTOBER 1914

ENGLAND

Most of the Royal Navy's work, patrolling, protecting the sea lanes, provided little outlet for Churchill's restless energy.

"Last night, at 8 o'clock when I was on my way upstairs to dress for dinner, a telephone message came from Churchill asking me to dine… He was in low spirits… oppressed with the opportunity of *doing* anything. The attitude of waiting, threatened all the time by submarine, unable to strike back at their Fleet, which lies behind the dock-gates of the Canal, Emden, or Wilhelmshaven, and the inability of the Staff to make any suggestions seem to bother him. I have not seen him so despondent before."[86]

CAPT. HERBERT RICHMOND, RN, ASSISTANT DIRECTOR OF OPERATIONS, ADMIRALTY.

The British Ambassador in Istanbul, Sir Louis Mallet, was told to make clear to the Turks that any incursion into Egypt would mean war.

"Your telegram of 23rd October gives the impression that Turkey considers sending an

---

83   *Leigh Chronicle and Weekly District Advertiser*, 27th November 1914.
84   Pte Fred Simkin, 1/5th Bn Manchester Regt, was killed in action on 4th June 1915. Commemorated on the Helles Memorial, he was the 18 year-old only son of Albert and Alice Prescott, of Beech Mount, Boothstown, Manchester.
85   *Hawera Star* (New Zealand), 24th April 1928.
86   Marder, *Portrait of an Admiral: The Life and Papers of Sir Herbert Richmond*, p. 121.

armed force over the frontier of Egypt as being in some way different from acts of war against Russia. You should disabuse the Turkish Government of any such idea, and inform them that a military violation of frontier of Egypt will place them in a state of war with three allied Powers."[87]

SIR EDWARD GREY, MP, BRITISH FOREIGN SECRETARY.

TURKEY

The American view of the situation contrasted starkly with that of Britain and Russia.

"Generally, things seem much more quiet here, and there seems to be less anticipation, or prospect, of Turkey going to war, as far as can be found out, from the opinion of officials, officers, and civilians, both Turkish and foreign, there is little possibility of war. The entire atmosphere seems better. From an authoritative source Turkey will not go to war unless Russia and the Allies are being beaten, from another = Turkey will not now go to war unless absolutely forced, but will not demobilise, but will keep in a state of readiness and precaution. From all I have seen and heard, I believe that Turkey now intends to keep the peace, as she realizes that it is for her own good."[88]

LT-CDR EDWARD MCCAULEY, USN, USS *SCORPION*, ISTANBUL.

EGYPT

"I am getting used to barrack life all right. I am now servant to the Rev. Denis Fletcher, senior curate at Rochdale Parish Church. Our band is playing to-night at the largest hotel in Cairo, the Continental. We played at a concert last night at the Church of England Institute, and there was a crowded audience.

"The regular soldiers who have been here had made the stage and painted the scenery. and some more are going to see the Pyramids. Some of our fellows have been photographed on camels, and they look all right. We have been to a little village called Heliopolis. There are similar amusements there to what one finds at the South Shore, Backpool [sic]. In fact, the different machines belong to the Blackpool firm."[89]

PTE JAMES TURNER, 1/6TH BN LANCASHIRE FUSILIERS, EAST LANCASHIRE DIVISION.

"The health of everyone is excellent. We have a few minor illnesses, but nothing serious. All the lads are growing fast, and they will not be recognisable when they return. Col.-Sergt. Allister is Col.-Sergt. Major of D Company, Col.-Sergt. J. Mason is Col.-Sergt. Major of C Company. Each Company consists of 260 men, and we commanders have

---

87 Scott, *Diplomatic Documents Relating to the Outbreak of the European War*, p. 1178.
88 Naval Attache's Reports, Office of Naval Intelligence, Unpublished Manuscript, US Naval War College, October 1914.
89 *Rochdale Observer*, 11th November 1914.

each five officers in our commands. Our hours of work are from 6 to 7, 8-30 to 12-30, and 4 to 5 p.m.; and it is hard training in this climate. However, we are all getting hardened, and everyone is tanned. Kindest regards to all."[90]

CAPT. JOHN GLEDHILL, 1/6TH BN LANCASHIRE FUSILIERS,
EAST LANCASHIRE DIVISION.[91]

INDIAN OCEAN

"The other transports are closing in. Four are in sight. The "Afric" is in front of us. She has turned around twice, and the crow's nest is seen to slide down the mast. A boat is at the stern, so there must be a man overboard. Life-belt drill to-day. 296 knots."[92]

2/LT GEORGE KELLY, 2ND BATTALION, AIF, HMAT *SUFFOLK*.

"The morning broke dull but fine, and very calm except for a bit of a swell. Firing was indulged in to-day. Captain Clayden lost his hat to-day while witnessing the shooting. Lt James also lost his. Fire drill was indulged in to-day for the first time. Am feeling very well."[93]

LT WILLIAM JANSON, WELLINGTON MOUNTED RIFLES, NZEF, HMNZT *ARAWA*.

25 OCTOBER 1914

AUSTRALIA

"Arrived at Albany at 6 this morning. There are about 12 other ships here. We anchored for a couple of hours, and had church and then steamed up to the wharf. The people gave us a great reception, but we were not allowed off. There was great scrambling for papers to see the news — most of the people brought along a paper to us. The "kids" on the wharf reaped a rich harvest of pennies. I don't know when you will get another letter, but I will write the first opportunity."[94]

PTE HAROLD FAWCETT, 2ND BATTALION, AIF, HMAT *SUFFOLK*.

INDIAN OCEAN

"Weather, no wind, very warm day, but a heavy swell. To-day at a quarter to 4 p.m.,

---

90 *Todmorden Advertiser and Hebden Bridge Newsletter*, 13th November 1914.
91 Capt. Gledhill was wounded during the Second Battle of Krithia on 6th May 1915.
92 Kelly, George Edward Eccleston, *Diary Kept by Lieut. G.E.E. Kelly, During the voyage from Sydney to Egypt as a Member of the Australian Expeditionary Forces*, p. 4, T. Dimmock Ltd., Printers, (Maitland).
93 *Hawera Star* (New Zealand), 24th April 1928.
94 *The Armidale Chronicle* (New South Wales), 11th November 1914.

our first burial at sea took place. It was Lance-Corporal Gillcrest[95] in the Ambulance Corps on the *Ruapehu*. At half-past 3 p.m. the *Ruapehu* steamed up between the two lines of transports, and at a signal all the boats stopped and the body was consigned to the deep, and at another signal the boats went ahead again. At a quarter past 4 p.m, parades were dismissed and sports on the deck were held."[96]

LT WILLIAM JANSON, WELLINGTON MOUNTED RIFLES, NZEF, HMNZT *ARAWA*.

26 OCTOBER 1914

ENGLAND

"Antwerp was a bitter blow to me, and some aspects of it have given a handle to my enemies, and perhaps for a time have reduced my power to be useful. From minute to minute one does not know that some fine ship will not be blown up by mine or submarine."[97]

WINSTON CHURCHILL, MP, FIRST LORD OF THE ADMIRALTY.

AUSTRALIA

More troopships arrived off Albany, Western Australia.

"We are anchored about 500 yards from the shore. It's a picture to see the bracken-covered hills, the dark rocks and snow-white sand. There are 19 troopships anchored in the inlet now. We expect to be here for two more days, and then a move forward. All the boys are anxious to get away… We have men in the company who will do credit to Australia in the field. It's wonderful to see them working as they do, including many who knew nothing at all about the work months ago."[98]

CPL FREDERICK GEE, 7TH BATTALION, AIF, HMAT *HORORATA*.[99]

27 OCTOBER 1914

IRISH SEA

A "fine ship" of Churchill's struck a mine that morning. *Audacious* had been the first of the super-dreadnoughts launched during his tenure at the Admiralty. As the RMS *Olympic* arrived on the scene, the leader of the ship's orchestra looked on.

---

95   L/Cpl William John (Jack) Gilchrist, New Zealand Medical Corps, died on 25th October 1914. Buried at sea, he is commemorated on the Otago Provincial Memorial. He was the 30 year-old son of the late William and Agnes Gilchrist, originally of East Gore, Southland.
96   *Hawera Star* (New Zealand), 24th April 1928.
97   Churchill, *The World Crisis 1911–1914*, p. 376.
98   *Bendigo Advertiser* (Victoria), 24th November 1914.
99   Gee, a former miner, was wounded on 25th April 1915. He embarked for home aboard the *Commonwealth*, leaving Suez on 21st January 1916, arriving at Melbourne on 29th February 1916. Gee was discharged from the Army due to the loss of his right eye on 13th August 1916.

"We sighted land at 10 a.m... The land was Tory island. An hour later we came below, when one of the stewards came to our quarters and said: 'You better get up on deck and see those two lovely warships.' The passengers soon got wind of the presence of war vessels and there was much uneasiness among them.

"As soon as the steward told us about the warships, we ran up on deck. The day was dark and cloudy and a stiff westerly gale was blowing. Off our starboard side we saw a big battleship down by the stern and heavy seas breaking over her. She was flying the code flag of the letter 'N,' which is a distress signal.

"As we approached, the other warship [HMS *Liverpool*], came over to us and at high speed... Hardly had the *Liverpool* cut across our bow when the order was given to man the starboard lifeboats. Before this, a call had been issued for volunteers. More answered than the boats could accommodate and when it came time for action the *Olympic*'s crew actually fought to get into the boats so eager were they to do something for their country and for the sailors on the doomed *Audacious*. When one of the boats hit the water they found in it a little bellboy 11 years old, who carried messages to and from the purser's office.

"Although the starboard lifeboats were manned, Capt Haddock suddenly changed his plan. Instead of dropping down on the port side of the pounding warrior, he decided to put about and approach on the starboard side. By so doing he made a lee, which enabled the 14 lifeboats dropped from the port side to accomplish a task that never could have been done if the original plan had been carried out.

"The seas were high and the men in the *Olympic*'s lifeboats had a hard pull. It took them 20 minutes to get over to the *Audacious*... I saw one boat come along the stern of the superdreadnought. It had five men in it. Just as it came into position for the designated men on the *Audacious* to jump into it a big wall of water pulled it up and slammed it upside down into the sea. Instantly four men came to the surface, and clutching at the beckets hauled themselves up on the keel. One man was missing. He was the officer in command. The men had not clung long to the upturned lifeboat when another boat came bounding by and picked them up.

"It was a fine piece of work, a beautiful pickup. On the deck of the *Audacious* the officers directed the work of sending off the crew. Men were picked for each post and told to jump on orders. There were about 900 on the *Audacious*, but only 280 were taken to the *Olympic*. About 400 were transferred to the other craft by the *Olympic*'s 14 lifeboats.

"Some of the small boats made three trips between the battleship and merchantman. About 200 men were left on board the *Audacious* to assist in the handling of lines and cables, while the *Olympic* made her futile efforts to take the *Audacious* in tow. Later these men were taken off to the *Liverpool* in the *Olympic*'s lifeboats...

"The *Olympic* dropped anchor off Lough Swilly at 8 p.m. [An hour later] a tremendous flash lighted up the entire ship... for [a] full 20 seconds afterwards burning fragments

[shot] upward...¹⁰⁰ Then there came a roar as if some mammoth boiler were letting off steam. It stopped as suddenly as it came."¹⁰¹

JAMES BEAMES, MUSICIAN, RMS OLYMPIC.¹⁰²

TURKEY

With war widely anticipated, British officials in Istanbul made preparations to leave. The German/Ottoman ships had vanished, not, like the *Audacious*, beneath the waves but across the Black Sea.

"The Ambassador is anxious to have all his arrangements cut and dried in case war should break out and he and his staff had to leave. It is not clear whether all British officials and servants here would have to leave with him. The question would depend principally on the attitude of the Turks, which cannot be ascertained until the time comes, if it does come. Meanwhile he has got authority from the Foreign Office to arrange, if possible, for any British officials desiring to stay in Constantinople, to do so. Could you ascertain how many... would want to stay... And could you warn everyone connected with the Consulate-General that they might have to leave at very short notice?"¹⁰³

SIR ANDREW RYAN, BRITISH CONSUL, PERA, ISTANBUL.

"I have reliable information that on the 22nd October Austrian Ambassador urged immediate war on Minister of Interior and Halill. Both these officials maintained that it would be wiser to wait until the situation in Egypt and Caucasus cleared before moving, and suggested it would be time enough to move in the spring. They were not sure that, if they went to war, Italy might not join the allies. Austrian Ambassador retorted that spring would be too late, and that it was essential to Germany and Austria that Turkey should declare herself with them at once. His Excellency was clearly greatly dissatisfied at their attitude.

"Enver Pasha, on the other hand, whom Austrian Ambassador saw subsequently, said that he was determined to have war, whatever his colleagues might desire. Turkish fleet would be sent into Black Sea, and he could easily arrange with Admiral Suchon to provoke hostilities.

"Fleet has, in point of fact, to-day gone into Black Sea, so it is impossible to foretell what is in store."¹⁰⁴

SIR LOUIS MALLET, BRITISH AMBASSADOR, ISTANBUL.

---

100 The explosion was probably the result of cordite charges and shells shifting as the ship sank. One man was killed aboard HMS *Liverpool* by the falling debris, PO William Burgess, RN. Buried at Lower Fahan (Christ Church) Churchyard, County Donegal, he was the 30 year-old son of Lawson and Matilda Burgess, of 12 Gainsborough Road, Plaistow, Essex.
101 *Springfield Weekly Republican* (Springfield, Massachusetts), 19th November 1914.
102 Married to Kezia Beames, one of their sons, Pte Frank Rupert Beames, 2nd Bn London Regt, was killed in action on 28th February 1917. He is commemorated on the Thiepval Memorial.
103 Ryan, Sir Andrew: Miscellaneous Correspondence, TNA FO 800/240.
104 Scott, *Diplomatic Documents Relating to the Outbreak of the European War*, p. 1180.

AUSTRALIA

"We arrived in King George's Sound, Albany, yesterday evening. I think we will be here for a few days. There is quite a fleet of transports here — about 23 — and more to come. We are not allowed to go ashore, so we might just as well be on the move as far as we are concerned. We are well looked after here so far as regards sickness. We have a hospital and two doctors, on board, and everything has to be kept very clean. The sea agrees with me, as I feel better every day. There is no 'sleeping-in' here in the morning. We are roused out at 5.50 and must be in bed at 10 o'clock at night."[105]

DVR EDWARD MORRIS, 1ST AUSTRALIAN LIGHT HORSE
REGIMENT, AIF, HMAT *STAR OF VICTORIA*.[106]

INDIAN OCEAN

"Roughest weather we have had, with heavy rain. Unable to do much drill, especially on the top decks; the horses all seem to be very well. On account of thick weather the warships are close in to us."[107]

LT WILLIAM JANSON, WELLINGTON MOUNTED RIFLES, NZEF, HMNZT *ARAWA*.

28 OCTOBER 1914

ENGLAND

The Cabinet decided to suppress the news of the loss of one of Britain's newest battleships, the *Audacious*, after striking a mine in the Irish Sea the previous day. Any hope that this could be kept secret was somewhat undermined by the event being photographed by passengers on the RMS *Olympic*.[108] It added to the pressure on Winston Churchill, who was having to deal with the hounding out of office of his First Sea Lord, Prince Louis von Battenberg because of his German origins.

"The sinking of the *Audacious* — one of the best and newest of the super-dreadnoughts, with a crew of about a thousand — is cruel luck for Winston, who has just been here pouring out his woes. After a rather heated discussion in the Cabinet we resolved not to make public the loss at this moment. I was very reluctant, because I think it is bad policy on the whole not to take the public into your confidence in reverses as well as in successes, and I only assented to immediate reticence on the grounds that

---

105 *The Wyalong Advocate and Mining, Agricultural and Pastoral Gazette* (New South Wales), 2nd December 1914.
106 Transferred to 45th Bn, AIF, Morris was killed in action on 9th June 1917. He is commemorated on the Menin Gate.
107 *Hawera Star* (New Zealand), 24th April 1928.
108 Some of those images appeared in the American press the following month.

(1) no lives were lost,[109] and (2) that the military and political situation is such that to advertise at this moment a great calamity might have had very bad results.

"Winston's real trouble, however, is about Prince Louis [von Battenberg] and the succession to his post. He must go, and Winston has had a most delicate and painful interview with him. Louis behaved with great dignity and public spirit, and will resign at once. Winston proposes to appoint Fisher to succeed him and to get [Sir Oliver] Wilson [VC] to come on also as Chief of the Staff, which I think will be very popular."[110]

HERBERT ASQUITH, MP, BRITISH PRIME MINISTER.

Meanwhile, the frustrations in London at the continued presence of the Germans in Istanbul simmered.

"You should warn Turkish Government that, as long as German officers remain on "Goeben" and "Breslau" and Turkish fleet is practically under German control, we must regard movement of Turkish ships as having a hostile intention, and, should Turkish gunboats proceed to sea, we must in self-defence stop them.

"As soon as Turkish Government carry out their promise respecting German crews and officers and observe the laws of neutrality with regard to "Goeben" and "Breslau," and free the Turkish fleet from German control, we shall regard Turkish ships as neutrals, but, till then, we must protect ourselves against any movements that threaten us."[111]

SIR EDWARD GREY, MP, BRITISH FOREIGN SECRETARY.

PENANG

The continuing depredations of the *Emden* would not have improved Churchill's mood nor eased others' fears about convoying men across the Indian Ocean.

"About one thousand yards away was an enemy warship… She was the Russian cruiser Yemtschuk… We took a leisurely course towards her in order to get within easy range. No movement on her deck, everybody asleep aboard, or unsuspecting. At three hundred yards we prepared to launch a torpedo… A terrible explosion. That first torpedo of ours lifted her off the water just as her gunners let go their first salvo. As a result her shells passed harmlessly over our heads."[112]

OBERLEUTNANT ZUR SEE DER RESERVE *JULIUS LAUTERBACH, SMS EMDEN.*

"We were all awakened with a start, and heard the sharp crack of guns. It did not take a second thought to know what it was, for as soon as we realised it was guns firing we knew it was the 'Emden.' Of course, we all turned out quick and rushed on deck, and were naturally a little alarmed to see that we were getting drawn into the direct line

---

109 As we have seen, one man had been killed by debris striking HMS *Liverpool*.
110 Asquith, *Memories and Reflections*, vol. 2, pp. 45–46.
111 Scott, *Diplomatic Documents Relating to the Outbreak of the European War*, p. 1182.
112 Julius Lauterbach quoted in Thomas, Lowell, *Lauterbach of the China Sea: The Escapes and Adventures of a Seagoing Falstaff*, p. 70, Doubleday, Doran & Company (New York) 1930.

of fire, the 'Emden' being between us and the Russian, a distance of about 300 yards away. We were struck several times from shells fired by the Russians which missed the 'Emden,' for the poor devils on the Zemchug were taken by surprise, and were no doubt firing without taking aim, one shot went clean through us, cargo and all, just a foot above water, another shell struck a sampson post aft and exploded, blowing the whole thing to pieces and cutting our main rigging, another shrapnel burst over us and severely wounded our second engineer in three places..., and piercing our decks with 27 holes."[113]

SECOND MATE DUDLEY KILBEE, SS CRANLEY.

"Our guns opened on her and then [we] turned and have her a second torpedo... The second torpedo must have landed squarely in her ammunition room. With an enormous explosion she rose right out of the water. Then an instant later she vanished from sight behind a curtain of smoke. When the smoke lifted nothing remained above the surface... but a few yards of the *Yemtschuk*'s mast sticking out of the water. The water was alive with Russian sailors, and in a few minutes native sampans were hurrying out to pick up survivors."[114]

OBERLEUTNANT ZUR SEE DER RESERVE JULIUS LAUTERBACH, SMS EMDEN.

"[W]e saw the *Emden* had gone about, and a few minutes after that we were out of the line of fire and then we saw the 'Zemchug' had been torpedoed, for there was a terrific explosion, and when the smoke cleared away, all that could be seen was the top of the 'Zemchug's' mast, and the *Emden* slowly steaming to sea again. When we saw the engagement was over, all merchant ships lowered their boats and pulled to the rescue of the Russians. Our boats picked up 25 of them, some of them terribly wounded."[115]

SECOND MATE DUDLEY KILBEE, SS CRANLEY.

"The [*Mousquet*] was now recklessly coming to the fray... She boldly cut loose at us with two torpedoes, and then, turning, tried to escape. But our gunners caught her with their second round; she shuddered and seemed to pause. Two more bursts of fire caught her, and over she went on her side. Then in a sliding dive she disappeared beneath the waves. We immediately put out our boats to pick up her crew, some of them badly wounded."[116]

OBERLEUTNANT ZUR SEE DER RESERVE JULIUS LAUTERBACH, SMS EMDEN.

"Suddenly the *Emden*... steamed away... because she saw the French torpedo-boat making for her. This gallant little boat [*Mousquet*] she blew to pieces about three miles out of the harbour, and as we had been dining with her officers two nights before, it made us feel pretty sick. After the *Emden* left, everyone with cars spent the morning bringing up the wounded Russians to the hospital. Some of them were

---

113 *Folkestone Express, Sandgate, Shorncliffe & Hythe Advertiser*, 6th February 1915.
114 Julius Lauterbach quoted in Thomas, *Lauterbach of the China Sea*, p. 70.
115 *Folkestone Express, Sandgate, Shorncliffe & Hythe Advertiser*, 6th February 1915.
116 Julius Lauterbach quoted in Thomas, *Lauterbach of the China Sea*, p. 71.

terrible sights, but they never murmured. Arthur once had as many as eleven in his car one journey. It was splendid the way they were rescued. One young fellow in the P. & O. Office saved 60 men himself by being prompt with the Company's launch."[117]

MARY ALICE VOULES.[118]

AUSTRALIA

"On shore at Albany. Took 105 minutes to write out telegrams from the boys to friends, and posted about 150 letters… The New Zealanders and four warships [came] in to-day. We visited about 20 transports. Saw the Kerr boys and Scott, who used to be with Jack Fletcher. Others called out, but I did not know them. Lectures by Bourke and Concannon. All letters and telegrams censored."[119]

2/LT GEORGE KELLY, 2ND BATTALION, AIF, HMAT *SUFFOLK*.

"We dropped anchor in Albany's outer harbour at 10.30 a.m. The morning broke fine with a heavy swell on; land was visible early in the morning. During the afternoon the usual boat race took place, the Artillery winning again. At 4 p.m. the skipper of the *Arawa* and Colonel Johnston, Colonel Meldrom, Elmslie, Walker, Taylor, Vet, Risk, James, Furby, Hardham, ship's second officer Lowack, Captain Kelsel (adjutant), and myself sailed down and rowed back from the Australian transport *Rangatiri*."[120]

LT WILLIAM JANSON, WELLINGTON MOUNTED RIFLES, NZEF, HMNZT *ARAWA*.

Amongst the men leaving Australia were British Army Reservists, heading home to rejoin their units.

"At 10 a.m.… we boarded a tug, in which we were taken to our troopship. When we were drawing near it, we saw it was the Aberdeen liner "Miltiades," and its number was 28. On coming alongside all the troops on board the "Miltiades" started singing "Old Soldiers never say die," etc., and shouts of "Pommies" were also heard.

"There was a band on board, composed of a big drum, a bassoon, and a brass instrument, but I suppose we were lucky to have even that. When we got on board we learnt that there were just over a thousand reservists on the ship. We had a concert that night, which was very much enjoyed by most of the troops."[121]

CPL ALEXANDER URE, RESERVIST, SEAFORTH HIGHLANDERS, HMAT *MILTIADES*.

---

117   *Strahearn Herald*, 16th January 1915.
118   Mary Voules died in the General Hospital, Penang, on 17th April 1917 after a fortnight's illness.
119   *Diary Kept by Lieut. G.E.E. Kelly*, pp. 4–7.
120   *Hawera Star* (New Zealand), 24th April 1928.
121   *Airdrie & Coatbridge Advertiser*, 19th June 1915.

29 OCTOBER 1914

RUSSIA

Where the Ottoman/German ships had gone after leaving Istanbul was answered in the morning. Odessa was first to be struck. As dawn broke, the *Goeben* approached Sevastapol on the Crimean peninsula.

> "In the morning twilight, land came in view ahead. We passed the lines of rocky cliffs over which a light mist still hung. We closed in to a range of 4,000 metres. Sebastopol lay before us…
>
> "On the bridge calculations were being made. The *Goeben*'s guns turned from their peace positions towards the land, and were sighted. The range figures were given through voice-pipes to the individual turrets. Everyone was waiting intently for the bell signal from the fire-control.
>
> "At this moment there came a flash of fire from the Crimean fortress. It began on the extreme left, a flickering line of flame which darted up spitefully and then ran along a line towards the right. Then came a second of deepest, uncanny silence. Then the thunder of the guns came over through the morning in a muffled roar…
>
> "For ten minutes this terrific conflict had raged; for ten minutes the devastating fire continued. Through the gun-ports, the gun crews could see figures running to and fro on shore. The deadly fire had begun its grim work of destruction. Flames rose on the high ground, licked their way greedily onward; a dense, dark smoke hung over the fortress works. Through glasses, shattered guns could be seen, hurled from their mountings."[122]
>
> WIRELESS OPERATOR GEORGE KOPP, SMS *GOEBEN/YAVUZ SULTAN SELIM*.

TURKEY

At first, some in Istanbul doubted the news of the attack on Russia. Some senior Ottoman politicians even claimed to have no knowledge of what had happened.

> "At a dinner where a few Englishmen, all well known to me, were present, a telegram giving the Odessa news created consternation. One of the intended diners had seen Jemal Pasha, the Minister of Marine, only two hours earlier, and he and others expressed the opinion that the telegram could not be true. All recognised that if it were it meant war. Accordingly one of those present was sent off at once by motor-car to see Jemal. Not finding him at his house, he followed him to a club, called him into a private room, and showed him the telegram. Jemal went green, expressed intense and genuine surprise or incredulity, and swore on the head of his daughter — an oath which no Turk lightly utters — that he, Minister of Marine though he was, knew nothing of the matter. He expressed his belief that Talaat Bey, Minister of the Interior,

---

122 Kopp, *Two Lone Ships*, pp. 95–97.

was in like ignorance. Talaat was immediately communicated with, and professed complete ignorance. Then Enver Pasha was rung up, and declared that he had just received a telegram to the like effect as that referred to. Whether or no he expressed complete ignorance of such an incident having been arranged, I do not know. The Grand Vizier emphatically repudiated any foreknowledge of the incident. Turkey had been forced into war."[123]

SIR EDWIN PEARS, BRITISH LAWYER, ISTANBUL

30 OCTOBER 1914

TURKEY

"This morning the Embassy received news from Petrograd via London that in addition to the incident at Odessa, Feodosia had been bombarded by the Turkish fleet, and that at Novorossiysk Turkish officers were sent ashore to demand the surrender of the town. In the latter case the officers were arrested and their ship left without further action.

"H.M. Embassy had already received yesterday morning news that an armed party of Bedouins, 2000 strong, had invaded Egypt and had reached a point over twenty miles inside the frontier. This news is not absolutely confirmed, but it is believed to be true.

"The authenticated incidents in the Black Sea have created a situation such that the Russian Ambassador has been instructed to ask for his passports and to leave Constantinople. It follows, as a consequence of the solidarity of the Entente Powers that the British and French Ambassadors will also leave Constantinople. A rupture of diplomatic relations between Great Britain and Turkey is therefore imminent… As soon as [that] takes place British subjects and British interests generally will at once be placed under the protection of the American Embassy.

"British subjects deciding to leave Constantinople under these circumstances must make their own arrangements. The situation is such that the Embassy considers it necessary to advise British subjects to leave. Nothing can at present be known as to whether any measures of expulsion will be resorted to by the Turkish Government."[124]

SIR ANDREW RYAN, BRITISH CONSUL, PERA, ISTANBUL.

"I called at the British Embassy. British residents were already streaming in large numbers to my office for protection, and fears of ill-treatment, even the massacre of foreigners, filled everybody's mind. Amid all this tension I found one imperturbable figure. Sir Louis was sitting in the chancery, before a huge fireplace, with large piles of documents heaped about him in a semi-circle. Secretaries and clerks were constantly entering, their arms full of papers, which they added to the accumulations already surrounding the Ambassador. Sir Louis would take up document after document,

---

123 Pears, Sir Edwin, *Forty years in Constantinople: The Recollections of Sir Edwin Pears, 1873–1915*, p. 353, Herbert Jenkins Ltd. (London) 1916.
124 Ryan, Sir Andrew: Miscellaneous Correspondence, TNA FO 800/240.

glance through it, and almost invariably drop it into the fire. These papers contained the Embassy records for probably a hundred years. In them were written the great achievements of a long line of distinguished Ambassadors… The records of other great British Ambassadors at the Sublime Porte now went, one by one, into Sir Louis Mallet's fire. The long story of British ascendancy in Turkey had reached its close."[125]

HENRY MORGENTHAU, AMERICAN AMBASSADOR, ISTANBUL.

The Ottoman's entry into the war had confirmed, had any doubt existed, where the real power lay: Enver Pasha.

"Just now I saw the Grand Vizier, who expressed to me his poignant regret for the attack of the Turkish fleet, affirming that it was entirely contrary to the orders of the Porte. He assured me that he would be able to set the Germans straight. To my answer that I had been instructed to leave and would have to obey orders, he responded that, understanding this, he would nevertheless address Petrograd directly in the hope of settling the affair. It is not his good-will, but his authority, that I doubt, and I believe his fall, and that of Djavid Bey, are not far distant."[126]

MIKHAIL NIKOLAYEVICH DE GIERS, RUSSIAN AMBASSADOR, ISTANBUL.

AUSTRALIA

As the allied diplomats and their nationals began to leave Istanbul, Australians and New Zealanders were preparing to leave Albany, though some less enthusiastically than others.

"On Friday, the 30th October, we left our position in King George's Sound for deep-water pier, situated just below the town, to get in supplies of water and vegetables… It was a grand sight to see the troopships in the long lines lying at anchor there, with the cruisers just out of the entrance. The wildflowers about the locality were very beautiful, there being scores of different varieties of all hues and shades. While we were out 15 men took French leave, and a special guard was sent out in the afternoon to capture them. They got 14 of them, and next day they were discharged with the exception of five, who had cells for a few days and fatigue for another three or four. The men who were discharged were sent off with just what they stood up in, and handed over to the military authorities."[127]

PTE RAY YOULDEN, 8TH BATTALION, AIF, HMAT *BENALLA*.

"We are still anchored at Albany. Heavy swell coming in. At about 4 a.m. we heard a gun fired, and on looking out of the porthole saw that the Australian warship *Melbourne* had her searchlight on a large steamer, the *Essex*, holding her up. Went on board the *Orari*. Horses seemed well. Lost two."[128]

LT WILLIAM JANSON, WELLINGTON MOUNTED RIFLES, NZEF, HMNZT *ARAWA*.

---

125 Morgenthau, Henry, *Secrets of the Bosphorus*, p.83, Hutchinson & Co., (London) 1918.
126 Scott, *Diplomatic Documents Relating to the Outbreak of the European War*, pp. 1431–1432.
127 *Maryborough and Dunolly Advertiser* (Victoria), 25th December 1914.
128 *Hawera Star* (New Zealand), 24th April 1928.

31 OCTOBER 1914

ENGLAND

The New Army's training remained in its early stages.

> "The training for recruits of the infantry is to consist of squad drill, company, battalion, and in large units of brigades and divisions. Though some divisions have been organized it may be said that the training of the large mass of men has not yet progressed beyond battalion, company or squad drill. Drills for the other arms in their functions is not further or not so far advanced. The training in the elementary drills are mostly conducted by old non-commissioned officers or soldiers, all of these available being urged to re-enlist. There is, of course, a great shortage of suitable material for drill instructors, non-commissioned officers and officers, and all having necessary military education to fill these ranks are obtained…

> "It is not now the intention to put any of the New Army in the field until it has had about six months' training, that is, not before about April 1st. The recruits will have at last four months' training in battalions and smaller units before the are trained in larger units."[129]

LT-COL. THOMAS C. TREADWELL, USMC, AMERICAN EMBASSY, LONDON.

EGYPT

Now that hostilities had finally begun, as if to reassure the European population, while reminding the Egyptians of their presence, a military procession was held through Cairo.

> "On Saturday, October 31, we had a route march. The procession was 4½ miles long, and took an hour and half to pass any particular spot. We marched past Commander of Forces in Egypt (General Maxwell), and were complimented for our splendid bearing. Large crowds watched us, the cheering was great, especially by the British people of Cairo, and the natives stared with open mouths and clapped."[130]

PTE ARNOLD MYER, 1/9TH BN MANCHESTER REGT, EAST LANCASHIRE DIVISION, CAIRO.

At the same time, the British were making approaches to Arab leaders.

> "Messenger has returned from Mecca with letter from Shereef Abdalla. Communication is guarded, but friendly and favourable. Desires 'closer union' with Great Britain, but expects and 'is awaiting written promise that Great Britain will abstain from internal intervention in Arabia and guarantee Emir against foreign and Ottoman aggression.' Shereef himself in a secret conversation with messenger, expressed himself more freely and openly, saying 'Stretch out to us a helping hand and we will never aid these oppressors.'"[131]

MILNE CHEETHAM, COUNSELLOR OF EMBASSY, CAIRO.

---

129   Naval Attache's Reports, Office of Naval Intelligence, Unpublished Manuscript, US Naval War College, October 1914.
130   *Wigan Observer and District Advertiser*, 28th November 1914.
131   Foreign Office papers, Grey, Sir Edward, TNA FO 800/48.

AUSTRALIA

Preparations for embarkation were continuing, with occasional interruptions.

"My instructions… were to wake the officers at 4 a.m. on 31/10/'14. I may tell you that I did not go to bed that night because by the time I had written my letters and finally completed my work it was 2 a.m. I slept until 3.30, however, and then taking my lantern I went round each tent and pulled the officers out. The varied expressions on their faces were very amusing. They abused me (as usual) and wanted to know if 'this was the real thing.' I assured them that it was. All assembled at the Colonel's tent, and in the grey dawn of the 31st October, 1914, the orders for the move were given. The officers dispersed to their companies, and soon a subdued buzz announced the fact that the men were roused. Without noise and without confusion, the camp fell as the rays of the sun topped the hill. Many remarks floated up to me as I stood at my place where my baggage was packed. Some were very amusing. "Is this a blanky 'dinkum go,' or are they pullin' our blanky legs," was the general tenor of the remarks. Everybody soon realised it was a 'dinkum go,' and the excitement grew. Blankets and other camp stores were rolled up and returned. Then breakfast. Dicksees [Dixies] were cleaned, and then No. 1 train personnel fell in and marched to the station. Nos. 2 and 3 trains followed at 20 minute intervals. Only three ladies saw the train out (Mrs. Johnston and family). No one realised we were going — even the men were still suspicious. We whizzed along until Perth was reached; here we halted for a few minutes, and then on again to Fremantle, where the train ran onto the wharf. The men were detrained and formed up by messes, and in half an hour the men of the first train were aboard. Then came No. 2 and 3 trains and in like manner they embarked. All was orderly — no fuss, very few people about. The constant practice on land stood us in good stead, and by 12 o'clock the s.s. *Ascanius* (Transport No. 11) pulled out into Gage Roads with 6 companies and headquarters of the 11th Battalion and the 10th Infantry aboard. She was followed at 1 o'clock by the s.s. *Medic* (Transport No. 7) at 1pm with Field Artillery, the A.M.C. Engineers, and 2 Companies of the 11th Battalion, and 2 Companies of the 12th Battalion aboard. Thus embarked the W.A. Contingent. Not a man was missing, and the praise of the staff was some recompense for the work put in by all ranks. Only the Quartermaster and myself were left behind. I met Mr. Mazel and Doctor Visalary on the wharf, also Miss Jackson. The Doctor stopped with me until I went out to the steamers in Gage Roads, where they had anchored. I found all going well on board. The men, under the N.C.Os, trained by us in camp, soon discovered the trick of rigging a hammock, and by 10 p.m. all ranks were content to sleep."[132]

LT & ADJ. JOHN PECK, 11TH BATTALION, AIF, HMAT *ASCANIUS*, FREMANTLE.

---

132   *The Grenfell Record and Lachlan District Advertiser* (New South Wales), 5th January 1915.

NOVEMBER 1914

"The state is at war with Russia, England, France."
Sheikh-ul-Islam

Proclamation of holy war, Istanbul, 14 November 1914.

1 NOVEMBER 1914

CORONEL

The German East Asia Squadron was traced to the eastern coast of South America. With the pre-dreadnought *Canopus* left behind because it was too slow and though the Admiralty denied him any reinforcement, Vice-Admiral Christopher Cradock decided to engage. Mindful of Troubridge's experience with the *Goeben*, he felt he had no choice despite commanding what was most definitely a force inferior to that of his opponents.

> "We must have stood up like silhouettes against the bright glow, while our opponents were hardly visible with dark clouds behind them.
>
> "At 7 p.m. the enemy opened fire. "What oh!" said we when their shots fell 200 yards over, "not a bad start!" We replied, and the action became general, the two German armoured cruisers firing on the *Good Hope* and *Monmouth*; the two light cruisers on us. Things began to get a bit warm, shells dropping ten yards short and drenching us with spray, the agonising part being that we couldn't spot our fall of shot owing to the light, though we were still magnificent targets for them.
>
> "At 7.45 p m. there was a roar like thunder, and a flame 300 ft. high lit up the heavens, and the poor old *Good Hope* had gone to her last moorings.
>
> "With a sinking feeling in the pit of my tummy I told the jolly marines (my gun's crew) that it was the German flagship, and went on steadily plugging away whenever the target was visible. By this time the second armoured cruiser had turned her fire on us, but it was growing dark and both sides were firing at the flashes of each other's guns.
>
> "At 8 p.m. we decided there was nothing to be gained by staying, so the *Monmouth* and ourselves drew off the latter reporting that she was making water forward and could only steam stern to sea. We told her to make off, and we endeavoured to draw the enemy away, but all to no purpose. We heard firing going on about half an hour later, and feared the worst."[1]

SUB-LT HAROLD HICKLING, RN, HMS *GLASGOW*.

> "At 8.58 the *Nurnberg* sank the *Monmouth* by bombardment at point blank range. The *Monmouth* did not reply, but she went down with her flag flying. There was no chance of saving anybody owing to the heavy sea."[2]

ADMIRAL MAXIMILIAN GRAF VON SPEE, SMS *SCHARNHORST*.

There were no survivors from *Good Hope* or *Monmouth*. Despite Churchill's refusal to accept any responsibility for what was the worst British defeat at sea in more than a century, many disagreed with him.

---

1 *Nottingham Evening Post*, 18th December 1914.
2 *Nottingham Evening Post*, 8th July 1915.

RUSSIA

The Turks' claim that the naval attack on Sevastopol and elsewhere was a response to Russian provocation was dismissed by their Foreign Minister.

> "I replied to the Turkish *Chargé d'Affaires* that I categorically denied that the hostile initiative was taken by our fleet. Further, that I feared that it is now too late, anyhow, to make any sort of negotiations. If Turkey had announced the immediate expulsion of all German soldiers and sailors, it might then still have been possible to enter into negotiations looking to reparation for the treacherous attack upon our coast and the damages caused thereby. I added that the communication presented by him in no wise affected the situation that had arisen."[3]

SERGEI DMITRYEVICH SAZONOV, RUSSIAN MINISTER OF FOREIGN AFFAIRS.

TURKEY

The failure of British diplomatic efforts was marked by a symbolic pyre.

> "All was now over. In the Embassy garden a huge bonfire was burning — the documents and records of British achievements in Turkey for over one hundred years were slowly burning before the eye of the Ambassador and his Secretaries. It was the funeral pyre of England's vanishing power in the Ottoman Empire.
>
> "At this juncture America took over British interests and a guard of sailors from the *Scorpion* was installed in our Embassy. Sir Louis presented his note at the Sublime Porte and arrangements were made for the English and French staff to leave by special train to Dedeagatch on the night of November 1st."[4]

BETTY CUNLIFFE-OWEN, BRITISH RESIDENT, ISTANBUL.

The Royal Navy did not take long to get in action with an attack on mine-layers at Smyrna.

> "We have been having some exciting experiences ... our boat and the *Scorpion* raided a fortified harbour, and sank a Turkish gunboat and a minelayer. They were anchored when we saw them first, and our Commander gave them ten minutes to raise steam or make a fight of it; but, before the ten minutes were up, they set fire to the forward part of the gunboat, and we opened fire and sent them both to the bottom."[5]

STO. 1 ARTHUR REDDING, RN, HMS *WOLVERINE*, SMYRNA.[6]

---

3   Scott, *Diplomatic Documents Relating to the Outbreak of the European War*, p. 1433.
4   Cunliffe-Owen, Betty, *Thro' the Gates of Memory (From the Bosphorus to Baghdad)*, pp. 61–62, Hutchinson & Co. (London) 1923.
5   *Grantham Journal*, 12th December 1914.
6   Redding was slightly wounded on 25th April 1915.

AUSTRALIA

"We left King George's Sound this morning. A glorious sight — one I may never see again. All of the troopships except two West Australian ships, came out in single file, attended by 5 war vessels. They say the escort will comprise 4 Japanese, 4 French, 4 British, and 2 Australian warships. It was fairly rough round Cape Leeuwin. Everywhere one looks there are troopships — a most impressive sight. Colonel Braund gave an address on Theosophy."[7]

2/LT GEORGE KELLY, 2ND BATTALION, AIF, HMAT *SUFFOLK*.

For some it was not a novel experience. To one British reservist, returning to rejoin his regiment, it was a familiar one: sights, sounds and smells evoking old memories.

"[W]e anchored until the morning of the 1st November. Directly opposite us were ten New Zealand troopships, all of the same colour — grey.

"On boarding the troopship your mind recalls previous times on other troopships, pleasant or otherwise, according as the life suits the tastes of the different individuals. For my part, I can't say it is altogether unpleasant, but at the same time I am not too much in love with it. It is a very lazy life. There is nothing to do and very little space to do it in; there is no freedom of movement; you can't get a walk for exercise to keep yourself in health and condition.

"The first thing you notice when you get between decks is the peculiar smell. In the passage leading past the galley (cook-house) there is a well-known odour which you would imagine was the concentrated essence of all the dinners that had ever been cooked there. It lingers in the passage and refuses to be shifted. In particular, there is a strong sensation of green peas, as though all the green peas that had ever been boiled there were evergreen, and flourished in immortal strength. In a day or so you get used to these smells, and you hardly know they ever existed.

"The troop decks are divided into messes with a long table and two forms for each mess; each table holds sixteen men, eight on either side, and there is not much room to move your elbows. The hammocks are slung from the roof on hooks, and they also are pretty closely packed together. In warm weather, however, a good many of the troops sleep on deck, where they can get some fresh air. But when we get to the west end of the Mediterranean Sea it will be too cold to sleep on top, so we will be glad to hang close together. Occasionally a man might lose his hammock and blankets, so he has to be content if he can pick up a pair of dirty blankets until he can manage to pick up someone else's.

"The food, generally, on a transport is not altogether up to the soldier's liking, but in this case, so far, it has been very good and there is plenty of it. The only fault is that

---

7   Kelly, George Edward Eccleston, *Diary Kept by Lieut. G. E. E. Kelly, During the voyage from Sydney to Egypt as a Member of the Australian Expeditionary Forces*, p. 7, T. Dimmock Ltd., Printers, (Maitland).

the soup for dinner is not very thick. We call it shadow soup, but we don't complain about it, because we are fortunate to get it, if it is only for a drink. We get plenty of butter and jam, tea, milk and sugar; meat or fish for breakfast, dinner, and tea; we also get a tiny drop of porridge for breakfast.

"There is some difficulty in the mornings and evenings trying to get a wash. There are only five basins, and a limited supply of fresh water – so that everyone tries to get there before the water is finished. That means the wash-house is packed full for an hour and a half every morning and evening.

"Washing clothes is anything but a pleasure; we have enough trouble to get our faces washed. To save me the inconvenience of washing, I have stopped wearing socks until such times as it gets too cold.

"We pass the time reading, sleeping, eating, playing cards or gambling.

"While we were in Albany harbour we were allowed to write letters. A small tug steamed round all the troopships collecting and delivering the mails once every day. We had to keep our letters open for inspection, because we were not allowed to give any information regarding troops, movements, or ships. The last mail was posted on Saturday, night, the 31st Oct., and I just managed to buy, and address, a post-card in time to slip into the mail bag as it was being tied up and lowered on to the mail tug.

"That night it was rumoured we were leaving early in the morning, so about 6.30 am. on the 1st of November, 1914, we saw two battleships coming out of the inner harbour, and some of us called out, "There is a move on, boys!" So we all got on the side rail of the ship to have a good view. I got right away up in the rigging, and I had a splendid view of a magnificent spectacle. We were on the extreme left, and all the others, with the exception of the ten New Zealanders, passed out in front of us. There were about 40, or perhaps more, troopships, and to see them filing out one by one, at a distance of nearly a mile apart, was a sight that will live long in our memories; and it was a spectacle anyone would be very fortunate to see, even once in a lifetime.

"I was enjoying the scene when two officers ordered me down from the rigging. I came down all right, but as soon as their backs were turned I was up again. I was soon ordered down again, but when I got a chance I climbed the mast, where I remained till breakfast-time. After breakfast I climbed the rigging again, and what a splendid sight it was to see the ten New Zealand ships filing out behind us in two rows.

"I then went up to the forecastle, where I mounted another high pinnacle, and watched the great armada forming in lines. From the position of our ship in rear of the right line – it was very difficult to make out how may lines there were, but there were three at least. Roughly judging, there was one mile between the ships in each line, and perhaps about two miles between each line. That is a rough guess, because it is very difficult for a lands man to judge distance on the sea. I could see the smoke of a cruiser in front, but one cruiser on either flank could be plainly seen.

"We were only about two hours' sail from Albany when there was a little excitement. A big shark or whale or some such docile bird — (I can't tell which, because I have not the knowledge to tell the difference between a crow and a red herring) — was seen close to the ship's side. From where I was I could see the shadow of it under the water, and now and again the fins would show above water. I suppose it was looking for some nice tit-bit, such as a tea biscuit, but we had nothing to spare on board; but as the shark did not appear to be going to worry us we thought we would keep the ladies for another more pressing case of emergency if required. Those sharks or whales are known to be very fond of old ladies, and they will follow a ship for days if they happen to find out there are any old ladies on board."[8]

CPL ALEXANDER URE, RESERVIST, SEAFORTH HIGHLANDERS, HMAT *MILTIADES*.

2 NOVEMBER 1914

SCOTLAND

David Beatty recorded the departure of two admirals and the advent of another. Prince Louis von Battenberg was forced out because of his roots; Cradock, killed by the Admiralty's negligence. But he had hopes for the return of Jackie Fisher as First Sea Lord.

> "Well, Prince Louis has gone… They have resurrected old Fisher. I wish he was ten years younger. He still has fine zeal, energy, and determination, coupled with low cunning, which is eminently desirable just now. He also has courage and will rule the Admiralty and Winston with a heavy hand…
>
> "Poor old Kit Cradock has gone, at Coronel, poor old chap. He had a glorious death. His death and the loss of the ships and the gallant lives in them can be laid to the door of the incompetency of the Admiralty. They have as much idea of strategy as the School Board boy, and have broken over and over again the first principles."[9]

REAR-ADMIRAL SIR DAVID BEATTY, RN, 1ST BATTLECRUISER SQUADRON.

AEGEAN SEA

Cheers greeted the news that the forts guarding the entrance to the Dardanelles were to be attacked the following morning.

> "[T]he Captain addressed us: "Well, my lads, the fact of the matter is, to-morrow at daybreak we are going to make a start by bombarding the outer forts of the Dardanelles. Cocoa will be served at four o'clock and every man to his station at half-past. We must not expect to get off without a scratch, as they have some very big guns there, but we must give them more than they give us."
>
> "This was received with hand claps, cheers and shrill whistles by the sailors until the Captain could not make himself heard above the din. It was really good to hear it.

---

8 *Airdrie & Coatbridge Advertiser*, 19th June 1915.
9 David Beatty, quoted in Chalmers, W. S., *The Life and Letters of David, Early Beatty. Admiral of the Fleet*, pp. 160–161, Hodder and Stoughton (London) 1951.

Every man was wishing for to-morrow to come. The time for which they had waited had come and they were quite ready."[10]

SIGNALMAN FREDERICK FRANCKLIN, RN, HMS *INDEFATIGABLE*.

INDIAN OCEAN

Life aboard a troopship during a long, uneventful voyage raised few smiles.

"Another attempt on my life to-day — by the doctor this time. Got one of my ears knocked a few days ago and it swelled on the inside. The doctor lanced it this morning. I reckon he took a delight in it. Now I am going about looking like a nun. Very ordinary day."[11]

PTE ROY DAVIES, CANTERBURY BN, NZEF, HMNZT *ATHENIC*.

"We had a little bit of physical exercise, which I had to do in my bare feet, because my feet are skinned and sore. The sea is a bit squally, and a few of us have the feeling that we will be sea-sick before the day is over. We got news by wireless today that Britain and Russia have declared war upon Turkey.

"It is a fine thing shaving when the ship is rolling. I just had one to-night, and I nearly fell over the wash basin. The fellow shaving next to me cut a great deep gash in the side of his jaw."[12]

CPL ALEXANDER URE, RESERVIST, SEAFORTH HIGHLANDERS, HMAT *MILTIADES*.

3 NOVEMBER 1914

ENGLAND

On the day of the first attack on the Dardanelles, the British Prime Minister relaxed.

"We have got down to three Cabinets a week and to-day is a day off, which is a great relief."[13]

HERBERT ASQUITH, MP, BRITISH PRIME MINISTER.

Kitchener, ever mindful of potential trouble within Britain's eastern empire, urged Lord Edward Cecil to offer concessions to Egyptians.

"The fervent situation in Egypt is one in which everything should be done to ensure that all the sympathies of the various classes of the people are with us.

---

10   *Thanet Advertiser*, 28th November 1914.
11   *Grey River Argus* (New Zealand), 24th December 1915.
12   *Airdrie & Coatbridge Advertiser*, 26th June 1915.
13   Asquith, *Memories and Reflections*, vol. 2, p. 47.

"I hope you are taking measures to render our administration as popular as possible by remission of taxation where the people are hard up and by relaxing any stringent financial regulations that bear heavily on officials or others.

"You can look upon this as an authority to you to risk considerable deficits… in order to conciliate all Egyptians during his period of crisis."[14]

LORD KITCHENER, SECRETARY OF STATE FOR WAR.

DARDANELLES

The Anglo-French ships approached the Dardanelles before dawn. The bombardment lasted just twenty minutes. All seemed impressed with the results. On the ships.

"Just before it got daylight all destroyers got an order to clear away from the forts and form up at a given point. When we arrived there, there were our big ships making towards the Dardanelles… We all lined up and set off full bore for the forts. We arrived within range and opened fire at 6 a.m…

"After about seven minutes' bombardment up shot a tremendous cloud of smoke, which plainly showed us that one of their magazines had blown up… What damage was done must have been considerable, as only three shots were fired from one side and about six from the other. They were all short. Apparently our ships out-range their guns. We only bombarded for a quarter of an hour… There were no doctors required with any of our ships. I heard a chap aboard say 'That will keep their sky pilot busy for the day'."[15]

LS CHARLES MASLIN, RN, HMS *BULLDOG*.

"We were fairly shifting those forts, for we could see them crumple under our heavy fire with our glasses, and there must have been terrible loss of life. Then all of sudden one of our lucky shots must have found its way to the enemy's shell magazine, for we saw a terrific explosion occur and a huge black greasy cloud of smoke appeared in the sky, almost blotting out the forts from our vision. Just after this our Admiral must have thought he had done enough, for we steamed away — and none too soon, for the enemy were just finding our range. Just as we turned, a well-directed shot screamed past our main-top and passed harmlessly ahead. Had we not turned we were told the shot would have caught us right amidships."[16]

SIGNALMAN FREDERICK FRANCKLIN, RN, HMS *INDEFATIGABLE*.

Anticipating an attack, the Turks had begun to remove ammunition out of the magazines in the vulnerable outer forts. It was bad luck that some of the exposed munitions were struck, killing 81 men in the process. But it helped foster a false impression of the effectiveness of naval gunfire against the forts guarding the Dardanelles.

---

14   Foreign Office. Private Offices. Grey, Sir Edward (Viscount), TNA FO 800/48.
15   *Burnley Express*, 28th November 1914.
16   *Thanet Advertiser*, 28th November 1914.

GREECE

The brief sortie to the Dardanelles might have provided encouragement to the Anglo-French. But others felt it had simply thrown away any chance of surprise.

> "This will be the end of the Dardanelles Expedition, for now the Turks will make it impregnable before the troops can arrive."[17]
>
> VICE-ADMIRAL MARK KERR, RN, COMMANDING ROYAL HELLENIC NAVY.

INDIAN OCEAN

Bad weather caused widespread sea sickness for many but not all.

> "Strong gale blowing and am not feeling too well — attended the lecture in the afternoon and evening, by Major Hart."[18]
>
> LT WILLIAM JANSON, WELLINGTON MOUNTED RIFLES, NZEF, HMNZT *ARAWA*.

> "A good number are sea-sick. I am not sick yet, but it won't take much to send me off. I keep up on the top deck as much as possible, because I don't think we feel the swinging so much on top as we do between decks, and it is also much fresher above. When anyone is sick, he is generally absent from the mess tables at meal times, and when he makes a reappearance he gets quite a reception from the others. The sea does not look very rough, but the boat is rocking a great deal, and some of the waves are coming on board, causing great merriment, especially if some of the troops get a drenching…
>
> "The boat was rocking very bad last night, but it did not disturb my sleep. I was like a bent cork-screw, twisted round a three-legged table and three chairs, which were all fixtures, so that the heaving of the ship did not shift me from my position. I don't think l am going to be sea-sick this trip, so I will be lucky."[19]
>
> CPL ALEXANDER URE, RESERVIST, SEAFORTH HIGHLANDERS, HMAT *MILTIADES*.

There were other difficulties experienced at sea. Lessons were there to be learned.

> "Two sentries have been sentenced to 21 days for sleeping on duty. A good thing this happened, because it impresses the boys with the seriousness of the offence. On active service the penalty would be death."[20]
>
> 2/LT GEORGE KELLY, 2ND BATTALION, AIF, HMAT *SUFFOLK*.

---

17   Kerr, Admiral Mark, *The Navy in My Time*, pp 188, Rich & Cowan Ltd. (London) 1933.
18   *Hawera Star* (New Zealand), 24th April 1928.
19   *Airdrie & Coatbridge Advertiser*, 19th June 1915.
20   *Diary Kept by Lieut. G.E.E. Kelly*, p. 7.

4 NOVEMBER 1914

INDIAN OCEAN

For those nearing the Equator the temperatures were rising and hygiene was enforced.

> "Beginning to feel the heat. Awnings out everywhere. A bath has been rigged near the front of the ship and companies take it in turn to bathe. If a man won't bathe he is thrown in clothes and all."[21]
>
> 2/LT GEORGE KELLY, 2ND BATTALION, AIF, HMAT *SUFFOLK*.

> "The ships are all in the same position, and you would actually think they were standing still. This morning after breakfast one of the New Zealand ships pulled out on to our right, and threw a target overboard attached to a cable, and had some shooting practice. Things went all right for a while until a bullet cut the cables and the target was left behind.
>
> "We have just heard that the German cruiser "Emden" has sunk a French torpedo boat and a Russian cruiser somewhere in the China Sea; also that the Germans have sunk a British cruiser in the Straits of Dover."[22]
>
> CPL ALEXANDER URE, RESERVIST, SEAFORTH HIGHLANDERS, HMAT *MILTIADES*.

GERMAN EAST AFRICA

An Anglo-Indian force was sent to take the port town of Tanga in German East Africa. What could go wrong did so.

> "Our convoy of transports, with *Goliath* and *Fox*, arrived off the coast of German East Africa on the 2nd of November. As soon as we anchored the captain of the *Goliath* gave the German commander twelve hours' notice to surrender, but the Germans sent word that if we wanted Tanga we would have to fight for it, and our General Officer Commanding then gave the order to disembark which we did [at 6.30 a.m., 3rd November 1914], getting wet through in the process, for we had to wade up to our necks in the sea before we could land.
>
> "We were met by a few German snipers, but we soon cleared them off. We then fixed bayonets and and opened out in skirmishing order. On our right we had the Kashmir Rifles and our left the 61st Pioneers. We then got the order to advance through a dense jungle, which was a very trying time for us, for you must remember we were not far from the equator. Well, anyway, we got through this without any opposition from the enemy, and came in sight the town of Tanga.

---

21 *Diary Kept by Lieut. G.E.E. Kelly*, p. 7.
22 *Airdrie & Coatbridge Advertiser*, 26th June 1915.

"It seemed so quiet and peaceful that I thought the Germans must have evacuated it, an impression which was soon dispelled, for as soon as we got into the town they let us have it. The Germans had fixed all the maxims in the windows of their houses. Men were falling all over the place, but still we pressed on. It was simply a death-trap, and the General gave the order to retire, which we in good order. We lost a lot of men in the retirement, and how I got out of it myself God only knows, I don't."[23]

L/CPL JAMES GREGSON, 2ND BN LOYAL NORTH LANCASHIRE REGT, 27TH INDIAN BDE.

Untried troops had been landed on a hostile, unfamiliar shore. After failing to move inland while the forces opposing them were light, they were unable to resist a disciplined counter-attack, which forced them back into their boats in some disorder. It did not augur well.

5 NOVEMBER 1914

INDIAN OCEAN

"At about 5.30 p.m. the *Osterley* (Orient liner) passed quite close to us, and the passengers crowded to get a sight of our fleet. We could easily hear their cheers, and our men crowded every available part of the ship to get a glimce [sic] at the first sign of life since leaving Albany, and to return the compliment with very hearty cheers. I thought it was a cheeky visit on the part of the skipper, and could not understand how they came to allow the ship to come so near. One in authority told me it was just cheek — catering for his passengers, I suppose; but he has since had to pay for it. The ship has been censored, and the skipper censured. No passengers will be allowed off until England is reached, and no mails. You can imagine what that will cost somebody."[24]

CPL FRANK SAUNDERS, 1ST DIVISIONAL TRAIN, AASC, AIF, HMAT *AFRIC*.

"The weather is getting warmer every day as we approach the equator, and the sea is hardly so rough. We paraded today with lifebelts on. If a shark were to catch any of us round the chest, the cork of the lifebelt would stick to its teeth, and it would have some difficulty in opening its mouth again…

"The first thing to break the monotony of our up-to-the-present uneventful voyage was the passing of the Orient Liner (mail) "Osterley," on her journey home from Australia. She was due to leave Fremantle on the Tuesday (3rd), so she passed us to-night, Thursday, 5th. There was great excitement as she passed us, quite close. She had on a great many passengers, and they were all shouting and waving, and needless to say our throats were sore in doing the same. We also sang our peaceful old war song as she passed — "Old soldiers never die," etc.

---

23 *Lancashire Evening Post*, 30th March 1915.
24 *Singleton Argus* (New South Wales), 12th December 1914.

"I was up in the rigging as usual, and in the excitement of the moment I forgot where I was. It was brought back to my mind when I happened to look down and saw the deck so far below. I was ordered down again by the same officer as before. I think he ought to know me pretty well when he sees me any time."[25]

CPL ALEXANDER URE, RESERVIST, SEAFORTH HIGHLANDERS, HMAT *MILTIADES*.

6 NOVEMBER 1914

ENGLAND

Kitchener often acted as if he were the Commander-in-Chief of the British Army rather than Secretary of State for War. He found it difficult to delegate; distrustful of subordinates outside a circle of close friends. This must have passed through Asquith's mind when told that Kitchener was considering replacing Sir John French with his old Chief of Staff, Sir Ian Hamilton.

"After lunch I found, on descending to the Cabinet room, Winston and Freddy Guest, the latter over a for a day on a secret mission from Sir John French,[26] a most disagreeable affair. It has been reported to French by some poisonous mischief-maker, that when K. [Kitchener] was at Dunkirk last Sunday he asked the French generals whether they were satisfied with Sir John, and even suggested, as a possible successor, Ian Hamilton. I do not believe there is a word of truth or even a shadow of foundation for the story, but it appears to have given great distress to Sir John French, and led to him to think that he had lost, or was losing, the confidence of the Government."[27]

HERBERT ASQUITH, MP, BRITISH PRIME MINISTER.

INDIAN OCEAN

"Weather calm and very warm. A.C. horse, No. 531, shot by Captain Taylor, vet. officer, and thrown overboard. Run today 253 miles. Saw flying fish for the first time. The body of flying fish seems to be green with white wings. *Maunganui* fell back again to-day, for what reason we do not know. "[28]

LT WILLIAM JANSON, WELLINGTON MOUNTED RIFLES, NZEF, HMNZT *ARAWA*.

"The sea is fairly calm to-day, and the weather is a bit hot and sultry. There is not much of a breeze… There is such a lot of gambling going on to-day that there is no room to walk anywhere on the decks… The most of us are now sleeping up on deck, it being quite close and warm at night. I get anchored every night on one of the deck

---

25 *Airdrie & Coatbridge Advertiser*, 26th June 1915.
26 French was the Commander-in-Chief of the British Expeditionary Force on the Western Front.
27 Asquith, *Memories and Reflections*, vol. 2, p. 49.
28 *Hawera Star* (New Zealand), 24th April 1928.

seats or forms. It is just narrow enough, and when the ship is rolling I have some difficulty in preserving my balance..."[29]

CPL ALEXANDER URE, RESERVIST, SEAFORTH HIGHLANDERS, HMAT *MILTIADES*.

7 NOVEMBER 1914

TURKEY

The Americans in Istanbul learned more about the attack on Russia that brought the Ottomans into the war.

> "From an eye witness of the attack by the Turkish torpedo boats, on Odessa, on the morning of Thursday, October 29th, I have the following: My informant was on one of the torpedo boats.
>
> "The Turkish Fleet left the Bosphorus on the morning of Tuesday, October 27th, with targets and apparently for target practice. When some distance out in the Black Sea the targets were sent back. The fleet then separated into three divisions. The division for Odessa, composed of three torpedo boats, separated from the others and proceeded towards that port. One torpedo boat fell out from accident to machinery or other cause. On approaching Odessa, very early on the morning of Thursday, October 29th, the Germans on the boats informed the Turkish officers and crews that war was to be declared by Turkey on Russia, or that the Russian Ambassador in Constantinople was to be given his passports at 3.30 or 4.00 a.m., and that at that time the torpedo boats would attack Odessa. My informant says the Turks were much surprised by this information. The information was not true, as they learned on their return to Constantinople.
>
> "When still dark the torpedo boats entered the harbor, and stood in behind the breakwater, entering by the Eastern entrance, and passing out by the western entrance... My informant reported that they had inflicted more damage on the port and shipping than I am inclined to believe was borne out by future investigation.
>
> "My informant also told me that the GOEBEN (SULTAN SELIM) and BRESLAU (MIDILLI) also bombarded Sevastopol at about the same time and placed mines off the port to prevent the exit of the Russian Fleet."[30]

LT-CDR EDWARD MCCAULEY, USN, USS *SCORPION*, ISTANBUL.

CYPRUS

The annexation of the island, already occupied by Britain for nearly 40 years, was marked by suitable ceremony.

---

29   *Airdrie & Coatbridge Advertiser*, 26th June 1915.
30   Naval Attache's Reports, Office of Naval Intelligence, Unpublished Manuscript, US Naval War College, November 1914.

"We marched down to the Town Hall (with our band) and lined the square. A proclamation in English, Greek and Turkish was then read, and after our band had played a few selections we marched back. To celebrate the occasion each man was given a pint of beer, and we had rather a better dinner at night. Of course we shall celebrate that day every year for ever. We are the only troops in the island, and it is quite possible we shall get an extra bar on the medal we shall get. It is awfully interesting to have taken part in such a historical event, and we are all proud to think that we were serving in the 8th Manchesters when they took over Cyprus."[31]

LT EDWARD HORSFALL, 1/8TH BN MANCHESTER REGIMENT, LIMASSOL.

"The men had ammunition and fixed bayonets and when they entered the square of the town the Greeks gave them a cheer... The Turkish flag was then pulled down and the Union Jack put up in its place. Some speech-making and selections from the band concluded the business and we marched away. When we got back we got a drink and a half holiday..."[32]

PTE JOHN WHELAN, 1/8TH BN MANCHESTER REGIMENT, LIMASSOL.

EGYPT

The moves to placate Egyptian domestic opinion ordered by Kitchener were reported to be taking effect.

"Everything going much better. Measures taken owing to your telegram and help afforded... financiers have agreed to remit part of this month's instalment of taxation. Results good and public feeling distinctly improved. Arrest of Turkish propagandists including Khedive's principal agents has had calming effect. Declaration of martial law well received, no excitement. General regret expressed at action of Turks but little sympathy for them manifested. I am confident we shall be able to keep country quiet."[33]

MILNE CHEETHAM, COUNSELLOR OF EMBASSY, CAIRO.

But not everyone was happy. The tradition of complaining about the food was kept up.

"We are having good food here, but not enough of it, and it is always stew for dinner. Some days we have stewed goat, and other days stewed camel, and I cannot eat either of them.[34] I have not had above one dinner since we landed, only what I have been

---

31 *Todmorden Advertiser and Hedben Bridge Newsletter*, 4th December 1914.
32 *Manchester Evening News*, 28th November 1914.
33 Foreign Office papers, Grey, Sir Edward, TNA FO 800/48.
34 An anonymous officer, probably Capt. Joseph Gledhill (see 18th December 1914), responded to this claim in a letter published in the *Rochdale Observer* on 19th December 1914: "The *Observer* of November 7th I see contained a letter complaining of the food we get — 'stewed camel and goat.' I should like to contradict that, as it gives people a very bad impression. The men never have camel, and the goat that is served occasionally is excellent. In addition to the Government ration of 1 lb. meat, 1 lb. bread, and ½ lb. fresh vegetables which each man has per day, there is also grant of 6d. per day per man which he may expend on extra rations. As a matter fact the men feed like turkey-cocks. Of course it is difficult get a great variety things out here, but they have everything that can be got."

able to buys myself. If I bought my dinner every day I should not have much money for anything else, so I do without dinner many a time unless I have saved a bit of bread from other meals for dinner. We only get a small quarter of bread for breakfast, and sometimes only two slices. We get a lot of tinned foods, such, a tin of herrings for six of is for tea, but we don't have that every day, sometimes we have dripping, and I like that. We walked over twelve miles the other day off four dog biscuits and a basin of tea for breakfast, and I had no dinner. At tea time they had the cheek to give us two more dog biscuits with jam and a bowl of tea. After that we had two hours' night marching."[35]

PTE WILLIAM LORD, 1/6TH BN LANCASHIRE FUSILIERS, EAST LANCASHIRE DIVISION.

INDIAN OCEAN

"Hotter than ever to-day and the sea more blue. Flying fish very plentiful. Lifebelt parade held at 9 a.m. today. The M.Rs. [Mounted Rifles] did shooting from the stem of boat. Temperature in shade 82. Sports to be held this afternoon. Hope to reach Colombo at 4 p.m. Sunday, 15th November. Run 249 miles. Lecture by Col. Johnston on British, German and French Artillery."[36]

LT WILLIAM JANSON, WELLINGTON MOUNTED RIFLES, NZEF, HMNZT *ARAWA*.

"There is a boxing contest on to-day, so I am taking advantage of everyone being out of the wash-house and bathroom to do some washing and have a good bath. From 12 noon yesterday till 12 noon to-day we covered 256 miles. That, up to the present, is the biggest distance we have covered in 24 hours. We are having sweepstakes nearly every day on the distance the ship covers."[37]

CPL ALEXANDER URE, RESERVIST, SEAFORTH HIGHLANDERS, HMAT *MILTIADES*.

8 NOVEMBER 1914

EGYPT

There was no lack of confidence following the news that war had been declared on the Ottomans.

"A native has just been round with a special edition of a paper containing news of the declaration of war on Turkey, and everybody in our room cheered wildly at the prospect of a scrap. From all accounts we are in for a bit of a dust-up, but if we can't beat the Turks with our hands in our pockets we shall want kicking."[38]

PTE FRED PRESCOTT, 1/5TH BN MANCHESTER REGT, EAST LANCASHIRE DIVISION.

---

35 *Rochdale Observer*, 7th November 1914.
36 *Hawera Star* (New Zealand), 24th April 1928.
37 *Airdrie & Coatbridge Advertiser*, 26th June 1915.
38 *Leigh Chronicle and Weekly District Advertiser*, 4th December 1914.

INDIAN OCEAN

As the weather remained oppressive, news of the worst British naval defeat in a century at Coronel did little to lift the mood.

> "Right up till and just after dinner it was red hot; then, all of a sudden, a change came up. On the horizon north west we could see heavy clouds and a white line of foam racing towards us, and in a very few minutes it was raining very heavily, and the afternoon here turned very cool. The rain lasted for about an hour and then the sun came out. Talk about muggy heat; it was something terrible. You have no idea what it is like in the tropics."[39]
>
> PTE RAY YOULDEN, 8TH BATTALION, AIF, HMAT *BENALLA*.

> "The HMS *Minitaur* [sic] left us this morning, the Australian warship *Melbourne* taking her place at the head of the line. News came through of the British disaster on the west coast of South America. Had a short lecture by the vet. officer, Captain Taylor. No other parade was held today other than the church parade."[40]
>
> LT WILLIAM JANSON, WELLINGTON MOUNTED RIFLES, NZEF, HMNZT *ARAWA*.

9 NOVEMBER 1914

ENGLAND

There were many more Muslims living within the British Empire than under the Ottomans. Asquith stressed that they had not wanted to fight the Turks; that the enemy was not Islam.

> "I wish to make it clear, not only to my fellow-countrymen, but to the world outside, that this is not our doing. It is in spite of our hopes and efforts and against our will. It is not the Turkish people, it is the Ottoman Government that has drawn the sword, and which, I do not hesitate to predict, will perish by the sword. It is they and not we who have rung the death knell of the Ottoman dominion, not only in Europe, but in Asia…
>
> "We have no quarrel with the Mussulman subjects of the Sultan. Our Sovereign claims amongst the most loyal of his subjects millions of men who hold the Mussulman faith. Nothing is further from our thoughts or intentions than to initiate or encourage a crusade against their creed. Their Holy Places we are prepared, if any such need should arise, to defend against all invaders and keep them inviolate.
>
> "The Turkish Empire has committed suicide, and dug with its own hands its grave."[41]
>
> HERBERT ASQUITH, MP, BRITISH PRIME MINISTER.

---

39 *Maryborough and Dunolly Advertiser* (Victoria), 25th December 1914.
40 *Hawera Star* (New Zealand), 24th April 1928.
41 Herbert Asquith quoted in Bluey, Ernest Charles, *The Dardanelles, Their Story and Their Significance in the Great War*, pp. 95–97, Andrew Melrose Ltd. (London) 1915.

INDIAN OCEAN

Early in the morning, an unidentified ship approached the Cocos Islands. It was *Emden*.

> "At 5.50 a.m. I was informed that a warship with four funnels was steaming for the entrance between Horsborough and Direction Island. Finding that the fourth funnel was palpably canvas, I found Mr. La Nauze[42] and instructed him to go immediately to the wireless hut, put out a general call that there was a strange warship in our vicinity, ask for assistance, and sign with our naval code. At the same time I went to the office and sent service messages, as previously instructed, to London, Adelaide, Perth, and Singapore. The *Emden*... came in at a great speed nearly as far as our outer buoy, where she wheeled and disclosed an armoured launch and two heavily manned boats under her counter. They were immediately slipped, and speeded straight for the jetty. Through a glass we managed to distinguish four machine-guns, two in the launch and one on the bow of each boat. I personally told Singapore that it was the *Emden*. So quick had been their movements, evidently with the hope of rushing our wireless, that the slip of the last-mentioned service message was passing through our 'autos' when they entered the office.
>
> "I returned to the wireless hut, where Mr. La Nauze informed me that the *Emden* and her collier, the *Buresk*, were endeavouring to interrupt him. I instructed him to continue the call, as the fact of forcing the two ships to use their strong Telefunken notes could only be regarded as a matter for suspicion if picked up by a warship. I stood at the corner of the hut to assume responsibility for the use of the wireless, until an officer and some half-dozen bluejackets ordered us to desist and leave. Armed guards ran to all buildings, and the office was taken possession of by force and the staff ordered out."[43]

DOVER FARRANT, SUPERINTENDENT, EASTERN EXTENSION TELEGRAPH COMPANY.

The *Emden*'s landing party soon got to work.

> "There was a fearful noise caused by the smashing of glass. Then in quick succession there were two loud explosions from the wireless, but although the mast swayed it did not fall until a third charge was used. Then down it came with a crash. Detachments of our visitors were next told off to search our rooms for papers, arms and ammunition. Then they opened all our boxes, and trunks, but did not attempt to loot. Their officer had forbidden them to do that. The first officer then proceeded to cut the cables near the landing point, just of the beach. He took two long lengths off and stowed them on his launch."[44]

ALAN LENTHALL, EASTERN EXTENSION TELEGRAPH COMPANY.

---

42   George La Nauze's brother, Capt. Charles Andrew La Nauze, 11th Bn, AIF, was killed in action on 28th June 1915. Buried in Shell Green Cemetery, the Mauritius-born former accountant was the 33 year-old son of Andrew and Grace La Nauze; husband of Lily R. La Nauze, of 116 Piesse Street, Boulder, Western Australia.
43   *The Argus* (Melbourne, Victoria), 29th November 1934.
44   *Gippsland Times* (Victoria), 11th January 1915.

La Nauze's transmissions had been picked up by HMAS *Sydney*.

> "[A]t 6.30 a.m... a wireless message was received from the Cocos Islands — "Strange warship at the entrance." I then ordered to raise steam for full speed at 7 a.m., and proceeded thither. I then worked up to 20 miles, and at 9.15 I sighted land ahead, and almost at the same moment the smoke of a steamer, which later proved to be H.I.G.M.S *Emden*, coming at me at a great rate."[45]

CAPTAIN JOHN GLOSSOP, RN, HMAS *SYDNEY*.

Their work done, the *Emden*'s men made to return to their ship, while the cable company's men prepared to eat.

> "Just as we were thinking about having breakfast a chap rushed in with the news that another cruiser was in sight. We broke all records in getting on to the roof to see who the stranger could be. When we discovered that it was a British ship we simply yelled and danced with delight, and waved our hats at the *Emden*'s landing party, who were, by this time, about a mile and a half out, and had discovered that their vessel had deserted them, the *Emden* having stood out to sea."[46]

ALAN LENTHALL, EASTERN EXTENSION TELEGRAPH COMPANY.

> "Looking for the *Emden*, we found that she hadn't waited for her boats... and very pretty she looked. She was burning picked Welsh coal, and there was scarcely the slightest sign of any smoke, while the other was enveloped in a black cloud through burning Australian coal. The *Emden* looked fine and clean, and immediately started the ball."[47]

DOVER FARRANT, SUPERINTENDENT, EASTERN EXTENSION TELEGRAPH COMPANY.

> "Then followed a wonderful sight — a sight, probably, I may never see again — a naval engagement on the high seas. Viewed from the vantage ground on which we stood — the roofs of our quarters — about as good a commanding position as the old Sydney cricket ground grandstand, it was a great spectacle."[48]

ALAN LENTHALL, EASTERN EXTENSION TELEGRAPH COMPANY.

The view closer to the action was less of a spectacle; it held fewer charms.

> "We sighted her at 9.30 a.m. and at 9.40 we opened fire... We could have stood off and sunk her without her touching us if we had liked, as our guns have a longer range than hers, but the captain for some reason or other chose to get near her, and so we were hit pretty often.

---

45 *Bendigo Advertiser* (Victoria), 9th December 1914.
46 *Gippsland Times* (Victoria), 11th January 1915.
47 *Birmingham Daily Gazette*, 8th January 1915.
48 *Gippsland Times* (Victoria), 11th January 1915.

> "It is a horrible sensation to hear the shells go whistling over one's head, and I stood on the bridge waiting to be picked off… One fellow on the bridge had his leg torn clean off by a shell, and he died immediately, poor chap."[49]

OS SIDNEY CAVE, RAN, HMAS *SYDNEY*.

> "The *Emden* got the range first and our after control was put out of action, also our forward range-finder. Our captain's seamanship and cleverness were very much in evidence. It was difficult to say the result after the first ten minutes. Then our boys commenced independent firing and picked their own ranges. That seemed to be the end of the *Emden*."[50]

ERA3 ALBERT ATTWATER, RAN, HMAS *SYDNEY*.

> "The *Emden*'s fire was accurate and rapid, but seemed to slacken suddenly. All the disasters occurred in this ship almost immediately. The *Emden*'s foremost funnel was the first to disappear entirely, and later the second went, and she was soon badly afire aft. Then the third funnel went, and I saw that she was making for the beach. She grounded at 11.20. I gave her two more broadsides."[51]

CAPTAIN JOHN GLOSSOP, RN, HMAS *SYDNEY*.

The *Sydney* was more heavily armed and significantly faster than the *Emden*. Ultimately, it proved a one-sided contest.

> "Our marksmanship at first was good, but soon the heavy British guns gained the upper hand, inflicting heavy losses among our gunners. As we were short of ammunition we were obliged to cease firing.
>
> "Though our steering gear was damaged by the enemy's fire we tried to get within torpedo range of the *Sydney*, but the attempt failed, as our funnels were destroyed, a fact which greatly influenced our speed. The ship was therefore run full speed on a reef on the northern side the islands."[52]

KAPITÄN ZUR SEE KARL FRIEDRICH MAX VON MÜLLER, SMS *EMDEN*.

With the *Emden* aground, Captain Glossop disengaged to track down the *Buresk*, one of *Emden*'s prizes employed as a collier.

> "I left her then to pursue a merchant man which had come up during the encounter. Although I had guns on this merchantman at different times during the action I did not fire. I came up to her at 12.10, and fired a gun across her bow, and hoisted the international code signal to surrender, which she did. I found out then that she was the British collier *Buresk*, with 18 Chinese crew, one English steward, one Norwegian cook, and a German prize crew of three officers and 12 men. The ship was fast sinking;

---

49   *The Daily Mirror*, 12th December 1914.
50   *London Evening Standard*, 15th December 1914.
51   *Bendigo Advertiser* (Victoria), 9th December 1914.
52   *Nottingham Evening Post*, 27th November 1914.

the sea cock was knocked out and damaged to prevent repairs. We took all on board, fired four shells into her, and returned to the *Emden*."[53]

CAPTAIN JOHN GLOSSOP, RN, HMAS *SYDNEY*.

"When we got back to the *Emden* we found their flag still flying and their stokers trying to get their forecastle gun, the only one workable, into position. We were forced to put twelve more rounds into her before she surrendered. Twelve men were killed in trying to haul their flag down. It was difficult, as all the decks were on fire. One chap dived from the forecastle, swam to the stern, climbed through a huge rent, and hauled down the flag. They flew a tablecloth as a flag of truce."[54]

ERA3 ALBERT ATTWATER, RAN, HMAS *SYDNEY*.

It did not take long for the news of the *Emden*'s destruction to reach the men on the transports.

"Yours truly had his ear glued to the Marconi room window, in company with anyone else who had the faintest notion of telegraphy. At 11 o'clock the message came from the *Sydney*: 'Am engaged with enemy's ship.' Then almost immediately afterwards: 'Enemy's ship beached to save her from sinking. Am following her transport.' Everybody yelled and hugged each other. 'Wonder if it's the *Emden*?' everybody said. Then came the news it was the *Emden*. Did we go mad? The whole fleet yelled at one another — 35,000 yelling men! How pleased we were that it was one of the Australian ships — a section of the great British Navy — that did the work."[55]

LT & ADJ. JOHN PECK, 11TH BATTALION, AIF, HMAT *ASCANIUS*.

Back on Cocos, the *Emden* party had returned to make preparations for a fight and/or their flight.

"The German officers were as courteous as before, but I was ordered to find all our men (29) and take them to the boat shed. All but five were mustered, the others having previously fled to other vantage points to view the fight... The Germans were very downcast at being ashore while their vessel was engaged in battle with a British cruiser, the first lieutenant particularly feeling the situation keenly. But they showed us some attention at intervals, the men helping Chinese "boys" to bring out sandwiches and drinks, which we shared with our temporary captors.

"The first lieutenant then said he was very sorry, but he had to demand supplies. He asked how much we had in stock, and on being told that we had a four months' supply he asked if he might have half. The superintendent said "Certainly," and our No. 6 (food) store was broken open and the contents loaded into the *Ayesha*. The officers again cautioned the men that there was to be no looting, and only half of each supply taken. The lieutenant then asked if we could give them any clothes, and we handed over a quantity of "ducks" and singlets, which were accepted gratefully.

---

53   *Bendigo Advertiser* (Victoria), 9th December 1914.
54   *London Evening Standard*, 15th December 1914.
55   *The Grenfell Record and Lachlan District Advertiser* (New South Wales), 5th January 1915.

"Being free again, we visited the scene of destruction... Offices, wireless, engine shed, workshop, carpenter's shop, in fact everything connected with the office, was one mass of ruins. The ice plant and condenser were spared by request, as affecting ourselves personally. Meanwhile our captors were busy making arrangements to fight a landing party should the British boat return, maxims being placed in suitable positions. Permission was given us to leave the island in the event of a fight for possession eventuating. One of our guards, a decent chap, asked me to post a letter to his mother in Germany if the *Emden* did not return, and this I promised to do — when the war was over. He sat in my room in full fighting kit and scrawled his note — the human touch which makes us all kin. Visits were made to the roof and we sat on the top, alternate German and British, using each other's glasses and exchanging news as well as we could and the effect of all they told me was that they did not want to fight Britain — that it was all the Kaiser's work; they had no quarrel with us...

"At 7 p.m. the *Ayesha* sailed. A final cheer we gave was replied to by three "hochs" from them, still another for us from aboard, and when the German flag was hoisted, three more. Their launch towed them out and one of our skiffs was seized in addition to their taking their two galleys. The *Sydney* at this time could just be seen by North Keeling in the dusk. Darkness came upon us, and with no lights, the electric light plant having been blown up, our quarters were very dismal, so we dined early and turned in, after the electrician, with two hours' hard work, had restored communication with the Durban and Singapore sections."[56]

ALAN LENTHALL, EASTERN EXTENSION TELEGRAPH COMPANY.

10 NOVEMBER 1914

INDIAN OCEAN

Having been delayed by picking up survivors, Glossop and the *Sydney* returned to look for the Germans.

"About 11.30 [a.m.] the *Sydney* was reported to be in sight. Another rush was made for the roofs. There we beheld the *Sydney* coming up from the north-east; with all her guns out, cleared for action. She anchored at the entrance and lowered two galleys, the first of which approached the shore displaying a white flag. The second boat stood off, the men in her piling arms while the first came to the jetty. They expected to find Germans in occupation, and meant to take us off before beginning hostilities. What a cheer we gave! Three after three. They cheered us in turn on learning that the Germans had left the previous night.

"Where is the *Emden*?" was the first question we asked.

---

56   *Gippsland Times* (Victoria), 11th January 1915.

"She will never worry you any more, sir," replied one of the men; she's hard and fast and blown to blazes on that island over there."

"We cheered our wildest again."[57]

ALAN LENTHALL, EASTERN EXTENSION TELEGRAPH COMPANY.

"We found that an *Emden*'s party, comprising three officers and 40 men had landed, and had seized and provisioned a 70 ton schooner, the *Ayesha*. They carried four maxims and two belts each. They had landed the previous night at 6. The wireless was destroyed.

"I landed a doctor and two assistants, and proceeded as fast as I could to the *Emden*. In view of the large number of prisoners and wounded, and the lack of accommodation in this ship, the captain of the *Emden* agreed that if I received his officers and men and all the wounded then for some time as they remained on the *Sydney*, they would cause no interference with the ships or buildings, and would be amenable to the discipline of the *Sydney*.

"I therefore set to work to transport them, a most difficult task, as I was on the weather side. I received the last man from the *Emden* at 3 p.m. I was then bound to go to the lee side to pick up 20 men who had managed to get ashore."[58]

CAPTAIN JOHN GLOSSOP, RN, HMAS *SYDNEY*.

Another Australian cruiser, sister to the *Sydney*, took over lead escort duties that day.

"Message from HMS *Minotaur* wishing us good-bye and good luck. Hope she is off to wipe out the five German warboats that beat ours near the South American coast. The H.M.A.S. *Melbourne* is now in charge of our convoy. We now know that the *Emden* passed within 20 miles of us on Sunday night. The convoy was so darkened that she could have passed within five miles without seeing us. There was a burial from the "Medic"[59] to-day."[60]

2/LT GEORGE KELLY, 2ND BATTALION, AIF, HMAT *SUFFOLK*.

11 NOVEMBER 1914

INDIAN OCEAN

The troopships continued their journeys. The majority who might have hoped for shore leave at Colombo were to be disappointed.

"Message received this morning that HMAS *Sydney* will rejoin us after she takes off

---

57   *Gippsland Times* (Victoria), 11th January 1915.
58   *Bendigo Advertiser* (Victoria), 9th December 1914.
59   Pte Frederick Courtney, 11th Bn, AIF, died on 10th November 1914 following pleurisy and pneumonia. The former school teacher is commemorated on the Chatby Memorial.
60   *Diary Kept by Lieut. G.E.E. Kelly*, p. 8.

the German prisoners. The fight took place at Cocos Islands. Another death on the "Medic"[61] to-day."[62]

2/LT GEORGE KELLY, 2ND BATTALION, AIF, HMAT *SUFFOLK*.

"All the troops are doing real well. We are to have our second inoculation to-day. Orders relating to our stay at Colombo have just been issued. I have been detailed for launch duty. There is to be no leave whatever, so I am one of the lucky few. We have some good fun at boxing. I have had them on [boxing gloves] several times. Result, a slightly blackened eye."[63]

LT ARTHUR COLMAN, 3RD COY AASC, AIF, HMAT *ORVIETO*.

12 NOVEMBER 1914

INDIAN OCEAN

Naval traditions were adhered to, as those crossing the Equator for the first time discovered, whether they liked it or not.

"A huge bath was rigged up, about 15 feet by 10 feet and 5 feet of water in it. Soon after dinner "Mr and Mrs Neptune" arrived on the scene and commenced operations. "Mr Neptune" and his assistant started the shaving saloon going. The shaving-brush was an extra large white-ash one, the soap, oily butter, soft soap and custard powders, a truly vile mixture to get on your face, while the razor was a wooden arrangement with a blade 15 inches long by 2 inches wide. After their shave the men had to go into the bath. Officers and all went in, irrespective of rank or clothes, and all those there was no room for in the bath had a huge fire hose turned on to them so that nobody should escape the drenching There were about three accidents during the business — a split head, sprained ankle, and one chap suffered from shock."[64]

PTE RAY YOULDEN, 8TH BATTALION, AIF, HMAT *BENALLA*.

"The *Sydney* will overtake us to-morrow, and our ship has to find room for the prisoners. It is great to notice the keen outlook our convoy has for strange ships. This morning a thin line of smoke appealed on the horizon. Immediately the Jap. dashed off. It was a British cruiser. Eventually she passed us and signalled: 'Good luck to you, Australians.'"[65]

LT ARTHUR COLMAN, 3RD COY AASC, AIF, HMAT *ORVIETO*.

That cruiser was probably the *Empress of Asia*. It was looking for the *Emden*'s landing party in the *Ayesha*.

61   Pte Charles Power, 11th Bn, AIF, died of pneumonia aboard the H.S. *Ascanius* on 11th November 1914. He is commemorated on the Chatby Memorial.
62   *Diary Kept by Lieut. G.E.E. Kelly*, p. 8.
63   *Sporting Judge* (Melbourne, Victoria), 23rd January 1915.
64   *Maryborough and Dunolly Advertiser* (Victoria), 25th December 1914.
65   *Sporting Judge* (Melbourne, Victoria), 23rd January 1915.

"During the afternoon we sighted a ship, and it passed us about 3 miles distant. We were all making wild guesses at what it might be. Some said it was a cruiser, some a battleship, others again said it was a mail boat. It was painted drab gray, and had 3 funnels, and I reckoned it was a Russian battleship. We were all wrong, however, because it turned out to be an auxiliary cruiser (*Empress of Asia*) commissioned to carry the mails. Another ship passed us tonight, but it was dark, so we did not see what she was like."[66]

CPL ALEXANDER URE, RESERVIST, SEAFORTH HIGHLANDERS, HMAT *MILTIADES*.

HONG KONG

For some, there was little else to do other than look out to sea for any sightings of strange ships. Steps were also taken to guard against any signs of discontent within Indian troops.

"Our job has been much like that of the British Navy: watching for the foe who never comes. Now that Tsingtao has fallen and the Navy have rounded up the German cruisers *Emden* and *Koenigsberg* there seems very little for us to do. We are sending home 250 R.G.A. by the ss. *Monmouthshire*, leaving to-morrow… We have handed over a good deal of our work to the Indian Artillery, and there is a general idea prevalent that in a few months' time all the regular troops stationed here will have left for home…

"In view of the participation of Turkey, you may be interested to know that the Indian troops are carefully taught that when a Mahommedan fights against the Indian's white brother he becomes, a bad Mahommedan. There need be no fear of the disloyalty of our native troops, and I don't mind predicting the absence of Turkey from subsequent maps of Europe."[67]

SGT ROBERT BRICE, 78TH COY, ROYAL GARRISON ARTILLERY, STONECUTTER'S ISLAND.[68]

13 NOVEMBER 1914

INDIAN OCEAN

More ships crossed the Equator but the fun stopped abruptly aboard one of them.

"We crossed the line at 7.30 a.m. amid heavy rain. A small land bird came on board in an exhausted condition, and as Ceylon is the nearest land, it must have come 400 miles at least. King Neptune had a great day. All officers and men and crew, whether they had crossed the line before or not, except the Colonel, Lieut. Heugh,[69] and the Sergeant-Major, were ducked before the ceremony began. At 2 p.m. Neptune and his

---

66   *Airdrie & Coatbridge Advertiser*, 3rd July 1915.
67   *Lynn News & County Press*, 26th December 1914.
68   Brice was discharged as no longer physically fit for service on 13th Dec. 1917 and died on 25th May 1918.
69   Lt David Heugh, 2nd Bn, AIF, died of wounds on the *Derfflinger* on 29th April 1915. He is commemorated on the Lone Pine Memorial. Mentioned in Despatches.

staff got to work on Captain Wallack.[70] I was next. I was carried down and examined by the "doctor," Major Gordon.[71] I was then lathered with paste, shaved, and tipped into a sail filled with water. Some of the officers got a rough time. We all got certificates from Neptune."[72]

2/LT GEORGE KELLY, 2ND BATTALION, AIF, HMAT *SUFFOLK*.

"Crossed the line to-day, but being a very wet day the Neptune business was only carried out unofficially. This is the first really wet day that we have had and all drill was suspended. To-day we took the lead of the fleet with two of the Australian fleet and a new warship *Hampshire* HMS, China Sea Squadron. A serious accident took place to-day during the Neptune business. Dr. [Ernest] Webb, in diving into the canvas tank, injured his back and it seems rather serious. A steamer passed us last might showing only two lights. Two doctors from the *Maunganui* came aboard this afternoon to attend to Dr. Webb."[73]

LT WILLIAM JANSON, WELLINGTON MOUNTED RIFLES, NZEF, HMNZT *ARAWA*.

14 NOVEMBER 1914

TURKEY

The Caliph's declaration of holy war against the Allies was proclaimed in Istanbul.

"All the Muslims in all countries… resort to Jihad… declare war against Russia, Britain and France and their helpers and supporters, who are enemies of the Islamic Caliphate… success depends on all Muslims to resort to jihad… the Muslims living under the sovereignty of Britain, France, Russia, Serbia, Montenegro and their supporters deserve severe suffering if they fight against Germany and Austria, who are helping the Ottoman government."[74]

SHEIKH-UL-ISLAM.

EGYPT

"It is a splendid sight to see all the shipping at Alexandria. The docks occupy a very fine position. The people engaged in coal carrying at the docks have to work ten hours a day without a stop for two shillings. They are running all the time. If they

---

70   Capt. Gordon Wallack, 2nd Bn, AIF, was killed in action on 19th May 1915. Buried in Lone Pine Cemetery, he was the 29 year-old son of Maj. Gen. Ernest T. Wallack, CB, CMG, and Alice May Wallack, of "Lucerne," Kingston Beach, Tasmania. His headstone bears the inscription: "BORN TO DIE 18TH NOVEMBER 1885 DIED TO LIVE 19TH MAY 1915."

71   Major Charles Gordon, 2nd Bn, AIF, was killed in action on 25th April 1915. Commemorated on the Lone Pine Memorial, he was the 45 year-old son of Robert and Annie Gordon; husband of J.M.E. Gordon, originally of Kingstown, Ireland.

72   *Diary Kept by Lieut. G.E.E. Kelly*, p. 9.

73   *Hawera Star* (New Zealand), 24th April 1928.

74   Sheikh-ul-Islam quoted in Hyde, Andrew, *Jihad: The Ottomans and the Allies 1914–1922*, p. 25, Amberley Publishing (Stroud) 2017.

should stop the overlooker gives them a few lashes with his dog whip. The police afford much amusement with their antics. They do not seem to lock men up or take their names, but throw stones at offenders. I saw one boy, after teasing a policeman with a whip, hit him with it and then run away."[75]

PTE HORACE TWEEDALE, 1/6TH BN LANCASHIRE FUSILIERS.[76]

INDIAN OCEAN

While not everyone appreciated getting a dunking as they crossed into the Northern Hemisphere, others were unimpressed with the food they were being served.

> "Some of the ship cooks went on strike because they were ducked yesterday, and as a result the Sergeants' mess had no breakfast."[77]
>
> 2/LT GEORGE KELLY, 2ND BATTALION, AIF, HMAT *SUFFOLK*.

> "At breakfast that day we had chops sent down. They were crook, so the men, immediately after having breakfast, got a stretcher and set a dead march going. They paraded all around the boat, finishing upon the promenade deck and lowering it over-board. A bugler sounded the last post and a firing party was also in attendance."[78]
>
> PTE RAY YOULDEN, 8TH BATTALION, AIF, HMAT *BENALLA*.

> "We were told that when the cruiser *Sydney* passed there was to be no demonstration, such as shouting and cheering, because they had wounded on board. The cruiser *Sydney* is one of our escort, and it was she who sunk the German cruiser *Emden* off the Cocos Islands."[79]
>
> CPL ALEXANDER URE, RESERVIST, SEAFORTH HIGHLANDERS, HMAT *MILTIADES*.

15 NOVEMBER 1914

CEYLON

The sights, sounds and smells of Colombo were unfamiliar to the majority of those pausing their journey there.

> "We dropped anchor at 2 p.m. in the roadstead, and spent most of our time looking through glasses at Colombo. We could see rickshaws, motors, and other vehicles racing about... We see inside the breakwater a mass of shipping, with British, Japanese and Russian warships. A five-funnel Russian cruiser [*Askold*] flying a white ensign

---

75 *Rochdale Observer*, 14th November 1914.
76 Pte Horace Tweedale, 6th Bn Lancashire Fusiliers, died of pneumonia on 17th October 1918.
    Buried at Hautmont Communal Cemetery, he was the 24 year-old son of Elisha and Maria Tweedale, of 69 Bent Gate Street, New Hey, Rochdale, Lancashire.
77 *Diary Kept by Lieut. G.E.E. Kelly*, p. 9.
78 *Maryborough and Dunolly Advertiser* (Victoria), 25th December 1914.
79 *Airdrie & Coatbridge Advertiser*, 3rd July 1915.

is anchored near us. A warm breeze is sweeping a delightful odour from the land. Complaints about food. Captain Kane[80] objected to cockroaches in his soup, and I offered to supply him with scorpions from mine. Campbell[81] would like to know if the porridge could be run through a strainer, as when the lumps were removed nothing remained. Tarrant[82] complained that the menu card showed a different name every meal for the same pudding."[83]

2/LT GEORGE KELLY, 2ND BATTALION, AIF, HMAT *SUFFOLK*.

"It is a beautiful place. I have not seen anything to compare with it as yet. The houses are of different colored stone, with colored tile roofs, and the trees which ebb out from between them show very dense foliage, and a very pretty green. Lawns run from these places nearly down to the shore… The lighthouse is a beauty. It has a revolving light, with three one-second flashes and three-second darkness. On Monday a canoe-load of natives came out, and when money was thrown overboard to them they did not take long to get it out. They are just like fish in the water. One of them climbed up on the boat and dived off the bridge for 4s. It was very funny hearing them name the coins as they got them – sexpen, she-len, two-shel, and so on."[84]

PTE RAY YOULDEN, 8TH BATTALION, AIF, HMAT *BENALLA*.

16 NOVEMBER 1914

CEYLON

After taking in the sights one New Zealander took advantage of the opportunity to talk to one of the *Emden*'s officers. There were moments of humour too but not everyone had the chance to enjoy themselves.

"Went ashore at Colombo again from 11 a.m. till 2.30 p.m. The sights I went into raptures over. Purchased silk worked blouse, also a bangle of moonstones. Had a motor car ride all around Colombo. Moved out of the harbour at 5 p.m. and anchored in the roadstead. We took on three officers and 30 men, prisoners from the *Emden*. Had an interesting talk in the smoking-room to the German officer Lieut. [Walter] Haas, and we learnt a good deal. The sights of to-day I shall never forget, no, never in my life. Word from the Colombo hospital that Dr. Webb was not expected to live. The

---

80   Capt. Francis Kane, AAMC, attd 2nd Bn, AIF. He was evacuated from the peninsula and admitted to hospital in Egypt, arriving on 19th May 1915. His health necessitated his return to Australia on 25th June 1915, where he was later discharged.
81   Capt. Irvine Campbell, 2nd Bn, AIF, died of wounds aboard HMT *Neuralia* off Malta on 2nd June 1915. He had been wounded at Gallipoli on 31st May 1915. Commemorated on the Lone Pine Memorial, he was the husband of Gertrude Ellen Campbell.
82   Lt Richard Tarrant, 2nd Bn, AIF, was wounded in action on 27th April 1915. He was wounded for a second time while serving on the Western Front on 7th August 1916, after which he was not recommended for further front line service.
83   *Diary Kept by Lieut. G.E.E. Kelly*, pp. 9–10.
84   *Maryborough and Dunolly Advertiser* (Victoria), 25th December 1914.

Empress of Russia, a steamer 20,000 tons was anchored in front of us, it was she who brought the *Emden* prisoners from Cocos Islands to Colombo."[85]

LT WILLIAM JANSON, WELLINGTON MOUNTED RIFLES, NZEF, HMNZT *ARAWA*.

"36 sailors and 14 officers of *Emden* came on board this afternoon. Strong guard. Big fat German officer watched as steward screwed porthole as tightly closed as possible with iron bar. Officer asked what for — pointed to his dimensions and asked did they think he'd get out through the porthole."[86]

SISTER CONSTANCE KEYS, AUSTRALIAN ARMY NURSING SERVICE, HMAT *OMRAH*.

"It is the good fortune of a soldier's life to see the world, but his misfortune to see it at a great distance away. We are lying here at Colombo at least 3 miles from the shore, hardly close enough to distinguish any movement of life ashore. Before our trip is finished we will have travelled ten thousand miles without seeing anything but water and an occasional glimpse of land many miles away."[87]

CPL ALEXANDER URE, RESERVIST, SEAFORTH HIGHLANDERS, HMAT *MILTIADES*.

17 NOVEMBER 1914

CEYLON

Not everyone was charmed by the presence of the *Emden*'s officers aboard their ship.

"We have 53 of the *Emden*'s crew, including the captain, the Kaiser's nephew, a doctor and a lieutenant. My fine deck cabin has been given to the captain, so I have to go below… The captain is a long, thin fellow, who wouldn't impress you as being clever, and the Kaiser's nephew is a sour, bad-tempered looking fellow. We have an interpreter here, and he does all the talking, assisted by young Casey,[88] … The captain and doctor of the *Sydney* say the sight aboard the *Emden* was sickening."[89]

LT ARTHUR COLMAN, 3RD COY AASC, AIF, HMAT *ORVIETO*.

"At noon we moved out into our place in the line, and passed between a Russian [*Askold*] and a British cruiser [HMS *Hampshire*]. The Russians stood to attention, and we cheered and we moved along. The British cruiser dipped her ensign. We had some oranges. They do not get yellow when ripe, remain green, and are sweet, but not up to our Australian oranges. We bade good-bye to Ceylon."[90]

2/LT GEORGE KELLY, 2ND BATTALION, AIF, HMAT *SUFFOLK*.

---

85 *Hawera Star* (New Zealand), 24th April 1928.
86 *The Australian Women's Weekly*, 19th April 1972.
87 *Airdrie & Coatbridge Advertiser*, 3rd July 1915.
88 Lt, later Major Richard Casey, Headquarters, 1st Australian Division.
89 *Sporting Judge* (Melbourne, Victoria), 23rd January 1915.
90 *Diary Kept by Lieut. G.E.E. Kelly*, p. 10.

INDIAN OCEAN

The naval battle fought between the *Sydney* and *Emden* had fascinated those who watched it from afar. For one of those spectators, things were different when they saw the results.

> "I was fortunate enough to visit the remains of the *Emden*… and was surprised at the havoc aboard, not to mention the most appalling sights and smell! After the very kindly manner with which we were treated by the landing party, one could almost feel sorry for these *Emden* men, who played the game so well."[91]

ALAN LENTHALL, EASTERN EXTENSION TELEGRAPH COMPANY.[92]

18 NOVEMBER 1914

EGYPT

> "It is rumoured that we are going to be sent back to England early in December, while some say that France will be our destination. I think that the war will be over by Christmas and that we shall all be at home by the latter end of January. As to whether this is so or not, we shall have to wait and see… One night we went out for a stroll and got lost. For a time we could not find anyone that understood our language and we were in a fix. Eventually we came across a white man who directed us back to the barracks… After this I swore that I would never leave the barracks again."[93]

L/CPL PETER CARR, 1/6TH BN LANCASHIRE FUSILIERS, EAST LANCASHIRE DIVISION.[94]

INDIA

> "General Birdwood, my Official Secretary in the Army Department…. told me that he had been offered by Kitchener the command of the Australian contingent, but that he realized that India having been bled white of officers, it would be extremely difficult if not impossible for me to find a competent officer to take his place, and that if I said the word he would stick to me and refuse the offer that had been made to him. I told him at once that he had put the question to me in a such a way that it was impossible

---

91 *Referee* (Sydney, New South Wales), 21st April 1915.
92 Alan Lenthall died, aged 29, in Nelson, New Zealand, after a long illness on 1st February 1917. According to an 'In Memoriam' notice published in the *Bairnsdale Advertiser and Tambo and Omeo Chronicle* (Victoria) on 7th February 1917, his illness had been "brought on partly from strenuous work in the Cocos Islands raid and the mutiny at Singapore."
93 *Rochdale Observer*, 18th November 1914.
94 Sgt Peter Carr, 1/6th Bn Lancashire Fusiliers, died of wounds on 3rd December 1917. Buried in Bethune Town Cemetery, the former cotton worker was the son of Sarah Ann Carr, of 142 Bury Road, Rochdale, and the late Luke Carr.

for me to refuse and that he was to accept the appointment… He was a splendid and very loyal officer to me in India."[95]

CHARLES HARDINGE, VICEROY OF INDIA.

INDIAN OCEAN

Their stop-over at Colombo ended, the troopships headed for Aden and the Red Sea. Boredom was relieved by conversation, the discovery of some unlikely connections and music. Of a kind.

> "We are now on our way again. First stop will be Aden. One of the *Emden*'s prisoners we have aboard once worked on the Fremantle trams, and another was a baker at Moonee Ponds."[96]

LT ARTHUR COLMAN, 3RD COY AASC, AIF, HMAT *ORVIETO*.

> "We left Colombo late last night, along with nine other ships. Had to rise early this morning owing to a shower of rain. We had some sports on board to-day, during which our band rendered us some very fine music. There was a change of instruments in the band to-day. We had a big drum, a tin platter served as a side drum, a tin whistle, a set of bagpipes, and a pair of clappers — the whole making a unique combination, which helped to relieve the monotony for a while."[97]

CPL ALEXANDER URE, RESERVIST, SEAFORTH HIGHLANDERS, HMAT *MILTIADES*.

19 NOVEMBER 1914

INDIAN OCEAN

The frustrations born of a long voyage emerged occasionally, not helped by the receipt of bad news. Others were feeling a little unwell, reactions to being inoculated.

> "A non-commissioned officer was de-graded to-day for insubordination.[98] The stripes were cut off his sleeve. It is a sad sight to see a man degraded, but it must be done for discipline sake, and non-commissioned officers should set the example."[99]

2/LT GEORGE KELLY, 2ND BATTALION, AIF, HMAT *SUFFOLK*.

---

95 Hardinge, Lord, *My Indian Years, 1910–1916. The Reminiscences of Lord Hardinge of Penshurt*, p. 106, John Murray (London) 1948.
96 *Sporting Judge* (Melbourne, Victoria), 23rd January 1915.
97 *Airdrie & Coatbridge Advertiser*, 3rd July 1915.
98 L/Cpl Harold Amos, 2nd Bn, AIF, was found guilty of "gross insubordination & indecent language to N.C.O." The war diary confirms his stripes were removed publicly and sentenced to 21 days confined to cells. In Egypt, he was found to be suffering from syphilis and recommended for discharge, as his services were no long required. He died on the return journey to Australia after blood poisoning supervened. Buried at sea, he is commemorated on the Chatby Memorial.
99 *Diary Kept by Lieut. G.E.E. Kelly*, p. 10.

"Feeling rather queer after yesterday's inoculating; unable to attend to my duties. At 3.3 this morning we passed an island with a lighthouse on it: it is a flashlight, I fainted at half-past eleven this morning after being inoculated, and feel very queer. Dr. Webb[100] died in the Colombo Hospital at 2.20 yesterday. News of this reached us by wireless."[101]

LT WILLIAM JANSON, WELLINGTON MOUNTED RIFLES, NZEF, HMNZT *ARAWA*.

20 NOVEMBER 1914

INDIAN OCEAN

"To-day is most perfect, with a beautiful breeze to clear the atmosphere… No drill to-day on account of the men being inoculated yesterday. The portion of the Australian fleet which we left behind at Colombo caught up to us at noon to-day, with the Japanese warship *Ibuki*. We passed the head of the Australian fleet on our way to Aden at 5 p.m. A death occurred on the *Maunganui*, and the burial will take place to-morrow."[102]

LT WILLIAM JANSON, WELLINGTON MOUNTED RIFLES, NZEF, HMNZT *ARAWA*.

"Still very warm. Slept on deck. Am on guard duty to-night. Majors Gellibrand[103] and Marsh[104] and Colonel Sellheim[105] to-night congratulated me on my signalling work at Colombo."[106]

LT ARTHUR COLMAN, 3RD COY AASC, AIF, HMAT *ORVIETO*.

21 NOVEMBER 1914

INDIAN OCEAN

The monotony of the voyage was brought to an abrupt halt for the men aboard two troopships in the early hours of the morning. Well, most of them.

"At about 4 a.m… I was awakened by someone pulling me out of bed, and arose to find a steamer just outside the port-hole of my cabin. I could see a man on the other ship in frantic haste pulling on his trousers. I could not realise what had happened

---

100   Lt Ernest Webb, New Zealand Medical Corps, died on 17th November 1914 (according to the Commonwealth War Graves Commission) as a result of breaking his neck diving into a makeshift swimming pool aboard HMNZT *Arawa* four days previously. Buried in Colombo (Kanatte) General Cemetery, he was the 32 year-old son of Herbert and Sarah Webb, of Mornington, Dunedin.
101   *Hawera Star* (New Zealand), 24th April 1928.
102   ibid.
103   Major John Gellibrand, Deputy Assistant Quartermaster General, 1st Australian Division.
104   Major Jeremy Taylor Marsh, Headquarters, 1st Australian Division.
105   Colonel Victor Conradsdorf Morriset Sellheim, Assistant Adjutant and Quartermaster General, 1st Australian Division.
106   *Sporting Judge* (Melbourne, Victoria), 23rd January 1915.

until the priest who had been sent down by the colonel to wake me, told me that our ship had collided with the *Shropshire*. Knowing me to be a sound sleeper, the colonel had sent him to wake me. If I had not done so, I am afraid I should have slept on. I climbed into my rifle belt, grabbed my revolver, and reported to the colonel on deck. Everything was very quiet and every man got to his station; men sat at their tables down below with life-belts on, and played cards and dominoes with the greatest unconcern… By the time the soundings had been made, it was ascertained we were not seriously damaged beyond the fact that a hole 25 ft long and six inches broad was torn in the prow, high up on the port side. It was rumoured that several men had fallen overboard from the *Shropshire*, but this proved to be incorrect."[107]

LT & ADJUTANT JOHN PECK, 11TH BATTALION, AIF, HMAT *ASCANIUS*.

"I was hardly asleep when there was a terrific crash, followed a few seconds later by another, and the violent ringing of the fire bells and the three blasts of our siren announced to us the horrible fact that we were in collision. The big steamer quivered, and shook from stem to stern, and we all scampered below for our life belts, for there are no boats for troops on a transport. The ten we carry will only hold the ship's complement and our officers. Rockets were sent up for assistance and the S.O.S. wireless call sent out on its merciful errand… Our men behaved splendidly, and there was no attempting to rush the boats or any sign of cowardice whatever. Half-an-hour or more passed, and our carpenters found that though we had sustained severe damage it was all above the water-line, and we were out of danger. Then we breathed freely, and half-an-hour later were dismissed…

"I'll never forget the scene at daybreak. The *Shropshire*, *Ascanius*, *Benalla*, *Argyleshire* and *Star of Victoria* all stationary and the surface of the dark sea covered with the bobbing lights of their life boats standing by us and two cruisers wheeling about flashing their searchlights everywhere. We, standing on parade, dozens and dozens of men being naked except for their lifebelts."[108]

DVR JOHN ROSS, 2ND BAC, AFA, AIF, HMAT *SHROPSHIRE*.

22 NOVEMBER 1914

EGYPT

Confidence was high at the ability of the garrison to deal with any potential Turkish invasion.

"One thing is quite certain… that is that we shall soon be moving from these barracks into the Citadel Barracks of Cairo. I believe that that means guards the whole day long, as many Germans and Turks are being kept prisoners there. We expect to finish our training at the end of another three weeks, and then we can hope for something definite. Nobody here worries much about Turkey and their threatened invasion.

---

107 *The Grenfell Record and Lachlan District Advertiser* (New South Wales), 5th January 1915.
108 *Hamilton Spectator* (Victoria), 29th December 1914.

In any case, there are 80,000 troops here of different kinds to resist them if they do attempt it, and I hear to-day that 40,000 Autralians [sic] are to be also dumped down. Although we all look forward to getting back to Todmorden again, we are — generally speaking — quite satisfied with our present existence, and spite of all sorts of annoyances which are bound to befall anyone who takes on this life."[109]

CAPT. ROBERT BARKER, 1/6TH BN LANCASHIRE FUSILIERS, EAST LANCASHIRE DIVISION.

INDIAN OCEAN

"The weather good and sea fairly calm. Our run to-day was 292 miles. Church service was held. I had the pleasure of taking the particulars of the German *Emden* prisoners as regards colour of hair, eyes, and complexion. Our four boats are still in the lead and keeping together. The 11 Australian are still to be seen astern. Our men are doing stables on the poop, as the signallers were inoculated yesterday."[110]

LT WILLIAM JANSON, WELLINGTON MOUNTED RIFLES, NZEF, HMNZT *ARAWA*.

"This being Sunday, we had a voluntary church service, so I took advantage of the deck being partly clear, and had a sharp walk up and down for exercise. After dinner I took my book and went up on deck, sat down and read, but, as per usual, I got sleepy, so I put the book under my head for a pillow, and stretching myself fell asleep. I wakened up in half an hour, and began thinking what a lazy life we were all leading. Right along the deck, as closely packed as could be, lay the troops, mostly sleeping, with a few reading. It would take a book to describe all the various garbs in which they were dressed, but they were suitable for lolling and lying about in warm weather."[111]

CPL ALEXANDER URE, RESERVIST, SEAFORTH HIGHLANDERS, HMAT *MILTIADES*.

23 NOVEMBER 1914

ADEN

Rumours circulated that the Australians and New Zealanders would be curtailing their journeys. Egypt was to be their new destination.

"We expect to be in England on December 17, and we will all be glad to set our feet on terra firma again. The voyage has been calm ever since we left Sydney. However, a troopship is a troopship, and not a pleasure ship. There is a rumour amongst us all that we may go to Egypt for the winter."[112]

PTE RALPH BRADSHAW, 4TH BATTALION, AIF, HMAT *EURIPIDES*.

---

109   *Todmorden Advertiser and Hebden Bridge Newsletter*, 11th December 1914.
110   *Hawera Star* (New Zealand), 24th April 1928.
111   *Airdrie & Coatbridge Advertiser*, 3rd July 1915.
112   *Windsor and Richmond Gazette* (New South Wales), 22nd January 1915.

INDIAN OCEAN

"There have been several deaths in the fleet since we left, but so far none on our boat, though we had to put two bad cases ashore at Colombo… We would have given anything for a run ashore there. One of our lads climbed down the stern hawser and got ashore in a native diver's boat. He is now doing 14 days' cells, which will stop any of us emulating his example at either Aden or Port Said. I got my second dose of inoculatlon [sic] for typhoid to-day, and it hasn't affected me a bit. I feel in splendid form, and am putting on condition. We have concerts every Wednesday and Saturday nights, and lectures during the hotter parts of the day on gunnery, while we all learn signalling, etc."[113]

DVR JOHN ROSS, 2ND BAC, AFA, AIF, HMAT *SHROPSHIRE*.

"Twenty-five men refused to be inoculated. Temperature of water 80; ship's run 289. Went down into the engine-room and stokehold and shovelled coal. "Oh, the heat" — soon had enough; 26 feet under water."[114]

LT WILLIAM JANSON, WELLINGTON MOUNTED RIFLES, NZEF, HMNZT *ARAWA*.

24 NOVEMBER 1914

ADEN

Aden appeared forbidding to some, welcoming to others but one salute to newly-arrived troop transports nearly ended very badly.

"We were watching the arrival of certain vessels… when the boom of a gun was heard. Simultaneously a big spout of water rose about 150 yards away from us and near to another transport. Then came a weird loud screaming moan, and a shell ricochetted about a mile. Some blanks followed. Somebody apparently fired a live shell in mistake and we nearly got it for breakfast."[115]

PTE JOHN BURGHOPE, 1/5TH BN QUEEN'S (ROYAL WEST SURREY REGT), RMS *ALAUNIA*.

"First sight of Arabia. Aden sighted at noon. Dropped anchor at 4.15 p.m. Aden is a barren mass of sharp granite peaks and rugged rocks with roads and tracks here and there leading up to forts on the peaks. We were told that regiments in disgrace are sent here on garrison duty. In the background is the low-lying desert of Arabia infested with lions. Several native villages are in sight. I don't know what they live on, except fish, which is plentiful. A searchlight playing from shore to-night."[116]

2/LT GEORGE KELLY, 2ND BATTALION, AIF, HMAT *SUFFOLK*.

---

113 *Hamilton Spectator* (Victoria), 29th December 1914.
114 *Hawera Star* (New Zealand), 24th April 1928.
115 *Surrey Mirror*, 1st January 1915.
116 *Diary Kept by Lieut. G.E.E. Kelly*, p. 11.

"I am lucky, as they chose me for duty on shore at Aden, where I am to purchase clothes for the Germans. Just a while ago I was taken to the *Emden*'s captain and Kaiser's nephew to get a list of what they require. They were most polite and said they would be much obliged to me for my work."[117]

LT ARTHUR COLMAN, 3RD COY AASC, AIF, HMAT *ORVIETO*.

INDIAN OCEAN

"On the 10th, Private Courney,[118] [sic] of A Company, on the *Medic*, died of pneumonia, and on the 11th, Private Power,[119] of G Company, on the *Ascanius*, died of pneumonia; both were fine boys, and their deaths cast a gloom over the ships, and especially over the West Australians. However, the battalion has now been completed to strength, because we discovered two men stowed away — one on the *Ascanius* and one on the *Medic*. The boy on the *Ascanius* was our Armourer-Sergeant in camp, but being a permanent armourer, the powers-that-be decided at the last moment he should not go, so he had to be discharged the day before we embarked. He hinted that I would probably see him again. Of course I had to ignore his insinuation. However, he turned up on board after we were well out at sea, and is once again our Armourer-Sergeant. The other man was a sergeant-major named Jelf. He was discovered after they got out to sea, and has now been enlisted in the place of Courteney. So you see that with luck we will land with a complete battalion."[120]

LT & ADJ. JOHN PECK, 11TH BATTALION, AIF, HMAT *ASCANIUS*.

25 NOVEMBER 1914

ENGLAND

Churchill expounded the possibilities offered by a successful assault of the Dardanelles but suggested that it could be presented as a diversion for operations elsewhere if adequate progress was not made.

"MR. CHURCHILL suggested that the ideal method of defending Egypt was by an attack on the Gallipoli Peninsula. This, if successful, would give us control of the Dardanelles, and we could dictate terms at Constantinople. This, however, was a very difficult operation requiring a large force. If it was considered impracticable, it appeared worth while to assemble transport and horse boats at Malta or Alexandria, and to make a feint at Gallipoli, conveying the impression that we intended to land

---

117 *Sporting Judge* (Melbourne, Victoria), 23rd January 1915.
118 Pte Frederick Courtney, 11th Bn, AIF, died on 10th November 1914 following pleurisy and pneumonia. The former school teacher is commemorated on the Chatby Memorial.
119 Pte Charles Power, 11th Bn, AIF, died of pneumonia aboard the *Ascanius* on 11th November 1914. He is commemorated on the Chatby Memorial.
120 *The Grenfell Record and Lachlan District Advertiser* (New South Wales), 5th January 1915.

ADEN

"Arrived at Aden 6 a.m. Lovely morning and sunrise. There are two lighthouses. There was no leave granted. The harbour was full of shipping. Troops from England were here on their way to Bombay. The balance of the Australian fleet arrived during the afternoon. Some of the transports required water and coal, but the *Arawa* is going to coal and water at Port Said. The *Sydney* left as we got here."[122]

LT WILLIAM JANSON, WELLINGTON MOUNTED RIFLES, NZEF, HMNZT *ARAWA*.

"The water is a deep green, and there are shoals of fish. As we entered the harbor hundreds of porpoises passed us. We coaled here, and while this operation was going on natives came out in their boats, bent on business. Talk about dates and Turkish delight! We all made ourselves ill, and as for tobacco, it was almost thrown at us. It is frightfully hot; we are told this is midwinter. I don't want to be here in the summer."[123]

PTE ARTHUR ASH, 10TH BATTALION, AIF, HMAT *ASCANIUS*.

26 NOVEMBER 1914

RED SEA

"Left Aden 6.30 p.m. New Zealand transports lead the fleet, which were all with us again. The *Hampshire* is our only escort. Saw land all the way today. Entered the Red Sea 3 p.m. today, the island of Pirim on our starboard bow, with lighthouses and forts on. Beautiful sunset. The weather is a good deal cooler towards evening. Five or six transports passed us today with no troops; they were evidently going for troops. About 10 p.m. the *Maunganui* sent up a rocket, the signal to alter course around a big rock in the Red Sea, but at that moment the steam valve or gauge broke on our boat, and she had to slow up until repaired. The *Maunganui* asked by lamp "What is the matter?"[124]

LT WILLIAM JANSON, WELLINGTON MOUNTED RIFLES, NZEF, HMNZT *ARAWA*.

---

121 Secretary's Notes of a Meeting of a War Council Held at 10, Downing Street, November 25, 1914. TNA CAB 42/1/4.
122 *Hawera Star* (New Zealand), 24th April 1928.
123 *The Advertiser* (Adelaide, South Australia), 12th January 1915.
124 *Hawera Star* (New Zealand), 24th April 1928.

(continued from previous page) there. Our real point of attack might be Haifa, or some point on the Syrian coast. The Committee of Imperial Defence in 1909 had recommended that a serious invasion of Egypt could best be met by a landing at Haifa."[121]

LT-COL. MAURICE HANKEY, SECRETARY, WAR COUNCIL.

> "All the other ships belonging to our fleet arrived during the night at Aden, so we all left together this morning about 7 a.m., escorted by the cruiser *Yarmouth*. We left the Japanese cruiser behind at Aden, which was one of our escort from Albany. We passed some Indian troopships to-day, but they appeared to be empty. We can tell troopships from other ships by a white square on the fore and aft part, with a number on it."[125]
>
> CPL ALEXANDER URE, RESERVIST, SEAFORTH HIGHLANDERS, HMAT *MILTIADES*.

27 NOVEMBER 1914

LEVANT

An American naval officer in Beirut reported a conversation he had had with an engineer said to have been involved in the development of the Dardanelles' defences.

> "These [forts] consist of two lines, the first at the mouth of the straits, the second some fifteen miles up the Dardanelles. The first line is two strong forts or series of forts... on either side of the mouth, covering thoroughly the entrance, with particular convergence of fire upon the main approach between the islands of Tenedos and Imbros. The second line of forts... is also located on high ground, fifteen miles up the passage, and has a converging fire down the straits, covering well into the area of operation of the first forts. Beyond the second line of forts the informant knew of no other.
>
> "The forts are of the Brialmont type, similar to those in Belgium, and are constructed of concrete and steel, the guns being placed in embrasures rather than in turrets or cupolas. The number of guns could not be ascertained, but they vary in size from the 6-inch rifle to a 12-inch mortar.
>
> "A modern system of searchlight, range finding stations, telephonic communications, etc., is installed.
>
> "The passage is of course mined, probably throughout its length, but the number and location of the mines is known only to the Turkish Army."[126]
>
> LT T. S. WILKINSON, USN, USS *SCORPION*, BEIRUT.

---

125 *Airdrie & Coatbridge Advertiser*, 10th July 1915.
126 Naval Attache's Reports, Office of Naval Intelligence, Unpublished Manuscript, US Naval War College, November 1914.

CYPRUS

The presence of a significant Turkish population concerned the garrison of the recently annexed island.

> "The situation here is awfully serious, and we are making all preparations to do what we can. It is a big job looking after this island, and there might be some really serious work to be done. Every man is confined to barracks (except those on duty outside), and in Egypt the position is the same. No man is allowed out even on business, unless he is armed and has an escort. There is no telling what Turkey will do, and you will see in the papers any of our doings. You mustn't get alarmed if one week no letter arrives. I might very possibly be out on patrol, and not return for days, or equally probable the letters might be censored, and this means delay. If you are imagining me undergoing all sorts of hardships, you are mistaken. Apart from the serious position of affairs, we should look upon this as a picnic. Christmas will soon be here, and no doubt we shall all feel homesick, for when all is said and done there is no place like England, and nobody will be keener to get into the old country again."[127]

LT EDWARD HORSFALL, 1/8TH BN MANCHESTER REGT, LIMASSOL.

RED SEA

As some looked forward to Christmas in England, others were making their own entertainment at sea.

> "Extra hot. Six engineers from one of our warships are on board, patching up our breakages. They reckon to have the gap repaired by the time we reach Suez. To-day the news went round that we are to be in England on Christmas Eve, so we shall eat our Christmas dinner there after all."[128]

PTE ARTHUR ASH, 10TH BATTALION, AIF, HMT *ASCANIUS*.

> "This ship is quite a Monte Carlo as far as gambling is concerned; some men winning and losing as much as £30 in a day. We have not heard our band playing for about a week. The last time it favoured us with a tune, it was composed of a drum, a penny tin whistle, a set of bagpipes, a tin meat platter, a very ancient and antique pair of cymbals or clappers, which might easily be mistaken for some old battle-worn shields belonging to some of the Roman warriors but that would make them more modern than they really appear."[129]

CPL ALEXANDER URE, RESERVIST, SEAFORTH HIGHLANDERS, HMAT *MILTIADES*.

---

127 *Todmorden Advertiser and Hebden Bridge Newsletter*, 27th November 1914.
128 *The Advertiser* (Adelaide, South Australia), 12th January 1915.
129 *Airdrie & Coatbridge Advertiser*, 10th July 1915.

28 NOVEMBER 1914

ATLANTIC OCEAN

The expected decisive clash at sea, the new Trafalgar, had not happened. The blockade of Germany, keeping the seas safe for allied vessels, was absolutely vital. But it attracted fewer headlines than the loss of the 'Live Bait Squadron,' the *Emden*'s adventures and the defeat at the Coronel. Public confidence had been dented and even some naval officers believed that the Admiralty, Churchill included, was failing. With memories of the calamitous Antwerp expedition still fresh, it was time to deal with von Spee's squadron. Fisher and Churchill ordered the battlecruisers *Inflexible* and *Invincible* south.

> "'King Neptune' and his suite came aboard... every man who 'had not crossed the line' before had be shaved with Neptune's hoop-iron razor and then tipped over backwards into conveniently placed canvas tank full of 'briny' 10 feet below. Officers as well went through the mill. There wasn't soul on the ship in a dry suit of clothes that day, for no sooner did a dry suit of clothes appear on deck than the wearer was promptly seized and bundled unceremoniously into one or more of the canvas tanks and severely ducked. Each man as he emerged from his 'baptism' then joined in the hunt for others who hadn't had some."[130]
>
> ERA4 ARTHUR LEEDALE, RN, HMS *INFLEXIBLE*.

It was not all fun and games. During the voyage, Cdr Hubert Dannreuther, *Invincible*'s gunnery officer, was able for the first time to practise firing at targets at 6,000 yards or more. He had joined the ship in 1912.

RED SEA

Rumour was confirmed by the announcement that the troopships' destination had been changed. The Australians and New Zealanders would not be going to England. Their destination was now Egypt.

> "Positively sweltering, and not a breeze. On the other hand in about a week we shall be in a freezing climate... A few minutes ago the bugle call sounded and we all rushed to the appointed spot. Our C.O., Colonel J. Lyon Johnston, awaited us with a smile. "Gentlemen," he said, "I have some news for you. Our course has been altered. We will disembark within three days, and proceed to Cairo in Egypt, and from there we will go straight to the front." I think it is about the last thing any of our fellows expected, but they received it cheerfully."[131]
>
> LT JOHN WILLIAMS, 11TH BATTALION, AIF, HMAT *ASCANIUS*.

---

130   *Harrogate Herald*, 24th March 1915.
131   *The Southern Record and Advertiser* (Candelo, New South Wales), 30th January 1915.

"Colonel Johnston read out a wireless message to the troops. It was to the effect that we are to disembark at an Egyptian port and proceed to Cairo to receive further training. The news occasioned great surprise, for although a large number of the men do not care where we go so long as we get off the boat, quickly, the majority wanted to see the Old Country first."[132]

L/SGT ALEC MARSHALL, 11TH BATTALION, AIF, HMAT *ASCANIUS*.

29 NOVEMBER 1914

EGYPT

Many of those with family connections to Britain were disappointed at not being sent there. The winter weather and conditions in the still developing camps might have changed some minds.

"The sole topic of conversation is our landing in Egypt. What is the reason? Shall we be kept there long? &c. Cairo, we are told, is a city with a population of about 600,000, has a beautiful winter climate, but is a particularly expensive place. Personally I would have much preferred England and the Continent."[133]

LT JOHN WILLIAMS, 11TH BATTALION, AIF, HMAT *ASCANIUS*.

30 NOVEMBER 1914

EGYPT

"Arrived at Suez at 1 a.m. What a scene. Natives in native boats crowding around, but the Colonel will allow none of them near us. They make all sorts of excuse. No avail. At 5 p.m. the sunset — and what a glory in colour. The sandhills and desert turned into a violet hue. Someone else can describe it; I can only gaze at the glory of it through the tiny clouds of a fine cigar."[134]

2/LT GEORGE KELLY, 2ND BATTALION, AIF, HMAT *SUFFOLK*.

---

132   *Kalgoorlie Western Argus* (Western Australia), 2nd February 1915.
133   *The Southern Record and Advertiser* (Candelo, New South Wales), 30th January 1915.
134   *Diary Kept by Lieut. G.E.E. Kelly*, p. 12.

DECEMBER 1914

"They are as fine a body physically as I have ever seen…
But do all Australians drink quite so much?"
Unnamed British Officer to Charles Bean

Australians at Mena Camp, near Cairo.

1 DECEMBER 1914

EGYPT

"Arrived at Suez 8 a.m. against a good head wind. Left Suez and entered the Canal at 1 p.m. The two flagships had evidently gone on. We passed a good many military camps, both native and British, along the banks. Camels and mules were numerous. Forty men and [an] 18-pounder were stationed on the look-out for snipers... It was dark at 5.30, so we lost the sights along the Canal. We hope to reach Port Said at daylight to-morrow. Each transport carries a powerful headlight as it goes through the Canal at night...

"We transferred our German prisoners (*Emden*) to the warship *Hampshire*. The canal is 80 miles long and we take up 66 miles with our transports."[1]

LT WILLIAM JANSON, WELLINGTON MOUNTED RIFLES, NZEF, HMNZT *ARAWA*.

"Suez reached at 9 a.m. It is a fair-sized town, and we can see a few trees. As trading boats were not allowed alongside, we could not procure any fruit. Left at 8 in the evening. The canal is only about two chains wide, and there is desert on both sides. Soldiers are patrolling all along the canal, and before we entered four machine guns were rigged up and 100 men posted on the top deck with loaded rifles — all for nothing."[2]

PTE ARTHUR ASH, 10TH BATTALION, AIF, HMT *ASCANIUS*.

2 DECEMBER 1914

MEDITERRANEAN

Interruptions in the monotony of voyages on troopships were not always welcomed.

"It was one of the worst nights I ever remember putting in, for more than half the troops, who numbered 2,500 odd, were down with ptomaine poisoning.[3] Oh, it was a night of horror. All round the deck the poor chaps were moaning and groaning. It was terrible to hear them, and I will never forget, it. You people in Australia can't possibly realise what an outbreak of ptomaine poisoning on a closely-packed troopship means. I worked under great difficulty, helping the poor fellows till midnight. I say great difficulty, for I was very queer, and kept dosing myself from the dispensary to keep going. After going from 8 p.m. till midnight, I went to bed, but not to sleep, for I was in too great pain, and at 1.30 was doubled up and in trying to reach the hospital for

---

1   *Hawera Star* (New Zealand), 24th April 1928.
2   *The Advertiser* (Adelaide, South Australia), 12th January 1915.
3   The 1st Australian Field Ambulance war diary records: "Severe epidemic of 'ptomaine poisoning' – Some 550 victims – chiefly 4th Bn." (AWM4 26/44/1).

treatment collapsed on the floor, and remained there till carried into the hospital, where a hypodermic injection of morphia sent me to sleep about 3 for a couple of hours."[4]

PTE OSCAR WARD, 1ST FIELD AMBULANCE, AAMC, AIF, HMAT *EURIPIDES*.

EGYPT

"There are a tremendous number of Australian troops coming to Egypt, 30,000 to be correct, and they are to be divided into three camps. I have been sent down with a detachment of the 9th Batt. Manchester Regiment, to be guard over the supplies for the troops at the Maadi Camp. Sergeant Braithwaite[5] has been sent to Mena Camp, at the foot of the Pyramids, on the same class of duty. Now Maadi is about ten miles from Cairo, and I came last Saturday afternoon. When I arrived there was nothing to be seen at the camp only the desert, marked out with flags, but to-day it is looking something like a camp, as the supplies are coming in daily, and in another day or two it will represent a white city. Sergt. Braithwaite is about eight miles from me, in another direction. I can see the Pyramids very plain during the day. We spent Saturday night sleeping in the open on the desert, as we had no tents, only one blanket and one oil sheet, and it was bitter cold. It is surprising how cold it goes during the night."[6]

SGT HARRY ILLINGWORTH, 1/9TH BN MANCHESTER REGT, MAADI CAMP.[7]

"We arrived at Port Said on December 2nd. The harbour was full of New Zealand and Australian boats, and about a dozen cruisers, French and English. Here we transferred the German prisoners to HMS *Hampshire*. While anchored in the harbour we saw three French hydro-aeroplanes landed. They belonged to a French battleship, their parent boat. In the evening we anchored outside Port Said."[8]

PTE CLAUDE HULBERT, 3RD FIELD AMBULANCE, AAMC, AIF, HMAT *RANGATIRA*.

"Arrived at Port Said at 1 a.m., and started to take coal. Kept awake all night by the noise the natives made while coaling. Went on board the *Orari* to see the horses; they had lost 17 up to date; only 1 of our squadron. Left Port Said for Alexandra [sic] at 3.3 p.m., amid cheering from every side, also from the three French warships which were there. There was no leave granted. Posted Lo's letter hurriedly. Our boat was the 7th to leave, the Australian flagship is in front of us. There are most interesting sights about Port Said."[9]

LT WILLIAM JANSON, WELLINGTON MOUNTED RIFLES, NZEF, HMNZT *ARAWA*.

---

4 *Windsor and Richmond Gazette* (New South Wales), 26th March 1915.
5 Sgt Noel Braithwaite, 1/9th Bn Manchester Regt, was killed in action on 7th June 1915. Commemorated on the Helles Memorial, he was the son of John and Mary Braithwaite, of Beech House, Ashton-under-Lyne, Lancashire.
6 An approximate date, as the letter was undated. *Stalybridge Reporter*, 26th December 1914.
7 Illingworth was killed in action on 5th June 1915. Buried in Redoubt Cemetery, he was the son of John Barraclough Illingworth, of 192 Oldham Road, Ashton-under-Lyne, Lancashire.
8 *Cheltenham Chronicle*, 16th January 1915.
9 *Hawera Star* (New Zealand), 24th April 1928.

3 DECEMBER 1914

EGYPT

It would not have surprised anyone that men long confined aboard ship would get on dry land, legally or otherwise. Some sights and sounds might have encouraged them to remain on board.

> "Arrived at Alexandria daylight to-day, after a fairly smooth trip. Had leave durig [sic] the afternoon. A good many men broke away from the ship during the evening. Alexandria is a great town. Of course, it has its slums as well."[10]
>
> LT WILLIAM JANSON, WELLINGTON MOUNTED RIFLES, NZEF, HMNZT *ARAWA*.

> "We started to go through the Suez Canal about 1 p.m. to-day. The canal is about 80 odd miles long, and I should say about 150 yards wide. At present it is guarded by Indian troops. They were throwing up sangars, redoubts and trenches as we were passing. There were three Indian soldiers playing the bagpipes as we passed one place. A good many of our fellows were very much surprised at seeing the Indian pipers, because they had not heard that the Indian regiments had pipe bands. There are a good many dredges working along the canal, and it appears to be mostly Frenchmen and Italians who are employed on them. The canal cuts through some lakes at parts. There is nothing to see on either side, but a lone stretch of yellow sand. Ships going through the canal require to have a pilot on board, also a searchlight placed in front to light up the way during the night. While at Port Suez we managed to purchase some oranges and apples from the natives who came alongside in small boats. That was the first taste of fruit we had since we left Australia five weeks ago, and although the oranges were hardly ripe still they were juicy and sweet, and they went down all right. The apples were small plum-shaped things and tasteless. They were being sold at 60 for a shilling, and the oranges a penny each. They can grow good oranges in Egypt, and sell them very cheap too, because I have bought them in Cairo, three for a piastre (1¼ d). We have just passed 2 ships and a French cruiser in the canal, and although it was dark there was a good bit of cheering."[11]
>
> CPL ALEXANDER URE, RESERVIST, SEAFORTH HIGHLANDERS, HMAT *MILTIADES*.

---

10   *Hawera Star* (New Zealand), 24th April 1928.
11   *Airdrie & Coatbridge Advertiser*, 10 July 1915.

4 DECEMBER 1914

ENGLAND

Britain's pre-war armed forces were effectively a colonial police force but participation in a world war required rather more than that. A shortage of volunteers was not a problem but these were far from being ready to fight.

> "In a great war, such as the present, the question of the supply of trained officers must always be a difficult one for any nation, owing to the large number of casualties, the need of officers to at once take in hand the training of the new levies… Even a country with the wonderful and thorough military system of Germany will have a shortage of trained officers in a great emergency like the present one. What then would be the case of a country with a small Regular Army deficient in officers[,] no reserve of officers, no men trained and available for officers' commissions, and no system of quickly training and supplying them in case of need? Such a country would be not only unable to replace the casualties in the officer list of its Regular Army in the field, with the result that this Army would rapidly deteriorate; but also would have no capable officers available to train and officer its large new levies…
>
> "A division of the British Army requires about 600 officers. The first new armies of one million men, which have already been raised require about 40,000 officers. The second million men, now in process of recruiting will require another 40,000 officers. Besides these, there are the officers required to replace the heavy casualties of the army in the field, and for new levies of Territorials.
>
> "Up to the present time about 50,000 additional officers have been provided; and for the total forces thus far provided for, and to replace casualties, some 100,000 officers have been, or will be required. That is, Great Britain within 6 months from the outbreak of war has required, or will require about 100,000 additional officers for her military forces."[12]
>
> MAJOR THOMAS C. TREADWELL, USMC, AMERICAN EMBASSY, LONDON.

EGYPT

> "Had early breakfast and marched to the *Orari* to unship our horses. Entrained it 11 a.m., and left Alexandria, for our camp at Zeitoum, arriving there at 8.30 p.m., about 10 miles from the pyramids. We had to sleep on the sands of the desert just as we arrived, with our greatcoats only. A night I shall never forget, sand snakes crawling about and the sand full of camel lice. However, we survived until the morning, then the hard work commenced of arranging the camp."[13]
>
> LT WILLIAM JANSON, WELLINGTON MOUNTED RIFLES, NZEF, ZEITOUN CAMP.

---

12 Naval Attache's Reports, Office of Naval Intelligence, Unpublished Manuscript, US Naval War College, December 1914.
13 *Hawera Star* (New Zealand), 24th April 1928.

"We have got a garrison church and the Rev. Dennis Fletcher, Rochdale, is conducting the services. Our barracks are the finest in the world, and are composed of a dozen large buildings. It will be grand when we are settled down; we can see the pyramids from the top balcony of our room. Our meals here are very good, and they put some kind of oil in the bread to keep it soft…

"We do not know how long we shall remain here or where we shall go next. We may in time go to France. We have all been supplied with fresh clothing which is more suitable for the hot climate."[14]

PTE ORMEROD CRABTREE, 1/6TH BN LANCASHIRE FUSILIERS, EAST LANCASHIRE DIVISION.

5 DECEMBER 1914

EGYPT

"Reached Alexandria at 7 a.m. A breakwater round the harbor is large enough to shelter all our ships. About 30 German ships captured by the British and the French are here. We are at anchor away from the wharf, and we are all eager to land."[15]

PTE ARTHUR ASH, 10TH BATTALION, AIF, HMAT *ASCANIUS*.

"We saw Alexandria early and got orders to disembark, so we were fairly busy. It was Sunday, too. We arrived at 11 a.m. and waited for the New Zealanders to complete landing. This is a very large place and it is over 2,000 years old so there are some very old buildings. It is a busy place, and there are plenty of yachts, motor pleasure boats, etc."[16]

SGT DOUGLAS SCOTT, 3RD AUSTRALIAN LIGHT HORSE REGT, AIF, HMAT *PORT LINCOLN*.

INDIA

How would India's Muslims react to war with the Ottomans? One observer did not expect trouble; he had not even noticed that holy war had already been declared.

"I am afraid there is some doubt whether the Mohammedans here would cause trouble over Turkey, but they are going the right way and showing good sense. The chief man at one of the biggest mosques in Calcutta, whom I saw myself, was confident that there would be no disturbance in India, and certainly not in Calcutta. If the Sultan or Caliph, as religious head of Islam, proclaim a holy war, all Moslems would be supposed to support him. Apparently the Sultan cannot do this by merely calling it a 'holy war.' There must be a conference and a consultation. At any rate no such 'holy war' has been proclaimed, and Indian Moslems do not seem likely to acknowledge it

---

14   *Todmorden & District News*, 4th December 1914.
15   *The Advertiser* (Adelaide, South Australia), 12th January 1915.
16   *Kapunda Herald* (South Australia), 29th January 1915.

as such, even if it were. The Germans will be disappointed if they counted on making trouble this way. If any trouble does come, one of the Mohammedan 'maswalas' told me, it would probably come from the hill tribes."[17]

HARRY GOUGH, "STATESMAN," CALCUTTA.

6 DECEMBER 1914

EGYPT

Those arriving in Egypt reflected on the past and looked forward to the future.

> "We are now at the scene of Nelson's great victory, and are to proceed on to Cairo, where Indian and Territorial troops are encamped, as the Turks under Gamal Pasha are advancing on Egypt, where they will meet with a warm reception. I am on tiptoe with expectation to see the wonders of this ancient land."[18]

PTE CHARLES ALEXANDER, 8TH BATTALION, AIF, HMAT *BENALLA*, ALEXANDRIA.

> "If we had gone Home we would have been amongst the Germans in the spring but I'm doubtful we'll ever get there. We are told that we are here to finish our training, and in case of a rising in Egypt. Of course, everything is all right yet, but you cannot depend on them, and if it should happen, well, you can call the Queensland Light Horse, the "Death or Glory Boys," because we'll never be taken alive to finish up on the ant-hills or such like."[19]

TPR STANLEY CHEERS, 2ND AUSTRALIAN LIGHT HORSE REGT, AIF, HMAT *STAR OF ENGLAND*, ALEXANDRIA.

7 DECEMBER 1914

MESOPOTAMIA

It was not always easy to assess how determined Turkish resistance had been after the shock of first action.

> "It's sheer luck, for if anyone were to look at us now they would think we were a pepper dredger, as we are full of holes. One of the masts, has been shot away, the funnel perforated, the decks are splintered, yet we are still floating…
>
> "One of the shells passed through an officer's body,[20] blew his heart clean out, and went on, blowing the wheel and the helmsman's hand away…. I pity the poor soldiers.

17 *Cheltenham Chronicle*, 5th December 1914.
18 *St. Arnaud Mercury* (Victoria), 6th February 1915.
19 *Tweed Daily* (Murwillumbah, New South Wales), 27th January 1915.
20 Lt-Cdr Frederick John Elkes, RNR, HMS *Ocean*, attached Armed Launch *Shaitan*, was killed in action on 7th December 1914. Buried in Basra War Cemetery, he was the husband of Annie Elizabeth Elkes, of 12 Mabel Grove, Lisburn Lane, Liverpool.

If this is war, I don't want to see or have any more. You can't realise what it like until you see it."²¹

AB WILLIAM BLAND, RN, HMS *ESPIEGLE*.

EGYPT

"We left Alexandria about one o'clock… Upon leaving Alexandria behind us we at once entered a land so intensely cultivated, that Westralians cannot help but make a comparison between it and their own land. The whole of the land that can be seen from the train between Alexandria and Cairo is best described as an enormous vegetable garden. Every available foot of soil appears to be contributing produce for the support of the population. It is a splendid example as to what can be done by systematic drainage and irrigation…

"The valley of the Nile must surely be one of the most thickly populated agricultural parts of the world. In some of the fields can be seen a small army of workers and their animals, in fact there are so many that in point of numbers they rival a Westralian country agricultural show or race meeting."²²

PTE EDWARD RICHARDS, 11TH BATTALION AIF, MENA CAMP.

8 DECEMBER 1914

FALKLAND ISLANDS

The *Invincible* and *Inflexible* had arrived the previous day. They began coaling at Port William at 7.20 a.m. and had not finished when German cruisers were sighted approaching the islands. Aground, the *Canopus* was employed as a guardship.

> "We went to general quarters… and opened fire with our 12-inch guns [at 9.19 a.m.]. Our other ships could do nothing, as the land was between them and the enemy, and most of them were coaling. We fired a lot of shots, and hit the *Gneisenau*. By this time the fleet had weighed, and was coming out at full speed. The Germans turned tail and fled, with the fleet at full speed after them. Of course we could not follow, as we were on the mud. Still, we had opened up the action, prevented the enemy from shelling the wireless station, and saved the fleet from being attacked while at anchor. All that day we waited eagerly for news."²³
>
> MIDSHIPMAN PHILIP DE CARTERET, RN, HMS *CANOPUS*.²⁴

After taking on enough coal, the battlecruisers gave chase.

---

21   *Surrey Advertiser*, 23rd January 1915.
22   *Bunbury Herald* (Western Australia), 23rd January 1915.
23   *The Sun* (Sydney, New South Wales), 12th March 1915.
24   Malet de Carteret was killed aboard HMS *Queen Mary* at Jutland on 31st May 1916. Commemorated on the Portsmouth Naval Memorial, he was the 18 year-old son of the barrister Reginald and Amy Malet de Carteret, of St. Ouen's Manor, Jersey, Channel Islands.

> "When we got out of harbour and started the chase, on a beautiful, but cold, day, all we could see of the enemy were five thick lumps of smoke away on the horizon. We didn't make straight for them, but headed off from the land, leaving them no option but to fight or surrender… The *Invincible*, *Inflexible*, and *Glasgow*, being faster ships, got a long way ahead of the remainder of our fleet, and a few minutes before 1 p.m. we started blazing away at the enemy light cruisers, which were in rear of the two armoured cruisers. The range was about 9 or 10 miles…
>
> "The *Invincible* and *Inflexible* now engaged the *Scharnhorst* and giving them "own back with interest" for our lost *Good Hope* and *Monmouth*. Clever manoeuvring helped us to dodge the enemy's shells… This ship got off very light, all things considered, for we were only fairly hit twice, though splinters riddled our decks. The *Invincible* was hit several times, but without serious damage…
>
> "The action was in three parts due to the enemy's manoeuvres, and during the intervals I went on deck for a glimpse of the now battered "Germs." About 4 p.m. word was passed round that the *Scharnhorst* was on fire and healing over heavily to starboard. A few minutes afterwards she sank, taking her crew to the bottom with her… The *Gneisenau* stuck it for nearly two hours after this, losing masts, funnels, one turret lifted clean over the side and other serious damage… Lying over at an angle of about 45 deg. she quietened down. Whether her colours were shot away or hauled down I don't know, but we, thinking she had surrendered, closed in to rescue her crew, but she fired another… and got a broadside in return. That broadside was her "coup de grace," for she turned quietly over and sank."[25]
>
> ERA 4 ARTHUR EMIN LEEDALE, RN, HMS *INFLEXIBLE*.

Of the 1,295 shells fired, an estimated 74 struck home. And many shells were defective. The problems affecting gunnery accuracy and shell quality would take time to address.

ENGLAND

In London, men at the Admiralty waited for news from 8,000 miles away.

> "Admiral Leveson came to my room and, calling me outside, told me of the receipt of the first of Sturdee's cables, "I am engaging German ships," and then we had to wait. Leveson and I both found we could not do any work. The excitement was too great. It must be remembered that we had our bad times. There had been a goodish number of losses… The morning on which the *Hogue*, *Aboukir* and *Cressy* were sunk was, for example, a bit of a facer. So was the day on which the tidings came of the loss of the *Audacious*; whilst the news of the fight off Coronel, though it had come in bit by bit was pretty heartbreaking. We were in a state, therefore, of peculiar tension; but presently the blessed news came in that we had sunk all the enemy's ships but one, and the relief was tremendous, the feeling of elation glorious."[26]
>
> CAPT. SIR DOUGLAS BROWNRIGG, RN, CHIEF WIRELESS TRANSMITTER CENSOR, ADMIRALTY.

25   *Harrogate Herald*, 24th March 1915.
26   Brownrigg, Douglas, *Indiscretions of the Naval Censor*, p. 22, Cassell and Co. Ltd. (London) 1920.

MESOPOTAMIA

The fighting was often hard-fought but the advance northwards continued. Turkish resistance was being overcome.

> "We had a very interesting time taking Kurnah. We could not get nearer than two miles except at the top of high water, but that was really near enough, we could range them easily, and see where our shots were going from the masthead, while they could not spot the fall of their shots at all. Also, we had an excellent view of the whole business, both the land and river fighting, from the masthead where I was stationed. The Turks were very strongly entrenched and had ten guns, all somewhat old, and their shells were not of the best, but they were well fought, and took a lot of silencing. The troops advanced on the trenches, and carried them without much trouble. They drove the Turks across the river, but then their trouble began, as they advanced right up to the river bank, and the Turks poured a very heavy fire on them from the tops of the houses opposite. There being no means of getting across the river at them, and we being unable to shell the houses from the ship, all except one were hidden from us (even from the masthead) by date trees, there was nothing for it but to retire.

> "The next attack was made two day later. We moved a bit higher up the river, and the troops, who had been reinforced, sent one force to attack the trenches, which the Turks had reoccupied, and another to endeavour to get across the Tigris higher up. The Turks were again driven across the Tigris, and late in the afternoon part of the force and two mule guns got across, about two miles up, and then orders were given to bivouac for the night, and continue the attack at daybreak. We had been kept pretty busy as four of their guns commanded the river. We soon silenced two, but the others continued to fire, though very wildly. As I said in my last letter, the launches had the hottest time, one was badly holed below water in the engine room by a shell, but managed to get back, and ran to the bank close to us before she sank. The captain of another, a RNR Lieutenant, who had been living board here for some time while the launch was being fitted out, got a nine pounder shell through his chest. Poor fellow, he had left the sea for good, as he thought, four years ago.[27] Besides him, the launches had one man killed[28] and two wounded, and we had two wounded. That night in the middle watch we saw a boat coming down the river, with much shouting and waving of lights, which proved to contain a deputation of Turkish officers, offering surrender unconditionally, so the attack planned for daylight was unnecessary.

> "Having sounded a channel very carefully, we managed to get up to Kurnah, though we were scraping along the bottom part of the way. Over a thousand Turks were taken prisoners, but no Arabs, they must have dispersed to their homes, as there was no doubt plenty of them had been fighting against us. There were said be 1,500 Arabs

---

27   Lt-Cdr Frederick Elkes, RNR, HMS *Ocean*, attached Armed Launch *Shaitan*, was killed in action on 7th December 1914. Buried in Basra War Cemetery, he was the husband of Annie Elizabeth Elkes, of 12 Mabel Grove, Lisburn Lane, Liverpool.

28   OS Edward Gibson, RFR, HMS *Ocean*, attached Armed Launch *Shaitan*, was severely wounded on 7th December 1914, dying on 9th December 1914. Buried in Basra War Cemetery, he was the son of Edward and Margaret Gibson, of 52 Shelly Street, Toxteth Park, Liverpool.

about 30 miles away that the Turks had hoped would fight for them, and I believe that is why the Turks surrendered instead of bolting, as they were afraid the said Arabs would turn on them. I am afraid it is all finished now, unless by any chance they decide to push on to Bagdad."[29]

LT-CDR ARTHUR SEYMOUR, RN, HMS *ESPIEGLE*.[30]

9 DECEMBER 1914

ENGLAND

Troubridge had been acquitted by Court Martial for failing to engage the *Goeben*. But senior figures in the Admiralty were not impressed.

> "The finding of the Court Martial appears to be correct on the evidence educed, but I am of opinion that its conclusions are wrong, both from the common-sense point of view and technically.
>
> "As regards the former, the instructions given to the Rear-Admiral that the First Cruiser Squadron and *Gloucester* were "not to get seriously engaged with superior force," taken in conjunction with the fact that he was twice informed that the *Goeben* was his objective, must, or should, have conveyed to him that his Squadron was not considered an inferior force, and that he was expected to attack her….
>
> "The whole action of the Rear-Admiral contrasts most strangely with that of the Captain of *Gloucester*, who, in spite of inferior speed and of vastly inferior power, clung tenaciously to *Goeben* until twice ordered back.
>
> "The finding of the Court Martial that the First Cruiser Squadron was an inferior to the *Goeben* and *Breslau* appears to be founded on exaggerated Gunnery expert advice combined with an omission to consider all the factors which bore on the case."[31]

REAR-ADMIRAL FREDERICK TUDOR, RN, THIRD SEA LORD.

Troubridge was never given a command at sea again. His career was effectively over.

EGYPT

"We landed to-day. The horses were fresh and took some holding on the wharf, but we ran them straight into the trucks off the ship so they were easy to truck. They will walk in anywhere now. We left Alexandria at 11 o'clock in the morning and got to Cairo at 5 in the afternoon. On our journey we saw thousands of acres of vegetables

---

29   *Coventry Standard*, 29th January 1915.
30   Lt-Cdr Seymour was awarded the DSO for his services during the Mespotamian campaign:
    "For excellent work throughout operations in Mesopotamia. During the attack on Nasiriyah on July 24th, 1915, Lt-Commander Seymour, who was in command of the armed launch "Shushan," fired the gun himself under very difficult conditions, and sank an armed Turkish patrol boat." (*London Gazette*, 19th November 1915.)
31   Tudor quoted in Lumby, *Policy and Operations in the Mediterranean 1912–14*, pp. 397–398.

and cotton fields along the Nile. It is the richest country I have seen in my life... We had to walk and lead the horses 10 miles after getting out of the train to Maadi. This is a garden of Eden itself. It is an English and French summer resort with lovely houses. Our camp is half a mile from here with not a tree or green blade to be seen except in the town, which is under irrigation from the Nile."[32]

SGT DOUGLAS SCOTT, 3RD AUSTRALIAN LIGHT HORSE REGT, AIF, HMAT *PORT LINCOLN*.

10 DECEMBER 1914

ENGLAND

With Coronel avenged, Churchill's mind turned to the possibilities of naval operations on the Belgian coast. Some feared a second Antwerp.

> "The tides are favourable from the 14th onwards, but firing would begin later each day. A gale would interrupt the naval operations.
>
> "Two battleships are all that can work off Ostend and Nieuport at one time. But arrangements would be made to replace any sunk or set on fire and to maintain the bombardment night and day as required. In addition three monitors, two gunboats, and six destroyers will be used. Total heavy guns twenty-six, of which nine are very heavy...
>
> "This force should be sufficient to support the advance of the Army on Ostend."[33]

WINSTON CHURCHILL, MP, FIRST LORD OF THE ADMIRALTY.

FRANCE

> "I fear Joffre and Foch will make difficulties. The preparations for a forward move, commenced as I told you when you were here, had even then proceeded farther than I thought, and I'm afraid we must carry this through now from our present position.
>
> "I am in close consultation with Foch and shall hear at once what view Joffre takes. But if he agreed to an immediate change of our position the forward move now projected (and for which troops have been moved into position) would have to be postponed for several days. He will hardly agree to this and I'm not altogether sure that, from a general point of view, he would be right in incurring the delay."[34]

FIELD MARSHAL SIR JOHN FRENCH, COMMANDER-IN-CHIEF,
BRITISH EXPEDITIONARY FORCE.

---

32   *Kapunda Herald* (South Australia), 29th January 1915.
33   Churchill, Winston, *The World Crisis 1915*, p. 53, Thornton Butterworth Limited (London) 1923.
34   Sir John French quoted in Churchill, *The World Crisis 1915*, p. 53.

11 DECEMBER 1914

EGYPT

The improvement in the physical fitness of the Lancashire Territorials was evident.

> "We finished our Brigade training… after a month's strenuous work. Friday was exceptionally hard, as we left barracks at 7 a.m., and did not return until 10-30 p.m. We were hard at it all day, and 14½ hours hard slogging on the desert in full marching order is no joke. Anyhow our boys marched into barracks singing, and you will be pleased to hear that not a single Todmorden lad fell out. We were highly praised by General Maxwell. I would just like to take them back to Todmorden in this condition, and march them through the town. The people would be astonished, as their improvement has been extraordinary… There can be absolutely no doubt but that we are fit to take the field at any time."[35]

CAPTAIN JOHN GLEDHILL, 1/6TH BN LANCASHIRE FUSILIERS, EAST LANCASHIRE DIV.

> "Most of us have by now been to see the Pyramids and the Sphinx… We have now spent enough time in Cairo to give a good idea of what it is like. After leaving the main gates of the barracks, you begin to get pestered with native hawkers and touts. Carriages, donkeys, matches, cigarettes and chocolate are bawled at you till you reach the car, which happily is not far. Once inside, you think you are free from their attentions, but not so. As the car moves towards Cairo they boldly mount the footboard, pressing their wares on you, until in desperation you cry 'himski,' and threaten them with violence. 'Yaller' is another word we have learnt, which means the same…

> "Egypt is a fairly healthy place, but the Chief Medical Officer of the Division has thought it necessary to publish some notes of warning to those desirous of keeping good health. Care must be taken to wash all fruit, and if possible it should be stewed. Over-ripe fruit should be avoided. Mediterranean fever is caused by drinking milk, and all milk used must be boiled. Typhoid, although practically unknown among the army of occupation, is very prevalent among visitors. The reason for our immunity is to be found in the inoculation which we have undergone. Bathing in the Nile is strictly prohibited, owing to the risk of disease. We have had letters from Todmorden which state that some of the 'Boys' here are not behaving themselves as they should. This is not so, as the 'Boys' only frequent those districts that are bad from idle curiosity, and not for the purpose that most letters suggest."[36]

PRIVATES JAMES CAVANAGH, SYDNEY JAMES COOK, MICHAEL GAVAGHAN, WILLIE EDMONDSON, L/CPL SAMUEL SHAW, 1/6TH BN LANCASHIRE FUSILIERS, EAST LANCASHIRE DIVISION.

---

35 *Todmorden Advertiser and Hebden Bridge Newsletter*, 1st January 1915.
36 *ibid.*, 11th December 1914.

Seeing the sights could be dangerous, as one man discovered to his cost.

"Drill. A man fell from the Great Pyramid. Badly injured and not expected to live.[37] Mena House is now a fine hospital. Young Dilley,[38] of Maitland, is a patient here and he took me through it."[39]

2/LT GEORGE KELLY, 2ND BATTALION, AIF, MENA CAMP.

"To-night we had leave to go into Cairo. Very few people in Cairo can talk English so it is all strange, but those who can talk will come up to you and help you. There are some lovely buildings here which put Adelaide in the shade. We were in Shepherd's [sic] Hotel, which covers about ten acres. They have all sorts of games in the street — cards, chess, and native games. They also smoke great long pipes, the stem of which is 3 feet long with a funny hole on it. They hire them for a smoke in the shops."[40]

SGT DOUGLAS SCOTT, 3RD AUSTRALIAN LIGHT HORSE REGT, AIF, MAADI CAMP.[41]

12 DECEMBER 1914

EGYPT

"Went to Cairo by motor-car along Pyramids' Avenue: a fine broad road running along a high embankment… In the modern part of Cairo there are splendid buildings but the natives' quarter is in ruin and desolation. Five minutes in it would drive our aldermen and sanitary inspector mad. Saw the Citadel, a great fort and barracks, built of sandstone. Cairo's most beautiful mosque is here, the mosque of Mahomet Ali. The minarets are the highest and most beautiful in Cairo. The Chamber of a Thousand Lights is in this mosque. The architecture is marvellous, and the arrangement of the lights is glorious. The Persian carpets are said to be worth thousands of pounds. I won't describe this wonderful place. It would take a Byron, a Ruskin, or our Joe Enright to do that. We next went through the bazaar and saw all sorts of things, a medley of relics, curious, etc. I saw a sword of the Crusaders' time, valued at £40… A good day's outing."[42]

2/LT GEORGE KELLY, 2ND BATTALION, AIF, MENA CAMP.

37   Possibly Dvr Arthur Beard, 7th Battery AFA, who was paralysed by an accidental fall from a pyramid, according his service record, on 12th December 1914. He died at the Military Base Hospital, Melbourne, on 10th July 1916.

38   Pte William Dilley, 2nd Bn, AIF, was wounded in action on 25th April 1915. He was returned to Australia, embarking on HMT *Wiltshire* on 31st August 1915, and was discharged as no longer physically fit for service on 7th February 1916.

39   Kelly, George Edward Eccleston, *Diary Kept by Lieut. G.E.E. Kelly, During the voyage from Sydney to Egypt as a Member of the Australian Expeditionary Forces*, p. 14, T. Dimmock Ltd., Printers, (Maitland).

40   *Kapunda Herald* (South Australia), 29th January 1915.

41   Sgt Douglas Scott, 3rd Australian Light Horse Regt, was wounded on 15th May 1915. He died aboard HMHS *Gascon* on 20th May 1915. Buried at sea, he is commemorated on the Lone Pine Memorial; the son of Thomas Scott, of Kapunda, South Australia.

42   *Diary Kept by Lieut. G.E.E. Kelly*, pp. 14–17.

"We stayed nearly a week at Alexandria. There were sixty-five of our boys arrested for leaving the ship without permission one night, and the following night another sixty. They were all fined or imprisoned, ranging from forfeiting the day's pay to four days' imprisonment and the loss of four days' pay. For once, I did not get into trouble."[43]

TPR ROBERT BICE, 1ST AUSTRALIAN LIGHT HORSE REGT, AIF, MAADI CAMP.

13 DECEMBER 1914

DARDANELLES

The old Turkish battleship *Mesudiye* was torpedoed and sunk in shallow water by the British submarine *B11*. What happened was recorded by an American naval officer in Istanbul.

"On December 13, the Turkish Battleship MUSSUDDIEH [sic], while lying in the Dardanelles, was sunk by an English, or French, submarine. The MUSSUDIYEH was moored, head and stern, in Sari Siglar Bay... with her head about N.N.W., and close in shore, about 300 to 500 yards. She was used as a floating battery, and her starboard 6-inch battery had been removed, undoubtedly for use in the shore batteries of the Dardanelles.

"I have this account from an eyewitness, absolutely dependable, but whose estimation of intervals of time may not be entirely accurate.

"It was at noon on Sunday, December, 13, with no wind, absolutely dead calm. The MUSSUDIYEH was struck on the port side and there was one big explosion. After about three minutes there were several shots, about eight in rapid succession, from the port battery of the MUSSUDIYEH, and the shells were seen to hit the water a short distance away. Then another two minutes, and she heeled slowly and steadily over to port, then sharply and stopped... with part of her starboard side and bottom showing above water, deeper down by the bow than by the stern, due either to the bow being in deeper water, or to the fact that the hole was made in the bow. Part of the starboard propeller was also above water. My informant estimates that there were 350 to 450 men on board. Many men were thrown into the water, and about 180 of these were saved. From 50 to 70 were rescued from the inside of the ship through the square ports not submerged. Many men were imprisoned, but about 50 were rescued during the three days following the sinking, by cutting away places for outlets, at least 50 were drowned."[44]

LT-CDR EDWARD MCCAULEY, USN, USS *SCORPION*, ISTANBUL.

The submarine's captain, Lt Norman Holbrook, RN, was awarded the Victoria Cross for the feat. His citation read:

"For most conspicuous bravery on the 13th December 1914, when in command of the Submarine B-11, he entered the Dardanelles, and, notwithstanding the very

---

43   *The Shoalhaven News and South Coast Districts Advertiser* (New South Wales), 30th January 1915.
44   Naval Attache's Reports, US Naval War College, 1914.

difficult current, dived his vessel under five rows of mines and torpedoed the Turkish battleship "Messudiyeh" which was guarding the minefield.

"Lieutenant Holbrook succeeded in bringing the B-11 safely back, although assailed by gun-fire and torpedo boats, having been submerged on one occasion for nine hours."[45]

14 DECEMBER 1914

EGYPT

As training got underway, some problems persisted. It was time to take action.

"I have had a lot of trouble over the inoculation question... there were 47 men, and, in addition, some on the *Arawa* who had not been done. I then said that I should make arrangements either to send back to New Zealand, or leave at the base, those men who still refused to be done. This had the effect of bringing in a few more, but yesterday there were still about 50 who absolutely persisted in their refusal. In this country, where everybody gets inoculated as a matter of course, it would be very wrong to risk keeping them, so I had them paraded last night and told them that, though I might respect the conscientious objections, and did not in any way wish to threaten them, the same time my first duty was to the force as a whole, and I could not risk the health of the vast majority of the men or an outbreak of enteric by keeping those who refused to submit to reasonable precautions. This had the effect of bringing in half a dozen more, but I am afraid there are still about 40 whom I am sending back by the *Athenic* along with a few of the worst characters. The sending back of the latter, and of those not inoculated, will have a very good effect on the discipline and efficiency of the force, as it will make everybody realise that they must conform to necessary orders, and also that we have no use for wasters."[46]

MAJOR-GENERAL SIR ALEXANDER GODLEY, COMMANDING NZ & A DIV., ZEITOUN CAMP.

Souvenir hunting was widespread, though that could involve grave-robbing.

"I was off yesterday afternoon and night, and spent my time going round the Pyramids and Sphinx and old Egyptian tombs. You have no idea how interesting it is. I saw the graves of Kings and Queens and soldiers. They were buried in caves, and at the head of each one is his or her picture cut in the rock. They must be thousands of years old, but still remain almost as good as the day they were cut. I secured a few little mementoes in the way of old Egyptian coins of King Pharoah's time; also a few pieces of alabaster stone and granite. I burst open a grave and took some stone from where a King's head had been placed thousands of years before."[47]

TPR ROBERT BICE, 1ST AUSTRALIAN LIGHT HORSE REGT, AIF, MAADI CAMP.

---

45 *London Gazette*, 22nd December 1914.
46 *Otago Witness* (New Zealand), 3rd February 1915.
47 *The Shoalhaven News and South Coast Districts Advertiser* (New South Wales), 30th January 1915.

HONG KONG

The presence of a ship associated with the *Emden* proved a draw to the curious.

> "We arrived here on November 30th, and came right into dock, and the work of repairs has been carried out without a single stoppage... We have since being here been the object of great interest, numerous visitors, including many soldiers, have been on board to see the effect the shell fire had on us..., which wounded our second engineer so severely that we were obliged to leave him behind in hospital, and we have learnt since arriving here that his left arm has been amputated. It seems marvellous that any us were left alive the way the shrapnels were bursting, but, luckily, only two were wounded, if the scratch I got can be called a wound. A small fragment of shell struck my left wrist and penetrated to the bone but it is now extracted, and soon healed up, but it has left a slight scar."[48]

SECOND MATE DUDLEY KILBEE, SS *CRANLEY*.

15 DECEMBER 1914

ENGLAND

> "That the attempt to preach a sort of Jihad should have proved a complete failure will not surprise anyone with a real knowledge of Eastern affairs... Further, there is something almost humorous in the idea that an invitation to war against the infidel could with any hope of success be made to appeal to the adherents of Islam when the Kaliph was himself in alliance with two infidel Governments."[49]

LORD CROMER, FORMER AGENT AND CONSUL-GENERAL, EGYPT.

EGYPT

> "I have had a look around part of Cairo. It covers a great space and needs a week of solid going to see all the city. We went into a restaurant and had a plate of fried eggs and bacon and a cup of coffee for three piastres (7½d). Wherever we went we met Australians, or New Zealanders, or Territorials. The city seems alive with them. I don't know definitely what our future movements will be, but think we will be here for two or three months undergoing a thorough system of training."[50]

CPL LEONARD FAWCETT, 3RD AUSTRALIAN LIGHT HORSE REGT, AIF, MAADI CAMP.

---

48   *Folkestone Express, Sandgate, Shorncliffe & Hythe Advertiser*, 6th February 1915.
49   *Staffordshire Sentinel*, 15th December 1914.
50   *Kapunda Herald* (South Australia), 29th January 1915.

16 DECEMBER 1914

ENGLAND

In the early morning ships of the German High Seas Fleet approached the English north east coast. They attacked the Hartlepools, Whitby but the heaviest damage was at Scarborough. One man was visiting the town to bury his recently deceased sister.

> "I was talking to my friends, about five minutes to eight, when we heard a kind of whistling noise, and then a loud report occurred about forty yards away. This abruptly terminated our conversation, and one of my friends remarked somewhat vigorously, 'The ——— Germans are here.' A shell had landed in a coal-yard, and it scattered the loose coal in all directions. I set off to run to where I was staying in order to get my wife away…
>
> "As I passed the first street a shell came and knocked the end house to pieces. I continued on my way, but so rapidly did the shells come that by the time I arrived at the next street another shell lifted the roof and chimney from a house… I got three streets further on, and was just turning into the house I was staying at when there was another loud crash. I glanced round and the third building, a shop, on the opposite side, was smashed in. Luckily, it was empty, but windows of other houses were blown out, including ours, and the room in which my sister lay dead. I hurried my wife out — she had to leave without her shoes fastened — and we made down the street towards Seamer."[51]

JAMES ATKINSON, SCARBOROUGH.

A forty-nine year-old porter, Leonard Ellis was the first to be killed there.

> "I saw Clare and Hunt's chemist shop opposite struck, and collapse. It was terrifying. The porter who had been employed at the shop for 33 years had his head blown off. He was standing in the shop doorway at the time. Another man[52] was killed near the shop."[53]

AMY ELT, SCARBOROUGH.

The result was panic, as people either sought shelter in their cellars or tried to leave the town. Those still slumbering that morning took time to realise what was happening.

> "For the moment I could hardly realise where I was, and even when awake I could not for a time imagine that shells were bursting over the town, but thought that target practice was being carried out. I eventually realised, however that guns were turned on the town, and I could not describe the sensation of sound as a shell burst, and then, just as the impression was dying away, came another crash… Immediately the sound of one died away another came, and to me it sounded like a bombardment

---

51 *Burnley Express*, 19th December 1914.
52 Harry Frith, a 44 year-old driver.
53 *Burnley Express*, 19th December 1914.

unimpeded by British ships… and as I went out of the hotel and into the town I saw several buildings which had been practically demolished. Whilst I was out, I saw one of the men who were employed in the signal station. He had his head bandaged, and his clothing was bespattered with blood…

"I made my way to the station to inquire for the time of the first train. Panic reigned everywhere, and the station was crowded with people eager to get away… The roads from Scarborough were all crowded with people, hurrying away… It was after ten o'clock when I got away, and though the bombardment was then over the station was still crowded and extra coaches had to be put on to the outgoing trains."[54]

JOHN KEERS, SCARBOROUGH.

"The damage, although very extensive in all parts of the town, we do not mind so much. It is the loss of life that hurts — and not one shot fired from here in return."[55]

AMY ELT, SCARBOROUGH.

'What was the Navy doing?' had often been asked since the war's outbreak but now the question was tinged with more anger than wonder. How could the Germans have been able to cross the North Sea, kill over 120 people and wound more than 440 others, and get away untouched? If news of the victory at the Falklands had released some of the pressure on the Admiralty, it returned now with interest.

17 DECEMBER 1914

EGYPT

After new arrivals had established themselves in their camps, training began in earnest. Following their long voyage, this could prove a challenging experience.

"The training, starting on Monday last in the desert, completely knocked the men up; some I had to take back to camp — one on a stretcher. I join the morning route marches, etc., to get into a hardy condition… The troops (almost every man save the sick), move off at 8 a.m., and return at 2.30, have dinner and rest. After this they need it, too.

"Twenty per cent. get leave each day; then 100 are on guard and picket duty out of each battalion (every day). We'll get into something like working order next week. Of course, it is a moot question how long we'll be here. Meanwhile I must close and go off to conduct a funeral."[56]

CAPTAIN CHAPLAIN WILLIAM MCKENZIE, ATTACHED 4TH BATTALION, AIF, MENA CAMP.

54   *Burnley Express*, 19th December 1914.
55   *ibid*.
56   *Echuca and Moama Advertiser and Farmers' Gazette* (Victoria), 9th February 1915. Pte Myles Cox, 4th Bn, AIF, died of pneumonia on 16th December 1914. Buried in Cairo War Memorial Cemetery, he was the 23 year-old son of Edward Standish Cox and Alice Victoria Cox, originally of Rylstone, New South Wales. His brother, Pte Edward Cox, also of 4th Bn, had succumbed to pneumonia in Alexandria on 13th December 1914. He is buried in Alexandria (Chatby) Military and War Memorial Cemetery.

18 DECEMBER 1914

EGYPT

Australians, including more recent ones, bade farewell to prisoners from the *Emden* and observed the installation of the new Khedive.

> "German prisoners left this morning on *Hampshire*. Sorry to see them go. Think they were very happy here. Jacob Geibil[57] — one of them — presented me with his capband."[58]
>
> SISTER CONSTANCE KEYS, AUSTRALIAN ARMY NURSING SERVICE, HMAT *OMRAH*.

> "We were here for the accession of the new Sultan of Egypt. It was a brilliant scene. The general impression gained from the educated Egyptian, with regard to the new protectorate, is that it is the best thing that Egypt could have hoped for. There is nothing to be feared from Egypt... I, although an Englishman, am proud to be serving as a colonial, with colonials."[59]
>
> PTE HUBERT SIZER, 9TH BATTALION, AIF, MENA CAMP.

> "British occupation of Egypt to-day. Sultan Hussein ascended the throne. A lot of Australian troops took part, but we did not see it. In Cairo in the evening. Most of the people pleased with the new Sultan, but some are not. The poorer people do not care who reigns so long as they get enough to eat."[60]
>
> 2/LT GEORGE KELLY, 2ND BATTALION, AIF, MENA CAMP.

The quality and availability of food was a concern for all. After a letter appeared in a local newspaper complaining about the quality of it given to Lancashire Territorials, one officer wrote to put the record straight. The individual responsible found himself in trouble.

> "I was surprised to see in your issue of the 13th November, a letter written by a member of the 6th Lancashire Fusiliers, in which he complains of the food issued to the men. He stated that stewed camel and stewed goat are issued, and that he had practically to purchase all his food out of his own pocket. I hasten to write and inform you that his statements are absolutely without foundation, and that when he was brought before the Commanding Officer for issuing false statements to the Press, he admitted that his letter was an absolute lie...
>
> "Owing to the large number of men in barracks, it is impossible with the present kitchen accommodation to roast the meat every day, and consequently each Company has to be content with two roasts per week, and stews are provided for the rest of the week. The meat issued is of the best quality, and consists of Australian mutton and Argentine beef, and everybody lives on this in Egypt. There is absolutely no necessity for any man to provide extra food for himself...

---

57   *Heizer* (Stoker) Jacob Geibel. Wounded on 9th November 1914, he became a prisoner of war at Malta.
58   *The Australian Women's Weekly*, 19th April 1972.
59   *The Telegraph* (Brisbane, Queensland), 22nd May 1915.
60   *Diary Kept by Lieut. G.E.E. Kelly*, p. 17.

"I enclose my diet sheet for the last week, which was drawn up by the Men's Messing Committee, and approved by me, and I am certain that this will show the people at home that their lads are not leading the miserable existence which the letter intended them to believe they were leading...

"Diet sheet for week ending 26th Nov., 1914:—

"Breakfasts:— Monday: Tea, bread and butter, sardines; Tuesday: Tea, bread, bacon and eggs; Wednesday: Tea and bread and treacle, porridge; Thursday: Tea and bread and butter; cheese and onions; Friday: Tea. bread, bacon and eggs; Saturday: Tea, bread and treacle, porridge; Sunday: Tea, bread and butter, cheese and onions.

"Dinners:— Monday: Roast beef, potatoes, vegetables, rice pudding; Tuesday: Stew, potatoes, bread, beans, tapioca pudding; Wednesday: Stew, potatoes, bread, peas, rice pudding; Thursday: Stew, potatoes, bread, tapioca pudding; Friday: Roast beef, potatoes, vegetables, rice pudding: Saturday: Stew. Potatoes, bread, tapioca pudding; Sunday: Stew, potatoes, bread, rice pudding.

"Teas:— Monday: Tea, bread and butter, tinned fruit; Tuesday; Tea, bread and butter, stewed figs; Wednesday: Tea, bread and butter, tomatoes, lettuce; Thursday: Tea, bread and stewed apples; Friday: Tea, bread and butter, marmalade; Saturday: Tea, bread and butter, tomatoes, lettuce; Sunday: Tea, bread and butter, salmon."[61]

CAPTAIN JOHN GLEDHILL, 1/6TH BN LANCASHIRE FUSILIERS, EAST LANCASHIRE DIVISION.

19 DECEMBER 1914

EGYPT

The arrival of more men in the country only increased the confidence those already there felt at their ability to deal with the Turks.

"We have a tremendous army out here now; you at home would be surprised; we have over 100,000 men, and more are coming. Fighting is not quite 40 miles away, and we shall soon be from moving forward from this base. I could open your eyes if I dared. We know the enemy's strength to a "t" and can put a big spoke in his wheels. I am afraid Mr. Turkey will be the most surprised old man in the East. If only he knew!!! Why, the troops here are enough to smash three Turkeys, let alone one. Never was such an English army landed here before, and not all Territorials, either. We shall give a good account of ourselves, I know."[62]

BDR NORMAN MCKAY, 18TH (EAST LANCASHIRE) BATTERY, 1/3RD EAST LANCASHIRE BRIGADE, ROYAL FIELD ARTILLERY.

---

61   *Todmorden Advertiser and Hebden Bridge Newsletter*, 18th December 1914.
62   *Preston Herald*, 19th December 1914.

20 DECEMBER 1914

ENGLAND

Churchill wrote to the Lord Mayor of Scarborough in the wake of the raid on the north east coat. His response might not have been what was expected.

> "We share your disappointment that the miscreants escaped unpunished. We await with patience the opportunity that will surely come.
>
> "But viewed in its larger aspect the incident is one of the most instructive and most encouraging that has ever happened in the war. Nothing proves more plainly the effectiveness of British naval pressure than the frenzy of hatred aroused against us in the breast of the enemy.
>
> "This hatred has already passed the frontiers of reason, it clouds their vision, it darkens their councils, it convulses their movements. We see a nation of military calculators throwing calculation to the wind, of strategists who have lost their sense of proportion, of schemers who have ceased to balance loss and gain."[63]
>
> WINSTON CHURCHILL, MP, FIRST LORD OF THE ADMIRALTY.

RUSSIA

More worrying news came from Russia, coupled with the concern that the real situation in the country was difficult to determine.

> "Delay in movements and rumours of Russian losses being very heavy made me feel anxious as to the situation here. My anxieties were confirmed by a talk with General Marquis de La Guiche, French Military Attaché, who told me he had heard that there was a great shortness in guns, munitions and rifles, especially guns and munitions…
>
> "I asked for an interview with the C.G.S., General Yanushkevich… [He] spoke out quite freely and frankly, telling me of the shortness of guns and munitions which delayed the Russian advance — that the G.O.C.'s of armies were bitterly disappointed at not being allowed to advance, but that it was obviously hopeless to do so under the circumstances.
>
> "It is a great pity that he never spoke out so freely before. However, it is no use crying over spilt milk, and all one can do now is to hope that they will keep us more in their confidence, instead of suddenly telling one of trouble after one had believed all was going well."[64]
>
> MAJOR-GENERAL JOHN HANBURY-WILLIAMS,
> BRITISH MILITARY MISSION, PETROGRAD.

63　*Birmingham Daily Gazette*, 21st December 1914.
64　Hanbury-William, Sir John, *Emperor Nicholas II As I Knew Him*, pp. 19–20, Humphreys (London) 1922.

EGYPT

Others were trying to take in the reality of their new surroundings.

> "Fancy so many Australians spending Christmas together in the shadow of the greatest of the "Seven Wonders of the World." Probably never since the days of Pharaoh have the Pyramids looked down upon such a host: I've only to look out of the tent opening as I write and I can see the two great Pyramids less than half a mile away. The Sphinx is hidden from view by a hill in the foreground…
>
> "On December 21 there is to be a great demonstration in Cairo, where the ceremony of hoisting the flag and the annexation of the country by Britain will take place… I was in Cairo yesterday and found the city getting ready for the proclamation. It is a great big city like Melbourne, only the people live closer together, and therefore the streets are more crowded."[65]

L/SGT ALEC MARSHALL, 11TH BATTALION, AIF, MENA CAMP.

21 DECEMBER 1914

ENGLAND

Churchill and Fisher were the subject of gossip between a Cabinet colleague and a newspaperman. The mood within Britain's political leadership was not positive.

> "Lunched with the McKennas. McK. [Reginald McKenna, Home Secretary] very sarcastic about the Navy. He cannot understand Fisher. Thinks he and Winston [Churchill] were manoeuvring for a spectacular battle on the old-fashioned lines."[66]

SIR GEORGE RIDDELL, NEWSPAPER PROPRIETOR.

22 DECEMBER 1914

ENGLAND

> "We discussed at length the Russian situation in Poland… According to K.'s estimate, if the Russians lost Warsaw the Germans might bring back forty divisions to reinforce the attack on the [Western] front."[67]

HERBERT ASQUITH, MP, BRITISH PRIME MINISTER.

---

65 *Kalgoorlie Western Argus* (Western Australia), 2nd February 1915.
66 Riddell, Sir George, *Lord Riddell's War Diary 1914–1918*, p. 47, Ivor Nicholson & Watson (London) 1933.
67 Asquith, *Memories and Reflections*, vol. 2, p. 50.

23 DECEMBER 1914

ENGLAND

"Winston is very sick about the whole affair [the German naval raid on the north east coast of England]. He is now interfering with the French campaign, and planning, it would appear, a second Antwerp escapade. But he sent a note across to [Lloyd George] in the Cabinet this week: 'The French are behaving odiously.' 'Which means,' said [Lloyd George] to me, 'that they refuse to alter the whole of their plans to add to Winston's personal glory'."[68]

FRANCES STEVENSON, LLOYD GEORGE'S SECRETARY.

EGYPT

At least one Australian said he was enjoying his training in the desert and his money was welcomed in Cairo.

"The desert is not half such a bad old place as one might imagine, there is much variety in it, and life there is fine and clean and healthy. One thing very noticeable about it at first is the stillness and silence. When marching at attention, nothing being heard but the rattle of the harness and swishing of our feet through the sand. We have been carrying out battle practice there for some time...

"The natives took it very quietly when Great Britain officially declared Egypt a British protectorate, there was no disturbance. The natives do not seem to regard us as invaders, we seem to be very popular where ever we go. Of course, the majority of us spend money very freely, so that the cafes and places of amusement are making more out of us than they would from the ordinary tourists in a good season. There seems to be as much chance of a scrap here as at Broadmeadows, but, of course, we all hope to go to France before long."[69]

PTE ROTHE BURTCHAELL, 5TH BATTALION, AIF, MENA CAMP.[70]

24 DECEMBER 1914

EGYPT

"I am of opinion, from letters received here, that you in England imagine we are either on the verge of, or actually, fighting. This is not so; we are, on the contrary, quiet up

---

68   Frances Stevenson quote in Taylor, A.J.P. (Ed.), *Lloyd George. A Diary*, p. 19, Hutchinson (London) 1972.
69   *Port Fairy Gazette* (Victoria), 1st February 1915.
70   A former station hand, Burtchaell was wounded on 26th April 1915, returning to the peninsula after treatment on 31st May 1915. Wounded more seriously in July, the former station hand returned to Australia, arriving at Melbourne on 25th August 1915.

to the present, although in the frontier things are unsettled. Whether we shall be involved therein in the future it is as yet difficult to state. All commodities in Egypt are very expensive, so that regulation pay does not go far. The climate conditions are at their best, and we have at present nothing to grumble at in that respect, and the men of Wigan and district are settling down splendidly to the diverse ways of oriental life. You can imagine what the flies are like when I tell you I have had to put some jam on the table to keep them quiet while I write."[71]

CPL SIDNEY PORTER, 1/5TH BN MANCHESTER REGT, EAST LANCASHIRE DIVISION.[72]

"Route march at 8.30 a.m. through farms and villages. The natives in different coloured garbs were picturesque, but the homes are ruins and rubbish everywhere as you march along. The water is driven through them by a system of pumping worked by wooden cogwheels driven by oxen, camels, or even natives. The water is raised in buckets on the wheels and emptied at the head of a drain. It then flows from drain to drain. The natives suffer from some eye disease and they are pitted with smallpox. The natives have crowds of donkeys which carry enormous loads of lucerne, sugar-cane, or cornstalks, and you can only see the lower part of the donkeys' legs. We returned to camp by the Cairo road after a march of 13 miles."[73]

2/LT GEORGE KELLY, 2ND BATTALION, AIF, MENA CAMP.

25 DECEMBER 1914

EGYPT

"Christmas Day. Church parade. To sing "Peace on Earth," etc., seems rather a farce with the nation in the deadly grip of war. Went to Cairo. Trams, lorries and cars were crowded. Went through the Museum, consisting of a marvellous collection of Egyptian relics of ancient rulers of ages ago. The mummies are wonderful. Queer to look on the face of a man who lived thousands of years ago; perfect faces, though dry and black. The stone carvings are wonderful and baffle description. It is Christmas night, and we wonder what they are doing in Maitland… Will we ever return?"[74]

2/LT GEORGE KELLY, 2ND BATTALION, AIF, MENA CAMP.

"The day here is just like an ordinary Sunday, with a few exceptions. We all attended our respective church parades, where special services had been arranged. In our company we had arranged for a good Christmas dinner. We purchased 70 fowls, allowing over half a fowl per man, besides which we were allowed one pudding, half a tin of fruits and a pint of beer each…

---

71 *Ormskirk Advertiser*, 24th December 1914.
72 Porter was discharged as no longer physically fit for service on 23rd September 1916 after contracting tuberculosis.
73 *Diary Kept by Lieut. G.E.E. Kelly*, p. 18.
74 ibid.

"I had an hour to myself so I went and climbed the big Pyramid, and had a look at the Sphinx. The view from the Pyramid, which is nearly 500 feet high, is wonderful, The beautiful valley of the Nile stretches for miles and miles before you, while on the other side there is nothing but sand for hundreds of miles. Tally McLeod[75] was with me, and we had our photos. taken with the Sphinx as a background. I'll send you one as soon as I get them. As far as we're concerned the day is almost over, as the men generally turn in before 9 p.m.

"There has been a big change in our training lately, and we're getting some very interesting new work to do. I believe there is to be another examination for 'N.S.O's. shortly, but I don't know whether those who passed the previous one have to go up again, or not."[76]

SGT SIDNEY PINKSTONE, 3RD BATTALION, AIF, MENA CAMP.

"Some of the boys received Christmas cards and letters from Tasmania. By jove I have often read about the excitement of soldiers receiving home letters; now I can understand the feeling. See us all crowd round a chap who is the lucky possessor of one, while he reads out what would interest us."[77]

PTE HUBERT WINDRED, 12TH BATTALION, AIF, MENA CAMP.[78]

MESOPOTAMIA

The Anglo-Indian army continued its progress along the Tigris.

"We are in Turkey, fighting against Turkeys. We have had five battles, and beaten them each time. Up till the time of writing we have captured over 1000 prisoners and 10 guns. The Turks tried to make it a holy war, but they got a shock. All we can see for miles is date palms. Dates we get plenty of at present. The country is very cold at nights and mornings. We feel it coming from India, where it is hot…

"We are spending our Christmas at this camp. Our camp is situated between two rivers — the Euphrates and the Tigris. These two rivers are mentioned in the Bible. This is supposed to be the old Garden of Eden."[79]

PTE DAVID KAY, 2ND BN NORFOLK REGT, 18TH BDE, 6TH (POONA) DIVISION.

---

75 Sgt Tallisker McLeod, 3rd Bn, AIF, was promoted 2/Lt on 28th April 1915; wounded on 19th May 1915; promoted Lt on 4th August 1915; and killed between 6th and 8th August 1915. Commemorated on the Lone Pine Memorial, the former jeweller was the 23 year-old son of Frank Donald and Marion Robertson McLeod, of The Groves, Cootamundra, New South Wales, originally of Geelong, Victoria.

76 *Cootamundra Herald* (New South Wales), 29th January 1915.

77 *Australian Town and Country Journal* (Sydney, New South Wales), 19th May 1915.

78 Windred, a former miner from Garrick, Tasmania, was wounded on 25th April 1915 or shortly afterwards, being admitted to 15th General Hospital, Alexandria, on 30th April. After treatment in England, Windred returned to Australia, leaving Portland on 24th June 1916.

79 *Perthshire Constitutional & Journal*, 27th January 1915.

INDIAN OCEAN

A good Christmas was had aboard the *Suevic* as it approached Egypt.

"We had thought they might give us something a bit out of the ordinary on Christmas Day, and they did too. It was not as good a Christmas dinner as some of the men would have had back home in sunny New South, but what man could grumble when he saw the mess orderlies staggering down the troop deck stairways with soup, roast fowl, roast pork, boiled potatoes, green peas, and steaming able-bodied plum puddings? The usual food for the evening meal is bread, butter, and jam. Sometimes there is also brawn or some other kind of cold meat, and when there is there are pickles to match it.

"The tea on Christmas night was deserving of two appetites a man. There were cold fowl, pork, and vegetables left over from dinner, in some instances, also pudding. The usual bread, butter, and jam were issued, and there were mince cakes in plenty as an added luxury. To cap all, for A squadron of the 6th Regiment, anyway, there was the huge cake presented to the squadron by Lieutenant A. R. (Roy) Hordern, who commands D troop. I don't know what it weighed, but it took two strong men to carry it in its box. It was two feet across at the base and about three feet high. A gorgeous iced affair it was, with an iced kangaroo on top and iced cherubs pursuing each other seraphically among colored Union Jacks around the desirable sides. It worked out at a good, solid slice a man, and each participant was duly grateful to Lt Hordern. This officer also had a pipe and a tin of tobacco presented to each member of his troop as a Christmas gift. Lt Hordern[80] himself is not with his troop, but is on another transport. His absence is regretted."[81]

TPR CLIFFORD HALLORAN, 6TH AUSTRALIAN LIGHT HORSE REGT, AIF, HMAT *SUEVIC*.

"We are still in the tropics, although it is cool to-day. Every morning at 11 o'clock we have a cup of tea and a gossip. The girls give Colonel Ramsey Smith a teapot of tea every morning. Every afternoon Matron Knowles, Sisters Pritchard, Mahon, Sweeney, and myself entertain him and Colonel Springthorp to afternoon tea in the saloon. You would laugh to see the doctors going in for egg and spoon races, potato races, sack races, and other events at the sports. Yesterday we crossed the Equator, and all the officers who had not crossed before were ducked in a huge tank of water, and then tarred. Dignity was nowhere. This morning there were three church parades. There are such a lot of musical people on board that the Concerts are really a pleasure. We have had our last vaccine infection and are all very glad. This afternoon the South Australian people, all the doctors and sisters, about 38 altogether, are to have afternoon tea together. This morning we all opened our Christmas parcels from

---

80   Lt Alfred Hordern, 6th Australian Light Horse Regt, was concussed by a bomb explosion on 24th May 1915, leaving him with blurred vision and a detached retina in his left eye. Returning to the peninsula on 2nd June 1915, he was evacuated once more due to diarrhoea. Later diagnosed as suffering from colitis and rheumatism, the former merchant returned to Australia aboard the *Thermistocles*, leaving Suez on 4th December 1915.

81   *The Sun* (Sydney, New South Wales), 3rd February 1915.

Dr. Halley. I had a tin of fancy biscuits, one tin and a box of chocolates, a big box of table raisins, and a great big slab, 12 in. by 6 in., of compressed crystalline figs. Everyone was treated the same. I hope they will hear of the pleasure they have given us. Tomorrow evening we expect to reach Colombo."[82]

SISTER EDITH MENHENNETT, AUSTRALIAN ARMY NURSING SERVICE, HOSPITAL SHIP *KYARRA*.

26 DECEMBER 1914

ENGLAND

The waters off the southern English coast were not tropical.

"Our work is very cold now, so we have been given warm clothing, and so we do not feel the cold winds so much. Every sailor has now a collar which goes round his neck, so on being thrown overboard he blows through a pipe in this collar, and of course the collar, being a kind of bladder, becomes inflated, and so keeps a man afloat for several hours."[83]

OS HORACE SYDNEY GRAY, RN, HMS *FORMIDABLE*.[84]

TURKEY

An American appreciation of the situation considered that the Caliph's call for all Muslims to rebel against Britain, France and Russia had been unsuccessful. It also reported an allied presence on Lemnos; a potential precursor to operations in the area.

"The attempt to bring about the "HOLY WAR" has failed utterly — that is, as a war like measure, as hoped for, and intended by, the Germans. It has caused some local disorders in the interior, I believe. As far as I can find out it has not even affected the Arabs in Turkey-in-Asia, who speak a different language, and have no great affection for his Turkish brother, especially disliking the Cinstantinople [sic] Turk…

"In a letter from a Greek living in the island of Lemnos, in the Aegean Sea, which is now Greek, having been taken from Turkey in the last war, it is reported that the allied fleet are using that island as a base; that there are French, English, Indian and Japanese troops there. The island would be a most convenient one for operations against the Dardanelles, but the report is open to doubt owing to Greece's neutrality. However, as Turkey still claims the island, it may have been taken over on this score. The confusion between Japanese and Indian troops… may… exist here.

---

82 *Kapunda Herald* (South Australia), 29th January 1915.
83 *Derby Daily Telegraph*, 15th January 1915.
84 The letter, acknowledging Christmas gifts, was undated. Gray, an 18 year-old former bottler, was killed with the loss of his ship on 1st January 1915. Commemorated on the Portsmouth Naval Memorial, he was the 18 year-old son of James Gray, of 15 Back Parker Street, Derby, and the late Caroline Gray.

"The feeling that the allied fleets are going to force the Dardanelles, and come up… to Constantinople, grows daily, but I cannot feel that that is their immediate attention. I believe that they are waiting for more troops for occupation, operations, and defense, and possibly for some landing of their troops elsewhere on the coast, and an advance against the Turkish troops, perhaps in the direction of… Constantinople, which is at present the only base of supplies for all the Turkish Armies operating in the field."[85]

LT-CDR EDWARD MCCAULEY, USN, USS *SCORPION*, ISTANBUL.

EGYPT

Families back in Britain were reassured that the risks of revolution, of holy war, were low. The same could not be said of climbing ancient monuments.

"I have read numerous letters from the towns of East Lancashire, including Wigan, and they have got a wrong idea altogether of the affairs of this country. They seem to think that this country is in quite an uproar through the supposed Turkish invasion. My dear friends in Wigan, it may be a bit of satisfaction to know that the troops and natives are as quiet and comfortable as any part of England. I have spoken men of the Mahommedan religion, and they are not at all in favour of the action of their co-religionists in Turkey, and they say it is suicidal for them to think of attempting to invade Egypt. They say unanimously that the condition of Egypt at the present time was never known to be better and they thank the British protection for it all. They have no wish to have the old days of slavery again brought, back. Every person who works here can now boast of a wage, no matter how small. They say it is better than working for nothing. If any Wigan people have got this idea of trouble in Egypt they can rest assured that the lads from Lancashire are as happy and comfortable as could be, and if the Turkish invasion is attempted I am sure that the lads from the Old Country will give a good account of themselves."[86]

PTE JOHN FARLEY, MANCHESTER REGT, ATTACHED STAFF, EAST LANCASHIRE DIVISION.

"Another victim has been added to the pyramids. One of the Manchester boys… fell from the top 400 ft and was killed instantly.[87] This makes the third victim since we have been here. Our company was on picket duty, and we saw the poor fellow fall. It awful to see a human being dashed to death."[88]

CPL ALFRED GLASSON, 8TH BATTALION, AIF, MENA CAMP.

---

85   Naval Attache's Reports, US Naval War College, 1914.
86   *Wigan Observer and District Advertiser*, 26th December 1914.
87   Pte James Donovan, 10th Bn Manchester Regt, was killed on 26th December 1914. Buried in Cairo War Memorial Cemetery, he was the 34 year-old son of Maria Law (formerly Donovan), of 223 West Street, Oldham, Lancashire, and the late Benjamin Donovan.
88   *Minyip Guardian and Sheep Hills Advocate* (Victoria), 2nd March 1915.

27 DECEMBER 1914

TURKEY

The head of the German military mission at Istanbul was confident that the defences of the Bosphorus and Dardanelles were secure.

> "A decisive enemy success could… only occur if a great troop landing were made in the Dardanelles simultaneously with a forced naval passage or just before. A landing following the naval passage could not depend on an artillery support of a fleet which had just forced its passage and had other problems at that time.
>
> "A sure chance of possessing Constantinople could at the most occur only by a simultaneous landing of strong Russian troops near the Bosphorus mouth in perfect cooperation with other allies.
>
> "But even against the Russian landing, the necessary measures had been taken. The Black Sea coast on both sides of the Bosphorus was defended by batteries and flying detachments while the 6th Army Corps located near, was ready to proceed against any troops. This Corps was particularly prepared for such a task by reason of many extensive exercises including night alarms.
>
> "So that already on December 27, 1914, I could properly telegraph the German H.Q. that the disquiet among the Military in Constantinople resulting from reports on the threatened danger to the Dardanelles was without foundation, and that necessary protective measures had been taken."[89]
>
> GENERAL OTTO LIMAN VON SANDERS, GERMAN MILITARY MISSION, ISTANBUL.

Less positive news regarding the fate of the Ottomans in the Caucasus was received in Istanbul.

> "I had a report from two different and independent sources, that the Russian Army had surrounded the Turkish 9th Division on the morning of the preceding day, at a short distance from Trebizond, as the Turks were advancing under the protection of guns from their men of war. Many Turks were killed and wounded, and the German officer in command of the division was taken prisoner. Coming… from independent sources, I am inclined to give it some credence."[90]
>
> LT-CDR EDWARD MCCAULEY, USN, USS *SCORPION*, ISTANBUL.

---

89   Von Sanders, Liman, *The Dardanelles Campaign*, p. 2, The Engineer School (Fort Humphreys, Virginia) 1931.
90   Naval Attache's Reports, US Naval War College, 1914.

28 DECEMBER 1914

TURKEY

"Men who have lived at the Dardanelles, and know the conditions there, believe that the Dardanelles can be forced. Talaat Bey, Minister of the Interior, also ad interim Minister of War, Minister of Marine, Minister of Finance, acknowledged the other day, that the Allied Fleet could come in if they were willing to lose a few ships…

"The Germans are getting pretty sick of things here, as they do not entirely approve of the way the Turks are running the war. There are many reports of a growing feeling of the Turks against the Germans. This I believe due to a variety of causes, among which can be stated that the Turks now feel that they were forced into this war against their will by the Germans and the Germanophile Turks, that there is a friction and jealousy in command and position, and the enlisted ranks of the Turks do not relish the treatment they have received at the hands of the German officers and others…

"Turkish victories in the Caucasus are being reported daily, but I do not believe them to be true. In fact I believe, from indications among Turkish officials, and the general feeling and conversation here, that the Turks are losing considerably in that direction."[91]

LT-CDR EDWARD MCCAULEY, USN, USS *SCORPION*, ISTANBUL.

EGYPT

"Orders were received that we were to finish training in Egypt, before proceeding to the front, and although disappointment was felt by many at not going right on to England, all were glad of the prospect of putting foot on shore again… A rather amusing incident occurred while we were coming through the canal, a British soldier on the bank called out, "Where are you from?" One of our lads promptly answered, "Ulverstone!" Judging by the look on "Tommy's" face, it is doubtful as to whether he ever located that important place on the map. After a stop at a port for several days, we left for Alexandria on December 8, arriving there the following day. After disembarking all entrained (with the exception of a few of us left for duty) for Maadi, a few miles out from the city of Cairo, where we are now in proper order, the first, second, and third Light Horse Regiments. The infantry and artillery are camped at Mena, a few miles from here, and close to the Pyramids."[92]

SQMS THOMAS DRAKE, 3RD AUSTRALIAN LIGHT HORSE REGT, AIF, MAADI CAMP.

---

91   Naval Attache's Reports, US Naval War College, 1914.
92   *Daily Telegraph* (Launceston, Tasmania), 27th January 1915.

INDIAN OCEAN

"After having been on the water for some little while the general opinion among the members of the portion of the Australian Light Horse with the Australian Imperial Expeditionary Force on this boat is that life on a troopship is not at all bad. Certainly it is much better than it was in the camps at Liverpool and Holdsworthy, where flies and dust, and food that would ruin the reputation of Paris House in one day, made for much discomfort. Here on board this steadily ploughing, man-and-horse-laden liner there is no dust, and flies are as scarce as unassuming Kaisers. As far as the condition of the sea is concerned, we might be making a trip from Circular Quay to Milson's Point, or at worst a joy trip to Manly…

"This is how the days go. Reveille, 5.30 am; coffee, 5.45; early morning stables and muck-out; 6; sick men and sick horses parade, 7; breakfast, 8; washing decks, 9; general parade and N.C.O's. Inspection, 10.30; midday stables, 11.30; dinner, 12.20 p.m.; general parade, 3; evening stables, 4.30; sick parade (men and horses), 4.30; tea, 6.30; guard mounting, 6.15; officers mess, 6.30: retreat. 6.45 pm; evening feed, 9; tattoo, 10; lights out, 10.15 p.m."[93]

TPR CLIFFORD HALLORAN, 6TH AUSTRALIAN LIGHT HORSE REGT, HMAT *SUEVIC*.

29 DECEMBER 1914

ENGLAND

With the war on the Western Front deadlocked, the casualty lists growing, alternative strategies fell under consideration. These were submitted to the Prime Minister; private initiatives not submissions on behalf of a General Staff.

"I have had two very interesting memoranda to-day on the War, one from Winston and the other from Hankey, written quite independently but coming by different roads to very similar conclusions. Both think that the existing deadlock in West and East is likely to continue… The losses involved in the trench-jumping operations now going on on both sides are enormous and out of proportion to the ground gained. When our New Armies are ready, as they will soon, it seems folly to send them to positions where they are not wanted and where, in Winston's phrase they will "chew barbed wire" or be wasted in futile frontal attacks.

"Hankey suggests the development of a lot of new mechanical devices, such as armed rollers to crush down barbed wire, bullet-proof shields and armour, smoke balls, etc. But apart from this both he and W. are for finding a new theatre for our New Armies. Hankey would like them to go to Turkey and in conjunction with the Balkan States

---

93   *The Sun* (Sydney, New South Wales), 3rd February 1915.

clear the Turk out of Europe. Germany and what is left of Austria, would be almost bound to take a hand. Winston, on the other hand, wants, primarily of course by means of his Navy, to close the Elbe and dominate the Baltic. He would first seize a German island, Borkum for choice, then invade Schleswig-Holstein, obtain naval command of the Baltic and enable Russia to land her troops within 90 miles of Berlin. This plan, apart from other difficulties, implies either the accession of Denmark to the Allies or the violation of her neutrality. There is here a good deal of food for thought. I am profoundly dissatisfied with the immediate prospect — an enormous waste of life and money day after day with no appreciable progress, and it is quite true that the whole country between Ypres and the German frontier is being transformed into a succession of lines of fortified entrenchments."[94]

HERBERT ASQUITH, MP, BRITISH PRIME MINISTER.

EGYPT

As Charles Bean worried about the risks to Australia's reputation, most Australians in Egypt faced more immediate challenges.

"The last week has been one of some anxiety to those who have the good name of Australia at heart. Cairo is one of the great pleasure resorts of the world, and a place where the soldiers in any neighboring camp can always have a reasonably enjoyable time during their hours of leave, provided they exercise the same amount of restraint as the ordinary tourist; but certain scenes have occurred and have become more common during the past few days which go a good way beyond that, and which are already affecting the reputation of Australia in the outside world… I was speaking the other day to one of the most distinguished men in the British army. "They are as fine a body physically as I have ever seen," he said. "But do all Australians drink quite so much?" The truth is that there are a certain number of men among those who were accepted for service abroad who are not fit to be sent abroad to represent Australia."[95]

CHARLES BEAN, AUSTRALIAN WAR CORRESPONDENT.

"I wish you people could see us tramping across the desert. Guess it would be a wise mother that would pick her own son, when when we come in after manoeuvres among the sands of the desert…

"On Sunday last, the 1st, 5th, 6th, 7th and 8th battalions had a fair grueller of a march. Over 500 men fell out on the road. To make matters worse we could not get a drop of water for them. The perspiration got in our eyes, and that with the dust, made it almost unbearable."[96]

CPL ALFRED GLASSON, 8TH BATTALION AIF, MENA CAMP.

---

94 Asquith, *Memories and Reflections*, vol. 2, pp. 50–51.
95 *The Advertiser* (Adelaide, South Australia), 22nd January 1915.
96 *Minyip Guardian and Sheep Hills Advocate* (Victoria), 2nd March 1915.

30 DECEMBER 1914

RUSSIA

With the Turkish offensive in the Caucasus ongoing, the British received a call for help.

> "To-day I was sent for by the Grand Duke Nicholas, who saw me with the Chief of the General Staff, my friend Prince Galitzin having told me beforehand of the nature of the interview.

> "I was told by the Commander-in-Chief that the position in the Caucasus was very serious, that the Turks were massing forces against the Caucasus army, and that though he could retain a Caucasian army corps, which was intended for this front, he had not done so, and had told the C-in-C of the Caucasus front that he must get on as best he could, but he felt sure that it would be for our mutual interests as Allies if we — that is, Great Britain — could render help by a demonstration of some kind which would alarm the Turks, and thus ease the position of the Russians on the Caucasus front. I answered that so far as I knew — and I had a pretty shrewd idea — our armies were not yet strong enough to spare sufficient men for a military expedition, but I asked him, in the event of its being possible, whether he thought a naval demonstration would be of any use. He jumped at it gladly."[97]

MAJOR-GENERAL JOHN HANBURY-WILLIAMS, BRITISH MILITARY MISSION, PETROGRAD.

EGYPT

Sir George Reid, former Australian Prime Minister, latterly the High Commissioner in London, visited the troops. Feelings were mixed at the prospect.

> "At time of writing we are just back in the main camp preparing for a march past Sir George Reid which takes place this afternoon. I don't know what sort of a figure we will cut, as we haven't had a wink of sleep since Monday night, last night we slept out on the desert, or rather tried to sleep. All were without blankets, and were chilled to the bone. We were glad to keep on the go for warmth. We were on outpost duty, and a miserable job it is."[98]

CPL ALFRED GLASSON, 8TH BATTALION, AIF, MENA CAMP.

> "Paraded under Generals Birdwood and Bridges. Sir George Reid, accompanied by General Maxwell, arrived, and made a wonderful speech, but missed the chance of his life by not addressing the soldiers of the baby nation of the world, Australia, from the Pyramid. What a reception he received."[99]

2/LT GEORGE KELLY, 2ND BATTALION, AIF, MENA CAMP.

---

97   Hanbury-William, Sir John, *Emperor Nicholas II As I Knew Him*, pp. 23–24, Humphreys (London) 1922.
98   *Minyip Guardian and Sheep Hills Advocate* (Victoria), 2nd March 1915.
99   *Diary Kept by Lieut. G.E.E. Kelly*, p. 19.

Australian troops parade under the Pyramids to be addressed by Sir George Reid.

31 DECEMBER 1914

ENGLAND

"Prince Louis of Battenberg has resigned. Fisher is to take his place. McKenna says he doubts whether Fisher will stand the strain for more than six months. He is considerably over seventy."[100]

SIR GEORGE RIDDELL, NEWSPAPER PROPRIETOR.

FRANCE

"This is in reality only a hurried line to wish you [Winston Churchill] all good luck for 1915; but as I am writing I want to tell you quite privately how far my plans have progressed towards the object we both have so much at heart, namely a powerful advance Eastward along the coast, supported by the Navy…

---

100   Riddell, Sir George, *Lord Riddell's War Diary 1914–1918*, p. 37, Ivor Nicholson & Watson (London) 1933.

"I do not like to tell you anything in detail until I am sure that the King of the Belgians will give his consent. But if my suggestions are accepted and the plan comes off, I can assure you there will be a land force of sufficient size to justify a vigorous Naval support and to give good promise of success."[101]

FIELD MARSHAL SIR JOHN FRENCH, COMMANDER-IN-CHIEF, BRITISH EXPEDITIONARY FORCE.

EGYPT

Sir George Reid touched on Charles Bean's stated risk to Australia's reputation in his address to troops at Mena.

"The youngest of these august pyramids was built 2,000 years before the birth of Christ. They have been silent witnesses of many strange events. Can they ever have looked down upon a more unique spectacle than this splendid array of Australian soldiers massed in their defence?... What brings your army here? Why do your tents stretch across this narrow parting of the ways, between worlds old and new? Is it a quest in search of gain, such as led your fathers to the Austral shore? Do you seek to invade and outrage weaker nationalities in a lawless raid of conquest? Thank God! your mission is as pure and noble as any soldiers undertook to rid the world of would-be tyrants...

"Do not forget the distant homes that love you. Remember Australia's good name and unstained honour, which she has given to your keeping in a supreme sense. A few wrong ones can besmirch the fair name of a whole army. The unworthy, if such there be, must be shunned, must be thrust out. Your first and best victories are those of self-control. Hearts of solid oak, nerves of flawless steel come that way. Lord Kitchener will send you to the front when you are fit... If any stains come on your bright new flags they must and will be stains of honour won by valour... Good luck! May God be with you each and all until we meet again."[102]

SIR GEORGE REID, AUSTRALIAN HIGH COMMISSIONER.

"Sir George Reid addressed the other half of the Division at Mena. I saw the old year out in Cairo, and wondered where I would see the next one swing into eternity — perhaps in eternity itself. How sentimental these holidays make us all."[103]

2/LT GEORGE KELLY, 2ND BATTALION, AIF, MENA CAMP.

---

101 Sir John French quoted in Churchill, *The World Crisis 1915*, p. 56.
102 *Western Mail* (Perth, Western Australia), 6th January 1938.
103 *Diary Kept by Lieut. G.E.E. Kelly*, p. 19.

Winston Churchill, First Lord of the Admiralty, one of the principal supporters of the plan to attack the Dardanelles.

JANUARY 1915

"... bombard and take the Gallipoli peninsula."
Lt-Col. Maurice Hankey, Cabinet Secretary

1 JANUARY 1915

ENGLAND

With the new year barely begun, the pre-dreadnought battleship HMS *Formidable* was struck by two torpedoes off Portland Bill from *U-24*. More men had tales to tell of how they survived the loss of their ship.

> "We did not think she would sink so rapidly, but gradually she began to capsize, and we climbed over the ship's side. I managed to get up to the keel but a lot who were following me slipped back into the water."[1]
>
> SIG ROBERT WILLIAM MILBURN, RN, HMS *FORMIDABLE*.

> "I never thought I should have reached the boat by jumping, but it was my last chance, and I jumped I never jumped before. When I fell into the boat I thought I should have died straight away. We were in the boat for 10 hours afterwards and you can tell what I felt like. I had only shirt and a pair of drawers, but no boots. I did welcome the coming of the [trawler] *Providence* because I was the worst off of the lot with being wounded so badly. They fairly threw me on the deck of *Providence*, for with the cold and exposure I was useless."[2]
>
> STO. 1 HAROLD SMITHURST, RN, HMS *FORMIDABLE*.

547 men would tell tales no more. Churchill once more put forward the plan to make a landing on the Belgian coast.

> "The battleship *Formidable* was sunk this morning by a submarine in the Channel. Information from all quarters shows that the Germans are steadily developing an important submarine base at Zeebrugge. Unless an operation can be undertaken to clear the coast, and particularly to capture this place, it must be recognized that the whole transportation of troops across the Channel will be seriously and increasingly compromised.
>
> "The Admiralty are of opinion that it would be possible, under cover of warships, to land a large force at Zeebrugge in conjunction with any genuine forward movement along the seashore to Ostend. They wish these views, which they have so frequently put forward, to be placed once again before the French commanders, and hope they may receive the consideration which their urgency and importance require."[3]
>
> WINSTON CHURCHILL, MP, FIRST LORD OF THE ADMIRALTY.

This plan was added to those already under consideration. Asquith continued to accept submissions by Cabinet colleagues, in Churchill's case after liaising directly with Sir John French, bypassing Kitchener completely.

---

1   *Newcastle Evening Chronicle*, 7th January 1915.
2   *Nottingham Evening Post*, 22nd January 1915.
3   Churchill, *The World Crisis 1915*, p. 58.

"I received to-day two long memoranda, one from Winston, the other from Lloyd George — the latter is quite good — as to the future conduct of the War. They are both keen on a new objective and theatre as soon as our new troops are ready, Winston of course for Borkum and the Baltic, Lloyd George for Salonika, to join in with the Servians, and for Syria."[4]

HERBERT ASQUITH, MP, BRITISH PRIME MINISTER.

EGYPT

The previous evening's celebrations took their toll on the men sent out to march on New Year's Day.

"This year was started rather well, and yet badly. We went out for a short march; in fact, it could not be called a march at all, for the distance we went was only about a mile. When we had gone out this distance we turned and formed up on the brigade ground where the Brigadier spoke to us as a man, on account of a lot of young fellows falling out before they had gone any distance, their excuse being that they were not well. The Brigadier putting this all down to cigarettes and drinking, which I think were the chief causes; the night before being New Year's Eve, a lot of them went to Cairo and came home the worst for drink. In the afternoon our section went on with our work, which consisted of range finding and how to draw up a range chart. We finished our day's work at half past two."[5]

PTE JULIUS JACOBSOHN, MG SECTION, 7TH BATTALION, AIF, MENA CAMP.

INDIAN OCEAN

Meanwhile, the uneventful voyages from Australia to Egypt paused occasionally for a burial at sea.

"On New Year's morning there was a burial off the *Borda*,[6] and in the afternoon we were joined by another boat off Fremantle."[7]

PTE GEORGE CLAPHAM, 14TH BATTALION, AIF, HMAT *ULYSSES*.

"New Year's Day, up early and greeting all. Did my washing, and had a bath, then lay round in sun and watched other transports. Some of our men got into the room where the soft drinks were kept, and had a good time. Getting hotter as we get nearer equator. I was inoculated the second time; arm very sore. Very anxious to know where we are going, and all letters censored by captain."[8]

PTE THOMAS WADDELL (WAD) CAMERON, 14TH BATTALION, AIF, HMAT *ULYSSES*.

---

4 Asquith, *Memories and Reflections*, vol. 2, p. 54.
5 *The Corowa Free Press* (New South Wales), 19th November 1915.
6 Pte Benjamin Acreman, 2nd Light Horse Field Ambulance, AAMC, embarked on the *Borda* at Brisbane on 16th December 1914. He died of typhoid on 1st January 1915. Commemorated on the Chatby Memorial, he was the son of Thirza Acreman, of Newmarket Road, Wilston, Brisbane.
7 *Gippsland Mercury* (Sale, Victoria), 13th April 1915.
8 *Kyabram Guardian* (Victoria), 19th March 1915.

AUSTRALIA

At Broken Hill, New South Wales, a railway excursion and picnic had been arranged for the holiday. They passed a local ice-cream seller's cart, from which flew a Turkish flag.

> "When the train steamed out from Sulphide-street everyone aboard seemed glad, hearty and cheerful... When nearly opposite the brick kilns... I caught sight of a red flag... I then saw the cart, and I said to my mate, 'It's rather late for an ice-cream cart to be going out to Silverton.' He said, 'Yes, I suppose it's some poor old beggar hoping to make a bit for himself'...
>
> "Just at this moment up bobbed two turbaned heads out of the pipe track trench, and there were two reports... But the shots missed. I saw one hit the sand, and the dust rattled up against the side of the engine. The other struck the under part of the brake-van next to the engine... We had no other thought than that they meant business, and we hastened to get the train out of range as quickly as possible."[9]
>
> W. STEWART BERRY, RAILWAY FIREMAN.

> "When our truck was directly opposite them (only about 30 yards away) I saw one of the Turks fire in the air. The other fired point blank at our carriage. I then thought it was time to duck. Just then a shot hit our truck, and the bullets flew all ways for a second or two. Even after the train stopped I did not realize the position... There were two killed in a truck ahead of us, and four were shot about four trucks behind us. The sight was dreadful in the truck behind us. One woman from Petersburg tried to shield her baby and was shot in the mouth. The baby was covered in blood. In this truck the blood was something terrible. It seemed like a slaughter house. One man following the train on a [motor] bike was shot dead."[10]
>
> FRANK WHITE, PASSENGER.

> "I... heard the noise, and by this time Broken Hill was in arms. I hurried home for my rifle, believing, as in fact everyone did that the Afghans, numbering about 45, were in revolt. When I got back to the street... seeing a motor car with four rifles, I hopped up and off we went to the battlefield or rather to the front. A spin of two miles brought us to the scene of operations and I could hardly believe my eyes. There were police and military, as well as civilians everywhere, while on top of a little hill were two Turks. They had natural fortifications in the shape of rocks, and it seemed to me that many of our men would be shot before we got them. Some commanding officer told us to go down on the flat and help in the attack on the flank, but by this time I could see that one of the Turks was either *hors de combat* or had escaped, the enemy fire having slackened off considerably... All this time there was a fierce fusillade going on from about 200 rifles, and I thought my luck was clean out, and that I would never got a shot. However... at that moment I got a view of the enemy's head. Hurriedly taking

---

9   *Barrier Miner* (Broken Hill, New South Wales), 8th January 1915.
10 *The People's Weekly* (Moonta, South Australia), 9th January 1915.

aim I fired and missed, got in two more shots, but before I could get in another, he dropped... When we got up to their position after the firing ceased, both were lying, one quite dead, and the other dying, and he expired shortly afterwards."[11]

THOMAS CLARKE, LOCAL RESIDENT.

On the bodies of the two men were found statements explaining their actions. One resented being fined for slaughtering a sheep in unlicensed premises; the other had served with the Ottoman Army. Both affirmed their loyalty to the Turkish Sultan.

"I am a poor sinner in the sight of the Almighty, and I supplicate His mercy. I am a poor resident of this country. I gave another name for my own purpose. One day the inspector accused me. On another occasion I begged and prayed but he would not listen to me. I was sitting brooding in anger. Just then the man Gool Badsha Mahomed came to me, and we made our grievance known to each other. I rejoiced, and gladly fell in with his plans, and asked God that I might die an easy death for my faith. Otherwise neither of us had any enmity against anybody for my faith and in obedience to the order of the Sultan and the order of the Koran, but owing to my grudge against the inspector it was my intention to kill him first. Beyond this there is no enmity against anybody, and we informed nobody. I swear this by God and on the Koran."[12]

MULLAH ABDULLAH.

"In the name of God all merciful and His Prophet, this poor sinner is a subject of the Sultan. My name is Gool Badsha Mahomed, Afghan Afridi. In the reign of Ahbdul Hamid Sultan, I have visited his kingdom four times for the purpose of fighting. I hold the Sultan's order duly signed and sealed by him. It is in my waist belt now, and if it is not destroyed by cannon shot or rifle bullets you will find it on me. I must kill your men, and give my life for my life by order of the Sultan. I have no enmity against anyone, nor have I consulted with anyone nor informed anyone. We bid to all the faithful farewell."[13]

GOOL BADSHA MAHOMED.

On or near the train, three people had been killed, a further six wounded, their ages ranging from 15 to 70. Another man was killed by Gool and Abdullah as they made their way away from the railway towards the location where they made their stand.

Back in Broken Hill, an angry crowd headed for the camp where local 'Afghan' camel drivers lived. Their path there blocked by police, they took out their anger on the local German club. It was burned to the ground, the local firefighters unable to reach the premises before it was well alight.

---

11   *The Casterton News and the Merino and Sandford Record* (Victoria), 18th January 1915.
12   *ibid.*, 14th January 1915.
13   *ibid.*

2 JANUARY 1915

TURKEY

In Istanbul there were rumours of a possible massacre of Europeans should the allied navies appear before the city.

> "I had an illuminating talk with Pallavicini [Austrian Ambassador]. He showed me a certificate given him by Bedri, the Prefect of Police, passing him and his secretaries and servants on one of [the] emergency trains. He also had seat tickets for himself and all of his suite. He said that each train would have only three cars, so that it could make great speed; he had been told to have everything ready to start at an hour's notice. Wangenheim made little attempt to conceal his apprehensions. He told me that he had made all preparations to send his wife to Berlin, and he invited Mrs. Morgenthau to accompany her, so that she, too, could be removed from the danger zone. Wangenheim showed the fear… that a successful bombardment would lead to fires and massacres in Constantinople as well as in the rest of Turkey."[14]

HENRY MORGENTHAU, AMERICAN AMBASSADOR, ISTANBUL.

ENGLAND

Kitchener could offer no military backing for an attack on the Dardanelles. But a naval 'demonstration' might be an option to support the Russians.

> "I do not see that we can do anything that will seriously help the Russians in the Caucasus… We have no troops to land anywhere. The only place where a demonstration might have some effect in stopping reinforcements going East would be the Dardanelles. We shall not be ready for anything big for some months."[15]

LORD KITCHENER, SECRETARY OF STATE FOR WAR.

EGYPT

The British Army had adopted the four company battalion structure. It was still being implemented at the beginning of 1915.

> "Instead of a battalion being composed of eight companies, as hitherto, it now comprises only four companies, but each of these new companies is twice the strength of the old companies. This is the system known as the platoon system adopted throughout the British army. Each company is in charge of a Major, or Senior Captain, who has a second captain to assist him. The companies are divided into four platoons of approximately sixty men, in charge of a Subaltern, who is known as the Platoon Commander.

---

14   Morgenthau, Henry, *Ambassador Morgenthau's Story*, pp. 187–188, Doubleday, Page & Co. (New York) 1918.
15   Lord Kitchener to Churchill quoted in Arthur, Sir George, *Life of Lord Kitchener*, vol. 3, p. 101, MacMillan & Co., (London) 1920.

In the new arrangement the old company in which the Northam boys were placed (G Company) has been amalgamated with H Company (made up of volunteers from the Southern and South-West districts of W.A.), and these two conjoined companies are now known as D Company. Major Denton[16] is their Company Commander, and in him the Northam representatives are fortunate in having a leader who is not only an excellent soldier, but one of the most popular and painstaking officers in the 11th Battalion. Capt. Jas. Croly[17] remains with them as second in command."[18]

PTE EDWARD RICHARDS, 11TH BATTALION, AIF, MENA CAMP.

3 JANUARY 1915

ENGLAND

Churchill cabled the commander of the Eastern Mediterranean Squadron, Vice Admiral Sackville Carden, about the feasibility of undertaking a naval attack on the Dardanelles unsupported by ground troops.

"Do you consider the forcing of the Dardanelles by ships alone a practicable operation?

"It is assumed older battleships fitted with mine-bumpers would be used, preceded by colliers or other merchant craft as mine-bumpers and sweepers.

"Importance of results would justify severe loss.

"Let me know your views."[19]

WINSTON CHURCHILL, MP, FIRST LORD OF THE ADMIRALTY.

EGYPT

As men toured the sites of ancient Egypt, rumours of a future attack by the Turks continued to circulate.

"We… came up to Ghiza station, where we waited for a train to take us out to our camp. Whilst waiting there a small but very nice looking little Arab boy came up to us and asked us if we came from Australia. He spoke English quite well. He knew his geography well too, and seemed to know quite a lot about Australia. He told us he went to school, and hoped to pass an examination this year to obtain a certificate to attend the secondary school. He has ambitions of becoming an engineer. I was

---

16   Major, later Lt-Col. James Denton, 11th Bn, AIF, was awarded the DSO for his performance during the landing at Anzac. The official citation was published in the *London Gazette* on 3rd July 1915: "During the operations in the neighbourhood of Gaba Tepe on the 25th April, 1915, for valuable services in obtaining and transmitting information to ships' guns and mountain batteries, and subsequently for holding a trench, with about 20 men, for over six days, repulsing several determined attacks."
17   Capt., later Major Arthur Croly, 11th Bn, AIF. Born in Dublin, Croly had previously served with the Royal Fusiliers. He was wounded on 28th April 1915.
18   *The Northam Advertiser* (Western Australia), 3rd February 1915.
19   Churchill, *The World Crisis 1915*, pp. 97–98.

quite taken with the little chap, who was only 12 years old… He wrote his name and address in my note book, and I wrote my name and Australian address on a piece of paper and asked him to write to me and let me know about his examination… You will have seen by the papers that Sir George Reid addressed us the other day. It is rumoured here that a Turkish invasion of Egypt is planned for next month."[20]

CAPT. CHARLES THOMPSON, 3RD BATTERY, AFA, MENA CAMP.

4 JANUARY 1915

ENGLAND

Churchill proposed to consult others about the potential for an attack on the Dardanelles. Not everyone in the Admiralty believed he was listening.

"I think we had better hear what others have to say about Turkish plans before taking a decided line. I would not grudge 100,000 men because of the great political effects in the Balkan peninsula; but Germany is the foe, and it is bad war to seek cheaper victories and easier antagonists."[21]

WINSTON CHURCHILL, MP, FIRST LORD OF THE ADMIRALTY.

"This morning Oliver sent for me & gave me a paper of the 1st Lord's to work upon. This concerns an attack upon Borkum — a favourite scheme of Bayly's[22]… It is quite mad. The reasons for capturing it are NIL, the possibilities about the same. I have never read such an idiotic, amateur piece of work as this outline in my life. Ironically enough it falls to me to prepare the plans for this stupendous piece of folly. Yet Sea Lords like Wilson (for he is in effect a Sea Lord) enter no protest. It remains with the Army, who I hope will refuse to throw away 12,000 troops in this manner for the self-glorification of an ignorant and impulsive man…

"The *Triumph*, firing against forts [at Tsing Tao] she could see, with spotting done for her on shore, made 13 per cent of hits. How many could we expect against invisible gun positions & with no spotting on shore?"[23]

CAPT. HERBERT RICHMOND, RN, ASSISTANT DIRECTOR OF OPERATIONS, ADMIRALTY.

EGYPT

Training continued but if it was not enjoyed then it could be enlivened by the introduction of the idea of a little competition.

---

20   *The Young Chronicle* (New South Wales), 12th February 1915.
21   Winston Churchill quoted in Morgan, Ted, *Churchill. The Rise to Failure: 1874–1915*, p. 494, Triad/Granada (London) 1984.
22   Vice-Admiral Sir Lewis Bayly was widely blamed for the loss of HMS *Formidable* and removed from his command.
23   Marder, *Portrait of an Admiral: The Life and Papers of Sir Herbert Richmond*, pp. 134–135.

"The reason for leave being stopped yesterday was that the battalion was going out at 2.30 this morning. Our section had the luck to get out of this, so we slept in for an hour later than usual, making it seven o'clock when we rose. After a little semaphore signalling and squad drill, we set to to finish off some targets which we commenced some time ago, and have not finished yet."[24]

PTE JULIUS JACOBSOHN, MG SECTION, 7TH BATTALION, AIF, MENA CAMP.

"Yesterday's paper seemed to indicate that we will have to go through a course of training in England before we are allowed to go to the front. The article read as though we Australians were not suitable owing to the wild boys not bowing down to discipline like our friends the Tommies. Of course it has aroused our ire to suggest that we are not as fit as the English Tommies to go to the front. I don't expect we have the discipline in our lines that they have, still we have chaps who are not afraid of hell itself (excuse the wild expression only one suitable). I am sure of this, that before going to the front we will have to go through a severe military examination, and I hope we will pass. It would be jolly hard on us if we were left and not have a chance of seeing anything of the war."[25]

DVR LEONARD JACOBS, 3RD FAB AMMUNITION COLUMN, AFA, MENA CAMP.

5 JANUARY 1915

ENGLAND

"Old Fisher seriously proposed, by way of reprisals for the Zeppelin raids, to shoot all the German prisoners here, and when Winston refused to embrace this statesmanlike suggestion sent in a formal resignation of his office. I imagine that by this time he has reconsidered it."[26]

HERBERT ASQUITH, MP, BRITISH PRIME MINISTER.

LEMNOS

Carden's response to Churchill did not exactly convey overwhelming confidence.

"With reference to your telegram of 3rd instant, I do not consider Dardanelles can be rushed.

"They might be forced by extended operations with large number of ships."[27]

VICE-ADMIRAL SACKVILLE CARDEN, RN, EASTERN MEDITERRANEAN SQUADRON (EMS).

---

24 *The Corowa Free Press* (New South Wales), 19th November 1915.
25 *The Adelaide Advertiser* (South Australia), 13th March 1915.
26 Asquith, *Memories and Reflections*, vol. 2, p. 54.
27 Sackville Carden quoted in Churchill, *The World Crisis 1915*, p. 98.

EGYPT

More rumours circulated that the Turks were planning to invade across the Suez Canal. Some of the men, tired of desert marches, looked forward to the prospect.

> "Platoon and company drill. Rumours of a large force of Turks nearing the Canal. We have hopes of finishing with toy soldiering."[28]
>
> 2/LT GEORGE KELLY, 2ND BATTALION, AIF, MENA CAMP.

> "We have had another stiff day on the desert to-day, carrying out field firing with ball ammunition, with very satisfactory results. We left camp at 7 a.m., returning at 5 p.m. There is not an ounce of fat on any one of us and our legs are all muscle and as tough as whip cord. We have just received word that several Italian steamers arrived at Alexandria yesterday crowded with refugees from Syria who state that the Turks have been devastating property, etc., in Syria. We expect to get on the road to Syria in a week. It may be only a rumour but if the Turks get near the canal we shall most certainly go to meet them there."[29]
>
> 2/LT CHARLES MASSEY, CANTERBURY BN, NZEF, ZEITOUN CAMP.

6 JANUARY 1915

EGYPT

If marching across the desert sand was hard, marching in a sandstorm was altogether different.

> "We had to march out with our faces to the wind, thus preventing us from seeing properly; then we had to work in this wind, which was really a Great Sahara dust storm. It was useless to wipe the sand out of your eyes, because you would not have your hands down from your face before, they would be full of sand again."[30]
>
> PTE JULIUS JACOBSOHN, MG SECTION, 7TH BATTALION, AIF, MENA CAMP.

> "For the first time since our arrival here it has been windy today, and the sand of the desert flying everywhere, a fair terror, and giving one some faint idea of what a sandstorm would be like. If we had a week of today's weather, I guess no one would want to remain any longer in Egypt… Egypt is a great land of mystery, and has some distinct interest. Nevertheless I fancy I should become awfully tired of being here for the sand literally gets on one's nerves. It is everywhere."[31]
>
> TPR JAMES MOSSMAN, AUCKLAND MOUNTED RIFLES, ZEITOUN CAMP.

---

28   Kelly, George Edward Eccleston, *Diary Kept by Lieut. G.E.E. Kelly, During the voyage from Sydney to Egypt as a Member of the Australian Expeditionary Forces*, p. 19, T. Dimmock Ltd., Printers, (Maitland).
29   *Lyttelton Times* (New Zealand), 25th February 1915.
30   *The Corowa Free Press* (New South Wales), 19th November 1915.
31   *Poverty Bay Herald* (New Zealand), 27th March 1915.

Some of the men would never leave Egypt.

> "I was one of the firing party to attend the funeral of one of our sergeants[32] on Monday. We had 16 miles train to the hospital, then marched for 7 miles to the cemetery. This occupied 2½ hours, as a great deal was at slow march, then from cemetery to train again and "home," beg pardon, I mean "camp" at 7.30. It was the Armourer Sergeant of our battalion whom we buried. He died through injuries received by a case of benzine exploding."[33]

PTE GEORGE TAYLOR, 9TH BATTALION, AIF, MENA CAMP.[34]

7 JANUARY 1915

ENGLAND

Much of the training on offer was being delivered by men with little more experience than those receiving it.

> "I am teaching musketry to my men, and it is a great relief to get away from squad drill, which was boring us both. I have four sections in the platoon of about fourteen men each (with N.C.O.'s the strength of a platoon is sixty), and I take one section at a time. The apparatus consists of a waterproof sheet, a sandbag, a rifle, and a target affixed to a box. All the lore I learnt at Hayling Island now stands me in good stead, and I don't think any soldier guesses that six months ago a service rifle was almost a mystery to me."[35]

2/LT JOHN ALLEN, 13TH (RESERVE) BN WORCESTERSHIRE REGT, LOOE, CORNWALL.

EGYPT

> "We had a day in our newly constructed mess rooms, which are about 150 by 56 feet, and are fitted up with tables and forms. We did work on the mechanism of the gun, and how to traverse fire properly. We had our first meals in these rooms to-day. When we had finished our day's work I sat down and wrote home to my people, and also to Ada. At eight o'clock that night the alarm was blown, and we had all to turn out in full marching order, much to our pleasure to find it as a false alarm, so we were able to get back into our tent again. "What was the reason of this?" did you say. Well, it

---

32 Sgt Joseph Moore, 9th Bn, AIF, died on 3rd January 1915. Buried in Cairo War Memorial Cemetery, he was the 42 year-old son of William and Elizabeth Moore; husband of Elizabeth Annie Moore, of Baroona Road, Milton, Brisbane. Originally from Lancashire, he had served in the British Army with the Loyal North Lancashire Regt.
33 *Maryborough Chronicle, Wide Bay and Burnett Advertiser* (Queensland), 23rd February 1915.
34 The former engine driver was killed in action at Proyart, near Peronne, France, on 11th August 1918. Buried in Heath Cemetery, Harbonnieres, he was the 33 year-old son of John and Zillah Taylor, originally of Maryborough, Queensland.
35 John Allen quoted in Montgomery, Ina, *John Hugh Allen of the Gallant Company. A Memoir*, p. 149, Edward Arnold (London) 1919.

just amounted to this. They wanted to see how quickly and quietly we could prepare should the occasion arise."[36]

PTE JULIUS JACOBSOHN, MG SECTION, 7TH BATTALION, AIF, MENA CAMP.

It was not all training. Some took the opportunity to meet local residents.

"We had a little Armenian boy with us who acted as our guide… He is a very fine little fellow and is going to write to New Zealand. He took us to his home. His father is an engineer and there is a daughter of about 18 years of age who was extremely nice. They entertained us with music and gave us good tea, and when we left we were told that we were always welcome. The houses here are very much different from those in New Zealand. They are generally two storeys high and are built of concrete. They always have balconies and it looks funny to us to see no fireplaces in the houses. The majority of the houses are very well furnished. Of course this is a great place for smoking. After tea they passed round the finest brand of Egyptian cigarettes. The young lady, Berrisa is her name, struck a match and lit the guests' cigarettes."[37]

PTE BLAIR CULLEN, OTAGO BATTALION, ZEITOUN CAMP.[38]

8 JANUARY 1915

ENGLAND

The War Council met at Downing Street at midday. After hearing from Kitchener that a German offensive in France & Flanders was imminent, the scope for allied action elsewhere was discussed.

"MR. LLOYD GEORGE laid stress on the great losses that would be entailed in any attempt to break through the German lines in France… He suggested that an attack on Austria might produce the desired effect… This plan… would enormously increase the military strength of the Allies by bringing in other nations…

"MR. CHURCHILL … urged, however, that we should not lose sight of the possibility of action in Northern Europe… Was there no possibility that Holland might enter the war on the side of the Allies?… If Holland could be induced to enter the war the advantages would far outweigh those of the Mediterranean…

"SIR EDWARD GREY said that [it] would be necessary to satisfy Holland that there was no prospect that she would share the fate of Belgium."[39]

LT-COL. MAURICE HANKEY, SECRETARY, WAR COUNCIL.

36 *The Corowa Free Press* (New South Wales), 19th November 1915.
37 *Mataura Ensign* (New Zealand), 17th March 1915.
38 Cullen, a former chemist, was killed in action on 2nd May 1915. Commemorated on the Lone Pine Memorial, he was the 21 year-old son of Peter Smith Cullen and Alexandrina Jane Cullen, of Mataura, Southland, originally of Gore, New Zealand.
39 Secretary's Notes of a Meeting of a War Council held at 10, Downing Street, January, 8, 1915. TNA CAB 42/1/12.

Churchill, *Formidable* in mind, next raised the possibility of an attack upon Zeebrugge. Other members of the Council sought the First Sea Lord's opinion on the matter.

> "MR. CHURCHILL asked the view of the Council as to whether the risk of a naval attack on Zeebrugge ought to be run in order to avoid ultimate risk from submarines.
>
> "SIR EDWARD GREY suggested that the opinion of the Admiral who would have to carry out the operation should be obtained.
>
> "MR. BALFOUR asked whether, if Zeebrugge was bombarded, the risk to transports and other ships in the channel would be materially reduced.
>
> "LORD FISHER thought not. In his opinion the results of a successful operation would not justify the danger involved."[40]
>
> LT-COL. MAURICE HANKEY, SECRETARY, WAR COUNCIL.

Though trained troops for fresh operations did not then exist, the War Council decided to keep its options open. But concluded the Admiralty should take care of the U-boats.

EGYPT

> "Some of the men have been playing up a bit, with the result that all leave was stopped from mid-day yesterday until midnight to-night. This was so as to give the pickets a chance to get any soldier who was in Cairo. Leave was stopped for the whole of the soldiers in Egypt...
>
> "To-day and yesterday we had a terrible dust storm. It was something awful. The dust would simply blind you, if you were not lucky enough to have a pair of smoked glasses. Luckily, I had a pair, which I purchased in Cairo. An Egyptian guide advised me to have a pair always on hand."[41]
>
> CPL HERBERT MARKWELL, DAC, AFA, 1ST AUSTRALIAN DIVISION.

9 JANUARY 1915

ENGLAND

Kitchener cabled General Sir John French to inform him that the proposed landings at Ostend and Zeebrugge had been rejected by the War Council. The ammunition required did not exist.

> "On a general review... the Council came to the conclusion that the advantages to be obtained from such an advance at the present moment would not be commensurate with the heavy losses involved... It must be borne in mind... that it is impossible at the present time to maintain a sufficient supply of gun ammunition on the scale which you considered necessary for offensive operations. Every effort is being made...

---

40 Secretary's Notes, War Council, January, 8, 1915, TNA CAB 42/1/12.
41 *The Maitland Daily Mercury* (New South Wales), 18th February 1915.

to obtain an unlimited supply of ammunition; but, as you are well aware, the result is still far from being sufficient to maintain the large number of guns which you now have under your command adequately supplied with ammunition for offensive purposes."[42]

LORD KITCHENER, SECRETARY OF STATE FOR WAR.

Nevertheless, operations beyond the Western Front were under consideration. French thought that was a bad idea.

"It was thought that ... German defences would be impassable for offensive movements of the Allies without great loss of life and the expenditure of more ammunition than could be provided. In these circumstances, it was considered advisable to find some other theatre where such obstructions to advance would be less pronounced, and from where operations against the enemy might lead to more decisive results.

"For these reasons, the War Council decided that certain of the possible projects for pressing the war in other theatres should be carefully studied during the next few weeks, so that as soon as the new forces are fit for action, plans may be ready to meet any eventuality that may be then deemed expedient, either from a political point of view, or to enable our forces to act with the best advantage in concert with the troops of others nations throwing in their lots with the Allies."[43]

LORD KITCHENER, SECRETARY OF STATE FOR WAR.

"Any attack on Turkey would be devoid of decisive result. In the most favourable circumstances it could only cause the relaxation of the pressure against Russia in the Caucasus and enable her to transfer two or three Corps to the West — a result quite incommensurate with the effort involved. To attack Turkey would be to play the German game and to bring about the end which Germany had in mind when she induced Turkey to join in the war, namely, to draw off troops from the decisive spot, which is Germany itself."[44]

FIELD MARSHAL SIR JOHN FRENCH, C-IN-C, BRITISH EXPEDITIONARY FORCE.

EGYPT

Some men found Egypt unpleasant; an unpleasantness not restricted to physical discomfort.

"We had two days [of] continuous [sandstorms] last week [—] a cold bleak wind you could almost lean against. It was impossible to see more than 100 yds at times and it was always possible to look straight at the sun without being dazzled. Everyone was coated about 1/16th inch thick with a fawn coloured powder. It got in our clothes, our food, our bedding and everything that was ours. I am now in town for a hot bath to wash the griminess of it away…

---

42   Kitchener quoted in French, Viscount, *1914*, pp. 308–309, Constable & Co. Ltd. (London) 1919.
43   Dardanelles Commission: Memoranda Produced as Evidence, TNA CAB/63/17.
44   French, Viscount, *1914*, p. 318.

"On Thursday I was in command of the Town Picquet... Duty consists most of seeking out the men (fortunately few) who are doing their best to bring a bad name on the force. To find these drunken sots & derelicts one has to go through the worst slums imaginable, streets unpaved with rotting, putrid vegetable matter in heaps together with other filth. Women of the street standing at nearly every door trying to entice you into their dens, some of which it was necessary to search owing to information given by native police which showed some wanted men had been seen there. Never before could I believe that vice could be indulged in on such a large scale."[45]

CAPT. GEOFFREY MCCRAE, 7TH BATTALION, AIF, MENA CAMP.

Others were able to take advantage of the improved facilities within the camps to further their training.

"We thoroughly stripped the gun and the lock, that is, took every possible part to pieces, and cleaned it up. In the afternoon we had a lecture on the machine gun and its principle object by Lt ———, of the Lancashire Fusiliers. In the evening S. Bowman[46] and I went to see Stan. Goulding,[47] who has been in hospital with bronchitis."[48]

PTE JULIUS JACOBSOHN, MG SECTION, 7TH BATTALION, AIF, MENA CAMP.

10 JANUARY 1915

EGYPT

Not everyone appreciated the opportunity to listen to more lectures.

"Now that our mess room is completed, we have now to submit to half-hourly or hourly lectures practically every night on some dismal subject. It would not be so bad if the officers could speak, and of this I am certain that they were not selected for their speaking ability. According to the Army orders, the rules of discipline must be read to the various companies every three months. Glad I am that I have not to listen to such trash every day."[49]

DVR MORTON COLLINGS, DIV. TRAIN, AASC 1ST AUSTRALIAN DIVISION, MENA CAMP.

Some Australians, however, had been able to offer training of a rather different sort to their new neighbours.

---

45   Letters from Geoffrey Gordon McCrae to his family, January 1915 to March 1915. AWM 1DRL/0427.
46   Pte Frank Bowman, MG Section, 7th Bn, AIF, a former salesman from Essendon, Victoria. Appointed L/Cpl, 1st May 1915; promoted Cpl, 6th June 1915; Sgt, 18th January 1916; 2/Lt, 21st March 1916; Lt, 5th September 1916; served with 2nd Machine Gun Company & 1st Australian Machine Gun Bn on Western Front; and returned to Australia, leaving Southampton, 3rd January 1919.
47   Cpl Stanley Gould, 7th Bn, AIF, died of wounds received at Anzac in Egypt on 1st May 1915. Buried in Alexandria (Chatby) Military and War Memorial Cemetery, he was the 19 year-old son of James and Florence Goulding, of 14 Wordsworth Street, Moonee Ponds, Victoria.
48   *The Corowa Free Press* (New South Wales), 19th November 1915.
49   *Cowra Free Press* (New South Wales), 6th March 1915.

"If the Australians have had no need to fight here, they have at least done excellent work in disseminating the English language. The results are highly satisfactory not to say startling. The newsboys… have mastered certain short offensive expressions thoroughly, and shout them out glibly in the main [s]treets. The Egyptian police, blissfully ignorant of their meaning, pay no heed whatever, and the young rascals are adding to their vocabularies daily."[50]

CHARLES BEAN, AUSTRALIAN WAR CORRESPONDENT.

11 JANUARY 1915

GREECE

The Greek Prime Minister submitted to King Constantine a British proposal for territorial concessions in exchange for Greek participation in the war.

"Till to-day our policy has consisted in the preservation of neutrality, in so far as our engagement with our ally Serbia has not required us to depart from it. But to-day we are called upon to take part in the war, not only in order to perform a moral duty, but in exchange for compensations, which, when realized, would make Greece great and powerful to a degree that the greatest optimists could not have contemplated a few years ago…

"My opinion that we should accede to the request that has been made us to take part in the war is also founded upon other considerations. By remaining impassive spectators of the conflict, we run other dangers than those already enumerated, which would be entailed by the eventual crushing of Serbia… But above all these dangers to which we should be exposed were we to enter upon the struggle, there rises a hope, and as I trust a well-founded hope, of saving a great proportion of Hellenism now in Turkey, and of creating a great and powerful Greece."[51]

ELEFTHERIOS VENIZELOS, GREEK PRIME MINISTER.

EGYPT

"Various rumors are floating round camp. One to the effect that we were going to Syria in four days started last week. So far no move has been made or likely to be made. They are keeping us here to deal with the Turks who are trying to invade Egypt. It is becoming monotonous training on the same old desert day after day. I would not mind another move to the front or something similar."[52]

PTE ERIC WARD, 1ST BATTALION, AIF, MENA CAMP.[53]

---

50   *Darling Downs Gazette* (Queensland), 20th February 1915.
51   Eleftherios Venizelos quoted in Kerofilas, Dr. C., *Eleftherios Venizelos. His Life and Work*, pp. 175–183, John Murray (London) 1915.
52   *National Advocate* (Bathurst, New South Wales), 17th February 1915.
53   Ward was buried by shellfire in his dugout on 29th June 1915 and invalided home. Returning to Australia on 4th July 1916, he was discharged as medically unfit on 16th September 1916.

"There is a great stir afoot, but we cannot get wind of anything official as yet. Anyhow we have to send an advance party away to Mena to-morrow, and the whole of the D.A.C. leave here on Tuesday morning at 5 a.m. From there it is rumoured we are going to Ismailia. I will be very sorry to shift camp from Ma'adi. We have had a fine camp and the residents have been most kind."[54]

CPL HERBERT MARKWELL, DAC, AFA, 1ST AUSTRALIAN DIVISION, MAADI CAMP.

12 JANUARY 1915
EGYPT

Lieutenant-General Sir William Birdwood took pains to communicate with the men under his command.

"In taking over command of the army corps composed of the Australian and New Zealand contingents I wish to tell my comrades of all ranks how proud I am at being associated with them in the great work which is before us all. We have been selected to fight for the honour and integrity of the British Empire. Before victory is assured much hard fighting will fall to us, fighting which will call for the highest degree, not only of discipline, but of self-denial and self-sacrifice…

"For myself, I pray God in all humility that I may prove myself worthy of the great trust which has been placed in my keeping, and that I may gain the confidence of my comrades, with whom I feel it to be an honour to be serving. If I can succeed in this, I well know you will make victory a certainty."[55]

LT-GEN. SIR WILLIAM BIRDWOOD, COMMANDING, AUSTRALIAN AND
NEW ZEALAND ARMY CORPS (ANZAC).

13 JANUARY 1915
ENGLAND

With other options ruled out by the lack of military resources, the War Council gave the go ahead to a purely naval attempt to force the Dardanelles.

"MR. CHURCHILL said he had interchanged telegrams with Vice-Admiral Carden, the Commander-in-Chief in the Mediterranean, in regard to the possibilities of a naval attack on the Dardanelles. The sense of Admiral Carden's reply was that it was impossible to rush the Dardanelles, but that, in his opinion, it might be possible to demolish the forts one by one. To this end Admiral Carden had submitted a plan…

"The Admiralty were studying the question, and believed that a plan could be made for systematically reducing all the forts within a few weeks. Once the forts were reduced

54   *The Maitland Daily Mercury* (New South Wales), 18th February 1915.
55   *The Press* (New Zealand), 24th February 1915.

the minefields would be cleared, and the Fleet would proceed up to Constantinople and destroy the "Goeben." They would have nothing to fear from field guns or rifles, which would be merely an inconvenience…

"The Admiralty should also prepare for a naval expedition in February to bombard and take the Gallipoli peninsula, with Constantinople as its objective."[56]

LT-COL. MAURICE HANKEY, CABINET SECRETARY.

"The War Council… spent the best part of three whole days — January 7, 8, and 13 — in surveying in the most comprehensive manner, and in the greatest detail, all our available resources in men and the calls which could be made upon on them… and accordingly on January 13 the Admiralty were ordered to prepare plans for an expedition, mainly naval in character, in February."[57]

HERBERT ASQUITH, MP, BRITISH PRIME MINISTER.

The detailed consideration Asquith described never extended to explaining how a navy was "to bombard and take the Gallipoli peninsula." By no means was everyone convinced that it could be done.

"On the 13th January, 1915, I was sent for by the First Lord (Mr. Winston Churchill) and he told me that HMS *Queen Elizabeth* was going to the Dardanelles, that the Navy was going to smash all the forts and go through to Constantinople and that I could go in command.

"I could not accept the offer as I knew it was an impossible task for the inefficient ships then in the Mediterranean to perform…[58] The idea that the battleships of the Mediterranean Squadron could reduce the forts and guns sprang from a sad want of knowledge. The authorities responsible for the mistaken idea were impressed by the success with which the German guns had reduced the Belgian forts, and concluded that in the same way ships' guns could reduce the Dardanelles forts. This deduction was due to a failure to realise the difference between firing on land and firing from a ship."[59]

ADMIRAL SIR PERCY SCOTT, RN.

Churchill made no reference to Admiral Scott's objections to the plan. But he did tell him not to speak about the operation.

"Sir Percy Scott has been cautioned as to secrecy. He is going out to assist in regulating the Director in *Queen Elizabeth*, but wishes to return from Gibraltar."[60]

WINSTON CHURCHILL, MP, FIRST LORD OF THE ADMIRALTY.

The First Sea Lord was not persuaded that a navy without an army in support could achieve the objectives set for it.

56 Secretary's Notes of a Meeting of a War Council held at 10, Downing St., January 13, 1915. TNA CAB/42/1/16.
57 Asquith, *Memories and Reflections*, vol. 2, pp. 88–89.
58 Scott, Admiral Sir Percy, *Fifty Years in the Royal Navy*, p. 281, George H. Doran Company (New York) 1919.
59 *ibid.*, p. 309.
60 Churchill, *The World Crisis 1915*, p. 112.

British submarine *B11* which sunk Turkish battleship *Mesudiye* on 13 December 1914.

"If the Greeks would land 100,000 men on the Gallipoli Peninsula in concert with a British Naval attack on the Dardanelles I think we could count on every success and quick arrival at Constantinople. A naval approach to Constantinople without any troops at all would occupy a month for the first shot fired at the mouth of the Dardanelles and would involve a loss of ships, and expenditure of ammunition and a wearing out of the heavy guns of the Fleet beyond approval and when the remains of the Fleet got to Constantinople it could do nothing else but carry out a futile bombardment with an accompanying massacre à la the bombardment of Alexandria! ... The real people to fight with us for Constantinople are the Bulgarians! — (one Bulgarian worth a thousand Greeks!)"[61]

ADMIRAL LORD JACKIE FISHER, RN, FIRST SEA LORD.

14 JANUARY 1915

ENGLAND

Churchill was now focused on the Dardanelles; his imagination fired by its possibilities.

"The attack on the Dardanelles will require practically our whole available margin. If that attack opens prosperously it will very soon attract to itself the whole attention of the Eastern theatre, and if it succeeds it will produce results which will undoubtedly influence every Mediterranean Power. In these circumstances we strongly advise that the Adriatic should be left solely to the French, and that we should devote ourselves

61   Letter from Fisher to William Tyrrell, Edward Grey's Private Secretary, 13th January 1915, TNA FO 800/107.

to action in accordance with the third conclusion of the War Council, viz., the methodical forcing of the Dardanelles."[62]

WINSTON CHURCHILL, MP, FIRST LORD OF THE ADMIRALTY.

Others learned of the gap between the First Lord's imagination and the realities of modern war.

"Eustace Fiennes, M.P., gave me an account of various incidents at Antwerp. He drove Winston from Dunkirk to Antwerp… When they reached Antwerp he visited the outskirts of the town and evinced great bravery in the face of the shower of shells which the Germans were pouring in. Later he said, "I am quite clear. This town must defended to the last; the infantry must fight street by street if necessary." He did not appreciate the effect of the German artillery."[63]

SIR GEORGE RIDDELL, NEWSPAPER PROPRIETOR.

EGYPT

Meanwhile the men in Egypt looked forward to doing more than chewing on sand.

"We did a ten mile route march to-day with all our war equipment, just what we take into the field nothing more or less and we found this a most friendly country. It gets up and throws itself at you. The only drawback is the quantity of soap and water one has to use to remove its caresses from one's person."[64]

CAPT. GEOFFREY MCCRAE, 7TH BATTALION, AIF, MENA CAMP.

"Letters are censored here now, so they evidently expect trouble… In the near future we expect to fight the Turks, and if we do not beat them we are all coming home again. We are nearly mad to get at them, and we watch the doings of our British Tommies for say what they like about the "Pommies," they can fight. And let me tell you that England is proud of Australia and the way she has responded to the call. Little does it occur to the mob at home that within a few days their countrymen here will be fighting for their lives. England needs every one, so tell the boys to pull themselves together and join us here… Only actual battle can improve us after the solid training we have had. We are getting well fed and living well indeed, now, so we cannot growl. There is everything on the grounds here, even picture shows."[65]

TPR ARTHUR HOLT, 3RD AUSTRALIAN LIGHT HORSE REGT, AIF, MAADI CAMP.

CEYLON

Bored by their long voyage some took unofficial advantage of their arrival at Colombo. Others yet to arrive, rolled up their sleeves.

62 Churchill, *The World Crisis 1915*, p. 112.
63 Riddell, Sir George, *Lord Riddell's War Diary 1914–1918*, pp. 51–52, Ivor Nicholson & Watson (London) 1933.
64 Letters from Geoffrey Gordon McCrae to his family, January 1915 to March 1915. AWM 1DRL/0427.
65 *Barrier Miner* (Broken Hill, New South Wales), 24th October 1915.

"[T]here was trouble; as leave to go ashore was refused to us. Some of the bolder ones decided to go over the side, and into the native fruit boats, in which they were rowed ashore. There must have been fully fifteen hundred who had their names taken; and as many more got back without. The chaps who were caught have been dealt with here. Our colonel gave his men seven days confined to camp; but others were fined up to £5. The same evening we steamed out in the Bay, where we anchored till next morning."[66]

PTE GEORGE CLAPHAM, 14TH BATTALION, AIF, HMAT *ULYSSES*.

INDIAN OCEAN

"For sheer calmness this little trip is just about beating the band. The sea has not, so far, got above a rolling swell, and a 16 ft. skiff, properly manned and navigated, could easily have made the voyage… It is mighty hot in the daytime, we being just about at the Equator now. Even so, many of the troopers consider they do not get sufficient exercise in leading horses, feeding, watering, and grooming them, mucking out their stalls, and cleaning the decks. So what do they do? They go down stoking at night-time for two or three hours, finishing up with a hot salt-water bath and a feed. It must be said that the amateur stokers look very well and happy. Perhaps the best-known of our trooper-firemen is the commander of the Second A.L.H. Brigade, Colonel Ryrie, who has been putting in a couple of hours every day in the stokehold in a weight-reducing effort. Perhaps he is getting rid of his superfluous. I don't know. Anyway, he still looks the ideal alderman amidships. He seems to be over strength for weight, but is still the sturdy, active fighting figure that Australian machine politics know well."[67]

TPR CLIFFORD HALLORAN, 6TH AUSTRALIAN LIGHT HORSE REGT, HMAT *SUEVIC*.

15 JANUARY 1915

ENGLAND

Stories of a Turkish invasion of Egypt were shared widely. Not a few dismissed them, as they dismissed the military competence of the Ottomans.

"There are… the usual boasts of a Turkish invasion of Egypt. The nearest point any Turkish troops have reached to that country is El Arish, and between that point and the Canal there are, as I have stated before nearly 160 miles of waterless desert. How they proposed to cross that with any considerable army is more than I know, more, indeed, than anybody knows."[68]

SAMUEL STOREY, CO-FOUNDER SUNDERLAND DAILY ECHO.

---

66   *Gippsland Mercury* (Sale, Victoria), 13th April 1915.
67   *The Sun* (Sydney, New South Wales), 7th March 1915.
68   *Sunderland Daily Echo and Shipping Gazette*, 15th January 1915.

EGYPT

The stay in Egypt was proving trying for animals as well as the men.

> "Although our horses landed in splendid condition they are now falling away considerably. Although we literally stuff them with feed, and they practically do no work, they are still falling away. The Egyptian chaff, called tibbin, is no good, and as no Australian chaff is available for some time, I fancy we will lose a few."[69]
>
> DVR MORTON COLLINGS, DIV. TRAIN, AASC 1ST AUSTRALIAN DIVISION, MENA CAMP.

> "Things are very quiet here at present. I have not been in Cairo for eight days now. I got very tired of the bustle. I expect we will be shifting down to the Canal any day, as we understand the Turks are coming down to take Egypt, but I guess they will get a great surprise if they do. If such is the case, everyone is prepared to play his part in the game, and I think they will find colonial mutton is harder to digest than they think."[70]
>
> PTE WILLIAM HAYES, CANTERBURY BN, NZEF, ZEITOUN CAMP.[71]

16 JANUARY 1915

EGYPT

> "There are twenty thousand men in our camp and ten thousand in the New Zealanders' and it is just like a fair sized town. We have all sorts of amusement and plenty of bands. We go out every day for a march which is very hard, for after you leave camp you can see nothing but sand, and when it is windy the sand nearly blinds you. At times we have had to stop the parades because it was so bad. Well, mother, it is not a bad life and I am keeping well; I weigh eleven stone two and I never felt better in my life."[72]
>
> PTE ALFRED VILLIERS, 9TH BATTALION, AIF, MENA CAMP.

17 JANUARY 1915

EGYPT

One battalion cook outlined a typical day's march.

> "I might tell you I am not too bright to-day. I rose at 4.30 a.m., prepared the breakfast of boiled eggs and tea for 228 men, and the bugle blew 'cookhouse' at 6.45 a.m. We

---

69 *Cowra Free Press* (New South Wales), 6th March 1915.
70 *Wanganui Herald* (New Zealand), 6th March 1915.
71 A Boer War veteran, Hayes, a former labourer, was admitted sick to No 2 Stationary Hospital, Mudros, on 26th June 1915; and, after a spell in Malta, to the 1st Southern General Hospital, Birmingham, on 2nd October 1915. He left England to return to New Zealand on 13th September 1915; discharged as no longer physically fit for service on 12th May 1916.
72 *The Evening Telegraph* (Charters Towers, Queensland), 12th March 1915.

had our meals and then had to prepare for the march. First of all we loaded our G.S. [General Service] waggon with all our kitchen utensils, and then dressed for parade. We fell in at 8.45 a.m., and got well on the march by 10 a.m. Our route was first of all along the main road to Cairo for about three miles. Then we turned off on to a side road, and gradually headed back to camp. We passed through three native villages… irrigation channels and fields of vegetation of all kinds, we had travelled about 13 miles, and got into the dreary desert, with sand feet deep. On this sand we camped for dinner, it being about 1 p.m. We started off about 2.30, the Light Horse leading, then the field artillery, our 4th Battalion Infantry, followed by the field ambulance. We waded along through the sand, sinking to our boot tops all the way for two miles. When we got to camp, all well satisfied. The cooks were dismissed from parade, and had to set to work to prepare a stew for tea. We soon had the 'dixies' simmering, and the tea well on the road. We were not in the best of humor, and every now and then an individual would stroll to the kitchens and ask how long would tea be. Of course, they all got some very sweet replies. The first we got was from a corporal, and before we had got our equipment off our backs he calmly asked how long will tea be. He got out of the cook house at the double, and following him was one of the most choice volleys I ever heard. At 6.45 p.m. we were ready, and the bugler blew the cook house call amidst a great cheer from the boys. They all got their tea, and we were at peace once again, and settled down to have our own snack. While we were having tea the eggs for breakfast the next morning arrived. No one spoke, and at last I said to the sergeant, 'I suppose those eggs will have to be cooked to-night.' His answer was, 'We have no other choice,' so I said, 'Just our luck.' I finished tea, and started to work again. The first job was to count the eggs out, three per man for 228 men a total of 684 eggs. Then to cook them. About 9 p.m. they were finished, and we rolled into bed, and very soon were not troubling about the eggs. The next I heard was at 6.30 a.m., and the words were, 'Come on, boys.' I turned over and rubbed my eyes, and said, 'Very good, sergeant.' We have six cooks in our tent. We sat up and looked at one another, and exclaimed, 'How do you feel?' We got up, and very soon got to work, and to-day we are going to give the boys curry and rice, to-morrow roast beef and cabbage and potatoes."[73]

PTE VERNON HOLLAND, 7TH BATTALION, AIF, MENA CAMP.[74]

The experience of Egypt was not all bad. This day witnessed a marriage.

"On duty all day, sanitary police. Private 10th Battalion married[75] — great scene. Writing letters and post cards in afternoon."[76]

PTE SYDNEY PENHALIGON, 3RD FIELD AMBULANCE, AAMC, MENA CAMP.

73 *Bendigo Advertiser* (Victoria), 20th February 1915.
74 Vernon Holland was killed in action at Helles on 8th May 1915. Commemorated on the Helles Memorial, the former insurance agent was the 24 year-old son of William Henry Vernon Holland and Amy Louisa Holland, of Piavella, Victoria, Australia.
75 L/Cpl Philip de Quetteville Robin, 10th Bn, AIF, married Nellie Irene Honeywell at Mena Camp, Egypt, on 17th January 1915. One of the battalion scouts, L/Cpl Robin is believed to be one of those who advanced furthest inland on 25th April 1915, together with the then Pte Arthur Seaforth Blackburn. Robin was killed on 28th April 1915 and is commemorated on the Lone Pine Memorial.
76 *The Queenslander* (Brisbane, Queensland), 21st April 1917.

18 JANUARY 1915

GREECE

The attitude of the Greek Government was divided. The Prime Minister supported Greek intervention in the war to secure territory in Asia Minor.

> "If… Germany and Turkey are victorious, not only will the 200,000 Greeks, who have already been driven from Asia Minor, have to renounce all hope of returning to their homes, but the number of those who will ultimately be expelled may assume alarming proportions. In any case, the triumph of Germanism will mean its absorption of the whole of Asia Minor.
>
> "Under these circumstances how can we let slip this opportunity which Divine Providence has given us, to realize our most daring national ideals, to create a Greece enfolding almost all the lands where Hellenism reigned supreme during its long history, a Greece comprising very fertile territories which would ensure our preponderance in the Aegean Sea?"[77]
>
> ELEFTHERIOS VENIZELOS, GREEK PRIME MINISTER.

EGYPT

Sand was not Egypt's only irritant. Problems with the mail frustrated many.

> "Although I am deeply grateful for your kind thoughtfulness to despatching the parcels of handkerchiefs and sox, I am sorry to state they have not arrived yet and I have despaired of ever getting them now.
>
> "By separate package I am forwarding an empty cigarette box. I smoked the cigarettes. They were presented to the soldiers by a firm in Egypt and would like the box kept among other things. I am also sending a few more photos. Have others reached you from here?"[78]
>
> L/SGT JOHN KERR, 5TH BATTALION, AIF, MENA CAMP.

19 JANUARY 1915

ENGLAND

Churchill wrote to Grand Duke Nicholas confirming the decision to force the Dardanelles. He asked for Russian co-operation.

---

[77] Eleftherios Venizelos quoted in Kerofilas, Dr. C., *Eleftherios Venizelos. His Life and Work*, pp. 185–189, John Murray (London) 1915.

[78] *Cobden Times and Heytesbury Advertiser* (Victoria), 20th February 1915.

"The Admiralty have considered with deep attention the request conveyed through Lord Kitchener from Your Imperial Highness for naval action against Turkey to relieve pressure in the Caucasus. They have decided that the general interests of the Allied cause require a great effort to be made to break down Turkish opposition in addition to the minor demonstration of which Lord Kitchener has telegraphed to you. It has therefore been determined to attempt to force the passage of the Dardanelles by naval force...

"The Admiralty hope that the Russian Government will co-operate powerfully in this operation at the proper moment by naval action at the mouth of the Bosphorus, and by having troops ready to seize any advantage that may be gained for the allied cause. It would probably be better to defer Russian action until the outer forts of the Dardanelles have been destroyed, so that if failure should occur at the outset, it will not have the appearance of a serious reverse. But it is our intention to press the matter to a conclusion."[79]

WINSTON CHURCHILL, MP, FIRST LORD OF THE ADMIRALTY.

It is not clear whether Churchill ever outlined why a Russian attack would benefit from military support but not that by the Anglo-French. If he did, the First Sea Lord was unaware of it.

"And now the Cabinet have decided on taking the Dardanelles solely with the Navy, using 15 battleships and 32 other vessels, and keeping out there three battle cruisers and a flotilla of destroyers — all urgently required at the decisive theatre at home! .... I don't agree with one single step taken."[80]

ADMIRAL LORD JACKIE FISHER, RN, FIRST SEA LORD.

EGYPT

"The regiment is, I am glad to say, in very good health, and the members are now all well trained and fit for anything they may be asked to do. We shall all be very pleased to receive our marching orders."[81]

MAJOR FREDERICK HURCOMBE, 10TH BATTALION, AIF, MENA CAMP.

20 JANUARY 1915

ENGLAND

One fear for the British, French and Russians alike was the potential impact defeat at the hands of an Islamic enemy would have on their own Muslim subjects. Churchill had thought of this.

---

79   Churchill, *The World Crisis 1915*, p. 119.
80   Jackie Fisher to John Jellicoe quoted in Hough, Richard, *The Great War at Sea 1914–1918*, p. 152, Oxford University Press (Oxford) 1984.
81   *The Express and Telegraph* (Adelaide, South Australia), 24th February 1915.

> "The attack on the Dardanelles should be begun as soon as the *Queen Elizabeth* can get there. Every effort will be made to accelerate her departure, so that fire can be opened on February 15. It is not desirable to concentrate the whole fleet of battleships required for the operation at the Dardanelles at the outset. This would only accentuate failure, if the forts prove too strong for us. *Indefatigable*, *Queen Elizabeth*, and three or four other British battleships, with the mine-sweepers and the *Ark Royal*, will be sufficient at the outset…
>
> "As soon as the attack on the Dardanelles has begun, the seizure of Alexandretta should take place. Thus if we cannot make headway in the Dardanelles, we can pretend that it is only a demonstration, the object of which was to cover the seizure of Alexandretta. This aspect is important from an Oriental point of view."[82]
>
> WINSTON CHURCHILL, MP, FIRST LORD OF THE ADMIRALTY.

Asquith, meanwhile, was hearing more about the growing friction between Churchill and his chief naval adviser.

> "Hankey came to see me to-day to say that Fisher, who is an old friend of his, had come to him in a very unhappy frame of mind. He likes Winston personally, but complains that on purely technical naval matters he is frequently overruled ("he out-argues me"), and he is not by any means at ease about either the present disposition of the fleets or their future movements. Though I think the old man is rather difficult, I fear there is some truth in what he says."[83]
>
> HERBERT ASQUITH, MP, BRITISH PRIME MINISTER.

EGYPT

The behaviour of some Australian soldiers in Egypt continued to cause concern — not least among other Australian soldiers.

> "They are sending about 600 men back to Australia out of the division — those that are not fit and those that have got cold feet, as the lads call it. It seems a jolly shame the way a lot of them have carried on since they landed here. One man arrived back in camp yesterday. He had been absent from camp without leave for 28 days, and he was sentenced to a district court martial, so do not know what they will do with him. One of the Terriers over on the Canal was away 21 days and he was tried, sentenced to death and was shot,[84] and that is what they want to do with some of this mob. There are about 10 per cent. in our battalion that are good for nothing, only targets."[85]
>
> PTE ANGUS SMITH, 10TH BATTALION, AIF, MENA CAMP.[86]

---

82   Churchill, *The World Crisis 1915*, p. 120.
83   Asquith, *Memories and Reflections*, vol. 2, p. 57.
84   No member of the Territorial Force was executed for desertion in Egypt at this time.
85   *Murray Pioneer and Australian River Record* (Renmark, South Australia), 11th March 1915.
86   Smith was wounded by shellfire on 9th October 1915. After treatment in Malta and Egypt, he left Suez on 28th January 1916, returning to Australia aboard the *Kanowna* to be discharged.

The attitude towards their commanders began to emerge at this time. Birdwood was liked. Major-General Sir Alexander Godley, commanding the NZ & A Division, not so much.

> "My, but ain't we glad we've got General Birdwood to keep our very own General Godley in hand! Things have been looking up ever since "Birdseed" took control. We got enough of our General coming over to do us all our lives… Generals Alex. Godley and Birdwood and their staff have just ridden past my tent, which is on "the main street" through the camp. Some one down the lines yelled out, "What-o, Alec!" I couldn't see Alec's face…
>
> "The boys are just back, covered in dust and sand and blinded with sweat. General Alec's latest order is that they have to oil their rifles. What rot! The sand clings badly enough without using oil to help to gather more sand."[87]
>
> CPL WILLIAM MILLS, AUCKLAND BATTALION, HELMIEH CAMP, HELIOPOLIS.

That evening, after reports were received of a Turkish advance towards the Suez Canal, camps that had settled down into dull routine became centres of activity.

> "We have just received word that we shall probably be in action to-morrow night. It is 8 o'clock now, and we are to get everything ready to-night. We entrain at 7 o'clock in the morning, and leave either for Suez, or Ismailia. All the band instruments have just been called in, and we have to take our places in the ranks, with the exception of 16 stretcher-bearers. The reinforcements are expected here on Wednesday, so we shall just miss them. I am in the best of spirits, and am not a bit nervous of the result, although there are supposed to be 80,000 Turks coming. It may be a false alarm."[88]
>
> L/CPL ARTHUR HAYBITTLE, AUCKLAND BATTALION, ZEITOUN CAMP.

21 JANUARY 1915

ENGLAND

Asquith pondered the options available for British military action. He seemed convinced at the military and political value of adding the forces of the Balkan states to the allies.

> "The main point at the moment is to do something really effective for Servia, which is threatened by an overwhelming inrush from the Austrians, reinforced by some 80,000 Germans. If she is allowed to go down things will look very black for us, and the prestige of the Allies with the wavering and hesitating States will be seriously impaired… There is a report that General Castelnau, who is one of the best French generals, is strongly of opinion that things there have reached a condition of stalemate, that neither side can do more than push a little here and retreat a little there. If so, it seems a criminal waste at such a critical time to put in new and good troops in that theatre.

---

87 *The Feilding Star* (New Zealand), 4th March 1915.
88 *ibid.*, 25th March 1915.

New Zealand Mounted Rifles disembarking in Egypt.

"There are two fatal things in war. One is to push blindly against a stone wall; the other is to scatter and divide forces in a number of separate and disconnected operations. We are in great danger of committing both blunders."[89]

HERBERT ASQUITH, MP, BRITISH PRIME MINISTER.

EGYPT

Asquith's uncertainty about where British forces were to go was shared by those camped near the Pyramids.

"We have been doing a lot of hard work, and are ready for any order. We are all looking to the day when we can say we are in a position to take a hand in the struggle. One does not know, of course, what possibilities there are in the east of the Mediterranean, but for myself, I anticipate being in France before a reply to this letter can reach me."[90]

L/CPL EDWARD BEAVIS, 4TH BATTALION, AIF, MENA CAMP.

AUSTRALIA

The Australian Prime Minister, William Hughes, responded to the stories of bad behaviour in Egypt.

"I read with a great deal of concern Mr. Bean's statement as to the conduct of some of our troops in Egypt. It is no doubt a very serious matter, but I should be more concerned if I believed such conduct fairly reflected the code of the Expeditionary Force generally. I don't believe it does for a moment. There are two things that may be considered. I speak as one who lived in a garrison town for some years, and I can hardly believe that even the most unruly of our troops have gone beyond what was quite common amongst British regular troops; and in that case, as in this, unruly spirits represented almost a negligible proportion of the whole. Secondly, some weight must be attached to the circles in which these young men find themselves after a long sea voyage, in all cases sufficiently irksome, and disembarkation in one of the most fascinating cities of the world. I shall, until there is good reason to alter my opinion, hold to the conviction that the Australian troops as a whole will do credit to Australia, not only on the battlefield, but generally."[91]

WILLIAM HUGHES, PRIME MINISTER OF AUSTRALIA.

---

89   Asquith, *Memories and Reflections*, vol. 2, p. 57.
90   *National Advocate* (Bathurst, New South Wales), 27th February 1915.
91   *The Advertiser* (Adelaide, South Australia), 22nd January 1915.

22 JANUARY 1915

EGYPT

Some Australians were keen to join the defence of the Suez Canal.

> "We are soon to get at the Turks. We were told to-day that any time would see us on the road to the Suez Canal; already two regiments of Territorials have gone there, and two batteries of our artillery and the infantry are standing to their guns, ready to move off at any minute. We are nearly dying to get at them. We are arranging our farewell concert to Maadi, so you see something is doing. Fred Gilling is at the canal with the engineers. We were told that 12 men were killed, but the heads won't tell us anything. We were told that it was a boiler explosion, but that is hard to believe. The officers here admit that there is fighting on the canal in small batches."[92]

TPR ARTHUR HOLT, 3RD AUSTRALIAN LIGHT HORSE REGT, MAADI CAMP.

23 JANUARY 1915

FRANCE

> "Augagneur informs us that the British Admiralty is organising, independently of ourselves, an operation in the Dardanelles. Now by virtue of Anglo-French naval agreements the right to command in the Mediterranean attaches to France and therefore our allies cannot do anything without previous reference to ourselves. It is decided that the Minister for Marine shall go to London in a day or two to discuss with Mr. Winston Churchill a proposal which must be closely studied before being accepted."[93]

RAYMOND POINCARÉ, PRESIDENT OF FRANCE.

EGYPT

Charles Bean, who never considered his reports of Australian misconduct to have been exaggerated, noticed a change after the camp facilities had expanded.

> "Partly because of the attractions that have lately been added to the camp at the Pyramids, partly because of certain stricter measures that have been taken, and partly because the camp training is really heavy, and the men come back from it fairly tired, the streets of Cairo are now far less crowded with Australians than they used to be. At one time there was a continuous stream of little carriage lamps and motor lights, reaching as far as a man could see, along the road from Cairo, and one could

---

92   *Barrier Miner* (Broken Hill, New South Wales), 24th October 1915.
93   Poincaré, Raymond, *The Memoirs of Raymond Poincaré, President of France, 1915*, p. 18, William Heineman Ltd. (London) 1926.

hear until after midnight a continuous roar — the clatter of hundreds of hoofs of cab horses on the asphalt of the Pyramids-road. You can go into Cairo almost any night now and not meet more than two or three taxis on the way. Australia is no longer in danger, I think, of losing her good name in this country."[94]

CHARLES BEAN, AUSTRALIAN WAR CORRESPONDENT, MENA CAMP.

The effect of artillery witnessed after one exercise was certainly sobering.

"I was out with the battalion yesterday. We attacked a line of trenches that had been dug the day before. When we started, at about 3000 yards, the artillery commenced to shell the trenches from our flank. First you see the flash from the gun and hear the whistling of the shelf as it rushes through the air. Then comes the report, and you turn to the target and see the shell burst either in the air above the target or on contact with the earth. Then comes the report from the shell-burst… After we had "driven the enemy out," we inspected the trenches and saw the damage done by the shells. We picked up pieces of jagged shell and hundreds of round bullets. I'm not anxious to get under fire from the big guns."[95]

CPL WILLIAM MILLS, AUCKLAND BATTALION, HELMIEH CAMP, HELIOPOLIS.

"It seems pretty certain that we shall go to France, but not at present, as it would be too cold for us. The boys are convinced that they are going to France, as they have had a lot of street fighting lately, and cannot understand what use that would be in fighting the Turks along the canal in the desert."[96]

2/LT CHARLES MASSEY, CANTERBURY BN, NZEF, ZEITOUN CAMP.

RED SEA

"We are passing through the Red Sea to-day, and the only cool spots are either in the freezing chambers or under the showers. We sighted a big boat this morning bound for Australia. She signalled us, but what it was I can't find out. Several of our chaps have died during the trip from pneumonia. One was buried at Aden,[97] and one poor fellow[98] directly after church parade to-day."[99]

PTE VEITCH LINDEMAN, 15TH BATTALION, AIF, HMT *CERAMIC*.

---

94 *The Capicornian* (Rockhampton, Queensland), 27th February 1915.
95 *The Feilding Star* (New Zealand), 4th March 1915.
96 *The Star* (Christchurch, New Zealand), 4th March 1915.
97 Pte Harold Robinson, 16th Bn, AIF, died of pneumonia on 20th January 1915. Buried in Maala Cemetery, Aden (modern Yemen), he was the 18 year-old son of James and Lila Robinson, of Noarlunga, South Australia.
98 Pte Philip Carleton, 15th Bn, AIF, is recorded as having died on the *Ceramic* on 24th January 1915. Buried at sea, the former clerk is commemorated on the Chatby Memorial, Egypt; the son of William Carleton, of Bowen Post Office, Bowen, Queensland.
99 *Clarence and Richmond Examiner* (Grafton, New South Wales), 17th April 1915.

24 JANUARY 1915

ENGLAND

While Maurice Hankey, the Cabinet Secretary, was convinced of the possibilities offered by attacking the Ottomans, he was concerned at the prospects for an unsupported naval attack. He, therefore, circulated copies of the 1906 report into the feasibility of an assault upon the Dardanelles to members of the War Committee. That study had this to say about an attack by sea alone:

> "Even if, as has been suggested, it were feasible to rush a number of His Majesty's least valuable ships past the batteries lining the Dardanelles and over the minefields which are believed to exist in the channel, their arrival off Constantinople would be no guarantee that the Sultan would be thereby brought to reason…

> "Without a large military force ready to land at a moment's notice, there would be no means of controlling the populace… The Turkish communications furthermore between Europe and Asia would remain virtually intact, and our squadron might find itself face to face with the necessity of again running the gauntlet of the Dardanelles under circumstances which might lead to its destruction."[100]

RUSSIA

News came through of the defeat inflicted upon the Turks in the Caucasus. This removed the need to attack the Dardanelles and came with a sting in the tail: the Russians were unable to participate in any co-ordinated attack from the Bosphorus.

> "The question of Russian co-operation over the Dardanelles business came up again and I had a long interview with the Grand Duke and Prince Koudacheff.

> "The former told me that the position in the Caucasus had been considerably eased by Russian successes, and he laid stress on the fact that he had made no suggestion as to the methods we should employ in rendering assistance to draw off the Turks from that theatre, and had never guaranteed any Russian co-operation, glad as they would of course be to give it should opportunity occur.

> "The Russian General Staff pointed out that their Black Sea Fleet, in view of the delay in building of their dreadnoughts, of the scarcity of their destroyers, and of lack of 'up-to-date' submarines, was only the equal of the Turkish Fleet (including, of course, *Goeben* and *Breslau*). Even, they added, that equality would only be reached when all the units could work together, and the absence of one or two of them would at once place the balance in favour of the Turks. The construction of their ships was such that they could only carry a four days' supply of coal. Coaling at sea was rendered

---

100 The Possibility of a Joint Naval and Military Attack Upon the Dardanelles, December 1906, TNA CAB 38/12/60.

extremely difficult by bad weather and the heavy seas which are met with in winter in the Black Sea. The nearest coaling base was 24 hours' sail from the entrance to the Bosphorus. However much they wished to co-operate with the British Fleet, their hands were tied."[101]

MAJOR-GENERAL JOHN HANBURY-WILLIAMS, BRITISH MILITARY MISSION, WARSAW.

EGYPT

After experiencing desert conditions for themselves, some men doubted the Turks would be able to attack the Suez Canal. But reinforcements continued to head that way.

"There is considerable rumour about the Turks marching in strong force across the desert toward us, but the heads do not seem to attach any great importance to it — even if they are there. Several armies have previously endeavoured to march through the long stretches of waste known as the desert; but all have failed — with the exception of the host led by a gentleman named Moses, quite a long while ago, and on that occasion, so I am led to believe, the trip occupied forty years. So that if one is coming now, with less guiding facilities than the historic Israelites had, the war may be over when they get here...

"We went out on a "night attack," and were lost for a time in the dense fog of the following early morning. (We have a chap named Burke and another named Wills in the company, and both were struggling manfully along. One of our fellows said we should never get lost because we had Burke and Wills with us. I told him if he looked searchingly around Melbourne he would find a monument erected to those very gentlemen for getting lost.)"[102]

L/SGT JOHN KERR, 5TH BATTALION, AIF, MENA CAMP.

"I had gone to town in the evening for a few things we needed, and while walking along one of the main streets we ran across one of our New Plymouth boys, and he greeted us with the news that we were off to the front first thing in the morning. This was the first we had heard of it. It just shows how secret our movements can be kept. Well, we rushed round and did our shopping, then jumped into a motor and off for our lives.

"When we arrived home the camp was in a great state, everybody rushing round packing kits and drawing ammunition and rations, and we eventually got to bed somewhere about midnight to be up again at 5 a.m."[103]

SGT WILLIAM OKEY, WELLINGTON BN, NZEF, ZEITOUN CAMP.

---

101 Hanbury-William, Sir John, *Emperor Nicholas II As I Knew Him*, pp. 36–38, Humphreys (London) 1922.
102 *Cobden Times and Heytesbury Advertiser* (Victoria), 20th March 1915.
103 *Taranaki Herald* (New Zealand), 7th April 1915.

25 JANUARY 1915

ENGLAND

'Jackie' Fisher continued to air his concerns about weakening the Royal Navy in home waters; the decisive theatre of operations.

> "We play into Germany's hands if we risk fighting ships in any subsidiary operations such as coastal bombardments or the attack of fortified places without military co-operation, for we thereby increase the possibility that the Germans may be able to engage our fleet with some approach to equality of strength. The sole justification of coastal bombardments and attacks by the fleet on fortified places, such as the contemplated prolonged bombardment of the Dardanelles Forts by our fleet, is to force a decision at sea, and so far and no farther can they be justified."[104]
>
> ADMIRAL LORD JACKIE FISHER, RN, FIRST SEA LORD.

EGYPT

Men continued moving towards the Suez Canal. Many welcomed the change.

> "All we were allowed to take in our valises was a change of underclothing, a shirt, and a couple of pairs of socks, and what with 150 rounds of ball ammunition in one's pouches it was quite as much as I wished to carry about.
>
> "However, at 10 a.m. we were on our way to Pont Limoun railway station, and we were off to the front, which happened to be a place called El Kubri, distant about five miles from Suez.... [It] was reached at 7 p.m. and the first night was spent under the blue sky and on the sandy desert."[105]
>
> SGT WILLIAM OKEY, WELLINGTON BN, NZEF, EL KUBRI.

> "8 p.m. — Word has just come through that we are off to the Suez Canal to-morrow to hold the canal against the Turks. The Turkish advance parties are already attacking the canal and we are leaving at twelve hours' notice. We are fed up with training here, and it will be a change to get into action. We shall be up all night to-night and we leave at 7 a.m. to-morrow. We are all writing in haste. The whole camp is in a terrific state of excitement. Bands are playing and everyone is singing patriotic songs, and all are eagerly looking forward to a fight with the Turks."[106]
>
> 2/LT CHARLES MASSEY, CANTERBURY BN, NZEF, ZEITOUN CAMP.

> "[W]hen the orders came through about the infantry moving, all was hustle and bustle. There was very little spare time for members of the A.S.C. and quite a number never got to sleep at all. Thirty waggons had to be sent off post haste to the Citadel in Cairo

---

104 Jackie Fisher quoted in Churchill, *The World Crisis 1915*, pp. 155–156.
105 *Taranaki Herald* (New Zealand), 7th April 1915.
106 *The Star* (Christchurch, New Zealand), 4th March 1915.

to draw ammunition, and then there were stores of all descriptions to be packed, after which everything had to be taken to the station and placed on board the train. I can assure you there was 'some' work going."[107]

TPR JAMES MOSSMAN, AUCKLAND MOUNTED RIFLES, ZEITOUN CAMP.

26 JANUARY 1915

ENGLAND

Sir Edward Grey put a positive gloss on the news that no Russian assistance would be forthcoming.

"This is the Russian reply about Dardanelles. It shows that, though Russia cannot help, the operation has her entire goodwill and the Grand Duke attaches the greatest importance to its success.

"This fact may be used with Augagneur to show that we must go ahead with it and that failure to do so will disappoint Russia and react most unfavourably upon the military situation, about which France and we are specially concerned just now."[108]

SIR EDWARD GREY, MP, BRITISH FOREIGN SECRETARY.

EGYPT

"[At] about 4 a.m., when they commenced their sniping, retiring again after about an hour's rifle fire, which did absolutely no damage. In fact, our Indian troops did not even reply. We just simply rolled over in our blankets and wished the Turks all sorts of "good luck" and were soon again in dreamland."[109]

SGT WILLIAM OKEY, WELLINGTON BN, NZEF, EL KUBRI.

"We have just received word that we shall probably be in action to-morrow night. It is 8 o'clock now, and we are to get everything ready to-night. We entrain at 7 o'clock in the morning, and leave either for Suez or Ismailia. All the band instruments have just been called in, and we have to take our places in the ranks, with the exception of 16 stretcher-bearers… I am in the best of spirits, and am not a bit nervous of the result, although there are supposed to be 80,000 Turks coming. It may be a false alarm."[110]

L/CPL ARTHUR HARBITTLE, AUCKLAND BATTALION, ZEITOUN CAMP.[111]

---

107 *Poverty Bay Herald* (New Zealand), 27th March 1915.
108 Sir Edward Grey quoted in Churchill, *The World Crisis 1915*, p. 157.
109 *Taranaki Herald* (New Zealand), 7th April 1915.
110 *The Feilding Star* (New Zealand), 25th March 1915.
111 Harbittle was wounded on 25th April 1915. Rejoining his battalion on 25th June, he was hospitalised again on 29th July and sent to Mudros. Discharged on 7th August 1915, the former painter was killed in action on 13th August 1915. Commemorated on the Chunuk Bair (New Zealand) Memorial, he was the 34 year-old son of Richard Frederick and Anne Elizabeth Haybittle, of 4 Bowen Street, Feilding.

> "The battalion fell in at 11 a.m. and marched to the Palais de Koubeh station, two miles away… Our full marching order was a fearful load and the men did not need to be told to take it off when we halted in a shady avenue of trees outside the station, waiting while supplies, etc., were loaded up. We entrained about 2 p.m.… stopped at a station called Zagazig, and then on again to Ismailia, seventy-nine miles and a half, arriving there at 6.45 p.m., the bright moonlight considerably assisted us in 'bivvying' for the night outside the station yard. The sand here is much softer than at Zeitun, and as the night was warm we slept comfortably. The Auckland battalion was alongside. The Wellington and Otago Battalions went to Kantara, some miles away. The Turkish advance posts are at present only ten miles out from Kantara, and about sixteen miles from Ismailia. We can hear the artillery guns from here, night and day, when things are reasonably quiet in camp. The main Turkish army is as yet some 150 miles away. We are only about a mile from Lake Timsah, through which the Canal runs. There are two gunboats on the lake at present."[112]

LT RAYMOND LAWRY, CANTERBURY BN, NZEF, ISMAILIA.

RED SEA

A former journalist with the 6th Light Horse watched firing practice at sea.

> "A couple of days ago our machine gun section got busy — busier than usual, that is. A gun was mounted right aft on the poop deck, the railing being taken down, and all made clear for firing practice. The target was an oil-drum, on which was tied a rope, run out 200 yards, or so, over the stern. At the drum end about 10 yards of the rope was thin steel hawser. As the drum rope was being paid out by several of the ship's crew, two or three of them, including the 'bosun, were nearly pulled into the water on a couple of occasions.

> "The drum was a difficult mark for the machine gun men to hit, as it performed all sorts of antics. It would drag along on or just below the surface for 10 or 15 yards, and would then leap into the air like a great flying fish (of which, by the way, we have seen many). To add to the difficulties of the marksmen there was the motion of the ship.

> "The marksmanship was, on the whole, very good. Some of the bullets which missed the target altogether, and others which cannoned off the steel hawser, skidded over the waves in great style. Many of us had never seen a machine gun working before, but it was easy for us to understand the deadliness of the weapon. The sound of the firing is a sort of peculiarly murderous thumping rattle…

> "After the machine gun practice several of our officers and a couple of squadron sergeant-majors, using revolvers and automatic pistols, took pot shots at boxes and bottles thrown overboard. It was easy to see that we have among us crack shots with these weapons."[113]

TPR CLIFFORD HALLORAN, 6TH AUSTRALIAN LIGHT HORSE REGT, HMAT *SUEVIC*.

---

112   *Lyttelton Times* (New Zealand), 25th March 1915.
113   *The Sun* (Sydney, New South Wales), 18th March 1915.

Indian machine gun team defending the Suez Canal.

27 JANUARY 1915

EGYPT

"This will be the first letter you have received from me written actually on the battlefield. We are now engaged with the Turks, although as yet we are not in the firing line, but acting as supports, about seven miles in the rear. An artillery duel has been going on all day, and two aeroplanes have passed over us (British scouts) this afternoon. It is a great compliment to us New Zealanders to be sent in preference to the English Tommies and Australians, and we are all anxious to see what we can do… An aeroplane has just come back with news which may mean us scrapping. Hoorah!"

CPL MOSTYN PRYCE JONES, CANTERBURY BN, NZEF, ISMAILIA.[114]

28 JANUARY 1915

ENGLAND

Asquith met Churchill and Fisher before the War Council meeting to confirm the course of action to be taken at the Dardanelles.

---

114   A former warehouseman, Jones, was wounded at Helles on 9th May 1915. He left Egypt aboard the *Tahiti* on 7th August 1915 to return to New Zealand, where he was discharged on 8th January 1916.

"They both came to see me this morning before the War Council and gave tongue to their mutual grievances. I tried to compose these differences by a compromise, under which Winston was to give up for the present his bombardment of Zeebrugge, Fisher withdrawing his opposition to the operation against the Dardanelles."[115]

HERBERT ASQUITH, MP, BRITISH PRIME MINISTER.

The minutes of the War Council recorded the discussion. Churchill mentioned Russian support for the assault upon the Dardanelles, if not its inability to participate; Carden's belief that the navy could force its way through, but not his "might."

"Mr. Churchill said that he had communicated to the Grand Duke Nicholas and to the French Admiralty the project for a naval attack on the Dardanelles. The Grand Duke had replied with enthusiasm, and believed that this might assist him. The French Admiralty had also sent a favourable reply, and had promised co-operation. Preparations were in hand for commencing about the middle of February. He asked if the War Council attached importance to this operation, which undoubtedly involved some risks?

"Lord Fisher said that he understood that this question would not be raised to-day. The Prime Minister was well aware of his own views in regard to it.

"The Prime Minister said that, in view of the steps which had already been taken, the question could not well be left in abeyance.

"Lord Kitchener considered the naval attack to be vitally important. If successful, its effect would be equivalent to that of a successful campaign fought with the new armies. One merit of the scheme was that, if satisfactory progress was not made, the attack could be broken off...

"Mr. Churchill said that the naval Commander-in-Chief in the Mediterranean had expressed his belief that it could be done. He required from three weeks to a month to accomplish it. The necessary ships were already on their way to the Dardanelles. In reply to Mr. Balfour, he said that, in response to his inquiries, the French had expressed their confidence that Austrian submarines would not get as far as the Dardanelles.

"Lord Haldane asked if the Turks had any submarines.

"Mr. Churchill said that, so far as could be ascertained, they had not. He did not anticipate that we should sustain much loss in the actual bombardment, but in sweeping for mines some losses must be expected. The real difficulties would begin after the outer forts had been silenced, and it became necessary to attack the Narrows. He explained the plan of attack on a map."[116]

LT-COL. MAURICE HANKEY, SECRETARY, WAR COUNCIL.

---

115 Asquith, *Memories and Reflections*, vol. 2, p. 59.
116 Maurice Hankey quoted in Churchill, *The World Crisis 1915*, pp. 163–164.

Asquith alluded to the prevailing atmosphere during the discussion.

> "When at the Council we came to discuss the latter, which is warmly supported by Kitchener and Grey and enthusiastically by A.J.B., [Balfour] old Jackie [Fisher] maintained an obstinate and ominous silence. He is always threatening to resign and writes an almost daily letter to Winston expressing his desire to return to the cultivation of his roses at Richmond. K. [Kitchener] has now taken up the role of conciliator, for which one might think that he was not naturally cut out."[117]

HERBERT ASQUITH, MP, BRITISH PRIME MINISTER.

Churchill recorded his version of events. Fisher had walked away from the table but had been talked round. Again.

> "Although the War Council had come to a decision with which I heartily agreed, and no voice had been raised against the naval plan, I thought I must come to a clear understanding with the First Sea Lord. I had noticed the incident of his leaving the table and Lord Kitchener following him to the window and arguing with him, and I did not know what was the upshot in his mind. After luncheon I asked him to come and see me in my room and we had a long talk. I strongly urged him not to turn back from the Dardanelles operation; and in the end, after a long and very friendly discussion which covered the whole Admiralty and naval position, he definitely consented to undertake it... We then repaired to the afternoon War Council meeting, Admiral Oliver, Chief of the Staff, coming with us, and I announced on behalf of the Admiralty, and with the agreement of Lord Fisher, that we had decided to undertake the task with which the War Council had charged us so urgently. This I took as the point of final decision... We had left the region of discussion and consultation, of balancings and misgivings. The matter had passed into the domain of action."[118]

WINSTON CHURCHILL, MP, FIRST LORD OF THE ADMIRALTY.

Asquith's recollection of the meeting went further.

> "I assert unhesitatingly that at this time the whole of our expert naval opinion was in favour of a naval operation. It is true that Lord Fisher disliked it. But his opinion, as he told me the same morning, was not based upon the technical or strategic demerits of a Dardanelles operation, but upon the fact that he preferred another and totally different objective in the Baltic. In that opinion, as Lord Kitchener appears to have pointed out to him, he was on the War Council in a minority of one. It was no doubt for that reason that he did not press his view. At the evening meeting the same day, in Lord Fisher's presence, and with his consent, it was announced that he had decided to undertake the operation."[119]

HERBERT ASQUITH, MP, BRITISH PRIME MINISTER.

---

117 Asquith, *Memories and Reflections*, vol. 2, p. 59.
118 Churchill, *The World Crisis 1915*, p. 165.
119 *Brisbane Courier* (Queensland), 4th August 1928.

EGYPT

One inexperienced British Indian Army officer recorded his routine along the Suez Canal and the death of a friend.

> "Our regiment (3rd Brahmans) has moved across the canal and some distance into the desert to make and occupy trenches ready to receive the Turks. I was able this morning, with the help of mule, boat and tram, to get to church in the town, where I heard Mass and received Holy Communion.

> "I have just ended my first week with my new regiment, and it has been one of the most enjoyable, and certainly the most interesting week of my life. Already I am picking up Hindustani quite easily; and you will be glad, I know, to hear that I have been given command of a company, which I did not expect till I could speak the language perfectly at any rate...

> "I am, of course, unable to give many details of my present work — mainly owing to its being active service — in fact, it looks like being very active indeed as the guns on our right have been very noisy to-day, and news has just come in, via the field telephone, of some action down on our right... Our hydroplanes did some rather cute bomb-dropping work over the enemy yesterday, which made them sit up.

> "As I am B.O. for the day (i.e. British Officer) I have to go round the outpost guards to-night, which means I cannot go to bed early, and so can devote the evening to writing to you... We are some distance away from the above town, and we are entrenched on the desert ready for our enemy, who are already beginning to make nuisances of themselves further down the line. Luckily, there is a convenient little shanty not too far from the trenches, one room of which we use as a mess — and so we are quite comfortable. But I must confess I hate this waiting — waiting — waiting.

> "You know the Patridges of Wimbledon. Young Patridge, who went out to Ceylon and was with the C.P.R.C. Contingent Company (i.e. Ceylon Planters Rifle Corps), took a Commission the same time as I did. He joined, however, as 'observer' in the Royal Flying Corps, and was attached to the French Aero Squadron working here. Last night he was taken out to make a reconnaissance. There must have been an accident, As they did not return to the headquarters, and this morning he and the pilot of the aeroplane were found dead not a mile from our trenches. Something evidently had gone wrong with the machine, obliging a descent on inconvenient ground when there was scarcely any light. He was buried this afternoon,[120] and it is nice to know that last Sunday he was at church with me and received Holy Communion. R.I.P."[121]

LT HENRY RAYMOND-BARKER, 3RD BRAHMANS, NEAR PORT SAID.

---

120   2/Lt Basil George Patridge, Indian Army Reserve of Officers, attd 2nd Queen Victoria's Own Rajput Light Infantry, was killed, mistaken for a Turk while returning to British lines on 27th January 1915. Buried in Port Said War Memorial Cemetery, he was the 23 year-old son of George E. T. and Florence Eva Patridge, of "Westleigh," Merton Park, Surrey.

121   *Wimbledon News*, 13th March 1915.

Troopships reaching Suez at this time had to barricade their decks.

> "At 4 a.m… we reached Suez. News had come along that the British troops had come in to contact with the Turks within five miles of the Canal: so you can guess we were all anxious to get along. Some of the New Zealand men were landed here. During the day the bridges of the boat were barricaded with sandbags. We on the Ulysses ran short of sand and used bags of flour and plates of iron."[122]
>
> PTE GEORGE CLAPHAM, 14TH BATTALION, AIF, HMAT *ULYSSES*.

> "We arrived at Port Said at 5 a.m., and anchored close to shore next to a French boat. Our bands tried to play the "Marseillaise," and failed so dismally that I felt ashamed of them. The French gave a rousing cheer… A French aeroplane and hydroplane have been flying over our boats all day, and the hydroplane came down and flew just clear of our masts."[123]
>
> PTE ARTHUR WATKINSON, 15TH BATTALION, AIF, HMAT *CERAMIC*.

29 JANUARY 1915

EGYPT

> "We were peacefully sailing up the Suez Canal on January 29th, when all of a sudden we heard the biggest roar of guns you could imagine. Bullets began to whistle over our ship. We were all ordered on deck, and our big guns started to speak. It was an awful cannonade. About 200 Turks had attacked us, but we soon scattered them. We did envy the New Zealanders being in the trenches. We had heard the day before we were attacked that Wairarapa boys had had a flutter with the Turks."
>
> PTE JOHN SWANSON, 8TH COY, AASC HMAT *CERAMIC*.

> "We entered the Canal at 9 a.m…, the *Ulysses* leading. At the entrance we passed the cruiser [sic] *Ocean*, and the tars gave three cheers for us as we went by. During our passage of the Canal we passed seven British and French gunboats and cruisers, and several of the mosquito fleet. As we steamed along we could see numbers of British and Indian troops. They were entrenched very strongly on either side of the Canal. I was told that outside the trenches there were land mines, spiked pits and barbed-wire entanglements. The latter, of course, we could see; and our officers gave us lectures on the different types of trenches and entanglements, as we went along. In the Bitter Lakes we passed a number of boats — some of them bound for Australia — and at four o'clock in the afternoon we anchored at Ismaila for the night."[124]
>
> PTE GEORGE CLAPHAM, 14TH BATTALION, AIF, HMAT *ULYSSES*.

---

122 *Gippsland Mercury* (Sale, Victoria), 13th April 1915.
123 *Warwick Examiner and Times* (Queensland), 22nd May 1915.
124 *Gippsland Mercury* (Sale, Victoria), 13th April 1915.

> "[T]he Australian and New Zealand troopships with the reinforcements passed through the canal, and we, standing on the banks, gave them great cheers, and they just about beat us in their hearty response. That same day the first submarine (*AE2*) I have ever seen went past, and really things did look busy while that thing was about."[125]

CPL ALAN SQUIRE, MG SECTION, WELLINGTON BN, NZEF, EL KUBRI.

30 JANUARY 1915

ENGLAND

Churchill was sent to liaise with the French about the plans for the Dardanelles.

> "We had a second meeting of the War Council before dinner on Thursday, and dispatched Winston to go and see French in order that both he and Joffre may realize the importance we attach to being able to send at any rate two divisions to help the Servians. Winston and Fisher have, for the time at any rate, patched up their differences, though Fisher is still a little uneasy about the Dardanelles."[126]

HERBERT ASQUITH, MP, BRITISH PRIME MINISTER.

FRANCE

> "Augagneur… showed me a confidential letter which Mr. Winston Churchill had written to him after the interviews he had with him, and the First Lord of the Admiralty undertakes to make no independent landing at Alexandretta. As to this Augagneur has invoked the spirit of our Agreement of 1912, but nevertheless he had some difficulty in obtaining his point; it may be that England, uneasy about the German and Turkish activities in Egypt, wishes to retain the right, if the moment should come, of making a dash at Alexandretta, or it may be that she specially hankers after that port. On the other hand Mr. Churchill has been emphatic as to France associating with herself in some degree with the operations which the Admiralty are diligently planning against the Dardanelles. Mr. Churchill, who asks that this matter should be kept a profound secret, says that the attack should take place about the 15th February, and that the Navy are prepared to dedicate, or really sacrifice, some old armoured ships… Augagneur, who has no great faith in the scheme, did not think it was his business to do anything to dissuade England, as it is England who is taking almost all the risk."[127]

RAYMOND POINCARÉ, PRESIDENT OF FRANCE.

---

125 *Hawera & Normanby Star* (New Zealand), 28th April 1915.
126 Asquith, *Memories and Reflections*, vol. 2, p. 60.
127 Poincaré, Raymond, *The Memoirs of Raymond Poincaré, President of France, 1915*, pp. 21–22, William Heineman Ltd. (London) 1926.

EGYPT

The Australian transport *Suevic* went aground during its passage of the Suez Canal.

> "We... managed to get off the mud by eight o'clock... Later on we passed Kantara, where the first shots of the Egyptian invasion were fired. This village is on the direct line of the caravan route from Palestine, via Ed Ariah, to Cairo. But the British force there was ready for all emergencies. There were a few Australians quartered here, and with characteristic irresponsibility they fired only one greeting at us, "Have you got any beer on board?"[128]

LT OLIVER HOGUE, 2ND AUSTRALIAN LIGHT HORSE BRIGADE, HMAT *SUEVIC*.

> "[A]s we were leaving, two aeroplanes came in sight, but went off across the desert. At Kantara, which was the point the enemy had attacked a couple of days before, we passed the *Swiftsure*, the flagship of the fleet in the Canal. She was working in conjunction with the infantry and the artillery. We also saw a large biplane at Kantara. Since we came through the Canal the Turks have made a general attack, and been thoroughly defeated. Of course, you will have read all about that in the papers... The Canal is a wonderful piece of work; and there are a large number of dredges constantly working in it, as the sand is always blowing in, and the wash of the boats cuts the banks away...

> "We reached Port Said at 3 p.m... and commenced to coal. Coaling is a very noisy operation here. Coolies carry the coal on their heads in baskets, and are singing all the time. We found it a good joke to throw a penny into a barge. The coolies would all drop their baskets, and there would be a great scramble for a few minutes, till the coin was found..."[129]

PTE GEORGE CLAPHAM, 14TH BATTALION, AIF, HMAT *ULYSSES*.

31 JANUARY 1915

ENGLAND

Asquith stressed the importance of drawing in the Balkan states and noted Churchill's enthusiasm.

> "Winston arrived soon after breakfast this morning, travelling, in his usual regal fashion, across the Channel in a Scout, and then in a special train... He had four interviews with Sir John French, and a pretty hard tussle. In the end they came to a sensible compromise... Of course he thinks... it is a pity to divide your forces instead of concentrating them. But this is one of the cases where policy overrides mere

---

128   *Nelson Evening Mail* (New Zealand), 25th March 1915.
129   *Gippsland Mercury* (Sale, Victoria), 13th April 1915.

strategy, and, if the effect of our stepping in to active assistance of Serbia is to bring in Rumania and Greece, it would mean in material terms an addition of perhaps 800,000 men to the Allies. Winston is for the moment as keen as mustard about his Dardanelles adventure."[130]

HERBERT ASQUITH, MP, BRITISH PRIME MINISTER.

EGYPT

The defenders of the Suez Canal were confident they could repel any attack upon Egypt. Others were conscious of divided sympathies within the country.

> "Everything during the past few months has been so very strenuous that it has been difficult to find time for anything outside military duties. One comes in after long field days, or desert days perhaps I should say, feeling dead beat with still other military duties to perform. Yesterday, we moved out of camp into barracks here at Abbasia [sic], which is about four miles from Heliopolis. It feels strange to sleep under a roof again after all these months, and the quarters here are so much more commodious than a tent that none of us will be very keen to go under canvas again if ever we are called upon to do so…

> "We have had some high winds which were extremely unpleasant, as one could not see for sand flying about, and one's food was gritty. It made it worse being in tents which cannot be closed so as to make them dust proof, and all our bedding and clothing got absolutely covered with sand. One day we carried out operations which lasted over 12 hours in a sandstorm. On the march back I walked two or three miles with my eyes practically closed, simply guiding myself by the tramp of the men's feet. For days afterwards men were going about with bloodshot eyes…

> "The papers report small skirmishes on the canal. When the news first came out there was great enthusiasm amongst the troops — prolonged cheering was heard in the lines at the thought of the chance of some fighting at last. The Turks will never pass the canal defences, but we hope to be taken from here to fight them in their own Empire — a modern crusade to Jerusalem or something of that sort. Our forces here, both British and Colonial are in such splendid condition that it would be a shame not to let us have a go at them somewhere. If I am not shot in the meantime I will write you again before long."[131]

LT GILBERT SPRAKE, 1/5TH BN EAST LANCASHIRE REGT, ABBASSIA.[132]

---

130 *The Brisbane Courier* (Queensland), 28th July 1928.
131 *Accrington Observer and Times*, 20th February 1915.
132 Sprake was killed in action on 4th June 1915. Believed to be buried in Twelve Tree Copse Cemetery, where he is commemorated by a special memorial, the former solicitor was the 28 year-old son of David and Isabella Sprake, of Whalley Road, Accrington, Lancashire.

"Of the population of Egypt (roughly) eleven million are Mahommedans and one million constitute the remainder, Christians (Koptics [sic], a form of Roman Catholic), for the greater part. Among the eleven million, naturally, there is great sympathy with Turkey, and there is a society in existence (decidedly anti-British) called the 'Nationalists,' whose ambition is to overthrow British sovereignty. Quite recently numbers of letters, urging an uprising, were stopped in the post office, and never reached their destination. Their meetings were forbidden, but the members started wearing black ties, but after a few had been punished in the police courts, the custom was stopped."[133]

L/SGT JOHN KERR, 5TH BATTALION, AIF, MENA CAMP.

INDIAN OCEAN

During its journey to patrol the German East African coast, HMAS *Pioneer* called in at the Cocos Islands. Some of the crew visited the wreck of the SMS *Emden*, finding some of its crew still aboard.

"We have had a fine trip. Our first port of call was Cocos Island, where we stayed all night and coaled ship and then we went to North Keeling Island to see the *Emden*. She is high and dry on the beach, only her stern is under water. To see her you would not believe anyone could have remained aboard her and lived. Her upper deck is a shambles, three funnels, her foremast and bridge are all blown down and her deck is full of shot holes. One of her battery guns was evidently hit by a shell from the *Sydney* and blown over the side only the pedestal remaining. There were also dead bodies aboard her. One man was lying in the ship's galley under a pile of pots and pans, another was staked to the ship's side by a big piece of broken steel-work — the smell was awful. One of our lads was pulling what looked like a good sea-boot from under a heap of rubbish and discovered a man's leg in it so of course he gave it up as a bad job.

"I got several curios in the shape of some post-cards written on in German and a big silver dollar... I have some nice shells and stones called cat's eyes, and have enough to make a bracelet."[134]

AB HERMANN STAHL, RAN, HMAS *PIONEER*.

---

133   *Cobden Times and Heytesbury Advertiser* (Victoria), 27th March 1915.
134   *The Sun* (Kalgoorlie, Western Australia), 2nd May 1915.

French and British warships off the Dardanelles.

FEBRUARY 1915

"We are very powerful, and I have no doubt we shall succeed."
Lt Edward Cromwell Colchester, RN, HMS *Irresistible*

1 FEBRUARY 1915

EGYPT

With the Turks having crossed Sinai, wild rumours circulated as its defenders were reinforced.

> "There is a lot of fighting on the canal. It is said to be… official that 400 Australians have been killed: they are infantry from Victoria and New South Wales. They shot two German officers here to-day. They were captured a couple of days ago with about 450 Turks and Bedouins. They were shot for inciting the Mohammedans to rise and fight against us. Many of the prisoners taken did not want to fight, and lots bear marks of bayonets in their backs showing that they were driven to fight."[1]

TPR ARTHUR ERNEST HOLT, 3RD AUSTRALIAN LIGHT HORSE REGT, MAADI CAMP.

> "Nos. 7 and 8 platoons received orders to be ready for the road by 2 o'clock. We were to take a blanket each and 24 hours' rations, our destination being the trenches over at Battery Point, across the canal from Ismalia, on the Arabia side… Once over the canal it is all sand, and the going was somewhat heavy. Battery Point is a very strongly-fortified position, and extends for some distance north and south of the crossing place. Our trenches, or rather the trenches which we were to occupy, were almost on the extreme right, or south, of the position, and the march over the sand along the banks of the canal was very trying, but this time for sure we never thought anything of the hardness of the ways, all our worries being as to whether we were at last to do something real."[2]

L/CPL GORDON KERENS, CANTERBURY BN, NZEF, BATTERY POINT.

2 FEBRUARY 1915

ENGLAND

> "I have been immensely impressed with the cumulative effect of the arguments presented in favour of military action in the Dardanelles at the earliest possible date.

> "These arguments have hitherto been presented singly at intervals in the discussions of three meetings. I feel bound to put them together in order to show how overwhelming a case they produce.

> "Here, then, is a summary: —

> "Russia is the nation upon which the Allies ought to rely mainly for the crushing blows required to bring about a decisive result. Russia has the largest reserves of men, and the eastern theatre of war gives sufficient space in which to employ great masses

---

1   *Barrier Miner* (Broken Hill, New South Wales), 24th October 1915.
2   *Otago Daily Times* (New Zealand), 6th May 1915.

without the inevitable siege warfare into which the fighting in the more limited western theatre degenerates.

"Russia, however, lacks warlike stores and equipment for her masses, and her needs can only be adequately supplied from Great Britain, France, and America.

"Hitherto these supplies have only trickled in through Archangel and Vladivostock. The former port is threatened with immediate closure by ice for some months, and the latter is at best an inadequate channel of supply.

"Unless the Dardanelles and Bosphorus can be opened up very soon, Russia will not receive the supplies indispensable to an offensive campaign, and the spring and summer may pass by without a decisive blow being struck at Germany or Austria.

"Hence the opening of a line of communications through the Dardanelles and Bosphorus is an indispensable military requirement in order to enable the Allies to take the offensive.

"In addition, there are two secondary strategical advantages to be obtained by operations in these waters if successful: —

"(1.) The attitude of the Balkan nations now hanging back will be definitely cleared up, and over 1,000,000 troops may be placed at the Allies' disposal (Roumania, 600,000; Bulgaria, 400,000; Greece, 200,000); more if Italy is drawn in.

"(2.) A line of communication up the Danube will be provided, bringing British sea power into the heart of the enemy's country, and enabling a British army, if desired, to operate against Austria....

"The reason why a military expedition is mentioned is that, as pointed out by the First Lord at the Sub-Committee, the navy can perhaps open the Dardanelles and Bosphorus to warships, which are more or less impervious to field gun and rifle fire, but they cannot open these channels to merchant ships so long as the enemy is in possession of the shores."[3]

LT-COL. MAURICE HANKEY, SECRETARY, WAR COUNCIL.

EGYPT

The Ottoman attack across the Suez Canal that afternoon was resisted but brought to a halt by a sandstorm.

"About 9 o'clock [a.m.], while the scouts were out, the Turks started shelling us with two big guns. My word, that sent a funny feeling through one! We could hear the boom of both guns, then the whistle of the shell, wondering all this time just where it is going to land. Luckily for us, they never got the range of our trenches, although two shells hit the gunboat, quite a number landed in the Canal, and about five landed fair and square into the signal station of Elferdan [sic], as close as 200 yards away from us. However, about 10 o'clock they stopped as suddenly as they had started, and that's

---

3 Committee of Imperial Defence. The War. Attack on the Dardanelles. Note by the Secretary. 2nd February 1915. TNA CAB 42/1/30.

the last we heard of them at Elferdan, with the exception of about ten prisoners that gave themselves up to us."[4]

PTE LEONARD WILLIAM HEMMINGS, CANTERBURY BN, NZEF, EL FERDAN.

"It is less than two hours since we were in action against the Turks. Just as we were going to have tea, a party of Turks was observed at about 4,000 yards' range. The time was 5.25 p.m. We opened fire on them forcing them to retreat hurriedly away from it. We have Ghurkas under our protection; also men from Punjab. We had only just enough light to be able to see… It has been blowing a sandstorm today here."[5]

CPL WILLIAM WRIGHT, 20TH BATTERY, 1/3TH (EAST LANCASHIRE) BDE,
ROYAL FIELD ARTILLERY, EAST LANCASHIRE DIVISION.

Though some had only recently arrived in camp, rumours about future action extended beyond Suez.

"By what I hear we might be going to the Dardanelles to tackle the Turks, and, if so, we will do our best when we get there. We do a lot of night drill, going out at dark and marching for miles over the desert, then have a fight at daybreak. There were 30,000 took part in one the other night, and it makes you pretty tired by the time you get back to camp; one only thinks of a good sleep."[6]

PTE ROBERT YOUNG, 14TH BATTALION, AIF, AERODROME CAMP, HELIOPOLIS.[7]

3 FEBRUARY 1915

EGYPT

Before dawn, around 3.00 a.m., the Turks brought their pontoons to the banks of the canal south of Toussoum.

"About 3.15 a.m. an English officer in charge of a Punjab Company, came up to me in a great state. He saw the enemy had brought boats and were crowding into them — in fact they were making a bold bid to cross the Canal. He asked me to go to his assistance.

"We were only too keen to get into it, so I sent a message to Major Brereton and took my platoon to assist the Punjabs. We got well down on the Canal side, took good cover and opened fire. There seemed to be hundreds and hundreds of them close to the bank, and we simply poured our fire into them. We must have riddled their boat, as one sank just in front of us crowded with Turks. There were swarms of them on the opposite bank, and although we could not see them we kept up a heavy rifle fire

---

4   *The Feilding Star* (New Zealand), 10th April 1915.
5   *Farnworth Chronicle*, 20th February 1915.
6   *Seymour Express and Goulburn Valley, Avenel, Graytown, Nagambie, Tallarook and Yea Advertiser* (Victoria), 21st May 1915.
7   Young, a former labourer, died of wounds at sea on 29th April 1915. Commemorated on the Lone Pine Memorial, he was the 31 year-old son of Emily Young, of Avenel, Victoria.

which must have done great damage. Fortunately for us some of our fire must have silenced one of their machine guns, as it ceased."[8]

2/LT ALEXANDER FORSYTHE, CANTERBURY BN, NZEF, TOUSSOUM.

"The orders were that we were to let them fill the boats and then open fire, and as fast as we could fire we did. No sooner was the boat of men wiped out than another lot would take their places. They tried this for over an hour, and then gave it up as the boats were no longer of any use; they were simply riddled with holes. The enemy then retired and took up positions along the bank in the sand hills, from which they poured a heavy fire into our bank, fortunately doing very little damage.

"When daylight came things became very interesting indeed. All our fellows showed wonderful coolness, and did some really good shooting with deadly effect. We could see Turks in every direction, some scattered about, and some in heavy bodies marching about taking up good positions. It was at this time that we saw the sight of our lives. Our artillery started in on them with shrapnel, and simply mowed the Turks down by scores. We were quite excited by now, and we went mad with delight at seeing those things happen, so we stood up and cheered wildly, although the bullets were pretty thick about us, and some very near. No one was hit, though, so no one minded."[9]

LT HARRY SAUNDERS, CANTERBURY BN, NZEF, TOUSSOUM.

The Lancashire Territorial artillerymen did indeed find some excellent targets.

"When it became lighter we thought we could make out the enemy crossing the Canal in boats, so we dragged our gun out of its pit and put it half way on to the bank, about 10 yards from the top, and opened fire nearly parallel with the canal. As it grew lighter we could see the enemy in all directions all over our left flank and front, and the boat loads going across. One of our maxims by this time had got on them as well, and their game was soon put a stop to. Later a torpedo destroyer sank the boats. By then we had got another gun out, and the major took on a reserve party hiding behind a crest. He soon put a shell into the middle of them and they fled in all directions. I also took them on, and between us we knocked them about. While this was going on ——— was up a tree directing the fire of our two other guns at the enemy in front, and kept them quiet. These were the real Turkish troops in blue uniforms, but those the major and I were engaged with were Syrians. While this was going on we were being sniped at, but fortunately none of our chaps was hit.

"We then saw… the Major [go] forward to his observation station. He spotted some trenches about 1,500 yards in front, which had been dug in the night and we opened fire on them. The enemy got a field gun to work, and all of a sudden four shells burst in front of us. We got four more rounds off and then came the most awful minute I have ever had. The enemy got our range and put four shots right into our guns, practically one on each gun. When the fourth shell had exploded just in front of No.

---

8   *Nelson Evening Mail* (New Zealand), 8th April 1915
9   *Otago Witness* (New Zealand), 31st March 1915.

2 gun, and spreading its bullets through our gun detachments like the three others, I thought everybody had been killed.

"I saw the layer on one of my guns fall over with a yell, and roll about. I thought he had been hit in the stomach, but it seems a bullet caught his helmet and went down his neck into his shoulder He then recovered his senses and ran to the tents. One of —'s gunners was caught with a bullet in the lungs, another was wounded in the ankle, and there was another badly hit. The remainder stuck to the guns jolly well, and about half a minute after this happened we got four more rounds off. The enemy's guns then went off us. Why, nobody can tell, but we were glad…

"Our four guns altogether fired about 320 rounds, so we all went to bed with awful headaches."[10]

LT THOMAS PARKE, 19TH BATTERY, 1/3RD (EAST LANCASHIRE) BDE, ROYAL FIELD ARTILLERY, TOUSSOUM.

The infantry also came under artillery fire before the attack fizzled out in the afternoon but not before the New Zealanders suffered their first battlefield fatal casualty.

"Up till this point we had had no shell fire directed at us; but we had not long to wait, and when they started it was hellish… Sergeant-major Williams,[11] who has seen a good deal of service, was sitting with me in my trench when a shell burst over us, and the base of the shell flew back and got him in the shoulder, inflicting a very severe cut in the fleshy part of the shoulder. We got him bandaged up, and sent him back to the hospital… You remember how the look of blood finishes me usually. Well, this chap was smothered in blood, and yet I was able to fix him up without feeling the slightest bit concerned. I must say I felt a bit rattled when he had been sent away, but it soon wore off."[12]

LT HARRY SAUNDERS, CANTERBURY BN, NZEF, TOUSSOUM.

"About 3 o'clock an order came for us to retire on headquarters. It was easier said than done, as we had to cross over an open space under fire. Two Punjabs were killed right away, and we did not feel very keen. However, we crossed in small groups, with the bullets whistling all round.

"I crossed with three men at the double. About three-quarters way across Ham, who was in front of me, dropped, shot in the neck, and Fowler, who came after me, got one through his hat. Ham we found aftewads [sic] got one bullet in the rifle also, so they were very plentiful. Ham was groaning badly, so I crept back and made him as comfortable as possible. He could speak and said he had lost the power of his arms and

10 *Lancashire Evening Post*, 12th March 1915.
11 Sgt Alfred Williams, Canterbury Bn, NZEF, was wounded in action on 3rd February 1915.
    He was a veteran of the Boer War, serving with the Rifle Brigade, and as a mounted infantryman in Somaliland. Later commissioned, Captain Alfred John Williams, 2nd Bn Wellington Regt, was killed in action on 1st October 1918. Buried in Flesquieres Hill British Cemetery, France, he was the 39 year-old son of Alfred Thomas and Jemima Williams, of 68 Pulford Street, Pimlico, London.
12 *Otago Witness* (New Zealand), 31st March 1915.

legs, so I knew the bullet must have lodged in the spinal column, and so paralysed him.[13] Shortly after the native stretcher bearers went out and brought him in."[14]

2/LT ALEXANDER FORSYTHE, CANTERBURY BN, NZEF, TOUSSOUM.

"I have not been on sick parade since we arrived in Egypt, but all cannot stand the same hard work. Lads who have worked in offices often fall out on our long marches. We are now having sandstorms during the day. The dust and sand are so thick that you cannot see the next man to you when marching, though he is only three feet away. It cuts the eyes, and the wind blows you to pieces. And the flies! It is a good thing we are going soon. They could not keep us here. A number of our brigades have already left, and we have news to-day that the 3rd Brigade is fighting somewhere in Turkey, and we have received orders to be ready to move off soon. We have been served with caps similar to those worn by the Territorials, but we will not wear them until we get to a cold place… Lately we have been having reveille at 3 a.m., owing to the heat, and before the wind rises, doing a 17 mile march on a mess tin of tea and four rounds of bread and jam. I am sure it is the Irish stew we get at night that keeps up our strength."[15]

PTE JULIUS CALMAN, 1ST BATTALION, AIF, MENA CAMP.[16]

4 FEBRUARY 1915

ENGLAND

Cabinet Secretary, Maurice Hankey was concerned about the plan to attack the Dardanelles with an unsupported naval force. In January he had circulated the 1906 feasibility study but to little apparent effect. He, therefore, commissioned Sir Julian Corbett to provide a commentary about previous attempts.

"The leading case, of course, is Duckworth's in 1807 — not pleasant reading, since it is the most conspicuous of the many failures of the "All the Talents" administration in war direction.

"The object was to assist our ally Russia, who was bearing the brunt of the struggle against Napoleon… To this end Duckworth was given… instructions that, if Turkey did not desist from hostilities with Russia, he was to demand the surrender of her fleet, and if they did not agree to this in an hour he was to take up a position which would enable him to destroy Constantinople and the fleet in the Golden Horn…

---

13   Pte William Ham, Canterbury Bn, NZEF, was severely wounded on 3rd February 1915, dying three days later. Buried in Ismailia War Memorial Cemetery, the former labourer was the 22 year-old son of William Ernest Ham, of Motueka, Nelson, New Zealand. He was the first NZEF battle casualty.
14   *Nelson Evening Mail* (New Zealand), 8th April 1915.
15   *Clarence and Richmond Examiner* (Grafton, New South Wales), 17th April 1915.
16   Evacuated from Gallipoli with diarrhoea on 27th October 1915, he was wounded on the Somme on 6th August 1916; and killed in action with 45th Bn in Belgium on 13th October 1917. Buried in Passchendaele New British Cemetery, he was the 21 year-old son of George and Alice Elizabeth Calman, of South Grafton, New South Wales.

> "Sir John Moore… at once put his finger on the weak point of the design. So strongly did he feel about separating the troops from the fleet… that General Fox sent him to meet Duckworth at Malta… "My opinion is," Moore wrote in his Diary, "It would have been well at present to have sent 7,000 or 8,000 men with the fleet to Constantinople, which would have secured their passage through the Dardanelles and enabled the Admiral to destroy the Turkish fleet and arsenal."…
>
> "The result was just what Moore anticipated… the Turks [turned] a deaf ear to Duckworth's overtures and menaces and [strengthened] their fortifications… not only at Constantinople, but also at the Dardanelles, till Duckworth… was forced to retired before his retreat was entirely cut off. As it was, he was only just in time; in repassing the new batteries his fleet suffered severely.
>
> "The result was a very severe blow to our prestige in the Near East, and, what was worse, our relations with Russia were poisoned. So far from strengthening the alliance, as was intended, the Czar began to give us up in despair, and a few months later signed the Peace of Tilsit with Napoleon."[17]
>
> JULIAN S. CORBETT, NAVAL HISTORIAN.

EGYPT

Periodic fighting continued along the Suez Canal but the threat had been staved off.

> "We… steamed slowly up the canal not far from Ismalia and we again encountered the Turks who had advanced right to the canal banks, and for an hour it was very lively. We fired shrapnel and canister at point-blank range, and the execution was frightful.
>
> "I am sorry to say that one of my mess-mates was shot and died an hour afterward. He was a Chief Yeoman of Signals.[18] We were joking about the possibility of it happening the night previous at supper time, but little did he think it was the last meal he was going to have. He leaves a wife and three children."[19]
>
> ERA3 HAROLD FORD, RN, HMS *SWIFTSURE*.

> "In our advance on the 4th we captured 80 camels and 40 Turks. The Turks, in numbers of 1 to 20, have been giving themselves up footsore, hungry, thirsty, and very glad to get in. Along the [Suez] Canal the enemy's loss was 4,000 [sic], and ours were 2 killed and 10 wounded."[20]
>
> PTE EDWIN VINCENT, CANTERBURY BN, NZEF, TOUSSOUM.

---

17 Committee of Imperial Defence. The War. The Dardanelles. Letter from Mr Julian Corbett to Colonel Hankey, 5th February 1915. TNA CAB 42/1/32.
18 Chief Yeoman of Signals Samuel John Smith, RN, HMS *Swiftsure*, was killed in action on 4th February 1915. Buried in Ismailia War Memorial Cemetery, he was the 36 year-old husband of Emily Elizabeth Smith, of Hazelwood, Hempstead, Chatham.
19 *The Orcadian*, 20th March 1915.
20 *Wells Journal*, 5th March 1915.

5 FEBRUARY 1915

ENGLAND

Training continued, as it continued everywhere, but occasionally enlivened by close harmony singing.

> "Don't think we live a thrilling life here. Much of our work is monotonous. I have lost all the pleasure I once had in shouting "At the halt! On the left! Form Platoon!" or in the novelty of being instantly obeyed and telling people to do things without saying "please." But there are occasions when it is all immense, to march out in front of your men and to hear behind you the wonderfully tuneful things they sing; to catch a glimpse of a soldier at church parade wholly absorbed over singing "Abide with me," and to think of the marvellous change in his life."[21]

2/LT JOHN ALLEN, 13TH (RESERVE) BN WORCESTERSHIRE REGT, LOOE, CORNWALL.

EGYPT

As the Turks pulled back from the Canal, men took time to record their recent experiences, while some made cautious incursions on the eastern side of its banks.

> "We had no casualties aboard of us. We silenced the enemy's guns after about two days fighting... We are just waiting for another smack at them. I hope they won't be long in coming again. I am sorry to say they are giving themselves up in bunches, about a dozen at a time; but there are still some of them giving trouble whom we can't get at. They are attacking about three places at once. Of course we have got plenty of ships here to see them off without the soldiers. We got two shells aboard of us,[22] but they did not do much damage, and soon made the necessary repairs to the ship."[23]

CARPENTER'S CREW THOMAS PERCIVAL, RN, HMS *CLIO*, SUEZ CANAL.

> "[A] reconnoitring party consisting of two of our platoons, the Indian Lancers, a company of Ghurkas, a part of the Camel Corps, and some of the Medical Corps went out to look for the enemy, who were believed to be entrenching on a hill about seven miles out. After a good deal of scouting and skirmishing they eventually came across about 50 Turks, who streaked for their lives when they saw us after them, but not before they had shot one of our Lancers through the head.[24] Our battery then got to work and shelled the hill."[25]

SGT WILLIAM OKEY, WELLINGTON BN, NZEF, EL KUBRI.

---

21   John Allen quoted in Montgomery, Ina, *John Hugh Allen of the Gallant Company. A Memoir*, p. 155, Edward Arnold (London) 1919.
22   The *Clio* was hit twice on the morning of 3rd February 1915.
23   *Port Glasgow Express*, 26th February 1915.
24   Sowar Muinuddin Khan, 1st Hyderabad Lancers, was killed in action on 5th February 1915. Commemorated on the Heliopolis (Port Tewfik) Memorial, he was the son of Husain Ali Khan, of Hyderabad, Deccan.
25   *Taranaki Herald* (New Zealand), 7th April 1915.

"Among the dead we found a German major.[26] Searching him, we found in his wallet a white flag, fitted with hooks ready to affix to either sword or bayonet. He is buried by himself, a wooden cross denoting the sport. The Turks were buried in heaps in the banks in the sand."[27]

PTE WALTER HOLLAND, 2ND (EAST LANCASHIRE) FIELD AMBULANCE, RAMC, EAST LANCASHIRE DIVISION.

The Australians, some of whom were unimpressed at missing out in the fighting at Suez, got a glimpse of the British commander of the still forming Australian & New Zealand Army Corps.

"Major-General Birdwood, of the British [Indian] army, and head over all the Australian troops in Egypt, has had a walk of inspection through the camp, and his few remarks concerning the men and horses were complimentary. He has none of the British officer's hauteur, of which I have read so much. Most Australians will work very hard and very faithfully for a good boss."[28]

TPR CLIFFORD HALLORAN, 6TH AUSTRALIAN LIGHT HORSE REGT, MAADI CAMP.

6 FEBRUARY 1915

ENGLAND

One American embassy official, soon to take up a post in Istanbul, took the opportunity to visit a German prisoner of war camp in southern England.

"I was yesterday able to visit the German prisoners at Southend both civilian and military, and to take away with me the most favourable impression. I was especially struck by the friendly relations existing between them and the English officers on duty, and the efforts by the latter when they properly could, to mitigate by human kindness a captivity which must necessarily be irksome. I spoke to many prisoners in German, and did not hear from them a single grievance though they had ample opportunity. It will be an agreeable duty for me to reassure our Turkish friends regarding any possible misapprehension they may entertain of the treatment of prisoners in England."[29]

LEWIS EINSTEIN, AMERICAN DIPLOMAT.

FRANCE & FLANDERS

As the momentum for a new campaign in the eastern Mediterranean developed, others began to express their concerns.

26   Major Wilhelm Karl von dem Hagen was killed on 3rd February 1915. He was buried by British troops two days later but now lies in Ismailia War Memorial Cemetery.
27   *Cheshire Observer*, 27th March 1915.
28   *The Sun* (Sydney, New South Wales), 21st March 1915.
29   Letter from Lewis Einstein to Sir Edward Grey, 6th February 1915. TNA CAB 800/107.

"Lloyd George and Montagu were here to-day on their way back from Paris. I understand there is still [a] question of sending troops to the Balkans. I find it very difficult to understand why the appearance of British and French soldiers in that part of the world should have so great an influence, and, unless something very decisive in that way will be gained by such a move, it appears to me to be a strategical mistake."[30]

FIELD-MARSHAL SIR JOHN FRENCH, C-IN-C, BRITISH EXPEDITIONARY FORCE.

EGYPT

Meanwhile, some detected a lack of enthusiasm for the daily routine of life in the camps.

"As I have not much news to tell you I will write a few lines on the day's work of our troops at Mena Camp, from the time we rise, until we go to bed again — that is, if the censor chap will be kind enough not to cut it out. Well, after a well earned night's rest, we are awakened in the morning at 6 o'clock by the melodious voice of our sergeant shouting, "Come on, lads! The reveille has gone; turn out!" Nobody seems to take any notice of him, but they all turn over and pull the blankets over their heads. Then after a few minutes we hear him bellow again. "Come on, you chaps, show a leg, the reveille has gone half an hour ago." A couple of us will peep out from under the blankets and rub our eyes. Someone will say, "By ———, ain't it cold? It is a cow to have to turn out at this time of the day." Then a corporal will put his head in the tent and shout, "Any sick here?" (he generally gets a couple from our tent). After a lot of grumbling we all turn out, and roll up our blankets, and get a wash in a "basin" which, I might say, is an old meat tin, and had to do 12 of us. Next we hear is the Non-Com. shouting, "Fall in, No. — Platoon." This generally takes about a quarter of an hour longer than it ought to, as we are all moping around, as if we can't help it. It is for roll call. After it is over, "cook-house door" blows, and we all line down to our mess houses for breakfast (which is not bacon and eggs but the same old stew). After breakfast we all get on to cleaning our rifles, and having a shave, which we are supposed to do every day (but don't). Then the bugle will blow, "Fall in." This generally takes us about ten minutes more than it ought to do, as some are only half way through a shave; others have not yet their rifles finished, or are away getting a bottle of water. At last we fall in, and have to be inspected. Some have to go before the skipper for not shaving. We are then marched on to the parade ground, the Colonel rides up on his gallant steed and shouts, "Battalion, shun!" Every man springs to it, and stands like a statue, then someone will put his hand up to knock a fly out of his eye, and the Colonel will go "crook" on him for moving. After he has had a good look at us, we get, "Form fours, right, quick march," and out we go about four miles through the sand to our parade ground for the day's drill, which consists of a lot of different items, such as an attack on an enemy that isn't there. Then we do a bit of bayonet fighting, rifle exercises, signalling, and lots of other things. At dinner time plenty of [Egyptians] come around selling tomatoes, cooked eggs, oranges and lots of other rubbish, which we all spend

---

30   Sir John French quoted in Arthur, Sir George, *Life of Lord Kitchener*, vol. 3, p. 108n., MacMillan & Co., (London) 1920.

our piastres on. We get back to camp about four o'clock or a little after. When we are dismissed, there is a dive for the cook-house for a dixie of tea, or up to the canteen for a pot of beer — that is, if we have the money to pay for it, for the proprietor won't give "tick." At 5 o'clock we have tea — more stew — but we lay into it, as we are generally pretty hungry at teatime. After tea one of us buys a paper ("The Egyptian Times"), and we discuss the war news. Some of us go to the pictures to put in an hour or so, but not very often, as the pictures are not up to much, and they show the same programme for about a week before they get another lot of films. We generally turn in about nine o 'clock."[31]

PTE ROBERT ROY GENNOE, 11TH BATTALION, AIF, MENA CAMP.[32]

7 FEBRUARY 1915

FRANCE

One senior British politician was looking beyond the fighting to the post-war settlement; an early indication of where Lloyd George's attention lay.

> "The French are very anxious to be represented in the [Mediterranean] expeditionary force. Briand thinks it desirable from the point of view of a final settlement that France and England should establish a right to a voice in the settlement of the Balkans by having a force there. He does not want Russia to feel that she alone is the arbiter if the fate of the Balkan peoples."[33]

DAVID LLOYD GEORGE, CHANCELLOR OF THE EXCHEQUER, PARIS.

EGYPT

The first of many similar letters to grieving families back in New Zealand described the death and burial of Pte William Ham. Others were happy to have the opportunity to write home.

> "We had been under heavy fire at very short range from 3 a.m. until 2 p.m., when he was unfortunately struck by a bullet in the neck. C.W. Forsyth did what he could for him, and immediately the stretcher bearers took him away. He received every attention possible, and was sent by train to Ismailia to the hospital there. I saw him on Friday, 5th, and he was very well cared for and everyone was most kind, and his nurse (a Melbourne lady) told me he was very good and patient. His case was hopeless from the first, but he suffered no pain. The funeral, at which the company, General ———, and many other high officers were present, took place this morning (Sunday) in the European cemetery here. I hope you will realise my sympathy for

---

31 *The Swan Express* (Midland Junction, Western Australia), 19th March 1915.
32 Gennoe, a former barber, was awarded the Military Medal (*London Gazette*, 16th November 1916) and was killed in action on 6th April 1917. He is buried in Morchies Australian Cemetery, France.
33 Lloyd George, *War Memoirs*, vol. 1, p. 243, Odhams Press Ltd. (London) 1938.

you all, and I feel his loss myself very much. It is a heavy responsibility for me having the lives of so many in a sense in my care. But you have some comfort that his death was in a good cause, and that we inflicted very heavy loss on the enemy and severely defeated them."[34]

MAJOR CYPRIAN BRERETON, CANTERBURY BN, NZEF, ISMAILIA.[35]

"It is two months now since we landed, and from all indications we shall be here another six weeks or two months yet. We have not had any fighting to do so far, although the Turks are being soundly thrashed a few miles away… We have been worked quite hard and have not been free on a good many evenings, when we are, we are generally too weary for letter writing. However, the bosses are slackening off a bit now and we are to be allowed more holidays. The task of carrying our heavy equipment so many miles a day over sand was too much for some of our lads so the authorities had to go a bit easier."[36]

PTE JOHN BRADFORD, 10TH BATTALION, AIF, MENA CAMP.[37]

8 FEBRUARY 1915

ENGLAND

Leading British and French politicians hoped to galvanise support in the Balkans by sending two divisions to Salonika, as the former, in particular, continued to bypass their General Staff.

"I had a rather interesting luncheon at Edward Grey's — Delcasse, Cambon, Kitchener, and Winston. Winston was very eloquent in the worst French anyone ever heard… We all agreed that the Serbian case is urgent and that we must promise to send them two divisions, one English and one French, as soon as may be to Salonika and force in the Greeks and the Roumanians. We must try to get the Russians to join if possible with a corps. Lloyd George told us he had got Sir John French's assent to this, but I have told K. [Kitchener] to send for him, and he is coming over tonight in one of Winston's destroyers."[38]

HERBERT ASQUITH, MP, BRITISH PRIME MINISTER.

EGYPT

A tentative probe beyond the banks of the Suez Canal found nothing but the detritus of a retreating army.

---

34   *The Colonist* (New Zealand), 29th March 1915.
35   Brereton was wounded at Gallipoli on 8th May 1915.
36   *West Coast Reporter* (Port Lincoln, South Australia), 7th April 1915.
37   Bradford was wounded on 27th April 1915. Returned to Australia on 26th November 1915, the former journalist was discharged as no longer physically fit for service on 10th March 1916.
38   *The Daily News* (Perth, Western Australia), 4th August 1928.

"At 2.30 a.m. the exciting moment arrived, the reveille sounded, and we hurriedly saddled up. We paraded and rode swiftly towards the canal, where we found a large Indian camp and two pontoon bridges... I tell you, with our horses under us, we had just a lovely sort of feeling. Four Maxim guns also went with us. We were joined at the other side by about 800 Mysore Lancers, splendidly mounted and fine-looking chaps they were, too, with about fifty camels. This all sounds exciting, but as a matter of fact we did no fighting: I'm simply describing what we experienced.

"Well, we soon got into the hostile country, scattering the enemy's scouts everywhere, so as to prevent any surprises, and went on for about 17 miles. Traces of the enemy were to be found everywhere, they having been there that morning. About 12.45 we halted to feed our horses, as we could not make the gees go further, for the country is very heavy deep sand. After a bit we returned, following the track the Turks took, and found all sorts of things — six dead camels, bits of clothing of all descriptions, tins, cartridge cases, and even unexpended cartridges. One shirt we found was absolutely saturated with blood. Everything seemed to shew the terrible state the enemy were in during their hasty retreat. All very interesting, but we were awfully disappointed that we were robbed of having a go at them... Still, to their credit let it be said, they did jolly well in getting across the desert, there being only one well on their way."[39]

PTE CHARLES MOORE, 2ND COUNTY OF LONDON
YEOMANRY (WESTMINSTER DRAGOONS).

AUSTRALIA

"At 6 a.m... we left Broadmeadows for the railway station in marching order, and carrying our sea kit bags, with changes of linen for the voyage. We had a warm march to the station of about 1½ miles, and entrained for North Melbourne. On arrival we detrained and marched to Victoria Docks and embarked on the troopship Pera, bound for Alexandria. We were allotted our troop deck and and a pannican of tea, and some bully beef and bread. We then stowed our kits and accoutrements in their places and settled down to work. I was appointed Q.M.S. for the voyage. My duty was to appoint mess-orderlies and tell off the proper number to each mess table. Then I had to draw from the store all utensils required for each mess table, which consists of 1 dixie with lid, 1 bucket, 1 tub, 2 meat dishes, 2 salt, pepper and mustard pots, 1 pickle jar, and 1 butter dish for each table; one plate, pannican, knife fork and spoon for each man. Then we drew our day ration, which consists of tea, sugar, mustard, salt and pepper, pickles and jam from the ship's stores. Butter we drew from the cold stores and the remaining food from the galley where it is cooked by the cook. Our breakfast consists of porridge and stew or chops and tea. Our dinner consists of soup, meat, and vegetables and pudding. The tea is composed of meat, bread and jam and tea; butter is on for all meals. The food on board is far better than that we had while in camp. While the orderly and myself were attending to the equipments for mess table, our mates were kept busy loading horses, and when they had to be led down

39 *Eltham & District Times*, 26th February 1915.

three decks it means work for brain and hands, for a horse sometimes objects to be led down a steep place, but after your mates use a little persuasion behind, he follows you at a run, and you have to travel to keep out of his road. When you have your horse on the right deck you put him in his stall and tie him up, which is different to tying a horse on land. We finished loading horses about 1 p.m. Then the boat hands prepared to cast off. No one was allowed on shore and the public had to keep well back from the ship until the last 15 minutes. Then they were allowed close to the ship side and a few farewells were spoken and not a few handkerchiefs were wet. We soon felt the ship moving, and as we gradually drew away. We watched the fluttering handkerchiefs, and thought of those near and dear to us. We all kept on decks and looked at the Victorian capital fading from sight, and thinking of how long it would be before we would see it again. We passed on down the bay and out of the heads with a smooth sea and then we saw the Victorian shore fall away and disappear from sight, and we were on the high seas bond for active service in foreign lands."[40]

CPL WILLIAM HOWELL, 8TH AUSTRALIAN LIGHT HORSE REGT, HMAT *PERA*.[41]

9 FEBRUARY 1915

ENGLAND

As politicians imagined great things from their great 'experiment' at least one of their professional advisors despaired of it all.

> "We have a War Council at No. 10, which Sir John French has come over to attend. The main question, of course, will be how soon, and in what form, we are to come to the aid of Serbia and whether and how far the French and Russians will join in. French will no doubt kick even at a single division being abstracted from his force, but he must be made to acquiesce in this. The two danger points at this moment are Serbia and Mesopotamia… I cannot help feeling that the whole situation in the Near East may be vitally transformed if the bombardment of the Dardanelles by our ships next week goes well. It is a great experiment."[42]

HERBERT ASQUITH, MP, BRITISH PRIME MINISTER.

> "Oliver told me a few days ago that Fisher does nothing. Winston proposes mad things & hangs grimly on to his silly Naval Division… Wilson opposes all suggestions made by anybody except himself… Winston, very, very ignorant, believes he can capture the Dardanelles without troops & that Borkum can be destroyed by bombardment. Strange fallacies!… It is hopeless trying to make war with men like these."[43]

CAPT. HERBERT RICHMOND, RN, ASSISTANT DIRECTOR OF OPERATIONS, ADMIRALTY.

---

40   *Stawell News and Pleasant Creek Chronicle* (Victoria), 29th April 1915.
41   Howell landed on the peninsula as a reinforcement to his regiment on 25th October but was invalided sick on 8th December. He returned to Australia, leaving Suez on 29th January 1916.
42   *The Brisbane Courier* (Queensland), 28th July 1928.
43   Marder, *Portrait of an Admiral: The Life and Papers of Sir Herbert Richmond*, p. 140.

EGYPT

Commanding officers took the chance to reflect on their experiences in the recent fighting.

"The Brigade has been at last in action, and is, I believe, the first Territorial Artillery to have that honour… There are no casualties in the 18th and 20th Batteries, and only four in the 19th Battery,[44] and they are all doing well and in no danger. Sergeant Gallagher[45] and Acting-Bombardier Hart[46] were the most serious. I have seen them in hospital, and they are quite comfortable. Gunner Jolley[47] and Gunner Morris[48] were the other two, and they were only slightly touched and are about again. They are all very pleased with themselves at being the first wounded gunners."[49]

LT-COL. CHARLES WALKER, 1/3RD (EAST LANCASHIRE) BDE, RFA, EAST LANCASHIRE DIV.

"The bulk of the fighting was done by the Indian troops, who fought with great coolness, and they were ably supported by some English Territorial guns and [an Egyptian] battery. The warships on the canal also did much damage and were greatly dreaded. The enemy took rather a bad knock all along the line, his total casualties being estimated at 3000 to 4000, including nearly 1000 prisoners and 600 odd dead buried by us. The Turks fell back, and a reconnaissance of six cavalry regiments and a camel corps (Colonel Johnston and I went out with them, although it went out over twenty miles) failed to find the enemy, two prisoners and two camels being the whole bag. Anyhow, we had a very pleasant ride."[50]

LT-COL. DOUGLAS MACBEAN STEWART, CANTERBURY BN, NZEF, ZEITOUN CAMP.[51]

10 FEBRUARY 1915

ENGLAND

Asquith was frustrated at news of a delay to the naval bombardment of the Dardanelles. He viewed it as the overture to a much wider Balkan campaign.

---

44 In addition to the four men wounded named below, Gnr. Lewis Grundy suffered from shock on 3rd February 1915. He was a cotton mule assistant; the son of John and Martha Grundy, of 17 Bee Fold Lane, Atherton, Lancashire.

45 Sgt John Gallagher was wounded on 3rd February 1915 and discharged as a result on 16th July 1915. He died on 15th June 1916. Buried in Bolton (Tonge) Cemetery, he was the 34 year-old husband of Annie Gallagher, of 22 Kirkby Road, Bolton, Lancashire.

46 Bdr Tom Hart was wounded on 3rd February 1915. Promoted Sgt, he was killed in action on 28th March 1918 serving with "B" Battery, 211th Brigade, Royal Field Artillery. Buried in Bienvillers Military Cemetery, he was the 24 year-old son of Thomas and Mary Hart, of 59 Harper's Lane, Bolton, Lancashire.

47 Gnr Ellis Jolley was wounded on 3rd February 1915. He transferred to the Royal Air Force on 20th April 1918 as a 2/Lt, posted to 228 Squadron on 14th November 1918. He left the service on 30th April 1919; the son of Isabella Jolley, of 57 Beaconsfield Street, Bolton, Lancashire.

48 Gnr Walter Morris was wounded on 3rd February 1915. He was a belt stretcher, son of James and Betty Morris, 9 Hey Club Row, Ainsworth, Lancashire.

49 *Farnworth Chronicle*, 27th February 1915.

50 *Lyttelton Times* (New Zealand), 23rd April 1915.

51 Stewart was killed near the Nek on 25th April 1915. Commemorated on the Lone Pine Memorial, he was the 38 year-old husband of Edith Macbean Stewart, of "Zeitoun," Merivale Lane, Christchurch.

"A telegram came this morning from Carden, the admiral, that the business out there, i.e., the Dardanelles, which was to have been begun next Monday, has had to be postponed for a few days as the requisite mine-sweepers could not be got together sooner. I hope it won't be delayed any longer, as it is all important as a preliminary to our de-marche in the Balkans. So far it has been a well-kept secret."[52]

HERBERT ASQUITH, MP, BRITISH PRIME MINISTER.

EGYPT

"I am still on guard at the powerplant. No fighting has taken place since last week. Prisoners are still dribbling in, and their tales are woeful. We have two of the Turkish pontoons intact. They are very light."[53]

SPR LOUIS AVERY, 3RD FIELD COY, AUSTRALIAN ENGINEERS, ISMAILIA.

11 FEBRUARY 1915

EGYPT

"To-morrow we are going for five days' brigade manoeuvres. The reinforcements have now arrived, and we are up to full battalion strength… We have had two deaths this week in our battalion. A corporal[54] died last night. It is very sad for the poor parents, but they have the consolation that they died for their country. The belongings of the lad who died on Monday,[55] were sold by auction amongst us. His issue clasp knife brought £1, his cap 15/ and a handkerchief 10/. I believe the money will be used to buy a headstone."[56]

PTE HAROLD FAWCETT, 2ND BATTALION, AIF, MENA CAMP.

Detachments from 7th and 8th Battalion Australian Infantry arrived too late to see action along the Canal. But the resilience of trench systems to shrapnel fire was noted.

"The Turks' shell fire was pretty good, and a shell went right through the funnel of one of our boats [HMS *Hardinge*] in the Canal, doing some damage. I had a look at some of the trenches before I left, and some were so riddled with shrapnel bullets you would think no one could escape, and yet I saw only three men were wounded; it seems marvellous how much it takes to kill a man. I also saw a lot of Turkish prisoners, and for the most part they were an unkempt crowd, but pretty well armed, many of their rifles being the very latest pattern."[57]

LT HERBERT LAYH, 7TH BATTALION, AIF, MENA CAMP.

52 *The Brisbane Courier* (Queensland), 30th July August 1928.
53 *Barrier Miner* (Broken Hill, New South Wales), 21st March 1915.
54 L/Cpl Ambrose Myall, 2nd Bn, AIF, died of pneumonia on 11th February 1915. The former mechanical engineer, originally of Wimbledon, Surrey, is buried in Cairo War Memorial Cemetery.
55 Probably Pte William Law, 2nd Bn, AIF, died of pneumonia on 6th February 1915. Buried in Cairo War Memorial Cemetery, he was the 20 year-old son of James and Emily Law, of Lambton, New South Wales.
56 *The Armidale Chronicle* (New South Wales), 20th March 1915.
57 *Coleraine Albion and Western Advertiser* (Victoria), 19th April 1915.

12 FEBRUARY 1915

EGYPT

The defeat of the Turkish attack on the Suez Canal and the condition of some of those taken prisoner led some to discount their military capability.

> "You will know by now we have had a bit of a scrap, but I assure you we would all prefer it to be with the Hun proper instead of the deluded Turk. An airman has just come in and said about 70,000 of them are massed about 50 miles from the canal, but they don't offer to come on. I think the last few days has taught them a lesson they will not forget in a hurry. It was pitiful to see them — half-clad, shoeless and not half-fed."[58]
>
> PTE DAVID WOODHEAD, 1/1ST HERTFORDSHIRE YEOMANRY, YEOMANRY MOUNTED BDE.[59]

> "I expect you have seen in the papers all about the Turkish invasion of Egypt. Our total losses the entire canal front were just over 100, mostly wounded, while theirs were probably over 3,000. They are now in full retreat, and the general opinion is that the invasion is over. If this is correct, it has been absolute fiasco from the Turkish or rather German — point of view...
>
> "We are now in a different camp — viz., on the east bank of the canal and on the shores of Lake Timsah. It is very dull now after the excitement of last week, and we would relish a move. We are all tired of Ismailia."[60]
>
> LT CLIFFORD ARBUTHNOT, 53RD (FRONTIER FORCE) SIKHS, 28TH INDIAN BDE, ISMAILIA.

13 FEBRUARY 1915

ENGLAND

The underestimation of Turks and questionable hopes for a grand Balkan alliance continued to cloud the thought of Britain's political leadership.

> "I have just been having a talk with Hankey, whose views are always worth hearing. He thinks very strongly that the naval operations should be supported by the landing of a fairly strong military force, and I think we ought to be able to do this without denuding [Sir John] French. If only these heart-breaking Balkan States could be brought into action the trick would be done with the greatest of ease. It is of much importance that in the course of the next month we should carry through a decisive operation somewhere, and this one would do admirably for the purpose."[61]
>
> HERBERT ASQUITH, MP, BRITISH PRIME MINISTER.

---

58   *Burnley News*, 6th March 1915.
59   Woodhead's own poor health led to his discharge as unfit for further service on 4th May 1915.
60   *Belfast News-Letter*, 3rd March 1915.
61   *The Brisbane Courier* (Queensland), 30th July August 1928.

EGYPT

Though missing out on the action itself, another Australian officer described his impressions of the front line. Impressed by the power of British naval gunfire, they repeated the strange story that the Turks had not brought their artillery across Sinai but dug up guns left for them by Germans some time before.

> "As the Turks had retired 60 miles in a week we were sent back to our base yesterday. It was all right while it lasted. The only regrettable thing was that the New Zealanders bore the brunt of it, instead of us… The Turks lost more than 1,000 killed and nearly as many wounded, besides about 1,200 prisoners. They came in to our outposts in groups with the white flag and gave themselves up. They said they were sick of it. They told us they got only one water bottle full for a week and rushed the water when we gave them a drink.
>
> "They were well armed, carrying a long German pattern Mannlicher (charger-loading), a fine rifle. The sighting was all in Turkish. They carried 300 rounds of ammunition each, in bandoliers round the chest and belt. They were dressed in what looked like British warmers, breeches and putties, but were getting very ragged and most of them had no boots on — only just shreds of leather that were once boots. They seemed glad to surrender…
>
> "I take off my hat to the shooting of the naval guns. They had blown the trenches to bits. The dead were everywhere, and the sand of the desert trenches was red with blood. The wounded were pitifully torn in all directions by shrapnel, and howled and moaned as we lifted them on to stretchers. I saw some of the most horrible sights one could wish to see. In one trench a doctor had been busy and his instruments and case were lying there when he was cleared out by John Bull's seamen gunners.
>
> "Just where we were was the El Ferdan Suez Canal house, and it came in for the bombardment as HMS *Clio* (which came 600 miles down the Yang-Tse-Kiang River, in China, to guard the Canal) was moored near by. We picked up the Turks' shells around our trenches; they were 18-pounder shrapnel and failed to burst or burst harmlessly on the ground, although they knocked holes through the Canal house walls… The 4.7 naval guns put the Turkish 18-pounders out of action in good style. How the Turks brought those 18 pounders across the dreary Sinai desert has yet to be told — it was a marvellous performance. We think that the Germans had them brought across in sections and buried before the war."[62]

LT JOHN PAUL, 8TH BATTALION, AIF, MENA CAMP.[63]

---

62   *The Chronicle* (Adelaide, South Australia), 20th March 1915.
63   Lt Paul was killed in action on 25th April 1915. Commemorated on the Lone Pine Memorial, he was the 21 year-old son of John Keating Paul and Nora Paul, of Fort Largs, Largs Bay, South Australia.

14 FEBRUARY 1915

ENGLAND

The Assistant Director of Operations at the Admiralty continued to bemoan the belief that an unsupported naval attack could achieve anything. No-one was listening.

> "The bombardment of the Dardanelles, even if all the forts are destroyed, can be nothing but a local success, which without an army to carry it on can have no further effect. So long as there is an army in the Gallipoli peninsula the Dardanelles Straits can never be a safe thoroughfare for trade, commanded as it would be by the field- and other guns entrenched on the heights. If we should confine our operations to bombardment, although we may pass into the Black Sea and so give Russia the command there, we shall have performed our task incompetently and shewed the Turks the real limitation of sea power divorced from military power. The most recent experience, corresponding with that of the past, has shewn that the capture of forts cannot be adequately carried out by ships alone. An army is also indispensable."[64]
>
> CAPT. HERBERT RICHMOND, RN, ASSISTANT DIRECTOR OF OPERATIONS, ADMIRALTY.

EGYPT

Charles Bean had received quite a backlash after reporting some bad behaviour amongst Australian soldiers in December 1914. The true voice of the Non-Conformist temperance campaigner thought matters were much worse than anything Bean had described.

> "The charges that I made were that, though the Australians were physically magnificent, and had many fine men among them, yet a big number of them were rowdy, immoral, and drunken; and... they were doing our Lancashire lads a lot harm by treating them to drink, and otherwise leading them wrong. These charges, I claim, are absolutely true. The behaviour of the Australians has been a scandal... Is it not right and proper that Rochdale parents should know of this fresh danger and temptation to their sons out here? It is difficult enough for these sons to go straight away from home influences: the Australians have made it more difficult. Am I not right to warn the parents, so that they may write to their sons and mention this? It is a plain duty to do so. I cannot see that considerations of patriotism should keep me silent. I yield to no one in my admiration for the glorious loyalty shown by our Colonies in this war. It is beyond all praise. But it can do no good to keep back the truth from our people at home. This policy of unlimited praise and no blame is a mistake in my opinion. People in England and in Australia should be told that a certain type of soldier sent to Egypt from Australia is a bad type without discipline and without much character.

---

64   Marder, *Portrait of an Admiral: The Life and Papers of Sir Herbert Richmond*, p. 142.

The discipline here will probably improve the type, but many of the men here now from Australia are doing the Empire no good, and creating a bad impression."[65]

REV. DENIS FLETCHER, RACHD, ABBAS HILMI BARRACKS, ABBASSIA.

15 FEBRUARY 1915

ENGLAND

Carden was sent an advisory note, suggestions rather than orders about the conduct of operations. Carden was not the man who needed to be told this.

> "The provision of the necessary military forces to enable the fruits of this heavy naval undertaking to be gathered must never be lost sight of; the transport carrying them should be in readiness to enter the Straits as soon as it is seen the forts at the Narrows will be silenced.
>
> "To complete their destruction, strong military landing parties with strong covering forces will be necessary. It is considered, however, that the full advantage of the undertaking would only be obtained by the occupation of the Peninsula by a military force acting in conjunction with the naval operations, as the pressure of a strong army on the Peninsula would not only greatly harass the operations, but would render the passage of the straits impracticable by any, but powerfully armed vessels, even though all the permanent defences had been silenced.
>
> "The naval bombardment is not recommended as a sound military operation unless a strong military force is ready to assist, or, at least, follow it up immediately all the forts are silenced."[66]
>
> ADMIRAL SIR HENRY JACKSON, CHIEF OF THE ADMIRALTY WAR STAFF.

While senior officers and politicians debated higher matters, those to be sent to carry out their plans had to deal with men dying even before going into action.

> "I didn't know how fond of one's men one becomes until last week one of my men died suddenly.[67] On Tuesday I saw him because some sergeant had mistakenly told him he was to be put into another platoon. He had been with me since I joined in November, and I had no intention of letting him go… On Wednesday morning he was taken to sick-quarters suffering from ptomaine poisoning, and at 6 that evening he died. The doctor[68] described it as a violent death. Even at this early stage of our service we had

---

65 *Rochdale Observer*, 6th March 1915.
66 Dardanelles Commission First Report, p.30. HMSO (London) 1917.
67 Pte Richard Roan, 13th (Reserve) Bn Worcestershire Regt, died on 4th February 1915. Buried in St. Martin-by-Looe (St. Martin of Tours) Churchyard and Cemetery, he was the 21 year-old son of Fanny Roan, of Tanhouse Lane, Malvern Link, Worcestershire.
68 Dr John Eustace Webb, RN (Retd.), acted as the temporary Medical Officer who gave evidence at the subsequent inquest into Road's death. Webb died at Looe on 6th August 1916, aged 53.

marched so much together, turned out to so many boring parades, and rejoiced at nearing home so often in company, that I felt as though I had lost a friend."[69]

2/LT JOHN ALLEN, 13TH (RESERVE) BN WORCESTERSHIRE REGT, LOOE, CORNWALL.

TURKEY

Many British and Commonwealth observers viewed the Germans as the real enemy; the Turks their dupes. One American journalist recently arrived in Istanbul offered a different perspective.

> "As seen from my present viewpoint, the entry of Turkey into the European war came about in this manner: The Ottoman government feared — and the German embassy succeeded in convincing it of this — that Turkey would have to engage in the war on the side of the Central States if she were not to be dismembered. The Entente governments, so runs the argument, would sooner or later have forced the Dardanelles for the purpose of making sure of Turkey. The waterway in question meant too much in the Entente scheme of things to have it remain the sole control of the Turks during the duration of the war. The Turks feared that they would never again oust the Western powers from Constantinople, once they had been installed. Though not enthusiastic adherents of the Central Powers, the Turks finally concluded that joining Germany and Austria-Hungary would be choosing the lesser of two evils."[70]

GEORGE ABEL SCHREINER, ASSOCIATED PRESS OF AMERICA, ISTANBUL.

EGYPT

The British commander in Egypt was well satisfied with the defeat inflicted upon the Turks but very hesitant to follow that up. Part of that caution was explicitly stated to be the desire not to be seen to suffer any kind of military reverse for fear of its impact on Muslim attitudes locally and in India.[71] Those fears would be stoked by events thousands of miles away this same day.

> "I feel the Turks must come on again after all their talk about a Holy War. We gave them a nasty knock, killing and wounding a great many more than I have reported, for every day bodies in hastily-dug graves are discovered; also many more were drowned in the Canal than we knew of. Yet they got away with their guns in fairly good order. As we knew that there was the Eighth Corps and part of the Third, Fourth, and Fifth Corps against us, and they only showed about 20,000 men, I did not think it safe to go out and meet them, for it was quite possible they were laying a trap for us, and I felt that anything in the nature of a reverse, or even a check, would have fatal results

---

69 John Allen quoted in Montgomery, Ina, *John Hugh Allen of the Gallant Company. A Memoir*, pp. 158-159, Edward Arnold (London) 1919.

70 Schreiner, George Abel, *From Berlin to Baghdad*, pp. 23-24, Harper & Brothers (New York) 1918.

71 The British were aware of an uprising planned by the Ghadr movement in India at this time. Though more nationalist than Islamist in sentiment, the organisers met in Lahore on 12th February 1915 to secure the support of disaffected elements within the Indian Army. The British had penetrated the organisation and mass arrests took place on 19th February 1915, two days before the date set for the rebellion.

in Egypt, for there is no doubt that the feeling here is pro-Turk and anti-English. It is odd, but nothing that we do or say is believed, whereas every Turkish or German lie is sucked in.

"It is satisfactory that our Moslems, both Indian and Egyptian, showed no disinclination to kill their co-religionists when they had the chance."[72]

GENERAL SIR JOHN MAXWELL, GOC BRITISH TROOPS, EGYPT.

Other men just wanted rid of the flies.

"Our work there [Suez] was all guard-mounting — 24 hours on, 24 hours sleep, and then on again the next day. Since about February 5th we have been just behind our line of resistance on the Canal as reserves, working like Trojans at trench digging, guards, outposts, washing clothes, etc. Occasionally we get time off for swimming.

"Aeroplanes hover over us every day. We hope to have a turn ourselves in the trenches shortly. I suppose you read of the little scraps that have taken place here. We have only had one casualty, and that unfortunate chap was shot by one of our own Indian outposts...

"I had always understood that the 'plague of flies' was removed from Egypt in Pharaohs days, but I am now of the opinion that most of them have found their way back again; we get simply swarms in the camp."[73]

SGT GEORGE COE, CEYLON PLANTERS' RIFLE CORPS, ISMAILIA.[74]

SINGAPORE

The Indian Army's 5th Light Infantry was unusual in being almost entirely Muslim. Scheduled to be shipped to Hong Kong the following day, some fearing they were actually being sent to fight the Ottoman Turks, around half the sepoys, 403 Rajput Muslims, mutinied.

"At about 3 p.m. a shot fired from the guardroom the 5th Light Infantry gave the signal for the rising. The men outside the guardroom collected there and took possession of the ammunition...

"At about 4.15 p.m. a body of some 100 men of the 5th Light Infantry appeared at the military hospital and prisoners of war camp, and surprised and overpowered the detachments of Singapore Volunteer Rifles and the Johore military forces.

---

72   Sir John Maxwell quoted in Arthur, Sir George, *Life of Lord Kitchener*, vol. 3, p. 113, MacMillan & Co., (London) 1920.
73   *Evening Star and Daily Herald*, 4th March 1915.
74   Granted a commission in the Border Regt, 2/Lt George Coe was killed in action with 1st Bn on the Somme, 1st July 1916. Commemorated on the Thiepval Memorial, he was the 30 year-old son of George John and Rosa Coe, of 2 Westgate Terrace, Long Melford, Suffolk.

> They killed Captain Gerrard,[75] commandant the prisoners of war; Second-Lieut. J. Love Montgomerie,[76] S.V.R.; Captain H. Cullimore,[77] second command, Johore military forces."[78]
>
> WILLIAM MAXWELL, ACTING SECRETARY TO THE HIGH COMMISSIONER FOR THE MALAY STATES.

Those who survived or evaded the attack managed to raise the alarm.

> "We were all lying in bed, having an afternoon nap prior to doing our evening's drill, when two officers rushed into our barracks, and one of them said — 'Our men have mutinied!'
>
> "We thought it a great joke, and took no notice of him, but when he said that two of his officers had been murdered, we were up like a shot, got on our kit, and made for the barracks as fast as we could in regular extended order, ready for the fray.
>
> "Along with five more, I was stationed on guard at the barracks until further orders, and we watched our men, about 80 strong, sweep up the hill to the officers' quarters, while the bullets were buzzing overhead. After they had gone, a Chinese boy came up to me and said there was 'Tuan' down on the road about a mile who had been shot… we brought in the mangled corpse of one of our men, who had only been with us two days.[79]
>
> "Arrived at the camp, we saw three or four of our men coming down the hill towards us, shouting 'Ammunition.' We got all the ammunition — 8000 rounds — and worked our way towards the Colonel's bungalow."[80]
>
> PTE JAMES MANN, MALAY STATES VOLUNTEER RIFLES.

New recruits for local volunteer units then undergoing training were close by. One of them immediately thought of an earlier mutiny, something that was never far from the imperial mindset.

> "I am afraid [it] was scarcely a riot, as a repetition of the scenes of the Indian mutiny was very narrowly averted. It would appear that the mutineers… split into two sections after obtaining possession of a large quantity of ammunition. One section

---

75   Capt. Percy Gerrard, Malay States Volunteer Rifles, was killed on 15th February 1915. Buried in Kranji War Cemetery, he was the 45 year-old Dublin-born son of Thomas and Elizabeth Gerrard, husband of Clare Gerrard, of 66d, Princes Square, Bayswater, London.

76   2/Lt John Montgomerie, Singapore Volunteer Rifles, was killed on 15th February 1915. Buried in Kranji War Cemetery, he was the 37 year-old son of David and Agnes Montgomerie; husband of Katherine Jane Esther Fairweather Montgomerie, of Pitheavlis Bank, 98 Glasgow Road, Perth.

77   Capt. Horace Cullimore, Royal Garrison Artillery, attd Johore Military Forces, was killed on 15th February 1915. Formerly of the Royal Marine Artillery, he was the 45 year-old son of Mary and the late George Cullimore, of the Phoenix Inn, London Road, Stroud, Gloucestershire; husband of Clara Anna Cullimore (b. Lausanne, Switzerland, 20th April 1875), of Midland Road, Wellingborough, Northamptonshire.

78   *Newcastle Journal*, 14th April 1915.

79   Possibly Pte William Leigh, Malay States Volunteer Rifles, who was killed while cycling to Normanton Camp on 15th February 1915. Buried in Kranji War Cemetery, he was the 25 year-old son of Joseph and Ellen Leigh, of Almondbury, Huddersfield, Yorkshire.

80   *Aberdeen Evening Express*, 1st May 1915.

proceeded to murder the Volunteers on guard over the German prisoners, while the others went out to the main roads and murdered all the white men they came across…

"The alarm was soon given, and thanks to telephones and an abundance of motor transport a body of men was assembled to deal with the situation. Unfortunately the trouble broke out on a holiday, which made things a bit difficult for most, but as regards the company I am in, advantage had been taken to mobilise us for training, and consequently within half an hour of the alarm we were awaiting orders to move on."[81]

PTE PERCY HILL, MAXIM COY, SINGAPORE VOLUNTEER ARTILLERY.

The wife of a local police inspector described what she saw. Two sepoys began to stop cars.

"While at tea in our verandah we suddenly heard shots, and everyone in the road below began to run… More shots, and the telephone ringing with the message, "a serious rising among the native regiment; men coming into town; get out the Sikhs." I went back to the verandah, and saw two natives with guns and bandoliers coming along the road beneath our house, shooting as they came. Bob[82] said, "I must shoot them," rushing downstairs and off to the Sikh Barracks, where the bugle immediately sounded "Fall in," and in no time I saw him leading out about twenty-five Sikhs. By this time the natives had under my eyes shot into a motor-car that came down the road all unsuspecting. I heard screams, and the car swerved into the drain and remained motionless. I knew its occupant had been killed. A warder rushed out from his quarters below our house, the native shot him down, and he lay across the road. On went the two natives, shooting and clearing the road, scattering everyone before them… It grew darker and men were racing up and down the hill with revolvers, and a crowd gathered by the motor in which poor Dr. Whittle[83] lay dead with seven bullets in him. His wife fell into the drain, and escaped to the Hospital. The warder[84] was wounded in the knee and jaw."[85]

ISABELLA HARRIOTT DEWAR.

Around 80 men with Lt-Col. Edward Martin, the 5th Light Infantry's unpopular commander, prepared his bungalow for the expected attack.

"When it got dark we barricaded the colonel's bungalow with cases and bags filled with gravel and placed these round the verandahs. We waited for the mutineers to attack, but it did not come off."[86]

PTE RICHARD TROWER, MALAY STATES VOLUNTEER RIFLES.

---

81 *Birmingham Daily Post*, 6th May 1915.
82 Isabella Dewar's husband, Police Inspector Arthur Robert Johnstone Dewar, a veteran of the Boer War, serving with the New Zealand Contingent.
83 Dr Edward Dennis Whittle was killed on 15th February 1915.
84 Warder J. 'Harry' Clarke died of his wounds on 16th February 1915. He was originally from Chalvey, Buckinghamshire.
85 *Stratford-upon-Avon Herald*, 9th April 1915.
86 *Surrey Mirror*, 27th April 1915.

Women and children on board the *Nile* while the Singapore Mutiny was being quelled.

Taking no chances, European women and children were immediately sent to places of relative safety.

"I can't describe… the horror of that evening; it makes me tremble. I seemed turned to stone watching these dreadful things, and having to wait and wait not knowing what danger my husband might not be in or when I should see him again. All the Chinese servants behaved well, but were terrified. Mr. Bourne, the Court Inspector, lately married, brought his wife to stay with me. She was so frightened, poor girl, just out from home. At 9.30 Bob returned, dead tired and very anxious, to tell us that the whole of the regiment of native infantry was out on mutiny. They had shot the European guard over the German prisoners, who had all escaped, and that orders were out that all ladies must go to one of the hotels or to Government House or the gaol. I choose the latter place, being near our house. Bob and I had a scratch dinner, and at 11 o'clock we went off with long chairs and cushions to the gaol with an escort of armed Sikh police."[87]

ISABELLA HARRIOTT DEWAR.

16 FEBRUARY 1915

ENGLAND

It was confirmed that the naval assault on the Dardanelles would begin on 19th February. The 29th Division was being sent but the attack would commence before they could arrive in the area.

"Mr Churchill informed the Cabinet that complete preparations have now been made

87   *Stratford-upon-Avon Herald*, 9th April 1915.

for the naval bombardment of the forts at & beyond the Dardanelles, which will commence on the morning of next Friday the 19th.

"Lord Kitchener, in conjunction with the Admiralty, has made arrangements for the despatch in the course of next week to Lemnos of the 29th Division, and for their reinforcement in case of need from Egypt, firstly by the Australian & New Zealand contingent."[88]

LT-COL. MAURICE HANKEY, CABINET SECRETARY.

"Not a grain of wheat will come from the Black Sea unless there is military occupation of the Dardanelles, and it will be the wonder of the ages that no troops were sent to co-operate with the Fleet with half a million soldiers in England."[89]

ADMIRAL LORD JACKIE FISHER, RN, FIRST SEA LORD.

Rosslyn 'Rosy' Wemyss, the man selected to prepare a base on Lemnos to receive the forces being sent to the Dardanelles, had been trying to make arrangements for an expedition in German East Africa.

"I was called to the telephone and curtly informed by some unknown person at the Admiralty that the East Africa expedition was off! My feelings can be better imagined than expressed! No Squadron — the arrangements for East Africa almost complete — and without a command! I repaired to the Admiralty at 3 p.m. in a towering rage, only to be told that the First Lord was not in the building and that nobody knew when he would be in, I was, however, quite determined not to leave without some sort of explanation, and after some delay at last waylaid Churchill in the passage and accompanied him to his room, but before I could open my mouth he informed me that the project was postponed, that it had that morning been decided to force the Dardanelles, that the island of Lemnos was to be made the base of operations, that he wished me to proceed out there at once — the next day in fact — that I should probably be the Governor of the island and that further orders were to follow me immediately, which by the way they never did!"[90]

REAR-ADMIRAL ROSSLYN WEMYSS, RN.

FRANCE

Asquith hoped attacking the Dardanelles would embolden the Balkan countries to join the allies. Poincaré thought nothing could be achieved before the operation had succeeded.

"M. Venizelos has been so dilatory in his reply that they have put off any idea of an expedition to Salonika, and as the Navy want the Army to help them in their attack on the Dardanelles forts, they propose that we should send as soon as possible a Division to Lemnos. This Dardanelles adventure, which Mr. Winston Churchill has

---

88 Meeting of Cabinet, 16th February 1915, TNA CAB 37/124/28.
89 Fisher quoted in Churchill, *The World Crisis 1915*, p. 189.
90 Wemyss quoted in Wemyss, Lady Wester, *The Life and Letters of Lord Wester Wemyss*, p. 197, Eyre and Spottiswoode (London) 1935.

conceived, will become a mirage in which the Allies will think they detect victory, and it would seem that the idea of renewing overtures to Sofia and Bucharest has been set aside, until the Narrows have been forced."[91]

RAYMOND POINCARÉ, PRESIDENT OF FRANCE.

EGYPT

There was no lack of confidence about the ability to deal with the Turks in some quarters.

"People look on the 'Terriers' as nothings, but I know differently, as I have served with the 1st East Lancashires in the Boer War, for which I received two medals with five bars. We have 200 Turkish prisoners here, and they look a poor lot. If they are all like these I say, 'Let them all come!'"[92]

PTE PATRICK SALMON, 1/5TH BN EAST LANCASHIRE REGT, EAST LANCASHIRE DIVISION.

One group of Australians enjoyed good food, good company and a good bath in Cairo.

"I went to town with George[93] and Bob Gray[94] and Will Latham.[95] We went to Rossmore House for dinner. We had the smoke-room to ourselves, and were able to eat to our hearts' content. Mohammed, the waiter, who generally looks after us, managed it. Before, we have always dined in the large dining room adjoining, and a fellow feels that the eyes of all the company present are on him, and he cannot have a decent "blow out." The first course was "devilled kidneys on toast" — O.K., with some beautifully tender mutton cutlets to follow. We had three each before we turned them down, and I think I ate nearly a fair feed of fried potatoes served with them. After this we had blanc-mange and jam. Then fruit (any quantity), finishing up with coffee. We always go to Rossmore House now. Miss Green, who runs the place, is very good to us, and always trying to make us feel at home. I had a hot bath before dinner, and as Bob remarked, "Removed the dust of ages." I had a real soak in the water as hot as I could bear it; then soaped down, and had a cold shower. It was scrumptious. We have to bath in a biscuit tin cut down in camp, so we enjoy the weekly wash in town. I haven't had a letter for some time now, but the others are in the same box, so we blame the P.O. officials."[96]

PTE WILLIAM BELSON, 9TH BATTALION, AIF, MENA CAMP.[97]

---

91 Poincaré, Raymond, *The Memoirs of Raymond Poincaré, President of France, 1915*, p. 41, William Heineman Ltd. (London) 1926.
92 *Burnley Express & Advertiser*, 6th March 1915.
93 Pte, later Sgt, George Reid Gray, 9th Bn, AIF.
94 Pte Robert Gray, 9th Bn, AIF, was wounded on 25th April 1915 and again on 7th September 1915. He left Egypt on 31st October 1915 for Australia, where he re-enlisted, serving later with the Australian Field Artillery.
95 Pte William Latham, 9th Bn, AIF, was wounded twice at Gallipoli, on 25th April and 24th June 1915 respectively. He was later commissioned, attaining the rank of Lt.
96 *Cairns Post* (Queensland), 8th May 1915.
97 Pte William Belson, 9th Bn, AIF, was killed in action on 25th April 1915. Commemorated on the Lone Pine Memorial, the former architect was the 22 year-old son of Charles and Mary Belson, of Malanda North, Queensland.

SINGAPORE

The British were fortunate that the mutineers lacked any clear leadership or plan, though those recovering from the shock of the previous day's events could be forgiven for not appreciating that. Split up into small groups, the sepoys tried to evade the growing forces ranged against them.

> "As soon as it was light the bullets began coming through the doors and windows, but no one was hit. We lay flat on the floors, and fired from the cover of the sandbags and boxes. About 8 o'clock we heard a maxim gun firing and increased rifle fire. It was the advance of the relief party and they consisted of bluejackets from HMS *Cadmus*, which happened to be in Singapore, R.G.A. men, Singapore Volunteers and some of the 36th Sikhs. Then we returned to Singapore, having been relieved about 9 o'clock. There was some more firing on the way to Singapore, but only long range."[98]
>
> PTE RICHARD TROWER, MALAY STATES VOLUNTEER RIFLES.

> "Every man's blood was up, and most of them had seen the bodies of murdered civilians, it being impossible to move them until we had driven the mutineers back. We moved off at 4.30 [in the] morning to attack the mutineers and rescue the Colonel and three other officers. Mr. Cotton and 80 men of the Malay States Volunteer Rifles, who were down for training, were surrounded by the rebels. It was not long before we came under fire. Within two hours we had rescued the Colonel and those with him, having shot down about 20 of the mutineers and wounded another 15. Our casualties were luckily very small. One man from HMS *Cadmus* was killed[99] and 6 wounded. Similar fights were going on in other parts of the island, as there were 600 mutineers, and they had scattered on our approach the previous night… we took 200 prisoners ourselves. Most of them gave themselves up on our approach as being loyal, but testing their rifles I found all had been fired through."[100]
>
> CAPT. STANLEY GREVILLE-SMITH, MIDDLESEX REGT, ATTD SINGAPORE VOLUNTEER RIFLES.

European civilians spent a fearful night; some learning what had happened to some of their friends.

> "We… passed a terrible night, shots being fired all round, warders drilling, prisoners yelling, and at five a.m. the children woke. We had not undressed or slept at all. The yard was full of warders, and at seven a.m. Mr. Codrington shouted to us to go to the old prison at once, so snatching up the babies and some cushions we flew downstairs, through those passages to the same long room that we had entered the night before. Here were warders at each window with rifles, and they expected an attack on the gaol…
>
> "I had a note from Bob to say he was well, and had to take a Sikh guard to protect the funerals of twenty poor victims, for, indeed, this number and more have perished. A

---

98   *Surrey Mirror*, 27th April 1915.
99   Sto. 1 Charles Anscombe, RN, HMS *Cadmus*, was killed on 16th February 1915. The 23 year-old son of Charles and Elizabeth Anscombe, of 101 Graveney Rd, Tooting, London, is buried in Kranji War Cemetery.
100 *Middlesex Chronicle*, 8th May 1915.

newly-married couple[101] — such a dear girl — were shot in their car and the chauffeur[102] also. Captain Izard,[103] the Archdeacon's cousin, and Major Galway,[104] of the R.G.A., and several civilians who met the natives on the road were killed. There are four distracted widows whose husbands were shot beside them in their cars. All the European guard on the German prisoners were killed but three. The report that all the Germans had escaped was false; only seventeen got away. The others refused to move."[105]

ISABELLA HARRIOTT DEWAR.

"1500 refugees, chiefly women and children arrived on board our ship… Many of the women were weeping for murdered relatives and others were nearly crazy with anxiety for their kinsmen who had pluckily volunteered to remain behind to aid in hunting down the murderers."[106]

DR WARREN MCNEIL, SS *NILE*.

Help from the Japanese, French and Russians was on the way. The immediate crisis might have passed but the fear of those in Singapore and the concerns of those in Dehli, Cairo and London regarding the loyalty of Britain's Muslim subjects, that Murray spoke of the previous day, would persist.

17 FEBRUARY 1915

TURKEY

As allied leaders feared the revolt of their Muslim subjects, so the Ottomans were all too keen to turn those fears into reality.

"Russia is our hereditary enemy, and Great Britain is the power which subjugated Islam… All the sorrowful eyes of Islam are today turned towards the Turks, who, since their appearance upon the historic scene, always have been the benevolent champions of the disciples of Mohammed and all Asiatics generally. The success of Ottoman arms can have but one result, namely the rising of all Mussulmans subject to the dominion of Russia, France or Great Britain. Persia is a living symbol of what Turkey would have become if we had not taken part in this present war."[107]

TALAAT BEY, OTTOMAN MINISTER OF THE INTERIOR, ISTANBUL.

101 Belfield Morth Woollcombe, an engineer employed by the Eastern Extension Cable Company, married Margaret Catherine Michie in Westminster on 19th August 1914. They sailed aboard the *Moldavia* from London on 22nd August 1914. They died in a car near the cable company's depot on 15th February 1915.
102 Hassan Kechil.
103 Capt. Francis Izard, Royal Garrison Artillery, was killed in a gharry on Outram Road around 4.20 p.m. on 15th February 1915. Buried in Kranji War Cemetery, he was the son of Emily Vallance, of 9 Promenade Terrace, Cheltenham, and the late Francis Izard.
104 Major Reginald Galwey, Royal Garrison Artillery, was shot in a rickshaw on Outram Road and died of his wounds later on 15th February 1915. Born in Spain, the son of William and Maria Galwey, of 84 Ashley Gardens, London, S.W., he is buried in Kranji War Cemetery.
105 *Stratford-upon-Avon Herald*, 9th April 1915.
106 *The Birmingham Age Herald* (Birmingham, Alabama), 25th April 1915.
107 *Newark Evening Star* (New Jersey, United States of America), 17th February 1915.

ENGLAND

The Great War alliance between Britain, France and Russia, three long-term antagonists, had not resolved the differences between them. The fate of Istanbul remained a controversial issue.

> "Kitchener takes rather a gloomy view of the Russian situation... The other question is the future of Constantinople and the Straits. It has become quite clear that Russia means to incorporate them in her own Empire. That is the secret of her intense and obstinate hostility to the idea of allowing the Greeks to take any share in the present operations. I do not know how it will be viewed in France or in this country. It is, of course, a complete reversal of our old traditional policy."[108]
>
> HERBERT ASQUITH, MP, BRITISH PRIME MINISTER.

The ad hoc, make-it-up-as-you-go-along British approach to the war caused more heart-searching, while Churchill was busy treading on other people's toes.

> "[T]he new scheme (Wortley says he expects one everyday) is to land the 29th Division on Lemnos where, it is argued, it can then be launched — sic! — in any required direction... The O.C. of that Division, having ascertained from the map-index where Lemnos was, naively enquired whether motor-transport (as laid down in the hand-book) should be taken! I said I was afraid if they took that sort of thing with them they would run the risk of sinking the Island!
>
> "Incidentally, Winston Churchill thinks that to have a Division at his back will help him to get through the Dardanelles. A scheme I view with some apprehension. I asked the other day what they mean to do when they get to Constantinople? Shell the town? How guard their communications and get supplies? I doubt whether it had been thought out very etraflico — all round."[109]
>
> CAPTAIN WYNDHAM DEEDES, KING'S ROYAL RIFLE CORPS.

> "Winston just now is absolutely maddening. How I wish Oc[110] had not joined his beastly naval Brigade! He is having a great Review of them today. He inspects the Brigade in a uniform of his own, which will cause universal derision among our soldiers! (Clemmy[111] said some time ago that inventing uniforms was one of Winston's chief pleasures and temptations.) He has just emerged from a fearful row with K. [Kitchener] by the skin of his teeth, and has now got himself into another...

---

108 *The Brisbane Courier* (Queensland), 30th July August 1928.
109 Wyndham Deedes quoted in Presland, John, *Deedes Bey. Sir Wyndham Deedes 1883–1923*, p. 147, MacMillan & Co. Ltd. (London) 1942.
110 Sub-Lt Arthur 'Oc' Asquith, RNVR, was wounded at Gallipoli while serving with the Hood Bn on 6th May 1915. He was to be wounded again, in France, on 19th October 1916.
111 Winston Churchill's wife, Clementine Churchill.

"Sir John [French] writes from St. Omer and tells K. that Winston has offered him 9000 of his Naval Brigade, ready to go into action at once in the trenches, and a squadron of armoured motor cars; and French asks if he may be allowed to remove the guns and use the motors, which shows what he thinks of the use of these expensive follies! K. is of course furious, and says to me he wonders what Winston would say if he, K., was always writing to Jellicoe, offering to do this and that. Of course, Winston is intolerable. It is all Vanity. He is devoured by vanity."[112]

HERBERT ASQUITH, MP, BRITISH PRIME MINISTER.

EGYPT

Life in Egypt was certainly losing its charm.

> "Too dreary is this hum-drum life,
> With guards and fat-i-gues,
> I wonder that our sergeant's bite,
> And deal in wild abuse.
>
> Even so many men are getting stale,
> And some are in the "clink,"
> Whilst several more are on the ale —
> Our only joy is drink.
>
> But putting all jokes aside,
> Our lads are doing well,
> And daily now we wait the word
> To sail for old Marsailles."[113]

PTE THOMAS RYAN, 2ND BATTALION, AIF, MENA CAMP.[114]

18 FEBRUARY 1915

ENGLAND

Early press reports of the events in Singapore described it as riot (though there was no damage to property) and its victims to have died as the result of accidents. Similar sensitivities were on display when the Cabinet was briefed about what had happened.

> "Lord Kitchener announced that the rioting at Singapore was well in hand."[115]

LT-COL. MAURICE HANKEY, CABINET SECRETARY.

---

112 Herbert Asquith quoted in Brock, Michael & Eleanor (Eds.), *Margot Asquith's Great War Diary 1914–1916*, p. 79, Oxford University Press (Oxford) 2014.
113 *National Advocate* (Bathurst, New South Wales), 17th February 1915.
114 Ryan was wounded 25th–30th April 1915, returning to Australia aboard the *Suevic* on 8th October 1915.
115 Note of Cabinet Meeting, TNA CAB 37/124/34.

EGYPT

"We are right in the desert at the Pyramids. And such a desert, as you cannot imagine! It rolls away as far as the eye can see to the west, in rocky undulations and sandy valleys. All over the low rises there are loose, apparently, water worn pebbles, some as large as an ostrich egg, but mostly small. These make good walking, and a great noise as you pass over them. The sandy patches are very heavy to march through, and sometimes extend for a mile or so. We are getting used to marching in it, but can always manage to get very tired. Only for the ever-visible Pyramids it would be very easy to lose oneself here, as everywhere, looks exactly similar, and it would be the last place on earth to get lost.

"Even within sight of our camp is the luxuriant Nile valley, and lately I have made several trips up the river, and through the areas irrigated by the Nile. The scenery is beautiful, and with water the natives can grow almost anything. The date palm is everywhere, and in the palm groves all kinds of fruits and vegetables are grown… Fruit and vegetables are plentiful, and we have lived well on oranges, mandarins and dates since we came here, and they have served to keep us in the best of health, and I am getting quite fat: my clothes are getting dead tight. I have also acquired a taste for sugar cane, of which there is tons here, and they say it is very healthy, too."[116]

SGT GEORGE GREEN, 6TH BATTALION, AIF, MENA CAMP.[117]

"Early breakfast. Field day to Tiger's Tooth. Erected dressing station and collected wounded and attended to same. Returned 5.30 p.m."[118]

PTE SYDNEY PENHALIGON, 3RD FIELD AMBULANCE, AAMC, AIF, MENA CAMP.

19 FEBRUARY 1915

ENGLAND

On the 108th anniversary of Admiral Duckworth's ultimately failed operation in the Dardanelles, the War Council met in London. This was the day the Anglo-French fleet would begin its bombardment of the forts guarding the Dardanelles. The main order of business was to consider how the advice of earlier studies, as well as historical precedent, no longer applied.

"[T]he Prime Minister read extracts from the [1906] General Staff Memorandum to the War Council, and the Secretary drew attention to the conclusion of the Committee of Imperial Defence… Subsequently copies of the Paper were distributed to all members of the War Committee.

---

116 *Woodend Star* (Victoria), 1st May 1915.
117 Green was killed in action on 25th April 1915. Buried in Lone Pine Cemetery, he was the 22 year-old son of George and Grace Green, of Macedon, Victoria.
118 *The Queenslander* (Brisbane, Queensland), 21st April 1917.

"It was generally held, however, that the Memorandum was not wholly applicable[119] to the conditions pertaining in 1915.

"(i.) Turkey in 1915, in addition to having to defend Constantinople, was at war on three other fronts — namely, the Caucasus, Egypt, and Mesopotamia.

"(ii.) Since the General Staff Memorandum was written, Turkey had suffered severe defeats in the Balkan wars, and had shown herself much less formidable as a military Power than had previously been assumed.

"(iii.) The conditions in the Balkans in 1915 were likely to be favourably influenced by a successful attack.

"(iv.) There had been considerable development of naval matériel since 1906.

"(v.) The fall of the Liège and Namur forts had led to the belief that permanent works were easily dealt with by modern long-range artillery, and this was confirmed by the fall of the outer forts.

"(vi.) The utilisation of aircraft had led to the hope that, in a comparatively confined space like the Gallipoli Peninsula, the value of naval bombardment, particularly by indirect laying, would be enormously increased.

"(vii.) The development of submarines led to the hope that the main Turkish communications to the Gallipoli Peninsula, which run through the Sea of Marmora, would be very vulnerable."[120]

LT-COL. MAURICE HANKEY, SECRETARY, WAR COUNCIL.

It is difficult to determine a purpose for this discussion. By the time it took place, the attack was already in progress. It was hardly to consider information with a view to coming to a decision, as that had already been made. But it is revealing of the extent to which evidence could either be imagined or moulded to fit with a predetermined conclusion. The belief that the fate of Belgian forts destroyed by fire from heavy howitzers on land had any relevance to an attempt to destroy forts by naval guns at sea would be shown for what it was. Nonsense. And there was not a single mention of mines.

TURKEY

Despite the weakness of the arguments in favour of attacking the Dardanelles, the strength of the forts did not reassure everyone.

"Pallavicini, the Austrian Ambassador, came to me with important news. The Marquis was a man of great personal dignity, yet it was apparent that he was this day exceedingly nervous, and, indeed, he made no attempt to conceal his apprehension. The

---

119 After the meeting of the Committee of Imperial Defence on 28th February 1907, at which Herbert Asquith, Sir Edward Grey, Lord Haldane, Lord Fisher and Sir John French had been present, concluded that: "The Committee consider that the operation of landing an Expeditionary Force on or near the Gallipoli Peninsula would involve great risk, and should not be undertaken if other means of bringing pressure to bear on Turkey were available." (Dardanelles Commission: Memoranda Produced as Evidence, TNA CAB 63/17.)

120 Dardanelles Commission: Memoranda Produced as Evidence, TNA CAB 63/17.

Allied fleets, he said, had reopened their attack on the Dardanelles, and this time their bombardment had been extremely ferocious."[121]

HENRY MORGENTHAU, AMERICAN AMBASSADOR, ISTANBUL.

DARDANELLES

One officer, a descendant of Oliver Cromwell, was conscious of the challenge they faced but was confident. Everyone wanted to see history being made.

"I can't say when you will get this, or how it will arrive, but when it does we shall be making history. We are all very excited, as this afternoon the fun commences. We shall not run any very great danger, but will have just to take our chance with the remainder. I would not miss this show for anything. It will take some weeks of strenuous work, and we shall be under fire nearly the whole time. We are very powerful, and I have no doubt we shall succeed in our undertaking… There is no need to feel alarmed about us at all; it would only be deuced bad luck to fetch us a cropper, and there are plenty of ships to pick us up if necessary. 'They' will get a dusting, I hope, if they know what's coming. I don't think we shall return much before the end of the war, and I am quite contented to be here until then. Anything is better than the North Sea."[122]

LT EDWARD CROMWELL COLCHESTER, RN, HMS *IRRESISTIBLE*.

"Well… we were just about to finish our patrol, and our relief was coming along; so was the big fleet to commence the bombardment. It was about 8 a.m., and we were growling because we wouldn't be there to see the bombardment, when there was a loud report and a splash a few yards from the ship's side. It was such a surprise, and we couldn't make it out when there was another splash, a little further out; then a shell screamed just over and fell about twenty yards away on the other side. By this we had found the Turks were firing at us, and it wasn't bad shooting either. However, they didn't get the chance to try it again, as we were going full speed out of it.

"We claim to be the first boat fired on a the Dardanelles bombardment, and that was our baptism… All day long we heard the guns booming, though we couldn't see anything as we were lying at the other side of Tenedos."[123]

STO. 1 WILLIAM CARRADICE, RFR, HMS *WEAR*.

HMS *Cornwallis* fired the first shot at 9.51 a.m., followed quickly by HMS *Triumph*.

"The morning dawned clear and calm, an ideal day for such an unideal task. Yonder, just out of range, the forts showed up in the sunlight. The great day had come. The men were ready. There were a few preliminary, precautionary measures to be taken. Any vulnerable part of the ship had to be protected by sand-bags. Hammocks had to

---

121 Morgenthau, Henry, *Ambassador Morgenthau's Story*, pp. 189–190, Doubleday, Page & Co. (New York) 1918.
122 *Cambridge Independent Press*, 26th March 1915.
123 *Consett Guardian*, 20th August 1915.

be packed here and there to prevent shell splinters striking men who must be exposed. As the bugles sounded "Action," a cheer resounded along the mess-decks, and in each ship men went to their battle stations…

"Great Scott! What is that? A noise which no cunning arrangement of vowels and consonants can convey. For a moment the whole ship trembles, and even we, buried here, hear the faint wail gradually dying away. The *Triumph* has fired the first gun! We shall get used to it in time, but the first shot "fired in anger" is curiously memorable.

"Now the supply parties have all their work to do in carrying out orders. Half stripped, yet sweating, they send up as ordered "10″ ordinary," "7.5″ lyddite," "Half charges cordite on the port side." To-day there is a heavy demand for 10″ shells, and more have to be hauled by pulley from the lower magazines. "Volunteers wanted!" shouts an officer, and immediately a dozen willing hands from the disengaged side of the ship are helping to trundle the heavy projectiles to the ammunition hoists."[124]

CHAPLAIN WILLIAM PRICE, RN, HMS *TRIUMPH*.

Admiral Carden ordered the bombardment to stop at 4.40 p.m. to allow its effectiveness to be determined. As the ships closed in to check their handiwork, silent forts opened fire.

"[T]hey hadn't returned our fire, so we were ordered in to about 5,000 yards so as we could get within range for our six-inch to fire.

"We gave No. 4 fort a good pummelling, and then passed on to No. 6 fort, the one the French had been firing at, and gave that a few rounds… The one we had been firing at all day (No. 4), and we thought had been silenced, opened fire as we turned, and another one on top of a cliff did likewise, so we were in between two fires. Shot and shell were screaming and snorting and splashing all over the place."[125]

MIDSHIPMAN GEORGE BALCOMBE, RN, HMS *VENGEANCE*.

The spotting aircraft could see what the ships could not, including howitzers firing from reverse slopes that the ships could not reach, but they were unable to communicate with the fleet. Their radios were not working.

PACIFIC

HMAT *Seang Bee*, conveying the 2nd Reinforcements to Egypt, had run aground on the Great Barrier Reef the previous day. The next high tide had not refloated the ship, so everyone was put to work to lighten the ship.

"After being on the reef for 24 hours, an order was given that all coal was to be dumped over the side, so as to lighten the forward portion of the ship. The work was hard and continuous. Officers joined in, and did their share of shovelling. It must have been a

---

124 Price, William Harold, *With the Fleet in the Dardanelles*, pp. 11–15, Andrew Melrose (London) 1915.
125 *Kentish Express and Ashford News*, 15th May 1915.

HMS *Queen Elizabeth* at Mudros.

novel experience for the officers to be sworn at by the men, who couldn't be expected to recognise them under the thick coating of coal dust and grime."[126]

PTE WILLIAM BOOTH, 3RD BATTALION, AIF, HMAT *SEANG BEE*.

During the enforced idleness, some of the men took advantage of the opportunity to bathe, though this was not without risk.

"The shallower the water the more beautiful does this garden on the floor of the ocean appear; and at times one is strongly tempted to risk a dive when some particularly fine piece of coral is seen. This, though, would be almost certain death, for by looking long and carefully enough here and there can be made out an ugly form, lying quietly on the bottom waiting for anything eatable that may happen along, and if that "anything" happened to be yourself — well, sharks are not, as a rule, respecters of persons.

"On one occasion a boatload went on to the reef for a swim in the shallows. Ere long a Jacko hove in sight and charged straight at 'em. His fin betrayed him, however, and, he arrived in time to see them peering over the side. It was, though, a very narrow shave.

"I regarded those swims as the best I have ever had and they had a great bucking up effect on all of us. The officers knocked out great enjoyment practising with their revolvers on the sharks. Now and then a triangular fin would glide along the (surface and presently as it came into the shallow water a huge monster could be plainly made out. Suddenly reports of revolvers would ring out from the saloon decks and great was the excitement if the creature leaped forward or rolled over and disappeared."[127]

PTE GERALD SUTTON, 13TH BATTALION, AIF, HMAT *SEANG BEE*.

The transport continued its journey the following day and the sharks left in peace.

20 FEBRUARY 1915

ENGLAND

Kitchener cabled General Maxwell in Egypt to tell him to prepare for the arrival of troops, sent to follow-up, not accompany, the naval forces in the Dardanelles.

"A naval squadron is proceeding to the bombardment of the Dardanelles, and during the first day they have silenced one fort and severely damaged another. In order to assist the Navy a force is being concentrated in Lemnos Island to occupy any captured forts. At present 2000 marines are on the island, to be followed about March 18 by 8000 more. You should have a force of approximately 30,000 of the Australian and New Zealand contingents under Birdwood prepared for this service. We shall send

---

126   *The Forbes Advocate* (New South Wales), 27th April 1915.
127   *National Advocate* (Bathurst, New South Wales), 24th April 1915.

transports from here to convey these troops to Lemnos, and they should arrive at Alexandria about March 9. You should, however, communicate with the Navy through Admiral Carden, commanding at the Dardanelles, as he may require a considerable force before that date, so that you may be able to send him what he most requires."[128]

LORD KITCHENER, SECRETARY OF STATE FOR WAR.

EGYPT

More training, this time at brigade level, was being undertaken.

"The 1st Brigade, consisting of the first four battalions, completed its brigade training with a four days' bivouac. We had a route march of 10 miles to the camping ground, where we were under active service conditions. At 5 o'clock the first afternoon we marched out in battle array to prepare for an expected attack. This meant entrenching ourselves. We dug far into the night and about five feet into the desert; and made the discovery that the desert is not always sand, for a couple of feet from the surface the shovel bounced off a hard shale, which necessitated the use of the pick. In our training, emphasis is laid on the construction of trenches, for to avoid heavy casualties in modern warfare an army must dig itself in; so if the Australians imitate the rabbit in its burrowing they will be well protected. The trench was completed about 2 a.m., and we were able to get forty winks before the hour of dawn arrived, when all stand to arms so as to be prepared for surprise attacks. An armistice was called between the hours of 6 and 9, when we breakfasted. An attack on our position was made by the 4th Battalion, and it was rather interesting to see an approaching "enemy" alternately in the open and under cover. By noon operations were over, and we returned to the bivouac ground. Here we washed out of our water bottle or mess tin for want of better conveniences, and at night slept soundly on the sand, covered with only a banket [sic], great coat, and a further unavoidable addition of a heavy dew.

"On Sunday night, at 10 o'clock, the whole brigade marched into the darkness. It was a case of follow the leader, for it was impossible to distinguish anything at more than 20 yards' distance. We were on the move all through the night, with occasional spells, when we would drop down to rest, and it was remarkable the short time in which some lads could raise a snore. At one spot we had to go in single file over a plank that led across an irrigation channel, and the pace became snail-like. The whole march, in fact went in fits and starts, something like the progress made by a nervous man walking bare-footed through a thistle patch. At 4.30 a.m. the brigade massed in close formation, and with bayonets fixed made a charge with the idea of surprising and routing a supposed enemy. The platoon I'm in brought up the rear in the bayonet charge, a position which will do me in any actual charge. During the bivouac we were fed well and always had plenty to eat, the daily menu consisting of stew, potatoes, tea bread, jam and cheese.

"The Brigadier, Col. McLaurin, addressing the troops before the return to Mena

---

128 Lord Kitchener quoted in Arthur, Sir George, *Life of Lord Kitchener*, vol. 3, p. 113, MacMillan & Co., (London) 1920.

Camp, expressed his pleasure at the success of the bivouac, and declared that the 1st Brigade was "fit for war." And the signs of the time seem to foretell, and Dame Rumor has it, too, that we shall soon make our departure from Egypt."[129]

PTE JOHN REID, 1ST BATTALION, AIF, MENA CAMP.

But even the most intensive training offered some the chance to take a breather.

"To-day our Port Lincoln section has a special job. We are doing brigade work — the third brigade (Q., Tas., W.A., and S.A. battalions), march out into the desert and attack certain positions, practising the advance by companies and sections, taking advantage of cover, etc. Our section is very often chosen for special work, and to-day we represent the enemy. We are divided into small parties of three or four men and occupy certain positions, signalling the effect which the attack is making on our positions. One of the English staff officers directs the operations and umpires. Tom Corcoran,[130] a cousin of mine named Edmonds,[131] and myself hold the last position. This is about a mile eastward of the big Pyramid and is quite lofty, being, in fact, a hill which had formed on top of an ancient temple. As we are not likely to be attacked for some time Tom and I left the other chap to keep a look out and descended to a room which has been excavated."[132]

PTE JOHN BRADFORD, 10TH BATTALION, AIF, MENA CAMP.

21 FEBRUARY 1915

ENGLAND

Some officers clearly knew exactly where they were going, probably expecting to be there before the letters informing their families and friends reached them.

"We are on our way again and are off on an expedition into the Mediterranean, perhaps to Constantinople. We are to be a landing party to the fleet and are trying to force the Dardanelles for a passage of Russian wheat. That is, as far as we are able to guess. We are likely to be away for a few months, and then are going to land in France and go up to the big show in Germany… I am promoted and am third in command of the battalion, and hope to get a battalion before the war finishes. To-morrow the King reviews us and then we are off."[133]

LT-CDR BERNARD FREYBERG, RNVR, HOOD BATTALION, ROYAL NAVAL DIVISION.

---

129 *Lachlander and Condobolin and Western Districts Recorder* (New South Wales), 24th January 1938.
130 Pte Thomas Corcoran, 10th Bn, AIF, later promoted Capt. and awarded the Military Cross for his gallantry on the Western Front in 1917. The former railway guard died of wounds on 30th May 1918. Buried in Borre British Cemetery, he was the 28 year-old son of Thomas and Margaret Corcoran, of Tantanoola, South Australia.
131 Pte George Edmonds, 10th Bn, AIF. The former farm labourer was wounded around the time of the landings at Gallipoli, being admitted to 17th General Hospital, Alexandria, on 30th April 1915.
132 *West Coast Recorder* (Port Lincoln, South Australia), 14th April 1915.
133 *Oamaru Mail* (New Zealand), 9th April 1915.

EGYPT

The sight of enemy prisoners fascinated many. Some spectators considered the best of them were still beatable.

> "We saw a few prisoners while we were at Ismailia, but they were a scraggy-looking lot… clothed in ordinary chaff bags, and looked as if they had not had a feed for a week, and had not had a wash for a month or more. The Bedouins and Syrians were a motley lot, too; they were dressed in all colours. This is not usual with the Bedouins, as they usually stick to black and white. All of this lot were equipped with Mauser rifles. Some had belts and some bandoliers, but both were of poor stuff, and the wonder to me that they did not break when they were filled with cartridges. The few Turks that were taken had a khaki uniform, and were better equipped in every way. They had strong leather belts and pouches and bandoliers, and looked very business-like. They also had Mauser rifles and bayonets. The bayonet is very long, and would give them an advantage in reach over us, but we make up for that with a good heart and head."[134]
>
> CPL HORACE HART, 8TH BATTALION, AIF, MENA CAMP.[135]

22 FEBRUARY 1915

ENGLAND

Bad weather was hampering the bombardment of the Dardanelles. But scepticism about the chances of a purely naval offensive against the Dardanelles was not restricted to the corridors of the Admiralty or War Office. One British journalist, and a former M.P., held out no hopes for its success.

> "A mere bombardment, however, from the sea can do no more than damage the outer forts, which would not enable our fleets to force the passage. This could only be done by landing large forces on one or both sides of the Strait and occupying Gallipoli at its other end. For such a serious operation there does not seem to be any present preparation. One of our new armies might be worse employed than in this offensive movement. Failing, however, any action on land the bombardment from the sea is not of prime importance.
>
> "It is an undoubted fact that a British fleet under Admiral Duckworth forced the Dardanelles [arrived] in front of Constantinople in a rather battered condition, but could do nothing of importance there for lack of troops, and had to fight its way back again through the Strait to the Mediterranean. Such a procedure is, I repeat,

---

134   *Otago Daily Times* (New Zealand), 6th April 1915.
135   Cpl Hart was killed in action on 8th May 1915. Commemorated on the Helles Memorial, the former builder was the son of Thomas Hart, of Dunedin, New Zealand.

impossible in these days of enormous guns and floating mines. If there be no army to sustain the fleet there can be no Allied victory here."[136]

SAMUEL STOREY, CO-FOUNDER SUNDERLAND DAILY ECHO.

EGYPT

If doubts existed elsewhere, the news that things were stirring was celebrated by others.

"On Monday evening, February 22, word came through that the 3rd Brigade were to leave Egypt at the end of the week. Wasn't there much excitement and great joy! Intense cheering right throughout the Brigade gave some idea of the manner in which the boys welcomed, the news."[137]

PTE GARNET RUNDLE, 12TH BATTALION, AIF, MENA CAMP.

23 FEBRUARY 1915

TURKEY

Reports received in Istanbul about the progress of the allied navies affected the public mood in the city.

"The city is full of all sorts of rumors… When news from the front is bad for the Turks it seems good to the Armenians. And vice versa. The Greeks don't seem to care who wins. To that extent they are true neutrals…

"My information is that the Turkish coast batteries at Kum Kaleh, Orchanieh, and Sid-il-Bahr are no more. I learned to-day that the Allied ships guns outranged them hopelessly. If that keeps up the British and French will be here before long."[138]

GEORGE ABEL SCHREINER, ASSOCIATED PRESS OF AMERICA, ISTANBUL.

ENGLAND

"I had a busy and rather tiresome morning. The gale is at last abating in the region of the Dardanelles, and the ships were going to resume this morning their pounding of the forts. Winston is sending off his Naval Division on Saturday to be at hand when the military part of the operations becomes to ripen."[139]

HERBERT ASQUITH, MP, BRITISH PRIME MINISTER.

---

136 *Sunderland Daily Echo and Shipping Gazette*, 22nd February 1915.
137 *Terang Express* (Victoria), 13th July 1915.
138 Schreiner, *From Berlin to Baghdad*, pp. 37–38.
139 *Brisbane Courier* (Queensland), 30th July 1928.

Kitchener instructed his protegé, Birdwood, to investigate the prospects for military action to assist the forcing of the Dardanelles.

"By the earliest possible opportunity you should proceed to meet Admiral Carden, to consult him on the spot as to the combined naval and military operations which the forcing of the Dardanelles is to involve; the result should be reported to me. It is important that you should learn from local observation and information the numbers of the Turkish garrison on the Gallipoli Peninsula, and what their composition is; whether it is considered by the Admiral that it will be necessary for troops to be employed to take the forts, and, if so, what force will be necessary; whether a landing force will be required of the troops to take the forts in reverse, and generally in what manner it is proposed to employ the troops. Will the Bulair Lines have to be held, and will any military operations on the Asiatic side be necessary or advisable."[140]

LORD KITCHENER, SECRETARY OF STATE FOR WAR.

DARDANELLES

The fascination of being part of history in the making was finite.

"Every second is interesting. It's lovely to watch the walls crumbling and to see the devils scuttling away. I suppose the worst of the winter is over with you now. The fellows in France will be glad of some sunshine. Things seem to be moving slowly. I hope this year will bring peace; I'm sick of all this life."[141]

LT EDWARD CROMWELL COLCHESTER, RN, HMS *IRRESISTIBLE*.[142]

EGYPT

Things were moving. Loud cheers greeted this announcement.

"Well, lads, I have just been informed by the Brigadier that two battalions of Australian Infantry will leave Egypt for good this week, and I am delighted to inform you that our battalion is one of the two. The whole of the Australian Forces will take their departure, but the honour of being the advance guard has been given to the 9th Battalion (Queensland), and the 11th Battalion (Western Australia). I don't know our destination, but I hope our Battalion will be still further honoured by being the first Australian force to set foot on the enemy's country. The hard training in Egypt has been heart-breaking, but you have stuck to it well, and I feel sure that right through the piece you will play the game."[143]

LT-COL. JAMES LYON JOHNSTON, 11TH BATTALION, AIF, MENA CAMP.

---

140 Dardanelles Commission: Memoranda Produced as Evidence, TNA CAB 63/17.
141 *Cambridge Independent Press*, 26th March 1915.
142 Lt Edward Cromwell Colchester, RN, HMS *Irresistible*, was killed in action on 18th March 1915. Commemorated on the Portsmouth Naval Memorial, he was the 31 year-old son of Edward Cromwell Colchester and Marguerite Branford Colchester, of Great Shelford, Cambridge.
143 *The Blackwood Times* (Greenbushes, Western Australia), 13th April 1915.

SINGAPORE

Two sepoys were reported to have undertaken some of the most widely reported killings in the first hours of the mutiny. These may have been the men, said to have been seen firing on civilians, who were selected as the first to be executed. Publicly.

> "Thousands of natives and Chinese and others looked on. The two were brought out handcuffed and stood up against the concrete prison wall. A firing party of 10 were standing 10 yards away… The proclamation was read in Chinese, Malay and English, and then it was "Present; fire." One could see the bullets hit, and the bodies, although against the wall, almost bounce. Five fired at each Indian. One immediately fell, the other did not fall, although I was sure he was dead. The firing party got the order to again fire, and this time 10 bullets, making 15 in all, went through him. After it was over they were carried away on stretchers, and we went and had a look at the wall; just two round holes in the concrete, about 6 in. or 8 in. in diameter, with blood everywhere."[144]
>
> STANLEY GOWEN, AUSTRALIAN MERCHANT.

24 FEBRUARY 1915

ENGLAND

The need for an army to participate in the offensive in the Dardanelles was discussed at the Committee of Imperial Defence. Once more, the concerns regarding a potential defeat at the hands of a Muslim enemy were not hard to detect. But a greater interest in what opposition such an operation might face would have been useful.

> "We are all agreed that, whatever else is done, the Bosphorus operation must be carried through to a successful termination. This may involve a pitched battle with Turkish troops in the neighbourhood of Constantinople; and, so far as I could gather from our latest discussion, we have no very precise information as to the number and quality of the Turkish troops with which, in such circumstances, we might have to deal. Evidently we must work with ample margins, for a check there might amount to disaster."[145]
>
> ARTHUR JAMES BALFOUR, COMMITTEE OF IMPERIAL DEFENCE.

Kitchener continued to shy away from any commitment to send an army to the Dardanelles, even airing the possibility that its defenders might evacuate the Gallipoli peninsula of their own accord. In the process he wrapped himself in a 'chicken and egg-esque' conundrum.

---

144 *National Advocate* (Bathurst, New South Wales), 5th June 1915.
145 The War. The Dardanelles and Balkans Operations. Memorandum by Mr. A. J. Balfour. TNA CAB 42/1/44.

"It is clearly essential that General Birdwood should get in into personal consultation on the spot with Admiral Carden. In concerting operations he should be guided by the following considerations. The object of forcing the passage of the Dardanelles is to gain an entrance to the sea of Marmora with the view of ultimately gaining possession of the Bosphorus and overawing Constantinople. The forcing of the Dardanelles is an operation to be effected mainly by naval means, and one which when successful will doubtless be followed by the retirement of the Gallipoli garrison. So far as our information of the situation goes, it does not appear to be a sound military undertaking to attempt a landing in force on the Gallipoli peninsula, the garrison of which is reported to be 40,000 strong, until the naval operations for reduction of the forts have been successful and the passage has been forced. The entrance of the Fleet into the Sea of Marmora would probably have the effect of rendering the Turkish position in the Gallipoli peninsula untenable and of enabling a force to occupy the peninsula if considered necessary. But to land with 10,000 men in face of 40,000 Turks while naval operations are still incomplete seems extremely hazardous. If it can be carried out without seriously compromising the troops landed for the purpose, there would be no objection to the employment of a military force to secure hold of the forts or positions already gained and dominated by naval fire so as to prevent their re-occupation or repair of damage by the enemy."[146]

LORD KITCHENER, SECRETARY OF STATE FOR WAR.

The War Council had backed the naval bombardments on the basis that they could be broken off, portrayed as a demonstration only, should little be achieved. Kitchener now thought that could never have been the case.

"Lord Kitchener said that he "felt that if the fleet would not get through the Straits unaided, the army ought to see the business through. The effect of a defeat in the Orient would be very serious. There could be no going back. The publicity of the announcement had committed us."[147]

MAJOR-GENERAL WILLIAM ROBERTSON, CHIEF OF
STAFF, BRITISH EXPEDITIONARY FORCE.

Events in Singapore can hardly be said to have created that mindset but they could only have reinforced the fears that lay behind it.

TENEDOS

Tasked with establishing a base at Lemnos, Wemyss met Carden to discuss the plans made to date.

"During my hurried interview with the First Lord he had briefly told me that an attack on the Dardanelles was impending, that the island of Lemnos was to be handed over to us, that I was to administer it as Governor and form a base at Mudros for the naval

---

146   Dardanelles Commission: Memoranda Produced as Evidence, TNA CAB 63/17.
147   *The Mercury* (Tasmania), 5th January 1926.

and military forces that were to take part in the operations, and he had informed me that instructions would follow. When, therefore, I visited Admiral Carden, I had hoped to obtain some more definite information and was proportionately disappointed that he had none to give me, beyond the fact that it was only the harbour and the town of Mudros, and not the whole island, that was to be in my hands. He gave me to understand that some 10,000 troops might shortly be expected, but of any plans for combined operations he appeared to be as ignorant as I was. There were already at Mudros a brigade of marines under the command of Colonel Trotman; they were embarked in the two transports *Braemar Castle* and *Cawdor Castle* in instant readiness to disembark and demolish the forts as they were silenced, and therefore could not be used by me for any other purpose. And so I entered into my kingdom with but vague ideas of what I had to prepare for, but always hopeful that I should shortly receive some further instructions, however indefinite they might be."[148]

REAR-ADMIRAL ROSSLYN WEMYSS, RN, GOVERNOR OF MUDROS.

EGYPT

The commander of the British Army in Egypt felt there was no prospect of a naval breakthrough at the Dardanelles. He asked for copies of earlier feasibility studies to be forwarded to him.

"Admiral Carden's telegram which I repeated to you strikes me as being so helpless that I feel no progress is likely to be made unless we, i.e. the military authorities, take the initiative. We may take it that there are about forty thousand Turks west, and thirty thousands east of the Hellespont who could probably be concentrated on either side at short notice. I make the following proposals considerably in the dark as I have no knowledge of the deep study which must have been made of the whole question of the forcing of the Dardanelles by the Imperial General Staff and the Navy for many years past, the result of which must be in the War office and [a] resumé of which I should much like.

"At first glance an obvious place for disembarkation of a force may seem to be Xeros Bay, but I understand the Gallipoli Peninsula is heavily fortified and prepared for defence everywhere, presenting practically a fort, advance against which from any quarter without many heavy guns would seem hazardous. Is there any possibility of a landing at Besika Bay, and an advance up the Asiatic side having a definite effect on the defences of the Gallipoli Peninsula. If so in view of the strength of the garrisons and defences on the Asiatic side are weaker than on the European. Meanwhile I recommend that troops be despatched from here and concentrated at Lemnos as shipping admits so as to be ready for eventualities."[149]

GENERAL SIR JOHN GRENFELL MAXWELL, GOC BRITISH TROOPS, EGYPT.

148   Wemyss, Wester, *The Navy in the Dardanelles Campaign*, p. 18, Hodder & Stoughton (London) 1924.
149   ANZAC General Staff War Diary, TNA WO 95–4280.

More men were hearing that things were moving.

> "There is a great deal of talk about moving from here, and there is no doubt about going in three or four weeks as far as I can make out, but our destination is uncertain. It may be France or Palestine. It came from the officers this time, who said that we would have to buck up as we would be called out in a very short time.... It is said that some of the infantry will most likely leave here this week... There was a little excitement in the camp last night on account of some of the Turkish prisoners escaping. Our inlying picquet was called out about 10 o'clock and about 30 shots were fired. They were all recaptured."[150]
>
> CPL LESLIE BRYSON, 6TH AUSTRALIAN LIGHT HORSE REGT, MAADI CAMP.

25 FEBRUARY 1915

ENGLAND

Contrasting views were expressed by the respective political heads of the British Army and Royal Navy. The one who thought swift success was within their grasp had never commanded men in battle.

> "When our new armies are prepared to take the field, it will be undoubtedly a matter of vital importance that they should be employed in the most effective manner, so as to secure decisive results, and that they should not be scattered on subsidiary operations.
>
> "Much depends upon the success of the Navy in forcing the Dardanelles, but we have not sufficient men available at present to attack the Turkish troops on the Gallipoli Peninsula. As the situation develops in the Near East, we shall be better able to judge how our troops could be best employed when ready."[151]
>
> LORD KITCHENER, SECRETARY OF STATE FOR WAR.

> "For us the decisive point, and the only point where the initiative can be seized and maintained, is in the Balkan Peninsula. With proper military and naval co-operation, and with forces which are available, we can make certain of taking Constantinople by the end of March, and capturing or destroying all Turkish forces in Europe (except those in Adrianople). This blow can be struck before the fate of Serbia is decided. Its effect on the whole of the Balkans will be decisive. It will eliminate Turkey as a military factor."[152]
>
> WINSTON CHURCHILL, MP, FIRST LORD OF THE ADMIRALTY.

---

150 *Goulburn Evening Penny Post* (New South Wales), 15th April 1915.
151 The War. Remarks by the Secretary of State for War on the Chancellor of the Exchequer's Memorandum (G-7) on the Conduct of the War. TNA CAB 42/1/45.
152 Churchill, *The World Crisis* 1915, p. 185.

FRANCE

"We are letting Sir Edward Grey know that the French contingent, under General d'Amade, will be ready to start on the 2nd March and will consist of two infantry brigades, a cavalry regiment, two groups of 75's and one group of 65's, 18,000 men and 5000 horses."[153]

RAYMOND POINCARÉ, PRESIDENT OF FRANCE.

DARDANELLES

"The… day's action of 19 February showed apparently that the effect of long-range bombardment by direct fire on modern earthworks is slight."[154]

VICE-ADMIRAL SACKVILLE CARDEN, RN, EASTERN MEDITERRANEAN SQUADRON.

But, with the weather improving, the bombardment was renewed.

"Daylight… found our fleet on their way towards the mouth of the world-renowned Dardanelles. That day spelt the fall of the outer forts. What a sight it was; they were the first really angry shots fired by our ship, and there seemed a strange feeling about it all. I can hardly describe it, one gets so used to firing guns in peace time; but the circumstances and the fact that we were under the enemy's fire ourselves lent a peculiar thrill to the whole experience."[155]

PTE WILLIAM MULLINS, RMLI, HMS *ALBION*.

"The "Queen Elizabeth," "Lord Nelson," and "Agamemnon" opened fire from a long range, and apparently soon began to put the forts out of action. It was impossible to see the effect on the forts themselves, as they were obscured from us by tremendous clouds of dust and smoke; but, since the ships began to creep nearer and nearer, it was pretty certain that they were meeting with little opposition.

"The big new ships kept well away, but the older vessels continued their approach, and at close range opened up broadsides time after time. It must have been hell inside those forts with shells raining in upon them every second… we learned that all the forts at the entrance had been silenced. Not one of the ships had been touched at all."[156]

DECK HAND JOHN GEORGE COWIE, RN, HMS *MAJESTIC*.

One of the ships had been touched. The family of a 16 year-old Boy sailor, the youngest member of HMS *Agamemnon*'s crew, received this news.

---

153 Poincaré, Raymond, *The Memoirs of Raymond Poincaré, President of France, 1915*, p. 47, William Heineman Ltd. (London) 1926.
154 Carden quoted in Forrest, Michael, *The Defence of the Dardanelles. From Bombards to Battleships*, p. 98, Pen & Sword Maritime (Barnsley) 2012.
155 *Western Daily Press*, 14th September 1915.
156 John Cowie quoted in Goodchild, George, *The Last Cruise of the "Majestic"*, pp. 70–71, Simpkin, Marshall, Hamilton, Kent & Co. Ltd. (London) 1917.

"I am very sorry to have to inform you that your son Walter [Mockett][157] was seriously wounded in action this morning, at about 10.30. He was at the time returning from taking a message to the captain, when a shell exploded near him, and fragments of it struck him in four or five places on the legs, and on his right thigh. He was at once carried below to the surgeons, and on examination it was found that the wounds, although bad, were not of a nature to endanger his life, and also I am glad to be able to say not so severe, as to require any amputation of the limbs, nor indeed to make it unlikely that he will recover completely... After the action he was sent to Hospital Ship *Soudan*, with the other wounded men, and where he will receive the very best nursing and attendance that can be had. We are very sorry that he should have had the bad luck to be hit, as he is the youngest member of our ship's company, and a very great favourite, although he has been with us only quite a short time. He will, as soon as possible, be taken to the RN Hospital at Malta, and will then, as soon as he is sufficiently recovered, probably be sent home, so that it may not be so very long before you have him again."[158]

CHAPLAIN WALTER SCOTT, RN, HMS *AGAMEMNON*.

It was time for landing parties to be prepared to complete the demolition of the guns in the forts at the entrance to the Straits. There was a problem, though.

"[N]o khaki clothing had been issued to marines of the fleet... A landing in whites or blues was out of the question, so orders were issued for one suit of whites to be dyed, and the nearest thing to a khaki dye was coffee!

"So coffee it was, and soon all our suits were dipped in a copper of this beverage and hung up to dry. Then all hands to overhaul our rifles and equipment, draw ammunition, and then to muster on the mess deck for a talk on the morrow's operation by our OC, Capt Panton,[159] RM."[160]

PTE FREDERICK POWELL, ROYAL MARINES, HMS *IRRESISTIBLE*.

LEMNOS

Mudros Bay was an excellent natural harbour. It lacked nothing except port facilities.

"I knew Lemnos well. Many a time had I visited the island when a lieutenant serving in HMS *Undaunted* and HMS *Astraea*, and later on when commanding HMS *Suffolk*. I had tramped many a mile over its hills after partridge; I had joined in many a picnic on the shores of its harbour, but never had I realized the poverty of the country or

---

157 Boy 1st Class Walter Mockett, RN, HMS *Agamemnon*, died on 5th March 1915 from wounds received on 25th February 1915. Commemorated on the Chatham Naval Memorial, he was the 16 year-old son of George and Alice Florence Mockett, of "Sunny Bank," Albert Road, Deal, Kent.
158 *Deal, Walmer & Sandwich Mercury*, 27th March 1915.
159 Capt. Henry Panton, RMLI, HMS *Irresistible*, was posted for duty with the island guard at Tenedos between 28th March to 1st September 1915; then at Salonika between 2nd October and 16th November 1915; and at Suvla as Naval Observation Officer 17th November to 7th December 1915. Transferred to HMS *Lancaster*, he served in waters of south western America before being promoted Major and appointed to command the naval defences of Port Stanley from 2nd February 1917 to 5th January 1919.
160 *The Argus* (Melbourne, Victoria), 22nd April 1944.

the entire lack of any of the ordinary adjuncts of civilization until the morning of February 25, when I landed to discover its resources. I wanted piers, cranes, water — there was nothing I did not want, and I found a complete absence of anything that could be of use to me."[161]

REAR-ADMIRAL ROSLYN WEMYSS, RN, GOVERNOR OF MUDROS.

26 FEBRUARY 1915

ENGLAND

The War Council heard the Admiralty's response to the news that the Army was considering plans to capture and hold the end of the Gallipoli Peninsula.

> "The Admiralty, however, had replied in the sense that the scope of the present operations was limited to a naval bombardment, except if it was necessary to land Marines to capture single works or observation stations for mines or torpedo stations; that the military forces were not intended to participate in immediate operations, but to enable him to reap the fruits of those operations when successfully accomplished…
>
> "We should be in a better position to judge the situation when the defences at The Narrows began to collapse. Another objection he had to the despatch of so large a force was that if serious and prolonged operations were contemplated sufficient ammunition was not available."[162]

LT-COL. MAURICE HANKEY, SECRETARY, WAR COUNCIL.

Kitchener cabled Birdwood in light of the discussion. His advice could have been clearer.

> "Forcing of Dardanelles is being undertaken by the Navy and as far as can be foreseen at present [the role] of your troops until such time as the passage has actually been secured will be limited to minor operations such as as final destruction of batteries after they have been silenced under cover of fire of battleships. It is however possible that there may be concealed howitzer batteries inland with which the ships cannot deal effectively and you might have to undertake special minor operations from within the Straits for dealing with these if called upon by Admiral Carden. Remember however that there are large enemy military forces stationed both sides of the Straits and you should not commit yourself to an enterprise of this class without ample reconnaissance and assurance of ample covering fire by the fleet."[163]

LORD KITCHENER, SECRETARY OF STATE FOR WAR.

Others, outside of the Admiralty and War Office, remained convinced that military co-operation was essential.

---

161   Wemyss, *The Navy in the Dardanelles Campaign*, pp. 19–20.
162   Dardanelles Commission: Memoranda Produced as Evidence, TNA CAB/63/17.
163   ANZAC General Staff War Diary, TNA WO 95/4280.

"The naval operations against the Turkish forts at the entrance of the Dardanelles have, with the more settled weather, been resumed, and the report is that these forts have been entirely put out of action. So far this is good news. I retain the opinion, however, that the Dardanelles cannot be forced without the co-operation of a land army."[164]

SAMUEL STOREY, CO-FOUNDER SUNDERLAND DAILY ECHO.

DARDANELLES

At 2 p.m. the orders were issued for demolition parties to land on the Asian shore from HMS *Vengeance* and on the European side from HMS *Irresistible*.

Lt-Cdr Eric Gascoigne Robinson, RN, led the party from the *Vengeance* ashore at 2.30 p.m., suffering the first casualties, one man killed[165] and two wounded.

> "The Turks were hidden in a churchyard [cemetery] just outside the town, and they opened fire on us as we passed by, and we had a little bit of excitement for a time, I can tell you. We were fighting for about an hour, and then ships joined in and started shelling their positions, and it became too hot for them. Then they retreated, and we started potting them."[166]

PTE NEIGHBOUR COULSTOCK, RMLI, HMS *VENGEANCE*.

Robinson's men had not attempted to camouflage their white uniforms, so he ordered them to stay put while he went on. What happened next was recorded in the citation for his Victoria Cross.

> "Lt-Cdr Robinson on the 26th February advanced alone, under heavy fire, into an enemy's gun position, which might well have been occupied, and destroying a four-inch gun, returned to his party for another charge with which the second gun was destroyed. Lt-Commander Robinson would not allow members of his demolition party to accompany him as their white uniforms rendered them very conspicuous."[167]

Their improvised camouflage did not protect all of those who landed on the European side; some were wounded but not by the Turks.

> "[O]ur ship, *Irresistible*, leading, followed by *Vengeance*, *Cornwallis*, and *Dublin*. Meanwhile, the detachment had donned its "khaki" suits, and a motley crowd we looked! The coffee had stained the suits all shades — dark brown, light brown, and yellow — and most suits had patches and creases of white where the coffee had failed

---

164 *Sunderland Daily Echo and Shipping Gazette*, 26th February 1915.
165 Sgt Ernest Turnbull, RMLI, HMS *Vengeance*, was killed in action on 26th February 1915. Commemorated on the Chatham Naval Memorial, the former fishmonger's porter, who had enlisted in London on 22nd March 1904, was the 28 year-old son of Mary Turnbull, of 40 Lambourn Road, Clapham. One of those wounded was struck by a bullet that had first passed through Turnbull.
166 *Surrey Mirror*, 26th March 1915.
167 *London Gazette*, 13th August 1915.

to make contact! Many were the smiles and rude remarks we had from the seamen and stokers that morning when we paraded!

"However, I'm sure we were far too excited to worry over such trifles — we all looked upon this trip as an adventure — just fun! Soon came the order to fall in, and, the old *Irresistible* having arrived off Sedd-el-Bahr, we soon tumbled into the cutters alongside and were on our way to the shore, towed by our picket boat.

"We cast off from the picket boat just before reaching the shore, and a few strokes of the oars grounded our cutters… the beach, sloping gently up to the walls of Sedd-el-Bahr fort, looked as peaceful as could be, with the exception of huge craters everywhere made by the shells of our guns.

"While the seamen were landing the demolition charges and getting things ready for their part of the show, we formed up, and with scouts out ahead off we went, a first and then a second line of skirmishers.

"What lay ahead? Slowly we moved forward, the scouts out in front making the all-clear signal, and soon we were up the slope and past the fort. The massive walls were in ruins, great gaps and craters, everywhere; two of the four guns — 14 in. — inside had been uprooted. The other two looked untouched, and upon these our demolition party was to operate. Well past the fort now, we halted some hundreds of yards farther on…

"Meanwhile the demolition party had reached the fort and commenced its task of destruction. Soon the explosives were packed, the fuses lit, and the demolition party safe behind cover. A tremendous and earth shaking crash, and the job was done. As we were lying flat in the open, debris rained upon us from above, two of our men getting nasty cracks from some of the pieces.

"Came the signal to retire, and slowly, with every caution, we all returned to the waiting boats, embarked, and were soon aboard again."[168]

PTE FREDERICK POWELL, ROYAL MARINES, HMS *IRRESISTIBLE*.

EGYPT

The stay at Mena Camp was drawing to an end for 3rd Australian Brigade but where were they to go?

"The West Australian infantry and artillery, the third field ambulance and the 10th battalion of South Australians have received orders to be ready to move out of camp tomorrow morning, as advance guard to the division; for France they tell us, but I should not be surprised if they alter it to Turkey when we get on the boat."[169]

PTE WALTER MORLEY, 11TH BATTALION, AIF, MENA CAMP.

---

168   *The Argus* (Melbourne, Victoria), 22nd April 1944.
169   *Kalgoorlie Miner* (Western Australia), 6th May 1915.

HMS *Agamemnon* at the Dardanelles.

27 FEBRUARY 1915

ENGLAND

The founder of *Jane's Naval Review* considered the minefields, not the forts, to be the main barrier to progress in the Dardanelles.

> "Danger lies rather in the mine fields laid in the narrows. These will have to be cleared under fire from field artillery and rifle fire (even supposing all forts to be silenced), unless a strong allied army operates along the Gallipoli Peninsula
>
> "Supposing, therefore, that (as assumed) an attempt to force the Dardanelles, and so reach and capture Constantinople is toward (the assumption may be wrong) it will be — as both Duckworth [in 1807] and Hornby [in 1878] described it in the past — 'a difficult and dangerous task.'"[170]
>
> FRED T. JANE.

DARDANELLES

The power of the bombardment, the noise, flames and smoke was impressive. Even coming under fire from areas where no guns were thought to be did not shake one observer's confidence.

---

170   *Land and Water*, 27th February 1915.

"It is a great sight see the enormous shells from the ships exploding on the forts, send up huge columns of smoke and often causing large landslips down the cliffs in front of the forts. Then the shells from the forts falling in the sea, sending up columns of water over 100 ft., except when hit, when a much larger column of smoke would spurt up and you wondered what damage was done and how many were killed. Fortunately not many hit though in the forts it must have been an inferno. The fact that they stood their ground shows what sort of people we were up against.

"Yesterday (Friday) our turn came and I can say truthfully say it was most exciting. The outer forts being finished, we had to silence the smaller (though not very small!!) ones, and find hidden forts, up to, but not including the enormous forts at the Narrows. We went in and started off by blowing up a bridge; that was not very exciting as there was no reply, but you could see men scurrying out of the way of our shells. Then we went for a fort which, however, did not reply, but while we were shelling it there suddenly appeared a lot of puffs from a place on the shore where we did not know there were any guns, and a few seconds later there were shells falling all round us; fortunately most of them fell in the sea, but some burst over us and showered us with fragments, though fortunately beyond a few ropes and some of our aerial being carried away, no damage was done.

"We had this sort of surprise on sundry other occasions, but marvellous to say nobody was hurt. In my opinion the shore people had rotten bad luck, as both their strategy and their shooting were good where the result produced was practically nil. We had one huge shell burst in the sea just under our port quarter, which lifted the whole ship and shook like dog does a rat; but even that must have burst just away from the ship's side as beyond bending in a plate and starting a few rivets, no damage was done."[171]

ASSISTANT PAYMASTER HAROLD CHESHIRE, RNR, HMS *MAJESTIC*.

EGYPT

The 3rd Australian Brigade was preparing to move but Charles Bean's comments about their conduct continued to rankle.

"I suppose you have seen the papers in which the article by Captain Bean, 'Australia's Fair Name,' was printed. I am pretty well acquainted with the general run of things here, and I cannot for the life of me see any justification for the remarks. The behaviour of the troops, so far as I can see, has been splendid...

"I am sure Australia is not in danger of losing her fair name. There are one or two wasters — you cannot expect to have 50,000 people of any sort without one or two mongrels who cannot behave themselves properly."[172]

L/CPL ERIC INGLIS, 10TH BATTALION, AIF, MENA CAMP.[173]

---

171   *Hastings and St. Leonards Observer*, 27th March 1915.
172   *The Express and Telegraph* (Adelaide, South Australia), 30th March 1915.
173   The former clerk was commissioned at Gallipoli on 4th August 1915. Later promoted Capt., he was awarded the Military Cross for gallantry on the Western Front.

MESOPOTAMIA

The British soldiers already in action with the Turks were very confident they had the beating of them.

> "We simply laugh at the enemy, for we are far superior to them in fighting, though they are much more numerous. But the worst is to come, and we are prepared for it. I am not wounded yet; but am in the best of health, and doing well under the conditions. Bert. Castle[174] and Olliffe,[175] are also well, and are still together. When the bullets whiz over our heads bob down a bit smart I can tell you, and once or twice have had some near goes."[176]

L/CPL PHILIP OLLIFF, 1ST BN OXFORDSHIRE & BUCKINGHAMSHIRE LIGHT INFANTRY, 6TH (POONA) DIVISION.

PACIFIC OCEAN

The last surviving vessel of the German East Asia Squadron, the *Dresden*, claimed its last prize. Sinking sailing ships was considered to be bad luck. The *Dresden*'s was running out.

> "We had left port about nine days when, at 2 30 in the afternoon, we sighted a smoke trail on the weather beam. By 4 30 we had been caught and were ordered to heave to. The German cutter was soon alongside. They boarded us, examined the ship's papers, and then allowed everyone to collect his belongings ready for departure. In the meantime, the Germans were busy pinching everything of value, stores, ropes, wires, pigs, chickens, paint, etc. to the *Dresden* in our lifeboats, and the Germans used their own boats to carry back the plunder. By 6 p.m. everything was finished. The Germans fired a charge of dynamite fore and aft of the ship, which blew two large holes in her. What had been a few moments before a fine ship, lazily rolling on the swell, with every inch of canvas set except the royals, vanished completely. I saw her go. The ship had been a home to me for the last four and a half years and had carried me safely through storm and calm, day and night, three times round the world."[177]

HENRY DESPICHT, 2ND MATE, SS *CONWAY CASTLE*.

28 FEBRUARY 1915

ENGLAND

Henry Woods, a former Grenadier Guards officer, had inspected the geography around the Dardanelles before the war. He had seen the forts and was aware of their strength.

---

174 Pte Albert Castle, 1st Bn Oxfordshire & Buckinghamshire Light Infantry, was wounded at Ctesiphon on 22nd November 1915.
175 L/Cpl Philip Olliff, 1st Bn Oxfordshire & Buckinghamshire Light Infantry, died as a prisoner of war at Mosul on 14th July 1916. Commemorated on the Basra Memorial, he was the 26 year-old son of Louisa and the late Eli Olliff(e).
176 *Bucks Herald*, 27th February 1915.
177 *Manchester Courier*, 24th April 1915.

"At the moment of writing it is difficult to say with any degree of certainty the number or the size of the guns now available for the defence of the Dardanelles. Certain it is, however, that the forts, to the maintenance of which the Turks have always devoted considerable attention, have been greatly improved during the last few years. At the time of the Turco-Italian war no stone was left unturned to prepare to repel the serious attack which it was feared might be made by the Italian fleet. Again, during the Balkan wars there is no doubt that the strength of these defences was considerably increased in order to be prepared for every eventuality. But judging from the manner in which the Germans, before and since the entry of Turkey into the present war, have completely taken over the control of the whole of the Ottoman Army, it is clear that we must expect that they have mounted every available weapon in the immediate neighbourhood of the Straits."[178]

HENRY CHARLES WOODS, BRITISH JOURNALIST.

TURKEY

Reports received in Istanbul mirrored those published elsewhere. The allied fleet was expected to break through.

"The Allied Fleets have been seriously bombarding the Dardanelles during the past week. Details are lacking, but, from preparations in, and around the city, and on the shores of the Marmara, it is expected that the English and French Fleets will force the straits…

"The latest reports from the Dardanelles are from people arriving here February 27, and 28. They report that all the women and children have left Chanak, or Dardanelles, and that all natives, living near the threatened area, are moving back into the country."[179]

LT-CDR EDWARD MCCAULEY, USN, USS *SCORPION*, ISTANBUL.

MEDITERRANEAN

Meanwhile, the Royal Naval Division was on its way to follow-up the Navy's expected success.

"The other day we, a few thousand of us, were told we sailed next day to make a landing. Off we stole that evening, thro' the phosphorescent Aegean… We rose at four, buckled on our panoply, hung ourselves with glasses, compasses, periscopes, revolvers, food and the rest, and had a stealthy large breakfast. That was a mistake! We paraded in silence under paling stars along the sides of the ship. The darkness of the sea was full of scattered flashing lights hinting at our fellow-transports, and the rest."[180]

SUB-LT RUPERT BROOKE, RNVR, HOOD BATTALION, HMT *GRANTULLY CASTLE*.

---

178   *Weekly Dispatch* (London), 28th February 1915.
179   Naval Attache's Reports, Office of Naval Intelligence, Unpublished Manuscript, US Naval War College, February 1915.
180   *The Sydney Morning Herald* (New South Wales), 23rd April 1927.

EGYPT

If the residents of Istanbul believed the Dardanelles would be forced quickly, doubts were emerging elsewhere.

> "I am very much in the dark as to the intentions and objects of the Fleet in forcing the Dardanelles, and await Birdwood's report with great interest. As I write, I hear the Fleet have forced the entrance, but they have the difficult part before them. There are seven lines of mines, the great part worked by electricity from the shore; the first line is just south of Kilid Bahr – Chanak, the entrance to the Narrows. This minefield is well protected by a series of strong forts on both sides of the Straits. There are also two mine-laying boats at Nagara whose mission is to let loose mines to float down with the current. The Admiralty seem to me to be over sanguine as to the capacity of the Fleet to force the passage without an expeditionary force. The Gallipoli Peninsula is very strongly organised for defence – all the bays on the northern littoral are defended, and from Maitos [sic] to Gallipoli it is an entrenched fort. Apparently there are any number of 15-cm. howitzers in prepared positions."[181]
>
> GENERAL SIR JOHN MAXWELL, GOC BRITISH TROOPS, EGYPT.

Others were not 'over sanguine' about the poor quality of the opposition they might face.

> "It has been said that the Turkish troops opposed to us on the canal were of poor quality, and others were Assyrians pressed into Turkish service. This is not the case… The prisoners indeed looked a sorry lot, but then one must bear in mind that the march across the Arabian desert was an arduous one, and was no doubt responsible for much of their "raggedness"…
>
> "We expect to leave here shortly, perhaps for Turkey, as our Allied fleets are smashing the Dardanelles forts. It will be fine to get to Turkey. It will mean another fighting bar on our medals, too… We have settled down again in our old camp, but there is a restless feeling everywhere of expectancy."[182]
>
> 2/LT CHARLES MASSEY, CANTERBURY BN, NZEF, ZEITOUN CAMP.

What was not in doubt was that momentum was building. Things were beginning to move.

> "We broke up camp at Mena… and marched to Cairo in the evening — a 10 mile walk, except for the few of us who were mounted — reached Cairo about midnight and got ourselves, our horses and baggage on to two trains and reached Alexandria at daybreak. The other three battalions of our 3rd Brigade also came down during the night. The boat allotted to us was the *Suffolk*, and we had some engineers on board with us, and a half company of B (Charlie Barnes' Company) was put on the *Nizam* with the transport."[183]
>
> CAPT. EDWARD BRENNAN, AAMC, ATTD 11TH BATTALION, AIF, MENA CAMP.

---

181 Sir John Maxwell quoted in Arthur, Sir George, *Life of Lord Kitchener*, vol. 3, pp. 118–119, MacMillan & Co., (London) 1920.
182 *Lyttelton Times* (New Zealand), 8th April 1915.
183 *Great Southern Star* (Leongatha, Victoria), 20th July 1915.

Badly damaged, the *Gaulois* exits the Dardanelles, escorted by a flotilla of French vessels, 18th March 1915.

MARCH 1915

"All I can remember seeing were chunks of deck flying up in all directions."

Lt James Montgomery, RNR, HMS *Irresistible*

1 MARCH 1915

ENGLAND

Asquith made a statement to Parliament about the Dardanelles.

> "It is a good rule in War to concentrate your forces on the main theatre, and not to dissipate them in disconnected and sporadic adventures, however promising they may appear to be. That consideration…has not been lost sight of in the counsels of the Allies. There has been, and there will be, no denudation or impairment of the forces which are at work in France and Flanders, and both the French and ourselves will continue to give them the fullest and, we believe, the most effective support…

> "The enterprise which is now going on, and has so far gone on in a manner which reflects, as the House will agree, the highest credit on all concerned, was carefully considered and conceived with very distinct and definite objects."[1]

> HERBERT ASQUITH, MP, BRITISH PRIME MINISTER.

He later recorded Churchill's confidence, buoyed by news of the offer of Greek support.

> "Winston is breast-high about the Dardanelles, particularly as to-night we have a telegram from Venizelos announcing that the Greeks, are prepared to send three divisions of troops to Gallipoli."[2]

> HERBERT ASQUITH, MP, BRITISH PRIME MINISTER.

DARDANELLES

American journalist George Abel Schreiner watched the afternoon's bombardment. An experienced artilleryman, he enjoyed explaining its effect to those less familiar with its power.

> "They steamed up close to the entrance [about 2.00 p.m.], raked the remains of Kum Kaleh and Sid-il-Bahr once more, and then occupied themselves with something or other near the village of Erenkoi and on a hill known as In Tepeh.

> "The fire was too far off to rouse our interest, though [Frederick] Swing insisted upon going to the tower platform… The flashes of the guns and exploding shells interested my friend very much, and I am afraid that a great deal of the kindergarten course in artillery I gave him was lost on him. This is the first time that my friend had seen such things. He seems torn between the curiosity of a child and the dread that things may happen, as I gathered when I gave him a lurid description of the effect of an exploding shell.

> "At 4.30 the Allied ships withdrew, that being tea-time for the British jackies. I was informed that the British adhere rigidly to their gastronomic habits."[3]

> GEORGE ABEL SCHREINER, ASSOCIATED PRESS OF AMERICA.

---

1   Hansard, Success of Allies Assured, House of Commons Debate, 1st March 1915.
2   Brisbane Courier (Queensland), 30th July 1928.
3   Schreiner, From Berlin to Baghdad, pp. 79–80.

One man was happy to escape the atmosphere in a gun turret on one of the ships.

"The way we banged at them was terrible. They were some time before they answered us, but when they did they were not slow in letting us have them. As you know, our gunnery is far superior to theirs, and when we got the range we just simply let fly, and every shell got home.

"The first day we had of it I was in one of the 6 in. gun turrets, so that in case of fire I should be near the fire hose, but they have moved me down below, and I am not sorry. What with the noise and the fumes after the guns had been fired, I had enough of it, although from there I could see what they were firing at, and could see the enemy return the compliment."[4]

CARPENTER'S CREW HORACE STOLWORTHY, RN, HMS *CORNWALLIS*.

EGYPT

More men left their camps to begin their journeys to Lemnos.

"Alexandria was reached at 1.30 a.m... then went on board the transport... but on inspecting her found her not nearly so well appointed from the rank and file's point of view at any rate, as our troop decks were very dirty and the hammocks of practically no use. We were her fifth lot of soldiers, so I suppose that accounted for much of her filth. The officers were much better off, as they had a cabin to themselves. Things were very quiet on the wharf, so a lot of the lads strolled up Alexandria... At three o'clock in the afternoon a picquet was sent up town to bring the stragglers back, after which no one was allowed ashore."[5]

PTE GARNET RUNDLE, 12TH BATTALION, AIF, HMT *DEVANAH*.

LEMNOS

As more members of the Australian & New Zealand Army Corps (ANZAC) arrived at Mudros, their training could focus on:

"(a) Disembarkation from Transports, landing on an open beach or at improvised floating piers; (b) Re-embarkation from an open beach or from improvised floating piers, to include embarkation from boats onto transports.

"Both should be practised frequently, by day & night. The Army Corps Commander is particularly desirous that this should be fully practised as much in what lies before us will depend on the training done now.

"(c) Occupation of positions cover both (a) & (b) by day & by night — in both cases without other reconnaissance beyond what can be done from a map or chart."[6]

LT-COL. ANDREW SKEEN, 24TH PUNJABIS, ATTD GENERAL STAFF, ANZAC.

---

4   *Spalding Guardian*, 2nd April 1915.
5   *Terang Express* (Victoria), 13th July 1915.
6   ANZAC General Staff War Diary, TNA WO 95/4280.

2 MARCH 1915

ENGLAND

Churchill stuck to his view that the Navy did not require military assistance. The Army would follow in its wake, not march in parallel. And that was no secret.

> "I have now heard from Carden that he considers it will take him fourteen days to enter the Sea of Marmora counting from March 2. I wish to make it clear that the naval operations cannot be delayed for troop movements."[7]
>
> WINSTON CHURCHILL, MP, FIRST LORD OF THE ADMIRALTY.

> "Today K. [Kitchener] and W.S.C. [Churchill] told us, with the air of importing a great secret, that their officers had decided to send to occupy Gallipoli 40,000 British and Colonial troops from Egypt, one French Division, one Naval Division from England, one Russian Brigade… So far as secrecy was concerned my brother-in-law John Fuller had heard this news from his friends at Brooks, where the servants were gossiping about it."[8]
>
> CHARLES HOBHOUSE, MP, POSTMASTER GENERAL.

One minister tried to calm any talk that the recent mutiny in Singapore was anything other than a little, local difficulty; a regimental matter.

> "It is important to make that clear, because it might easily be supposed that an outbreak of this kind in a Moslem regiment might conceivably be due to some spread of what is known as a Jehad feeling — a sympathy with the altogether unauthorised declaration of a Holy War which some who profess to represent Islam have endeavoured to make but which has been not merely not accepted but absolutely denounced as false by those who are best qualified to represent the true Mahomedan feeling… the disturbance undoubtedly was of a purely local and special character connected with regimental matters."[9]
>
> MARQUESS OF CREWE, THE LORD PRIVY SEAL AND SECRETARY OF STATE FOR INDIA.

ATLANTIC OCEAN

> "Got up and had a delicious salt bath, then ate splendid breakfast. Read in the morning. After lunch sat with Henry (a lieutenant), Asquith and others on the boat deck. Watched dolphins sporting about in the water… Far out we could see the other transport, which had the rest of the Brigade on board, signalling to us. We replied. Made a discovery to-day that we are definitely going to the Dardanelles."[10]
>
> CHAPLAIN HENRY FOSTER, RN, 2ND (NAVAL) BRIGADE, HMT *GRANTULLY CASTLE*.

---

7   Churchill quoted in Ballard, Brig. Gen. Colin Robert, *Kitchener*, p. 301, Faber & Faber (London) 1930.
8   David, Edward (Ed.), *Inside Asquith's Cabinet. From the Diaries of Charles Hobhouse*, p. 224, John Murray (London) 1977.
9   *Hansard*, The Regimental Riot At Singapore, 2nd March 1915.
10   Foster, Rev. H. C. Foster, *At Antwerp and The Dardanelles*, p. 55, Mills and Boon (London) 1918.

DARDANELLES

Though the weather was deemed too rough to send any demolition parties ashore, another bombardment was carried out. The *Canopus*, last heard of aground in the Falklands, took part.

> "We opened fire with our 12 in. guns and found the range first time, but got no reply for a while. We manoeuvred about to bring the 6 in. guns into play, and the reply we then got was fast and furious; shells were soon flying around and over us. Eventually the fort got our range and let us have a few shells on board, but not enough by a long way for us to have a piece each for a curio — not even enough for the officers to have a piece each. They did manage to blow off our mainmast half way up, and the top part came down with a run; we also had a couple of holes in our funnels, our boats splintered a bit, and a few ropes smashed here and there. One large shell landed nicely on the quarter-deck and went through the plank and steel deck into the ward-room, through the next steel deck and nearly through the next making a tidy mess. Luckily it struck that part of the deck immediately over the bulkhead, and apart from scattering the electric-light and a few steel plates did little damage. I am glad to say we did not have a man scratched. Thank God we all came out as we went into action, with the exception of a few headaches from the reports of our own guns. We were bombarding for more than four hours — it may have been five, as I did not take the exact time — and a number of their guns were destroyed, but I cannot say for certain the amount of damage done."[11]
>
> STO. 1 GEORGE ADAMS, RN, HMS *CANOPUS*.

According to Schreiner, the firing had been very poor.

> "The big event seems to be drawing nearer. To-day, for the first time, the ships of the Allies ventured well into Erenkoi Bay… A little before noon appeared four of the British line ships, members of the *Majestic*, *Victoria*, and *Agamemnon* classes, accompanied by two cruisers and a herd of smaller fry. After a while the cruisers and most of the small vessels withdrew, and the bombardment began…
>
> "The British ships were… not stingy with their ammunition. The same can be said of the Turkish howitzers on the Anatolian and Gallipoli hills. Things became rather hot in the lower stretch of the bay… The British gunners were working hard. Great tongues of yellow fire leaped from the turrets of the three ships. Huge clouds of reddish smoke seemed to spring out of nothing in the next instant, and seconds later a great geyser of earth and powder fumes would break into view somewhere on the Anatolian shore.

---

11   *Western Daily Press*, 19th April 1915.

> "Fort Dardanos especially was severely punished, as the British must have thought. But that was a fallacy. From the tower I could see quite plainly that much of the British fire was far too high… The fire of the ships was very poor. I concluded that the high sea was responsible for that."[12]

GEORGE ABEL SCHREINER, ASSOCIATED PRESS OF AMERICA.

TENEDOS

> "I took the General [Birdwood] over to Tenedos to see Admiral Carden. The weather was still bad, but a bombardment was going on. We had a conference with the Admiral; he was evidently coming to the conclusion that the enemy's concealed guns and mobile batteries were rendering mine-sweeping impossible, and that they would have to be cleared away before any real progress could be made with the work of attacking the inner forts. To accomplish this troops would have to be landed upon the Peninsula and the question arose with how much opposition would such a landing meet? These matters would have to be settled elsewhere than at Mudros or Tenedos."[13]

REAR-ADMIRAL ROSSLYN WEMYSS, RN, GOVERNOR OF MUDROS.

EGYPT

As more ships left Alexandria bound for Lemnos, not everyone completed the journey.

> "11 a.m. we moved off. The pier was practically deserted. Our band started playing, and in five minutes there was a huge crowd watching us depart. They seemed to come out of thin air and from "round the corners." Going out we passed the U.S.A. cruiser *North Carolina*. Her band played "God Save the King," and someone aboard her signalled "good bye and good luck." Our band then played their national tune, and followed with "Yankee Doodle." Soon we were once more in the Mediterranean, and then, for the first time learnt our destination (the favorite winning) — Lemnos Island, 40 miles south-west, of the Dardanelles. A rather heavy sea was running, and although this boat behaved splendidly, quite a few found their spell on land had made them bad sailors. I didn't. A rush was made for sleeping room up on deck, and I managed to secure a position."[14]

PTE GARNET RUNDLE, 12TH BATTALION, AIF, HMT *DEVANAH*.

> "I camped on the port side promenade deck… Not feeling comfortable I picked up my blankets and went downstairs about 1.30 a.m. On going to the place where my section slept near the upright bars protecting the open hatch I felt something liquid sticking to my feet thro' my sox. In bending down to find out what it was by the gleam of the light I saw blood issuing from a man's throat. I immediately ran about 10 yards and put my blanket away and returned and placed the man's black kit bag further under

---

12 Schreiner, *From Berlin to Baghdad*, pp. 81–83.
13 Wemyss, *The Navy in the Dardanelles Campaign*, p. 28.
14 *Terang Express* (Victoria), 13th July 1915.

his head as I thought it would lessen the flow of blood. Then [I] wrapped my handkerchief round his kneck [sic]. I could not tie it up as the blood was flowing all over my hands and the handkerchief... was too small. After doing this I ran as quickly as I could to warn the officer on watch as I did not know where the medical officer slept... After this I returned to where the man lay and waited till Capt. Dukes arrived... I recognised the the man from the first as Pte J. Davenport[15]... I thought that the man had attempted suicide. There was not the slightest indication of anyone else being concerned in the affair as far as I could see."[16]

PTE NORMAN WEBB, 9TH BATTALION, AIF, HMT *IONIAN*.[17]

3 MARCH 1915

ENGLAND

"The mention in the official report of landing parties from the *Vengeance* and *Irresistible*, which completed the work of demolishing the forts at the entrance to the Straits is an interesting foreshadowing of what is to come. The last time a British fleet passed through the Dardanelles was 1878, and Sir Geoffrey Phipps Hornby, who was in command, reported to the authorities that unless the Gallipoli Peninsula was in friendly hands great difficulty would be experienced, even if the fleet succeeded in passing through, in keeping the route open to supply ships, colliers, and other craft auxiliary to the work the fleet. That difficulty exists no less today, and must be overcome."[18]

HUBERT FERRABY, NAVAL CORRESPONDENT.

DARDANELLES

Though bad weather once more delayed proceedings, confidence of success remained high. But those aboard one ship had learned reasons to be cautious.

"We have had very slight casualties up to now, and those only occurred on shore whilst a party were engaged in blowing up the remains of some of the forts. We still have a hard task before us, but everybody is hoping and expecting that the squadron will be off Constantinople within the next week or so."[19]

LS CHARLES CLEMENTS, RN, HMS *VENERABLE*.

---

15   Pte James Holton (alias Davenport), 9th Bn, AIF, died on 7th March 1915 as a result of self-inflicted injuries. Buried at Alexandria (Chatby) Military and War Memorial Cemetery, he was the 34 year-old son of Fanny Holton, of London.
16   NAA: B2455, DAVENPORT J H.
17   Webb, a former clerk, was killed in action on 25th April 1915. Commemorated on the Lone Pine Memorial, he was the 25 year-old son of Sidney and Ann Webb, originally of Paddington, Queensland.
18   *Dundee Evening Telegraph*, 3rd March 1915.
19   *Richmond and Twickenham Times*, 3rd April 1915.

"Although the *Triumph*, when bombarding at close range, had on more than one occasion such success that enemy guns were actually seen to be slewed round or knocked upward, the reduction of land fortifications by ships' guns is generally regarded as a matter of doubtful success, however close the range may be. The experience of the *Triumph* at Tsingtau was valuable. She had learned to distrust silenced forts. Often guns would reopen fire after hours of silence had given the impression that they were out of action. In the Dardanelles, only by putting ashore demolition parties could the complete destruction of the guns be assured. And this is hazardous work."[20]

CHAPLAIN WILLIAM PRICE, RN, HMS *TRIUMPH*.

LEMNOS

Landing parties could achieve limited, local successes but, after speaking to Carden, Birdwood concluded much more was required.

"I anticipate, that if required to land by the Navy, we shall not be able in taking concealed guns or howitzers to restrict movements to minor operations, as any guns are sure to be in strong positions and very numerous, and would be covered by strongly entrenched infantry, who in some places would doubtless be able to command coast fort guns which might have been reduced by the Navy. With the exception perhaps of two Mounted Brigades, and one of these lately arrived from Australia is but little trained. I will certainly try to dominate the Eastern side from the Gallipoli peninsula, as I am particularly anxious to avoid, if possible, placing any troops on the Asiatic side; for not only do I fully realize the danger of placing there a more or less isolated force, but I know from personal observation that the country is big and difficult, and even a whole Division would soon lose itself."[21]

LT-GEN. SIR WILLIAM BIRDWOOD, COMMANDING ANZAC.

MEDITERRANEAN

"Save for a rough sea and sighting one of our watch dogs well out to starboard there was nothing exciting. Everybody was looking for rest, and cards were about the only recreation — that is, when not on parade. Sure, they keep us at it. We had a couple of hours this morning, getting the sand out of our rifles, and a little physical drill during the afternoon. We were going to have some boat drill, too, but the sea was too rough and the decks too wet."[22]

PTE GARNET RUNDLE, 12TH BATTALION, AIF, HMT *DEVANAH*.

---

20   Price, William Harold, *With the Fleet in the Dardanelles*, pp. 122–123, Andrew Melrose (London) 1915.
21   Dardanelles Commission: Memoranda Produced as Evidence, TNA CAB 63/17.
22   *Terang Express* (Victoria), 13th July 1915.

4 MARCH 1915

GERMANY

The Straits' defences did not include any submarines. That was going to change.

> "The Turks want a submarine, but the Austrians have refused, as being too dangerous. We have offered them a crew."[23]
>
> GROSSADMIRAL ALFRED VON TIRPIRZ, SECRETARY OF STATE, GERMAN IMPERIAL NAVAL OFFICE.

DARDANELLES

Landings were made on the shores north and south of the Straits. Those heading for the northern side got into their boats at 9.00 a.m.

> "We took a zig-zag course from the ship we left, and when we got within a thousand yards of Seddul Bahr, our ships opened fire to cover our landing. The men were landed from the destroyers into picket boats, and they were towed ashore. We had a good landing, the ships kept up firing for 15 to 20 minutes, then stopped, and started up the hill into the village. As soon as we reached the top of the hill, snipers opened fire on us from some houses, which were difficult for us to get at, but we kept them at bay with the aid of our ships' guns, until the demolition party had done their work in the forts.[24] Our seaplanes were flying round giving signals to our ships where the enemy was. We started to advance through the village, but we had orders to retire, as the enemy was in strong force... Our losses were one non-commissioned officer and two privates killed, and one private wounded, and one sailor killed and one wounded on the picket boat. They were working a machine gun. We retired under heavy shell fire by picket boats to a destroyer, then by zig-zag course to our ship. When we got on board we were for our tea. That was our first experience in warfare."[25]
>
> PTE WILLIAM POLLARD, PLYMOUTH BATTALION, RMLI.

At 9.10 a.m. a party from the Plymouth Battalion and the *Nelson* reached the Asian shore. It, too, met with strong opposition.

> "This landing was the signal for sharp and well directed rifle fire to be opened by the enemy upon the jetty and approaches from it to the village. It soon became apparent that the buildings in the vicinity of the FORT entrance, afforded well-concealed shelter for the enemy's riflemen. Throughout the day, however, no success was achieved in pushing through the Northern end of KUM KALE village...
>
> "At 2.45 p.m. I thought the situation warranted my pushing on to FORT 4, and the

---

23 Admiral von Tirpitz quoted in Chatterton, E. Keeble, *Dardanelles Dilemma. The Story of the Naval Operations*, p. 245, Rich & Cowan Ltd. (London) 1935.
24 Two Nordenfeldt guns were destroyed.
25 *Cheshire Observer*, 24th July, 1915.

Reserve of one Platoon was ordered to support this movement. An Advanced Guard was moved forward in charge of Major A. E. Bewes, followed by the demolition part under Lieut. W. L. Dodgson RN. This advance progressed as far as the LAGOON…. At this point we were dominated by the heights of YENI SHEHR village, and our scouts were very soon held up by well directed rifle fire from FORT 4, and fire trenches which we now saw on the NORTHERN slope of YENI SHEHR hill. This fire increased in intensity at 3.45 p.m. and it became abundantly clear that the ground was too strongly held to be captured by the small force at my disposal."[26]

LT-COL. GODFREY MATTHEWS, PLYMOUTH BATTALION, RMLI.

The landing had failed. A withdrawal was effected but one of the beach party was killed.

"I was a messmate of Alec's,[27] and the other signalman that landed with him, so am better able to tell you the real facts. We were two of many that were landed, and it was during this time that Alec asked me, should anything happen to him, to write and tell you. There is no man living who could have his duty more bravely or more eagerly than did your brother. I know this, because I overheard the officer in charge praising him. It was indeed a sad blow to me to hear that he had been shot by the enemy, and both I and all his messmates sympathise with you deeply in your great loss. No man, I am sure, would wish to die a more noble death, for to be killed in defence of one's country's honour is the greatest of all."[28]

SIGNALMAN ALBERT MARSLAND, RN, HMS *NELSON*.[29]

Cumming was one of 47 casualties: 20 killed, 24 wounded and three men reported missing. The Turkish response had been far more aggressive than the previous month. The disappointment weighed heavily on those watching offshore.

"Great difficulty was experienced in withdrawing the attacking force, the enemy having gained possession of a cemetery up near Mendere Bridge, commanding the ground over which our men had to fall back after their repulse. Eventually destroyers went close in and covered the retirement. At dusk the destroyers *Scorpion* and *Wolverine* ran in and landed parties to search the beach from Kum Kale to the cliffs below Fort 4 for wounded stragglers. The former brought off two officers and five men who had been unable to reach the boats… It was the first time I had watched men being killed and wounded — from a position of perfect safety — unable to do anything to help — and it was a most distressing experience."[30]

COMMODORE ROGER KEYES, RN, CHIEF OF STAFF, EMS, HMS *INFLEXIBLE*.

26   Royal Naval Division General Staff War Diary, TNA WO 95/4290.
27   OS Alexander Cumming, RN, HMS *Lord Nelson*, was killed in action at Kum Kale on 4th March 1915. Commemorated on the Helles Memorial, he was the 18 year-old son of William Cumming, of 15 King Street, New Elgin.
28   *Northern Scot and Moray & Nairn Express*, 3rd April 1915.
29   Signalman Albert Marsland, RN, HMS *Nelson*, was killed in action on 28th June 1915. Buried in Lancashire Landing Cemetery, he was the 23 year-old son of Alfred and Margaret Marsland, of Manchester.
30   Keyes, Sir Roger, *The Naval Memoirs of Admiral of the Fleet Sir Roger Keyes. The Narrow Seas to the Dardanelles 1910–1915*, pp. 203–204, Thornton Butterworth (London) 1934.

LEMNOS

"On March 4 the vanguard of what eventually proved to be a large army arrived in the form of 5,000 Australians. Of these, 1,000 were landed and encamped close to the town, the others remaining afloat in their transports, principally on account of the lack of water."[31]

REAR-ADMIRAL ROSSLYN WEMYSS, RN, GOVERNOR OF MUDROS.

"After a good trip of about three days, we arrived at Lemnos, which we found to be a very mountainous island, with a beautiful harbor, which was entered through a gap between the hills and which could not be seen from the sea. We proceeded through the winding entrance, and on reaching the inner harbor found there a large number of vessels riding peacefully at anchor, none of which could be seen until we were practically alongside them in harbor. The change from Egypt to Lemnos was much appreciated by all on board, the weather being very cold and bracing, and the majority of the troops saw for the first time snow-clad mountains. On the voyage from Egypt we had passed numerous islands on which snow could be seen, and from Lemnos could be seen in the distance [the] snow-covered peaks glistening in the sun."[32]

LT-COL. JAMES LYON JOHNSTON, 11TH BATTALION, AIF, HMT *NIZAM*.

"Going to our position at the far end of the harbor, we passed several warships and many smaller boats, torpedo destroyers, submarines, mine sweepers, and colliers. We had the honor of being the first transport to arrive. The first impression of Lemnos Island is good. The harbor is practically oval, and from our position we could not see the ocean, as the entrance is long and has big hills on both sides. The island itself is very hilly, but beautifully green, and the slopes are much cultivated. Big windmills are on all the high crests."[33]

PTE GARNET RUNDLE, 12TH BATTALION, AIF, HMT *DEVANAH*.

EGYPT

"Parading daily in the toilsome heavy sand proved exceedingly arduous... Lately, however, we have been doing little training, and are inclined to believe that our preliminary task is at an end, and that we are waiting for orders to proceed to a more active form of service... we are to proceed to the Dardanelles, as Dame Rumour will have it."[34]

L/CPL WILLIAM FRY, 6TH BATTALION, AIF, MENA CAMP.

---

31 Wemyss, *The Navy in the Dardanelles Campaign*, p. 29.
32 *The Sun* (Kalgoorlie, Western Australia), 11th June 1915.
33 *Terang Express* (Victoria), 13th July 1915.
34 *Kapunda Herald* (South Australia), 16th April 1915.

5 MARCH 1915

MEDITERRANEAN

"[W]e entered the Straits of Gibraltar about 4 a.m. We did not go up on deck, but from the port-hole I could see the coast of Southern Spain quite plainly. A line of destroyers was stretched across the Straits, and an examination ship came up and asked us who we were before we were allowed to proceed."[35]

CHAPLAIN HENRY FOSTER, RN, 2ND (NAVAL) BRIGADE, HMT *GRANTULLY CASTLE*.

DARDANELLES

Schreiner again observed the shelling of the forts. Once more, the spectacle was impressive but the actual effect negligible. Naval guns were not designed for this purpose.

"War drew appreciably closer to-day. Some of it came in the form of direct fire from the Bay of Erenkoi, and more from across the peninsula of Gallipoli. The Allies stationed a few of their line ships and cruisers in the Aegean, off Ariburnu, and took under fire from there the batteries of Kilid-il-Bahr, across the strait from us…

"At about 3.45 [p.m.] an Allied hydroplane hove into view above Kilid-il-Bahr. The observer was to report on the damage done. What he reported I don't know, of course. I hope he told the truth — that the effect so far had been nil… The many earth columns that rose, the vivid flashes of explosions, and the ever-gaining blaze made an impressive picture. Fuad Bey was sure that no stone was left on top of another… I told him he was very much mistaken…

"Allied ships kept up their fire diligently, but… the ammunition used… was not suited for the work in hand. The shells were made to do their best in armor-penetration. But there is no armor worth mention along the Dardanelles. The parapets and traverses are of sand. The shells of the Allies do not go far in that. Each little grain acts as a brake upon the side of the projectile, whose programme is in that manner quickly checked.

"When the retarded fuse finally springs the fulminating charge the shell has not penetrated deep enough to have the explosion do much damage. Shells that would go through steel armor as through a piece of cheese are nearly worthless against the sand protecting the emplacements."[36]

GEORGE ABEL SCHREINER, ASSOCIATED PRESS OF AMERICA.

---

35 Foster, *At Antwerp and The Dardanelles*, p. 56.
36 Schreiner, *From Berlin to Baghdad*, pp. 87–94.

GABA TEPE

"We [were] firing across the strip of land into the big forts well up the Dardanelles, but we did not do much damage, as we could not get the right ranges because the seaplane had some mishap and came down, injuring the airman."[37]

STO. JOHN LAPSLEY, RNR, HMS *QUEEN ELIZABETH*.

"A seaplane went up to direct the fire, and while getting to his height, before going over the forts, something went wrong, and down came the machine; what a time it seemed to be in falling, and my heart seemed to stop beating. With a big splash they landed in the water and were picked up; both occupants were seriously injured though neither were killed,[38] which seemed a miracle, as they had fallen 3,000 feet."[39]

STO. 1 WILLIAM CARRADICE, RFR, HMS WEAR.

"[T]he *Queen Elizabeth* was ordered to carry out indirect firing at the Chanak Forts, from a position to the south of Gaba Tepe. *Prince George* was ordered to discover any shore batteries and prevent them from interfering with the *Queen Elizabeth*. Troops were observed as well as some trenches, which, from the absence of freshly turned soil did not appear to have been recently dug.

"At 12.38 a shore-gun opened fire at *Queen Elizabeth* and *Prince George* and *Inflexible* replied on the place from which the enemy appeared to be firing. At 12.45 a shrapnel shell burst in the ward-room of the *Queen Elizabeth*, only a very short time after the last person had left; if it had arrived five minutes earlier it would have caused many casualties."[40]

SURGEON OCTAVIUS ANDREWS, RN, HMS *PRINCE GEORGE*.

EGYPT

"I think we are moving shortly, but we do not know when. We are expecting hurried orders any time. The 9th, 10th, and 11th infantry left last Monday for the Dardanelles… I only wish it were we who had gone. We are sick of this place, and will be getting sanded if we are here much longer."[41]

TPR ARTHUR HOLT, 3RD AUSTRALIAN LIGHT HORSE REGT, AIF, HELIOPOLIS.

---

37 *Linlithgowshire Gazette*, 9th April 1915.
38 Flight-Lieutenant Walter Garnett, RNAS, and Flight-Commander Hugh Williamson, RNAS, were flying in a Sopwith 807 when its propeller failed. They were both injured in the crash-landing into the sea; Williamson severely. After transferring to the Royal Flying Corps, Garnett was killed in an unexplained flying accident at the Central Flying School on 21st September 1916. He is buried in Upavon Cemetery, Wiltshire.
39 *Consett Guardian*, 20th August 1915.
40 Andrews, Octavius William, *Seamarks and Landmarks. Being Leaves From the Log of Surgeon Captain O. W. Andrews*, p. 274, Ernest Benn Limited, (London) 1927.
41 *Barrier Miner* (Broken Hill, New South Wales), 24th October 1915.

6 MARCH 1915

ENGLAND

It had been hoped that a grand Balkan alliance would emerge once operations against the Dardanelles commenced. How this was ever considered likely is unclear.

> "Russia, despite all our representations and remonstrances, declines absolutely to allow the Greeks to have any part in the Dardanelles business or the subsequent advance on Constantinople, and the French appear inclined to agree with her. On the other hand, the Greeks are burning to be part of the force which enters Constantinople and yet to avoid committing themselves to fighting against anybody but the Turks and possibly the Bulgarians. They won't raise a finger for Serbia, and even want all the time to keep on not unfriendly terms with Germany and Austria."[42]

HERBERT ASQUITH, PM, PRIME MINISTER.

Fred Jane, founder of the eponymous naval journal, while confident of success and having very little respect for the Turks, still considered swift progress dependent upon their surrender.

> "The official details now published of the preliminary operations in the Dardanelles indicate very clearly the immense relative superiority of forts to ships... however, the forts were not "first class," and in addition thereunto, being Turkish, are most unlikely to have been in any high state of efficiency. Yet they survived the first heavy bombardment, and were only finally reduced after over seven hours' firing from the British ships *Queen Elizabeth*, *Agamemnon*, *Irresistible*, *Vengeance*, *Albion*, and *Cornwallis*, and the French ships *Gaulois*, *Suffren*, and *Charlemagne* — all ships making excellent practice against an indifferent reply.

> "From this we can get a clear inkling of the magnitude of the task on which the Allied Fleet is engaged, and — unless Turkish resistance suddenly collapses — progress is likely to be slow and tedious."[43]

FRED T. JANE, NAVAL CORRESPONDENT.

MEDITERRANEAN

> "In the distance we could see the African coast, with Algiers nestling in among the hills, with its dazzling buildings of white stone; and, farther away still, the mountains with their snow-clad peaks.

> "In the evening, when all was still there was one officer, who was often to be seen pacing the deck alone. No one ever thought of disturbing him, as we knew by instinct that he wished to be alone. It was Rupert Brooke."[44]

CHAPLAIN HENRY FOSTER, RN, 2ND (NAVAL) BRIGADE, HMT *GRANTULLY CASTLE*.

---

42   *The Brisbane Courier* (Queensland), 30th July August 1928.
43   *Land and Water*, 6th March 1915.
44   Foster, *At Antwerp and The Dardanelles*, pp. 56–57.

DARDANELLES

"Bombardment from 9.30 to 4.30 — breakfast to tea, with a short pause for lunch.

"Four British vessels of pre-dreadnaught [sic] types steamed boldly into Erenkoi Bay, milled for position, and took under desultory fire the batteries of Erenkoi, Dardanos, and those at Kilid-il-Bahr. The effect of the bombardment was *nil* again.

"Fort Anadolu Hamidieh drew fire to-day for the first time. We were on the tower platform at the time. Half a dozen shells buried themselves in the yard of the battery. The British ships in Erenkio Bay let it go at that.

"The shells fell not more than eight hundred yards away from us, and some of the splinters hit the wall and platform of the venerable pile on which we roost in the day. We put up a brave front to the signal-men who share the platform with us, but were not sorry when the British withdrew…

"Well, the great event cannot be far off. I understand that the Allied fleet is growing each day. The German aviators in the Ottoman service go nightly on patrol to Tenedos, Imbros, and Lemnos. Their reports are not very encouraging to the men in charge of the defense of the Dardanelles and the city that lies beyond them."[45]

GEORGE ABEL SCHREINER, ASSOCIATED PRESS OF AMERICA.

EGYPT

"We expected to receive word to move this week, but I don't know where we will make for when we get a move on. It is not a great "tale" this time, as I heard the Colonel say that he expected word to move any day, and it was also read out in orders that everything was to be ready to be packed up at the shortest notice. Perhaps we will go next week, and it will most likely be to Turkey. Some of the Australian troops have already shifted."[46]

TPR LESLIE BRYSON, 6TH AUSTRALIAN LIGHT HORSE REGT, AIF, MAADI CAMP.

INDIAN OCEAN

"Corporal te Moanui[47] of A Company departed from this life to-day. Farewell Kia tipuna."[48]

2/LT AUTINI PITARA KAIPARA, MAORI CONTINGENT, HMNZT *WARRIMOO*.

---

45 Schreiner, *From Berlin to Baghdad*, pp. 95–96.
46 *Goulburn Evening Penny Post* (New South Wales), 15th April 1915.
47 Cpl Mikaera Te Moananui, New Zealand Maori Contingent, died of pneumonia on 6th March 1915. Commemorated on the Auckland Provincial Memorial, he was the son of Tihitapu Te Moananui, of Paeroa.
48 *Gisborne Times* (New Zealand), 21st May 1915.

7 MARCH 1915

RUSSIA

Anxious to prevent risking a separate peace being concluded, the British signalled their agreement to Russia taking control of the Bosphorus and Dardanelles; a region they had striven for decades to keep them out of.[49]

> "Will you please express to Mr. Grey the profound gratitude of the Imperial Government for the complete and final assent of Great Britain to the solution of the Straits and Constantinople in accordance with Russian desires."[50]
>
> SERGE SAZONOV, RUSSIAN MINISTER FOR FOREIGN AFFAIRS.

DARDANELLES

> "I must record that the fire of the Allied ships is improving. Much damage would have been done to the Turkish emplacements to-day were it not that all of them lie on "soft" land — meadow soil of a light, loamy character.
>
> "From 1.10 to 2.30 the cannonade was terrific. Shot followed shot. The air was rent with the roar of explosions, echoing and reverberating from the steep hills of Gallipoli to the mountains of Anatolia… There is no doubt that we are living here on a volcano. The Turks and Germans are beginning to realize the situation, and it is worrying them."[51]
>
> GEORGE ABEL SCHREINER, ASSOCIATED PRESS OF AMERICA.

EGYPT

> "The 3rd brigade has taken its departure from here, but as to its destination we are totally in the dark. Our brigade, comprising the 5th, 6th, 7th, and 8th battalions, is reported as being the next to go, but when or where to I can't say. Even if we knew, all our letters have to pass through the hands of the censor, and we would not be allowed to give any information about it."[52]
>
> PTE JOHN WADESON, 7TH BATTALION, AIF, MENA CAMP.

> "We are still wondering when we will leave Egypt, and are all more or less anxious to do so — anyway, that is how the troopers of the 2nd A.L.H. Brigade look at things. Wild yarns circulate almost continuously. One minute an excited trooper rushes into his tent, and yells to his mates, "Boys, we're going to Turkey in two days from now!"

---

49 Whatever was agreed at this time, Kitchener, for one, anticipated that any post-war settlement would lead to future disputes between the erstwhile allies. See 16 March 1915.

50 Serge Sazonov quoted in Ponsonby, Arthur, *Falsehood in War-Time*, p. 120, George Allen & Unwin Ltd. (London) 1928.

51 Schreiner, *From Berlin to Baghdad*, pp. 96–97.

52 *Kyabram Free Press and Rodney and Deakin Shire Advocate* (Victoria), 23rd April 1915.

The next minute the story is that Kitchener wants all the A.L.H. to be ready with horses and rifles in France, "at the first sign of spring." Following this is the rumor that we are to get more training in England before going to the front; but it is part of the Army game to let the trooper and private know as little as possible."[53]

TPR CLIFFORD HALLORAN, 6TH AUSTRALIAN LIGHT HORSE REGT, AIF, MAADI CAMP.

INDIAN OCEAN

"[D]eceased was let down to the sea at noon.[54] The four troopships stopped as the body was lowered. After the body had gone, a firing party of 16 fired three rounds and the massed buglers sounded the Last Post. The troops were filled with sorrow and some wept for their lost friend. A sight I will never forget… All the men of the troopships were lined up for the ceremony. The firing party gave an impressive funeral drill in honor of the dead."[55]

2/LT AUTINI PITARA KAIPARA, MAORI CONTINGENT, HMNZT *WARRIMOO*.

8 MARCH 1915

MALTA

"We saw the harbour at its best, in glorious sunshine, and anchored off the Fish Quay about 3.30. An hour after out arrival a French battle-cruiser was sighted coming into the harbour, and a splendid spectacle she made as she steamed in. We stood at the salute, while our band came up on deck and played "The Marseillaise"; then her officers returned the compliment. Permission to go ashore was granted to officers, but not to the men, greatly to their disgust."[56]

CHAPLAIN HENRY FOSTER, RN, 2ND (NAVAL) BRIGADE, HMT *GRANTULLY CASTLE*.

DARDANELLES

The day's poor visibility restricted allied fire; so, too, did an order to limit the expenditure of ammunition.[57] The Royal Navy was running short of shells, while Turkish howitzers continued to cause problems.

"On March 8, the first time the *Queen Elizabeth* was inside the Straits, we led her in, looking for floating mines and submarines. When we got to somewhere near Dardanus the "Q.E." began to play. We had a lovely view of the operations. Presently

---

53   *The Sun* (Sydney, New South Wales), 14th April 1915.
54   Cpl Mikaera Te Moananui, New Zealand Maori Contingent, died of pneumonia on 6th March 1915. Commemorated on the Auckland Provincial Memorial, he was the son of Tihitapu Te Moananui, of Paeroa.
55   *Gisborne Times* (New Zealand), 21st May 1915.
56   Foster, *At Antwerp and The Dardanelles*, pp. 58–59.
57   Corbett, Sir Julian S., *Naval Operations*, vol. 2, p. 194, Longmans, Green and Co. (London) 1923.

there was a whine, which grew to a shriek, and a salvo of big shells dropped so close, that the water splashed aboard. They were intended for the "Q.E." but as we were quite close to her we nearly caught them. Again that awful whine, and another salvo dropped at out stern, and pieces of shell rattled on to our deck. We were wondering where the next lot would land, as it happened, it came quite close to the "Q.E." We were ordered out; the "Q.E." followed us out."[58]

STO. 1 WILLIAM CARRADICE, RFR, HMS *WEAR*.

As many civilians who could left Chanak that morning. Schreiner witnessed it before heading across the Narrows to see what damage had been done at Kilitbahir.

"F. [Frederick] Swing and I were hauled out of our beds at three o'clock in the morning by a tremendous babble under our windows. It was dark yet. A great crowd was surging about the quay, however... Every now and then a *mahonie*, loaded to overflowing with men, women, children, and baggage, would set sail for the Central Dardanelles. We learned that off Nagara several steamers were waiting for the refugees... Those who wanted to go to Constantinople could do that, though the advice was being given that it might be best to visit relatives in the interior...

"In the afternoon my friend and I took a *mahonie* to go to the works of Kilid-il-Bahr. We wanted to see what damage the bombardment had done. No British ships were in sight, though the weather was good. The fact that we had to sail around the mine-field made the trip agreeably long...

"The effect of the Allies' fire was not great by any means. The damage done was small — ridiculously small. Not a single gun had been damaged. A small barracks, several sheds, three living-houses and large number of outhouses had been destroyed. A large barracks and five other buildings were slightly damaged. Two gunners dead and seven wounded were the total of the casualties.

"It was rather different beyond the precincts of the coast-defence establishment. The southern part of the town of Kilid-il-Bahr had suffered severely. Shells had set the buildings afire, and an inspection of the burned-over area showed that most of the shells of the Allies' guns had struck there."[59]

GEORGE ABEL SCHREINER, ASSOCIATED PRESS OF AMERICA.

Forts were small targets; the actual guns themselves, smaller still. The sea was not a stable firing platform, resulting in around only 2–3% of allied shells finding their mark.

SMYRNA

"On March 8th night sweeping was ordered. The sweeping is always done by two boats, and the "Okino" and the "Beatrice" were partners. When the sweep was finished the "Beatrice" slipped her sweep wire, leaving her partner to heave it in while

---

58   *Newcastle Evening Chronicle*, 18th August 1915.
59   Schreiner, *From Berlin to Baghdad*, pp. 107–111.

she proceeded back to the fleet, according to orders. That was the last task that the "Okino" did. After the sweep wire had been hove board the "Okino" steamed full speed towards the fleet, but within five minutes the ship and crew were blown up either through contact with a mine or it may be a stray shell struck her. Out of the 15 hands, ten were killed or drowned."[60]

CAPT. HARRY JAMES, HMT *BEATRICE* AND STANLEY FRY, MATE, HMT *RENARRO*.

LEMNOS

As more troopships arrived at Mudros, the men mostly remained aboard, there being no facilities to accommodate them on the island.

"Rumour says that the Turks and Germans are strongly entrenched on the Dardanelles... Fresh troopships are arriving daily and are bringing Royal Marines and French troops. Rumour again says that 25,000 French are to come, so evidently the powers that be intend making a determined attack on Turkey."[61]

PTE JAMES AITKEN, 11TH BATTALION, AIF, HMT *NIZAM*.

"[The] battalion went ashore for the first time, rowing over in the ship's cutters. It took about two hours to get us all ashore. We then did a route march, and as it had been raining heavily during the night, we were boot-top deep in mud nearly all the way. Twas our first real experience in mud since joining, as Pontville and Mena were both too sandy to give us this pleasure. All the same, it was good to walk on fertile land once more, and the eight miles we walked were full of pleasure."[62]

PTE GARNET RUNDLE, 12TH BATTALION, AIF, HMT *DEVANAH*.

EGYPT

"The Fleet are very sanguine, I think far too much so. It is quite on the cards that they will have no great difficulty with the coast defence guns and batteries, but to deal with movable armaments is another question. I am all for leaving the Gallipoli peninsula severely alone — let it be full of Turks and Germans and bottle them in by getting hold of the Bulair lines. There are practically no roads, no supplies, no water and no inhabitants anywhere that our troop are likely to operate, therefore all has to be carried. This points to pack, but we are all organized for wheel and have no train."[63]

GENERAL SIR JOHN MAXWELL, GOC, BRITISH TROOPS, EGYPT.

"We appear to be getting near to our departure, as most of the infantry have left Mena, and our supplies are being carted to the railway. I think to-day week will see the last

---

60   *Western Times*, 29th April 1915.
61   *Western Mail* (Perth, Western Australia), 6th January 1938.
62   *Terang Express* (Victoria), 13th July 1915.
63   Sir John Maxwell quoted in Arthur, Sir George, *General Sir John Maxwell*, pp. 173-174, John Murray (London) 1932.

of this place, and about time too, as the sand has sent quite a number of our men to the hospital. The majority are suffering from deafness, sore ears, eyes and throat. The wind and dust were so bad to-day that all parades were cancelled, and we spent the time hugging our tent poles to save our happy homes being broken up. To-day week we went through another musketry course and I topped the aggregate score in the squadron and in all probability it is for the regiment."[64]

CPL FREDERICK ELWORTHY, 1ST AUSTRALIAN LIGHT HORSE REGT, AIF, HELIOPOLIS.

9 MARCH 1915

DARDANELLES

Having observed British and French ships manoeuvring in Erenkeui Bay it was decided to lay a new minefield parallel to the shore there.

"I was personally double-checking the engine room and then got the boilers ready for smokeless sailing, the torpedo-man, Bettaque,[65] and the Turkish mine-laying crew cleared the mines ready for launch. Two German NCOs and stokers were at my disposal for the operation of the engines and boiler. This was to guarantee that my commands were executed quickly and correctly. At 5 o'clock in the morning I had the anchor raised. The weather was good for this operation. A light mist lay on the water, which gradually turned into a steady rain. With an average of 140 revolutions, the minelayer made its way from Nagara along the Asian coast.

"Since it was still dark and several minefields had to be negotiated, great caution was needed. However, the Turkish Mine Captain knew the critical points exactly, and so *Nusret* arrived safely at its destination. Throughout the voyage, [engine] revolutions were maintained according to my orders. This enabled me to sail completely smokeless, although the Turkish Eregi coal is very unsuitable for this purpose. At 07.10 hrs I had us turnabout and bound for home; simultaneously, I had the mines laid at 15 second intervals by Hafis Nasimi, the Turkish Mine Captain. Overall, 26 mines were laid in the general direction of SW-NE."[66]

MARINE-OBERINGENIEUR ARNHOLDT REEDER, NUSRET.

LEMNOS

"Imagine my feelings when this morning there appears on the scene a French General who informs me that he is the precursor of a French Army and has apparently been told that I will supply him with all that they need. Truly the ways of those in authority

---

64 *The Gundagai Times and Tumut, Adelong and Murrumbridge District Advertiser* (New South Wales), 16th April 1915.
65 *Offizierstellvertreter* Rudolf Bettanque was killed during mine-clearing operations on 7th May 1916.
66 Caption, Australian War Memorial photograph, AWM P04411.003.

are beyond conception. This wretched island is evidently supposed to be a land flowing with milk and honey and water."[67]

REAR-ADMIRAL ROSSLYN WEMYSS, RN, GOVERNOR OF MUDROS.

"Practised disembarkation. Went ashore, marched through two villages; very interesting. Traders ashore; goods very high priced. Eggs four for 6d., bread 6d. loaf, &c. Paid £1/10/."[68]

PTE SYDNEY PENHALIGON, 3RD FIELD AMBULANCE, AAMC, AIF, HMT *MALDA*.

EGYPT

While there was speculation about where the men who had trained in Egypt would end up next, others were in no doubt. And they would know.

"We are anxiously awaiting orders to leave there, but I do not know where we are bound for. We hear all sorts of rumours; some say we are to go to Turkey, and others to Europe, but I am inclined to think we shall go to England and from there to France."[69]

SGT SAMUEL WOODS, CANTERBURY BN, NZEF, ZEITOUN CAMP.

NEW ZEALAND

"A French and probably a British force will operate in the Dardanelles, and a base will he established in Northern Africa. It is a most important move, and the result will be of great importance in the struggle. You can look for very stirring news from the new centre of operations in the near future."[70]

WILLIAM MASSEY, PRIME MINISTER, NEW ZEALAND.

10 MARCH 1915

ENGLAND

It was confirmed that Britain's last Regular division, albeit one brought up to strength with some infantry and supporting arms drawn from the Territorial Force, would be sent east.

"Lord Kitchener, being then more reassured as regards the position in other theatres, informed the War Council that he was prepared to release the 29th Division for the Dardanelles."[71]

MAJOR-GENERAL WILLIAM ROBERTSON, CHIEF OF STAFF, BEF.

---

67 Wemyss, *The Navy in the Dardanelles Campaign*, p. 29.
68 *The Queenslander* (Brisbane, Queensland), 21st April 1917.
69 *Ellesmere Guardian* (New Zealand), 24th April 1915.
70 *Evening Star* (New Zealand), 10th March 1915.
71 *The Mercury* (Tasmania), 5th January 1926.

DARDANELLES

The attempts to clear the minefields with unarmoured, converted trawlers continued that night. They were horribly vulnerable to fire from the shore.

> "They let us get right up… and as we turned round to take our sweeps up, one of our number was blown up. Then they peppered us from each side, for one-and-a-half to two miles. We heard cries for help. I said, 'We shall have to do the best we can and go back and pick up.' There was no waiting, no saying 'Who shall go?' As soon as I called for volunteers, three jumped in. I kept the vessel as close as I could to shelter them. I did not think any would come back alive, but no one was hit, and I said, 'Now we'll get the boat in.' Just as we got the boat nicely clear of the water, along came a shot and knocked it in splinters. I shouted, 'All hands keep under cover as much as you can!' and I got on the bridge and we went full steam ahead. I could not tell you what it was like, with floating and sunken mines and shots everywhere. We got knocked about, the mast almost gone, rigging gone, and she was riddled right along the starboard side. One of the hands we picked up had his left arm smashed with shrapnel. That was all the injury we got. When we got out, the Commander came alongside and said, 'Have you seen any more trawlers?' I said 'Yes, we've got the crew of one aboard, the *Manx Hero*.' We were the last out, and I can tell you, I never want to see such a sight again!"[72]
>
> SKIPPER ROBERT WOODGATE, RNR, HMT *KOORAH*.

> "[I]f any one deserves the V.C. it is Captain Woodgate, a native of Beer, East Devon and his crew on the "Koorah." The trawlers "Gwenllian" and the "Manx Hero" were sweeping partners in the Dardanelles, when the "Manx Hero" was blown up. The "Koorah" and the other sweepers were on their way down to the fleet as the explosion took place. Hearing cries for help, Captain Woodgate turned his ship round, and, with the truest traditions of British pluck and seamanship, decided to save his fellow sweeper. He called for volunteers to man the small boat. It was thrown over the ship's side, and into it jumped the boatswain, Joseph Abbott, of Brixham, and two deck hands, Thomas Thompson and Robert Strachan. Away they rowed with shot and shell pitching thick around their tiny craft, while Capt. Woodgate stood at the helm keeping his ship as near the boat as was prudent. The crew of eleven were rescued and put aboard the "Koorah." Some skippers would have cut the small boat adrift with the hail of shell showering on the vessel, but Woodgate ordered it to be hove aboard. The tackle was fastened to the bowring, but scarcely had the boat moved deckwards than a shell crashed into her and shattered her to matchwood. The "Koorah" was now the last sweeper, and with powerful searchlights flashed upon her, there seemed very little chance for her to get through, as she was the centre of the fire from the forts. Again sound judgment saved the "Koorah." Observing the faults of the marksmen, Woodgate steered the trawler towards the northern shore, and came through the fire

---

[72] Robert Woodgate quoted in Wood, Walter, *Fishermen in War Time*, pp. 164–165, Sampson Low, Marston & Co. Ltd. (London) 1918.

zone safe. The daring deed took nearly ninety minutes to accomplish. Capt. Woodgate is regarded by his fellow sweepers as the hero of the operations, and if he does not get the V.C. or some special recognition there will be disappointment."[73]

CAPT. HARRY JAMES, HMT *BEATRICE* AND STANLEY FRY, MATE, HMT *RENARRO*.

LEMNOS

If luck played a part in the mine-sweepers' survival, little was enjoyed by one submarine.

"*AE2* was coming into harbour… after a day's patrol duty, when she had the misfortune to run on some rocks. I was asleep down below, when I was awakened by a grinding smash. The poor little ship trembled for a moment, and then lurched right over to starboard. All the crew started to clamber up through the conning tower, and you may depend upon it that I made a move in that direction as soon as possible. When we arrived on deck we received orders to take our heavy sea boots off and prepare for a swim. It was a bitterly cold night, in fact, all around us the hills were covered with snow. Some searchlights were played on us from shore to light the water between us and the shore if the necessity came to take to the briny. Thank goodness help arrived and took off the greater part of the crew to a place of safety. If they had had to swim for it not many would have reached the shore, as there was a nasty sea breaking on the rocks, which would probably have smashed them to bits. I did not leave the ship, as I am about the best swimmer on board, so had more chance to save myself, besides the skipper wanted some of us to remain behind to try to get the boat afloat again. We did succeed after six weary hours in the cold. We all went aft, and managed to secure a wire hawser, which came from a destroyer. We all got soaking wet while doing this which made us ever so much colder. When all was ready we went full speed astern, and the destroyer tugged and off we came. The poor little ship was badly damaged."[74]

AB ALBERT NICHOLS, RAN, HMAS *AE2*.

11 MARCH 1915

ENGLAND

More doubts emerged in the Admiralty concerning the ability to take control of the Straits without military assistance. Turkish resistance to naval landing parties had increased and was only expected to harden further.

"Admiral Carden's Report… on the progress of operations in the Dardanelles shows he has made good progress in the Dardanelles, but that his operations are now greatly retarded by concealed batteries of howitzers, and that their effects are now as formidable as the heavy guns in the permanent batteries. He also states that

---

73   *Western Times*, 29th April 1915.
74   *Morning Bulletin* (Rockhampton, Queensland), 19th June 1915.

demolition parties are essential to render the guns useless. The enemy's military forces have prevented this work from being effectually completed at the entrance, and they will be in even a better position to prevent it further up the Straits...

"To advance further with a rush over unswept minefields and in waters commanded at short range by heavy guns, howitzers and torpedo-tubes, must involve serious losses in ships and men, and will not achieve the object of making the Straits a safe waterway for the transports."[75]

ADMIRAL SIR HENRY JACKSON, ADMIRALTY WAR STAFF.

LEMNOS

"We reached Lemnos… about six o'clock in the evening…., and were struck with the spaciousness of this beautiful natural harbour. The entrance is by a narrow strait of not more than half-a-mile across, and, at the first approach, you might be led to believe this was the extent of the harbour, but soon you would discover you mistake. In Mudros Bay, so called from a village of that name with a modern church, the whole Mediterranean Fleet could lie in comfort, protected alike from storms and submarines."[76]

CHAPLAIN HENRY FOSTER, RN, 2ND (NAVAL) BRIGADE, HMT *GRANTULLY CASTLE*.

DARDANELLES

"This afternoon I had a long talk with Admiral-General von Usedom Pasha, the purpose of which was to get some information on relations between the Turks and Germans here.

"I learned that these relations are not entirely frictionless. The Germans say that are doing all the work. The Turks claim all the credit. Von Usedom Pasha said that he did not mind this, but that some of the officers under him were less liberal in that respect. He was content with being of account to his fatherland anywhere. Whether that was at the Dardanelles or in the North Sea would never worry him.

"There were one or two things which the Turks and Germans had overlooked so far. The white minarets and houses, the great towers of the kalehs at Tchanak Kaleh and Kilid-il-Bahr, and the white stone revets of some of the batteries, have in the past made excellent targets for the Allied gunners. That having been ascertained, these surfaces are now being cross-hatched with black paint."[77]

GEORGE ABEL SCHREINER, ASSOCIATED PRESS OF AMERICA.

---

75 Admiral Jackson quoted in Churchill, *The World Crisis 1915*, p. 212.
76 Foster, *At Antwerp and The Dardanelles*, pp. 59–60.
77 Schreiner, *From Berlin to Baghdad*, p. 117.

EGYPT

"I am very pleased to say that I have now done six months in the army (less a fortnight), and have not been up before the O.C. once. Some chaps have books half filled with defaults. There are times when the officers yap at you so much that you have to bite your tongue to prevent yourself from telling then your opinion of them. Of course if you told them they would have you in the guard tent... I believe we will go to the Dardanelles."[78]

PTE THOMAS BERRY, 13TH BATTALION, AIF, HELIOPOLIS.[79]

"I hope you people don't attach much weight to the articles which have appeared in the papers about the supposed bad conduct of some Australians in Egypt. One cannot expect to find every man perfect in such a big body. There are bound to be a few 'wasters' among them... Twelve of us girls left Mena House this afternoon to take up duty at Heliopolis with the 1st Australian General Hospital for awhile."[80]

STAFF NURSE GRACE BURNS, AANS, 2ND AUSTRALIAN GENERAL HOSPITAL.

12 MARCH 1915

ENGLAND

"I have just been reading the Admiral's report of the operations so far. They are making progress, but it is slow, and there are a number of howitzers and concealed guns, which give them a good deal of trouble. I think the Admiral is quite right to proceed very cautiously. Winston is rather for pushing him on."[81]

HERBERT ASQUITH, MP, PRIME MINISTER.

As General Sir Ian Hamilton told it, his appointment was confirmed after the briefest of meetings.

"I was working at the Horse Guards when, about 10 a.m., K. [Kitchener] sent for me... 'We are sending a military force to support the Fleet now at the Dardanelles, and you are to have Command.'...

"I asked no question, packed up my kit, ordered my train, started that night. Not another syllable was said on the subject. Uninstructed and unaccredited I left that night for the front; my outfit one A.D.C., two horses, two mules and a buggy."[82]

GENERAL SIR IAN HAMILTON, C-IN-C, MEF.

---

78 *The Young Chronicle* (New South Wales), 20th April 1915.
79 Berry, a former clerk, was seriously wounded on at Gallipoli on 25th April or shortly afterwards and admitted to hospital in Alexandria on 30th April 1915. Berry returned to Australia on 4th September 1915 to be discharged.
80 *The Register* (Adelaide, South Australia), 13th April 1915.
81 *The Brisbane Courier* (Queensland), 30th July August 1928.
82 Hamilton, Sir Ian, *Gallipoli Diary*, vol. 1., pp. 1–2, Edward Arnold (London) 1920.

Hunter-Weston, the newly-appointed commander of 29th Division, prepared by discussing his task with key figures at the Admiralty.

> "Official date of taking over Command of the 29th Division. I left Leamington [at] 9 a.m. Busy day at the War Office... Had a chat with Sir Francis Hopwood, Civil Lord of the Admiralty. Had a long chat with Winston Churchill, First Lord of the Admiralty at 5 p.m. on the situation in the Dardanelles."[83]
>
> MAJOR-GENERAL SIR AYLMER HUNTER-WESTON, COMMANDING 29TH DIVISION.

LEMNOS

The difference between a harbour and a port was not appreciated by some, it appears.

> "On the 12th the Franconia, the first of the transports carrying the Royal Naval Division, some 8,500 strong, arrived. Major-General Paris, R.M.A., who commanded them, was on board with his staff. He told me that his troops had been embarked in such a manner that it was sufficient for a man to be in one transport to be certain that his greatcoat was in another; his whole force would have to be disembarked, re-organized and re-embarked again before they would be ready for service. He was under the impression that this could be done at Mudros, as apparently were the authorities at home; but it only needed a few words of explanation from me to convince him of the absolute impossibility of such a proceeding. We had neither wharves, nor cranes, nor even landing piers and an insufficiency of boats, and the only course to take was to send the whole force to Egypt, where, at Alexandria or Port Said, this complicated work could be carried out."[84]
>
> REAR-ADMIRAL ROSSLYN WEMYSS, RN, GOVERNOR OF MUDROS.

EGYPT

Camp facilities were well developed by this stage but the tenants were keen to leave.

> "Our camp at present is quite a small township, with numerous shops and restaurants, bathhouses, photographers, afternoon tea places, laundries, reading rooms, and so on... Every battalion has everything it needs. We even have a wet canteen... his canteen is open only at certain times of the day. In the day time we have big attack practices. At night, also, there are attacks and marches out into the desert, returning about 9 a.m. for breakfast, after having had no sleep."[85]
>
> PTE GEORGE FRANKLIN, CANTERBURY BN, NZEF, ZEITOUN CAMP.[86]

---

83 Diary entry, 12th March 1915, Maj. Gen. Hunter-Weston, British Library, MS 48364.
84 Wemyss, *The Navy in the Dardanelles Campaign*, p. 29.
85 *Lyttelton Times* (New Zealand), 4th May 1915.
86 Promoted Cpl, Franklin was wounded on 5th June 1915 but remained with his unit. After falling ill, he was admitted to hospital on 9th July 1915. While being transferred from Mudros to Malta he died of peritonitis aboard HMT *Andania* on 18th August 1915. Commemorated on the Lone Pine Memorial, he was the 28 year-old son of George James Franklin and Mary Jane Franklin, of 523 Madras Street, St. Albans, Christchurch.

"Whoever it was that said that New Zealand was "God's Own Country" spoke a true word. I'm not surprised at Moses and the Israelites wanting to leave Egypt, and you can take it from me it was not Pharaoh that made them trek. It was mosquitoes, ants, scorpions, beetles and other alligators…

"Cairo I have explored thoroughly, climbed Cheops Pyramid twice, paid the Sphinx a visit, also the buried temple and tombs, also consider myself able to write a book on mosques, give lectures on Biblical scenes and places, tell lies about the victorious battles on the canal, etc. I have dug up skeletons at Mataria, eaten spaghette [sic] at Greek restaurants, drunk beer and talked politics with the highest in the land at Heliopolis House Hotel, have sat in the coronation chair of the Sultan Hassien (on whom be peace) at the mosque, of that name. In fact, I've done Cairo, and want a change."[87]

PTE JOHN PENN, NZAC, ZEITOUN CAMP.

13 MARCH 1915

ENGLAND

Kitchener warmed Birdwood not to undertake any landing in force before reinforcements and Sir Ian Hamilton arrived.

"In answer to your question, unless it is found that our estimate of the Ottoman strength on the Gallipoli Peninsula is exaggerated and the position on the Kilid Bahr Plateau less strong than anticipated, no operations on a large scale should be attempted until the 29th Division has arrived and is ready to take part in what is likely to prove a difficult undertaking, in which severe fighting must be anticipated."[88]

LORD KITCHENER, SECRETARY OF STATE FOR WAR.

He also briefed Hamilton concerning the coming campaign. The Army's role remained subsidiary to the Navy; ill-defined; an advance to contact on a large scale.

"The Fleet have undertaken to force the passage of the Dardanelles. The employment of military forces on any large scale for land operations at this juncture is only contemplated in the event of the Fleet failing to get through after every effort has been exhausted…

"Having entered on the project of forcing the Straits there can be no idea of abandoning the scheme. It will require time, patience, and methodical plans of co-operation between the naval and military commanders. The essential point is to avoid a check, which will jeopardise our chances of strategical and political success…

"Owing to the lack of any definite information we must presume that the Gallipoli Peninsula is held in strength and that the Kilid Bahr plateau has been fortified and

---

87 *Taranaki Herald* (New Zealand), 30th April 1915.
88 Kitchener quoted in Churchill, *The World Crisis 1915*, p. 214.

armed for a determined resistance. In fact, we must presuppose that the Turks have taken every measure for the defence of the plateau, which is the key to the Western front at the Narrows, until such time as reconnaissance has proved otherwise...

"In order not to reduce forces advancing on Constantinople, the security of the Dardanelles passage, once it has been forced, is a matter for the Fleet... The occupation of the Asiatic side by military forces is to be strongly deprecated...

"As it is impossible now to foretell what action the Turkish military authorities may decide upon as regards holding their European territories, the plan of operations for the landing of the troops and their employment must be left for subsequent decision."[89]

LORD KITCHENER, SECRETARY OF STATE FOR WAR.

That evening things took a dramatic turn. Room 40, part of Naval Intelligence, using the codes provided by Russians the previous August, intercepted a German cable.

"For Admiral Usedom.[90] H.M. the Kaiser received the report and telegram relating to the Dardanelles. Everything conceivable is being done here to arrange the supply of ammunition. For political reasons it is necessary to maintain a confident tone in Turkey. H.M. The Kaiser requests you to use your influence in this direction. The sending of a German or Austrian submarine is being seriously considered."[91]

ADMIRAL GEORG VON MÜLLER, CHIEF, NAVAL CABINET.

Though the message showed concern about ammunition supply, did that necessarily mean there was already a critical shortage? There seemed little doubt in the Admiralty.

"Lord Fisher took the message, read it aloud and waved it over his head. 'By God,' he shouted, 'I'll go through tomorrow!' Mr Churchill, equally excited, seized hold of the letter and read it through again for his own satisfaction. 'That means,' he said, 'they've come to the end of their ammunition.' 'Tomorrow,' repeated Lord Fisher, and at that moment I believe he was as enthusiastic as ever Mr Churchill had been about the whole Dardanelles campaign. 'We shall probably lose six ships, but I'm going through.' The 1st Lord nodded. 'Then get the orders out.' And there and then Lord Fisher sat down at Mr Churchill's table and began to draft out the necessary orders."[92]

CAPTAIN WILLIAM HALL, RN, DIRECTOR OF NAVAL INTELLIGENCE.

This had not been the only indication that the Turks' ammunition stocks were low. Press reports, some appearing only days after their entry into the war, had stated that munitions were scarce generally. A decoded German cable represented rather higher grade intelligence but it is hard to escape the conclusion that this was an instance of confirmation bias. It was believed because it was wanted to be believed. Even individuals

---

89 Lord Kitchener to Sir Ian Hamilton quoted in Arthur, Sir George, *Life of Lord Kitchener*, vol. 3, pp. 122–123, MacMillan & Co., (London) 1920.
90 Inspector-General of Coast Defences and Mines, Dardanelles
91 James, William Milbourne, *The eyes of the Navy; a Biographical Study of Admiral Sir Reginald Hall, K.C.M.G., C.B., LL.D., D.C.L.*, pp.62–63, Methuen (London) 1955.
92 Hall quoted in in Beesly, Patrick, *Room 40. British Naval Intelligence 1914-18*, p. 81, Hamish Hamilton (London) 1982.

holding extremely low opinions of the fighting qualities of the Dardanelles' defenders added caveats to their contentions.

> "We have a certain number of ships which we can spare for these operations, ships which we could lose without jeopardising our naval superiority. This — coupled with the fact that the enemy are not a brainy folk — makes the Dardanelles effort possible. But... nothing has ever happened to negative the old proverb that one gun on shore is worth a dozen such guns afloat. We cannot be too careful in avoiding false deductions for successes in the Dardanelles."[93]

FRED T. JANE, NAVAL CORRESPONDENT.

DARDANELLES

With the failure of the mine-sweeping operations to date, a fresh effort was made, replacing reservists with Royal Navy personnel. Schreiner watched them come.

> "Of a sudden one of the projected rays fixes itself on some object near the Gate of the Dardanelles. In the same moment the object emits a yellow flash — and then another...

> "Guided by their search-lights, the Turkish gunners take the Allied vessels under fire. The black hollow between the Gallipoli and Anatolian heights begins to echo and reverberate the crash of artillery and exploding shells...

> "The Turkish search-lights showed its low and glistening hull, three funnels, two masts, turrets, and gun-barrels, like high-lights in a smudge drawing. Now and then the flashes of the ship's own guns would reveal other details. Its gunners were not saving ammunition. They were sending it to the sites of the Turkish projectors as fast as it was possible to serve the pieces...

> "Later a cruiser advanced and began to bombard the Turkish projectors, whose lights served better in the heavy atmosphere to reveal their location than to illume the waters of Erenkoi. A lively cannonade between ship and anti-mine battery. It ended another attempt to get rid of the mines.

> "So far the Allies have made five serious attempts to sweep the channel. None of them has been a success. The few mines they have been able to get have been replaced by the Turks with Russian mines which have floated into the Bosphorus after having been liberated with malice aforethought by the Russians near the entrance to that strait."[94]

GEORGE ABEL SCHREINER, ASSOCIATED PRESS OF AMERICA.

---

93 *Land and Water*, 13th March 1915.
94 Schreiner, *From Berlin to Baghdad*, pp. 119–122.

On the receiving end of most of this attention was the cruiser, *Amethyst*.[95]

> "When we gets abeam of the light the sight-setter of my gun got the range — 2,500 yards for the light, and so burst off as hard as we could. Well, they opened fire, and the first shot splashed underneath the gun, and wet us all. Of course at this marksmanship we all grinned, and the next shot we fired put a searchlight out. Well, then they started both sides. It was like hell let loose.

> "Along comes a shot and puts our lights out. Another smashes the steering gear, and the next killed two out of my gun crew, tore the leg off another, splinters in leg and arm on the other, and me, lucky, was hit in the shoulder. I had just gone for another projectile; had the gun been loaded I should not be writing this. When I was hit all I remember was a great yellow flash and a roar."[96]

BOY 1ST CLASS GEORGE DIXON, RN, HMS *AMETHYST*.

> "There were dead bodies lying all around us, and the deck was streaming in blood. It was pitch dark, and the wounded were crying out in their agony. I had a most marvellous escape, for I had just been relieved at a gun, and had got about three fathoms away, when a shell came and killed all the gun's crew; thus I escaped without a scratch."[97]

OS DAVID TUMBLETY, RN, HMS *AMETHYST*.[98]

The volume of fire directed at the *Amethyst* did not suggest the defenders had serious concerns about their ammunition supply.

LEMNOS

> "Telegrams pouring in in basketsful, messages by wireless telegraphy, orders from England (generally contradictory), requests from all parts of the Mediterranean, demands for the possible and impossible from every quarter, this will give you some idea of what I have to deal with at all hours of the day and night. Merchant captains and military officers seem to vie with each other in seeing which can be the most tiresome, and all the time I know that one small exhibition of temper on my part at the wrong moment may do infinite harm. So you can imagine that I sometimes have a trying time."[99]

REAR-ADMIRAL ROSSLYN WEMYSS, RN, GOVERNOR OF MUDROS.

---

95   Almost all of the 27 dead and 43 men wounded that night were from the *Amethyst*.
96   *Sheffield Evening Telegraph*, 27th April 1915.
97   *Berkshire Chronicle*, 11th June 1915.
98   Tumblety was killed in action on 18th May 1915 and is commemorated on the Portsmouth Naval Memorial. He was 17.
99   Wemyss, *The Navy in the Dardanelles Campaign*, pp. 34–35.

14 MARCH 1915

LEMNOS

Strengthening resistance to shore parties and the failure of the minesweepers convinced Carden that the Navy needed help. This was not the response of a man responding to an urgent order to attack; someone persuaded that the defenders were short of ammunition.

> "In my opinion military operations on large scale should be commenced immediately to ensure my communication line immediately fleet enters Sea of Marmora.
>
> "The losses in passing through Narrows may be great; therefore submit that further ships be held in readiness at short notice and additional ammunition be despatched as soon as possible."[100]
>
> VICE-ADMIRAL SACKVILLE CARDEN, RN, EAST MEDITERRANEAN SQUADRON.

EGYPT

The charms of the country had clearly paled after months being sand-blasted, propositioned and marched across the desert. Some of the training had been tough.

> "We also feel inclined to shed bitter tears when we try to dig into it, see the sand trickling in almost as fast as we shovel it out. On Thursday we rose at three a.m., and went forth to weild [sic] the pick and shovel again, returning to camp at 12 o'clock. There was a strong wind blowing, and we were enveloped in clouds of flying sand. Saturday was a general holiday, and I spent the time having a good rest."[101]
>
> PTE FREDERIC MUIR, 1ST AUSTRALIAN BATTALION, AIF, MENA CAMP.[102]

> "We are simply sick of Cairo, sick of Mena, sick of the Pyramids, sick of the natives and their habits — sick of everything that sounds or appears Egyptian. The whole surroundings have become a bore. Cairo is nothing but a low down, brothel-infested hell!"[103]
>
> SGT NORMAN PINKSTONE, 3RD BATTALION, AIF, MENA CAMP.

> "Our first field day was done with ball ammunition, and we advanced to the attack under the support of shrapnel fire from the New Zealand Artillery. It was novel to hear the shells whizzing over our heads and see them bursting ahead over the targets on the position we were attacking; and when the firing line was formed and fully developed one got a good idea of what it is like in the real thing. The firing line was

---

100 Cardin quoted in Churchill, Winston, *The World Crisis 1915*, p. 218, Thornton Butterworth Ltd. (London) 1927.
101 *South Coast Times and Wollongong Argus* (New South Wales), 30th April 1915.
102 Muir, a former law student was wounded on 25th November 1915 and died three days later aboard HMHS *Glenart Castle*. Commemorated on the Lone Pine Memorial, he was son of Alice O'Donnell, of Unanderra, New South Wales.
103 *Cootamundra Herald* (New South Wales), 23rd April 1915.

about two miles long, and when we and the machine-guns were all banging away there was some noise, believe me. The poor old targets had a very bad time, and were simply shot to pieces. That was the only time we have used ball ammunition, all the other times have been with blank, as there has been both an attacking and defending party in these affairs. The day before yesterday, I think, was about the biggest operation. Our whole division, about 20,000 men all told, attacked a lot of Territorials, and as they had taken up a very strong position we had a very strenuous engagement. We are ready for service now and expect to go at any time, and, believe me, we won't be sorry to leave this. Tramping about the desert is very heavy going, and gets very monotonous, and the dust and sandstorms are far from a delight."[104]

CPL CECIL GOSS, 15TH BATTALION, AIF, AERODROME CAMP, HELIOPOLIS.[105]

"To-day is a brute — clouds of dust and dirt everywhere, and as a result sore eyes are frequent. Our death roll keeps mounting up. I believe since arriving in Egypt we have buried 260 men. Copies of the apology made by Bean, the war correspondent, have been circulated through the various camps, and no doubt he was made to eat apple pie over the lying and slanderous statement he made to the Press of Australia."[106]

DVR MORTON COLLINGS, NO 1 COY, DIV. TRAIN, AASC, AIF, MENA CAMP.

15 MARCH 1915

SEA OF MARMORA

While sailing to the Dardanelles aboard the Yuruk, Enver Pasha told the American Ambassador that even if the Anglo-French fleet broke through it could achieve nothing.

"We have plenty of guns, plenty of ammunition, and we have these on terra-firma, whereas the English and French batteries are floating ones. And the natural advantages of the straits are so great that the warships can make little progress against them. I do not care what other people may think... Indeed, I do not know just what these English and French battleships are driving at. Suppose that they rush the Dardanelles, get here into the Marmora, and reach Constantinople, what good will that do them? They can bombard and destroy the city, I admit, but they cannot capture it, as they have no troops to land. Unless they do bring a large army, they will really be caught in a trap... It seems to me to be a very foolish enterprise."[107]

ENVER PASHA, MINISTER OF WAR.

---

104 *Taranaki Herald* (New Zealand), 26th April 1915.
105 Goss was wounded at Gallipoli in 17th May 1915. He rose to the rank of Captain and was decorated with the Military Cross for gallantry on the Western Front.
106 *Cowra Free Press* (New South Wales), 24th April 1915.
107 Enver Pasha quoted in Morgenthau, Henry, *Secrets of the Bosphorus*, p.133–134, Hutchinson & Co., (London) 1918.

EGYPT

An officer defended one of his men after hearing about ugly rumours from home.

"Knowing the interest you took in the first volunteer force that left Mackay under myself for active service abroad, I feel sure will contradict the unjust and cruel statement that has been circulated in Mackay and district regarding one of its members, i.e., Albert Graffunder.[108] This, I need hardly say, is an utter falsehood, without a vein of truth in it. In fact, Graffunder was made a Corporal a week after joining, and is Senior Corporal of my Squadron, and is looked upon by the whole regiment as one of the best Non-commissioned Officers we have. There is not a mark against him, and he is the next for promotion to Sergeant. This, I think, speaks for itself as to the esteem in which he is held by his officers and comrades. Trusting you will give the above publicity and thus remove a slur cast upon one who is doing his duty for his King and country."[109]

CAPT. GILBERT BIRKBECK, 2ND AUSTRALIAN LIGHT HORSE REGIMENT, AIF, AERODROME CAMP, HELIOPOLIS.

"We have had some very solid days since we came back from Ismailia. We have been having divisional training this last fortnight, and we have had two big days this week to put in. We do anything from eighteen to twenty-five miles on these days, part of the distance being an attack. I can tell you 'doubling' over this sand in the heat is not very nice. We shall not be sorry when this divisional training is over."[110]

PTE ROBERT WATSON, AUCKLAND BATTALION, ZEITOUN CAMP.[111]

INDIAN OCEAN

"Another glorious day. The breeze is very strong and cooling and it reminds me of the voyage from New Zealand to Australia. We are getting down to our work again as the sea is calm as could be... The band is playing, the men are singing, and everything is bright. What a beautiful sunset as the sun dies in the west! There is nothing around us by the changing sea. The breeze is simply sweet as most of us lounge about in the lightest of garb."[112]

2/LT AUTINI PITARA KAIPARA, MAORI CONTINGENT, HMNZT *WARRIMOO*.

---

108   Cpl Albert Graffunder, 2nd Australian Light Horse Regt.
109   *The Daily Mercury* (Mackay, Queensland), 27th April 1915.
110   *Dominion* (New Zealand), 10th May 1915.
111   Watson was killed in action on 26th April 1915. Commemorated on the Lone Pine Memorial, he was the son of Martha Watson, of 17 Volcanic Street, Mount Eden, Auckland.
112   *Gisborne Times* (New Zealand), 21st May 1915.

16 MARCH 1915

ENGLAND

Looking forward, Kitchener issued a warning that a successful occupation of the Bosphorus and Dardanelles would not be the end of the matter.

> "Assuming that the war is brought to a successful conclusion, a partition of a certain part of the Turkish dominions will doubtless have to be undertaken. Russia, we know, will secure Constantinople together with the control of the Dardanelles and the Bosphorus. It is safe to assume that the claims of Syria, which the French have put forward for so many years, will have to be satisfied to a considerable extent. With Russia established on the Dardanelles and with France in possession of Syria, the strategical and economic situation in the Levant cannot fail to be profoundly modified, and the position of Egypt will be considerably affected. It must not be forgotten that, after the conclusion of peace, old enmities and jealousies which have been stilled by the existing crisis in Europe, may revive. We have, in fact, to assume that, at some future date, we may find ourselves at enmity with Russia, or with France, or with both in combination, and we must bear this possibility in mind in deciding how, when the time for settlement comes and the question of partition of Turkey in Asia arises, our interests can best be safeguarded."[113]

LORD KITCHENER, SECRETARY OF STATE FOR WAR.

TENEDOS

Wemyss was called to see Carden off Tenedos.

> "I proceeded there in a destroyer and found him ill and obliged to give up the command. The situation thus created was a delicate one, for his departure would leave me the senior officer, since Rear-Admiral de Robeck, his second in command of the squadron operating against the forts, though older and senior to me in the Service, was actually my junior on the Rear-Admiral List. Here was I, organizing the base, an arduous task inevitably bound to suffer from a change of command, whilst de Robeck was in the middle of a complicated operation, in full possession of and knowing its most intricate details of which I was completely ignorant; yet surely the senior officer's place was with the squadron at the front. I discussed the situation from every point of view with him and with Commodore Keyes, his Chief of the Staff, and eventually made up my mind that no other course was open to me except to return to Mudros to carry on my work there, leaving the operations in de Robeck's hands."[114]

REAR-ADMIRAL ROSSLYN WEMYSS, RN, GOVERNOR OF MUDROS.

---

113   Alexandretta and Mesopotamia, TNA CAB 42/2/10.
114   Wemyss, *The Navy in the Dardanelles Campaign*, pp. 34–35.

LEMNOS

"We went ashore… determined to visit, if possible, the three Turkish villages which were said still to be in existence on the island of Lemnos. He [Sub-Lt Arthur Tisdall] had been told off to make maps and observations of the road to Castro, the capital of the island. When his work was finished, we walked four or five miles to a Greek village, and came across the schoolmaster, who seemed to be a most important person. Tisdall's knowledge of Ancient Greek was a valuable asset, and the schoolmaster insisted upon us going into the school. He them made the children sing a patriotic song to us, which was about driving out the Turks, and re-establishing Greek and Christian civilisation."[115]

CHAPLAIN HENRY FOSTER, RN, 2ND (NAVAL) BRIGADE, HMT *GRANTULLY CASTLE*.

DARDANELLES

During his visit to the Straits, the American Ambassador learned why it was so difficult for a fleet to destroy forts on shore.

"The land for nearly half a mile about seemed to have been completely churned up; it looked like photographs I had seen of the battlefields in France. The strange thing was that, despite all this punishment, the batteries themselves remained intact; not a single gun, my guides told me, had been destroyed…

"I was much puzzled by the fact that the Allied fleet, despite its large expenditures of ammunition, had not been able to hit this Dardanos emplacement. I naturally thought at first that such a failure indicated poor marksmanship, but my German guides said that this was not the case… a rapidly manoeuvring battleship is under a great disadvantage in shooting at a fixed fortification."[116]

HENRY MORGENTHAU, AMERICAN AMBASSADOR.

EGYPT

"At 4 p.m. to see Sir W.B. [William Birdwood] told me he as off tonight to see Sir Ian Hamilton at Lemnos — that Sellheim[117] said there were 100000 best Turkish Troops in Gallipoli and vicinity — Arty [artillery] was to go to Alexa[ndria] probably this week so as to be ready to embark. Operations in Dardanelles… to get amm[unition] to Russia. Conference at 8 p.m. on night attacks."[118]

MAJOR-GENERAL SIR WILLIAM THROSBY BRIDGES, COMMANDING
1ST AUSTRALIAN DIVISION, MENA CAMP.

---

115 Foster, *At Antwerp and The Dardanelles*, p. 62.
116 Morgenthau, Henry, *Ambassador Morgenthau's Story*, pp. 213–215, Doubleday, Page & Co. (New York) 1918.
117 Colonel Victor Conradsdorf Morisset Sellheim, Officer Commanding Australian Intermediate Base, Egypt.
118 Transcript of diary of William Throsby Bridges, p. 6, 1915, AWM 2DRL/0469.

"All our heavy baggage is being sent away from here, so we expect to follow shortly. We were out at the Rifle Range on Saturday, and the best shots were to be picked out as snipers… We buried one of our company at the English cemetery to-day.[119] He was buried with full military Honours the body being carried on a gun carriage."[120]

PTE ARTHUR WATKINSON, 15TH BATTALION, AIF, AERODROME CAMP, HELIOPOLIS.[121]

17 MARCH 1915

LEMNOS

A meeting of senior naval and military officers was held aboard HMS *Queen Elizabeth* ahead of the attack next day. Vice-Admiral John de Robeck, Carden's successor, led the discussion.

> "Admiral de Robeck told the soldiers that he could silence the fortress guns, but the mobile artillery was his chief difficulty, as it interfered with the clearing of the minefield; he expressed, however, confidence in his ability to force a passage through the Straits without voluntary military assistance on a large scale, and he intended to do so. If our army could land at Bulair, the Turks on the Peninsula would be cut off. He said the Turks were working like beavers every night, none were seen during the day, but new trenches and wire appeared every morning. All possible landing places were being protected by trenches and wire, and ships' guns would not be able to give the troops much support — an opinion which he based on their failure to help the Marines on the 4th March."[122]

COMMODORE ROGER KEYES, RN, CHIEF OF STAFF, EMS, HMS *QUEEN ELIZABETH*.

> "I went to attend a Council of War… on board the *Queen Elizabeth*, in which de Robeck, now a Vice-Admiral and my senior officer, had hoisted his flag. As a result of the conference it was decided that another attack on the forts was to be made the next day to cover a further attempt to sweep the mine field, and that on its upshot would depend the decision as to the extent and manner of the employment of the Army of which Ian Hamilton was to be Commander-in-Chief."[123]

REAR-ADMIRAL ROSSLYN WEMYSS, RN, GOVERNOR OF MUDROS.

The outgoing commander of naval operations submitted his report. It did not strike an optimistic note.

---

119 Pte Leslie Clegg, 15th Bn, AIF, died of pneumonia on 9th March 1915. Buried in Cairo War Memorial Cemetery, he was the 38 year-old brother of Edith Clegg, of Altrincham, Cheshire.
120 *Warwick Examiner and Times* (Queensland), 22nd May 1915.
121 Watkinson, a former draper, was wounded at Quinn's Post on 29th May 1915. Rejoining his unit on 29th October 1915, he was commissioned in France on 19th August 1916. Watkinson was reported missing, later confirmed to be a prisoner of war, at Bullecourt on 11th April 1917.
122 Keyes, Roger, *The Naval Memoirs of Admiral of the Fleet Sir Roger Keyes. The Narrow Seas to the Dardanelles 1910–1915*, p. 223, Thornton Butterworth Ltd., (London) 1934.
123 Wemyss, *The Navy in the Dardanelles Campaign*, p. 40.

"The rest of the day's action on the 19th February showed apparently that the effect of long-range bombardment by direct fire on modern earthwork forts is slight. Forts 1-4 appeared to be hit on many occasions by 12-inch common shell well placed; but when the ships closed in all four guns opened fire. And on the second day, although a heavy and prolonged fire was poured into the forts 70 per cent. of the heavy guns were found to be in a serviceable condition when the demolition parties landed."[124]

VICE-ADMIRAL SACKVILLE CARDEN, RN, LATE EAST MEDITERRANEAN SQUADRON.

18 MARCH 1915

ENGLAND

Asquith was handed a message expressing Hamilton's reservations about the prospects for an unsupported naval operation. The issue was about to be resolved.

"K… showed me a very interesting telegram from Ian Hamilton, who got to the Dardanelles on Tuesday night [16th March 1915]. The Admiralty have been over-sanguine as to what they could do by ships alone. Every night the Turks, under German directions, repair their fortifications, and the channel is sewn with complicated and constantly renewed minefields. The French general, d'Amade arrived at the same time as Ian Hamilton, and they are going to make a really thorough and, I hope, scientific survey of the whole situation."[125]

HERBERT ASQUITH, MP, BRITISH PRIME MINISTER.

DARDANELLES

The Anglo-French fleet approached the entrance to the Dardanelles around 10.00 a.m. This was what the earlier bombardments had been building towards. This was the attempt to force the Dardanelles. Could ships without soldiers succeed?

"It was a glorious morning as the… *Inflexible*, Lord Nelson and *Agamemnon* steamed away from the anchorage north of Tenedos led by the most powerful warship at that time in the world, the *Queen Elizabeth*. There was only just enough wind to make the queer-looking Greek windmills go round. Everything was now favourable to the fleet for we had had days of bad weather on end… The French squadron under Admiral Guepratte followed half an hour later."[126]

LT HARRY BENNETT, RN, HMS *CANOPUS*.

---

124   *Dardanelles Commission. First Report*, p. 31, HMSO (London) 1917.
125   *The Brisbane Courier* (Queensland), 30th July August 1928.
126   *The Sydney Morning Herald* (New South Wales), 19th March 1936.

"Long range gun fire was carried out by the "Agamemnon," "Lord Nelson," "Queen Elizabeth," and the "Inflexible" each being detailed off to engage a fort…

"There is always danger from shell fire in these operations, especially from "Howitzers," which are troublesome things… It seemed as if we must have been struck times without number. It put me in mind of being in a corrugated iron building at which crowd of school boys had been given permission to throw as many stones as they possibly could at it."[127]

PO ERNEST BUCKERFIELD, RN, HMS *AGAMEMNON*.

"The *Inflexible* was very unlucky… A shell from a six-inch or eight-inch howitzer put one of her 12-inch guns out of action at 12.16, and at 12.29 a four-inch projectile fired from a field gun near Erenkeui, hit a signal yard and burst on the roof of her fighting top, killing three men and wounding two officers and three men, the two former mortally; only one man escaped. At 12.47 the Narrows found the *Inflexible*'s range. A heavy shell, probably 14-inch, fell close alongside her, and though it did not strike her, it burst just below the surface and caused a leak in a couple of compartments on the port side aft. Almost simultaneously a 9.4-inch shell made a jagged hole in her starboard side above the water line, and a few minutes later a 9.4-inch shell went through her foremast, and bursting in the Navigator's deck cabin, caused a severe fire which destroyed all communication with the foretop. Her picket boat was also hit at the same time by a heavy shell, but the crew managed to bring her alongside and escape unscathed before she sank. The flames from the fire shot up all round the foremast and scorched and smoked the wounded, and prevented their removal for some time."[128]

COMMODORE ROGER KEYES, RN, CHIEF OF STAFF, EMS, HMS *QUEEN ELIZABETH*.

"Worse was to follow, for shortly after this she [*Inflexible*] hit a mine in the fore torpedo flat. That tore an enormous hole in the side and the whole of the submerged tubes crews, 27 in number, were either killed or drowned. She was immediately ordered to retire outside and get to shallow water…

"When the French ships arrived to engage they were flying enormous Tri-colours, bands playing, and all the rest of it. Guepratte was a well-known fire-eater. He steamed his squadron through the British at 16 knots right up to within 5000 yards of the forts then turning, led his ships past. The fire of the French was rapid and sustained. Opening and closing the range to put the shore gunners off, Guepratte poured in a withering fire that registered ashore…

"The reports from the British sea 'planes, however, were extremely disappointing, for, though it appeared as if though nothing could possibly live ashore, the enemy still had all their guns manned, and were firing away merrily. As the action went on the whole of the sky became tinged with orange; there were several violent explosions around Chanak, that surely must have been magazines going up, but still those enemy guns

---

[127] *Derbyshire Advertiser and Journal*, 30th April 1915.
[128] Keyes, *Naval Memoirs 1910-1915*, pp. 234-235.

spoke. At 1.30 the French squadron was ordered to withdraw, receiving the congratulations of the Commander-in-Chief, Admiral de Robeck, as they steamed down the straits. This constituted the first phase of the attack."[129]

LT HARRY BENNETT, RN, HMS CANOPUS.

"At 1.54 the *Suffren* was leading "B" line out, the *Bouvet* being immediately astern of her. I happened to be looking at them, to see if they had been much knocked about, the *Suffren* had just passed, and the *Bouvet* was almost abreast of us, when I saw a great column of smoke shoot up, which I thought was the burst of a heavy shell striking her, followed by a tremendous explosion, which looked as if her magazine had blown up, she heeled over, still going very fast, capsized and plunged out of sight, with incredible swiftness. Within a minute of the explosion there was nothing to be seen but a few heads in the water. Five officers and about 30 men, who were engaged in the fire control and upper deck batteries, were picked up by the *Wear* and our picket boats; 639 of her company, and her gallant Captain Rageot, lie entombed in her."[130]

COMMODORE ROGER KEYES, RN, CHIEF OF STAFF, EMS, HMS QUEEN ELIZABETH.

"Some officers and sailors, remnants of this fine crew, clung to everything that could support them… But the *Bouvet* soon turned turtle, revealing her keel; the propellers beating the air in a final spasm…

"In less than a minute (45 seconds exactly) the valiant *Bouvet*, ripped apart by a floating mine, capsized and disappeared forever, taking the greater part of the officers and crew with her, twenty officers out of twenty-five and six hundred and forty petty officers, quartermasters and ratings out of seven hundred. Most of the survivors were rescued by the pinnaces from the *Prince George* and *Agamemnon*."[131]

REAR-ADMIRAL ÉMILE GUÉPRATTE, SUFFREN.

"We immediately rushed towards her, but ere we had time to wink, she had gone. When we arrived on the spot, there was nothing to denote that a battleship had been there, except a little wreckage and a space churning water. There were a few boats there, and they had picked up the survivors. We steamed round and round, but no more bodies rose.[132]

STO. 1 WILLIAM CARRADICE, RFR, HMS WEAR.

"Both the French and British damaged ships arrived at the entrance to the straits at the same time, the *Inflexible*, steaming fast with torrents of water pouring over her side, and with all her crew mustered aft in case she should founder, while the *Gaulois* came out slowly firing all her guns into Asia, well down by the bows and with the quarterdeck covered with chairs, as though they were going to have a reception. But the most amazing thing of all was to see a port hole open, and, mark you, in a sinking

---

129   *The Sydney Morning Herald* (New South Wales), 19th March 1936.
130   Keyes, *Naval Memoirs 1910–1915*, p. 236.
131   Guépratte, Émile, *L'Expedition des Dardanelles 1914–1915*, p. 73, Payot (Paris) 1935.
132   *Newcastle Evening Chronicle*, 18th August 1915.

ship and only about a foot above the sea, with a French matelot's head stuck out of it! Escorted by two destroyers she was run ashore on a sandy beach at Rabbit Islands, north of Tenedos."[133]

LT HARRY BENNETT, RN, HMS *CANOPUS*.

"About three p.m. we were ordered up to closer range with three other vessels, and were soon firing away ourselves.

"We were not left in peace for very long, for we suddenly discovered that 14 inch and 11 inch shells were beginning to fall unpleasantly near. As soon as possible we got our fore-turret to bear on the port firing at us, and replied to their fire. It is a strange thing, but as long as the ship is firing back, one does not worry about the enemy's fire. It was rather unpleasant to see the flash of the guns in the fort, then a pause of about 12 seconds, during which time one was wondering where the shells would land.

"Suddenly, was an enormous splash alongside the ship, accompanied by the bang of the shell as it exploded, and immediately water was falling in cascades over the vessel. Even up in the foretop, where I was stationed, we were drowned out. However, as soon as we replied to the fire, it was all right; each one had his duties to perform, and so had no time for waiting for the fall of the enemy's shells.

"The Turks did some very pretty shooting indeed, and the marvel is how we were not struck. Shells were falling right alongside of us and kept our decks almost awash with water. Up aloft it was unpleasantly wet, and made spotting through glasses a matter of difficulty. Things were now getting so warm we were ordered to retire out of range. This we did under sternway, and were soon out of range of the fort."[134]

LT JAMES MONTGOMERY, RNR, HMS *IRRESISTIBLE*.

"We had a furious bombardment until 4.30 p.m., when the crash came and I was nearly thrown on the deck. I knew we had struck a mine, and the order came for everyone on upper deck. There was naturally a rush of men proceeding on deck, and I yelled out 'No panic,' as I happened to be in a passage through which they passed in going, but there was none, and the men behaved well."[135]

ASSISTANT PAYMASTER REGINALD HAYWARD, RNR, HMS *IRRESISTIBLE*.

"Up aloft it seemed as if the masts would be shaken out of her. The vessel immediately listed over to starboard and then laid quiet. There was no doubt what had happened, "Irresistible" was apparently not irresistible where mines were concerned, for, judging by the smoke and steam coming up through the engine-room skylights, the mine had struck her in a vital spot. We soon arrived down on deck to find the men pouring from below in obedience to the order, "Everybody up from below, close watertight doors, magazines, &c."[136]

LT JAMES MONTGOMERY, RNR, HMS *IRRESISTIBLE*.

133 *The Sydney Morning Herald* (New South Wales), 19th March 1936.
134 *Maidenhead Advertiser*, 26th May 1915.
135 *Somerset Standard*, 16th April 1915.
136 *Maidenhead Advertiser*, 26th May 1915.

"We had just been hit again forward by another big shell that shook us all up, and then before we had time to recover ourselves, a tremendous shock was felt. Fully half a dozen of our crew were thrown violently over, and when order had been regained there was the old *Irresistible* heeling over to port at an angle of fully 45 degrees, and our gun pointing in the air, for all the world like an anti aircraft gun.

"Except for the men who had been thrown down picking themselves up, nobody had moved in our casemate. All we could do was to wait orders, looking, meanwhile, into each other's white, set faces. But we were not to remain long waiting. Orders came along to clear the casemate, everybody to get on deck. So, opening the casemate doors, we trooped out, wondering what would be the next thing to happen."[137]

PTE FREDERICK POWELL, ROYAL MARINES, HMS *IRRESISTIBLE*.

"It was really laughable to see everyone, as soon as they arrived on deck, puffing and blowing up their swimming-collars, and then going about with what looked like large sausages round their necks! We soon found that the damage was serious, for the starboard engine-room had practically disappeared, and the port was half full of water, so that we could not move the engines. As soon as the enemy on shore saw we were disabled they opened a heavy fire on as with field guns and howitzers, and made things rather warm on deck.

"About twenty minutes after the explosion a torpedo-boat destroyer came alongside, and into her the boys and ordinary seamen were ordered. About this time some of us had a very narrow shave, for a shell burst within a few yards of us, killing the ordinary seaman next to me. One look was sufficient to satisfy me that the man was finished. It is strange how one can think of these things at such a time; but as I moved from the spot I debated in my mind about making sure the man required no help. My reasoning was, one shell having landed in that spot, the odds were against others landing there. So I went back, and, while looking for the dead man's identification disc, got hurt myself, thus proving my reasoning to be wrong.

"The second shell burst beneath the wooden deck on which I was standing. All I can remember seeing were chunks of deck flying up in all directions, with spurts of flame coming through the holes, while a violent concussion under my feet seemed to jar me to the marrow. The next thing I realised was being helped into the destroyer by some of the men who would insist that I was badly wounded. I remember quite well feeling annoyed and protesting in a far away voice that there was nothing the matter with me. Later, when one of the men asked if he might wipe my face with such a dirty towel, l realised why they thought I was wounded. (I say no more)."[138]

LT JAMES MONTGOMERY, RNR, HMS *IRRESISTIBLE*.

---

137 *Hampshire Telegraph*, 1st October 1915.
138 *Maidenhead Advertiser*, 26th May 1915.

"The *Ocean* was ordered to take her in tow, but when it was realised that she could not possibly float for long, these orders were cancelled."[139]

LT HARRY BENNETT, RN, HMS *CANOPUS*.

"Suddenly a man standing right aft pointed to the entrance of the Straits, and we could see, far away, a destroyer or boat of some sort speeding towards us, smoke flying from her funnels — a welcome sight indeed."[140]

PTE FREDERICK POWELL, ROYAL MARINES, HMS *IRRESISTIBLE*.[141]

"We… could see the *Irresistible* had been struck, so off we went at full speed. When we got up to her we could see she was doomed, and all her crew were on deck waiting for us, and you should have heard them cheer when we got to them. It took nearly half an hour for us to take them all on board, and we steamed away through a hail of shells with 620 of them besides our own crew of 70, and we never had our paint scratched. Don't you think that was a miracle? But I must tell you we have a very brave captain and if ever a man deserves a V.C. it is Capt. Metcalfe.[142] There were no other destroyers there at the time and the poor *Irresistible* was powerless."[143]

STO. 1 GARNET DAWSON, RN, HMS *WEAR*.

"We could do nothing with the ship as the engine rooms were swamped out. Shells began to fall round the ship and after two or three minutes the Turks had the range beautifully. I was on the quarter-deck and two shells pitched about five yards from me, which laid out a large number of our men, and it was not a pleasant sight. A destroyer then came to the rescue, and the men cheered as she came in the distance. After some time, when most of the men had got on the destroyer and the wounded as well, the order came for all of us to leave, but as I was aft and most of the shells were pitching in a certain place, I took off my boots (of course there was a knot in one of the laces) and dived in and swam to the destroyer and was pulled up by a rope. I need not have gone into the water, but I should have had to have gone forward to get in the destroyer and I should have run the risk of being hit by lyddite which would not have been pleasant."[144]

ASSISTANT PAYMASTER REGINALD HAYWARD, RNR, HMS *IRRESISTIBLE*.

"By this time all hands had been sent on board the T.B.D. Shell was flying around everywhere, but only the "Irresistible" was being hit. It was miraculous how the T.B.D.

---

139   *The Sydney Morning Herald* (New South Wales), 19th March 1936.
140   *Hampshire Telegraph*, 1st October 1915.
141   Powell served briefly on HMS *Queen Elizabeth* before seeing service on land on the peninsula as well as on the Western Front. He went to live in Australia and joined the Royal Australian Navy after the outbreak of the Second World War. Employed training recruits, he was killed in the accidental explosion of a smoke bomb when on a range on 26th November 1945.
142   Captain Christopher Metcalfe RN, HMS *Wear*, was awarded the DSO, the citation read: "On the 18th March, after the "Irresistible" struck a mine, he took the "Wear" alongside and rescued nearly the whole of the crew under a very heavy fire which caused several casualties — a very fine display of seamanship" (*London Gazette*, 16th August 1915).
143   *Retford, Gainsborough & Worksop Times*, 2nd July 1915.
144   *Somerset Standard*, 16th April 1915.

escaped. There were 560 men on her upper deck when she left. From other ships we heard afterwards that it looked as if least half our company must have been laid out. The Captain, Commander, and about six officers remained behind, hoping to get the ship taken in tow. Twelve volunteers to remain were asked for, and immediately dozens of men started to scramble back. We finally left the old ship under heavy fire, the T.B.D. in a rather top heavy condition."[145]

LT JAMES MONTGOMERY, RNR, HMS *IRRESISTIBLE*.

"As we got close several men jumped into the water, and I was assisting pulling them aboard; others walked aboard over the focs'le. What a sight! I shall never forget it. The ship was sinking, and the crew was lined up ready to leave. Shells were bursting everywhere, and the fumes were choking. Still there was work to be done, and it wasn't the time to stop and study. The wounded were all passed aboard, and the crew came as fast as possible. Shrapnell [sic] was bursting overhead, shells were throwing the water over the decks, shells were bursting in *Irresistible*. The noise was deafening. The safety valves were open and steam was blowing off. Yet everything was carried out without panic. By the time we had got the 610 survivors aboard there was hardly room to breathe. At last we reached the *Queen Elizabeth*, and discharged our living cargo on one side, while her fifteen-inch guns were booming from the other."[146]

STO. 1 WILLIAM CARRADICE, RFR, HMS *WEAR*.

"We were hastened below as the *Queen Elizabeth* was still firing and was in range of guns. After a time she moved out of the Straits, and I was lent some warm clothing, and after a cup of tea and a cigarette I felt 'merry and bright,' except when I thought of the loss of our dear old ship and the casualties, which were extraordinarily small considering. I cannot understand now how it was more were not killed. Poor old Colchester,[147] who was a fine chap, was shot through the mouth by shrapnel and killed almost instantaneously. As always happens in these affairs, 'the one shall be taken and the other left.' The *Irresistible* lasted some time, but there was a bad list to starboard when I left, and there was no sign of her after dark."[148]

ASSISTANT PAYMASTER REGINALD HAYWARD, RNR, HMS *IRRESISTIBLE*.

"Our steamboat was blown up, we were hit in the foremast, in the forefunnel, our wireless gear shot away in two places, and our quarter-deck ruined. One shot nearly waltzed the after turret into the ditch, a shell carried away a voice pipe over my head (I was in the after-turret). The gun-room was wrecked by a 14 in. shell, we got hit all along the mess decks with small shell, 6 in. or 10 in., and we had one gun's crew wounded, and the gun partially damaged. We had a shot in our upper-deck, which carried away a deck pillar, then scattered a box of ammunition, and finished off the

---

145 *Maidenhead Advertiser*, 26th May 1915.
146 *Newcastle Evening Chronicle*, 18th August 1915.
147 Lt Edward Cromwell Colchester, RN, HMS *Irresistible*, was killed in action on 18th March 1915. Commemorated on the Portsmouth Naval Memorial, he was the 31 year-old son of Edward Cromwell Colchester and Marguerite Branford Colchester, of Great Shelford, Cambridge.
148 *Somerset Standard*, 16th April 1915.

engine room hatch, and there was another shot which burst by the fore-conning tower and did'nt [sic] leave much but remnants of one of our 12 guns. We were going up to tow the "Irresistible" away, and on our way up we got a mine, good and solid on our starboard side just by the after-turret. The turret jumped and shivered and remained standing, otherwise it would have been good-bye to "yours truly," and the rest of the turret's crew."[149]

MIDSHIPMAN DENIS GODDARD, RN, HMS OCEAN.

"My station in action was down the after steering engine-room... It was an awful shock to us down below, although several of my pals were down there, and most of them were Bristol reservists, the same as myself. We were having a chat together when we heard an awful crash. We all kept cool, but could tell the ship was listing over to starboard. The order came "All hands on deck," When I reached the upper deck I saw the destroyers coming alongside of us. Our captain was quite cool, and stood on the fore bridge smoking his pipe watching us leave the ship. The Turks were not satisfied to see our ship sinking, for as we were boarding the destroyers they were firing all around us, but not one man was hit by their firing... I think the officers and crew of the destroyers deserve the V.C. for the way they manoeuvred under heavy fire as we were being rescued. I think there was only one man hit out of the whole ship's company. We saved the ship's pet, Jack Johnson, our black cat, I am pleased to say."[150]

STO. 1 RALPH BENNETT, RFR, HMS OCEAN.

"Just after we got inside the Straits the *Ocean* struck a mine, so we stood by her. She had a good number of the *Irresistible*'s crew on board, a large number of the remainder having been taken off by the Wear. While we were standing by the *Ocean* she took a list of about 15°, and looked as if she might capsize at any moment, so we went alongside her port side and took off about 500 of her crew. Other destroyers went alongside the other side also. It was rather a nasty situation, as she was under fire from a battery of fairly big guns — howitzers, I think — and was turning slowly round and round in the current. Salvos of five or six heavy shell, about 8- or 10-inch, were arriving frequently. Evidently, one of these salvos having just missed us, burst under water and lifted the little Chelmer bodily into the air... The damage done to the ship was eighteen feet of the bottom blown in, and the centre boiler-room flooded...

"Just before the explosion a yeoman of signals belonging to the *Irresistible* or *Ocean* had been sent aft to ask the senior officer on board if he would care to use the captain's cabin. We were anxious to get the officers down below, to leave more room on the upper deck. Apparently the yeoman reached this officer at the same time as the explosion; but the message was faithfully delivered. He sent back the reply,

---

149 *Folkestone, Hythe, Sandgate & Cheriton Herald*, 12th June 1915.
150 *Western Daily Press*, 4th June 1915.

"Thank the Captain for his kindness; but I will stay on deck and see it out to the end." This officer had already abandoned the *Irresistible* to come to the *Ocean*, only to be mined again."[151]

LT-CDR JAMES CLARK, RN, HMS *MOSQUITO*.

"At last our boat was full, and we put off and turned round. We then had the most exhilarating ride ever anyone had. Steaming out of the Straits at 28 knots under heavy shell fire — not small shells, but 14-inch shells from the Chanak Forts. Besides this, the wind was the wrong way. That means the shells would pitch near us before we heard them coming. When we were out in the open sea we went alongside the HMS *Agamemnon* pre-dreadnought. We went aboard her, and had some food and hammocks given us. They did not keep us long waiting, but assembled on the quarter deck the next day, and told off for different ships and stations. I was sent to the *Majestic*."[152]

BOY 1ST CLASS GORDON ROSS, RN, HMS *OCEAN*.

"Now it was not my luck to be one of the first to get away, although I would have liked to, and I never heard anyone sing 'Tipperary,' and all those other heroic songs. I will say the ship's boys were not panic-stricken, but it was case of escaping under a murderous cross-fire.

"It was getting dusk — about 6 o'clock — and we could see their guns flashing and several forts on fire. That night their searchlights burned at Chanak, which leaves no doubt as to the navy's great task."[153]

GNR FREDERICK ASTLES, RMA, HMS *OCEAN*.

Turkish gunners, as the men aboard *Irresistible*, *Ocean*, *Mosquito* and the rest could attest, had not conserved their ammunition. With minefields intact, torpedo batteries untouched and the forts still functioning, supported by the *Goeben* if required, the defenders, though shaken, remained in place. They would be available for any renewed action the following day. The same could not be said of a third of the allied capital ships that had attacked that morning.

Somewhere, perhaps someone in the Admiralty that night, recalled the words contained in the 1906 study: "Even if, as has been suggested, it were feasible to rush a number of His Majesty's least valuable ships past the batteries lining the Dardanelles and over the minefields which are believed to exist in the channel, their arrival off Constantinople would be no guarantee that the Sultan would be thereby brought to reason."[154]

---

151 James Clark quoted in Dorling, Taprell, *Endless Story. Being an Account of the Work of the Destroyers, Flotilla-Leaders, Torpedo-Boats and Patrol Boats in the Great War*, pp. 55–56, Hodder & Stoughton (London) 1932.
152 *Sheffield Evening Telegraph*, 14th July 1915.
153 *Chester Chronicle*, 3rd July 1915.
154 The Possibility of a Joint Naval and Military Attack Upon the Dardanelles, December 1906, TNA CAB 38/12/60.

HMS *Ocean* sinking, 18th March 1915.

EGYPT

Men living in a desert, somewhat drier than the Dardanelles, wondered when they would be leaving; recent arrivals tried to make the most of things, even if it was unappealing.

> "Still our letters are headed 'Egypt.' When are we going to leave this land of sand? is the burning question. All this week we have been drilling hard on a new line of training. Can't understand this sudden rush of drill. For over three weeks we did practically nothing and only left the gun park twice. Now the "heads" seem to be trying to get as much training into the day as possible. Not that we mind the work — we rather appreciate it. But why this burst of keenness to teach us new drill?"[155]
>
> GNR. CEDRIC PATERSON, 8TH BATTERY, AFA, AIF, MENA CAMP.[156]

> "We have just been issued with our new service caps — something like officers', only of inferior quality — to supersede our old hats. From any point of view you like the old hats are the best. Of course our original ones have seen their best days — as you can imagine, but we even yet prefer them. There is something essentially Australian about the old "slouch"; it keeps — or helps to keep — the burning sun off your face; in wet weather it affords some little protection; you can sleep in it, roll on it, pack it up, and it will come up smiling every time. But these caps are neither fish, fowl, nor good red herring. They keep neither sun nor rain off, are uncomfortable, heavy and bear too much the appearance of a Territorial headpiece to be appreciated by the Colonial."[157]
>
> L/SGT JOHN KERR, 5TH BATTALION BATTALION, AIF, MENA CAMP.

---

155   *The Daily News* (Perth, Western Australia), 28th May 1915.
156   Paterson, a former station overseer, was wounded at Gallipoli on 10th May 1915.
157   *Cobden Times and Heytesbury Advertiser* (Victoria), 24th April 1915.

"We heard a lot of talk on the boat of bad behaviour of Australian troops in Egypt, but they seem a well-disciplined, efficient lot to me. The camp is well catered for. Each arm of the service has its own wet canteen — English beer at one piastre a pint (2d½). I don't like the beer much. The chief advantage of the climate is that the beer gets cold with cold nights and is still cool at midday, despite the heat of the day."[158]

L/CPL JOHN PLAYNE, 10TH AUSTRALIAN LIGHT HORSE REGT, AIF, MENA CAMP.

19 MARCH 1915

ENGLAND

A sombre meeting heard of the losses incurred during the failure at the Dardanelles.

"MR. CHURCHILL read a series of telegrams from Admiral Carden and Admiral de Robeck on the subject of the progress in the Dardanelles... The most serious feature of these telegrams was the sinking of the "Irresistible," "Ocean," and "Bouvet," the running ashore of the "Gaulois," and the disablement of the "Inflexible." He also mentioned that the XXIXth Division was due to arrive at Alexandria on the 2nd April."[159]

LT-COL. MAURICE HANKEY, SECRETARY, WAR COUNCIL.

In private, Hankey suspected he knew why all those who advised against an unsupported naval attack had been ignored.

"Personally on the first day [the] proposal was made I warned the P.M., Lord K's Chief of Staff, Lloyd George and Balfour that [the] Fleet could not effect passage without troops, and that all naval officers thought so. I also begged Churchill to have troops to co-operate, but he wouldn't listen, insisting that [the] Navy could do it alone. In my belief Churchill wanted to bring off [a] coup by [the] Navy alone to rehabilitate his reputation, which was damaged (unjustly) by [the] Antwerp affair."[160]

LT-COL. MAURICE HANKEY, SECRETARY, WAR COUNCIL.

DARDANELLES

Churchill expected the attack to be renewed immediately. He was to be disappointed.

---

158 *The Albany Advertiser* (Western Australia), 21st April 1915.
159 Secretary's Notes of a Meeting of a War Council held at 10, Downing Street, March 19, 1915, TNA CAB/42/2/14.
160 Maurice Hankey, quoted in Roskill, Stephen, *Hankey. Man of Secrets*, vol. 1, p. 168, Collins (London) 1970.

> "From what I saw of the extraordinarily gallant attack made yesterday, I am being most reluctantly driven to the conclusion that the Dardanelles are less likely to be forced by battleships than at one time seemed probable, and if the Army is to participate, its operations will not assume the subsidiary form anticipated."[161]
>
> GENERAL SIR IAN HAMILTON, COMMANDER-IN-CHIEF, MEF.

> "This is written after the latest attack on the forts… Comments are prohibited, but we are progressing slowly and safely, and more cannot be expected. The place is very strongly fortified, and, being a narrow waterway, offers considerable difficulties in the way of manoeuvring, so it will necessarily be a slow job."[162]
>
> ERA 3 JOEL MOSS, RN, HMS *CORNWALLIS*.

It would be a slow job. The Army would be employed in the lead role from now on. Some still disagreed but others could not know any of that.

> "After breakfast I surveyed the entrance of the strait from a jetty in front of the hotel. No ships were in sight. At eight we were in Fort Tehemenlik. But the danger flag, a white field with three red disks in it, was not up.
>
> "Nine o'clock came and still no Allies. At ten no smoke even could be seen behind Tenedos, The same state of affairs prevailed at eleven. Noon came and the coast was still clear. The afternoon passed and all was well. But they may come to-morrow."[163]
>
> GEORGE ABEL SCHREINER, ASSOCIATED PRESS OF AMERICA.

TENEDOS

The *Nusret*'s new minefield in Erenkeui Bay had not been detected. How had the mines that caused so much damage got there?

> "How our ships struck mines in an area that was reported clear and swept the previous night I do not know, unless they were floating mines started from the Narrows!

> "I was sad to lose ships and my heart aches when one thinks of it; one must do what one is told and take risks or otherwise we cannot win. We are all getting ready for another 'go' and not in the least beaten or downhearted. The big forts were silenced for a long time and everything was going well, until *Bouvet* struck a mine. It is hard to say what amount of damage we did, I don't know, there were big explosions in the forts!"[164]
>
> VICE-ADMIRAL JOHN DE ROBECK, RN, EASTERN MEDITERRANEAN SQUADRON.

---

161 Sir Ian Hamilton quoted in Bacon, Sir R. H., *The Life of Lord Fisher of Kilverstone, Admiral of the Fleet*, vol. 2, p. 221, Hodder & Stoughton, 1929.
162 *Otago Daily Times* (New Zealand), 1st June 1915.
163 Schreiner, *From Berlin to Baghdad*, p. 148.
164 John de Robeck quoted in Hamilton, *Gallipoli Diary*, vol. 1, p. 40.

"I... visited de Robeck at Tenedos and found him naturally enough somewhat depressed at the turn of events. He spoke of disaster, a term I begged him not to use, and after conferring with him on the steps necessary to take as a consequence of the battle of the day before I left him more cheerful than I had found him.

"The experience we had undergone pointed to the following argument: the battleships could not force the Straits until the mine field had been cleared — the mine field could not be cleared until the concealed guns which defended them were destroyed — they could not be destroyed until the Peninsula was in our hands, hence we should have to seize it with the Army. Any main operations must therefore be postponed until such time as preparations for a combined attack could be made."[165]

REAR-ADMIRAL ROSSLYN WEMYSS, RN, GOVERNOR OF MUDROS.

LEMNOS

"Landing practice. The wicker rafts were made at the time of the Fashoda business for the invasion of England! Nothing could be more complicated. They inspire no confidence in spite of the real ingenuity of the arrangement.

"I am charged with a mission together with Captain Goetz and Lt Petiot. We have to explore the country, or rather a specified sector, in order to decide on the spot for a bivouac... A fine old man has the goodness to conduct us to all the springs, one by one, at which his lambs drink."[166]

MO JOSEPH VASSAL, 6TH COLONIAL REGT, BRIGADE *COLONIALE*, 1ST DIVISION, CEO

EGYPT

"We are awaiting events anxiously, for we think it probable that our destination depends upon the result of the bombardment of the Dardanelles. It is eight years since I was in Constantinople, and I'd rather like to go there again, for curiosity's sake... When the young Turks captured the town I was in it, and I would like to be at its second capture. It would be a unique experience."[167]

PTE SAMUEL MARSHALL, 3RD BATTALION, AIF, MENA CAMP.[168]

20 MARCH 1915

AMERICA

The view of the operation from across the Atlantic: under-way, under-prepared and under-resourced.

---

165 Wemyss, *The Navy in the Dardanelles Campaign*, pp. 41–42.
166 Vassal, Joseph, *Uncensored Letters from the Dardanelles*, p.12, William Heinenman (London) 1916.
167 *Molong Express and Western District Advertiser* (New South Wales), 1st May 1915.
168 A former railway worker, London-born Marshall was wounded shortly after the landing at Gallipoli, being admitted to hospital in Egypt on 30th April 1915.

"Necessity for serious land operations in co-operation with the battleship bombardment of the Dardanelles is now being suggested in London. The delay in adopting this course, as well as the desultory nature of the bombardment, indicate that the Allies have gone into the Constantinople adventure without preparing in advance for all eventualities.

"It seems apparent that the campaign has been regarded from the first as experimental and as presenting possibilities of difficulty or success not predictable at the start. There are reasons which suggest that it has been judged necessary in Paris and London to attempt to conquer the Dardanelles with a minimum of effort… so does the reluctance of the Allies to use a large field force against the land defenses."[169]

JOSEPH WARREN TEETS MASON, AMERICAN JOURNALIST, UNITED PRESS.

ENGLAND

"I embarked on the S.S. *Arcadian* for the seat of war. My destination, I learned, was to be the Dardanelles, and the campaign, I surmised, was likely to be more romantic than any other military undertaking of modern times… After the usual orderly panic consequent on the loading of a troopship we glided from the quay, our only send-off being supplied by a musical Tommy on shore, who performed with great delicacy and feeling "The Girl I left Behind Me" on a tin whistle. The night was calm and beautiful, and the new crescent moon swung above in the velvet sky — a symbol, as I thought, of the land we were bound for."[170]

CAPT. JOHN GILLAM, ARMY SERVICE CORPS, HMT *ARCADIAN*.

LEMNOS

"Wind and rain, big seas running 1620 nearly ashore. About 30 lifeboats broke loose and ran ashore. No parades all day. Had a sleep all afternoon."[171]

PTE SYDNEY PENHALIGON, 3RD FIELD AMBULANCE, AAMC, AIF, HMT *MALDA*.

EGYPT

Many fell ill during their journey to and stay in Egypt. One man described the experience of sunstroke, necessitating a six weeks' stay in hospital.

"I did not fall down, as is usually the case, but a deadly tiredness came over the whole of my body, and I went below and started to play patience at the table, when suddenly I got very cold, and then I went and got my greatcoat and laid down. I was cold until dinner time, and then had a plate of soup. In about a second I was as hot as I had I been cold, so I lay down again. About 5 o'clock I got worse, so I thought I would go up

---

169   *Evening Public Ledger* (Philadelphia), 20th March 1915.
170   Gillam, John Graham, Major, *Gallipoli Diary*, p. 23, George Allen & Unwin Ltd. (London) 1918.
171   *The Queenslander* (Brisbane, Queensland), 21st April 1917.

and see the doctor. I just managed to get to the hospital, and that was all. The doctor took my temperature, and then said curtly to the orderly, "Take him to the hospital." I was told after that my temperature was 104.8. I had a very bad time for four days, and was partly unconscious all the time. My head was aching tremendously, and I had a pain right down at the end of my spine that gradually spread from hip to hip…The fourth and last night I was delirious. I was very hungry, and when the doctor came round I asked him for a piece of bread and butter, and when he said "No" I roused at him, and asked him how he expected me to live if he did not give me anything to eat. You must remember that I was delirious, and had eaten little or nothing of any description for four days, and was not at all responsible… When I left Australia I was a little over 11 stone, and when I got out of bed I was 8 stone, with legs as thin as a wasp, and all ribs sticking out. I had nothing to eat for over a fortnight but two pints of milk and two egg flips a day, but they gradually increased my diet as I got stronger."[172]

SGT JOSEPH SPARKS, 15TH BATTALION, AIF, AERODROME CAMP, HELIOPOLIS.[173]

"There are about 300 acres covered by tents and there must be close on 16,000 horses. One thing which upset the camp was the death of Frank Parker,[174] as two days before he was down here talking and laughing as happy as could be. The next we heard was that there was no chance for him. Nearly all our troop went to the funeral, and a lot of others as well as B squadron. I could not go as I had a bad foot, but it is better now."[175]

TPR OSCAR HASSELL, 10TH AUSTRALIAN LIGHT HORSE, AIF, MENA CAMP.[176]

CEYLON

"After leaving Wellington for Australia it took us fully 14 days from New Zealand to Albany, and from Albany to Colombo 18 days, and on our way to Egypt we are making ourselves as if we were at home. We used to enjoy ourselves, and have games of all sorts. Boxing is the most interesting game of all, for, telling you the truth, no one in the whole Maori contingent could beat me for boxing for my own weight. Well, uncle, this letter was written to you in Colombo garrison hall in my spare time. We had four days out in Colombo. We used to give the people in Colombo a haka of all kinds, just to give us exercise if we get to Egypt. I will write to you again. It takes two or three weeks from Colombo to Egypt, and then to the barracks… Other than that I could not tell you for sure whether we are going to the firing line or not…

172 *Cairns Post* (Queensland), 3rd May 1915.
173 Sparks fell ill aboard HMAT *Ceramic* on 3rd February and discharged from the Military Hospital, Alexandria, on 16th March. Commissioned, 2/Lt Sparks was wounded on 19th May, leading to the amputation of his left hand aboard the hospital ship *Galeka* on 21st May. The former clerk left Suez on 5th July to return to Australia, arriving at Melbourne on 6th August 1915.
174 Major Francis Parker, DSO, (*London Gazette*, 19th April 1901) 3rd Brigade Australian Field Artillery, died of meningitis in Mena House Hospital at 10.05 p.m., 17th March 1915. Buried in Cairo War Memorial Cemetery, the former barrister was the 38 year-old son of the Hon. Sir Henry Stephen Parker, K.C.M.G., and Lady Parker; husband of Dorothy Winifred Parker.
175 *The Albany Advertiser* (Western Australia), 19th May 1915.
176 Hassall was killed in action at the Nek on 7th August 1915. Commemorated on the Lone Pine Memorial, he was the 24 year-old son of Ethel Hassall and the late Albert Young Hassall, originally of Albany, Western Australia.

"Another thing I want to tell you is that our officers are very good to us. Our captain of the Maori contingent is Captain W. Pitt,[177] brother of C. Pitt. He is very kind to us, and we were kind to him."[178]

PTE MIKA PURU, NEW ZEALAND MAORI CONTINGENT, GARRISON HALL, COLOMBO.

21 MARCH 1915

DARDANELLES

"The damage done by the bombardment of four days ago is hardly what I had expected… There is a small dent in Turret No. 1. Turret No. 3 was struck by a shell fragment near the gunport. As a result of that the gun could no longer be elevated or lowered. But a little work with a steel saw fixed that. No. 5 turret is slightly damaged near the base.

"In Fort Rumeli Medjidieh two guns are temporarily out of action. In Rumeli Hamidieh a gun is dismounted. Fort Techemenlik mourns the temporary loss of a gun. One of the casemates there was demolished. In Anadolu Hamidieh a 35.5-cm. has been torn from its anchorage and its carriage has also been badly mauled. Here, too, a casemate caved in.

"The most remarkable thing is the Turkish list of casualties — twenty-three Turks and Germans dead, and seventy-eight wounded. Many of these are civilians… But the Allies are still expected back… New guns are arriving every day… More ammunition is also arriving."[179]

GEORGE ABEL SCHREINER, ASSOCIATED PRESS OF AMERICA.

EGYPT

"To-morrow is to be a big inspection of the whole of the Second Division of troops in Egypt, we are to be inspected by the General. Last week, we had the misfortune of burying one of my tent mates, although he did not have the good luck to fight the Germans, we believe he did his duty and was tendered a soldier's burial. On this occasion, I had the opportunity of being one of the pall-bearers. I must say that it was a big relief to me when it was over."[180]

PTE SEFTON GLASTONBURY, 16TH BATTALION, AIF, AERODROME CAMP, HELIOPOLIS.[181]

---

177 Capt. William Tutepuaki Pitt, a Boer War veteran. Born at Gisborne on 11th June 1877, he was the son of Major Chowell Dean Pitt & Maata Te Owai.
178 *Poverty Bay Herald* (New Zealand), 19th April 1915.
179 Schreiner, *From Berlin to Baghdad*, pp. 148–149.
180 *Yorke's Peninsula Advertiser* (South Australia), 30th April 1915.
181 The former tram conductor was wounded in the first week of the Gallipoli campaign. Rejoining his unit on 28th July, he was evacuated to 2nd Australian Stationary Hospital, Lemnos, on 13th August suffering from a nervous breakdown. After treatment in England, Glastonbury was returned to Australia aboard the Star of Australia on 19th January 1916.

INDIAN OCEAN

"[W]e... passed some more passenger boats as well as the auxiliary cruiser *Empress of Russia* sailing due east at the Arabian Sea. At 4.15 p.m. we passed a town on the Arabian shore called Mocha, about six or seven miles from the sea. It looks a beautiful town with big buildings (bigger than Auckland). At 4.30 p.m., the heat of the sun is getting strong again and the sky looks hazy for the atmosphere is thick."[182]

2/LT AUTINI PITARA KAIPARA, MAORI CONTINGENT, HMNZT *WARRIMOO*.

22 MARCH 1915

LEMNOS

"At a meeting of Admirals and Generals held on board the *Queen Elizabeth* in Mudros harbour... the whole situation was reviewed, and the pros and cons for an immediate attempt at landing troops on the Peninsula were debated. Of the desirability for instant action there was never a doubt, but the chance of surprise — to my mind an absolute necessity for the success of such an enterprise — had vanished. The enemy was thoroughly on his guard and to throw the troops immediately available on shore in face of an almost certain stubborn defence without carefully prepared plans and with the transports unorganized was out of the question."[183]

REAR-ADMIRAL ROSSLYN WEMYSS, RN, GOVERNOR OF MUDROS.

EGYPT

"There are just thousands of soldiers here and the inspection to-day was one of the biggest I have seen. There were I believe over twenty thousand New Zealand and Australian troops and Sir Arthur McMahon took the salute. I shall never forget what a grand sight it was. By the way we have seen the battleship "Ocean" which was sunk the other day. She passed through the Canal while we were there. There was also a submarine which passed through with her deck just out of water; the first I had ever seen."[184]

PTE FREDERICK LOCKER, WELLINGTON BN, NZEF, ZEITOUN CAMP.

"I must tell you of my experience with a young French lady, she advertised in the paper "will teach French to Australian soldier in exchange of being taught English." Well next leave I wandered in the direction of the address given in the advertisement, but what did I see, 20 of our boys after the job, well to my surprise she picked me out, and

---

182 *Gisborne Times* (New Zealand), 21st May 1915.
183 Wemyss, *The Navy in the Dardanelles Campaign*, pp. 42–43.
184 *Patea Mail* (New Zealand), 24th May 1915.

we started on our lessons, which continued for a week, when she spoilt the game by wanting me to marry her. Well I have not been there for any more French lessons."[185]

BUGLER MILTON THORNTON, 3RD BATTALION AIF, MENA CAMP.[186]

23 MARCH 1915

ENGLAND

There was growing frustration in London at the lack of progress in the Dardanelles. They wanted the Navy to attack again before waiting for the army; before the defences were strengthened further. The men on the spot had already concluded that was impossible.

"We had a longish Cabinet this morning. The news from the Dardanelles is not very good. There are more mines and concealed guns than they ever counted upon, and the Admiral seems to be in rather a funk. Ian Hamilton has not yet sent his report, but the soldiers cannot be ready for a big concerted operation before April 14. I agree with Winston [Churchill] and K. [Kitchener] that the navy ought to make another big push so soon as the weather clears. If they wait and wait until the army is fully prepared they may fall into a spell of bad weather or find that submarines, Austrian or German, have arrived on the scene."[187]

HERBERT ASQUITH, MP, BRITISH PRIME MINISTER.

LEMNOS

"At meeting to-day with Generals Hamilton and Birdwood the former told me Army will not be in a position to undertake any military operations before 14th April. In order to maintain our communications when the fleet penetrates into the Sea of Marmora it is necessary to destroy all guns of position guarding the Straits. These are numerous, and only small percentage can be rendered useless by gunfire. The landing of demolishing party on the 26th February evidently surprised enemy. From our experience on the 4th March it seems in future destruction of guns will have to be carried out in face of strenuous and well-prepared opposition. I do not think it a practicable operation to land a force adequate to undertake this service inside Dardanelles. General Hamilton concurs in this opinion."[188]

VICE-ADMIRAL JOHN DE ROBECK, RN, COMMANDING
EASTERN MEDITERRANEAN SQUADRON.

"We got rather a shock when we arrived, as, although we had heard of the loss of the

---

185 *Camden News* (New South Wales), 22nd April 1915.
186 Thornton, a former tailor, was killed in action on 19th May 1915. He is buried in 4th Battalion Parade Ground Cemetery, he was the 20 year-old son of Thomas and Mahala Thornton, of Argyle Street, Camden, New South Wales.
187 *The Brisbane Courier* (Queensland), 31st July 1928.
188 John de Robeck quoted in Churchill, Winston, *The World Crisis 1915*, p. 232, Thornton Butterworth Ltd. (London) 1923.

Irresistible, *Ocean* and *Bouvet* while we were at Malta, we still thought that things were going swimmingly and that we should be through the straits before very much longer… However, when we got to Lemnos we discovered that in reality all that had been done was to knock out the two forts at either side of the entrance, and that the loss of the three ships on March 18 was quite a severe setback for us. This rather damped our spirits, as we had actually thought on our way out that if we wanted to come in for any of the show at all we should have to hurry up or we should find the fleet at Constantinople before we arrived on the scene."[189]

LT FRANK STEPHENSON, RN, HMS *IMPLACABLE*.

EGYPT

"We had a sham fight…, this Brigade against the 1st Brigade, who are at Heliopolis, 14 miles away, we started out at 8 o'clock and met them in the Moquattom Hills about half way. I had my brigade occupying a seven miles front and conducted the operations by flag signals, heliographs, telephone and wireless. It was a great success, and was a drawn battle in my favor. General Walker, who was chief umpire, made very favorable criticisms on the work done. We then entertained the officers of the other brigade at dinner as they came on to our camp and bivouaced for the night."[190]

COLONEL GRANVILLE DE LAUNE RYRIE, 2ND AUSTRALIAN LIGHT HORSE BRIGADE, MAADI CAMP.

RED SEA

"A glorious morning, lovely calm sea, sweet-scented breeze and atmosphere. Welcome clear sky which makes us well and bright. The sea is calm as a pond as her precious travellers on ships pass us every hour. At 6 p.m. we passed 'The Twelve Apostles Islands.'"[191]

2/LT AUTINI PITARA KAIPARA, MAORI CONTINGENT, HMNZT *WARRIMOO*.

24 MARCH 1915

TURKEY

"Enver requested me to wait for him in my office. He came soon afterward and asked if I were willing to take command of the Fifth Army to be organized for the defense of the Dardanelles. I assented at once and informed him that the troops now there would have to be reinforced at once as we had no time to spare."[192]

GENERAL OTTO LIMAN VON SANDERS, COMMANDING 5TH OTTOMAN ARMY.

---

189   *The Press* (New Zealand), 22nd April 1938.
190   *The Queanbeyan Age and Queanbeyan Observer* (New South Wales), 21st May 1915.
191   *Gisborne Times* (New Zealand), 21st May 1915.
192   Liman von Sanders, Otto, *Five Years in Turkey*, p. 57, United States Naval Institute (Annapolis) 1927.

ENGLAND

> "Winston came to talk about the Dardanelles. The weather is infamous there & the Naval experts seem to be suffering from a fit of nerves. They are now disposed to wait till the troops can assist them in force, which ought to be not later than about April 10th. Winston thinks & I agree with him, that the ships, as soon as the weather clears, & the aeroplanes can detect the condition of the forts & the positions of the concealed guns, ought to make another push: & I hope this will be done."[193]
>
> HERBERT ASQUITH, MP, BRITISH PRIME MINISTER.

Churchill signalled de Robeck. It was as if the loss of five of the fifteen capital ships deployed on 18th March had never happened. The Turks had not fled at the sight of a flag.

> "You must not underrate the supreme moral effect of a British fleet with sufficient fuel and ammunition entering the Sea of Marmora, provided it is strong enough to destroy the Turco-German vessels. The Gallipoli Peninsula would be completely cut-off if our ships were on both sides of the Bulair Isthmus. It seems very probable that as soon as it is apparent that the fortresses at the Narrows are not going to stop the fleet, a general evacuation of the peninsula will take place; but anyhow, all troops remaining on it would be doomed to starvation or surrender. Besides this there is the political effect of the arrival of the fleet before Constantinople, which is incalculable, and may well be absolutely decisive."[194]
>
> WINSTON CHURCHILL, MP, FIRST LORD OF THE ADMIRALTY.

MALTA

Hunter-Weston next sought out those with knowledge of conditions at the Dardanelles. He began with Limpus, previously the head of the British Naval Mission at Istanbul.

> "After mooring, I went on shore… to see Lord Methuen (the Governor) & then went to see Admiral Superintendant [sic] of Dockyard at Malta, Vice-Admiral Limpus who had been in Turkish Service. A long talk with him[195]…
>
> "As you know, I have always thought that if the Turks are capably directed & have the necessary material of war, they should be able to prevent our ships getting through the straits, & they should also prevent our landing troops on the Peninsular, for they have had ample warning & they have lots of men. They should by now have made the whole Peninsular into a strongly entrenched camp, with hidden trenches all around the coast, with lots of men to hold them. They should have howitzers in hidden positions covering every possible landing place. If they have done this, & our folks out there should know by this time, the passage of the Dardanelles is not a feasible operation & we shall have to venture home again & be employed elsewhere… It is, as

---

[193] Herbert Asquith quoted in Brock, Eleanor & Brock, Michael (Eds.), *Letters to Venetia Stanley*, p. 506, Oxford University Press (Oxford) 1982.
[194] Churchill, Winston, *The World Crisis 1915*, p. 236, Thornton Butterworth Ltd. (London) 1923.
[195] Diary entry, 24th March 1915, Maj. Gen. Hunter-Weston, British Library, MS 48364.

I have said to you always, a big gamble & if there is any chance of pulling off the big stakes it is well worth trying. If I don't see a good chance, I shall counsel caution. If I see a good chance, I shall try & work it out carefully & push it through with determination & vigour. But the decision does not lie with me."[196]

MAJOR-GENERAL SIR AYLMER HUNTER-WESTON, COMMANDING 29TH DIVISION.

LEMNOS

"Physical training in morning, bandaging exercises and lectures afternoon. Sergeant Clark on insensibility and hemorrhage. S.M. on work of a field ambulance in the field."[197]

PTE SYDNEY PENHALIGON, 3RD FIELD AMBULANCE, AAMC, AIF, HMT *MALDA*.

EGYPT

"Yesterday we went out at 3 p.m. and were out in trenches all night. We took our tea but came back to camp at 7 a.m. for breakfast. Needless to say we are getting tired of this work and long for the action of actual fighting. Strange to say the men will put up with anything if they are gong to move, but are very tired of this."[198]

PTE RICHARD BASSETT, 8TH BATTALION, AIF, MENA CAMP.

"It is grand to hear the shells (they use live shells) screeching over your head, but it is at the same time awful to see the damage done by the shells to the targets placed a mile or two away. The New Zealand artillery are attached to our division, and they are regarded as very good shots, and they deserve it, too. There are not many dummy guns standing after they have had a few shots."[199]

PTE JOHN FINCH, 14TH BATTALION, AIF, AERODROME CAMP, HELIOPOLIS.[200]

25 MARCH 1915

ENGLAND

"Grey and I had a really interesting conversation about the whole international situation. Winston is very anxious that if, when the war ends, Russia has got Constantinople, and Italy Dalmatia, and France Syria, we should be able to appropriate some equivalent share of the spoils — Mesopotamia with or without Alexandretta, a sphere in Persia, and some German colonies, &c. I believe that at the moment Grey and I are the only two men who doubt and distrust any such settlement… Taking Mesopotamia, for

---

196 Letter from Hunter-Weston to his wife, 24th March 1915, British Library, MS 48364.
197 *The Queenslander* (Brisbane, Queensland), 21st April 1917.
198 *Colac Reformer* (Victoria), 29th April 1915.
199 *Bendigo Advertiser* (Victoria), 12th May 1915.
200 The former labourer was evacuated from Gallipoli suffering from neuritis on 20th July 1915. Admitted to No. 1 General Hospital, Heliopolis, on 27th July and was invalided to Australia on 3rd September 1915.

instance, means spending millions in irrigation and development with no immediate or early return, keeping up quite a large army in an unfamiliar country, tackling every kind of tangled administrative question worse than any we have ever had in India, with a hornets' nest of Arab tribes."[201]

HERBERT ASQUITH, MP, BRITISH PRIME MINISTER.

"The attack on the Dardanelles is suspended, whether because of the bad weather or for some other reason one cannot say. We have been told that all conceivable difficulties had been provided for, but I regret to have to doubt this. Rather it seems that the Admiralty had formed the opinion that success could be achieved by a sea attack alone. From the first I have thought this impossible."[202]

SAMUEL STOREY, CO-FOUNDER SUNDERLAND DAILY ECHO.

TENEDOS

Despite enjoying panoramic views from Mount Elias, some observers lacked perspective.

"To-day is the seventh successive day that it has blown such a gale as to make bombardment impossible… To the soldiery and the ignorant masses of the Turkish population the prolonged interruption of the bombardment is encouraging. Being too stupid to realise that the weather makes any difference, they joyfully assume that the Turkish guns have driven the enemy away, and their leaders foster the illusion."[203]

G. WARD PRICE, BRITISH WAR CORRESPONDENT.

LEMNOS

"Sorting mail till 10.30. Physical training in morning. Bandaging and fixing fractures. Lectures afternoon by Sergeant Short on hygiene. Sergeant Clark on hemorrhage, S.M. on intelligence corps."[204]

PTE SYDNEY PENHALIGON, 3RD FIELD AMBULANCE, AAMC, AIF, HMT *MALDA*.

MALTA

"Up at 7 & to see the Admiral Superintendent — Vice Admiral Limpus at 9 a.m… Learnt from him what landing arrangements had been made at Lemnos &… what intelligence as to Turkey he could give… After lunch, I had a talk with Lord Methuen. Neither Methuen nor Limpus are in favour of Dardanelles operations now."[205]

MAJOR-GENERAL SIR AYLMER HUNTER-WESTON, COMMANDING 29TH DIVISION.

---

201   Asquith, *Memories and Reflections*, vol. 2, pp. 82–83.
202   *Sunderland Daily Echo and Shipping Gazette*, 25th March 1915.
203   *Daily Herald* (Adelaide, South Australia), 4th June 1915.
204   *The Queenslander* (Brisbane, Queensland), 21st April 1917.
205   Diary entry, 25th March 1915, Maj. Gen. Hunter-Weston, British Library, MS 48364.

EGYPT

"After riding round the trenches back to camp at 7 a.m. Office morning. Saw McCay and told him I thought Semmens[206] unfit to command Battn. He agreed and promised to see him and say he thought him too unwell to retain command. Conference at 4.30 re last night's operation. Left camp 8.45 p.m. to see MacLaurin's night assault."[207]

MAJ.-GEN. SIR WILLIAM THROSBY BRIDGES, COMMANDING 1ST AUSTRALIAN DIV., MENA CAMP.

SINGAPORE

The response to the 'riot' by 5th Light Infantry reached its bloody conclusion.

"The last batch of twenty-two were shot by the Volunteers, who made up a party of over 100. Unfortunately, I was in the firing party — a distinctly unpleasant experience. One does not mind so much shooting at a man when you know you have an equally good chance of being shot, but it goes against the grain to shoot a man up against a post."[208]

PTE PERCY HILL, MAXIM COY, SINGAPORE VOLUNTEER ARTILLERY.

26 MARCH 1915

ENGLAND

"Defence, defence, defence all the time. Offence nowhere, or, where offence is conducted, as at the Dardanelles, it is so inadequately gone into, & the 1st Lord's personal vanity occupies so large a place in the arrangements that the operation is either a fiasco or is most wasteful in lives or material — or both.

"The 1st Lord would not hear of the Naval Division going to the Dardanelles. He could not understand that the capture of the Dardanelles was a joint operation, or else he wanted all the glory for himself. Result, an attempt to do with the Fleet what is beyond the power of the Fleet."[209]

CAPT. HERBERT RICHMOND, RN, ASSISTANT DIRECTOR OF OPERATIONS, ADMIRALTY.

GALLIPOLI

"On the morning of March 26 we landed at the port of Gallipoli where the headquarters of the Third Corps had been for some time, and established temporary headquarters there."[210]

GENERAL OTTO LIMAN VON SANDERS, COMMANDING OTTOMAN FIFTH ARMY.

---

206 Lt-Col. James Michael Semmens, 6th Bn, AIF, was sent home to Australia on 13th April 1915.
207 Transcript of diary of William Throsby Bridges, 1915. AWM 2DRL/0469.
208 *Birmingham Daily Post*, 6th May 1915.
209 Marder, *Portrait of an Admiral: The Life and Papers of Sir Herbert Richmond*, p. 148.
210 Liman von Sanders, *Five Years in Turkey*, p. 57.

TENEDOS

"I went to Tenedos for the day in the *Doris*. I was very anxious to see how the sweepers were getting on. We also inspected the aerodrome, which we had been making for some days with the assistance of Greek refugees, in preparation for the aeroplanes which the First Lord had hurried out, on receipt of our complaints as to the inefficiency of our seaplanes. An excellent aerodrome had been made by the removal of vineyards and crops, stones and rocks, and rolling out a flat surface with oil drums filled with cement."[211]

COMMODORE ROGER KEYES, RN, CHIEF OF STAFF, EMS.

LEMNOS

Admiral de Robeck resisted Churchill's pressure to launch an immediate naval attack. While agreeing success could still be achieved, he temporised. Citing Hamilton, he argued that it made sense to wait as the military forces assembled.

"I do not consider the check on the 18th was decisive, and I am still of the opinion that a portion of the Fleet would succeed in entering the Marmora. Nothing has occurred since the 21st to alter my intention to press the enemy hard until I am in a position to deliver a decisive attack.

"On 21st I was prepared to go forward irrespective of the Army, as I fully realised that this matter must be carried through to a successful issue regardless of cost, and also because, in view of the military opinion expressed in your 70[212] and which, if persisted in, would in wise assist the Navy in their task. I did not anticipate the possibility of military co-operation in forcing the Straits, though I have always been of opinion that the decisive results would be best obtained by a combined operation rather than a naval or military force acting alone.

"On 22nd, having conferred with General Hamilton, and heard his proposals, I learnt that the co-operation of the Army and Navy was considered by him a sound operation of war, and that he was fully prepared to work with the Navy in the forcing of the Dardanelles, but that he could not act before 14th April. The plan discussed with General Hamilton, and now in course of preparation pending your approval of my 256, will effect, in my opinion, decisive and overwhelming results."[213]

VICE-ADMIRAL JOHN DE ROBECK, RN, EASTERN MEDITERRANEAN SQUADRON.

EGYPT

"We are all getting a bit sick of doing nothing, we would like to go to the front and do a bit of fighting. We have had no word lately about going. The old Turks came back

---

211   Keyes, *Naval Memoirs 1910–1915*, p. 267.
212   Admiralty telegram, 26th February 1915.
213   Keyes, *Naval Memoirs 1910–1915*, pp. 267–268.

at the Canal the other night and perhaps the authorities think they had better not take away too many troops from Egypt while they are poking about, because it would never do to let them get anything like a footing as those rotten Egyptians would rise in a minute if they thought the Turks had a show."[214]

COL. GRANVILLE DE LAUNE RYRIE, 2ND AUSTRALIAN LIGHT HORSE BDE, MAADI CAMP.

"Real excitement prevailed on the evening of Friday, March 26, when it was known that the Maori contingent was to arrive. Fatigue parties soon had their tents erected on ground immediately opposite our regiments' lines. Thousands and thousands of the fellows trooped down to the Helmeih station to welcome [them], and practically every band turned out for them. They were certainly given a most enthusiastic reception, much more so than the Third Reinforcements, who arrived about 11 the same evening."[215]

L/CPL LESLIE SOLE, WELLINGTON BN, NZEF, ZEITOUN CAMP.[216]

"Night after night we go out and bivouac in the desert, with the star-lit canopy of the heavens as our only roof — alone in the vast silence of the desert, Strict silence is the order of night marches, and as we tramp noiselessly along the sand, the slight jingle of an isolated equipment is the only noise heard to let you know that a whole division or 10,000 men are in close proximity and on the move. The work we do is considered good, and on observing at daylight the change effected one would wonder at it. Where before lay the unbroken line of the desert you now see stretch after stretch of trenches, manned by khaki-clad figures, with the parapet in front and the parados behind sprinkled with sand to make them invisible from a distance. It is wonderful the work a battalion of men can do in the dark with pick and spade when expecting sudden attack. You might be lying asleep, rolled in your blanket, when suddenly a distant shot announces that your outposts are in touch with the enemy. Like lightning a whispered command comes along the sleeping lines, and in a moment thousands of human ants are digging themselves in at a furious rate, since we do not know the strength of the enemy or how long our screen and patrols can keep them in check. The trenches dug, they are manned, and soon the grim silence of the desert is broken by the crack of a thousand rifles and spurts of flame leap from the trenches like a thousand fireflys, whilst now and again the artillery guns speak, and the boom goes rolling and reverberating over the desert, to be answered in kind the next moment by the opposing artillery. It is very interesting; and as near the real thing as possible."[217]

CPL SYDNEY BOLITHO, 6TH BATTALION, AIF, MENA CAMP.[218]

---

214 *The Queanbeyan Age and Queanbeyan Observer* (New South Wales), 21st May 1915.
215 *Taranaki Herald* (New Zealand), 20th May 1915.
216 Sgt Leslie Sole, Wellington Bn, NZEF, died of wounds aboard the transport *Southland* on 9th May 1915. Commemorated on the Lone Pine Memorial, he was the 24 year-old son of Thomas Gore Sole and Alice Sole, of Brown Street, New Plymouth, Taranaki.
217 *The Bendigo Independent* (Victoria), 5th May 1915.
218 Promoted Sgt on 12th May, the former clerk was wounded on 18th July. After treatment in Malta, he was transferred to England, where he was later attached to the Australian Army Pay Corps. Poor health led to Bolitho's return to Australia on 5th November 1917.

27 MARCH 1915

ENGLAND

"I have been somewhat severely criticised in the past for insisting in these Notes that forcing the Dardanelles must necessarily be a very difficult and dangerous operation, and not the mere "naval parade" which so many people were inclined to imagine that it would be. Now that losses have been sustained, there is a tendency for the undue public optimism of yesterday to be replaced by an equally undue pessimism. It is necessary, therefore, to emphasise the fact that whatever public opinion in the matter may be, our Admiralty most certainly did not enter upon these operations without carefully counting the cost or without being prepared for, and anticipating, losses. Nor was it under any delusions as to the relative fighting values of ships and forts. Consequently, though we have had one set-back, and may yet experience others, there is every reason to believe that Constantinople will ultimately be reached."[219]

FRED T. JANE, NAVAL CORRESPONDENT.

"We are now under canvas, and I can tell you it is mighty cold, though it should be getting warmer each day now. I feel the cold very much here, especially the snow storms… Today we had a sham fight, two divisions against the other two. We left camp at 6 a.m., and returned at 5.30 p.m., covering 25 miles in all. By the time you receive this, we will be on our way from England, but whether it will be for France or the Dardanelles we do not know. I feel in a different world altogether being so far from home."[220]

CPL GEORGE FLEMING, 11TH BN KING'S (LIVERPOOL) REGT, WATT'S COMMON, ALDERSHOT.

LEMNOS

"Went with mail to steamer, then to Mudros with kit bag for Dr. Merwin. Sailed back. Rest of men gone ashore loading wounded on ships."[221]

PTE SYDNEY PENHALIGON, 3RD FIELD AMBULANCE, AAMC, AIF, HMT *MALDA*.

EGYPT

"I had a general view of the sandy soldiers serving King and Empire. The boys are well and happy and tucker is good. Major-General Godley welcomed us and spoke highly of the conduct of the troops. He gave us good advice about the temptations which we must fight in Egypt."[222]

2/LT AUTINI PITARA KAIPARA, MAORI CONTINGENT, HMNZT *WARRIMOO*.

---

219   *Land and Water*, 27th March 1915.
220   *The North West Post* (Formby, Tasmania), 9th June 1915.
221   *The Queenslander* (Brisbane, Queensland), 21st April 1917.
222   *Gisborne Times* (New Zealand), 21st May 1915.

"Sir Ian Hamilton inspects us to-morrow, and as he is supposed to be one of Kitchener's right hand men we hope that he also expresses a favorable opinion. If it blows like it is doing at present, he will only be able to see very few of us."[223]

SGT NORMAN MIGHELL, 15TH BATTALION, AIF, AERODROME CAMP, HELIOPOLIS.[224]

"There are very few indeed who will be sorry when the dust of Egypt is shaken off their feet, and may it be soon... No doubt the sand makes hard marching, but it is better than mud. The latter is better dodged, and all are thankful that they did not have to face the rigours of an English winter."[225]

L/CPL CHARLES IVE, OTAGO BATTALION, ZEITOUN CAMP.[226]

"In order to distinguish one batallion [sic] from another we have each been ordered to wear colors — ours is red and-black, and as I gaze on my arm my mind readily goes back to many stirring scenes on the Cobden recreation ground and on the East Melbourne reserve...

"I have several coins which I got from prisoners off the "Emden." I value them highly, and would like to send them home, but am afraid to risk it. Did some Egyptian coins reach you that I sent?"[227]

L/SGT JOHN KERR, 5TH BATTALION, AIF, MENA CAMP.

28 MARCH 1915

ITALY

"I went round to the Turkish Embassy, and there read through the Turkish official account of the operations of March 18th. It was an extremely interesting document, very soberly written, and as it was only intended for their own representatives, and not for the world at large, it could be considered accurate. It served to confirm the opinion held by many that we had under-estimated our task, and that the attack of March 18th had never stood any chance of succeeding."[228]

ELLIS ASHMEAD-BARTLETT, WAR CORRESPONDENT, ROME.

223　*Cairns Post* (Queensland), 24th May 1915.
224　A former articled clerk, Mighell, was wounded, a gunshot wound to his jaw, on 26th April 1915. After treatment in Egypt and England, he sailed from Portland to return to Australia on 7th November 1915.
225　*Southland Times* (New Zealand), 8th May 1915.
226　Ive was wounded at Gallipoli, sent to Egypt aboard the transport *Dongola* on 3rd May 1915. Transferred at Alexandria to the hospital ship *Letitia* on 7th May, the former journalist was then sent England, admitted on 20th May 1915 to the 1st Southern General Hospital.
227　*Cobden Times and Heytesbury Advertiser* (Victoria), 19th May 1915.
228　Ashmead-Bartlett, Ellis, *Uncensored Dardanelles*, pp. 26–27, Hutchinson & Co. (London) 1928.

Turkish prisoners captured during the fighting at Suez marched through Cairo.

EGYPT

Hunter-Weston outlined more of his concerns about the coming campaign. These were not shared universally.

> "The Turkish Army, having been warned by our early bombardments and by the landings carried out some time ago, has concentrated a large force in and near the Gallipoli Peninsula. It has converted the Peninsula into an entrenched camp; has, under German direction made several lines of entrenchments covering the landing places with concealed machine gun emplacements and land mines on the beaches; has put in concealed positions guns and howitzers capable of covering the landing places and operations with their fire. The Turkish Army in the Peninsula is being supplied and reinforced from the Asiatic side and from the Sea of Marmora and is not dependent on the Isthmus of Bulair…

> "With a reasonable prospect of success therefore the forcing of the Dardanelles should be undertaken whatever the cost. But without a reasonable chance of ultimate success the operations should go no further than they have.

> "To go away after coming so far would have a bad effect, would cause much talk and some laughter, and would need great moral strength and courage. To incur a local defeat here on land would however, be a serious disaster to the Empire…"[229]

> "Throughout this war none of the combatants has ever succeeded in breaking quickly through even indifferent entrenchments. The usual result has been stalemate. Success has only been attained after long and careful preparation and after the expenditure of an enormous amount of High Explosive Gun ammunition both from quickfirers and howitzers. We are very short of gun ammunition and are particularly short of High Explosive Shell. There appears therefore every prospect of getting tied up on

---

[229] Appreciation of situation in the Dardanelles according to information received up to the 25th March, 1915, the date of leaving Malta, Maj. Gen. Hunter-Weston, British Library, MS 48364.

an extended line across the Peninsula, in front of the Turkish Kilid Bahr plateau trenches — a second Crimea... We shall have a precarious and insufficient landing place and shall be entirely cut off if stormy weather intervenes."[230]

MAJOR-GENERAL SIR AYLMER HUNTER-WESTON, COMMANDING 29TH DIVISION.

"Near by our camp here is a big prison, where are accommodated over 600 Turkish prisoners of war. I took a stroll down there to-day to have a look at the beggars, and I reckoned I saw the riff-raff of Europe. They are far better off as prisoners than they were as Turkish soldiers. While I was there three fresh ones (or rather new ones) arrived under escort. They looked a sorry sight indeed. Two of them wore khaki puttees over a pair of dilapidated boots, whilst the third wore the remains of a pair of leather leggings, which looked suspiciously like as though portion of them had provided a meal, as they had a very chewed and frayed appearance. I saw them served out with a dungaree uniform each, and more important still, plates and mugs, so I suppose they'll reckon they are in clover now. They also got a blanket each, so I suppose they'll be more comfortable at night than they were with an old grey overcoat each."[231]

TPR JAMES BROWNHILL, 6TH AUSTRALIAN LIGHT HORSE REGT, AIF, MAADI CAMP.

29 MARCH 1915

ENGLAND

"In our loose national habit of not thinking things out, to be cocksure is counted as manly. It is not. Forcing the Dardanelles against modern mines and guns inspires awe in the mind of students of history... cocksureness as to the success of our fleet, at all events on its first or second attempt, is misplaced. We shall get through — of that there is no doubt — but the cost of getting through may darken many British homes, and quench the light of many women's eyes."[232]

ARNOLD WHITE, JOURNALIST AND AUTHOR.

"Doubtless means are being got ready for a larger and wider attack, which it is hoped may be more successful. Nobody now doubts that the passage cannot be forced nor Constantinople assailed without troops as well as sailors, and it is a pity that this view was not taken by our Government from the first."[233]

SAMUEL STOREY, CO-FOUNDER SUNDERLAND DAILY ECHO.

EGYPT

Hunter-Weston continued to air his reservations about the prospects for the landings.

---

230 Note with regards to the opinion of Major-General Hunter-Weston, as to the utilisation of our forces in the Eastern Mediterranean, March 1915, 28th March 1915, British Library, MS 48364.
231 *Wellington Times* (New South Wales), 3rd May 1915.
232 *Oxfordshire Weekly News*, 31st March 1915.
233 *Sunderland Daily Echo and Shipping Gazette*, 29th March 1915.

"At the conference I was the only one who expressed doubt of success although, as you know, I have the reputation of being something of a thruster. I do not see how we can win sufficient territory to make a secure base; I do not see how we can count upon such a measure of success as will make us independent of the vagaries of the sea[234]…

"I had a talk with Braithwaite at his office. Tommy took him a fair copy of my appreciation. Sir Thomas Cunningham had a chat with me before dinner, & dined with me at Staunton's table at the Majestic Hotel."[235]

MAJOR-GENERAL SIR AYLMER HUNTER-WESTON, COMMANDING 29TH DIVISION.

"Hunter-Weston had written me a letter from Malta (just to hand) putting it down in black and white that we have not a reasonable prospect of success. He seemed keen and sanguine when we met and made no reference to this letter: so it comes in now as rather a startler. But it is best to have the black points thrust upon one's notice beforehand — so long always as I keep it fixed in the back of my mind that there was never yet a great thought or deed which was not cried down as unreasonable before the fact by a number of reasonable people!"[236]

GENERAL SIR IAN HAMILTON, C-IN-C, MEF.

"To-day we were inspected by and marched past General Sir Ian Hamilton, and, in spite of adverse circumstances… The day was the reverse of perfect, the rays of a scorching sun beating down like the heat from an Australian bush fire; while a fierce desert wind, laden with sand, dust, and particles of stone, blowing in our faces the whole day long, made things anything but pleasant. Still, the united brigades of the force rose magnificently to the occasion — not a man stirred, every bayonet fixed like clockwork, and each arm sloped in absolute harmony; in the march past the step was faultless, and it needed no congratulatory word from the war-worn general or his staff to assure us we had done well…

"Rumor has it that Sir Ian Hamilton is going to take charge of the Australians and lead us to the front. We hope it is correct, and if it is so we will be proud indeed to be found in action with such gallant leadership."[237]

L/SGT JOHN KERR, 5TH BATTALION, AIF, MENA CAMP.

"This day's training will be under Battalion arrangements but should be carried out by Companies. The training will begin by instruction in rapid distribution of tools, computation of the task, telling off of working parties, and siting and marking out of trenches. The digging will proceed by reliefs, the whole personnel participating, and the aim should be for each Company to thoroughly complete, to full DEPTH, with communication trench, observation posts and recesses complete, sufficient fire trench for 100 rifles; machine guns sections will similarly construct emplacements, etc.

---

234 Hunter-Weston quoted in Montgomery-Cunninghame, Sir Thomas, *Dusty Measure. A Record of Troubled Times*, p. 212, John Murray (London) 1939.
235 Diary entry, 29th March 1915, Maj. Gen. Hunter-Weston, British Library, MS 48364.
236 Hamilton, *Gallipoli Diary*, vol. 1, p.62.
237 *Cobden Times and Heytesbury Advertiser* (Victoria), 19th May 1915.

"All trenches to be filled in before leaving the site. The Battalion will march at 7 a.m. and move direct to this training area."[238]

CAPT. CHARLES DARE, 14TH BATTALION, AIF, AERODROME CAMP, HELIOPOLIS.

30 MARCH 1915

ENGLAND

Carden had been replaced by de Robeck as the commander of the Eastern Mediterranean Squadron on 16th March. Poor health was given as the reason for the change. That was not what Carden told a friend, according to one source.

> "I have recently had a conversation with a gentleman who is a personal friend of and has seen Vice Admiral Carden in London[239] since the latter's return from command of the operations at the Dardanelles.
>
> "Vice Admiral Carden stated that he was in perfect health despite the Admiralty announcement that he was "incapacitated by illness," and that the rumour, prevalent in London, that he had been wounded, was also untrue. He stated that the reason for his relief was follows —
>
> "After he had sent the *Amethyst* through the narrows[240] on March 13th and that vessel had returned with only minor injuries due to gunfire, the Admiralty desired him to take his fleet through; in his judgment such a course was not justified as he considered the risks greater than the chances of success; he was then given the option of carrying out the Admiralty's desires or of being relieved of his command; he chose the latter alternative, and was accordingly succeeded by Admiral de Robeck on March 16. The action of March 18 in which the *Irresistible*, *Ocean* and Bouvet were lost was an attempt to carry out the Admiralty's desire to force the narrows at once.
>
> "Vice Admiral Carden considers that the result of the attempt on the 18th vindicates his judgment, and stated that he was urging his case at the Admiralty on those grounds, and claimed the right of reinstatement in his command."[241]

NAVAL CONSTRUCTOR LEWIS B. MCBRIDE, USN, AMERICAN EMBASSY, LONDON.

Carden had failed to impress the likes of Jackson, Birdwood and Wemyss. Did Churchill remove a man who did not share his conviction that an all-out effort would win through or exhaust the defenders' ammunition in the process? If Carden had taken a principled stand, one vindicated by subsequent events, it is a stance he failed to mention later.

---

238 14th Bn Australian Infantry War Diary, AWM4 23/31/6.
239 Carden arrived back in London on 6th April 1915.
240 The *Amethyst*, though widely reported to have done so, never reached, let alone passed through, the Narrows.
241 Naval Attache's Reports, Office of Naval Intelligence, Unpublished Manuscript, US Naval War College, March 1915.

DARDANELLES

"The Allies seem to have lost all interest in the Dardanelles. Their ships have gone to parts unknown. A few of the older tubs have stayed, however, and day before yesterday they came to the entrance and peppered something on the heights behind Kum Kaleh...

"Life in a bombarded town is not altogether pleasant. Everything still smells of dead fire. Of inhabitants only the dogs and cats are left. Poor things!"[242]

GEORGE ABEL SCHREINER, ASSOCIATED PRESS OF AMERICA.

EGYPT

Hunter-Weston met Hamilton and discussed the reservations he had raised in his earlier letter.

"To Savoy Hotel to see Sir Ian Hamilton. Expressed my pessimism. Back to Majestic Hotel; paid my bill & went to The Savoy & lunched at same table as Sir Ian Hamilton. After lunch went with de la Chapelle in new 6 cylinder Studebacker [sic] Car, my car as G.O.C. 29th Div., to call on General d'Amade, commanding the French Division. We began talking in English but soon dropped into French & we talking for over an hour. He agreed with my views."[243]

MAJOR-GENERAL SIR AYLMER HUNTER-WESTON, CMDG 29TH DIVISION.

A large scale attack exercise was held. One of the 'wounded' described his part in it.

"On Tuesday [30th March 1915] our brigade did an attack and defence on a canal. It was inter-brigade work, the object being to co-operate with the Field Ambulance. We left camp at 6.30 a.m. and got back at 2 p.m. In approaching, the enemy, I was part of a flank guard, and we struck the enemy (which was one company from each battalion) where they weren't supposed to be. They caught our small party of about 20 men in the open. Our only hope was to get some cover about 100 yards ahead. So we rushed it. By the time we got there we were all declared "wiped out," by our umpire. I wouldn't like to say that was why we went for that particular piece of cover, but the fact remains that we sat down in the shade of some date palms and watched the others attack. There's no point in rushing about under an Egyptian sun when a little brainwork will get one out of it.

"The umpires were riding about distributing tickets which told how a man was wounded and what to do with him. We couldn't get hold of any of these, as the umpire on our particular spot said we were dead and didn't need a ride home in the ambulance waggons. When word came down the line for the brigade to assemble we made our way back to the rallying place.

---

242 Schreiner, *From Berlin to Baghdad*, p. 150.
243 Diary entry, 30th March 1915, Maj. Gen. Hunter-Weston, British Library, MS 48364.

"Whilst waiting for the rest to come up we saw the stretcher-bearers at work. I saw a party of three men passing us, and asked them who they were and where they were going. One chap said, 'Oh, I'm unconscious, and this chap was shot in the leg, and this chap has a scalp wound. We got tired of waiting for the stretchers, so we're on our way over to the ambulance waggons'."[244]

CPL WILLIAM MILLS, AUCKLAND BATTALION, HELMIEH CAMP, HELIOPOLIS.

31 MARCH 1915

ENGLAND

The Dardanelles sucked in more resources, as it evolved from a purely naval to a combined operation involving the Army. This so alarmed Fisher that he wrote to Churchill.

"With the departure of the last batch of reinforcements of destroyers, submarines, and mine-sweepers for the Dardanelles we have reached a point when the general situation must be carefully reviewed, more especially with regard to the margin of superiority over the German Fleets which we retain in home waters — at their selected moment...

"I consider that we have now descended to the bare minimum of superiority in home waters, and that to dispatch any more fighting ships, of any kind, to the Dardanelles operations would be to court serious losses at home.

"We have been fortunate of late in that we have had no submarine losses of moment in the North Sea, but danger is always present, and a turn of the wheel may give the enemy chances of attrition which he will take... Consequently I desire now to state my definite opinion that we must stand or fall by the ships now out in the Mediterranean or on their way there. If the concerted operations about to be undertaken are not successful, and we incur heavy losses either of great or small ships, we cannot afford to send any more from home waters to complete the work... nor even if more naval force is required to extricate our Army...

"I therefore urge that, before the final plunge is taken and the troops landed, the Dardanelles operations should again be carefully examined from this standpoint by the War Council. We should have before us the considered report of Sir Ian Hamilton, with the remarks of the Army Council thereon; we should reconsider the position in view of the lack of damage hitherto done to the guns in the forts by the previous bombardment; we should consider the possible arrival of enemy submarines, and, in a word, all the factors which might result in the next operation being indecisive or unsuccessful so as to call for greater naval force...

244   *The Feilding Star* (New Zealand), 19th May 1915.

> "We can recover from an indecisive or even an unsuccessful result of these operations in the Dardanelles; we can recover from an abandonment of the operations, should this be necessary; but we could never recover from a reverse to our main fleet in the decisive theatre at home. It would be ruin. Our existence depends on our unchallengeable naval supremacy."[245]

ADMIRAL LORD JACKIE FISHER, RN, FIRST SEA LORD.

LEMNOS

> "Had there not been such a thing as mines, undoubtedly we should have reached Constantinople before now, and outside Abdul's harem. We are having lovely weather compared with that at home, and should it continue I should say you should read a good bit of news from here shortly."[246]

LEADING SIGNALMAN GEORGE DURRANCE, RN, HMS *LORD NELSON*.

EGYPT

Excerpts from the British combined operations doctrine were circulated, specifically those regarding "embarking from boats,"[247] to the 29th Division. The document laid out some general principles that should be followed in amphibious warfare.

> "When operations oversea are contemplated by the Government, it will be necessary for the naval and military authorities to advise as to the forces to be employed for the attainment of the object, having regard to the information available concerning the enemy, the topography and resources of the proposed theatre of operations, the anchorages, landing places and harbours, and the districts inland. A detailed scheme will also be required for the organization and mobilization of the expedition; and plans must be prepared for its embarkation and disembarkation and, as far as possible, subsequent operations…
>
> "In certain circumstances it may be necessary, owing to the naval risks involved in their execution, to change or modify the plans originally drawn up for the disembarkation and subsequent operations. Such alterations will, however, usually involve delays, and will thus reduce the possibility of inflicting a surprise, which is all important. It will, therefore, as a rule, be best to carry out the original plans in spite of the risks which must be incurred."[248]

Those reading the document for the first time might well have reflected on the difference between the doctrine and their own circumstances.

---

245 Lord Fisher quoted in Bacon, Admiral Sir R.H., *The Life of Lord Fisher of Kilverstone. Admiral of the Fleet*, vol. 2, pp. 225–227, Hodder & Stoughton (London) 1929.
246 *Grantham Journal*, 24th April 1915.
247 29th Division General Staff War Diary, TNA WO 95/4304.
248 *Manual of Combined Naval and Military Operations. 1913*, p. 10, HMSO (London) 1913.

Australians aboard a battleship, bound for Gallipoli.

Anzac Cove, the afternoon of April 25, with guns of the 2nd Brigade, Australian Field Artillery, being brought ashore.

APRIL 1915

"The air was simply alive with death."
Capt. John Milne, 9th Battalion, AIF

1 APRIL 1915

MALTA

"I went across to the Old Port to call on Admiral Limpus. The Admiral had been for some years head of our Naval Mission in Turkey, and he naturally knew the Turks and the defences of the Dardanelles better than anyone else. One would have thought that as he was Senior Naval Officer in the Mediterranean, he would have been given command of the fleet operating against the Straits… Limpus was extremely sceptical about the prospects of the Expedition. He declared that the attack on March 18th ought never to have been made as the forts and defences were far too strong. "Now," he added, "we have given the Turks warning that we intend to strike, and they will be ready for us on the Peninsula itself."[1]

ELLIS ASHMEAD-BARTLETT, WAR CORRESPONDENT.

EGYPT

"Divisional Scheme. Breakfast 6.30. Assembled Comdrs at 7.45. Gen Birdwood Director told me he thought plan of operation shld [should] be changed to land on Asiatic side — S[outh] of Besika Bay. Wagstaff[2] out to dinner afterds [afterwards] gave me orders embarkation was to begin on Sunday."[3]

MAJOR-GENERAL SIR WILLIAM BRIDGES, COMMANDING
1ST AUSTRALIAN DIVISION, AIF, MENA CAMP.

"You called me silly and mad when I enlisted, but you no doubt see now that I took the right step. I only did what any other young fellow physically fit should be doing now. I told you it would be a pleasure trip, but only to humour you. I knew all the time the seriousness of my step, but now that you see it was inevitable I have no hesitation in stating the true state of affairs at the time of my enlistment. I couldn't, if I were at home, see all the elderly men of the town drill while I carried on as if nothing had happened like a lot of young men are doing. Troops from everywhere are arriving here lately: Maoris, English, French. Soon we will be mixed in bloody battle. We are almost certain to participate in the operations in the Dardanelles and the conquest of Turkey. If successful in that we will probably move on to the western seat of war in France and help in the invasion of Germany."[4]

PTE THEODORE JAHNS, 2ND BATTALION, AIF, MENA CAMP.[5]

---

1   Ashmead-Bartlett, *Uncensored Dardanelles*, p. 27.
2   Major Cyril Wagstaff, Royal Engineers, attached General Staff, Australian & New Zealand Army Corps.
3   Transcript of diary of William Throsby Bridges, p. 9, 1915, AWM 2DRL/0469.
4   *The Tamworth Daily Observer* (New South Wales), 19th June 1915.
5   Pte Theodore Jahns, 2nd Bn, AIF, was reported missing, later presumed to have been killed in action on 2nd May 1915. Commemorated on the Lone Pine Memorial, he was the 21 year-old son of Henry Theodore Jahns and Kate Jahns, of Darling Street, Tamworth, New South Wales.

"It was very funny the other night. We were out for a night attack, and I and two other chaps were sent out as scouts; and we got lost. It was very dark when we were sent out. We were told to come back to the main body in an hour's time, but when we found out that we were lost, we lay down and went to sleep. When we woke up the next morning we found out we were about 5 miles off the place where we were told to pick up the main body, and 15 miles from camp. So we set off for the camp; we got into camp about dinner time, and you can guess we did not waste any time in getting our dinner. We had had nothing to eat since 5 o'clock the evening before… I don't know when we are going to move from here; we are all feeling anxious to get into it. The officers won't tell us anything, but by the hints they throw out I think we will soon be into it."[6]

PTE ANGUS PETTIGREW, 14TH BATTALION, AIF, AERODROME CAMP, HELIOPOLIS.[7]

"We are on the point of leaving Egypt now. The infantry have already gone, and the last of these went to night. Our destination is not yet made known to us, but everything points to the Dardanelles or at least seems so. If such is the case, I think we are in for a fairly hot time, but we are quite prepared for whatever comes. I have had visits from some of the Essendon boys last week, and this week I have been to repay the visits. On Sunday, I went over to the 5th Battalion to see some of the chaps that I know viz.: Leslie Williamson,[8] Rupert Middleton[9] (Sgt), Jim Workman[10] and others…

"On Tuesday, I went to the 7th Battalion, where most of the Essendon chaps are, and we had a fine old talk in one of the bright clear moonlight nights that we get here. Ted Cobbin[11] accompanied me; by the way, he is keeping as well as I am, and I tell you that is well. We are great chums. To-morrow is Good Friday. I do not see too many hot cross buns coming this way. We are to have a holiday. Easter has come upon us quite suddenly this year. I attended a funeral of one of our artillery comrades[12] yesterday. The proceedings were most impressive. You have no idea how a loss such as this affects the strong men here. He was buried in the Old Cairo Military Cemetery, with the usual "last post" sounded and salutes. I have seen the locust plague in its true

---

6   *Shepparton Advertiser* (Victoria), 3rd June 1915.
7   Pettigrew was killed in action on 29th April 1915. Buried in Courtney's and Steel's Post Cemetery, Anzac, where he is commemorated by a special memorial, he was the 19 year-old son of John and Annie Pettigrew, of Congupna Road, originally of Shepparton, Victoria.
8   Pte Leslie Williamson, 5th Bn, AIF, a former draper, was killed in action on 25th April 1915. Commemorated on the Lone Pine Memorial, he was the 21 year-old son of Henry and Louise Williamson, of 1 Chatham Street, Newmarket, Victoria.
9   L/Sgt Rupert Middleton, 5th Bn, AIF, was taken aboard the hospital ship *Gascon* on 25th April 1915 suffering from a sprained ankle and concussion. The former signwriter returned to Australia on 30th August 1915.
10 Pte James Workman, 5th Bn, AIF, was wounded 6th–11th May 1915. He was evacuated to England on the hospital ship *Sicilia* on 3rd July 1915.
11 Shoeing Smith Edward Cobbin, 5th Battery, 2nd Bde Australian Field Artillery, a former blacksmith from Ascot Vale, Victoria.
12 Dvr Charles Newman, 2nd Brigade Australian Field Artillery, died of pneumonia on 30th March 1915. Buried in Cairo War Memorial Cemetery, he was the 33 year-old son of George Cole Newman and Marian Newman, of 19 Montifore Street, Coburg, Victoria Australia.

light. Clouds upon clouds of the insects fly daily over the camp, and nearly smother the crops of the surrounding district."[13]

CPL CYRIL BURDEU, 5TH BATTERY, AFA, AIF, MENA CAMP.[14]

While reservations about the prospects for the operation had been shared both publicly and in private, confidence born of complacency, the hope that things would somehow work out for the best, pervaded this British intelligence summary.

"It is the general opinion that the Turks will offer an energetic resistance to our landing, but when once we are firmly established on the Peninsula, it is thought possible that this opposition may crumble away, and that they may turn on their German masters. The average Turk has always been most sympathetic to the British Nation, and it is known that many look with envy on the prosperity which Egypt enjoys under British rule."[15]

29TH DIVISION GENERAL STAFF WAR DIARY.

2 APRIL 1915

EGYPT

"Drove Arthur Tufnell & de la Chapelle out to Victoria College where I saw d'Amade who told me he had seen Hamilton & told him his opinion of his (Hamilton's) plan; but that Hamilton refused to listen."[16]

MAJOR-GENERAL SIR AYLMER HUNTER-WESTON, COMMANDING 29TH DIVISION.

If some of Hamilton's commanders were unhappy, so, too, were many of the men under their command. Months in desert camps, long marches in deep sand, sand that penetrated everything and critical comments about bad behaviour did not improve their morale. The Good Friday holiday offered a chance for a break in the routine. But leave was restricted, a further cause for grievance, and all that frustration turned to violence.

"I met Paul and the other two at the usual place of recreation, the skating rink. Then we went and had tea. We had just sat down and ordered our tea when there was great excitement the other side of the street. On making inquiries, we found that the "red caps," i.e., the military police from the English [sic] army had shot several lads. Troops from both Australia and England were fighting them with chairs, sticks, crockery, and whatever came to hand, from restaurants and shops. The red caps were eventually driven off, after three had been injured."[17]

PTE JULIUS JACOBSOHN, MG SECTION, 7TH BATTALION, AIF, MENA CAMP.

13   *The Essendon Gazette and Keilor, Bulla & Broadmeadows Reporter* (Moonee Ponds, Victoria), 13th May 1915.
14   Cpl Cyril Burdeu, 5th Battery, 2nd Brigade Australian Field Artillery, was killed in action on 10th May 1915. Buried in Beach Cemetery, he was the 22 year-old son of Arthur Penrose Acteson Burdeu and Annie Jane Burdeu, of 40 Pascoe Crescent, Essendon, Victoria.
15   TNA WO 95/4304.
16   Diary entry, 2nd April 1915, Maj. Gen. Hunter-Weston, British Library, MS 48364.
17   *The Corowa Free Press* (New South Wales), 17th December 1915.

"I did not see the necessity of keeping the men hanging about, and gave them the option of having leave until "Last Post" or going back to camp. To my surprise, less than half of them decided to take leave, the others preferring to return to camp. Having fixed up the passes of those who wished to stay in Cairo, I put the others on a tram with an N.C.O. in charge, and the officer and I made our way to Shepheard's, where we found the others just about to go in to dinner.

"We did full justice to the dinner, and were sitting in the lounge enjoying our coffee and cigars when the A.P.M. of Cairo arrived and called for all Australian and New Zealand officers present. He told us that a riot had broken out in the Haret el Wasser, and that the Australians and New Zealanders were burning the street down. He called upon us to go and quell the riot, and there was nothing for it but to obey, much as we felt inclined to curse him for what gave promise of being the "end of a perfect day." So we duly sallied forth in pairs to see what we could do.

"On arriving at the scene of the trouble we found the street crowded with soldiers. Australians predominated, although there was a fair sprinkling of New Zealanders and British troops. In the centre of the street a bonfire of furniture and bed clothes was burning, despite the efforts of an Egyptian fire brigade to extinguish it. Their efforts were considerably hampered by the fact that their hoses had been punctured in places by bayonets or cut with knives, and there were too few of them to attempt to patrol the hose lines, even if the size of the crowd had not made such action impossible.

"Most of those present were simply spectators, but some gangs were actively employed looting the houses and tossing furniture and effects into the streets. The presence of the British military police (redcaps) only tended to incense the mob, and after a while they were withdrawn. With the help of a number of patrols of Australians, New Zealanders, and Tommies, plus a squadron of the Westminster Dragoons, the street was eventually cleared and order restored. When we saw that the authorities were well on the way towards getting the matter in hand we decided that our presence was no long necessary, and once again we made a bee-line for Shepheard's. We found it guarded by troops from the East Lancashire Division, who were drawn up round it with fixed bayonets. We had some difficulty in getting through the cordon, as their instructions were to permit no Australians or New Zealanders to enter. Eventually, however, we found an officer to whom we stated our case, and he passed us through to enjoy once more our interrupted taste of the "flesh pots of Egypt"."[18]

CAPT. & ADJ. HORACE VINEY, 3RD AUSTRALIAN LIGHT HORSE REGT, AIF, MAADI CAMP.

Exactly what sparked the trouble was unclear and rumours spread quickly to fill the void.

"It was started in a lowdown part of the city by a New Zealander getting stabbed in the eye. No Britisher will stand the knife, so a crowd of them got to it properly, slung a couple of [Egyptians] through a third-storey window, and followed up with chairs, tables, and even a piano, if you please! Then the military police came on the scene, and, being too frightened to make an arrest, one drew his revolver and shot one of

18   *The Adelaide Chronicle* (South Australia), 9th March 1933.

the soldiers dead and wounded another. They got to this policeman with a chair, and the others continued the shooting. During the melee the mob sat fire to buildings everywhere, and had furniture burning in the streets. Two regiments of Territorials were called out, with bayonets fixed. No sooner did they enter the street than chairs, tables, bedding, beds, and everything were thrown from upstairs windows upon them. A few minutes of that and the Terriers unfixed bayonets of their own accord, and the officer then ordered them to shoot on the mob. Then he told them to refix bayonets and charge. Of course, they did it! I don't think the officers know what we are. Rather than do that anyway they joined in and helped smash shop windows and so forth. It eventually ended about 11 o'clock at night. Ask yourself had those regiments fired on the crowd what would Australia think of them, and would Australians fight alongside the "Tommies" afterwards? Not while there was a Terrier in Egypt would the boys give in to them. But, as they said, we are the roughest they have ever seen, and only too true they found it out."[19]

TPR ARTHUR HOLT, 3RD AUSTRALIAN LIGHT HORSE REGT, AIF, MAADI CAMP.

"About 8.10 I had just finished cleaning my bayonet, when someone called out, "Fall in, everyone — rifles and bayonets." I armed and then rushed down and turned our men out of the canteen and messroom. We mustered about 40 men in our company, two sergeants, two corporals, and an officer.

"Then word came up for 25 men and two non-coms. per company. There being two sergeants there, that meant that the two corporals would be left out. That didn't please me — I didn't want to miss seeing a riot. So I pushed my frame in with the privates, making myself as inconspicuous as possible. Next thing was the issuing of 50 rounds of ammunition per man. We marched down to Helmiah station and entrained, arriving at Cairo at 9.20. That wasn't bad going, considering that 100 men were raised in the dark on a general-leave day. On the way in we passed several tramcars bearing some of our New Zealand men back to camp. They told us that everything was over and good-naturedly called us "scabs" and "blacklegs."

"We marched to the barracks of the Westminster Dragoons. The Dragoons wear red-roofed caps, and they supply the military police for Cairo city. They are very well hated, these "redcaps," by the Colonials. Then, having got our orders, we marched up to Shepherd's [sic] Hotel — the haunt of the officers. Here we found a strong picket of English Terriers who were guarding the place from an expected raid by the mob. Then we started in to clear the town. All the fun was over, and all we did was to run in every uniformed man we saw. The soldiers must have gone mad. Down in one quarter every window that was smashable from the street was smashed. Down the centre of one street ran a pile of half-burnt furniture and bedding. A piano had been thrown from the second storey of one house, and it was badly spilt over the pavement. Sort of "broken melody."

---

19   *Barrier Miner* (Broken Hill, New South Wales), 24th October 1915.

"When we had collected a fair number of prisoners we would take them down to the barracks and then go back and make another start. Some of the men were in an awful state with torn clothing and blood-stains. We were kept busy until about 2.30 a.m., and then we returned to the barracks and slept on some straw that was strewn on the ground in the open. I was so weary that I slept until 6.30, in spite of the cold. I can tell you, I put away a substantial breakfast.

"I have heard several explanations as to the cause of the rioting, but I don't know the true reason. I was told that when the Australians were given leave they were told that it would be their last leave. They decided to paint Cairo crimson — and, by Jove, they did! It was an Australian Day, all right, although it must not be thought that our boys were not in it —they were.

"At the start of things a bottle was thrown out of a second storey window at a mounted 'red-cap,' and it killed him. His mate drew his revolver and shot the thrower. Then they shot at the crowd. One hears all sorts of rumours as to how many were killed and wounded, but I don't think we will ever know for certain. Personally, I don't believe any of the "redcaps" were killed. Then the authorities called on every New Zealander for picket work. But what could our chaps do, unarmed? I believe there were some pretty serious moments when the mob approached the picquets. By the time we got there the sober rioters were gone and all danger was over. But the damage had been done."[20]

CPL WILLIAM MILLS, AUCKLAND BATTALION, HELMIEH CAMP, HELIOPOLIS.

"There are terrible doings in Cairo. 20,000 troops are on leave — it being Easter. They have got into the lowest part of the town & are rioting. We do not know all the truth of the story yet but 3 men shot down. I have been brought into hospital & they say the Citadel is full. The authorities have had to call out two regiments of British to fire on our own boys. About 20 have been injured I believe."[21]

SISTER ALICE ROSS-KING, 1ST AUSTRALIAN GENERAL HOSPITAL, AANS.

3 APRIL 1915

FRANCE & FLANDERS

Anglo-French intentions at the Dardanelles had long passed rumour or speculation. Everyone knew what was to come.

"Colonel Hankey, Secretary of the Committee of Imperial Defence, arrived to see me. He is over in France for 3 days. He states Lord Kitchener is more hopeful as regards the ammunition. As to the Dardanelles operations I asked why the naval bombardment had taken place before the military part of the operation was on the spot. He quite agreed with my view, and said 'the operation had been run like an

---

20 *The Feilding Star* (New Zealand), 19th May 1915.
21 Diary relating to the First World War service of Sister Alice Ross-King, AANS. AWM PR02082.

American Cinema Show' — meaning the wide advertisement which had been given every step long before anything had actually been done."[22]

GEN. SIR DOUGLAS HAIG, COMMANDING FIRST ARMY, BRITISH EXPEDITIONARY FORCE.

GREECE

At a meeting in Athens, the British Military Attaché aired his concerns about the coming attack on the Dardanelles. He was familiar with the reservations previously expressed by some senior officers.

> "I made clear the grave doubts which I had concerning a successful issue, and of the wisdom of the plan in its wider aspects. I explained the reluctance of the leaders of the Expeditionary Force to raise, of their own accord, the question of a change, and their view that any proposed alteration should come to them through diplomatic channels…
>
> "Though my views were represented, nothing was done. The only visible result was a reproach from my Russian colleague for supporting a scheme which might have been brought the Greeks to the neighbourhood of Constantinople!"[23]

SIR THOMAS MONTGOMERY-CUNNINGHAME, BRITISH MILITARY ATTACHÉ TO GREECE, ATHENS.

LEMNOS

> "We steamed into Mudros Bay at 3 p.m., and there our gaze fell on one of the most magnificent spectacles the world has ever seen — the greatest Armada of warships and transports ever assembled together in history. Here for the first time I saw the mighty *Queen Elizabeth*, our latest and greatest battleship, carrying eight of the new 15-inch guns, shepherding a long line of pre-dreadnought battleships, beginning with the *Lord Nelson*, *Agamemnon*, *Swiftsure*, and *Triumph*, and followed down the tide of times by the *London*, *Prince of Wales*, *Canopus*, *Cornwallis*, *Majestic*, *Goliath*, and many others."[24]

ELLIS ASHMEAD-BARTLETT, WAR CORRESPONDENT.

EGYPT

> "The Maoris and third reinforcements arrived last week, and together with us they marched past General Sir Ian Hamilton, and created a favourable impression. The General remarked that we were a most magnificent lot of men.
>
> "We expect to be fighting under him next week somewhere in Turkey, as we are moving to Alexandria and taking a boat to an unknown destination. No horses are

---

22   Sheffield, Gary and Bourne, John (Eds.), *Douglas Haig War Diaries and Letters*, pp. 113–114, Weidenfeld & Nicholson, (London) 2005.

23   Montgomery-Cunninghame, Sir Thomas, *Dusty Measure. A Record of Troubled Times*, pp. 212–213, John Murray (London) 1939.

24   Ashmead-Bartlett, *Uncensored Dardanelles*, p. 28.

to be taken, but I think the mounted men will act as infantry. My opinion is that we are bound for Gallipoli Peninsula or Alexandretta. We are to go with the First Munster and First Scottish, and another crack corps of Fusiliers, so we shall be with veterans.... I expect all the batteries will be going, and we are told that there is a tremendous lot of the famous French 75 guns waiting at Alexandria. Of course we are to co-operate with the French and Russian forces, and it is reckoned that there will been army of 200,000 troops."[25]

L/CPL ARTHUR HAYBITTLE, AUCKLAND BATTALION, ZEITOUN CAMP.[26]

4 APRIL 1915

LEMNOS

"Very quiet day. Most of [the] officers went to hospital ship Canada. Orders to move about Friday. Church parade. Shifted position in bay — came alongside. Steward[27] died."[28]

PTE SYDNEY PENHALIGON, 3RD FIELD AMBULANCE, AAMC, AIF, HMT *MALDA*.

EGYPT

More units received their orders to move. Their destination, if not revealed to them, was not difficult to fathom.

"We broke camp to-day; it was a marvellous transformation scene. One moment we see a city of tents; the next, they are gone, and the undulating, billowy stretch of desert once more comes up in unbroken sameness to the foot of the Pyramids — the monuments of a Dead Past, that would gaze with open-eyed amazement on our present day preparation for civilised slaughter. At 6 p.m. we began our 10-mile march to Cairo, and on arrival there boarded the train for Alexandria."[29]

PTE EDWIN FLOODY, 6TH BATTALION, AIF, MENA CAMP.

"The Colonel had a yarn to the noncoms, this morning, and told us that we were leaving on Wednesday night. Where to, he wouldn't tell us. But he told us we were going on the Lutzor [Lutzow], and that most probably we would have to land in the face of opposition! The "where" is not hard to guess."[30]

CPL WILLIAM MILLS, AUCKLAND BATTALION, HELMIEH CAMP, HELIOPOLIS.

25   *Waikato Times* (New Zealand), 19th May 1915.
26   Haybittle, a former painter, was killed in action on 13th August 1915. Commemorated on the Chunuk Bair (New Zealand) Memorial, he was the 34 year-old son of Richard Frederick and Anne Elizabeth Haybittle, of 4 Bowen Street, Feilding.
27   Steward R. Rogers, SS *Malda*, died on 4th April 1915. He is buried in East Mudros Military Cemetery. Born in Liverpool in 1880.
28   *The Queenslander* (Brisbane, Queensland), 21st April 1917.
29   *Jamieson and Woodpoint Chronicle and Upper Goulburn Advertiser* (Victoria), 18th June 1915.
30   *The Feilding Star* (New Zealand), 19th May 1915.

"Lately we have been without tunics, and with only belt and braces. On Thursday, however, we put on tunics and full marching order, and although our waterbottles were filled we were not allowed to drink any during the whole of a hot and dusty 15 miles march. It was merely to test us, and see if we were fit for active service. Friday was a whole holiday, but I did not go out, as I am sick of Cairo.

"To-day we have been served with a lot of new clothing, and from all accounts we will be leaving here some time during this week. If we go to the Dardanelles as we expect, we will not have much opportunity to write, so please don't be disappointed if you do not hear from me so often."[31]

L/CPL NORRIS TUNNICLIFFE, CANTERBURY BN, NZEF, ZEITOUN CAMP.

5 APRIL 1915

ENGLAND

It had been hoped that the operations in the Eastern Mediterranean would galvanise support, drawing Balkan nations into the allied camp. It had not happened.

"A lot of Foreign Office stuff came in yesterday and to-day, but nothing very encouraging. The Italians are still holding out for their one and a half pounds of flesh, but I do not mean to give them up, particularly now that it is clear that Rumania will be inclined to hang back until they very definitely come in. The delay in the Dardanelles is very unfortunate. Visible progress and, still more, a theatrical coup in that quarter would have goaded all the laggard States into the arena." [32]

HERBERT ASQUITH, MP, BRITISH PRIME MINISTER.

"At to the time of writing (Monday night) there have been no further developments in the Dardanelles… Pending the clearance of the shores by a land force, it is difficult to see how progress afloat can be aught but very slow and tedious, especially since every spell of bad weather gives opportunity to the enemy to make good his damages."[33]

FRED. T. JANE, NAVAL CORRESPONDENT.

TURKEY

Britain had long been an ally of the Turks, an effective partner to block Russian westward expansion beyond the Black Sea. Germany's rise had brought erstwhile rivals, the British, French and Russians, into a (not always easy) alliance. Conscious of territorial ambitions at their expense, the Turks looked elsewhere for support.

"We rejected the Triple Entente offer to guarantee Turkey's integrity for thirty years because its acceptance would have been detrimental to Turkey's sovereignty.

---

31 *Nelson Evening Mail* (New Zealand), 31st May 1915.
32 Asquith, *Memories and Reflections*, vol. 2, p. 72.
33 *Land and Water*, 10th April 1915.

"Turkey's experiences with promises made by the Powers forming the Triple Entente have not been favourable to the promotion of confidence. Had the past actions of these Powers been different, the page in history which is now being written might read differently.

"We were tired of the hypocrisy which actuated the Entente Powers when dealing with Turkey, so we did what provocation forced us to do and went to war.

"The Turkish people want a chance to work out their destiny. Our start, six years ago was good; much has been accomplished already everywhere in material and intellectual progress here…

"Turkey has been misrepresented hitherto and misunderstood, hence she has lacked the sympathy to which she has been entitled. Hitherto we were a pawn in Europe's politics, whose interests were wholly unconsidered.

"We are fighting for a chance for Turkey to exist for the sake of Turkey.

"The claim that Turkey is bound to pass under the sway of Germany is absurd.

"Have Austria-Hungary and Italy passed under the sway of their powerful ally in the alliance?"[34]

SAID HALIM PASHA, OTTOMAN GRAND VIZIER.

"There has been large movements of troops in the vicinity of Constantinople and thru the streets, at different times, during the last two weeks, and it is believed they are being sent to the Dardanelles, but this is not certain. Several hundred troops have been seen marching thru Tophane into Stamboul at about 8.00 p.m., on two occasions during the past week."[35]

LT HERBERT S. BABBITT, USN, USS *SCORPION*, ISTANBUL.

LEMNOS

"We left Egypt on March 1, and now are waiting to have a go at the Turks at the Dardanelles, which will have taken place before you receive this, so we will no longer be tourists, as the "Bulletin" truthfully calls us; but the powers that be know best what to do with us, and will use the "boys" when they are ready. We expect a very warm time of it when we do start."[36]

CAPT. LESLIE MULLEN, 12TH AUSTRALIAN BATTALION, AIF, HMT *DEVANHA*.

EGYPT

As their stay in Egypt was drawing to a close, it was believed that bored men could become good soldiers; all that pent-up frustration would find its outlet.

---

34   *Liverpool Echo*, 5th April 1915.
35   Naval Attache's Reports, Office of Naval Intelligence, Unpublished Manuscript, US Naval War College, April 1915.
36   *The North Western Advocate and the Emu Bay Times* (Tasmania), 26th May 1915.

"There was rather a bad riot the other night in that street just beyond Shepheard's, where all the whores live. I can't quite discover the reason of the outbreak; it was confined entirely to the soldiers, the civil population wisely kept out of it. It appears that some soldiers wanted to pay off old scores with a lady who had sent them to hospital, and that in so doing one of them got hit or stabbed, and covered with blood which excited the rest, so they set to work to wreck the establishment, pulling out all the furniture, etc., piling it up in the street, and setting fire to it, and also the house. The Military Police and picket tried to stop them, but by this time a huge crowd of soldiers had collected, not a few being drunk, and set upon the police and pickets… The former drew their pistols and fired, mostly in the air; this further excited the crowd, and one or two Australians drew Brownings and retaliated. For a time things looked very ugly, as the soldiers attacked and looted other shops and brothels, but a strong picket and a squadron of Yeomanry appeared on the scene, and very soon normal conditions were restored. Four men were wounded, but none dangerously, with pistols, and some six or eight were knocked about; I wonder we have not had more of these rows. I cleared out 500 venereals to Malta; Methuen did not appreciate the compliment, but I was getting badly blocked, and these cases were not doing well under canvas.

"The Australians and New Zealanders are just about fed up with Egyptian sand and training; the real thing will do them all the good in the world; they are, I believe, fine fighters and shoot straight, the only question being whether their officers are good enough to keep them in hand."[37]

GENERAL SIR JOHN MAXWELL, GOC BRITISH TROOPS, EGYPT.

"When different units get together in a camp the amount of thieving, technically called skirmishing, is beyond belief to anyone unaccustomed to camp life. At present we have two mules that do not belong to us. One wandered into our camp and a man who claimed it as belonging to his unit was told he had to prove his statement before he would be allowed to remove it, which he failed to do… I am a fairly good skirmisher myself, and when a wagon pole, for which I was responsible when unloading at the docks, did not turn up, I had two in its place in no time. We afterwards found that neither of them would fit any of our wagons. The cook has been handicapped in his work by having no table, but to-day he has one about 12 feet long which he tells me he got "over the road" last night when it was dark."[38]

LT GEORGE DAVIDSON, 89TH (1ST HIGHLAND) FIELD AMBULANCE,
RAMC, MEX CAMP, ALEXANDRIA.

"Reached Alexandria at 3.30 a.m… We were informed, on arrival, that we would board the transports at 10 a.m. During the interval… I had a look round the town. It is a beastly place — nothing to see and nowhere to go… They have one advantage here — a splendid port and a busy one. There are a lot of German ships interned here;

---

[37] Sir John Maxwell quoted in Arthur, Sir George, *General Sir John Maxwell*, pp. 176–177, John Murray (London) 1932.

[38] Davidson, George, *The Incomparable 29th and the River Clyde*, pp. 15–16, James Gordon Bisset (Aberdeen) 1920.

some of them are beauties, too. I suppose they will start using them soon. Eventually 10 o'clock came round, and we boarded the s.s. *Galeka* (a Union Castle liner)."[39]

PTE EDWIN FLOODY, 6TH BATTALION, AIF, HMT *GALEKA*.

"This will be a very hurriedly written letter, for we have just had word that we are to move on very soon, and everything is excitement and absolute chaos.

"We have no knowledge of where we are to go, but popular opinion seems to be that we will land somewhere near the Dardanelles. To-day is Monday, and we expect to leave on Tuesday night, and there are fully 30 reinforcements to be equipped before then."[40]

SGT NORMAN MIGHELL, 15TH BATTALION, AIF, AERODROME CAMP, HELIOPOLIS.

"We have received marching orders, and will probably leave here on the seventh. You would laugh if you could see me now. They did not issue any candles to-day, but l have found one in my kit, and am lying with just my head and shoulders in the tent and the writing pad on my kit… I cannot give you any details of life here, because I have lots to write to to-night, as I do not know when I will get another chance. I can only add that I am living as straight as a die, and am ready for whatever fortune cares to send me."[41]

PTE JOHN ZIEGLER, CANTERBURY BN, NZEF, ZEITOUN CAMP.[42]

6 APRIL 1915

EGYPT

"Today Sir Ian Hamilton, accompanied by General d'Amade, inspected the 29th Division… It went off very well, but my Division being so scattered & some not yet off the ships, I could not have a spectacular review like the French had. The men are certainly very fine. And the officers all up to the highest peace standard, & up to full strength. The criticism I make is that the Lieutenant Colonels in some instances are too old. And that the Division is not a well oiled machine, as were the old Regular Divisions. If I had had it in camp as a whole for three weeks, it would be a much better fighting machine, but as it is, it is uncommonly good & will, I am sure, give a very good account of itself."[43]

MAJOR-GENERAL SIR AYLMER HUNTER-WESTON, COMMANDING 29TH DIVISION.

---

39 *Jamieson and Woodpoint Chronicle and Upper Goulburn Advertiser* (Victoria), 18th June 1915.
40 *Cairns Post* (Queensland), 24th May 1915.
41 *Lyttelton Times* (New Zealand), 24th June 1915.
42 Ziegler died of wounds aboard the hospital ship *Sicilia* on 18th June 1915. Commemorated on the Lone Pine Memorial, the former accountant was the 27 year-old son of Frederica Ziegler, of Shakespeare Street, Greymouth, New Zealand.
43 Extract from letter, 6th April 1915, Maj. Gen. Hunter-Weston, British Library, MS 48364.

> "We were penned up aboard on board while the boat was tied to the pier and goods were being loaded. Well… we had not much do do, so we passed our time away at the piano and at cards. Whilst we were not at these pastimes we were either buying fruit, nuts, chocolates, etc., over the side of the boat, or watching the French, English, or Indian troops dodging about from place to place. The French uniform, as far as we could see, is very peculiar and gaudy, the trousers being bright red, a long swallow-tail coat of royal blue, trimmed with brass buttons, and a cap peaked somewhat like a porter's, and a larger and squarer peak. The rifle they use is like our old type, the old Le [sic] Enfield, and they use a bayonet about two feet long."[44]
>
> PTE JULIUS JACOBSOHN, MG SECTION, 7TH BATTALION, AIF, HMAT *GALEKA*.

> "I had a pretty bad time last week. First of all, I got the influenza, and just as I was getting over that I got the sand colich, and, my word, that turned me up properly. Well, dear uncle, I don't think I will be in Egypt any more when you receive this letter, because we are moving off about 11/4/15 for the Dardanelles, and as we have to land under the protecting fire of our own naval guns I am expecting a particularly lively time for a while…
>
> "Good-bye, good-bye, and God bless you, uncle, for what you have taught me; it is doing me good now, for I have great trust in the Lord — trusting, always trusting in Him to bring me safely back."[45]
>
> L/CPL ARTHUR WALKER, 8TH BATTALION, AIF, MENA CAMP.[46]

7 APRIL 1915

TURKEY

The Sultan spoke to American journalists, telling them of his confidence in the face of the coming assault. German officers, too, expressed confidence in their ally.

> "I am convinced the Dardanelles cannot be forced. The brave conduct of the Turkish troops in the recent operations permits me to conclude that, although the allies use every means at their disposal, they will be unable to achieve their purpose…
>
> "It has been said that it was luck that made our victory on March 18 so complete and great, but we in the Turkish have a saying, 'luck is infatuated with the efficient.'
>
> "It appears unjust to me that the allies want to force the Dardanelles and take Constantinople just to import foodstuffs from Russia. But our army and coast defense force have shown their ability and willingness to do their duty. I am speaking here not alone of the Turkish defenders of the Dardanelles, but also those Germans who so efficiently and bravely co-operated with them.

---

44   *The Corowa Free Press* (New South Wales), 17th December 1915.
45   *The Horsham Times* (Victoria), 25th June 1915.
46   Walker, a former farm labourer, was killed in action on 29th April 1915. Buried in Shell Green Cemetery, he was the son of William Walter Walker, of Dimboola Post Office, Dimboola, Victoria.

"I would thank you if you would say for me that my admiration for the German troops in the east and the west is so great that it is impossible for me to express in words my high opinion of their valor and efficiency."[47]

SULTAN MEHMED V, ISTANBUL.

"Turkey to-day is better prepared than ever. She has one and a quarter millions of well-trained men, besides hundred thousands ready for any emergency. As the Entente is politically interested in exaggerating news, which is only partly favourable, the destruction of the outer forts of Sed-ul-Bahr and Kum Kaleh was made to appear a great victory.

"How little Constantinople was alarmed by the new attacks can be seen from the fact that the Sultan remained in his palace a short distance from the sea. The attack had not the least effect on the population."[48]

FIELD MARSHAL COLMAR FREIHERR VON DER GOLTZ, ISTANBUL.

ENGLAND

The confidence expressed in Istanbul was not shared by some in London or France.

"It is as well to recall that when [attacking the Dardanelles] was first mentioned to the Cabinet McK. [Reginald McKenna] asked on whose authority it was undertaken. W.S.C. replied 'after consultation with the 'C.I.D.'[49] and on McK. [McKenna] observing that the Cabinet had no responsibility for it, Churchill declared that he was willing to take the whole responsibility himself. It is evident that the Navy has hitherto failed completely so far in its task, and though it has been reinforced by 4 battleships and a squadron of destroyers, and aeroplanes and very doubtful venture… W.S.C. reported that the Grimsby trawlers, who are mine sweeping, did not object to the risk of being blown up by mines, but declined entirely to face shellfire, and have had to be withdrawn."[50]

CHARLES HOBHOUSE, MP, POSTMASTER GENERAL.

FRANCE & FLANDERS

"We discussed the Dardanelles. He [Sir John French] showed me a letter he had just received from Johnnie Hamilton, dated from Lemnos. Johnnie had returned from a cruise in the Bay of Saros, he had gone so near the coast that he had seen the rows and rows of trenches. He had gone to the mouth of the Dardanelles, where he was at once shelled. This all sounds like going through! Sir John agrees now that to employ troops for this enterprise is madness. I think I must see if I can't do something to stop

---

47   *Richmond-Dispatch Times* (Richmond, Virginia), 8th April 1915.
48   *Nottingham Daily Express*, 8th April 1915.
49   Lt-Col. Maurice Hankey, Secretary to the Committee of Imperial Defence.
50   David, Edward (Ed.), *Inside Asquith's Cabinet. From the Diaries of Charles Hobhouse*, p. 234, John Murray (London) 1977.

it. As I always say, I am not clear that, even if successful, it would be a good thing to do. Winston [Churchill] ought to be, and must be, *dégommé* over this."⁵¹

LT-GEN. SIR HENRY WILSON, PRINCIPAL LIAISON OFFICER, BEF.

EGYPT

Hunter-Weston had drawn up an appreciation of the feasibility of the attack upon the Dardanelles after consulting quite widely, sharing his concerns with his peers. But now, armed with the knowledge that it was going ahead and the belief that nothing could not be achieved, he felt it worth the attempt.

> "Our enterprise if successful may bring the end of the War more nearly in sight. The troops are grand & will do the trick if it be now possible; but after so much warning, the odds against us are very heavy. However, nothing is impossible & I'll do all that in me lies to pull it off."⁵²

MAJOR-GENERAL SIR AYLMER HUNTER-WESTON, COMMANDING 29TH DIVISION.

> "We… are expected to sail on the 10th, in our old ship the "Marquette"… north, presumably for Gallipoli, but some say Smyrna, to join in what will be a most bloody affair — so we have been warned by Lord Kitchener who, in an address to our Infantry Battalions, has said that the work before us will be hard in the extreme, and that he had reserved our Infantry as the finest Battalions in the Army for this arduous job, and told them that they must be prepared to face great hardships and great sacrifices."⁵³

LT GEORGE DAVIDSON, 89TH (1ST HIGHLAND) FIELD AMBULANCE,
RAMC, MEX CAMP, ALEXANDRIA.

> "We had to pack our gear yesterday ready to go. Our Colonel told us this morning that the name of the boat we are to go on is "The Ascot," also that they anticipate we will have to land under fire, and that we will be fighting right away. He told us, too, that we had to take two days ration with us, and our bottle of water would have to last two days, that we would have to be extremely careful with it… So you see we are going to get to it straight away, and by jove the chaps are anxious to be at it."⁵⁴

PTE THOMAS BERRY, 13TH BATTALION, AIF, AERODROME CAMP, HELIOPOLIS.

> "Just a few lines to let you know that we are leaving Egypt to-morrow. Before you get this postcard we will be fighting… the Turks, so you can take it easy that I will do my best and play the game. You will get no more letters for a good while, only censored post cards."⁵⁵

PTE ROBERT CURRIE, CANTERBURY BN, NZEF, ZEITOUN CAMP.⁵⁶

---

51   Sir Henry Wilson, quoted in Callwell, Maj-Gen. Sir C.E., *Field-Marshal Sir Henry Wilson Bart., G.C., D.S.O. His Life and Diaries*, vol. 1, pp. 221–222, Cassell & Co. (London) 1927.
52   Extract from letter, 7th April 1915, Maj. Gen. Hunter-Weston, British Library, MS 48364.
53   Davidson, *The Incomparable 29th and the River Clyde*, pp. 24–25.
54   *The Young Chronicle* (New South Wales), 18th May 1915.
55   *West Coast Times* (New Zealand), 1st June 1915.
56   The battalion left for Alexandria on 9/10th April, sailing for Lemnos on 11th April aboard HMT *Lutzow*.

8 APRIL 1915

ENGLAND

"[W]e have openly announced that we are going to take the Dardanelles with our Army, and as our preparations at Mudros & elsewhere can have left little doubt that we are going to do so, it is there that the Turks have now made their main defence."[57]

CAPT. HERBERT RICHMOND, RN, ASSISTANT DIRECTOR
OF OPERATIONS, ADMIRALTY WAR STAFF.

MEDITERRANEAN

"Since the Covering Force is likely to be required to get into tows from transports by night, men must be practised in doing so. Eight units will be used — four on each side of each ship, — of which seven in some cases may be Jacob's ladders. No lights of any kind will be allowed. Men must be gradually accustomed to getting into boats fully equipped, carrying 200 rounds of ammunition, 3 iron rations, and pack. Rifles should be slung while getting into boats, and unslung as soon as the men are in the boats."[58]

MAJOR-GENERAL WALTER BRAITHWAITE, CHIEF OF
GENERAL STAFF, MEF, HMT *ARCADIAN*.

"The forces for the present expedition against Turkey have concentrated in Alexandria, and are at present over 100,000 strong, mostly British but also largely French. To-day the pioneers of this huge force have set sail, and as far as I can gather our boat was the second to go out. We are doing 14 knots and in two or three days should reach our journey's end. The day is beautiful and the Mediterranean its deepest blue.

"I have been having a talk with the captain of the "Ausonia". He has only 64 tons of water on board, while he should have had ten times that amount. There are no pipes laid to the docks and the whole of the shipping has to depend on six water lighters which carry 60 tons each. At present these are totally unable to supply the huge number of transports in Alexandria."[59]

LT GEORGE DAVIDSON, 89TH (1ST HIGHLAND) FIELD AMBULANCE,
RAMC, HMT *AUSONIA*.

EGYPT

"This appreciation of the situation will interest you greatly to read. It gives the views I expressed to Sir Ian Hamilton & his Chief of the General Staff (Gen. Braithwaite). They must of course be kept absolutely secret, though as they do not express the views of the Commander of the force or of the Government, they are not of any value to the enemy. Having expressed my views & so done my duty as Counsellor, I now

---

57 Marder, *Portrait of an Admiral: The Life and Papers of Sir Herbert Richmond*, p. 149.
58 MEF General Headquarters, General Staff, TNA WO 95/4263.
59 Davidson, *The Incomparable 29th and the River Clyde*, pp. 23–24.

wholeheartedly accept whatever place may be decided on by the Commander and shall do my "damnest" to bring this (the Government's) plan to a successful issue; thus doing my duty as a man of action & a subordinate leader."[60]

MAJOR-GENERAL SIR AYLMER HUNTER-WESTON, COMMANDING 29TH DIVISION.

"Nearly all the transports have to go to Alexandria to enable the units to be re-sorted and reorganised before a disembarkation can be attempted on the Gallipoli coast. Stores, men, guns, horses, and mules have been shipped piecemeal from England, and, as there are no quays or cranes at Mudros, Alexandria is the nearest port available."[61]

ELLIS ASHMEAD-BARTLETT, WAR CORRESPONDENT.

"Goodness knows where we are going, but it seems practically certain that Turkey is the place. I think there will be some very hard fighting there, as the Turks are very good on the defensive, and especially in their own country. We will probably land under the cover of the warships at the Dardanelles. I expect you will have read all about it before this letter reaches New Zealand. We are all ready to go and very keen to have another real slap up against the Turks. Our troops are really a fine body of men now, and will do justice to our little Dominion."[62]

SGT (FRANCIS) CLIVE UPTON, CANTERBURY BN, NZEF, ZEITOUN CAMP.[63]

"We were just preparing to return home [Mena].... when one of the officers came galloping down and told us that all gear had to be returned to stores immediately. At first we treated the matter as a joke, but from the excited look on the officer's face we saw that it was all right... There was very little sleep for anybody that night as the whole camp was excited. We had to pack our web kits with absolute necessaries and discard all the gear we could, as we were only allowed what we could carry on our backs. This load, in most cases, consisted of a great coat, two pairs of socks, one set of underclothing, one pair of boots, one flannel shirt, one khaki shirt, a hold-all (containing knife, fork, spoon, razor, a brush and comb), a housewife and a couple of shirts. All this had to be carried, and it made a very fair load. One blanket and waterproof per man were also carried... This may be the last letter I will ever write to you, but I feel pretty confident I will come out all right."[64]

PTE HAROLD WILLIAMSON, 1ST BATTALION, AIF, HMAT *MINNEWASKA*.[65]

60 Extract from letter, 8th April 1915, Maj. Gen. Hunter-Weston, British Library, MS 48364.
61 Ashmead-Bartlett, *Uncensored Dardanelles*, p. 30.
62 *Ashburton Guardian* (New Zealand), 20th May 1915.
63 Upton, a former grain buyer, was wounded at Gallipoli, being taken aboard the *Dongola* on 3rd May 1915, and evacuated to Egypt. Commissioned, 2/Lt Francis Clive Ramsden Upton, 2nd Bn Canterbury Regt, NZEF, was killed in action on 2nd October 1916. Commemorated on the Caterpillar Valley (New Zealand) Memorial, he was the 26 year-old son of Thomas Evererd Upton, of Ashburton, Canterbury.
64 *The Yass Courier* (New South Wales), 20th May 1915.
65 Williamson, a former carpenter, was killed in action on 19th May 1915. Buried in Shrapnel Valley Cemetery, he was the 25 year-old son of Thomas and Esther Williamson, of Motor Garage, Yass, New South Wales.

"We leave for Dardanelles to-morrow — the place where the lead pills fly in the air; the fun begins. Our Light Horse won't be with us, as they are not much use on the Continent in this war, so the men are trying to join the infantry now."[66]

PTE (THOMAS) WADDELL (WAD) CAMERON, 14TH BATTALION,
AIF, AERODROME CAMP, HELIOPOLIS.[67]

9 APRIL 1915

LEMNOS

"The Dardanelles is not a fit and proper place for peaceful policemen. The Germans and Turks refuse to move on when requested, and often become very violent. The worst part of it is they often get down below the hills somewhere and pelt away at you with their howitzers, and you hear the music approaching. 'Whiz-z-z-z,' then a splash, sometimes a hit. You have the greatest difficulty to find them out — when we do they get it with plenty of energy behind… We do not mind the forts. We can hammer them one by one as we go along, but we have on either side of us thousands of troops pelting away with field guns and howitzers, and it is not at all kind of them. When a shell comes singing past one's head I, for one, automatically go to 'trembling stations.' I think all do the same."[68]

YEOMAN OF SIGNALS CHARLES GOULD, RFR, HMS *CANOPUS*.[69]

AEGEAN

After expressing concerns about the doubts raised by his commanders, Hamilton's confidence reasserted itself, assured that old problems would be dealt with by novel methods.

> "Never till to-day has solicitude become painful. This is the fault of Birdwood, Hunter-Weston and Paris. I read their "appreciations of the situation" some days ago, but until to-day I have not had the unbroken hour needed to digest them…
>
> "Hunter-Weston's appreciation, written on his way out at Malta, is a masterly piece of work. He understands clearly that our true objective is to let our warships through the Narrows to attack Constantinople. "The immediate object," he says, of operations in the Dardanelles is to enable our warships, with the necessary colliers and other unarmoured supply ships — without which capital ships cannot maintain themselves — to pass through the Straits in order to attack Constantinople"…

---

66   *Kyabram Guardian* (Victoria), 11th May 1915.
67   Cameron, a former grocer, was killed in action on 8th August 1915. Commemorated on the Lone Pine Memorial, he was the brother of Flora Brown, of 49 Moore Street, Fitzroy, Victoria.
68   *Hampshire Telegraph*, 9th April 1915.
69   Gould was one of five of *Good Hope*'s crew left on Isla Auchilu to establish a wireless station four days before the ship was lost with all hands at Coronel.

"Broadly, he thinks that we are so short of ammunition and particularly of high explosive shell that there is every prospect of our getting tied up on an extended line across the Peninsula in front of the Kilid Bahr trenches. Should the enemy submarines arrive we should be "up a tree"...

"Paris's appreciation gives no very clear lead. "The enemy is of strength unknown," he says, "but within striking distance there must be 250,000." He also lays stress on the point that the enemy are expecting us — "Surprise is now impossible — The difficulties are now increased a hundredfold. To land would be difficult enough if surprise was possible but hazardous in the extreme under present conditions." He discusses Gaba Tepe as a landing place; also Smyrna, and Bulair...

"The truth is, every one of these fellows agrees in his heart with old Von der Goltz, the Berlin experts, and the Sultan of Egypt that the landing is impossible. Well, we shall see, D.V. [God Willing], we shall see!! One thing is certain: we must work up our preparations to the nth degree of perfection: the impossible can only be overborne by the unprecedented; i.e., by an original method or idea."[70]

GENERAL SIR IAN HAMILTON, C-IN-C, MEF, HMT *ARCADIAN*.

More prosaically, others looked for different solutions to the more immediate problems they faced.

"After lunch I had a conversation with my new friend, the captain of the "Ausonia"... The most interesting piece of news I got out of him was that our destination was Lemnos, but that he expected that it was merely as a rendezvous for the whole force, and was only 48 miles from Sedd-el-Bahr, on the south point of Gallipoli. His view is that we will land a short way north of that... He is in distress over his shortage of water as none is to be had in the small islands. This shortage of water got me into trouble with the O.C. the troops on board at general parade this morning. Many of the men had not shaved for two days, and some looked untidy and unwashed, but all put this down to their being denied water to slake their thirst, which must come before washing and shaving, but the order was "see that it does not happen again." I advised one particularly hirsute chap to lower his shaving brush into the sea to-morrow at the end of a string."[71]

LT GEORGE DAVIDSON, 89TH (1ST HIGHLAND) FIELD AMBULANCE, RAMC, HMT *AUSONIA*.

EGYPT

"We are very keen to get to work. It is a ruthless, blood-thirsty business, but one has only to remember the Belgian atrocities and the massacre of Christians by the Turks during the past few years to feel that one's cause is just and righteous. With this feeling one can face the greatest personal danger undaunted. We are sons of an Empire of which the foundation stone is liberty, and which is dedicated to the proposition that

---

70 Hamilton, *Gallipoli Diary*, vol. 1, pp. 86–94.
71 Davidson, *The Incomparable 29th and the River Clyde*, p. 26.

no man can stand on its shores and be a slave. We are prepared to give our lives, if necessary, that that nation may long endure, and, as you say, a simple, clear faith is a great help in facing the ultimate crisis of life. So we shall all of us do our little bit at the Dardanelles. I always considered that this war would last for years, and would call forth all England's manhood before we could win, and on looking back and weighing the possibilities I am sure I took the right course."[72]

CPL FREDERICK HALL-JONES, AUCKLAND BATTALION, ZEITOUN CAMP.

"I issued 24 hours' iron rations to every officer and man of the company, made final preparations for departure, and entrained and left Helmieh at 10 p.m. There was a noticeable difference in the behaviour of the men compared with their demeanour on our last entraining, when we went to Ismailia. On that occasion, it was Bedlam let loose. This time, everyone was quiet, and did his work without noise or grumbling."[73]

SGT CLIVE SWEARS, AUCKLAND BATTALION, ZEITOUN CAMP.

"We landed in [Alexandria] five days ago, and are at present encamped at a place called ——— almost on the verge of the desert. The camps are on land, hard baked with the sun, but taking everything into consideration we are not badly off… There is a rumour that we won't be long here. This certainly isn't our destination."[74]

PTE GEORGE SIDEY, 1/5TH BN ROYAL SCOTS (LOTHIAN REGIMENT), MUSTAFA PASHA BARRACKS.

LEMNOS

"Arrived at Lemnos after a calm, uneventful trip. This island has an area of about 12 square miles. In entrance and harbor it is similar to Albany. Hilly and green everywhere. Villages here and there built of stone, and numerous windmills. Found that the 3rd Brigade (which included the W.A. Battalion) had been here for about one month. There are 38 transports here already, and more coming in every day. Fourteen battleships, five submarines… one hydroplane… and four desroyers [sic] are here also. Dardanelles about 45 miles away; divisional signallers have to keep communications, with ships and carry despatches, so I am having a lot of journeys to different ships.

"Men practice daily on the disembarking, with full kit on. Information of disposal of Turkish troops elaborate and complete. Have been supplied with maps of Gallipoli Peninsula, complete as regards details, soundings, and positions of enemy forts. Enemy's force estimated at 110,000 men."[75]

CPL ROBERT HUNTER, 1ST DIV. SIGNAL COY, AUSTRALIAN ENGINEERS.

---

72 *Colonist* (New Zealand), 30th June 1915.
73 *New Zealand Herald*, 24th June 1915.
74 *Berwickshire News*, 18th May 1915.
75 *The Evening Star* (Boulder, Western Australia), 26th June 1915.

10 APRIL 1915

MEDITERRANEAN

> "The unexpected happened last night, when we pulled out of port as soon as the guns were put on board. Some of the chaps risked going ashore, and nine were left behind. This offence is very serious… We considered ourselves a bit crowded on board the *Medic*, but it was nothing to this boat; we are simply packed like sardines, but everyone is perfectly happy because we know we shall have some excitement."[76]
>
> DVR ALFRED PATERSON, 8TH BATTERY, AFA, HMT *ATLANTIAN*.

LEMNOS

Hamilton led a discussion of how the landings were to be conducted. Supported by his Chief of Staff, Braithwaite, the Navy was represented by de Robeck, Wemyss and Keyes.

> "I would like to land my whole force in one — like a hammer stroke — with the fullest violence of its mass effect — as close as I can to my objective, the Kilid Bahr plateau. But, apart from lack of small craft, the thing cannot be done; the beach space is so cramped that the men and their stores could not be put ashore. I have to separate my forces and the effect of momentum, which cannot be produced by cohesion, must be reproduced by the simultaneous nature of the movement."[77]
>
> GENERAL SIR IAN HAMILTON, C-IN-C, MEF, HMS *QUEEN ELIZABETH*.

Hamilton outlined his novel methods in a memorandum to Kitchener. He stressed his determination to press on as a means of overcoming problems. God was expected to play a part. But would divine intervention be required to prevent them being defeated in detail?

> "I arrived here this morning and I have just cabled you a brief account of my discussion with the Vice-Admiral, and of his full agreement with our scheme.
>
> "The more I ponder over the map and consider the character, numbers, and position of the enemy, the more I am convinced that the very essence of success must lie in upsetting the equilibrium of the Turk by the most rapid deployment of force possible over a fairly wide extent of country, combined with feints where troops and launches cannot be spared for an actual serious landing.
>
> "My main reliance will be on the Twenty-ninth Division, the covering force of which will be landed at dawn at Seddel-Bahr, Cape Helles, and, D.V., in Morto Bay. I put in a special "D.V." to the Morto Bay project because the transports will there be under fire from the other side, and whether they can stick it or not is rather a question. Still, they must try. Also, no doubt, they will be under long-range fire from field-guns and

---

76   *Northern Times* (Carnarvon, Western Australia), 17th July 1915.
77   Hamilton, *Gallipoli Diary*, vol. 1, p. 96.

perhaps howitzers from behind Achi Baba. To help these fellows along, subsidiary landings in boats will be made along the coast in small groups from Tekke Barnu up to opposite Krithia. Even a few men able to scramble up these cliffs should shake the first line of defence which stretches from Old Castle northwards to the coast. The Australians meanwhile will make a strong feint which will, I hope, develop into a serious landing operation north of Gaba Tepe. Braithwaite has marked out a good circular holding position, stretching from about Fisherman's Hut round to Gaba Tepe, and if they can maintain themselves there, I should hope later on they may be able to make a push forward for Koja Dere. Whatever this does, it will tend to raise anxieties in the minds of the men opposed to the Twenty-ninth Division, and will prevent the plateau being reinforced. I fear we must expect casualties from guns in concealed positions, both on the sea and whilst this is being done. But that is part of the hardness of the nut.

"Meanwhile the Naval Division will move up and make a simultaneous feint somewhere opposite Bulair, which will keep the Gallipoli people on tenterhooks at least for a time.

"These are my plans in broad outline. I do not want to talk about the difficulties, for I try to keep my mind fixed on my own objective, feeling sure that if I can stick to that and carry it through with vigour, the enemy will not be able to do all the wonderful things which theorists might expect."[78]

GENERAL SIR IAN HAMILTON, COMMANDER-IN-CHIEF, MEF.

EGYPT

"Well, we are leaving at midnight, and all our gear is packed and we are having a rest before we move as I do not suppose there will be much chance of sleep later. We are now sure of going to the Dardanelles, so I suppose things will be more exciting than anything we have yet had. In future I will be unable to write you letters on account of the censor, but I shall post you by the field service cards as often as they are served out. I believe there is an enormous fleet of troopships in the harbour so it must be a big force that is going."[79]

PTE FREDERICK LOCKER, WELLINGTON BN, NZEF, ZEITOUN CAMP.[80]

"It was a most impressive sight to see regiment after regiment, followed by medical corps, army service corps, artillery, engineers and signallers, marching gaily to the station. They were all happy. And what a reception they received from the mounteds and the Army Service Corps men attached to the latter! Cheer after cheer went up for

---

78   Sir Ian Hamilton to Lord Kitchener, quoted in Arthur, Sir George, *Life of Lord Kitchener*, vol. 3, pp. 130–131, MacMillan & Co., (London) 1920.
79   *Patea Mail* (New Zealand), 24th May 1915.
80   Locker was wounded on 29th April 1915. After treatment in England, he returned to New Zealand aboard the *Ruahine* on 17th November 1915 and was discharged.

the boys as they filed along. It is hard to say when or where we will meet again, for I am certain that none of us realise what we have got before us, or know when we will see our homes again.

"The mounteds do not know when they will leave, and, as the company of the army service corps I am in is attached to them, I have a few more weary days to spend in Egypt. I would have loved to get away with the first lot. I am certain that, as soon as we land and get settled down, there will be plenty left for us to do, for this is going to be a big struggle before it is finished."[81]

L/CPL JOSEPH COOKSLEY, 3RD COY, NZASC, ZEITOUN CAMP.[82]

"I spent [the] evening with some "cobbers" in the 15th Battalion (Jack Merrell,[83] Billy White,[84] "Taffy" Nicholls,[85] and "Bunny" Owens[86]), and saw them off when they marched out at 11 p.m. to entrain. There was not a single man there who would have swapped his pack for a pair of [Light Horseman's] leggings. The boys went off in high spirits, joking about their chances of getting through, or otherwise."[87]

CPL CHARLES SCOTT, 2ND AUSTRALIAN LIGHT HORSE REGT, HELIOPOLIS.

11 APRIL 1915

AEGEAN

"Had a cold, windy night, but nice to-day. Passed several islands of the Aegean Archipelago to-day. Passed another boat, with a pontoon in tow. 3 p.m., passed island of Patmos, could see the white buildings plainly. 3.30 p.m., lecture by General Birdwood. Our job, briefly, is to clear the Gallipoli Isthmus of small land forts, etc., so that the fleet can enter the Narrows. He said it was one of the hardest undertakings of the war, but he had complete confidence in the Australians."[88]

CPL HOWARD BOTH, 1ST DIV. SIGNAL COY, AUSTRALIAN ENGINEERS, HMAT *MINNEWASKA*.

---

81 *Lyttelton Times* (New Zealand), 24th May 1915.
82 Promoted Sgt, Cooksley, a former teamster, was killed by shellfire on the Western Front on 18th June 1917. Buried in Trois Arbres Cemetery, Steenwerck, France, he was the 26 year-old son of the late Henry Driver Cooksley and Elizabeth Cooksley, originally of Kaiapoi, Christchurch.
83 Pte Frederick Merrell, 15th Bn, AIF, a former farm labourer, was awarded the Military Medal for his gallantry at Zonnebeke on 26th September 1917. He returned to Australia as a Lieutenant on 1st May 1919.
84 Probably Pte William Whyte, 15th Bn, AIF, a former traveller from Denmark Hill, Ipswich, Queensland. His service at Gallipoli ended after being evacuated with contusions to his back in June 1915. He left England to return to Australia on 20th November 1918.
85 Pte Roy Nicholls, 15th Bn, AIF, was killed in action 9th/10th May 1915. Commemorated on the Lone Pine Memorial, the former draper was the son of Bessie Nicholls, of Auchenflower, Brisbane, Queensland.
86 Pte Robert Owens, 15th Bn, AIF, was evacuated from Gallipoli on 30th August 1915 suffering from bronchitis. He returned to Australia after being treated in Malta and England on 19th January 1916 due to bronchitis and neurasthenia.
87 *Queensland Times* (Ipswich, Queensland), 5th June 1915.
88 *The Areas' Express* (Booyoolee, South Australia), 20th August 1915.

LEMNOS

"After lunch I met Lt.-Col. Rooth of the Dublins, who gave me some authentic information concerning the proposed military landing on Gallipoli. The covering party for the whole expedition is to be our 86th Brigade. The Munsters are in the S.S.T. "Caledonia," (B ii) lying alongside our ship. The Lancashires are there also. All these, along with our stretcher bearers, land together from cutters, and the date fixed is in all probability Wednesday, April 14, or the following day at latest. A very warm reception from the enemy on shore is expected, as I gather from the way the Dublin officers talk. It is also said that we will have to make a dash for it under the cover of night."[89]

LT GEORGE DAVIDSON, 89TH (1ST HIGHLAND) FIELD AMBULANCE, RAMC, HMT *AUSONIA*.

EGYPT

"Hamilton's force is embarking as hard as it can, the Australians, 29th Division, French and Naval Division, at Port Said are practically off; New Zealanders are beginning to-day. Alexandria harbour was so full of transport, it was almost impossible for ships to move.

"I have let Hamilton have Cox's brigade, and a brigade of Mountain Artillery, also some officers and much material he asked for.

"D'Amade — who has much aged — has asked for, and got, 200 horses, of which 150 are Artillery. I am helping the French all I can with their hospital arrangements, which are very difficult. Sixty of the 'Légion Étrangère' (mostly Italians) deserted; I hope we will be able to catch them."[90]

GENERAL SIR JOHN MAXWELL, GOC BRITISH TROOPS, EGYPT.

12 APRIL 1915

ENGLAND

Some very well connected observers were far from happy at the coming assault upon Gallipoli. And they had no doubt who was responsible for it.

"The Dardanelles is the greatest horror of the lot. Antwerp was bad enough in all conscience, but the Dardanelles is worse... I believe the man who originated the idea of the Dardanelles scheme was Hall, who is now DNI [Director of Naval Intelligence]. Perhaps you may recall him if I tell you that he goes by the name of 'Blinker' Hall.[91] For some time he has been urging the forcing of the Dardanelles as a Naval operation quite feasible provided always that it was undertaken simultaneously with the landing

---

89   Davidson, *The Incomparable 29th and the River Clyde*, p. 28.
90   Sir John Maxwell quoted in Arthur, Sir George, *General Sir John Maxwell*, pp. 178, John Murray (London) 1932.
91   Admiral Sir William Reginald 'Blinker' Hall was the Director of Naval Intelligence between 1914 and 1919.

of a large military force. At last he persuaded the Admiralty to study the question seriously, and they did, and came to the conclusion that it was a feasible thing and not only feasible, but not very difficult, provided all the ships wore crinolines[92] and, especially, that a military force should be landed at the same time. Winston gets to hear of this and having no military knowledge whatever, but being at the same time a man of dramatic instincts who looks upon every phase of this war as an opportunity for him to appear before the footlights, he rushes the thing through. Crinolines would take six weeks to make; then no crinolines. Floating mines could be stopped in some mysterious way of his own, which he never divulged. A large force would take two months: another delay in the dramatic coup. In vain Jackie [Fisher] and his Board protested against the supersession of the crinolines and the military force. He carried this precious War Committee of the Cabinet with him, and I am told on authority that I can absolutely trust, that he gave the Committee to understand that on this question the Board of Admiralty were with him. You know the rest of the story… Altogether it is a most hopeless business and entirely due to the hotheadedness of a man who ought never to have been in charge of that Department."[93]

HOWELL GWYNNE, EDITOR, *THE MORNING POST*.

TURKEY

"I am told the Turkish population is beginning to think the Russians will get in here soon, and that some of the better class are secretly spreading a propaganda to open the Dardanelles to the British and French, and let them in first, if occupation by the Russian does actually become imminent, as they fear, above all else, an occupation by the Russians, and expect no mercy in such a case."[94]

LT HERBERT S. BABBITT, USN, USS *SCORPION*, ISTANBUL.

LEMNOS

A joint military naval/military meeting was held to consider details of the planned landings. Captain George Hope, commanding HMS *Queen Elizabeth*, led the meeting aboard the transport *Arcadian*.[95] Commander Edward Unwin expressed his concerns about the viability of landing men in open boats.

"I have only this to say, Sir, that as the beaches on which our men are to land are defended, it seems to me fatal to land them from open boats. The boat grounds and becomes a sitting target, full of men without cover, scrambling over one another trying to get out. The troops will be hampered by their packs and at their very worst.

---

92   Anti-torpedo bulges.
93   Howell Arthur Gwynne quoted in Wilson, Keith (Ed.), *The Rasp of War. The Letters of H. A. Gwynne to The Countess Bathurst 1914–1918*, pp. 78–79, Sidgwick & Jackson (London) 1988.
94   Naval Attache's Reports, Office of Naval Intelligence, Unpublished Manuscript, US Naval War College, April 1915.
95   The date of the meeting is uncertain but believed to be 11th/12th April 1915. However, Hunter-Weston, who was present, had not arrived at Lemnos aboard HMT *Andania* until the evening of 11th.

The enemy has only got to concentrate on the boat and scarcely a man will get out alive. Panic and confusion is certain!"[96]

CDR EDWARD UNWIN, RN, HMT ARCADIAN.

Hope then asked what Unwin for his solution to the problem.

"Why, run an old ship ashore full of troops, cut a hole in her bow, then make a gang-way with lighters to the shore, and the troops can dash ashore along it, covered by machine-gun fire from the ship. Also, the ship can bring in tons of ammunition and stores — the more weight you put in the stern, the higher up it will bring her bow — and what is more, she can carry large supplies of water, and would be a distilling station, and clearing-station for wounded ready made. Of course, if the enemy have field-guns ashore, doubtless they could defeat her, but I understand they have no artillery."[97]

CDR EDWARD UNWIN, RN, HMT ARCADIAN.

Concerns about the supply of ammunition for the fleet prompted guidance to be issued about its expenditure in support of the initial landings. The real fight was expected to be the renewed assault on the forts guarding the Dardanelles.

"Fire from covering ships is to be used to assist the operations of our troops, the primary objective being the enemy's artillery…

"The secondary objective is the shelling of the enemy's troops, in trenches or in the open. The shelling of enemy's trenches, except to cover an infantry attack or for other specific object, is, as a rule, a waste of ammunition…

"It is not desirable for me to definitely limit the expenditure of ammunition of covering ships during the landing and subsequent advance of the troops, but the following figures are issued as a guide of what I consider the maximum expenditure allowable on the day on which the troops are landed, without prejudice to future operations.

12" to 9.2" inclusive	…	20	rounds per gun.
7.5"	…	80	" "
6" Q.F. or B.L.	…	100	" "
4" Q.F.	…	100	" "
12 pdr. Q.F.	…	100	" "

"This allowance should not be exceeded without urgent military necessity."[98]

VICE-ADMIRAL SIR JOHN DE ROBECK, RN, EASTERN MEDITERRANEAN SQUADRON.

Not everyone depending upon the effectiveness of the naval barrage, those tasked with undertaking the landings, was filled with confidence at their prospects.

---

96 Unwin quoted in Usborne, C.V., *Smoke on the Horizon. Mediterranean Fighting 1914–1918*, pp. 145–146, Hodder & Stoughton (London) 1933.
97 Unwin quoted in Usborne, *Smoke on the Horizon*, pp. 146–147.
98 MEF General Headquarters, General Staff, TNA WO 95/4263.

"Orders were issued yesterday that we were to practice disembarking to-day in preparation for the landing on Gallipoli. The different units had to line up in the stations allotted to them, ours luckily being on the saloon deck where we will get use of the accommodation ladder instead of the rope ladder as first proposed. Except for our rations, which had not been issued, we had on our full marching order loads — revolver, water-bottle, ammunition, haversack, field glasses, map case, Burberry and ground sheet. When we land we will have about 5 lbs. of rations in addition...

"They say the whole of Gallipoli swarms with Turks, and the whole coast is covered with trenches and barbed wire entanglements 6 feet high. They talk as if it meant absolute annihilation of our small covering force of about 5000. The whole remainder of the Expeditionary Force, I presume, will lie out at sea till the coast is clear should we succeed in clearing it, but it is very evident every man I have spoken to has practically no hope of ever returning."[99]

LT GEORGE DAVIDSON, 89TH (1ST HIGHLAND) FIELD AMBULANCE, RAMC, HMT *AUSONIA*.

"General Birdwood was on board this afternoon. He said "How are you boys faring?" "All right Sir," "Anxious to get ashore?" "Yes sir." "Well you will land in a few days, and then get all the fighting you want." So we shall soon be in the thick of it all, and you can guess how eager we are to be there after this long wait and many doubts as to whether we should ever be used except for garrison work. The work ahead should much more interesting than if we had gone to France; and if we succeed in reaching Constantinople, we will feel as if we had helped to do something. I am on guard for the first time to-night; previously it had been picket for me."[100]

DVR ALFRED PATERSON, 8TH BATTERY, AFA, HMT *ATLANTIAN*.

"Went ashore. Bought Greek store. Proprietor spoke English. Had quite a good stock, including whisky, brandy, champagne, etc. Asked him if local folk bought that. Answered — "No, I knew the English were coming, and provided for same." Our coming to this island was a secret, so what ho, the spies!"[101]

LT ARTHUR COLMAN, AASC, AIF, HMAT *KATUNA*.

"Combatant officers and men are forbidden to divert their attention from the enemy in order to attend wounded men. A wounded man unable to bind up his own wound may be assisted by the nearest private soldier, who as soon as the wound is bound up (with the wounded man's own first field dressing) will at once rejoin his unit, taking forward the wounded man's ammunition.

"Both in advance and retreat wounded are to be left where they lie till removed by stretcher bearers and are in no case to be moved by combatants. Turkey like ourselves has agreed to the Geneva Convention and wounded falling into the enemy's hands will be under the protection of the Red Crescent, which corresponds to our Red Cross...

---

99   Davidson, *The Incomparable 29th and the River Clyde*, pp. 28–29.
100 *Northern Times* (Carnarvon, Western Australia), 24th July 1915.
101 *Sporting Judge* (Melbourne, Victoria), 11th September 1915.

"All ranks are to be warned that, when landing, they will have to wade waist-deep in water. Arrangements should therefore be made to keep haversacks containing rations from getting wet. With the web equipment it has been found that they can be carried on top of the pack secured on top of the pack secured to the spare ends of the pack attachment straps."[102]

LT-COL. CLAUDE PERCEVAL, AA & QMG, 29TH DIVISION.

EGYPT

There was no secrecy about the preparations at Lemnos. As there were no proper port facilities there either, those ships whose cargoes had been improperly loaded in Britain had to be sent to Alexandria.[103] But the problems did not end there.

"At 11am… I received instructions from then O.C. Base Supply Depot, Alexandria, to take charge of the loading of HMT "Eddystone". I was given a list of Supplies to be shipped thereon. These comprised 30 day's [sic] rations for 60,000 men and 17,000 horses…

"The loading was done by gangs of Arabs provided by a contractor. He engaged men at the ship's side at the beginning of each shift. Each foreman engaged his own gang. This process took anything up to half an hour of absolute pandemonium, and often stretched into the hours during which work was supposed to be in progress. Moreover, it was generally found after work had commenced that two or three unauthorized men had slipped into the gang, and there was further stoppage and pandemonium till these were ejected. Arrangements should have been made to engage the men before working hours, and away from the ship. The police should have kept all unauthorized persons outside the docks.

"The quality of the labour provided was very bad, and as a result the cargo is very badly stacked. The Winchmen were particularly bad, and a great deal of breakage and delay resulted. During the greater part of the time there was only one depot unit in charge of the loading, and it was impossible to prevent pilfering, which took place on a wholesale scale."[104]

2/LT ALBERT COX, 232ND DEPOT UNIT OF SUPPLY, ARMY SERVICE CORPS, ALEXANDRIA.

"We are taking two accessory hospitals in the vicinity. One will accommodate at least 450, and the other 550 patients. One is for contagious cases, the other for certain infectious and general cases. The main hospital is being fitted for 1000 patients, rising under pressure to 1200, so that, we are laying ourselves out altogether for a possible demand of 2200 patients. That is a hospital six times the size of the Melbourne

---

102   1/5th Bn Royal Scots (Lothian Regt) War Diary, TNA 95/4325.
103   Writing post-war, Major-General Sir George Aston, Royal Marines, went so far as to state: "…for a landing on a hostile coast, the holds of troop transports must be packed very carefully, so that what you want to land first is to be found on the top. Neglect of this lesson made the success of the Dardanelles operation impossible, instead of very improbable." (Aston, Sir George, *Memories of a Marine. An Amphibiography*, p. 241, John Murray (London) 1919.)
104   232nd Depot Unit of Supply War Diary, TNA WO 95/4358.

Hospital, and I think, larger than any other hospital in the world. We hope that it will not be wanted to this extent, but General Ford and General Williams have determined after courteous consultation with us, that it is better to spend some money on equipment and be ready, than to experience the risk of being overwhelmed with sick and wounded. To-night I have heard the cheers — as I have every night this week — of men passing to the front. We have a short respite in which to get our house in order, and we hope that we shall be agreeably disappointed and not overwhelmed with sick and wounded. If they do come, however, we will have the satisfaction of knowing that we have done everything that men can do to provide for them."[105]

MAJOR JAMES BARRETT, AAMC, 1ST AUSTRALIAN GENERAL HOSPITAL, HELIOPOLIS.

"The camp is quite deserted to what it was a week ago; for the last seven days troops have been pouring out of Cairo. Great trains of khaki-clad troops are for ever coming and going. What does Cairo care from what corner of the Empire they come — it is sufficient to know that they are here. You see a battalion or two strike camp, followed by a battery of artillery. You ask the officer in command where they are going. His orders are to entrain for Alexandria at a certain time; beyond that he knows nothing. This perpetual movement of troops is a great and mysterious business."[106]

LT FRANK HARTNELL, WELLINGTON BN, NZEF, ZEITOUN CAMP.[107]

13 APRIL 1915

OFF GALLIPOLI

What could be seen during an offshore reconnaissance of the potential landing sites worried some of those looking on.

"[T]he examination by the G.O.C. and members of the Corps and First Australian Divisional Staff, of the coast line of the Gallipoli Peninsula, from the Gulf of Saros to Cape Helles… fully satisfied me of the difficulties of effecting a "landing" on the exposed and narrow beaches below steep and rugged cliffs, particularly of guns and horses, except just to the north of Gaba Tepe, where the Olive Grove Valley opens out with a gentle slope to the sea. I noticed, however, how well it could be swept by enemy fire, and I thought I could detect barbed wire in plenty on the beach…

"The information gained on this reconnaissance… naturally gave rise to serious doubts and grave apprehension of the possibility of failure, and even disaster; and in any case, very serious losses, in the minds of men competent to properly appreciate the situation."[108]

BRIGADIER-GENERAL JOSEPH HOBBS, 1ST AUSTRALIAN DIV. ARTILLERY, HMS *QUEEN*.

---

105 *Daily Standard* (Brisbane, Queensland), 22nd June 1915.
106 *Taranaki Herald* (New Zealand), 1st June 1915.
107 Hartnell was severely wounded at Chunuk Bair on 8th August 1915.
108 *Western Mail* (Perth, Western Australia), 26th April 1934.

The man appointed to command the covering force noted the barbed wire near Gaba Tepe.

> "If that place is strongly held with guns, it will be almost impregnable for my fellows."[109]
> COLONEL EWEN SINCLAIR-MACLAGAN, COMMANDING 3RD BDE, HMS *QUEEN*.

LEMNOS

The Navy's stated primary target was the Turks' artillery. The problem was no-one knew where it was.

> "The enemy holds the Kilid Bahr plateau in strength, and is believed to have a number of troops concentrated in the neighbourhood of the Anafarta villages and Maidos. There may be Divisions (20,000 men) distributed in these areas. Gun emplacements have been located at Kaba Tepe and Nibrunesi Point, but repeated air reconnaissances have failed, as yet, to disclose any guns…
>
> "There has been no opportunity to reconnoitre for forming up places and rendezvous; it will be necessary to move troops forward as soon as they land, taking advantage of any cover to get units together. Until the beach control personnel is disembarked, special arrangements must be made to organize the forming up and control of the troops as they come ashore."[110]
> MAJOR-GENERAL WALTER BRAITHWAITE, CGS, MEF, HMT *ARCADIAN*.

> "Just a few lines, as we are very near the fighting place now — only about 40 miles off. We expect to sail any moment to start operations. From what I can gather landing will be a difficult thing and full of risk, but it should be exciting, and more especially if the Turks show any resistance. We have no idea where we are going to land. We only arrived at this place this morning, and a large number of ships are congregated here. You can depend that when we do start we will be all over them in a very few days, as the crowd of men here are all old trained soldiers of the very best quality, who will soon wipe everything that comes before them off the face of the earth."[111]
> CAPT. DOUGLAS LINDSAY, 1/5TH BN ROYAL SCOTS (LOTHIAN REGT), HMT *DONGOLA*.

TENEDOS

Hunter-Weston examined the latest aerial reconnaissance photographs.

> "Went to Tenedos in Cruiser "Dartmouth" with Wolley Dod[112] & Street.[113]

---

109   Sinclair-Maclagan quoted in Bean, C.E.W., *The Story of ANZAC: From the Outbreak of War to the End of the First Phase of the Gallipoli Campaign, May 4, 1915*, p. 222, Angus & Robertson Ltd. (Sydney) 1941.
110   MEF General Headquarters, General Staff, TNA WO 95-4263
111   *Edinburgh Evening News*, 7th May 1915.
112   Colonel Owen Wolley-Dod, Lancashire Fusiliers, GSO1, General Staff, 29th Division.
113   Major Harold Street, Royal Field Artillery, attached General Staff, 29th Division.

Landed & interviewed Comdr Samson[114] at Aerodrome. Saw result of air reconnaissance & arranged for copies to be sent."[115]

MAJOR-GENERAL SIR AYLMER HUNTER-WESTON, COMMANDING 29TH DIVISION.

The following day he took a look at the peninsula for himself.

14 APRIL 1915

LEMNOS

"On "Dartmouth" proceeded at daylight to S. end of Gallipoli Peninsular. Cruised round Tekke Burnu, Cape Helles, Seddel Bahr, Morto Bay to Eski Hissarlik. Then up West Coast to Gaba Tepe & back to Tenedos. Saw no enemy but some smoke from camps behind N. spur of Achi Baba, near Krithia & also further South. The South end of the Peninsular showed heavy entrenchments & much wire. Got back to Mudros & went on to "Queen Elizabeth" to arrange principle operations by the fleet."[116]

MAJOR-GENERAL SIR AYLMER HUNTER-WESTON, COMMANDING 29TH DIVISION.

"It is desirable to practise the disembarkation of the detachment detailed to land from the Collier [*River Clyde*] when operations commence. I am to request that you will issue the necessary orders to carry this out; the detachment should parade, dressed, equipped, and carrying everything that they will have to at a later date.

"Arrangements have been made to have the Collier alongside Transport B.2 by 7 am on the 17th inst. to take troops from that ship.

"The P.N.T.O. [Principal Naval Transport Officer] is arranging to transfer details from transports B.1 and B.3.

"The Naval Officer Commanding the Collier is selecting the landing place; the O.C. troops will receive all instructions from him."[117]

LT-COL. WILLIAM DE LANCEY WILLIAMS, HAMPSHIRE REGT, ATTD MEF GENERAL STAFF.

"Orders are out for the usual drill to-morrow which now always consists of boating, landing, and climbing rope ladders swinging about in mid-air.

"After dinner I had a long talk with one of the ship's officers who had been in the navy for years, and is now attached to this boat to look after things naval. The charge ashore of the covering party he considers a vast mistake, and his idea is that the authorities have just discovered this too, and are reconsidering its advisability. A few machine-guns could wipe us all out before we get ashore. We are to be covered by

---

114   Commander Charles Rumney Samson, No. 3 Squadron, RNAS.
115   Diary entry, 13th April 1915, Maj. Gen. Hunter-Weston, British Library, MS 48364.
116   *ibid*.
117   MEF Headquarters, General Staff War Diary, TNA WO 95/4263.

the navy, but what is the use of big guns against individuals planted everywhere in trenches. However it is not for us "to reason why".[118]

LT GEORGE DAVIDSON, 89TH (1ST HIGHLAND) FIELD AMBULANCE, RAMC, HMT *AUSONIA*.

"We, on board the *London*, are kept very busy rehearsing the landing of the 11th Battalion of the Australian Infantry under Colonel Johnstone [sic], which is to be disembarked from our ship. Every day parties of men fully equipped are brought on board and practised in climbing ladders to and from the boats. To facilitate the rapid disembarkation of a great number, wide wooden ladders have been made on board, up and down which two fully equipped men can climb at a time. Rope ladders were also experimented with, but these have turned out unsatisfactory, on account of the sagging, and have been abandoned. These wooden ladders, together with the ship's gangways, enable 500 or 600 men to embark from, or disembark into, the boats and steam pinnaces in a very few minutes with a minimum of delay. The crews of the boats are kept busy all day landing the troops on the shores of the bay, and bringing them off again at night. On all the ships similar rehearsals are practised."[119]

ELLIS ASHMEAD-BARTLETT, WAR CORRESPONDENT.

"We were all glad to leave [Egypt], though we've all got reminders of it. We all have bugs, and, by jove! they bite. Great big bugs, all colours; and besides, the blankets have lice on them. Cheery, eh! We are only allowed one shirt on and one in our packs, so they get dirty, and water for drinking is scarce, so we can't wash much. If you saw me now! What with the sun and the muck, I'm quite brown. Then we are not allowed to shave upper lips, so I have a bit "mouser," to add to this the bugs on a dirty shirt, and you have the complete soldier."[120]

PTE GEORGE SIDEY, 1/5TH BN ROYAL SCOTS (LOTHIAN REGT), HMT *DONGOLA*.

MESOPOTAMIA

What kind of a fight were the Turks expected to put up? Those already in action against them described the fighting as fierce but "British pluck" had won through.

"We have just had a three days' battle with the Turks at Shaiba, April 14th, about 12 miles from here. To get to Shaiba we had to march through five miles of mud and water, which was up to our knees, and in some places it was over our knees. When we got out of the water we had to march about five miles to camp, through a sandy desert, and when we got there, I, like the rest, was completely knocked up. The water there was very salty for drinking purposes, and I am not sorry we are back in Busra again. At Shaiba we remained in the trenches two days, and the next day we marched out to fight them. They were superior in numbers, but after a hard fight we beat them well, making them do a "Turkey trot." They left some guns and ammunition behind them,

118 Davidson, *The Incomparable 29th and the River Clyde*, p. 32.
119 Ashmead-Bartlett, *Uncensored Dardanelles*, pp. 36–37.
120 *Berwickshire News*, 18th May 1915.

also many wounded and dead. The Norfolks lost 32 killed and about 106 wounded; this is not so bad considering we were weak, being in the trenches two days under a very hot sun, and they were superior in numbers. I did very little in the fight, because an officer was wounded near me, and I was called upon to help to carry him back to hospital, which I can assure you was a very warm job. Some of the Turkish bullets are like lumps of lead, others are sharp-pointed and round-nosed, all making a very bad wound. I was not sorry when it was over, as we were all done up. It is very hard fighting on a desert; water is very scarce, and sand is plentiful."[121]

PTE HENRY LEEDER, 2ND BN NORFOLK REGT, 18TH BDE, 6TH (POONA) DIVISION.

"Our poor regiment got it pretty hard; we had forty killed or died through wounds and nearly three hundred wounded, and anyone who got out of it without a scratch could shake hands with himself. The Turks were much stronger than we were, but you know the old British pluck did them. They could not make it out. There were two white regiments — Norfolks and Dorsets — and when the charge was given these two regiments and the natives raced for the first lot of trenches and cleared them in a grand style, and that was enough for the Turks, and they retired like sheep, and our boys touched them up and helped them on. Our force captured all their camp ammunition, and several guns."[122]

PTE CHRISTY SMITH, DCM,[123]
2ND BN NORFOLK REGT, 18TH BDE, 6TH (POONA) DIVISION.

15 APRIL 1915

LEMNOS

"[W]hile enjoying a cup of tea at a table of Engineer officers, we heard what is evidently the latest proposal about the invasion of Gallipoli. Instead of landing us from troopships we all go on battleships, which seems to us to be an improvement. We are also likely to land at three if not four different points at the same time. This new plan will likely take a few more days to develop, so that we may expect a few days' grace yet. We have very exact maps of Gallipoli on a large scale, with full accounts of all the possible landing places and the interior, with soundings round the whole peninsula, the nature and the amount of water to be expected at various points, etc."[124]

LT GEORGE DAVIDSON, 89TH (1ST HIGHLAND) FIELD AMBULANCE, RAMC, HMT *AUSONIA*.

"The detachments went ashore to-day. Believe the party is to be the landing party. I am left out. Didn't I go mad! I've got to stay on board in charge of the cable section

---

121 *Lynn Advertiser*, 18th June 1915.
122 *Lynn Advertiser*, 10th September 1915.
123 Smith was awarded the decoration for his bravery that day: "For gallant conduct at Barjisiyah (Turkey in Asia) on 14th April 1915, in constantly carrying urgent messages and bringing up ammunition. He was noticed on several occasions for his courageous behaviour, and has been subsequently wounded." (*London Gazette*, 5th August 1915.)
124 Davidson, *The Incomparable 29th and the River Clyde*, p. 33.

horses. They will not be able to land, on account of the country, for a while. Well, as long as I get ashore later on I cannot growl. Somebody has got to do it, I suppose, and it is my luck."[125]

CPL HOWARD BOTH, 1ST DIV. SIGNAL COY, AUSTRALIAN ENGINEERS, HMAT *MINNEWASKA*.

"It is strange how the training and dicipline [sic] makes you forget that there is any risk in this — you only live for the day — how we shall behave under fire is yet to be proved; and I don't believe that any of us will be able to say that we weren't afraid at first going into action. There are about twenty men-o'-war, 15 of these British. Everyone expresses confidence in the British navy, and prays that the Germans will send their boats out into the North Sea to try conclusions."[126]

DVR ALFRED PATERSON, 8TH BATTERY, AFA, HMT *ATLANTIAN*.

EGYPT

"The immensity of the preparations for moving a large army into Turkey, following the forcing of the Dardanelles, requires a great deal of time. We are only a very small cog in a very great machine, and all our wishing to get away will not move us before the other parts are ready. Most of the Australian divisions have gone on, leaving only line of communication units behind. Our turn will come as soon as some space in which to organise has been won from the enemy's territory."[127]

CAPT. EDMUND MILNE, RAILWAY SUPPLY DETACHMENT, 11TH COY AASC, ABBASSIA CAMP.

INDIAN OCEAN

As anxious as some were to leave Egypt, not everyone got that far.

"[T]his morning about 5.30 a.m. I heard the report of a fire-arm but thinking it was on the upper deck I took no notice of it. At a little past seven o'clock Lieut. Young said to me "Go and see Mr. McLay as there is something wrong with him." I went in and saw Mr. McLay lying half way across the settee covered in blood.[128] I saw no revolver. I did not move or touch the body. I then proceeded to the hospital for the troops' doctor. I attended Mr. McLay. I did not take him tea or coffee because he appeared to be absent from his room."[129]

EDWARD DRUMMOND, ASSISTANT STEWARD, HMAT *WILTSHIRE*.

"I am a lieutenant attached to the 7th Batt AIF on board transport A18 SS *Wiltshire* at sea. I know the deceased 2 Lieutenant Stanley Gordon McLay of 7th Battalion. I last

---

125 *The Areas' Express* (Booyoolee, South Australia), 20th August 1915.
126 *Northern Times* (Carnarvon, Western Australia), 24th July 1915.
127 *The Young Chronicle* (New South Wales), 4th June 1915.
128 2/Lt Stanley McLay, 7th Bn, AIF, died on 15th April 1915. An enquiry held on the troopship concluded that he had died as the result of an accidental gunshot wound, as there was no evidence that he was intent on taking his own life. Commemorated on the Chatby Memorial, he was the son of James McLay, of Welshman's Reef, Victoria.
129 McLay's army service record NAA B2455: MCCLAY S.

saw him last night (at 9.30 p.m. 14th inst). He was then suffering from sea sickness. I then said "How do you feel?" He informed me that he did not feel very well. I noticed nothing unusual about his manner. I next saw him between 7 & 7.30 a.m. this morning. I went to his cabin to see how he was. I found the door open fastened on the hook, with [an] electric light on. I went in and found him lying on his back across the settee. I noticed blood on his arm which I took to be [the] result of vomiting and without further examination I called the steward who was in the passage and went for the doctor… I was not conversant with his private affairs. With the exception of feeling a little worried about his duties as ship's quartermaster he appeared to be quite normal."[130]

LT ERIC YOUNG, 7TH BATTALION, AIF, HMAT *WILTSHIRE*.[131]

16 APRIL 1915

LEMNOS

As embarkation practice continued, Hamilton heard more from those who did not share his confidence in the coming campaign.

"Mr. [Gerald] Fitzmaurice, late dragoman at the Embassy at Constantinople…, says the Turks will put up a great fight at the Dardanelles. They had believed in the British Navy, and, a month ago, they were shaking in their shoes. But they had not believed in the British Army or that a body so infinitely small would be so saucy as to attack them on their own chosen ground. Even now, he says, they can hardly credit their spies, or their eyes, and it ought to be easy enough to make them think all this is a blind, and that we are really going to Smyrna or Adramiti. They are fond of saying, "If the English are fools enough to enter our mouth we only have to close it." Enver especially brags he will make very short work with us if we set foot so near to the heart of his Empire, and gives it out that the whole of us will be marching through the streets of Constantinople, not as conquerors, but as prisoners, within a week from the date of our making the attempt."[132]

GENERAL SIR IAN HAMILTON, C-IN-C, MEF.

"We had disembarkation practice in earnest to-day, climbing down over the ship's side on rope ladders, into the boats. We then rowed to the shore, and spent the day most enjoyably on land, returning to the ship as darkness set in."[133]

PTE CECIL YORKE, CANTERBURY BN, NZEF, HMT LUTZOW.

---

130   McLay's army service record.
131   Young was evacuated from Gallipoli due to shell shock on 15th July 1915. Promoted Capt. and transferred to 59th Bn, AIF, he was awarded the Military Cross for bravery on the Western Front.
132   Hamilton, *Gallipoli Diary*, vol. 1, p. 114.
133   *Lyttelton Times* (New Zealand), 24th July 1915.

MEDITERRANEAN

Just after passing Skyros, the unescorted transport *Manitou*, taking an artillery brigade and an infantry working party, 20 officers and 626 men all told, to Lemnos spotted a ship approaching. Originally taking it to be a British warship, it proved to be the Ottoman destroyer *Demir Hissar*.

> "We… were held up by a Turkish torpedo-boat. She gave us 10 minutes to get out of the ship. This was not much good, as there were not nearly enough boats to hold us all. Two capsized while they were being lowered. This was a terrible sight, and I won't write about it; it would give you nightmare[s] for weeks. One boat fell right into the water — the davits snapped clean away like pipe-stems. After all the boats had gone, most of the men left took to the water, some on planks and some on broken pieces of the ship.
>
> "All this time the torpedo-boat was loosing off torpedoes at us. She let off three altogether, and not one hit us. I had a beautiful view of the last torpedo. The torpedo-boat came right on to within 150 yards, then turned broadside towards us. I saw a little whiff of smoke, and, simultaneously the torpedo plunging into the sea, leaving a white trail behind it, and coming straight for us. This was the most horrible sight of all. You should have seen the fellows jumping off into the water from all parts of the ship. Hart and I went over to the other side of the boat preparing for a sudden disappearance if this one struck her. We waited and waited — and nothing happened! All this time the wireless operator had been sending S.O.S. for all he was worth, and all his Majesty's ships and transports for 100 miles around were answering. The door of the wireless box was ajar, and the noise of the instruments was like about a thousand women all knitting socks for the soldiers at once.
>
> "You can imagine how bucked up we were when at last a trail of smoke showed over the horizon to the southward, then another. They got thicker and blacker every moment and soon we could see the hulls of two of our cruisers racing towards us as hard as they could, with mountains of foam at their bows. It was just at this moment, as Hart and I stood by each other on the deck, that the Turk fired his third and last torpedo, and we thought the cruisers would be too late. But when he missed for the last time, he put on speed and cleared out for all he was worth to the eastward. Two of our torpedo-boat desstroyers [sic] were coming up by now and they were after him like a shot. We heard later that the Turk was compelled to beach his boat on a small island. Our two torpedo-boat destroyers then broke up his hull. A little mine-sweeper and two transports had come up and they picked up about 300 of our men altogether and restored them to us. Well we are all out of it now, except some of the poor fellows who were in too great a hurry to start with, but it was a pretty close thing."[134]

LT RODERICK MACDONALD, 147TH BDE ROYAL FIELD ARTILLERY, HMT *MANITOU*.[135]

---

134 *Poverty Bay Herald* (New Zealand), 4th August 1915.
135 Transferred to the Royal Air Force, Capt. Macdonald, 53rd Training Depot Station (Dover), was killed in a flying accident when his aeroplane, Sopwith Camel E9970, spun into the ground from 300 feet. Buried in Dover (St. James's) Cemetery, he was the son of Margaret Ellen Macdonald, of Norman House, St. Margaret's, nr. Dover, Kent.

"From the bridge I made my way as quickly as I could to my cabin where I put on my left belt and, taking a flask of brandy and some chocolate from my haversack, put them in my pocket. Regaining the deck, I was just in time to see a second torpedo leave the tube. It came straight towards me and I remember instinctively looking up to see whether all was clear overhead or whether any rigging intervened to obstruct my free passage skywards. I held on tight to something... Now!

"But again nothing happened. I moved quickly along the deck. Most of the boats were already in the water. I leaned over the rail between two jerking davits and looked down. The boat was suspended a few feet above the water, its further progress stopped by men men who, stepping out from the deck below me, were grasping the falls of rope and passing themselves down hand over hand. As I watched, more men reached out to the ropes and swung themselves clear of the ship. The strain was too much and, with a report like a gun, the davits on either side of me snapped off short in their sockets and fell inwards and downwards. For a fraction of a second I saw agonized faces upturned in the boat... The boat splashed into the water. One after the other, men, still grasping the ropes in their hands, fell head-long in rapid succession on their comrades below and then, just as the struggling mass was overturning, the heavy davits with a sickening thud crashed fair and square.

"I heard my name being called insistently and, some distance away I recognized one of my N.C.O's standing up in the stern of a boat. He implored me to leave the ship but I waved my hand to show him I was not coming and turned away. Exactly why I cannot say, but from the very start I had subconsciously decided to remain on the ship as long as there was something to stand on. Perhaps my mind harked back to a summer afternoon on the Severn years ago when, as a small school boy, I had stayed too long in the water, been seized with cramp and only rescued with great difficulty. In any case I think it was the certain knowledge that in very cold water I could not hold up for long, that decided me to delay the final struggle as long as possible. I was certainly not swayed by any vain hopes of rescue before the ship went down.

"Making my way forward I joined a small group of officers who were busying themselves in throwing overboard everything that they could lay their hands on that would float and it was whilst we were engaged in this task that the torpedo boat was suddenly discovered to be on the move again. She made off in a direction at right angles to the *Manitou* on the port side, and then turned and stopped broadside on at a distance of perhaps a mile. What was she up to now? Would she open fire with the gun which we had noticed on her deck and which looked like a 6-pounder, fire more torpedoes or what? How long she remained over there I cannot say, but as we were momentarily expecting something to happen, it seemed a long time — perhaps a quarter of an hour or thereabouts. Up in the bows we discussed the situation the while we fortified ourselves for our long swim by sipping brandy from our flasks and munching chocolate. Some were for going overboard and swimming for it, arguing that to remain on the ship when she went down would not give one a fair chance. One of my subalterns came up and asked if I "minded his going over." We watched him and others let themselves down by a rope hanging from the bows and swim away from the ship. The transport which had followed us out from Alexandria — one of the P. & O. fast steamers of the Brindisi–Port Said route as it turned out — came up and seeing us in trouble, passed us at top speed. This wise but apparently unfriendly act was, I believe, in obedience to instructions issued by the Admiralty as a result of torpedoing of the "Cressy," "Hogue" and "Aboukir"...

"A shuffle on the deck behind me, and turning round I beheld my subaltern very wet and shivering.

"What on earth are you doing here?"

"'Well, sir, I found the water so d...d cold that I thought I would come back, so I swam to the ladder and here I am'! This did not sound promising! I took my boots off; slacks would only be in the way and hinder me, so off they came too; socks and pants, nothing to them up now — off with them! Finally, in the buff from the waist downwards I felt prepared. Prepared, but of confidence in myself to swim for more than a few minutes I had none. And the curtain rang up for the final scene.

"'She's coming back again' called out someone, and all eyes were at once turned in the direction of the torpedo boat which was seen to be approaching us at full speed with smoke belching from her funnel. Well, we were in for it this time! Whatever it was that had gone wrong before, we felt would not go wrong again. All that time she had spent over there had obviously been employed in overhauling her beastly torpedo gear, and having satisfied herself that all was now functioning properly, she was coming in to finish us off. We watched her fascinated and in silence. Closer and closer she came and when well within her old range of 60 to 80 yards, round she swung and disclosed the same old six [crew] at the torpedo tube. A flutter of white foam below her stern as the engines were reversed. Now she's ready. Splash! Torpedo number three was on its way.

"'We're for it' Damn it! why was my mouth so dry? 'Hold on, you lads' said someone, and then the same someone began counting slowly — 'One, two, three.... four.... five.... six.... SEVEN! A breathless pause, and then — 'By God, she's missed us again!'

"Whatever the cause of this third failure, the old "Manitou" still rode the waters of the Aegean inviolate. Slowly turning, the torpedo boat passed across our bows within biscuit throw, her crew standing on deck and gaping in speechless amazement and mortification at the ship which had had held so utterly at their mercy. With black smoke pouring from her, she rapidly dwindled to a mere speck on the horizon and was then lost to view. Our ordeal was over."[136]

MAJOR ALAN THOMSON, 147TH BDE ROYAL FIELD ARTILLERY, HMT *MANITOU*.

---

136   Thomson, Colonel A. F., 'Torpedoed? An incident of the Gallipoli Campaign' in *The Journal of the Royal Artillery*, vol. LVII, July 1930, pp. 260–262, Royal Artillery Institution (Woolwich) 1930.

Although all three torpedoes missed their target, fifty-one men died while trying to leave the *Manitou*; many not keen to remain aboard a ship containing a large amount of artillery ammunition while it was under attack.

The *Demir Hissar* was chased by a number of Royal Navy vessels, beached itself on the Greek island of Chios and its crew were interned.

17 APRIL 1915

ENGLAND

After arriving home from Istanbul, one Englishman painted a picture of a city on the verge of revolution; of an army unfit to fight. It was what many wanted to believe to be true.

> "I have come into contact with several highly-placed officials, and they have told me that the British are certain to get through the Dardanelles. This is borne out by the precautions taken in Constantinople. All the museums have been emptied, and the collections taken to Konia, in the interior of Asia, which will probably be the seat of the Turkish Government, if they are beaten in Europe. All the valuables from banks, and the personal effects of rich Turks have been removed.

> "The Turks consider the situation hopeless, and I think that within two months she will conclude a separate peace with the Allies.

> "At the Haiderpasha railway station which is on the direct route to Konia, a train with steam up waits night and day to take away the Sultan in case of necessity. I have passed the train frequently, and the Sultan's coach is attached…

> "The Germans themselves will be in danger of massacre if the British Fleet arrives. Already there are loud murmurs of discontent against the policy which has led to the war. People are saying that Turkey made the greatest mistake of her life when she allowed herself to be drawn into war on the side of Germany.

> "About 90 per cent. of the Turkish soldiers are wearing patched clothing, and there is very little uniformity of dress. A regiment of soldiers arriving from the Caucasus were so exhausted that some of the men staggered and fell in the street. The officers kicked and beat them, and one soldier was carried away leaving a trail of blood behind him."[137]
>
> WILLIAM EATON, FORMERLY CONSTANTINOPLE TELEPHONE COMPANY.

137   *Stalybridge Reporter*, 24th April 1915.

MALTA

"We hope to be afloat again tomorrow (April 18) and that it will not be long before we are ploughing the ocean again. I do get tired of remaining in harbour, and am ever so much more happy when the salt spray is splashing in my face, and the salt laden air is blowing my hair in all sorts or manners. I love the deep blue ocean with all my heart because it seems to be in sympathy with my loneliness. Will you believe it, in Malta alone I have spent £20 in one month, just visiting the places of interest, and I do not throw money away, I can tell you?"[138]

AB ALBERT NICHOLS, RAN, HMAS *AE2*.

DARDANELLES

The British submarine *E15* ran aground during its attempt to force a passage through the Narrows. Stuck under the guns of Fort Dardanos at Kephez Point, it was a sitting duck.

"We tried to get through the Dardanelles submerged, but had the ill-luck to run aground when we got a good way up. The forts on both sides opened fire on us, and it was something awful while it lasted. We had six killed, nine wounded, and sixteen got off with nothing more than a severe shaking and soaking. It was surprising that there was not more killed, the shells were flying all around us for least half an hour."[139]

AB JOHN LOCKERBIE, RN, HMS *E15*.

An American diplomat was given some gruesome details about the fate of the crew.

"The Turks did not at once realize that the E 15 had run aground. They opened fire, and one of the first shots struck the conning tower and cut the commanding officer in two,[140] the lower part of his body falling at P.'s[141] feet. Another shell burst in the ammonia tank and fumes asphyxiated six sailors; the others jumped overboard. When the Turks saw these swimming they went to their rescue at no little risk, for the current was running strong. The English dead were buried on the beach; but as soon as Djevad Pasha, the commandant, heard of this, he gave orders that they be reinterred in the British cemetery, and a service said over their remains."[142]

LEWIS EINSTEIN, AMERICAN EMBASSY, ISTANBUL.

---

138 *Morning Bulletin* (Rockhampton, Queensland), 19th June 1915.
139 *Dumfries and Galloway Standard*, 23rd June 1915.
140 Lt-Cdr Theodore Brodie, RN, the commander of HMS *E15*, was killed on the submarine's conning tower on 17th April 1915. Buried in Chanak Consular Cemetery, he was the 31 year-old son of George Gordon Brodie and Louisa Mary Brodie, of Woodlands, Cheltenham.
141 Lt Clarence Palmer, RNVR, was formerly the British Vice-Consul at Chanak. Taken prisoner, there was a suggestion that he was to be executed as a spy but he survived his captivity.
142 Einstein, Lewis, *Inside Constantinople. A Diplomatist's Diary During the Dardanelles Expedition April — September, 1915*, pp. 2–3, John Murray (London) 1917.

LEMNOS

"Had breakfast at six, paraded at seven and stood on deck till 10.45 waiting our turn to cross to a collier that is to be used in the Gallipoli attack. The intention is to run her ashore at full speed, ploughing into the sands, when her load of 2000 men are to get overboard as best they can on to floating gangways. By a long circuitous route we all got into our places, and were packed close on the various decks which have had large square openings cut through the iron plates of the sides of the ship, and from these and the upper deck we have to decamp as quickly as possible."[143]

LT GEORGE DAVIDSON, 89TH (1ST HIGHLAND) FIELD AMBULANCE, RAMC, SS *RIVER CLYDE*.

"Physical exercise, troops formed up in full marching order for practice in disembarkation but at 12 noon message received stating practice cancelled owing to insufficient boats. Men practised in descending and ascending the rope ladders in full marching order."[144]

CAPT. CHARLES DARE, 14TH BATTALION, AIF, HMT *SEANG CHOON*.

18 APRIL 1915

DARDANELLES

Concerned that the *E15* might be brought into use by the Ottomans — fearing, too, that its secret papers might fall into Turkish hands — immediate attempts were made to destroy it. Shelling it from 12,000 yards was discounted, so the *Majestic* and *Triumph*'s picket boats were sent to sink the submarine.

"The ... submarine *B11* went to blow her up, so that the Turks would not make use of her. Anyhow *B11* came back, and told us that she could not see the grounded submarine on account of a dense fog, so the *Triumph* and *Majestic* dashed up the Dardanelles to try and sink her with gun fire, but we were unsuccessful owing to the Turkish forts opening a terrific fire on us when we came out. We all cheered our captain for his daring dash under fire.

"At night we launched our picket boat, and the *Majestic* launched hers, and two torpedoes were attached to each boat. There were volunteer crews on each boat, and they left at 10-30. The *Majestic*'s picket boat was sunk with gun fire from batteries on shore, and the *Triumph*'s picket boat was left to do all the work. They hit the submarine with both torpedoes."[145]

PTE WILLIAM SHORE, RMLI, HMS *TRIUMPH*.

---

143 Davidson, *The Incomparable 29th and the River Clyde*, p. 34.
144 14th Bn Australian Infantry War Diary, AWM4 23/31/6.
145 *Stalybridge Reporter*, 17th July 1915.

"The ... picket boat, along with the [*Triumph*'s] picket boat, did a very fine piece of work. The submarine *E15*, having run around near Chanak, it became highly important that she should be sunk, otherwise the Turks might have got her off and used her against us, much to our disadvantage. So at midnight two small boats went up the Straits to the Narrows, with the searchlights from the Narrows playing down upon them, making the quite visible for the Turks to fire at, which they did with much vigour. Our picket boat fired two torpedoes into the submarine, and thus made her *hors de combat*, but while so doing the picket boat was hit by a shell in the stern. One man was killed, and the boat was lost. The [*Triumph*'s] picket boat took our picket boat's crew off, and thus they returned minus one boat, but victorious."[146]

SBA HARRY BRADSHAW, RN, HMS *MAJESTIC*.

LEMNOS

Hamilton issued his orders to the French. They were to land on the Asian side of the straits to prevent any Turkish forces there firing on the British landings at Helles.

"In order to assist the preliminary operations of the troops landing in the Gallipoli Peninsula, I have assigned to the force under your command the task of effecting a landing near Kum Kale, with the object of engaging the attention of any hostile troops which may be near the entrance to the Dardanelles on the Asiatic shore....

"The information received does not point to any considerable numbers of the enemy being met with by your landing party. With a view, however, to safeguarding the landing the troops in Morto Bay [S Beach], it is important to prevent the enemy from placing field batteries in the neighbourhood of Kum Kale, whence they could bring fire to bear on transports anchored in the bay.

"The landing near Kum Kale is intended to be in the nature of a diversion, and it is not desirable to extend the scope of the operations further than is necessary for clearing the region between Kum Kale and Yeni Shehr and west of the Mendere Chai."[147]

GENERAL SIR IAN HAMILTON, C-IN-C, MEF.

3rd Australian Brigade received its instructions for the landing. These stressed the need to move inland quickly and to secure the area's northern flank.

"The objective of the covering force ... and the subsequent action of the Army Corps is defined in these extracts, that is: — to secure a position covering the Kaba Tepe — Fisherman's Hut Landing Place; the landing of the Army Corps; and an advance to sever the enemy's North and South communications in the vicinity of Mal Tepe.

"In your instructions to the covering force, you should keep in mind the advantage of landing on a broad front and the necessity for occupying as rapidly as possible the covering position laid down as [an] objective in Force Orders.

---

146 *Farnworth Chronicle*, 19th June 1915.
147 MEF Headquarters, General Staff War Diary, TNA WO 95/4310.

"In view of the reported presence of guns... and of troops and guns in the Peren Ovasi Valley, the covering force will have to advance and occupy the ridge running first East from Gaba Tepe and then North East...

"To assist it in this task, the rest of your Division is being landed immediately after the covering force, and should be disposed with a view to securing the above line and the Northern Flank in the direction of Fisherman's Hut.

"When this line has been secured, you will be guided by the situation as to whether you make a further advance, or consolidate your position until the landing of the bulk of the Army Corps permits the development of an advance towards its objective — Mal Tepe."[148]

BRIG.-GEN. HAROLD WALKER, GENERAL STAFF, ANZAC, HMT *MINNEWASKA*.

"As you see, we are still on board the troopship, and anchored in a bay at the Island Lemnos, just near the Dardanelles. I am braving the Censor this time, and think I can get this through... We have been practising disembarking and re-embarking. It was a treat to get on a place where there was no sand... The bay is full of transports and warships, about 70 all told, and we are waiting for a few more troops before we make a move. We have all sorts of ships of war here, from the great *Queen Elizabeth* (which was only launched in December last), to the smallest submarines, three or four airships and a couple of balloons. We hear that a move is to be made to-morrow (Monday). I hope so, as we have been here about ten days already, and life on board ship is not all that could be desired. We are in for a pretty rough time when we land, against the Turks, but have no doubt of the ultimate result."[149]

CAPT. FLEM CAMPBELL, 2ND BATTALION, AIF, HMAT *KARROO*.[150]

EGYPT

"It looks as though the Light Horse will have to wait until the "sloggers" have effected a landing, before it will be possible for us to land our horses. We shall get the chance before long now, I feel sure...

"We go through all kinds of fights against a friendly foe, in the shape of one of the other regiments, and the work is very severe... Needless to say, as we are supposed to be under fire, about the only men who have any idea as to where they are going, are the men in the lead. The others simply tear along in a pall of thick dust, raised by the thudding hoofs of the leaders, and as gutters and mounds and ditches are frequent, it's not all play. Our chaps are mostly all aboard, though, but the other day, Alick

---

148   Australian & New Zealand Army Corps, General Staff War Diary, TNA WO 95/4280.
149   *The Gosford Times and Wyong District Advocate* (New South Wales), 18th June 1915.
150   Fleming died aboard HMHS *Neuralia* on 2nd June 1915 from wounds received on 31st May. Commemorated on Lone Pine Memorial, he was the husband of Gertrude Campbell, of 'Miamba', Hill Street, Scone, New South Wales.

Chrystal,[151] of Wellington, had a rather nasty smash, by his horse rearing and falling back on him. The beggar jolly nearly fell on top of me, as I was right behind him when he fell. He got badly squashed, up; but now he is pretty well recovered again, after a couple of weeks in hospital."[152]

TPR JAMES BROWNHILL, 2ND AUSTRALIAN LIGHT HORSE REGT, MAADI CAMP.

MESOPOTAMIA

Hard fighting had taught British soldiers already in action against the Turks to respect their opponents, very different from the ragged condition described elsewhere.

"When you say there is not much fight left in 'Johnny' you are wrong. I can assure you no one was more surprised than I to run up against such stubborn resistance, because, like yourself, I was under the impression that the Turks were a secondary consideration, and that we were here to scrap the local Arabs…

"Despite the terrible pasting the Turk gets I cannot help but admire him as a most stubborn fighter. I believe in giving credit where it is due, and when you take into consideration his inferiority in equipment, etc., and how he managed to live in such a desert, the least you can say is that he put up a good fight."[153]

DVR FREDERICK COLLEY, 'S' BATTERY, ROYAL HORSE ARTILLERY.

19 APRIL 1915

ENGLAND

Kitchener forwarded a report of the recent fighting in Mesopotamia. The Ottoman defence had been skilful, stubborn but, ultimately, overcome.

"Following extract may be of interest to you, describing the recent fighting at Basra: —

"The Turkish troops were well disciplined, well trained and brave. Their machine guns had been well concealed, and were used with great effect, and their trenches were admirably [situated]… Turks had no idea of being shot out of their trenches, and had to be turned out by a charge of the whole line with the bayonet. If pluck and determination of our troops, both British and Indian had not been of the sternest, and if they had not been handled with initiative and decision, battle would not have been won. Trenches were finally captured about 4.30 pm, and being so well concealed, brunt of taking them fell on the Infantry.

---

151   Tpr Alexander Raymond Chrystal, 2nd Light Horse Brigade Headquarters, went to Gallipoli with his unit on 15th May. He was taken ill and evacuated to Lemnos on 27th August 1915. Found to be suffering from jaundice and the effects of a fractured pelvis, the former grazier returned to Australia, leaving Suez on 20th January 1916.
152   *The Forbes Advocate* (New South Wales), 25th May 1915.
153   *Gloucester Journal*, 29th May 1915.

"The Turks were so severely handled that they retired nineteen miles during the night, and later information indicated that the next day they continued their retirement."[154]

LORD KITCHENER, SECRETARY OF STATE FOR WAR.

DARDANELLES

"I will tell you of an incident that occurred last night and the early hours of this morning. One of our submarines went up the Dardanelles two days ago, got into difficulties and went ashore. Up to the present I have not heard anything definite as to the fate of the crew. Attempts have been made by aeroplane to destroy the submarine so that she should not fall into the hands of the enemy and possibly be used against us. Two ships' steamboats were sent from 'Majestic' (my ship) and 'Triumph' with torpedoes, each manned by volunteer crews with one exception. Now this job was an extremely dangerous one. I cannot give you an idea as to what it was really like, there were hundreds of guns, large and small, firing at these two little boats and lots of searchlights trained on them. However, they carried out the job successfully, our boat torpedoing the submarine, but not before the stern had been blown off the small boat and so seriously injuring one man that he died before reaching the ship. The singular thing about it was the man who was killed was not a volunteer. So brave a deed has not, in the opinion of many, been accomplished since the war began. We buried at sea to-day the unfortunate man,[155] and a more impressive scene you could not have witnessed!"[156]

AB WILLIAM SCOTT, RN, HMS *MAJESTIC*.[157]

LEMNOS

"Conference at 10 a.m. on the *Queen Elizabeth* — called by Sir Ian Hamilton — The 23rd St. George's Day fixed for landing — All admirals including Guepratte and all Generals includg D'Amade attended. Adml Thursby was ready as soon as he got the *Triumph* and others back. Adml Weymss could not be ready for 4 days mainly owing to lack of picquet boats. Aftds [Afterwards] called on French Admiral. Aftn [Afternoon] conference of Brig and C.O. explained my operation and disembarkation orders."[158]

MAJOR-GENERAL SIR WILLIAM BRIDGES, COMMANDING 1ST AUSTRALIAN DIVISION.

---

154 MEF Headquarters, General Staff War Diary TNA WO 95/4310.
155 Armourer Thomas Hooper, RN, HMS *Majestic*, died of wounds on 18th April 1915. He is commemorated on the Plymouth Naval Memorial.
156 *County Down Spectator and Ulster Standard*, 7th May 1915.
157 Scott was killed when his ship was torpedoed on 27th May 1915. Commemorated on the Plymouth Naval Memorial, he was the 31 year-old son of William and Agnes Scott, of 8 Albert Street, Bangor, Co. Down.
158 Transcript of diary of William Throsby Bridges, p. 6, 1915, AWM 2DRL/0469.

"Didn't sail last night after all. A list of the landing party has just been posted up. There is much disappointment amongst the men whose names have not been posted up. It will be most dangerous work, still, I think, not a man on board would remain were he given the chance to be amongst the first to land. We hear that the Turks have barbed-wire entanglements right to the water's edge; and we shall have to cut our way through these. I think I mentioned that the 8th battery is to be the first to land — a great honor. It will be something to boast of should we return safely."[159]

DVR ALFRED PATERSON, 8TH BATTERY, AFA, HMT *ATLANTIAN*.

"All sick men to be ashore to-day by sunset. We are evidently about to move. Two hospital ships have been manned, and medical officers in charge appointed. The Dardanelles are said to be about eighty miles from here. Disembarkation practice continues daily…

"General Birdwood has sent us a letter, in which he says that we are about to undertake "one of the most difficult tasks any soldier can be called upon to perform, and a problem which has puzzled many soldiers for years past… We are going to have a real hard and rough time of it until, at all events, we have turned the enemy out of our first objective." "Hard times," the general says, "none of us mind, but to get through them successfully we must always keep before us the following facts." He then refers to possible difficulties of transport, and reminds the men that they "must not think their wants have been neglected if they do not get all they want." Speaking of the necessity for the men individually to conserve food and water, he remarks: "Men are liable to throw away their food the first day out, and to finish their water-bottles as soon as they start marching. If you do this now we can hardly hope for success, as unfed men cannot fight, and you must make an effort to try and refrain from starting on your water-bottles until quite late in the day. Once you begin drinking you cannot stop, and a water-bottle is very soon emptied." The letter concludes with some further practical advice, and urges the troops not to waste their ammunition at any time by indiscriminate firing."[160]

PTE CECIL YORKE, CANTERBURY BN, NZEF, HMT *LUTZOW*.

20 APRIL 1915

DARDANELLES

One of the *E15*'s survivors described the good treatment they were receiving as prisoners of war.

"Our poor old boat is a proper wreck. We had five killed; our captain was the first. We have just been told we are to be removed to Constantinople, where we shall be much safer than at this place. We have been very well treated. The food is strange, but we are

---

159 *Northern Times* (Carnarvon, Western Australia), 24th July 1915.
160 *Lyttelton Times* (New Zealand), 24th July 1915.

getting used to it. We have all sorts of visitors to see us. They bring us cigarettes and do not seem able to do enough for us. Very different from what you hear at home."[161]

CERA SAMUEL TODD, RN, HMS *E15*.[162]

"On the 20th April we spent a long day in the Straits bombarding hidden batteries on the Asiatic side, and at night, between 11.30 and midnight, when the Turks thought all was quiet, we suddenly opened fire on a redoubt and trenches in rear of the lighthouse on Cape Helles, where it was known that the Turks were busy working under cover of darkness. When the guns were duly laid on the position, the searchlights were switched on. It was not possible to know the result, but it was thought that the fire could not have failed to be effective."[163]

SURGEON OCTAVIUS ANDREWS, RN, HMS *PRINCE GEORGE*.

LEMNOS

Training and the sharing of rumours were the chief pastimes of most men at Mudros.

"The troops are practising disembarkation on every opportunity. The naval authorities supply steam launches, cutters, pinnaces, life boats &c., and in turn the troops from the different boats are towed ashore. Everything is made as real as possible, and the officers impress upon the men the seriousness of the movement... A full pack is carried, and consists of 200 rounds of ammunition, change of clothes, overcoat, mess tin, waterproof sheet and rifle. The whole weighs nearly 70 lbs., and when one climbs down a swinging rope ladder impeded with such gear the operation presents difficulties. The other day a soldier lost his grip and fell into the sea, but was rescued with the aid of a boathook, but minus his rifle. When it is considered that the landing will probably be under a heavy rifle fire it will be readily seen that the future is full of possibilities."[164]

A/SGT WILLIAM FRY, 6TH BATTALION, AIF, HMT *GALEKA*.

"We were told to-day we were not to be the first to land. For some reasons plans have been changed. Needless to say there is much disappointment, and we want to know why we have had our hopes raised only to have them dashed to the ground at the eleventh hour... We don't know what is going on outside this little patch of water. No reliable news reaches us. Only rumors, rumors, rumors which are contradicted a few hours afterwards. There are about 140 boats here now, and the French troops have not arrived yet. 80,000 of them are yet to turn up — they land on the Asiatic side of the Dardanelles."[165]

DVR ALFRED PATERSON, 8TH BATTERY, AFA, HMT *ATLANTIAN*.

---

161   *Gloucestershire Echo*, 26th May 1915.
162   The letter is undated, sent from hospital in 'Chanak.' Todd, a former engine fitter from Edinburgh, survived his captivity and returned to service with the Royal Navy.
163   Andrews, Octavius William, *Seamarks and Landmarks. Being Leaves From the Log of Surgeon Captain O. W. Andrews*, pp. 292, Ernest Benn Limited, (London) 1927.
164   *The Mount Barker Courier and Onkaparinga and Gumeracha Advertiser* (South Australia), 9th July 1915.
165   *Northern Times* (Carnarvon, Western Australia), 9th October 1915.

"We have been practising disembarkation during the past four weeks, in anticipation of the great move to the Dardanelles. I consider we are most fortunate in this move, as it is considered that it will have a marked bearing on the issue of the war…

"For the past month we have done very little except boat crews' work, but, nevertheless, we recognise that although we have waited eight months, this waiting has played a very important and necessary part in the great game, and now that the time has come to strike, we are going to strike hard. The impression is that we leave here at midnight, and the attack will be made at daylight. It is thought there are probably 150,000 to 200,000 Turks to meet us, but the attacking force will be sufficient for them. The men are all very eager."[166]

PTE ARTHUR WEYMOUTH, 3RD FIELD AMBULANCE, AAMC, HMT *NIZAM*.

21 APRIL 1915

LEMNOS

"Admiral Thursby met me on board the *Minnewaska* with several of his officers. I had assembled my own Australian and New Zealand officers, and together we went fully into all the final details of the landing operations, in the hope that we should be able to make the attempt within 24 hours."[167]

LT-GEN. SIR WILLIAM BIRDWOOD, COMMANDING
AUSTRALIAN & NEW ZEALAND ARMY CORPS.

Hunter-Weston contemplated his part in attempting "the impossible."

"The common herd being wise after the event, will say either: "How easy, why think twice about so obviously correct an operation," or will say: "Criminal idiots to attempt the impossible." Personally I should not have undertaken this operation, the chances of success being so small & the consequences of defeat so disastrous. I expressed that view to the C. in C. but he decided otherwise. That being so, I am wholeheartedly for carrying the plan to a successful issue & I intend to pull off the outside chance & make a success of it. I tell the Division that we will succeed. I believe we shall succeed & I think I have made the Division full of confidence of success after hard fighting, which latter they are all on for. They are out for blood & will take a lot of stopping."[168]

MAJOR-GEN. SIR AYLMER HUNTER-WESTON, COMMANDING 29TH DIV., HMT *ADANIA*.

After setting down his thoughts privately, he sent a message to the men under his command.

"To each man of the 29th Division, on the occasion of their first major going into action together.

---

166 *The Mercury* (Hobart, Tasmania), 14th June 1915.
167 Birdwood, Lord, *Khaki and Gown. An Autobiography*, p. 255, Ward, Lock & Co. Ltd. (London) 1942.
168 Extract from letter, 21st April 1915, Maj. Gen. Hunter-Weston, British Library, MS 48364.

> "The Major-General commanding congratulates the Division on being selected for an enterprise the success of which will have a decisive effect on the war.
>
> "The eyes of the world are upon us, and your deeds will live in history.
>
> "To us is now given an opportunity of avenging our friends and relatives who have fallen in France and Flanders. Our comrades there willingly gave their lives in thousands and tens of thousands for our King and country, and by their glorious courage and dogged tenacity they defeated the invaders and broke the German offensive.
>
> "We, also, must be prepared to suffer hardships, privations, thirst and heavy losses by bullets, by shells, by mines, by drowning. But if each man feels, as is true, that on him individually, however small or however great his task, rests the success or failure of the expedition, and therefore the honour of the Empire and welfare of his own folk at home, we are certain to win through to a glorious victory.
>
> "In Nelson's time it was England, now it is the whole British Empire, which expects that each man of us will do his duty."[169]
>
> MAJOR-GENERAL SIR AYLMER HUNTER-WESTON,
> COMMANDING 29TH DIVISION, HMT *ANDANIA*.

Sinclair-Mclagan, the commander of the Australian covering force, also took the opportunity to share a final message to his men ahead of their landing.

> "It is necessary that you should understand that we are about to carry out a most difficult operation, viz., landing on an enemy coast in the face of opposition. Such an operation requires complete harmony of working between the navy and army and unhesitating and immediate compliance with all orders and instructions.
>
> "You have been selected by the divisional commander as the covering force, a high honour which we must all do our best to justify. We must be successful at any cost. Whatever footing we get on land must be held on to and improved by pushing on to our objective, the covering position which we must get to as rapidly as possible, and once obtained must be held at all costs and even to the last man.
>
> "In an operation of this kind there is no going back. We shall be reinforced as the navy can land troops, and meantime 'Forward' is the word, until on to our position, when 'Hang on' is what we have to do, until sufficient troops and guns are landed to enable up to push on.
>
> "We must be careful not to give the enemy a chance of any kind; no smoking or lights or noise from midnight onwards till after daylight. Take every chance of reorganising (under cover if possible). Attacks must be as rapid as the ground will allow. You will probably have to drop your packs; but carry tools forward as far as you can, it may mean saving many lives later in the day. Until broad daylight the bayonet is your weapon, and when you charge do so in as good a line as possible; one or two pieces of good bayonet work now may stand us all in good stead later on.

---

169 Hunter-Weston quoted in Foster, Rev. H. C. Foster, *At Antwerp and The Dardanelles*, pp. 76–77, Mills and Boon (London) 1918.

"Every man must keep his eyes skinned and help his officers and N.C.O.'s to the utmost by reporting quickly things seen. Look out for your flanks. After taking a charger out, shut the cartridge pocket. Once ashore don't be caught without a charger in the magazine. Look after each cartridge as if it were a ten pound note.

"Good fire orders, direction, control, and discipline will make the enemy respect your powers, and give us all an easier task in the long run. Wild firing will only encourage the enemy. Keep your food and water very carefully. We don't know when we shall get any more...

"We must expect to be shelled when in our positions, but remember, that is part of this game of war, and we must stick it no matter what the fire. One thing I want you to remember all through this campaigning work is this, and it is very important; you may get orders to do something which appears in your positions is the wrong thing to do, and perhaps a mad enterprise. Do not cavil at it, but carry it out whole-heartedly and with absolute faith in your leaders, because we are after all only a very small piece on the board. Some pieces have often to be sacrificed to win the game, and after all it is to win the game that we are here."[170]

COLONEL EWEN SINCLAIR-MACLAGAN, 3RD BDE, 1ST AUSTRALIAN DIVISION.

Events would show the extent to which Sinclair-Maclagan would abide by his own orders. Meanwhile, Ashmead-Bartlett claimed to have already written off any chances for success.

"This afternoon I went on shore for the last time and had a farewell walk through the little village and out into the country accompanied by the parson from the London. On returning to my launch I found Sir Ian Hamilton on the quay and had a talk with him. He seemed to be extremely confident, in excellent spirits, and even told me a funny story about some Australians. I asked him how he reckoned his chances, and he replied that he thought they were very good. He said his Intelligence Department considered the Turks had about 35,000 men on the Gallipoli Peninsula. Personally I was far from sharing his confidence and ventured to remark, on saying good-bye, "General, the task ahead is one of the most difficult that has ever been undertaken, and the Expedition can only succeed if you have sufficient troops to push right inland at the start, and if the Government keeps you well supplied with reinforcements."

"By this time I had become convinced that the Expedition was almost certainly doomed to failure."[171]

ELLIS ASMEAD-BARTLETT, WAR CORRESPONDENT.

In the background, others were having to deal with reality of the improvised arrangements and the complete absence of adequate infrastructure for the whole enterprise.

"With reference to the use of the "Umsinga" as an issuing Store Ship, I have to point out the situation that has arisen from her not having been to Alexandria to discharge stores required there and to be properly arranged as a ship to issue from at sea.

---

170   *Western Mail* (Perth, Western Australia), 13th January 1938.
171   Ashmead-Bartlett, *Uncensored Dardanelles*, pp. 37–38.

"In her present condition she cannot fulfil the purpose for which she was sent out to be used, as there is so much cargo on her that it is out of the question to make detailed issues of any of the stores on board, and it is absolytely [sic] assential [sic] before she leaves Lemnos that a very considerable quantity of stores, not less than 200 tons and [if] possible more, should be removed off her. I understand it is desired that this ship should follow the 29th Division, and another one the Aus: & N.Z. Army Corps; at present no second store ship is available, but if the "Anglo-Egyptian" could be got alongside the "Umsinga" in this harbour, there is some hold space on her, and it might be practicable to so arrange stores that the difficult operation of making detailed issues off a ship at sea could be carried out in a fairly satisfactory manner.

"It is however essential that before either ship can be used for this purpose that they should be alongside each other under favourable conditions for at least 3 days before either can be got ready to leave this harbour. Until it is tried, it is not practicable to say how long it will take, and it may require longer, but unless these 2 ships can be worked under these conditions before leaving Lemnos, there is a grave risk of serious difficulty arising over the issue of Ordnance stores."[172]

COL. PERCY BAINBRIDGE, DEPUTY DIRECTOR ORDNANCE SERVICES, 29TH DIVISION.

22 APRIL 1915

ENGLAND

Churchill betrayed signs of stress as the date of the landings neared.

"Winston spoke freely regarding the Dardanelles Expedition. He again referred to the imminence of a great battle. He said the Turks had now fortified positions formerly unfortified and that the lives of many of our men might be lost in consequence. He said that his calculations had in a measure been put out by the mobility of the Turkish guns, which enabled them to train readily on the ships which were confined in a narrow sea. Originally it had been thought that the attack might be successful by sea alone, but this had proved impracticable. This war could not be won by sitting still, as some people thought. Offensive operations were necessary, but before deciding on a plan which might, and probably would, result in the loss of many valuable lives, Winston had given the whole subject most anxious consideration. If the operation is successful, its effects will be most important. They are worth the risk. It is better to risk lives in this way than to allow the war to drag on indefinitely. If the operation is unsuccessful, Winston recognises that the effect on his career may be serious. "They may get rid of me," he said. "If they do, I cannot help it. I shall have done my best. My regiment is awaiting me." He seemed calm, but no doubt is feeling the strain."[173]

SIR GEORGE RIDDELL, NEWSPAPER PROPRIETOR.

---

172 29th Division General Staff War Diary, TNA WO 95/4305.
173 Riddell, Sir George, *Lord Riddell's War Diary 1914–1918*, p. 80, Ivor Nicholson & Watson (London) 1933.

LEMNOS

Churchill was not the only one organising his thoughts at this time. Some, perhaps seeking to reassure themselves, chose to accentuate the positive.

> "The Peninsular, which before our guns warned the Turks of their danger, was miserably defended has now been made as strong as German brains & Turkish hands can make it. But our troops are glorious & will, I am confident, carry all before them. The Australians & New Zealanders impress me as magnificent troops, & as for the 29th Division, it is undoubtedly the finest Division that any country has at present in the field. It is indeed the only Division that has not been knocked to pieces & that is still a Division of Regulars. It is not a fully trained Division… had little opportunity to for practice in cooperation. The machine therefore sometimes creaks a little at the joints. But it has come on wonderfully in the last week or two & as to its great fighting power there can be no doubt."[174]

MAJOR-GENERAL SIR AYLMER HUNTER-WESTON, COMMANDING 29TH DIVISION.

> "Lord Kitchener sent us the other day an account of the fighting at Busorah [Basra], preparing us for what was before us. The Turks had fought desperately, were well trained, and well led, and could only be turned out of their trenches at the point of the bayonet."[175]

LT GEORGE DAVIDSON, 89TH (1ST HIGHLAND) FIELD AMBULANCE, RAMC, HMT *AUSONIA*.

> "To-morrow is England's great day, and within a few hours of its close one of the greatest undertakings in the history of the world will either be a success or a dismal failure. If it is successful, Australia's name will be before the world for a long time, and the 3rd Infantry Brigade, commanded by Col. E. G. Sinclair-MacLagan, D.S.O. will have won laurels which will live for ever."[176]

MAJOR CHARLES BRAND, BRIGADE-MAJOR, 3RD BDE, 1ST AUSTRALIAN DIVISION.

> "The prevailing opinion amongst the Tommies is that the landing will be a soft job, with Queen Bess and her sisters pounding the land defences with shells. …The feeling prevails that when once the landing is effected [the Turks] will cave in, and that will have a great influence on the duration of the war. But a Scotsman said to me to-day, "Remember, Kitchener said 'A three years' war.'"[177]

CAPT. JOHN GILLAM, DIVISIONAL TRAIN, ARMY SERVICE CORPS, 29TH DIVISION.

> "I have a special job as soon as we land. I am one of the eight wire-cutters to the company. They called for volunteers, and I went as one. We will either get a chance of distinction or an early exit from all troubles, the main of which is carrying our

---

174 Extract from letter, 21st April 1915, Maj. Gen. Hunter-Weston, British Library, MS 48364.
175 Davidson, *The Incomparable 29th and the River Clyde*, pp. 38–39.
176 *The Observer* (Adelaide, South Australia), 12th June 1915.
177 Gillam, *Gallipoli Diary*, pp. 28–29.

equipment. Our spies tell us that the wire entanglements between where we land and Constantinople are something marvellous."[178]

PTE COLIN GLASGOW, 12TH BATTALION, AIF,[179] HMT *DEVANHA*.

MESOPOTAMIA

More news from those already in action against the Turks echoed earlier accounts: the fighting had been hard but Turkish resistance had been overcome and they were forced into retreat. But those in Mesopotamia looked to the coming assault upon the Dardanelles to bring things to a conclusion.

"You will have seen from the papers that we have had some hard fighting here lately. About 15,000 regular Turks and 10,000 Arabs came down to mop us up last Sunday week. After three days of hard fighting we managed to turn the tables, and our small force of 5,000 or 6,000 sent these gents in hasty retreat. It was big odds against us, and we had a pretty tough time of it, but we won handsomely in the end, and as far as can be ascertained there are now only dead Turks within 100 miles of us, the live ones finding discretion the better part of valor. Our losses were pretty heavy, but not excessive. My own regiment had one British officer killed and two wounded. We had two native officers killed and three wounded, and some 20 men killed and over 80 wounded. This is a big list when our strength is only about 610. I am glad to say I came through the thick of it untouched. The ordeal was not pleasant, but at the time one does not think much of these things. In fact, one is hardly conscious of bullets flying in thousands when one is busy dressing the wounded. In the big battle of Bargesick Woods on April 14 I was in the firing line all the time. Men were killed and wounded on all sides. We are now hoping that the Dardanelles will be forced and end this part of the show."[180]

CAPT. JOHN HARPER-NELSON, INDIAN MEDICAL SERVICE, ATTD 119TH INFANTRY (THE MOOLTAN REGT), 17TH BDE, 6TH (POONA) DIVISION.

23 APRIL 1915

LEMNOS

"The Scheme is audaciously bold, and I think we have done all we can to help to make it a success. But the authorities at home! They seem to think that it is a picnic party for all the assistance they have given us. Of course the initial mistake was bombarding before we had an army to land. Had we had troops to pour in after the first bombardment the whole thing would have been finished and done in a very short time. But

---

178 *The North Western Advocate and the Emu Bay Times* (Tasmania), 8th July 1915.
179 Pte Colin Glasgow, 12th Bn, AIF, was killed in action on 25th April 1915. Commemorated on the Lone Pine Memorial, he was the son of Mabel Fanny Myer (formerly Glasgow) and the late Lewis Henry Glasgow, of Casino, New South Wales.
180 *The Adelaide Advertiser* (South Australia), 24th June 1915.

we hadn't. The ignorance of the Admiralty is nothing less than criminal. Then this hastily devised plan of sending troops out to Lemnos without an organization or staff to do it. The whole of the troops had to be unpacked and re-packed again, and this had to be done in Alexandria. Then the whole thing had to be reorganized by me out here, and I had no staff! By beseeching and telegraphing we got 12 officers out from England, and we should have had 56! and everything else in the same proportion... However, that of it is all finished, and now we are going to utilize what we have created. But alas! we have given the Turks or rather the Germans, time to prepare and the landing will be a very different thing now to what it would have been a month ago.... Never in the history of the world has such an expedition sailed — never has a big campaign been so hastily organized and got together, and never has such an undertaking got so little consideration given it from home. I believe we shall succeed simply because everybody is determined that it *must*. There is no alternative."[181]

REAR-ADMIRAL ROSSLYN WEMYSS, RN, HMS *EURYALUS*.

"On the 23rd the troopships began to get a move on, and the send-off we gave to the Australians and Fusiliers was worth coming out here to hear. The most comical part of the business was when we got under weigh, and passed a Russian cruiser, our band played a Scotch tune, and broke into an Irish air as we passed a French battleship."[182]

STOKER PETTY OFFICER CHARLES COOK, RN, HMS *IMPLACABLE*.

"Boys! We have been instructed along with the 9th and 10th Battalions to form the covering party for the Australian landing on the Gallipoli Peninsula. 'A' and 'C' Companies will go from here on HMS *London*. 'B' and 'D' Companies on destroyers and the landing will be effected in the way that we have been lately practising. The position of honour has been assigned to us in being thus chosen as vanguard for one of the most daring enterprises in history.

"Boys! the General informs me that it will take several battleships and destroyers to carry our brigade to Gallipoli; a barge will be sufficient to take us home again! Once ashore, we must make good our footing. The navy guarantees to land us, but it refuses to take us off again!"[183]

LT-COL. JAMES LYON JOHNSTON, 11TH BATTALION, 3RD BDE, 1ST AUSTRALIAN DIVISION.

"General Hunter-Weston gave an address to all the senior officers of his Division, in which he stated, amongst many other things, that the weight and moral, as well as material effect, of the naval preliminary bombardment would astonish the world, and especially the Turk to such an extent that his resistance would be paralysed. In fact I think we, most of us, gathered from his lecture that we could hardly hope to come to grips with the Turks much before we had reached Constantinople."[184]

BRIG.-GENERAL WILLIAM MARSHALL, COMMANDING 87TH BRIGADE, 29TH DIVISION.

181 Wemyss quoted in Wemyss, Lady Wester, *The Life and Letters of Lord Wester Wemyss*, p. 215.
182 *Cambridge Independent Press*, 4th June 1915.
183 *Western Mail* (Perth, Western Australia), 13th January 1938.
184 Marshall, Sir William, *Memories of Four Fronts*, p. 53, Ernest Benn Ltd. (London) 1929.

> "A different atmosphere pervades our ship to-day, a feeling of strain and anxiety is more or less on every mind, not that it would be apparent to an outsider except in a case or two. Bad news has leaked in all the time from the navy and our airmen, all the time this is getting worse, such as the account that Gallipoli swarms with well-armed Turks, wire entanglements of great breadth and height everywhere, and, of course, trenches… and our having to face their well-defended positions in open boats is not altogether comforting, and naturally all feel a bit anxious…
>
> "The particular part of the coast on which I land with the 89th Field Ambulance is a short way west of Sedd-el-Bahr, landing in the collier "River Clyde," on which there will be a force of 2100. I have already spoken about this boat. From what is going on I will be surprised if we do not leave Lemnos to-night."[185]

LT GEORGE DAVIDSON, 89TH (1ST HIGHLAND) FIELD AMBULANCE, RAMC, HMT *AUSONIA*.

> "The first boat of the fleet leaves, named the *River Clyde*, an old tramp steamer, painted khaki. She contains the Dublin and Munster Fusiliers. Fore and aft on starboard and port the sides are cut away, but fastened like doors. She will be beached at "V" Beach, and immediately that is over, her sides will be opened and the troops aboard will swarm out on to the shore. Good luck to those on board! She slowly passes the battleships, and turning round the boom, is soon out of sight."[186]

CAPT. JOHN GILLAM, DIVISIONAL TRAIN, ARMY SERVICE CORPS, 29TH DIVISION.

> "I shall be glad to get on land again, though we are expecting to have a very hot time when we do land. The Turks have everything nicely ready for us. Our regiment has been picked out as one of the first to land. This is really an honour that I suppose we should be proud of; so we shall have to do the best we can and trust to luck. You can bet I shall do my best to pull through safely. You must all enjoy yourselves and not bother about me."[187]

PTE CHARLES ALDERTON, 1ST BN LANCASHIRE FUSILIERS, 86TH BDE, 29TH DIVISION.[188]

As the troopships and warships made their way towards the peninsula, the Australian submarine *AE2* was heading for the Dardanelles.

> "Aboard the *Queen Elizabeth* I was received by the Chief of Staff, Commodore [Roger] Keyes. If I still believed it might be possible to dive through the Dardanelles, I would be permitted to try. He took me to the Admiral, who was kindness itself. Without belittling the difficulties he simply asked how we proposed to overcome them. He found it difficult himself to believe the feat was possible, but its military value would be so great it must be tried. If we got through the other boats would immediately be sent to follow. Finally, wishing us luck, he concluded: "If you succeed there is no calculating the result it will cause, and it may well be that you will have done more to finish the war than any other act accomplished."

---

185 Davidson, *The Incomparable 29th and the River Clyde*, pp. 40–41.
186 Gillam, *Gallipoli Diary*, p. 29.
187 *Cambridge Independent Press*, 7th January 1916.
188 Alderton was killed on 25th April 1915. Commemorated on the Helles Memorial, he was the 25 year-old son of Charles and Lucy Alderton, of 42 Great Eastern Street, Cambridge.

"I hurried back to *AE2*, where we immediately commenced to take in provisions and prepare for sea. Two hours later we were threading our way out through the crowded harbour. A practice dive at sea, and so to a convenient anchorage to await the fall of night."[189]

LT-CMDR HENRY STOKER, RN, HMAS *AE2*.

24 APRIL 1915

ASIA MINOR

"On April 24th we had a grand maneuver exercise with the 11th Division on the Asiatic side, which included the possibility of an enemy landing in Little Basika Bay. Late that afternoon I returned to Gallipoli."[190]

GENERAL OTTO LIMAN VON SANDERS, COMMANDING OTTOMAN 5TH ARMY.

LEMNOS

"Towards evening the… Admiral sent for me… There was… to be a minor alteration to the orders. Instead of attempting to pass Chanak without being seen by the enemy, we were to attack and sink, if possible, any mine-dropping ships found in the Narrows — if we got there. The reason was obvious. The morrow, Sunday 25th, was the day of disembarkation of our attacking army. The transports were to approach the shore at daylight, and while the landing of troops was carried on, the fleet would engage the forts and batteries. For this purpose some battleships would be operating in the entrance of the Strait, and therefore it was to be expected that many floating mines would be launched in the Narrows. So *AE2* must endeavour to hamper the movements of any mine-dropping ships and, in the words of the Chief of Staff, "Generally run amuck" off Chanak."[191]

LT-CMDR HENRY STOKER, RN, HMAS *AE2*.

HEADING TO HELLES

"When we awoke… we were lying off the isle of Tenedos, surrounded by battleships, cruisers and destroyers. To our dismay a stiff breeze sprung up, and at one time it looked as if operations would have to be postponed; at about one o'clock, however, the wind dropped, and it grew wonderfully calm. At 3 p.m. a trawler came alongside to take some of the Dublin Fusiliers to the s.s. "River Clyde"…

"A few hours later, two companies of the Dublins that remained were taken off in a trawler to one of our battleships for the landing. Father Finn went with these latter

---

189 Stoker, Commander H.G., *Straws in the Wind*, p. 100, Herbert Jenkins Ltd. (London) 1925.
190 Von Sanders, Liman, *The Dardanelles Campaign*, p. 12, The Engineer School (Fort Humphreys, Virginia) 1931.
191 Stoker, *Straws in the Wind*, pp. 105–106.

companies. He shook my hand warmly as he said "Good-bye," and gave me a small medal of "Our Lady of Mount Carmel," such as he had been distributing earlier in the day to his men. "Take this and wear it," he said, "and may it bring you good fortune, and take you safely home." And with a wave of the hand, he ran down the gangway on to the trawler."[192]

CHAPLAIN HENRY FOSTER, RN, 2ND NAVAL BRIGADE, HMT *AUSONIA*.

"Arrived on coal boat [*River Clyde*] at 6.30. Place in stern fitted up for officers' supper; two lime barrels and a few rough boards form table: whisky: tinned meat: biscuits: 2200 of us on board: all happy and fit. We start in two hours: only 12 or 13 miles to go: then anchor 1½ miles from land and wait for daylight and bombardment; then at proper moment rush in… Some sleeping or pretending; others smoking; I doing latter and sitting on board after trying to snooze with head on a big box and less high one in small of back; but too uncomfortable for anything, so whipped out my "bookie" and scribbled."[193]

LT GEORGE DAVIDSON, 89TH (1ST HIGHLAND) FIELD AMBULANCE, RAMC, SS *RIVER CLYDE*.

"When we left L[emnos] we went as far as [Tenedos] where two companies and headquarters changed over to a collier for the landing. She was not comfortable."[194]

CAPT. & ADJ. GEORGE REID, 2ND BN HAMPSHIRE REGT, 88TH BDE, 29TH DIVISION.

"Half Battalion Royal Fusiliers and Headquarters went on board HMS *Implacable* about 7 p.m., from which ship we had been practising getting into boats, and so on; the other Half Battalion, under Brandreth,[195] spent the night on two Fleet minesweepers. At about 10.30 p.m. the brigade and warships all sailed for the peninsula, arriving there by night."[196]

LT-COL. HENRY NEWENHAM, 2ND BN ROYAL FUSILIERS, 86TH BDE, 29TH DIVISION.

"About 4 p.m… we started getting the troops on board, 750 altogether, just about double our complement, so you can imagine what a squash there was. We all gave up our cabins to the officers of course, as it was the last decent night's rest they would have for some time. We left Tenedos at 9.30 p.m. with four tows of boats astern and with two trawlers and two fleet sweepers (ships of about 300 tons or so) following us, each with one tow of boats astern of it, the idea being that we were to land the lot aboard of us first and then the boats would go back for the rest of the troops of the

---

192   Foster, *At Antwerp and The Dardanelles*, pp. 79–80.
193   Davidson, *The Incomparable 29th and the River Clyde*, p. 41.
194   *Stratford-upon-Avon Herald*, 28th May 1915.
195   Major Lyall Brandreth, 2nd Bn Royal Fusiliers, was killed in action on 6th June 1915. Commemorated on the Helles Memorial, he was the 46 year-old son of the late Admiral Sir Thomas Brandreth, KCB; husband of Clare Rosabelle Briggs Brandreth, of "Fairmount," Hillcrest Road, Hythe, Kent.
196   Lt-Col. Henry Edward Berkley Newenham quoted in Creighton, Oswin, *With the Twenty-Ninth Division in Gallipoli*, pp. 55, Longmans, Green & Co. (London) 1916.

covering party who would have embarked in the fleet sweepers from the transports by then."[197]

LT FRANK STEPHENSON, RN, HMS *IMPLACABLE*.

"Another bright day. Some transports and battleships leaving harbour. Issue extra days' rations to troops on board, which makes four days' that they will have to carry. Their packs and equipment now equal sixty pounds. How they will fight to-morrow beats me. I tried a pack on and was astonished at its weight. We have left harbour and are steaming for the scene of the great adventure. Hope we shall not meet a submarine or drifting mines. Have spent the evening with some young officers of the Essex. They all seem a trifle nervous, yet brave and cheery."[198]

CAPT. JOHN GILLAM, DIVISIONAL TRAIN, ARMY SERVICE CORPS, HMT *DONGOLA*.

"Tomorrow morning the bombardment of the South end of the Gallipoli Peninsular will commence & my fine fellows will be landed at 5.30 a.m. if the fates permit. A most dangerous & difficult enterprise but a glorious one & I believe we shall be successful. The Australians (followed by the New Zealanders) land first before us some dozen miles further north. The Naval Division yet further North & the French away to the East. By evening tomorrow the papers of History will have been marked heavily for good or ill for the good of England…

"The view tonight at sunset was beautiful to see, with the 'Queen Elizabeth' steaming away right into the Western red. The entrance to the Dardanelles & the height of Achi Baba, the Mecca of my morrow's hopes, just seen in the centre of the view to the north…

"I have no anxieties, no troubled fears as to tomorrow. My mind is absolutely at rest. I have done my part. Have issued & explained my orders. I have 'enthused' my officers & men. The rest is in God's hands, & I go to sleep now with my mind at rest."[199]

MAJOR-GENERAL SIR AYLMER HUNTER-WESTON, COMMANDING 29TH DIVISION.

HEADING TO ANZAC

Aboard HMS *Prince of Wales*, Sinclair-Mclagan, in conversation with Bridges, his divisional commander, did not exactly exude confidence. And he was not alone in that.

"I do thank you for the great honour of having this job to do with my Brigade. But if we find the Turks holding the ridges in any strength, I honestly don't think you'll ever see the Third Brigade again."[200]

COLONEL EWEN SINCLAIR-MACLAGAN, 3RD BDE, 1ST AUSTRALIAN DIVISION.

---

197 *The Press* (New Zealand), 23rd April 1938.
198 Gillam, *Gallipoli Diary*, p. 30.
199 Diary entry, 24th April 1915, Maj. Gen. Hunter-Weston, British Library, MS 48364.
200 Sinclair-Maclagan quoted in Bean, *The Story of ANZAC*, vol. 1, p. 244.

"As the *Minnewaska*, conveying the First Divisional Artillery Staff and the First Infantry Battalion, steamed slowly out of Lemnos Harbour on the evening of April 24, 1915, I confess my feelings were not too cheerful. I fully realised all that could happen in getting the troops and guns ashore, and how difficult it might be to establish a footing there. The fact that the day of the "Landing" was also the silver anniversary of my wedding did not help me much — certainly there would be a great fireworks display, but there was also the probability of my wife becoming a widow."[201]

BRIG.-GEN. JOSEPH HOBBS, COMMANDING 1ST AUSTRALIAN DIV. ARTILLERY.

Some of the men, if they had any fears, chose not to share them with their friends.

"In the afternoon Brigade Major Brand addressed our Battalion, when he told us in quiet terms that the Third Brigade had the honour of being selected as the landing party, to be reinforced later in the day by the Second and First Brigade, comprised [of] the 9th, 10th, 11th, and 12th Battalions, and of these our Battalion was chosen as the first to step ashore. Brigade Major Brand said that to successfully take this position there would need to be a great sacrifice of men, and he wanted us to fill up the gaps in our ranks smartly as fast as they should occur…

"We were shown a plan of the attack and given our respective duties, and dismissed till later. I thought the lads would naturally feel a bit glum when they realised that many of their number would never see daylight after April 25. However, such was not the case, as every one was singing, playing cards, or otherwise amusing themselves. In fact, one would think from the joviality of the throng that they were going to a picnic excursion."[202]

PTE CHARLES PEINIGER, 9TH BATTALION, 3RD BDE, 1ST AUSTRALIAN DIVISION.

Sailors aboard HMS *Queen* gave their guests a warm welcome. One recognised an old friend.

"The Scouts… — I am in them, good luck! — were given the order to prepare to leave the troopship at 10:45 a.m., and to carry three days' rations and water. We paraded at 10.30, and immediately started to embark into the torpedo boat destroyer which was to take us to the cruiser [sic]. The cruiser 'up-anchored' at noon, and proceeded slowly out to sea at the head of a line of five others. The sailors on board the cruiser could not do enough for us, they gave us tobacco, cigarettes, and pipes, till we were overloaded. A number of us were taken by the bluejackets and given dinner on their own accounts. I had the jolliest time for months, with soup, and, of course, choruses, and a sailor's dance or two thrown in, and a gramophone. By good luck I was able to get a hot bath, the first for a month. The chief gunner at the big gun turrets (12 in. guns) took us over the whole of the working of the big guns, and by jove it was marvellous!"[203]

PTE GEOFFREY PRESTON, 9TH BATTALION, 3RD BDE, 1ST AUSTRALIAN DIVISION.

---

201 *Western Mail* (Perth, Western Australia), 26th April 1934.
202 *The Brisbane Courier* (Queensland), 28th June 1915.
203 *The Brisbane Courier* (Queensland), 23rd June 1915.

"I... had the pleasure of meeting a Stourbridge man, P. Young,[204] of Wollaston, who emigrated to Australia, and since the outbreak of war joined the Australian Imperial Volunteers. Our meeting was an unexpected one... We had to land the Australian troops [and they] came on board... and we made them as happy and comfortable as possible, sharing tobacco and cigarettes, and giving each other mementoes, such as military badges, cap ribbons, with the ship's name on, knives, etc. Then we had to say good bye and good luck."[205]

L/STO. CHARLES CHILLINGSWORTH, RN, HMS QUEEN.

"At 3 p.m. our boats brought the 500 men of the 11th Australian Infantry on board for the last time. Numbered squares had been painted in white on the quarter-deck, and on each of these a company fell in. The men were then dismissed and made their way forward to the mess decks. The hospitable British tars handed over their limited accommodation to the newcomers, who were to bear the brunt of the attack. At 5 p.m., our force, the Second Division of the fleet, consisting of the *Queen*, *Prince of Wales*, *London*, and *Majestic*, with four transports bearing troops, and the covering ships *Triumph*, *Bacchante*, and *Prince George*, slowly steamed out of the bay. As we passed through the long lines of waiting transports, our bands played the national anthems of all the Allies, and deafening cheers greeted our departure."[206]

ELLIS ASHMEAD-BARTLETT, WAR CORRESPONDENT.

"On Saturday a portion of Battalion headquarters and A and C. Companies embarked on HMS *London* and together with the other battleships set sail for the scene of action. There was not the slightest indication of excitement amongst the men who were going out to take part in one of the most difficult operations in warfare — to effect a landing under hostile fire on an enemy's coast. They were all as cool and collected as if going on an ordinary parade. We all got settled down on the *London*, where we were treated most royally by officers and men. They did everything it was possible to do to make us comfortable during the journey, and all of us who were privileged to make that trip on the *London* will always remember with gratitude the great kindness extended to us by all ranks. While on the *London* we met Fleet Surgeon McMillan, who is a brother of Dr. McMillan, of Kalgoorlie, and he could not do enough for us especially those of us from the Goldfields, Having been treated thus sumptuously we turned in to have a few hours rest before the hard work we all knew to be ahead of us."[207]

LT-COL. JAMES LYON JOHNSTON, 11TH BATTALION, AIF, HMS LONDON.

"The West Australian men were transferred from the transport *Suffolk* to the battleship *London* on Saturday morning, April 24, and taken to the Dardanelles that night. Whilst on the battleship the navy fellows treated us like long-lost brothers, and fed us with the best that the boat could produce. The last feed I had before the battle

---

204　Pte Percival Young, 9th Bn, AIF, a sawyer originally from Stourbridge, Worcestershire. He served on the peninsula until evacuated due to dysentery on 12th August 1915.
205　*County Express for Worcestershire and Staffordshire*, 22nd May 1915.
206　Ashmead-Bartlett, *Uncensored Dardanelles*, p. 44.
207　*The Sun* (Kalgoorlie, (Western Australia), 11th June 1915.

in the morning was a huge plateful of meat pie somewhere about midnight. I was blown up like a poisoned pup, and had a difficulty to fasten my equipment belt. We were to land with our full equipment on, so, with 250 rounds of ammunition, we had about 70 lb. on our backs."[208]

PTE LAWRENCE KUHLMANN, 11TH BATTALION, AIF, HMS *LONDON*.

"[H]alf our battalion embarked on destroyers and were taken to the HMS *London* — a big pre-dreadnought with four 12 inch guns — and started out — all the battleships and cruisers in line — the transports (with the remainder of the troops) following… We buzzing about like bees. The *Queen* cruised about the Aegean Sea all the afternoon and at dark started slowly up the Gulf of Saros — the gulf on the west of the Gallipoli Peninsula (the Dardanelles being on the east). The officers of the *London* were awfully good to us — they all gave up their beds to us and fed us up like fighting cocks. If you ever hear anyone say anything derogatory about the Navy in future just plug him and explain it from me. They really are the finest lot of men I have ever met."[209]

CAPT. EDWARD BRENNAN, AAMC, ATTD 11TH BATTALION, AIF, HMS *LONDON*.

"Well, we have finished with Lemnos now, and I will make an end. I am writing in the twilight, and a keen chill wind is making this exposed seat on the boat deck a little too uncomfortable. Love[210] and myself have been in the best of trim since we began soldiering, having dodged the colds and chill of Mena and Lemnos.

"In a few hours we shall be fighting a more tangible foe than ill-health. Let us hope our luck will hold good on the peninsula of Gallipoli. At 9.30 p.m. we go on board the Destroyers, and will be taken on them as near our landing place as the depth of water will allow. Then we take the rowing boats, and then ashore. We shall have an imposing reception. For the rest one must take his chance, while doing his little bit."[211]

L/CPL MAURICE O'DONOHUE, 11TH BATTALION, AIF, HMAT *SUFFOLK*.[212]

"After waiting all these months, at last we have made a decided move — I guess that a good many of the 11th Battalion have seen the sun rise for the last time. — We have a nice little load to carry with us — which weighs eighty pounds, so I guess we have our work cut out without fighting. Things are moving — so I must stop. If this should happen to be the last, which I hope not, you can take it as a goodbye letter. We are under steam! Hoo Roo!"[213]

PTE JAMES CARRINGTON, 11TH BATTALION, AIF, HMAT *SUFFOLK*.[214]

208   *Kalgoorlie Miner* (Western Australia), 31st July 1915.
209   *Great Southern Star* (Leongatha, Victoria), 20th July 1915.
210   Pte William Love, 11th Battalion, AIF, was wounded on 25th April 1915. The former teacher was invalided to Australia, leaving England on 17th March 1916.
211   *The Northam Advertiser* (Western Australia), 14th July 1915.
212   O'Donohue, a former teacher, was killed in action on 25th April 1915. Buried in Baby 700 Cemetery, he was the son of James O'Donohue, of Boundary Road, Kalgoorlie, Western Australia.
213   *Western Mail* (Perth, Western Australia), 20th January 1938.
214   Pte James Edward Carrington, 11th Bn, AIF, was killed in action on 25th April 1915. Commemorated on the Lone Pine Memorial, he was the 33 year-old son of James and Hannah Carrington, of 25 South Street, Fremantle, Western Australia.

"When we got well out to sea, and the troops were all brought up to the foredeck, and our colonel (R. W. Owen) gave us some good, sound, solid advice, and also read a letter which the Brigadier (Col. M'Laurin, of the 1st Infantry Brigade) had sent him the previous day. It read as follows: 'As your Brigadier I ask each and every one of you to do your duty for your King and your Commonwealth of Australia. Think of the honor and glory this will be to Australia if we are successful in our attack upon the Dardanelles. I know that every man will, therefore, do his duty and strive to win the attack, and gain a great victory that will ever remain in the history of Australia. I wish you all good luck.' Then our Colonel, whom we all love as a son loves his mother, spoke as follows: 'Men, for the past eight months we have been training vigorously for the time when we should be called upon to answer the call, and be fit, as every one of you are, for war. In that space we have each had time enough to know each other. I know you and you know me. You have never failed me in anything yet, and I, doing my part, have never failed you. Therefore, both you and I can go into action, each knowing that we will act as one man. You have four good company commanders. Take notice of what they say, and act upon their advice, because you know they are acting on my commands. Do this, and I know that the third battalion will in no way be far behind. I am proud of all my men and I hope you have a good opinion of me. Good-bye and good luck.'

"It would be impossible to describe the enthusiasm that was put into those mighty cheers that were called for the 'dad' and our colonel. I think the boat must have rocked a bit, for you can imagine cheers coming from a thousand men at once and each doing his utmost to roar louder than his mate."[215]

PTE JOHN COLLESS, 3RD BATTALION, AIF, HMT *DERRFLINGER*.

"At last we make our final move, and very soon we will have begun to do what we came away for, and have waited so long to do. While you are in church to-morrow thinking of us, we may be needing all your prayers, as it is either going to be a hard fight or an easy walk in; but everything is ready and everyone quietly confident of success. It is going to be Australia's chance, and she makes a tradition out of this that she must always look back on. God grant it will be a great one. The importance of this alone seems stupendous to Australia, while the effect of success on the war itself will be even greater. The battalion (seventh) is up in the front again with an important job before it, and there seems no reason why it should not do well."[216]

LT ALAN HENDERSON, 7TH BATTALION, AIF, HMT *GALEKA*.

"The "Clan" anchored to the north of Lemnos until after dark. This was the eve of the attack, everybody was in high spirits, and the orderly officers (Lieutenant McLeod[217] and myself were detailed for this duty during the whole voyage) had a hopeless task in getting quietness after "Lights Out"; somebody imitated a rooster crowing and it was taken up by both troop decks, and for a couple of hours we had to put up with

---

215   *Mudgee Guardian and North West Representative* (New South Wales) 28th October 1915.
216   *The Argus* (Melbourne, Victoria), 24th June 1915.
217   2/Lt Maurice McLeod, 8th Bn, AIF, was killed in action on 25th April 1915. Buried in Shell Green Cemetery, he was the 21 year-old son of Mary Jane McLeod, of 405, Gregory Street, Ballarat, Victoria, Australia, and the late Sydney Stears McLeod.

dogs (all sizes and sorts) barking, hens cackling, as well as the natural call of horses, cattle, pigs, sheep, turkeys, in fact every call that could be thought of, and all done most realistically. It was funny beyond words."[218]

CAPT. WILLIAM YATES, 8TH BATTALION, AIF, HMAT *CLAN MACGILLIVRAY*.

"We moved last evening into the outer harbour, our band playing… the decks and rigging swarming with men cheering and being cheered by the battleships and transports, which we passed on each side. On one big ship was painted large the legend, 'To Constantinople.'

"Later we had what is termed 'collision drill.' At the 'alarm' everybody lined up on deck, and those for whom there was no boating accommodation were briefly addressed by the officer commanding the troops on board. He emphasised the small chance there was of any life being lost in view of the number of ships which were with us, and which, in case of accident, would be able to reach us within five minutes. The colonel mentioned that the ship's boats, as soon as they were full, would push off and get away, so relieving any congestion there might be. The remainder of the men he urged, above all, to keep cool and not lose their heads in the possible contingency of our being torpedoed or mined during the night…

"Orders have been read to us concerning the treatment to be accorded the inhabitants of the country when we land. In dealing with the women, special care has to be taken. Should we have any instructions we must communicate them if possible through the men folk. If we have occasion to enter a house we must knock first to give the women time to veil themselves; and on no occasions where it is not all absolute necessity must we speak to, or take the slightest notice of, any women at all. Our platoon commander has also told us how to ask for water in Turkish (ya arkadash). Translated, the words mean: 'O comrade, for God's sake give me water'."[219]

PTE CECIL YORKE, CANTERBURY BN, NZEF, HMT *LUTZOW*.

EGYPT

Others were preparing to receive men returning from the Dardanelles after the fight began.

"Since l wrote you last the hospital has continued to expand, though we thought it to have finished, and I am again seeking for more buildings. The 2000 limit will soon be passed, and it will be much larger still in all probability. The necessities of the case simply drive us and there is no alternative…

"The Commander-in-Chief visited the Red Cross depot this afternoon, and expressed unbounded satisfaction with this operation. At the same time he pointed out that with the oncoming of summer, we had better get only the lighter things at present. We shall pass the warmer clothing to the front as fast as possible. Although Egypt is warm we are informed the Dardanelles are exceedingly cold."[220]

MAJOR JAMES BARRETT, AAMC, 1ST AUSTRALIAN GENERAL HOSPITAL, HELIOPOLIS.

218   *Camperdown Chronicle* (Victoria), 10th August 1915.
219   *Lyttelton Times* (New Zealand), 24th July 1915.
220   *Daily Standard* (Brisbane, Queensland), 22nd June 1915.

25 APRIL 1915

DARDANELLES

The Australian submarine *AE2* entered the straits at 3.00 a.m.

> "It was absolutely dark, still, and dead calm as *AE2*… crept slowly along on the surface. With broken clouds shutting out such light as a moonless sky even yet contrives to give, the searchlights seemed more powerful than before. As we neared the white cliffs one felt forced to edge away from the light and nearer and nearer the European shore.
>
> "The long beam of light swept slowly along over the water, searching from southern shore towards the entrance, and then along the gloom under the steepness of the northern shore. Each time, as it touched *AE2* with brighter and yet brighter finger, one held for the instant one's breath, lest the steady sweep, arrest for a moment, would show a suspicion of our shadowy presence… But as the minutes passed by and custom eased the eerie feeling caused by the passing light, a necessitous boldness forced us farther along, now at dead slow speed on one engine."[221]
>
> LT-CMDR HENRY STOKER, RN, HMAS *AE2*.

Spotted, the submarine came under fire.

> "Mighty close was the bang of that gun, and mighty close to my head the broken swish of the shell as it hurtled past. With too much thought for the eyes of the watchers by the searchlights we had edged to within a mile of the European shore, and had been sighted by the look-outs of a battery of guns near Suandere River. Within a minute we were submerged, with above us the darkness preventing sight through the periscope, but a faint glimmer of light in the eastern sky giving promise of the quickly approaching day."[222]
>
> LT-CMDR HENRY STOKER, RN, HMAS *AE2*.

ANZAC

The plan was for the 3rd Australian Brigade, the covering force, to land on a front from Gaba Tepe in the south to the area near Fisherman's Hut in the north. Once ashore, the men were to move inland quickly, the high ground to the north securing the flank of the planned move towards Third Ridge. Mal Tepe, with its views overlooking the Dardanelles, was the ultimate objective for the first 24 hours. Contact with the enemy was not required before that plan began to unfold.

> "About midnight half of our lot was transferred to a destroyer and the other half to the battleship "Queen" and the rest of the Battalion was also transferred from transports to battleships and destroyers… Looking round our destroyer you could see our lads

---

221   Stoker, *Straws in the Wind*, pp. 107–108.
222   *ibid.*, p. 108.

fully equipped talking softly to each other, while some were in the forecastle drinking cocoa with the sailors, while others slept soundly without a care in the world."[223]

PTE CHARLES PEINIGER, 9TH BATTALION, 3RD BDE, 1ST AUSTRALIAN DIVISION.

"At 12.30 — It is a perfect moonlit night. We are passing the north east point of some island — probably still Lemnos. On the end of the island a pinpoint light is flashing three times every few seconds. Ahead of us is a single tiny sternlight always dead ahead. Away to the left far out on our port bow are two other lights — one after the other. Astern is another ship — I can see the faint glow of some cabin or galley light — for the rest she is simply a dark shape, no side lights or masthead lights. All the five of us probably more — are ploughing ahead stoutly through the darkness. One great thing in our favor — the night is… calm. The breeze of the day has died down. The sea under the moon is like satin with the soft velvet of the island rising out of it on our right. We are heading almost due N.E."[224]

CHARLES BEAN, AUSTRALIAN WAR CORRESPONDENT, HMT *MINNEWASKA*.

"We steamed up to within about a mile of the coast and anchored. The front three ships had troops on board, and the back three assisted with boats. Each ship provided two tows, consisting of a steamboat and three pulling boats, so that there were twelve tows altogether. We anchored off the coast shortly after midnight, and lowered all our boats. We received the order to disembark the troops about 1 a.m. The night was very still, and the moon was just setting. We filled up all our boats, one tow being on each side of the ship."[225]

MIDSHIPMAN HON CHARLES GIFFORD, RN, HMS *LONDON*.

"The *Bacchante*'s light (she had been sent on ahead to indicate the position) was sighted at 1 a.m… Soon afterwards the ships were stopped and the boats hoisted out, while the troops were given a good hot meal and a tot of grog. It was a perfectly still night with hardly a breath of wind, every sound was magnified, so much so that it seemed impossible that the noise of the boat-hoist engines could not be heard for miles away. We eagerly scanned the direction of the shore, the loom of which could just be seen, to see if we could detect any movement, but all was still. At last all the boats were in the water and orders given for the troops to be placed in them. This was carried out so quietly and expeditiously that I did not realize it had begun and sent to know what was the delay. My flag-lieutenant soon returned and informed me that nearly half the men were in the boats, I would not have believed that the operation could have been carried out so quietly that I could not hear them, although on the bridge only a few yards away."[226]

REAR-ADMIRAL CECIL THURSBY, RN, HMS *QUEEN*.

---

223 *The Evening Telegraph* (Charters Towers, Queensland), 16th June 1915.
224 *The Mildura Telegraph and Darling and Lower Murray Advocate* (Victoria), 9th July 1915.
225 *Grantham Journal*, 29th May 1915.
226 Thursby quoted in Wester-Wemyss, Lord, *The Navy in the Dardanelles Campaign*, p. 89, Hodder and Stoughton (London) 1924.

"After we got on the destroyer it was perfect silence and no smoking. No lights to be shown on deck under any consideration. Of course we managed to sneak a few smokes by getting below in the sailors' quarters... We had just finished a smoke when they called us from below to fall in, as we were getting close. The conversation while we were below was of the Turks. One of our chaps, Snowy Thompson,[227] from Pirie, says I'll bet you there is not a Turk for a hundred miles; all these precautions for nothing."[228]

PTE EDWARD WATHERSTON, 10TH BATTALION, 3RD BDE, 1ST AUSTRALIAN DIVISION.

"2.30 [a.m.] On deck again. The moon is almost down now. Our Third Brigade has to land in the little interval between the moonset and the dawn. They must be getting near there now — ten miles ahead of us perhaps — we are steaming just north of a high coast line — it must be Imbros. There are clouds on the high velvet black hills. Other very high mountains which must be the Island of Samothrace to the north."[229]

CHARLES BEAN, AUSTRALIAN WAR CORRESPONDENT, HMT *MINNEWASKA*.

"At 3.30 [a.m.] we were cutting the water at a terrific rate, heading straight for the Turkish coast somewhere... The next half hour was the worst of the war. The silent uncertainty was awful. Everything was silent except for the roar of the boats skimming the water. The silence could be felt as we neared the shore."[230]

PTE ALFRED WHALAN, 9TH BATTALION, 3RD BDE, 1ST AUSTRALIAN DIVISION.

"3.30 [a.m.] We are now clearing the last point of Imbros. The moon is down and it is much darker. I cannot see the mainland beyond, although I know it is there — the distance is only twelve miles. Far on our right on the point of Imbros Island or on some ship stationed in the channel between us and the land are two white lights, one above the other, and a little aslant as if on a mast. A colonel of the army corps staff is leaning over the rail beside me. A light somewhat suddenly behind some land far away to the right ahead of us appears a searchlight. It must be somewhere in the Dardanelles, south of the peninsula. It swoops in a scared way to right and left and left and back again and suddenly disappears. It cannot be one of the lights on our warships — it is too far up the straights for that. It must be one of the searchlights from one of the Turkish forts up the strait...

"It is just on 4 [a.m.] Wonder if they have heard anything or is it that they are sweeping the straits as an ordinary nightly precaution. Another searchlight further up the strait — we can only see the haze of this one also searching round like the startled eyes of some frightened animal. There goes the first searchlight again. Anyway they can neither of them reach anything on this side of the peninsula. Just at this moment dawn had begun to break. The fringe of light is now on our port side. The ship must have turned suddenly southwards. One can see the land — a high ridge of hills there also and reaching away on either side of us. We are moving in between two flanking

---

227 Pte James Thomson, 10th Bn, AIF, was wounded on 25th April or shortly afterwards; admitted to 1st Australian General Hospital, Cairo, on 29th April. He returned to Australia after developing heart disease.
228 *Port Pirie Recorder and North Western Mail* (South Australia), 7th July 1915.
229 *The Mildura Telegraph and Darling and Lower Murray Advocate* (Victoria), 9th July 1915.
230 *Molong Express and Western District Advertiser* (New South Wales), 19th June 1915.

"ships, merchant ships evidently stationed there to give us the position. There is not a sign of anything moving on those hills at all — nor the least sound. And yet it is well past four — the time when our Third Brigade ought to be rushing out of their boats and somewhere up the grey slope that we can see ahead. They hoped at least to get to shore unseen — they were not to fire a shot if possible till daylight — to trust to the bayonet till then alone. It is still too dark to see what I am writing but the dawn is slowly growing. A line of officers is gradually lining the rail under the bridge — a ship's officer or two amongst them as excited at the rest. Down on the forecastle the men are beginning to cluster to the sides."[231]

CHARLES BEAN, AUSTRALIAN WAR CORRESPONDENT, HMT *MINNEWASKA*.

"I was out of bed at 4 a.m. and on deck. The dawn was just breaking about 4.10 a.m., as I was leaning over the ship's rail. All at once I was brought to my senses by a flash and awful boom from a warship close. Within a few minutes it was a living hell — for a dozen warships had started on their day's work. The noise was something awful — we could not hear ourselves speak."[232]

DVR CHARLES HEMMING, NO 1 COY, AASC, AIF, HMT *MINNEWASKA*.

"Each ship provided two tows, consisting of a steamboat and three pulling boats, so that there were twelve tows altogether… When the steamboats had towed the boats into shallow water, they slipped them."[233]

MIDSHIPMAN HON CHARLES GIFFORD, RN, HMS *LONDON*.

"I altered course to the north to get away from Gaba Tepe (as soon as I could see it) as I knew it was heavily defended. Had we gone to the beach there we would have been enfiladed from Gaba Tepe, got into barbed wire and pits at the beach, and probably totally annihilated."[234]

MIDSHIPMAN JOHN METCALF, RNR, HMS *TRIUMPH*.

"As we approached the shore those of us who knew the position of the bay in which it had been arranged we should land, realised that something had gone wrong. We were making the shore opposite some high cliffs instead of at the bay. There was a little momentary confusion amongst the boats which closed in from all sides… The boats from the different battleships got mixed up — some which should have been on the flanks were in the centre, others vice versa. This was undoubtedly the most anxious time we had had, and in the semi-darkness was enough to upset any nerves."[235]

LT-COL. JAMES LYON JOHNSTON, 11TH BATTALION, 3RD BDE, 1ST AUSTRALIAN DIVISION.

"We got into the cutter from the *Queen*, and the four boats containing A Company, in common with the rest of the storming party, were taken in tow by a steam pinnace

---

231 *The Mildura Telegraph and Darling and Lower Murray Advocate* (Victoria), 9th July 1915.
232 *Nepean Times* (Penrith, New South Wales), 7th August 1915.
233 *Grantham Journal*, 29th May 1915.
234 Letter, John Metcalf to Jim Bostock, 14th March 1973.
235 *The Sun* (Kalgoorlie, Western Australia), 11th June 1915.

and taken as close as possible to the shore. Then she cast off and a few strokes of the oars put the bow ashore.

"Lieutenant Chapman[236] was right forward, and hopped over, and was followed quickly by the rest. Somewhere about 17 men were out of the boat before the first rifle shot rang out... Lieutenant Chapman was the first man ashore by a small margin."[237]

PTE JAMES SPIERS, 9TH BATTALION, 3RD BDE, 1ST AUSTRALIAN DIVISION.

"Lieutenant Chapman [called] 'All out!' and we immediately hopped over the side. I saw the Lieutenant's silhouette as he hopped overboard and I followed him, in water up to our waists, and we scrambled ashore... I was the second man to land."[238]

PTE JAMES BOSTOCK, 9TH BATTALION, 3RD BDE, 1ST AUSTRALIAN DIVISION.

"We reached the beach about 4.30 [a.m.], and half of our chaps got out of the boat without anything happening. We were just beginning to think that the Turks were not there, when one shot was fired. Then shots came from all directions. I was lucky to start with one shot only hitting my cap and throwing me back in the water. I was not very long scrambling out again. We lay at the foot of the hill for about ten minutes, and the bullets did fly around. We could see several men hit while coming in other boats."[239]

PTE WILLIAM FISHER, 9TH BATTALION, 3RD BDE, 1ST AUSTRALIAN DIVISION.

"One [Midshipman] in our boat was such a slim, smooth-cheeked kinchim, that if you had seen him in the streets you would have wondered why the truant inspector was not doing his business properly, Shrapnel was falling like rain, but this kid stood up in the stern sheets and called out in a squeaky voice that did not reach half-way up the boat: 'Pull, men, pull,' and the men did pull. The sight of that cool stripling made us forget the shrapnel a bit."[240]

SGT WALTER AYLING, 11TH BATTALION, 3RD BDE, 1ST AUSTRALIAN DIVISION.[241]

"Now it was we realised the difficulty caused by the close packing. We found we could not use the oars freely, but the boys dipped the oars down and bent to it with a will; they were, however, fearfully hampered, and consequently we moved at a snail's pace, but move we did, while the water around us was being churned to foam by bullets. Fortunately their fire was high, but all the same some of the bullets found a mark. At last the boat grounded... Bucking off our heavy packs, bayonets were fixed. In the meantime I had yelled for officers and was answered by Lts. Rockliff[242] and

---

236 Capt. Duncan Chapman, 9th Bn, AIF. He was killed as a Major with 45th Battalion on 6th August 1916; buried in Pozieres British Cemetery, Ovillers-La Boisselle, France.
237 *The Telegraph* (Brisbane, Queensland), 18th January 1934.
238 *ibid.*, 24th January 1934.
239 *The Brisbane Courier* (Queensland), 11th June 1915.
240 *The Sun* (Sydney, New South Wales), 20th August 1915.
241 Ayling was awarded the DCM (*London Gazette*, 3rd July 1915): "On 25th April 1915, during operations near Kaba Tepe, for gallantry in commanding his platoon after his officer had been wounded. When compelled to retire he carried the wounded Officer with him, and on obtaining reinforcements again led his platoon to the attack."
242 Lt, later Major William Hudson Rockliff, 11th Bn, AIF, a former school teacher from Perth, Western Australia.

Macfarlane.[243] On enquiry, I found that they had about 100 men between them; so, being by virtue of my rank in charge, I ordered Macfarlane to move his platoon to the right to join up with the remainder whom I could see charging up the precipitous slopes of the hill."[244]

CAPT. JOHN PECK, 11TH BATTALION, 3RD BDE, 1ST AUSTRALIAN DIVISION.

"I was in the scouts' boat, pulling No 2 oar, and Colonel Weir, Captain Lorenzo,[245] and Lt Talbot Smith (the latter in charge of the scouts) were in the next seat. Our chaps bent down as far as possible in order to present a smaller target for the bullets, but for some time the officers continued to sit erect. A bullet passed between the colonel and me, and others nearly found their mark, so somebody leant against the colonel, thereby compelling him to bend over, and this action probably saved him from getting hurt... I could not bend down as far as I should have liked to on account of the oar, but I forgot all about the bullets by putting all my strength into the strokes. Suddenly the boat grounded... I am afraid I didn't trouble about shipping my oar, but tumbled into the water and half ran and half floundered to shore."[246]

PTE JOHN BRADFORD, 10TH BATTALION, 3RD BDE, 1ST AUSTRALIAN DIVISION.

"There were 33 scouts, and Lieutenant Talbot Smith was in our boat... We got down on the beach, and Lieutenant Smith[247] gave us orders to fix bayonets... He then jumped up and called "Come on, lands, they are only a few patrols!" and dashed up the cliff."[248]

PTE JOHN SUTHERLAND, 10TH BATTALION, 3RD BDE, 1ST AUSTRALIAN DIVISION.[249]

The noise of the first engagement was heard in the fleet watching offshore.

"The first sign — 4.37 — low down either on the line of sea or shore a signal lamp has flashed. We cannot say if it is on some small boat close in or on the Beach itself. 4.38 — For the first time listening eagerly, I can catch faintly from the direction of the shore a distant knocking — first an isolated knock or two, then a second's wait, then knock, knock, again knock, knock, twenty or thirty knocks. It stops for a few seconds and comes again continually like the knocking of an axle box heard very far off, very faintly through the bush. To my mind there is no mistaking that sound whatever. It is the first time I have heard the sound but I have not the slightest doubt as to what it is. It is the first distant echo of rifle firing — first a few shots then heavier and continuous

---

243 2/Lt, later Capt. Archibald MacFarlane, 11th Bn, AIF. The former analyst was killed in action at Pozieres on 22nd July 1916. Commemorated on the Villers-Bretonneux Memorial, he was the 27 year-old son of John and Martha MacFarlane, of 190 Townshend Road, Subiaco, Western Australia.
244 *The Grenfell Record and Lachlan District Advertiser* (New South Wales), 27th July 1915.
245 Capt., later Lt-Col. Francis Lorenzo, DSO, 10th & 49th Bns, AIF.
246 *West Coast Recorder* (Port Lincoln, South Australia), 11th August 1915.
247 Lt Eric Talbot Smith, 10th Bn, AIF, was severely wounded on 25th April and died in Egypt on 30th April 1915. Buried in Alexandria (Chatby) War Memorial Cemetery, he was the son of Sydney Talbot Smith and Florence Talbot Smith, of Kensington Park, Adelaide, South Australia.
248 *Daily Herald* (Adelaide, South Australia), 24th September 1915.
249 Sutherland, a former electrician, was wounded on 25th July 1915. Rejoining his unit in February 1916, he died of wounds in Belgium on 4th October 1917. The son of James and Martha Sutherland, of Gilbert Street, Gilberton, Adelaide, South Australia, is buried in Lijssenthoek Military Cemetery.

"—I call the attention of the ship's officer next to me to it. He listens and can hear it too. He has no more doubt than I have. Some of the others cannot catch it at first, but within a few minutes there is no question. Heavy firing is going on somewhere on the slope of that hill. Perhaps it is on the beach. Looking hard one cannot see the least sign of a flash. So they must have been discovered early. I wonder how things are with them. If only we could see — the light is growing fast into broad daylight."[250]

CHARLES BEAN, AUSTRALIAN WAR CORRESPONDENT, HMT *MINNEWASKA*.

"Suddenly a faint sound of rifle firing was heard and I knew the beach had been reached. I immediately signalled the destroyers to advance, and in they dashed at full speed. Dawn was now breaking and we could gradually make out white patches which were our boats on the beach; one or two on the left seemed wrecked and broken to pieces. Howitzers and field guns were now pouring in a hot fire and shells seemed to be coming from every direction. As the light increased the position could be more clearly defined; the men had landed from the destroyers and were dashing up the slopes to join their comrades who were already well up the hills. The noise made by rifle and machine-gun fire was growing in volume, punctuated by the louder explosions of shells from the heavy guns. The destroyers had stood in until their keels nearly touched the beach and so rapidly had the disembarkation proceeded that 4,000 men had been thrown ashore on a front of over a mile in just over half an hour."[251]

REAR-ADMIRAL CECIL THURSBY, RN, HMS *QUEEN*.

"To prevent the accidental discharge of a rifle before landing, which would have given warning to the enemy, the order had been given that rifles were not to be loaded until after the landing was effected, and that as far as possible, the bayonet was to be relied on. This, like all other orders, was strictly carried out, and a good story is told, although I have not been able to verify it, that one of the men jumped off the boat into very deep water, and when he came up again he had his bayonet fixed."[252]

LT-COL. JAMES LYON JOHNSTON, 11TH BATTALION, 3RD BDE, 1ST AUSTRALIAN DIVISION.

"I saw a bit of a knob of ground just on the beach, so reckoned that would be a good place to make for. You can just see me going for that shelter. When I got out of the water that beach seemed very lonely, I can tell you, but the water behind me was packed with swearing Australians. Very quickly there were half-a-dozen of us under this bit of knob. We were getting rid of our packs and filling our magazines and fixing our bayonets."[253]

PTE LAWRENCE KUHLMANN, 11TH BATTALION, 3RD BDE, 1ST AUSTRALIAN DIVISION.

"Day was now breaking, and from our position we could see a long way. A formidable fort [Gaba Tepe], about a mile to the south of our landing spot, began to talk, but the *Bacchante* moving close inshore engaged at once, and we were interested spectators of the duel, which did not last long, the *Bacchante* proving the victor. A casual look

---

250 *The Mildura Telegraph and Darling and Lower Murray Advocate* (Victoria), 9th July 1915.
251 Thursby quoted in Wester-Wemyss, *The Navy in the Dardanelles Campaign*, p. 90.
252 *The Sun* (Kalgoorlie, Western Australia), 11th June 1915.
253 *Kalgoorlie Miner* (Western Australia), 31st July 1915.

through the glasses at this fort and surroundings showed us what elaborate preparations the enemy had made for our reception. Merciful Providence and the navy co-operated to guide us to a safer spot, for had we landed a mile further south I am quite sure not a man of us would have been left. Wire entanglements, that would almost imprison a small bird, cover the hill and beach, and stretch far our under the water, whilst concealed gun position… could search the water and beach to a wide radius."[254]

L/SGT ALEX MARSHALL, 11TH BATTALION, 3RD BDE, 1ST AUSTRALIAN DIVISION.

"My boat was rather unfortunate, as we hit one in front and stopped, so out I jumped, being the first in the bow and I landed in the water up to my neck. I had, in the meantime, removed my pack which was very heavy, and held it above my head to keep it dry. Anyway my men followed me, and we reached terra firma after a bit of trouble. There were no orders given, as the men were told previous to landing exactly what to do, and my word they did it. We fixed bayonets at about 4.45 a.m., and made up the cliffs."[255]

LT ROBERT CHAMBERS, 9TH BATTALION, 3RD BDE, 1ST AUSTRALIAN DIVISION.

"You ought to have heard the cheer when they gave us the word to charge. You could have heard it miles if you could have stopped to listen. Some were saying (or roaring), 'Come on, Australia!' and others 'Australia for ever!' and some 'Come on, boys!' and 'Give it to 'em, boys!' but the funniest part of it was hearing chaps talking to themselves, or at least at the Turks, and calling them all the Saturday night bar fight words ever heard…

"I nearly lost my rifle the first shot. I forgot to put it close to my shoulder, and it kicked nearly out of my hands."[256]

PTE FREDERICK RICHARDSON, 10TH BATTALION, 3RD BDE, 1ST AUSTRALIAN DIVISION.

"With one accord all fixed bayonets and charged the hill, a terribly rough and steep one, the men having to pause several times to regain breath; but with dreadful yells we rushed on, cleared the trenches… One of the first to fall was Captain Annear,[257] our second in command, who was shot dead as soon as we reached the summit, Lt Macdonald[258] being wounded in the shoulder at the same time."[259]

L/SGT ALEX MARSHALL, 11TH BATTALION, 3RD BDE, 1ST AUSTRALIAN DIVISION.

"I confess to being really and thoroughly funked then. I lay down and hugged the earth tightly, and tried to collect my terrified senses. Men were already dead, dying, and helpless all round. With a sudden access to fury or courage, perhaps both, I stood

---

254 *Sunday Times* (Perth, Western Australia), 25th July 1915.
255 *The Telegraph* (Brisbane, Queensland), 17th June 1915.
256 *The Advertiser* (Adelaide, South Australia), 7th July 1915.
257 Capt. William Annear, 11th Bn, AIF, was the first Australian officer to be killed during the campaign. Commemorated on the Lone Pine Memorial, originally from Ballarat, he was the 40 year-old son of the late John and Annie Annear.
258 Lt David Macdonald, 11th Bn, AIF, was wounded on 25th April. Rejoining his unit on 2nd June, he was killed in action on 28th June 1915. Buried in Shell Green Cemetery, he was the 25 year-old son of William and Catherine Macdonald, of "Caithness," Reginald Street, Cottesloe, Western Australia.
259 *Sunday Times* (Perth, Western Australia), 25th July 1915.

up and made into the country. I hadn't gone 25 yards before I found it was absolutely necessary to take cover again… The air was thick with bullets and shrapnel shellfire. In my second position I stopped till the firing died down again, and again rushed forward. The machine gunners collected together, but the fire was still too hot to permit of one even raising a head…

"Then the Turks ran before the bayonets of our advanced companies, and in a few minutes we had the first line of their trenches. The country is simply a series of steep bluffs. I stumbled over a heap of dead (bayoneted) Turks."[260]

PTE JOHN WHITE, MG SECTION, 11TH BATTALION, 3RD BDE, 1ST AUSTRALIAN DIVISION.

"When I got ashore and started attending the wounded the first man I dressed was… Lieut. D. McDonald. He was shot through the shoulder, and wanted badly to go on with his men, but after a lot of persuasion I got him to get into the lifeboat and go off to the hospital ship."[261]

PTE EDWARD LANGOULANT, 3RD FIELD AMBULANCE, AAMC.

"[5 a.m.] Now at last as we move in we can see just on this side of the beach a swarm of small boats — small boats everywhere. Each seems to be going its own way — rowing, not in tow. Some seem to be stationary in clumps as if they might be in difficulties — it is hard to tell at this distance. 'I don't like the way they are all scattered about' says a staff officer near me."[262]

CHARLES BEAN, AUSTRALIAN WAR CORRESPONDENT, HMT *MINNEWASKA*.

"Ned Bird[263]… and I were the two picked to run the first telephone line, and we were very proud of it, I can tell you… We each carried a mile of wire across our shoulders. Getting out of the boat in fairly deep water nearly up to my middle, someone threw Ned's rifle and he lost it in deep water. He came to me nearly frantic about his rifle. He ran along the beach and grabbed the first dead man's, and away we started after our Brigadier. We left our base operator on the beach and took to the cliffs among the prickly bushes and shrubs. The perspiration poured out of us."[264]

SPR HARRY MARSHALL, DIVISIONAL SIGNAL COY, AUSTRALIAN ENGINEERS, 1ST AUSTRALIAN DIVISION.

"After gaining the top we were confronted by a deep valley… When we appeared on the ridge we were met by a perfect hurricane of lead, it was confounding, and death to rise up even a few inches from the ground. The only chance to move was when there was a slight lull of the firing in your direction, bullets seemed to be flying everywhere,

---

260　*Daily Herald* (Adelaide, South Australia), 12th June 1915.
261　*Sunday Times* (Perth, Western Australia), 23rd January 1916.
262　*The Mildura Telegraph and Darling and Lower Murray Advocate* (Victoria), 9th July 1915.
263　Spr Edgar Bird, 1st Div. Signal Coy, Australian Engineers, was killed in action on 3rd June 1915. Buried in Shell Green Cemetery, the former farmer was the 25 year-old son of Mary and the late John Bird, of Lexton, Victoria.
264　*The Bundberg Mail and Burnett Advertiser* (Queensland), 15th June 1915.

so if you moved you were a 'goner.' We had no artillery landed at so early an hour, so the Turks had nothing to fear from our big guns."[265]

PTE CLAUDE OVERELL, 9TH BATTALION, 3RD BDE, 1ST AUSTRALIAN DIVISION.

Turkish fire also targeted those approaching the shore, meaning some men never reached it.

"The disembarkation proceeded quietly but quickly. As soon as one boat was filled it shoved off and pulled for the shore, and made room for another. By this time the light of day was flooding the scene, and as soon as they could be distinguished, the destroyers became the object of fire. Bullets began to fly unpleasantly near, and every now and then a louder whistling proclaimed a shell, which had passed over us and splashed into the water. Soon "boom," when a heavy gun was fired, and we knew that the warships were drawing in. Just then I heard a crack behind me, and felt a numbing pain in my right side and a knock on the head. Putting my hand up I felt the warm blood, and almost immediately after I began to feel dizzy. Telling the next fellow to me that I was hit, I slipped down to the cover of the engine hold, and Will [Hese] who was next to me, took off my shoulder straps and equipment. Then two sailors helped me down to their mess deck. Fortunately an A.M.C. man was on board, and he quickly had me fixed up with my field dressing. I did not feel much pain, and was quite conscious, but I felt a bit dizzy."[266]

PTE ERIC RATCLIFF, 12TH BATTALION, 3RD BDE, 1ST AUSTRALIAN DIVISION.

Sinclair-Maclagan was concerned about the possibility of a Turkish counter-attack coming from the south and redirected forces to meet it.

"Just after we ("C" Coy) landed on that well peppered beach, and whilst we were 'assembling,' the brigadier told me to move up to some ridges on our extreme right — one platoon being detached first to get hold of a hilltop (I never had that platoon again, they fought in with other units during the day). Well, the other three platoons got to the ridges desired — hung on there for awhile, and did some reconnoitring further to the south; we then swung round a bit towards the east, and gained another ridge, had no losses so far — beyond loss of breath climbing those hills! (There were several hit en route to the shore, and one, I think, was killed). Then after we had done a bit of long-range firing from the last ridge mentioned, just trying to get the range to some suspected trenches, we decided to push on a bit further east and inland."[267]

CAPT. JOHN WHITHAM, 12TH BATTALION, 3RD BDE, 1ST AUSTRALIAN DIVISION.

"At this moment the enemy began to retire, with all our boys in hot pursuit, shouting 'Imshi,' which means in Egyptian 'Clear out or get.' — We had no time to look for prisoners. Snipers picked a good few of our men off. As you know, I was through the South African campaign, but never faced anything like it out there."[268]

SGT JAMES CHRISTIE, 12TH BATTALION, 3RD BDE, 1ST AUSTRALIAN DIVISION.

265 *Daily Standard* (Brisbane, Queensland), 16th July 1915.
266 *Daily Telegraph* (Launceston, Tasmania), 15th June 1915.
267 *The Critic* (Hobart, Tasmania), 2nd July 1915.
268 *The Albany Advertiser* (Western Australia), 24th July 1915.

Others headed inland towards what became known as Walker's Ridge towards the Nek and Baby 700. The 12th Battalion, the brigade reserve, instead of being concentrated, was now committed piecemeal over a wide front.

"After a few moments' consultation with Captain Lalor it was decided that I should take my platoon up the left flank and support the 11th Battalion, while he, with what men he could gather together, advanced on my left.

"By this time the Turks had been driven back, and except for some snipers who were concealed in the scrub on the steep hills, we were not under direct fire for the time being. We had to cross some terribly rough country, with deep gullies, the sides of which were almost perpendicular in places. Reaching the top of one of these gullies I found Captain R. W. Everett, of the 11th Battalion, and he asked me to go to his left flank and tell Captain Lalor that there was a machine gun in that direction, which was troubling them. When I came up with Captain Lalor I delivered my message, and we then pushed up a gully which was very rough and very thickly covered with scrub."[269]

LT EDWARD BUTLER, 12TH BATTALION, 3RD BDE, 1ST AUSTRALIAN DIVISION.

"Many of us crossed a narrow connecting ridge, which saved descending the deep gully. Had the Turks been cooler shots, they could have picked us off one by one like flies… when we reached the final heights not a soul was to be seen save three dead men whom our fellows had shot down."[270]

L/CPL WILLIAM HASTINGS, 11TH BATTALION, 3RD BDE, 1ST AUSTRALIAN DIVISION.[271]

"Heavily loaded with a machine gun I struggled on while the air sang. I was halfway down, the steep slope when someone on the crest behind knocked some stones down. One of them struck me and knocked me unconscious. When I came to, a Red Cross man had hauled me back and bandaged me with my field dressing…

"On my way back to our dressing station I had a nerve-wracking time. I was helping a chap who had dislocated his collarbone. I had his rifle and equipment to bring along, too, and it was an awful job. We were in a storm of shrapnel and stray bullets nearly all the way. Poor chap; his nerves were completely shattered, and every time a shell burst the look in his eyes was agonising."[272]

PTE JOHN WHITE, MG SECTION, 11TH BATTALION, 3RD BDE, 1ST AUSTRALIAN DIVISION.

"He [Lt Roberts][273] was wounded in the neck with shrapnel, which seemed to make him rather faint. I offered to assist him back under cover of the hill where the "Red Cross" were at work, but he would not hear of going back. He just tied a pocket

269   *The Sun* (Sydney, New South Wales), 22nd August 1915.
270   *The Daily News* (Perth, Western Australia), 22nd June 1915.
271   Wounded on 27th April, the former bank clerk was commissioned. 2nd Lt Hastings was killed at Pozieres 22nd–25th July 1916. The 38 year-old former bank clerk is commemorated on the Villers-Bretonneux Memorial; the son of William and Harriett Hastings, of Gosford, New South Wales.
272   *Daily Herald* (Adelaide, South Australia), 12th June 1915.
273   Lt John Roberts, 9th Bn, AIF, was killed in action on 25th April 1915. Buried in Walker's Ridge Cemetery, where he is commemorated by a special memorial, he was the 33 year-old son of the late Arthur and Jessie Roberts; husband of Sarah Kennedy (formerly Roberts), of Hancock Street, Ipswich, Queensland.

handkerchief round his neck, and gripping his revolver shouted "Come on boys." We all followed him into some thick pine scrub… he had not gone ten paces into the scrub before he was killed."[274]

PTE FRANK HOLLOWAY, 9TH BATTALION, 3RD BDE, 1ST AUSTRALIAN DIVISION.

"Sergeant Graham[275] was shot in the arm and thigh later on, and died very quickly, and a young chap named Sergeant Fowles[276] was shot dead. This looked bad for those with stripes or anything that made them look like officers, so I pulled mine off pretty quick."[277]

L/CPL JOSEPH CHILDERS, 9TH BATTALION, 3RD BDE, 1ST AUSTRALIAN DIVISION.

"At 5.3 a.m. the fire intensified, and we could tell from the sound that our men were firing. It lasted until 5.28 and then died down somewhat. No one on board knew what was happening, although dawn was gradually breaking, because we were looking due east into the sun, slowly rising behind the hills, which are almost flush with the foreshore, and there was also a haze. Astern at 5.26 we saw the outline of some of the transports gradually growing bigger and bigger as they approached the coast. They were bringing up the remainder of the Australians and New Zealanders."[278]

ELLIS ASHMEAD-BARTLETT, BRITISH WAR CORRESPONDENT, HMS *LONDON*.

Two 10th Battalion scouts, Arthur Thorburn and Philip Robin, penetrated furthest inland, reaching Scrubby Knoll. They found few defenders but later saw formed Turkish reinforcements approaching their isolated position.

"Up till now I had seen no one that I knew, as all the battalions of our Brigade were completely mixed up. Just as I started down into the valley, however, I met Phil Robin[279] and Micky Smith and together we pushed on after the enemy. Travelling across this valley was a decidedly lively time as the scrub was full of snipers and every little while a bullet would come closer than was pleasant. However we got to the top of the ridge in safety and there found several other chaps but no one in charge. Just at that moment, however, Captain Herbert[280] came up and so Phil Robin reported to him. He had decided to entrench there and so sent Phil and me out to watch a valley on his front and flank while he did so, and this, by the way, was about the only bit of

---

274 *Western Star and Roma Advertiser* (Queensland), 20th November 1915.
275 Sgt Thomas Graham, 9th Bn, AIF. His death in action is officially recorded as taking place on 2nd May 1915. Born in Bromsgove, Worcestershire, he was the brother of Winifred Backhouse, of 83a Berkley Street, Liverpool.
276 Sgt Herbert Fowles, 9th Bn, AIF, was killed in action on 25th April 1915. Commemorated on the Lone Pine Memorial, he was the 22 year-old son of John Kentwell Fowles and Agnes Ellen Fowles, of "Menahous," Wooloowin, Queensland.
277 *Maryborough Chronicle, Wide Bay and Burnett Advertiser* (Queensland), 22nd November 1915.
278 *The Referee* (Sydney, New South Wales), 23rd June 1915.
279 L/Cpl Philip de Quetteville Robin, 10th Bn, AIF, was killed in action on 28th April 1915. Commemorated on the Lone Pine Memorial, he was the 30 year-old son of Rowland and Mary Robin, of 28 Edwin Terrace, Gilberton, Adelaide, South Australia. He married Nellie Honeywill, who was working as a nurse in England and travelled from London to Cairo for the wedding, on 17th January 1915. After a brief honeymoon, Nellie returned to London. Nellie Robin and her infant son died soon after she had given birth on 22nd November 1915.
280 Capt. Mervyn Herbert, 10th Bn, AIF, was wounded 26th/27th April 1915 and invalided home. He rejoined the battalion at Lemnos on 26th November 1915.

scouting I got to do on the first day. We stayed out there until driven in by the enemy who were coming to the attack in force, and Phil and I then got into the trench. In a few moments, however, Captain Giles[281] asked us to get a message through for reinforcements and Phil sent me back to bring some up. On returning some hour or so later with them I got into the trench about fifty yards from Phil."[282]

PTE ARTHUR THORBURN, 10TH BATTALION, 3RD BDE, 1ST AUSTRALIAN DIVISION.[283]

Meanwhile one 10th Battalion officer considered his exhausted men had advanced as far as they could.

"If I could only have got the men to dig holes for themselves when we took the ridge our losses would have been fewer, but we were all knocked out by the awful hill climb… It is a wonderful thing that we ever got into this position. As a matter of fact, the 3rd Brigade took and held it without the direction of any of our generals, in fact, Major Brand and myself were the only senior officers there at the time. The others did not arrive until we had started to dig ourselves in."[284]

MAJOR FREDERICK HURCOMBE, 10TH BATTALION, 3RD BDE, 1ST AUSTRALIAN DIVISION.

Shortly after arriving on the peninsula, Sinclair-Maclagan had sent Charles Brand, his Brigade-Major, to take control of the situation and to report what he found.

"Quarter of an hour after the advanced units landed the brigadier and staff set foot on shore. It was then he realised that the original plan had gone somewhat awry. Prompt action was necessary to prevent the enemy forestalling us on Plateau 400 (Lone Pine position) and the ridges sloping towards Gaba Tepe. To me, as brigade major, was entrusted the task of seeing this was done. Colonel MacLagan selected his first headquarters on the spur which now bears his name. Half an hour later he moved up to the head of Wire Gully, an off-shoot from Shrapnel Gully, in the vicinity of Plateau 400…

"The semi-darkness, thick undergrowth, the steepness and criss-crossing of gullies and ravines rendered direction and control most difficult. With daylight the situation became clarified, but still far from satisfactory. Slight though the opposition was at first, the Turks realised the danger of the invader gaining a foothold. Gradually the opposition became greater and casualties mounted up… My report stated it was hopeless to try to reach and effectively hold the covering position originally assigned. The brigadier also came to the same conclusion."[285]

MAJOR CHARLES BRAND, BRIGADE-MAJOR, 3RD BDE, 1ST AUSTRALIAN DIVISION.

The disorganisation that Brand described, the intermixing of units was very evident; and the casualties amongst officers or anyone showing initiative, could have only compounded

---

281   Capt., later Major Felix Giles, DSO, 10th Bn, AIF.
282   Letter from Arthur Seaforth Blackburn to his brother Charlie, 3rd June 1915, AWM2018.785.56.
283   As Lt Arthur Blackburn, he was the first member of the 10th Battalion to be awarded the VC, at Pozieres in July 1916 during the Battle of the Somme.
284   *The Advertiser* (Adelaide, South Australia), 24th June 1915.
285   *Daily Advertiser* (Wagga, Wagga, New South Wales), 25th April 1935.

Sinclair-Maclagan's lack of faith in his ability to secure the original objectives. So, at 5.30 a.m., just one hour after Chapman set foot on the peninsula, despite the limited (if effective) opposition encountered thus far, he took the decision to consolidate a position on Second Ridge. Did Sinclair-Maclagan, as Hurcombe alludes to, simply endorse the actions of his subordinates or had Brand had the same effect on his commanding officer as Wray had on Troubridge when he turned away from the *Goeben* back in August? Whatever the case, despite the fact that major Turkish reinforcements did not reach Third Ridge before 8.00 a.m., Sinclair-Maclagan failed to push those under his command to take their objective; failed to follow orders he had conveyed to his men; and ended any forward momentum. The attacking force became the defenders from that moment.

"Owing to the extraordinary circumstances obtaining, it was impossible to keep the various units of the brigade intact, but although all the battalions got mixed up, the men played the game splendidly and attached themselves to the nearest officer or N.C.O., and carried on. At one stage of the proceedings, quite early on Sunday morning, I found myself in command of two of our own companies and four from other battalions, and it was with these, under orders from the Brigadier, that the rush for the main position occupied was first made, and in consequence of this and of the fact that tactical considerations necessitated my sending my four platoons in four different and widely-separated directions. I lost my own company, or the greater part of it."[286]

MAJOR EDMUND DRAKE-BROCKMAN, 11TH BN, 3RD BDE, 1ST AUSTRALIAN DIVISION.

"The only way to make headway was in the rushes with our bayonets fixed. I managed to be in one of these. After that, I was just observing some fire from the battleship, which was trying hard to put a battery on Gaba Tepeh out of action, when I got hit. I was stunned for a minute from the knock of my glasses and then I suddenly found my arm was across my back and no feeling in it, so I got a man to put the first field dressing on… I was then suffering some pain, so I took my morphia tablet, which the M.O. gave us before starting. It lessened the pain. I only had the one as I gave the other to my platoon sergeant earlier. He was shot across the back, and had his spine injured."[287]

LT ROBERT CHAMBERS, 9TH BATTALION, 3RD BDE, 1ST AUSTRALIAN DIVISION.[288]

"Had some great excitement getting to the next ridge, the ground suddenly fell away from a point about 40 yards in advance of where we had been lying on the crest, and 'twas almost a steep drop to the gully below. Well, that sheer drop was done in quick time by every fellow who commenced the journey, for that machine gun had the range of that cliff face (it was a chalky, gravelly cliff face) just beautifully, and to add to one's comfort, shrapnel began to burst overhead; 'twas really ludicrous (after one had got safely down himself, of course) to watch the others tumbling or rolling and sliding down. I suppose this was some time between 9 and 10 a.m. We reorganised

---

286 *The West Australian* (Perth, Western Australia), 28th July 1915.
287 *The Telegraph* (Brisbane, Queensland), 17th June 1915.
288 Promoted Capt., Chambers, a former draftsman, was killed in action on the Somme on 21st August 1916. Buried in Regina Trench Cemetery, Grandcourt, France, he was the 23 year-old son of Claude and Evelina Chambers, of 172 Merthyr Road, New Farm, Queensland.

those sections which we had got down in the gully, under cover of an under-feature, or small ridge between the two main ridges, and moved on towards the scrubby ridge. I remained on the intervening ridge in order to get together the remaining sections and platoons — but they only came on in driblets — other companies being mixed up with them; one platoon commander (Holland[289]) was hit about that time — on the ridge I had just left — and his men got mixed up with men of another battalion. For some time my job was getting hold of small parties of men coming from the rear, organising them and feeding them up the hill in front; my company sergeant-major and the usual small staff of messengers and signallers were with me."[290]

CAPT. JOHN WHITHAM, 12TH BATTALION, 3RD BDE, 1ST AUSTRALIAN DIVISION.

"The first authentic news we received came with the return of our boats. A steam pinnace came alongside with two recumbent forms on her deck and a small figure, pale but cheerful and waving his hand astern. They were one of our midshipmen, just 16 years of age, shot through the stomach… From them we learned what had happened in those first wild moments."[291]

ELLIS ASHMEAD-BARTLETT, BRITISH WAR CORRESPONDENT, HMS *LONDON*.

"The enemy now had light enough to use his field guns from Gaba Tepe, and shelled the boats heavily. Gaba Tepe was at once engaged by the *Triumph* and *Bacchante*, but the guns were so well placed that they continued in action at intervals during the whole landing. This shell fire enfiladed the beach and caused many casualties in the boats. Those casualties caused further delay in the disembarkation, as wounded men were left in the boats, and even put in the boats from the beach. When the boats returned to the transports it was necessary to take the wounded on board, and, as provision had not been made for this, increasing delays took place with each tow or string of boats."[292]

MAJOR WALTER CASS, BRIGADE MAJOR, 2ND BDE, 1ST AUSTRALIAN DIVISION.

6th and 7th Battalions of the 2nd Brigade were aboard HMAT *Galeka* off the coast near Fisherman's Hut. The delays led to a change of plan.

"The naval tows, which were to meet our transport at 4.45 a.m., did not turn up, and as the enemy's guns were getting our range, and shells were bursting over the ship, Commander Somerville,[293] the Naval Transport officer, instructed me to get the men ashore in ship's boats."[294]

LT-COL. HAROLD ELLIOT, 7TH BATTALION, 2ND BDE, 1ST AUSTRALIAN DIVISION.

289   Lt Thomas Holland, 12th Bn, AIF, wounded on 25th April 1915. Rejoined his unit 20th May 1915. Later promoted Major, he suffered a nervous breakdown in England and left for Australia on 31st October 1917.
290   *The Critic* (Hobart, Tasmania), 2nd July 1915.
291   *The Referee* (Sydney, New South Wales), 23rd June 1915.
292   Walter Cass quoted in Schuler, Philip, *Australia in Arms. A Narrative of the Australasian Imperial Force and Their Achievement at Anzac*, pp. 109–110, T. Fisher Unwin Ltd. (London) 1916.
293   Lt-Cmdr, later Admiral of the Fleet Sir James Somerville, RN.
294   *Bendigonian* (Bendigo, Victoria), 1st July 1915.

The men in four boats, containing 140 men of 7th Battalion's B Company, began to row.

"When we had got within about 400 yards of the shore, we were greeted with a hail of bullets, and it being our first time under fire, you can imagine how uncomfortable we were. Lieut. Scanlon,[295] who was in charge of the boat, yelled out: "All except the rowers, keep your heads down," an order which needed no repeating. So the target for the enemy was minimised. One by one the rowers were shot, but as soon as one of them was unable to continue rowing, another man would sit up and take the oar from his fallen comrade. I might say that during the whole time Lieut. Scanlon stood up in the boat, encouraged the rowers and men, and gave the direction to the man at the tiller thus we were able to reach shore."[296]

CPL ALFRED YOUNG, 7TH BATTALION, 2ND BDE, 1ST AUSTRALIAN DIVISION.

"The first to go was a fine fellow in Vernon Brooks,[297] who was rowing, and cried out, "They've got me," and dropped to the bottom of the boat dead. Next was poor old Joe Cowan,[298] another rower, who got one through the stomach, but was lucky enough to recover."[299]

L/CPL JOHN KIELY, 7TH BATTALION, 2ND BDE, 1ST AUSTRALIAN DIVISION.

"One of my men was hit first through the neck, then I was next to get it through the chest. It nearly knocked me out of the boat. I quivered like a rabbit, but pulled myself together at once. I was steering the boat at the time, so had to let go with my hands and use the tiller with my foot. I managed to keep the nose for the shore. The poor chaps that were rowing stuck bravely to it, and so did the others. You could see a look of revenge on their faces. As soon as we bumped the shore they scrambled out except myself, three killed and seven wounded. I tried hard to get out, but I got as far as amidships and there I had to stay in the bottom of the boat."[300]

LT ALBERT HEIGHWAY, 7TH BATTALION, 2ND BDE, 1ST AUSTRALIAN DIVISION.

"The boat which I occupied was the last to get to the beach, and we witnessed the others given a regular "hell." The Turks' trench was 100 yards from the shore, so you can imagine what a splendid target we were; and being so crowded we could not use a single rifle ourselves."[301]

CPL RUPERT LEE, 7TH BATTALION, 2ND BDE, 1ST AUSTRALIAN DIVISION.

"As I was climbing out of the boat a bullet hit me in rather an awkward place — on

---

295   Lt, later Lt-Col. John Scanlan, DSO & Bar, 7th & 59th Bns, AIF. Wounded at Helles on 8th May 1915.
296   *The Essendon Gazette and Keilor, Bulla and Broadmeadows Reporter* (Moonee Ponds, Victoria), 15th July 1915.
297   Pte Vernon Brooks, 7th Bn, AIF, was killed on 25th April 1915. Commemorated by a special memorial at No. 2 Outpost Cemetery, he was the 22 year-old son of Robert Thomas and Agnes Brookes, of 3 Fern Ave, Windsor, Victoria.
298   Two men called Cowan were wounded with 7th Bn on 25th April: Pte Albert Walker Cowan and Cpl Charles Davenport Cowan.
299   *Echuca and Moama Advertiser and Farmers' Gazette* (Victoria), 21st October 1915.
300   *The Corowa Free Press* (New South Wales), 18th June 1915.
301   *Numurkah Leader* (Victoria), 25th June 1915.

the left buttock. We then sprinted across the beach and took cover behind some small sand heaps, leaving some more of our men on the beach. They had got struck in crossing the beach. During the sprint I received my second present in the calf of the leg, but it did not disable me… The poor chap next to me was shot three times in as many seconds — the first time in the abdomen. The second shot shattered his knee and the third broke his arm. His agony was awful and in his great pain he implored me to shoot him. Just then a fourth bullet got him, and that ended his suffering."[302]

CAPT. HERBERT LAYH, 7TH BATTALION, 2ND BDE, 1ST AUSTRALIAN DIVISION.

"Next was Sgt Collins… He made an attempt to get to the bank, but got five shots into him before he got to the water's edge. I then made a dash, falling into the water behind the iron side of the dinghy. By this time others had woken up to this cover, and we shoved the boat into the bank (long ways along the bank), and tried to fire, which was not a success. Sergeant Collins, with his bad wounds, was still giving us orders, and we could not give him any assistance as it meant death. When we fixed bayonets and were about to charge we saw the Turks in full retreat."[303]

L/CPL JOHN KIELY, 7TH BATTALION, 2ND BDE, 1ST AUSTRALIAN DIVISION.

"After an interval of about half an hour our little party (16 in number) charged the trenches of the enemy, who at the sight of the bayonets, fled, and we then occupied some "tailor made" trenches. In a short time not a Turk was to be seen. This was confirmed by a patrol, which we sent out."[304]

CPL RUPERT LEE, 7TH BATTALION, 2ND BDE, 1ST AUSTRALIAN DIVISION.

"One was a sad case, that of Powley (Wangaratta, Victoria), who was in our platoon. Charlie[305] lay on the bank dying with a bullet wound in the head, his brother Jack[306] was with us after looking for him and found him in a dying condition and could do nothing for him. Wounded were lying all over the bank. Just before the firing ceased we got Sgt Collins under cover, and made him comfortable until the medical men arrived and fixed up all the bad cases."[307]

L/CPL JOHN KIELY, 7TH BATTALION, 2ND BDE, 1ST AUSTRALIAN DIVISION.

"He was shot when we were landing on the beach, and I was by his side. No words can tell how broken-hearted I am. He was shot through the head, and I knelt down beside him to say, 'Good-bye,' but Charlie did not answer me."[308]

PTE JOHN POWLEY, 7TH BATTALION, 3RD BDE, 1ST AUSTRALIAN DIVISION.

302 *Bairnsdale Advertiser and Tambo and Omeo Chronicle* (Victoria), 19th June 1915.
303 *Echuca and Moama Advertiser and Farmers' Gazette* (Victoria), 21st October 1915.
304 *Numurkah Leader* (Victoria), 25th June 1915.
305 Pte Charles Powley, 7th Bn, AIF, was killed in action on 25th April 1915. Commemorated by a special memorial in No. 2 Outpost Cemetery, he was the 21 year-old son of Edward Hall Powley and Arabella Jane Powley, of Boweya, Victoria.
306 Pte John Powley, 7th Bn, AIF, was wounded on 25th April 1915. A second wound on 8th/9th August 1915 led to his return to Australia, leaving Suez on 31st October 1915.
307 *Echuca and Moama Advertiser and Farmers' Gazette* (Victoria), 21st October 1915.
308 *Glenn Innes Examiner* (New South Wales), 17th June 1915.

> "The Turks kept up a continuous fire on the boats, although they knew they were full of dead and wounded. While I was lying there one just took the skin off my shoulder, and another ripped my haversack. Dozens of them went within an inch of me. When the bullets cut through the sides of the boat you could smell the burning paint. A shell came screaming towards us once. It touched the side of the boat and went into the water without exploding, but sent up a huge spray that nearly drowned us. About three hours afterward, the Red Cross came to our assistance, and our wounds were doctored. I was for seven hours in a cramped position in that boat before I got on to the hospital ship."[309]

LT ALBERT HEIGHWAY, 7TH BATTALION, 2ND BDE, 1ST AUSTRALIAN DIVISION.[310]

> "All this time concealed batteries were shelling the boats pulling back from ship to shore and from shore to ship. If anyone deserves praise it is the navy boys who worked these boats unceasingly all day, and never got a chance to retaliate. Is it a wonder that several grabbed rifles from the wounded men on the beach and tore up the cliff to the firing line to avenge their comrades?"[311]

SPR HARRY MARSHALL, DIV. SIGNAL COY, AUSTRALIAN ENGINEERS, 1ST AUSTRALIAN DIV.

Colonel James McCay, commanding 2nd Australian Brigade, conferred with Sinclair-Maclagan after landing. He was persuaded to divert his men away from their objective of Sari Bair and Hill 971, to turn south, to deal with the perceived threat from the direction of Gaba Tepe that most concerned Sinclair-Maclagan.

> "On reaching the beach there was a certain amount of confusion. Men from all four battalions of the 2nd Brigade began landing at the one time, to find on the beach many men from the 3rd Brigade who had gone forward. Because of the landing being made a little farther north than was anticipated or intended, the 3rd Brigade had gone to the left flank, and the 2nd Brigade, after a hurried consultation between the two brigades, moved to the right flank… But practically all semblance of company and battalion formation was lost."[312]

MAJOR WALTER CASS, BRIGADE MAJOR, 2ND BDE, 1ST AUSTRALIAN DIVISION.

> "7.17. Someone says there are men on the skyline. And through the telescope I can see them — a few at this point — more of them further along the skyline — why there are crowds of them. Some are standing up — others moving over the hill — others sitting down apparently talking. Are they Turks or Australians? The Turks wear khaki, but the attitudes are extraordinarily like those of Australians — something of the stockyard fence about them. Behind them — I think on a nearer ridge — a long line of men is quietly digging on a nearer hill. Time and time again I have seen the Engineers digging in the desert at Mena in just such a line. Surely those are the round disc-like tops of our men's caps.

---

309 *The Corowa Free Press* (New South Wales), 18th June 1915.
310 Evacuated to Egypt, Heighway's wounds led to his return to Australia, leaving Suez on 5th July 1915.
311 *The Bundberg Mail and Burnett Advertiser* (Queensland), 15th June 1915.
312 Walter Cass quoted in Schuler, *Australia in Arms*, pp. 109–110.

"There can be no question of it. Everybody knows it now. Those men are Australian — and whilst we are looking for them on the nearer ridges and especially that shoulder rising from the beach on the right they were right back there on the further hills. I can't say what a load that has lifted off one's mind. Well done, boys — great work."[313]

CHARLES BEAN, AUSTRALIAN WAR CORRESPONDENT, HMT *MINNEWASKA*.

"On shore all was busy. Companies, battalions, brigades were landing how and where they could... I found that the battalion, instead of being in support, had gone straight into the firing line, and the 5th, which was apparently late, was now in support, but was just going forward to reinforce. We were climbing up hill, into valleys, up again, through tangled undergrowth and low scrub, almost too steep in places to climb, and where we could advance only in single file. The bullets which were meant for our firing line were now zip-zip-ing over the hill on to us, but doing no damage. I took my platoon just over the ridge, extended, and came into the firing line, on the right of the 6th, and the left of the 5th, up with the 12th...

"For nearly two hours we were lying here; now a few men would up and rush a few yards, then one or two would attempt to get ahead. Then, crawling and creeping, we still struggled to get near enough to use the bayonet. Every time more men came up to us, or any of us got up to run forward, it was the signal for a burst of bullets. All this time, too, shells were bursting over us and in rear. At one time four men and I had been lying in the one place for a considerable time unable to move; the fire ceased momentarily, and we up and dashed on about 10 yards. We were no sooner down than a shell burst just where we had been lying, and made a hole big enough to bury a small horse. I think we all looked at each other with a sickly sort of grin. Our line was very thick here; but our casualties were awful. Every few yards there was a dead or a wounded man... A man right next to me had a bullet through the peak of his cap. The bullet came downwards and buried itself in the earth right under his chin. Another fellow on the other side got one in the legs, and was groaning. I told him to get to the rear, but before he could move he got another in the side. A man named Roche,[314] an old man-o'-war's man, who was in the China war, just groaned and sobbed, a little cough, and died. He got one through the head. His death affected me very much. He was a good soldier; he was sticking close to me on the field. I felt he was a tower of strength."[315]

LT EDWIN SPARGO, 6TH BATTALION, 2ND BDE, 1ST AUSTRALIAN DIVISION.

"Major Fethers sent me to reinforce the firing line, and when I was about half way to it I got the first hit through the left arm above the elbow. That made me feel as cool as a cucumber, but I did not get more than a couple of hundred yards ahead of this spot when I thought the world had come to an end — a shell exploded a few feet from my head, and I promptly went to sleep. That was about 9 o'clock."[316]

LT EDWARD McVEA, 5TH BATTALION, 2ND BDE, 1ST AUSTRALIAN DIVISION.

313   *The Mildura Telegraph and Darling and Lower Murray Advocate* (Victoria), 9th July 1915.
314   Pte David Roche was killed on 25th April. Originally from Cork, Ireland, the former sailor is commemorated on the Lone Pine Memorial.
315   *Terang Express* (Victoria), 11th June 1915.
316   *The Sydney Mail* (New South Wales), 23rd June 1915.

"Battalions were mixed up hopelessly, owing to the mix-up in landing; but notwithstanding that, wherever the men found themselves — whether with strange officers, or N.C.O.'s, or by themselves — they worked splendidly together. All knew the objective, and pushed on towards it, which after all is the golden principle of attack stripped of its verbiage. Some of the boys in their enthusiasm, went too far forward — a thing easy to do in this involved country, even when equipped with a map and compass as I was; and it was during a little expedition of mine on the far-flung left flank induced by the sight of men going too far out, that I met my 'pill.' This was at 9.15 a.m., about a mile inland, when, after leaving Rockliff in charge. I set out to find my headquarters. My search had been fruitless (it has since transpired that I had outstripped them), and I sat down to consider the situation when I noticed men about half-a-mile in front. Thinking they were our own people, I set-out after them, caught them up, and found them to be a mixed force. After gathering them together, a messenger rushed by calling for reinforcements for Ryder,[317] of the 9th battalion. I decided to help him out, and with that object started forward. The fire became very heavy, and I could see that with our force we could not hold out for long. Ryder sent me a message to say that he was going to get back on to a hill to my right rear and accordingly I waited to cover the movement. As men drifted back, I drew them into my fold, and finally had about one hundred men with Lts. Newman[318] (11th battalion), Haig, and Loutit[319] (10th battalion). Our friends the enemy soon discovered us, and vented his wrath upon us with machine guns and rifle fire; but as we kept down it passed over us, much to their disgust. Meanwhile… we could see reinforcements of our side forming up; and as our position covered the second objective of the 11th battalion. I was about to write a message asking for reinforcements when I felt something like a salute from the business end of a mule land on my right arm. Newman was watching me write, and did not realise that I was hit until I told him — quietly — as I did not want the men to know. He would not believe me until the blood began to gush out. I had no pain, as the arm went quite 'dead' immediately; and I had no difficulty in making my way back two miles to the Medical Officer… The enemy were hailing shrapnel in the gullies and on the beach as we (the wounded) passed along, and many wounded men 'went home,' together with their stretcher-bearers. As an ex-artillery man I could not help admiring the artillery work of our enemies, which was splendid. Of course I've not the slightest doubt that the guns were run by Germans… At 12 noon, after a furious argument — culminating in a direct order from Col. Howse, V.C.,[320] Assistant Director of Medical Services to go aboard — I shouldered my few belongings and embarked on the Hospital ship *Glasgow*."[321]

CAPT. JOHN PECK, 11TH BATTALION, 3RD BDE, 1ST AUSTRALIAN DIVISION.

317   Capt. John Ryder, 9th Bn, AIF, was wounded on 25th April 1915. He returned to Australia aboard the *Ascanius*, leaving England on 17th March 1916.
318   2/Lt John Newman, 11th Bn, AIF. Later promoted to Major, he was awarded the DSO for his conduct during the Third Battle of Ypres.
319   Lt Noel Loutit, 10th Bn, AIF. He was evacuated from the peninsula on 4th August 1915 after falling ill. He ended the war as a Lt Colonel and decorated with the DSO & Bar for his actions on the Western Front.
320   Col. Neville Howse, VC, Assistant Director Medical Services, 1st Australian Division. He was the first Australian to be awarded a VC after rescuing a wounded man under fire at Vredefort on 24th July 1900 during the Boer War. He ended the war knighted and a Major-General.
321   *The Grenfell Record and Lachlan District Advertiser* (New South Wales), 27th July 1915.

"There were a lot of wounded Turks about but as there were so many of our own wounded I had not much time to look at them, besides they had no field dressings like our men carry. I gave some morphia to a few of them, but most of them spat it out... On the plateau I met my A.M.C. Sergeant, and it was very fortunate, as two can do better than one, especially with fractures and bad hemmorrhage [sic] cases. We fixed a couple of shattered legs and went on down into the big gully — along that for a bit and up on to the top of the main ridge, which our fellows had just taken. The first wounded man up there that I struck was Peck[322] our adjutant — he had a bullet through the shoulder — just missed the bone."[323]

CAPT. EDWARD BRENNAN, AAMC, ATTD 11TH BN, 3RD BDE, 1ST AUSTRALIAN DIVISION.

"We... scrambled down the other side, only to be confronted by a steeper hill. This was also cleared, and we then halted to reform under cover. As soon as this was done we moved over this hill and here we saw several tents of the Turks. These were given a wide berth as it was feared they were mined, and worked over to the third ridge, when the enemy gave us a heavy fire... However, we occupied the third ridge at about 7 a.m., and orders came from the brigadier that this position must be held at all costs, and to dig in, so with entrenching tools, picks and shovels, we proceeded to throw up cover."[324]

LT HERBERT ORBELL, 12TH BATTALION, 3RD BDE, 1ST AUSTRALIAN DIVISION.

"I hadn't got a hole big enough to hide a bottle of West End when they counter-attacked us, but I kept pegging away with my old entrenching tool every chance I got, and I finished up by getting a fairly snug place, though I used to get stiff as a board lying in one position. I made it large enough to kneel up and fire at last."[325]

PTE FREDERICK RICHARDSON, 10TH BATTALION, 3RD BDE, 1ST AUSTRALIAN DIVISION.

"Reinforcements were then being called for on our right flank, and Major Hurcombe[326] was bringing ammunition to us, telling us to hold on as the artillery had landed. The times were exciting, even for those like myself, who had been at the game before. It put me in mind of January 6, 1900, when the Boers made their fierce attack on Ladysmith. Soon after I got back, I felt a stinging pain in my left foot, but thought no more about it, and believing it was cramp — until we changed our position, so as if possible to upset their range. I could not walk, owing to having been shot through the foot. The wound must have been what I felt previously. It was far safer to remain where I was. I hung on until about 3 p.m., when I was obliged to crawl back out of it."[327]

PTE CYRIL HORSFIELD, 10TH BATTALION, 3RD BDE, 1ST AUSTRALIAN DIVISION.

---

322   Capt. & Adj. John Peck, 11th Bn, was wounded on 25th April. He ended the war as a Lt Col. Awarded the DSO and CMG.
323   *Great Southern Star* (Leongatha, Victoria), 20th July 1915.
324   *Daily Telegraph* (Launceston, Tasmania), 8th September 1915.
325   *The Advertiser* (Adelaide, South Australia), 7th July 1915.
326   Maj. Frederick Hurcombe, 10th Bn, AIF, was hospitalised on 2nd July suffering from a nervous breakdown.
327   *The Express and Telegraph* (Adelaide, South Australia), 11th June 1915.

"Just ahead of us was a bit of ground which seemed to afford some natural cover. I decided to get there. I called up the men near me, and ordered a dash. Four or five of us were up and off. I had gone about 15 [?] yards when crash — a sledge-hammer struck me in the region of the heart. I said, "That's through the heart, I'm dead," was perfectly satisfied and commenced to die. I made no elaborate swan-song, but thought of the bad luck it was to little mother. All of a sudden I found I was better. My dying was only a temporary faint. I thought I must be only bruised. I put my hand under my coat and felt a little warm blood. A man next to me named Williams, an Irishman, I think, opened my tunic, and there was a small neat hole. I was again certain I was shot through the heart, and wondered more why I wasn't dead. Williams picked me up, but no sooner had we moved than zip-zip went a shower of bullets. We got down again. I ordered him to leave me alone, and reminded him that it was a crime to leave the firing line to help a wounded man, and in his best brogue he cried, "That's no ——— good to me; I'm going to get you back. I think I fainted again, for I found him with his legs twisted round mine, dragging himself on his back. We argued and cursed in desperation. I got up, and to get rid of him bolted with all the strength I had left, and, strange to say, got across that fire-swept zone absolutely unharmed."[328]

LT EDWIN SPARGO, 6TH BATTALION, 2ND BDE, 1ST AUSTRALIAN DIVISION.[329]

"On the top of a hill we found some men from every battalion, who said there more ahead, so we set off. We only got about 30 yards when down we went flat, for such an inferno greeted us as would be difficult to imagine. We continued to crawl forward, for the machine guns were firing over the tops of the bushes. We progressed in this manner for a while, and the firing redoubled in intensity, and shrapnel exploded closer. At this point I funked it, and calling out to my mate, I found that he was of the same mind, namely, that it was an absolute cert that if we moved an inch from our bush we would be cut into mince meat. We both agreed that we were well gone in any case, but if we stayed where we were we might do some good, and might have a million to one chance of hanging out till it got dark. Do you think it was cowardly? Our firing line was supposed to be ahead of us (I now have my doubts), though we were well ahead of our platoon. If there were men ahead of us it was our place to go on. On the other hand, I really believe that it was absolutely certain death to go from our bush, for we would have been riddled, in a second, and be no more good, but a soldier should never think of himself.

"Sometimes I think I did right, and at other times I am worried. We had been there about five minutes when my mate gave a yell and cried out that he was hit. Keeping flat I swung round and found he had been shot through the leg with shrapnel. I ripped open his field dressing and applied it, but I seemed to be all thumbs, and it was all soaked before I got it half on. I then wound his puttee round it, but the blood still poured through. All the time he kept calling to me to never mind him, but to look after myself. I swung back to my bush and picked up my trenching tool to dig in

---

328 *Terang Express* (Victoria), 11th June 1915.
329 Spargo was killed on 7th August 1915. Commemorated on the Lone Pine Memorial, he was the son of James and Jessie Spargo, of 173 Tooronga Road, East Malvern, Victoria.

when a bullet struck it in front of my head. Almost immediately another grazed the place where one is popularly supposed to get it when running away. A second or two after I got it bang. I roared louder than a bull. My poor wounded mate worked his way over to me, got my equipment off and ripped open tunic and shirt. By this time I had recovered somewhat, and finding I was not blown up started to drag myself to the rear, I remember going about 60 yards and sitting up behind another bush, feeling much stronger, but thinking that nothing mattered much. Then I got on my feet and waddled, back over the hill."[330]

PTE JAMES COUGHLAN, 6TH BATTALION, 2ND BDE, 1ST AUSTRALIAN DIVISION.

"The 3rd Battalion's turn came about 8 a.m., and A Co. went first. I was O.K. until I saw the bodies of four poor beggars on the destroyer covered over with a tarpaulin. The blood was running out from under it, and it quite upset me. Didn't get my nerve back until we got into the rowing boats, and then I was O.K."[331]

PTE HERBERT SMYTHE, 3RD BATTALION, 1ST BDE, 1ST AUSTRALIAN DIVISION.

"It was just breaking dawn, and as we looked towards the sound of the firing we were faced by almost perpendicular cliffs, about 200 ft. above sea level, and, as we were of opinion that most of the fire was coming from this quarter, it was evident that this was the direction of our attack. Therefore, after a minute or two, having gained our breath, we started to climb. Soon I came upon Colonel Clarke and Lt Patterson, and together, on our hands and knees, we climbed to the top of the first ridge. Up to this time I had not seen a sign of a Turk, but as we moved a little to our left we discovered a trench overlooking the beach, and, fixing bayonets, we received the order to go for it, but unfortunately the Turks had no desire to wait for us; when they saw the bayonet they cleared out in great disorder, leaving much ammunition and some equipment.

"Our men opened fire on them, and several of the enemy were wounded. They retired on to the forward slope of a rise, about 1000 yards to the rear and to the left, and here took up a position in the scrub. We pursued them, and opened fire at about 350 yards. Colonel Clarke, who was about 25 yards to my right, called for a signaller, and commenced to write a report for British headquarters, but was shot through the heart and died at once. Private Davis was also killed here, and Major Elliott, going to the rescue of the colonel, was shot through the shoulder and the elbow, fracturing his arm. Lt Patterson now took charge of the party, and gave orders to hold on until more men came to hand. Very shortly after this a patrol sent out reported that the enemy had retired over the hill. Captain Burt then came up with more men."[332]

LT IVOR MARGETTS, 12TH BATTALION, 3RD BDE, 1ST AUSTRALIAN DIVISION.

"I managed to get my chaps on a spur half-way up, with the loss of only one, and we stopped to help some others on our left, who had some rather exposed country to cross. They had just made one beautiful rush, which cost 'em about ten men, and we were able to enfilade the Turks' trench beautifully at about 600 yards. We cleared 'em out, bagged

330   *Omeo Standard and Mining Gazette* (Victoria), 25th June 1915.
331   *Jerilderie Herald and Urana Advertiser* (New South Wales), 9th July 1915.
332   *The Examiner* (Launceston, Tasmania), 4th August 1915.

four or five out of a squad a bit nearer on our left; and then continued our climb. We got to the top about 6.30, just in time to see some of the 12th Battalion bowled over. I saw both Col. Clarke,[333] of Tasmania, and Major Elliott[334] hit. The Colonel was killed instantly.

"By 7.30 or 8 we had driven the Turks out of all their trenches, within three-quarters of a mile of the beach and were able to take stock a bit. I found I had been landed a long way from the rest of the battalion, and was among a lot of the 12th, 9th and 10th about 200 in all. Captain Lalor was the senior officer left. We were about half a mile to the left [on Battleship Hill] of the rest of the brigade, but as it was an important ridge, we decided to stay as long as possible. As it turned out, it was as well we did. We were all mixed up as I told you. While Lalor's[335] people dug themselves in a little, I and about 50 of the tenth went forward to cover them. Almost as soon as we got to a position the fun started. A little after 9 their counter attack commenced strongly and at 10.30 we who were covering, had to get back. From then on it was plain hell."[336]

CAPT. ERIC TULLOCH, 11TH BATTALION, 3RD BDE, 1ST AUSTRALIAN DIVISION.

"After a stiff climb we got to the top of the cliff, and then found that our colonel, Lt-Col L. F. Clarke, D.S.O., of Tasmania, had been killed, and Major C. L. Elliot (West Australia) wounded…

"Rushing forward a little, we got under cover, and attempted to form up the different companies. Captain Lalor, Captain L. E. Burt, Lts P. J. Patterson, and I. S. Margetts, of the 12th, and Captain E. W. Tulloch, of the 11th, an old Ballarat College boy, were amongst those present. Word was sent along to us to prepare for a counter attack, and Captain Lalor, who took command, decided to hold a spur just ahead of us. We had just got to it, and I was writing a message for him to the brigadier, when the Turks were seen coming over some rising ground (Hill 971) on our left front. Captain Lalor instructed me to advance and hold them back while he dug in."[337]

LT EDWARD BUTLER, 12TH BATTALION, 3RD BDE, 1ST AUSTRALIAN DIVISION.

"I was the only colonel who went past the first ridge with his battalion. Col. Lee, commanding the 9th, sprained his ankle;[338] Col. Johnston, of the 11th, was nearly drowned getting from the boat to the shore; and Col. Clarke, of the 12th, was killed on the first ridge."[339]

LT-COL. STANLEY WEIR, 10TH BATTALION, 3RD BDE, 1ST AUSTRALIAN DIVISION.

333 Lt-Col. Lancelot Clarke, DSO, 12th Bn, was killed in action on 25th April 1915. Buried in Beach Cemetery, he was the 57 year-old son of the late Joseph Johnston Clarke and Charlotte Elizabeth Clarke; husband of Beatrice Fox Clarke, of Holebrook, Tasmania.
334 Major Charles Elliott, 12th Bn, was wounded on 25th April. He assumed command of the battalion on his return on 19th September, relinquishing it after the withdrawal to Lemnos on 8th December. Elliott was wounded again on 10th August 1918, ending the war with a DSO & Bar and the *Legion d'Honneur*.
335 Capt. Joseph Lalor, 12th Bn, was killed in action on 25th April 1915. Buried in Baby 700 Cemetery, he was the 30 year-old son of Joseph Peter and Agnes Lalor, husband of Hester Lalor, of Elgin Street, Hawthorn, Victoria, Australia.
336 *The Daily News* (Perth, Western Australia), 16th June 1915.
337 *The Sun* (Sydney, New South Wales), 22nd August 1915.
338 Weir was being diplomatic. Harry Lee's nerve had failed him shortly after landing. He did not take any part in the action. A medical board on 5th April 1916 concluded that he was "unfitted for the strain of Active Service." NAA: B2455, LEE HARRY WILLIAM.
339 *The Register* (Adelaide, South Australia), 4th August 1915.

"As we were mixed up with men of other battalions, Captain Burt decided that, owing to a slight lull in the fighting, it would be better to re-form. When this was completed we moved up a little in advance of our first position, and Captain Lalor, who had just come up, decided to hold the position and dig in, but afterwards decided to advance to the ridge over which the enemy had retired, so we pushed on. It was soon discovered that the enemy was in strong force, and were attempting to get round on our left flank, consequently that flank retired, and we had to follow suit, but yet again we pushed up when reinforcements arrived from the Second Battalion. I was then ordered to line the ridge overlooking the deep ravine running up from the beach, but soon afterwards was sent up to reinforce the front again. When I got up I found that ammunition was short, so went back to try to get some sent up, and also to find out where the rest of my men were. I joined Lieutenant Patterson, and after resting a few minutes under shrapnel I again joined the firing line, doubling forward through the scrub, and frequently falling, owing to the thickness of the undergrowth. The fire came very hot from a ridge on our left flank, and a party of the Second Battalion occupied it. I believe they had a machine gun. We could see people moving on our left flank, but did not shoot at them, as we were informed that Indian troops were on our left. This we afterwards found out must have been a ruse of the enemy."[340]

LT IVOR MARGETTS, 12TH BATTALION, 3RD BDE, 1ST AUSTRALIAN DIVISION.

"[A]t 8.20, by common consent, we came to a halt. We waited to get some definite orders through, and meantime many were wandering about asking such questions as "Where is such and such a company or battalion? Where was Major So-and-so?" &c. When they did find their mates they got into little groups and discussed the casualties so far. The scene looked like some ordinary day's work on Lemnos. Our company O.C. came along and hustled us up a bit, and said — "You had better start to dig in, as you will be getting shrapnel over here in half an hour's time." After ten minutes' spasmodic work with entrenching tools he said — "Take it easy, lads, and have a smoke-oh. We're sending down to the beach for tools." Work ceased, and the extraordinary sight of men drying their socks and jerseys could be seen. There might not have been a Turk within twenty miles of us, for all the chaps seemed to care."[341]

PTE EDWARD WHITE, 9TH BATTALION, 3RD BDE, 1ST AUSTRALIAN DIVISION, AIF.

"Our brigade landed about 8 a.m. I was [with] the brigade and regiment headquarters staff... We moved up a gully and watched the remainder land... We were kept in reserve while the remainder reinforced the 3rd Brigade.

"Soon the horrors of war came upon us, as hundreds of wounded men were being taken down the gully to the beach."[342]

SSM JOHN SLOAN, 4TH BATTALION, 1ST BDE, 1ST AUSTRALIAN DIVISION.

---

340   *The Examiner* (Launceston, Tasmania), 4th August 1915.
341   *The Age* (Melbourne, Victoria), 26th June 1915.
342   *Casino and Kyogle Courier and North Coast Advertiser* (New South Wales), 3rd July 1915.

"We landed in a bad place, and it's just as well. The Turks were expecting us at another place, and had we gone there we would never have got ashore. They had guns and machine guns, splendid trenches, obstacles, and even barbed wire entanglements and mines in the water to welcome us with. Where we actually did land was not very strongly guarded and we sort of surprised them, and we had got ashore and established ourselves before they could bring sufficient troops to prevent us...

"After a short rest in a ravine we pushed on. Talk about hill! we had to simply pull ourselves up by the undergrowth in places. None of them were very high, but they were all very steep, and we had to stop for a spell every little while. It was during one of these short spells that we had our first casualty. A bullet got Sergeant Cavill[343] in the neck and killed him. The bullets from the fighting in front were flying around pretty thick. You could hear in every direction the sharp crack as they passed. Finally we got on top of a hill with a pretty good trench in it. The fact that it was a Turkish trench didn't worry our consciences in the least. We just took possession of it and inwardly thanked the Turks for saving us the trouble of digging one. Unfortunately it had no field of fire, so we got up on to the crest of the hill and tried to pick out some of the Turks who were now potting at us... But we could not see a sign of them as the whole country is covered with scrub about 3 ft high... We dug ourselves in, so that we were safe from rifle fire. It's lovely work lying as close to the ground as a snake and trying to dig yourself a trench at the same time."[344]

PTE HERBERT SMYTHE, 3RD BATTALION, 1ST BDE, 1ST AUSTRALIAN DIVISION.

"9.40. Left the ship waving good bye to those we are leaving behind... As we come in now one can see the hills our Third Brigade went up this morning. It is an astonishing feat. The place is just like a section of the Blue Mountains in N.S.W. To north and south on either side there is a plain up which the guns of the navy can sweep I suppose for some way. Between these plains rises a sheer ridge. [In] front of us at one point high above our heads it curls to a point in an absolute precipice with what looks like a knife edge on top [the Sphinx]. Even the other hills are half precipitous, some of them covered with a low scrub and grass...

"The destroyer has stopped and three or four boats have come alongside... We scramble into the bows of a whaleboat alongside — she is quickly filled with men, four seamen get to the oars. A row for a hundred yards or so — the boat grounds about ten yards from the beach. We jump out; it is not an easy job with a heavy pack. A splash through water about to the top of our gaiters and we are are on the shingle. About twenty yards to our left the naval signalmen are erecting a wireless pole. There are one or two wounded men being carried along the beach — troops being formed up everywhere and marched off constantly. Someone says the first two boats suffered heavily — there is no sign of our dead."[345]

CHARLES BEAN, AUSTRALIAN WAR CORRESPONDENT.

---

343 Sgt Walter Cavill, 3rd Bn, AIF, was killed on 25th April 1915. Commemorated on Lone Pine Memorial, he was the husband of Elizabeth Cavill, of Portland, New South Wales.
344 *Jerilderie Herald and Urana Advertiser* (New South Wales), 9th July 1915.
345 *The Mildura Telegraph and Darling and Lower Murray Advocate* (Victoria), 9th July 1915.

"All our section were with the Brigadier, each one doing something or reading signals. We were working direct with the gunboat by helio[graph], also by wireless. By this time our boys had a portable wireless plant buzzing on the beach under the cliff. We could telephone back and the wireless jerked the waves out to sea. We were controlling the ship's fire from our position on the hills."[346]

SPR HARRY MARSHALL, DIV. SIGNAL COY, AUSTRALIAN ENGINEERS, 1ST AUSTRALIAN DIV.

"Well as soon as we reached the beach, we were in the thick of it, wounded and dead lying all over the place. We soon had a dressing station rigged up and going full swing. As soon as we had the wounded on the beach fixed up, we scrambled up the cliffs into the trenches at the top which our boys were holding and started bandaging there. But we soon ran out of bandages, and had to use handkerchiefs, shirts and putties; in fact, anything at all. We had to carry the wounded down on our backs, as it was too steep to use the stretchers, and it's not a joke getting down a mountain-side with a 12-stone man on your back I can tell you. Things were getting very hot in the place where I was working, and I thought the Turks were going to drive us into the sea, but the New Zealanders reinforced us about 12 o'clock. They passed where I was in single file, and I caught a glance of a few of the Invercargill boys, but not many recognised me and no wonder, as I was covered in blood."[347]

PTE DOUGLAS PORTER, 4TH FIELD AMBULANCE, AAMC.

"Many of the men landed in six feet of water. Colonel Stewart and I, with more luck than the rest, got hold of a small boat and by this means got ashore quite dry. The shells at this time simply poured down, and men began to drop. Calling our men together we marched along the beach. There was a constant stream of wounded. The sight of the dead strewn about, some lying on the shingle, some half in and half out of the water, the boats half-full of wounded, the passing along of orders — all this made us realise for the first time the grim realities of war. It is necessary to explain that only the First and Second Canterbury Companies, together with the Headquarters personnel, landed at this time. The Twelfth and Thirteenth, being on another ship, were to come along later.

"Our orders were to climb a spur running at right angles to the beach and reinforce the Australians, who were on a plateau 900 feet above the sea. The track leading to it was so very narrow that troops could work their way up only in single file. I was left near the landing-place, with the signallers and other Headquarters men, to pick up the two remaining companies and take them up after the others. To be left among the dead and dying was not to my liking, so leaving some men to watch the boats arrive I scaled the ridge, but on the way met Stewart and Captain Sarginson coming back. Stewart said that the track was blocked, and the men would take some time before they got to the top."[348]

MAJOR ALBERT LOACH, CANTERBURY BN, NZEF, NEW ZEALAND BDE, NZ & A DIVISION.

346 *The Bundberg Mail and Burnett Advertiser* (Queensland), 15th June 1915.
347 *Southland Times* (New Zealand), 16th February 1916.
348 *Lyttelton Times* (New Zealand), 29th July 1915.

"I landed with the rest of our company about 11 a.m., and we advanced up the cliff. It was pretty stiff going. It had to be negotiated in single file up a track which had been cut by the Engineers. As soon as we arrived on top of the hill, we were under a perfect hail of bullets, though they were going very high. Then the artillery started dropping shrapnel, which was more upsetting than the rifle fire, but after a few shells, we became used to it.

"All this time we were advancing towards our firing line. Our side was getting a pretty bad time, as we had only rifles against their artillery."[349]

SGT CLIVE SWEARS, AUCKLAND BATTALION, NEW ZEALAND BDE, NZ & A DIVISION.

"No sooner had we touched the shore than we were rushed up the hill, having first thrown off our packs, and made straight for the firing line. Just as we reached the top of the hill the shrapnel was turned on us. It was awful, the poor beggars were downed in dozens. A dozen of our chaps were put out of action by one shell, one being killed (Sergeant Bruce[350]), and the remainder wounded — Clive Swears, Sergt. Robertson,[351] Lieut. Wooley,[352] Private Seed,[353] and seven other chaps whom I did not know — and I was almost in the centre of the lot, and was not even scratched."[354]

CPL VINCENT HOLLIS, AUCKLAND BATTALION, NEW ZEALAND BDE, NZ & A DIVISION.

"In taking a short cut across the top of a ridge and endeavouring to dodge under the hail of bullets in my hurried advance, I fell about 20 ft and sprained my ankle. Not being able to walk, I was taken off."[355]

PTE ARTHUR PHILLIPS, AUCKLAND BATTALION, NEW ZEALAND BDE, NZ & A DIVISION.

"After watching the scene for about thirty minutes I applied for leave to go up and see if all was well, but I was ordered to stay where I was, to keep in touch with the Colonel and await the arrival of the Twelfth and Thirteenth Companies, which were late in landing. The scene at this stage was worthy of description. Out on the Aegean Sea transports were constantly arriving. Between them and the beach was a row of warships belching forth their deadly missiles. Destroyers, laden with men, and towing barges were hurrying to and fro, the arriving barges being filled with men alive and well, while those retiring were filled with wounded, dying and dead.

"On the beach all was bustle. Troopers were landing the ammunition, mules and donkeys were being beached, chaplains were burying the dead and medical officers were attending the wounded. There were dreadful sights amongst the latter. Some

---

349 *New Zealand Herald*, 24th June 1915.
350 Sgt Harry Bruce, Auckland Bn, NZEF, is recorded by the CWGC as dying on 27th April. Commemorated on the Lone Pine Memorial, he was the 24 year-old son of Arthur and Helen Bruce, of Kakahu, Geraldine, Canterbury.
351 Possibly Sgt-Major Alexander Robertson, Auckland Bn, NZEF. Admitted to hospital in Alexandria on 30th April, he returned to New Zealand on 15th July 1915.
352 Lt, later Capt., George Woolley. He was invalided home on 5th November 1916.
353 Pte Henry Seed, Auckland Bn, NZEF.
354 *Thames Star* (New Zealand), 8th July 1915.
355 *The Star* (Christchurch, New Zealand), 18th June 1915.

men had gaping wounds, others had a hand, arm or leg blown off or shattered. Men lay all over the place. Some were to live, some to die."[356]

MAJOR ALBERT LOACH, CANTERBURY BN, NZEF, NEW ZEALAND BDE, NZ & A DIVISION.

"A wireless station had already been erected. Wounded men were being brought down the hills in stretchers every now and then, or were limping or being helped along by their comrades to where the Royal Army Medical Corps was at work. We fresh troops, however, did not stay on the shore much longer than was necessary to take off our packs. These we stacked in piles according to units, and then at once proceeded across country towards the firing line.

"I was one of a party of about thirty men attached to the Canterbury machine gun for duty until the limbers could be brought ashore. Almost all of us previously had had a three weeks' course on the gun, so that we could replace casualties. But our particular work on this first day of battle was to carry reserve ammunition, of which each had two boxes."[357]

PTE CECIL YORKE, CANTERBURY BN, NZEF, NEW ZEALAND BDE, NZ & A DIVISION.

Work to build piers got underway to make landing drier for everyone.

"I had a very ticklish job… our boat carried a lot of timber to build a landing stage to enable the troops to land, and the boat's officers called for a volunteer party to row ashore and tow the timber after them… I was detailed to get twelve men and take the timber ashore. I called for volunteers and was simply rushed, so had to take the first twelve. The timber was 6 in by 6 in., and about 12 ft long, and tied in bundles of about six or eight pieces. This was thrown overboard and all the bundles were tied together at one end. A steam pinnace from one of the war boats then came alongside and took us off our boat, and also took the timber in tow. They took us to within 400 yards of the shore, and then stopped and tied the rope that the timber was on to a buoy that was anchored there. We were then ordered to get on the timber, which we did immediately, and then the boat cleared off and left us there. I called out to the officer in charge for orders what to do, and his reply was: 'Get the timber ashore!' While he was telling me I was trying to balance myself on a bundle of timber which started to revolve, and after playing at treadmills for five or six revolutions of the timber, I lost my balance and fell into thirty fathoms of deep blue sea. It did not take me long to scramble up again, and I was thankful I could swim. When I got back I looked round for some means of getting ashore, and I discovered that attached to the buoy was a rope, with the other end on shore; so I tried to secure this rope, which was under water. In trying to lean over to get the rope I over-balanced and again had another drenching, but as I was thoroughly wet I did not mind. I eventually secured the rope and then started a long tug-of-war. It was hard work, but the boys I had were of the best. I had one on each bundle to distribute the weight, and two besides myself on the centre bundle. The man on the front end to guide the rope was one of

---

356 *Lyttelton Times* (New Zealand), 29th July 1915.
357 *ibid.*, 23rd July 1915.

our shoeing smiths and worked splendidly; but, when we landed he turned to me and said, 'Sergeant, I was scared!'

"We got the timber ashore under very heavy fire. The shrapnel was bursting all around us and hitting the timber everywhere, but not one of the boys was touched. As soon as we had made it secure on the narrow beach I made the boys take cover under the hill, as none of us were armed in any way whatever, and, went to find the colonel to report to him. By this time the beach was simply crowded with the wounded and dead, and there were some awful sights. When I got back to my boys I found them all fully armed, and with plenty of ammunition, and as there was no chance of us joining our column again for some time, we rushed up the hill to do our little bit."[358]

SGT SYDNEY NAPPER, 2ND BDE AMMUNITION COLUMN, AFA, 1ST AUSTRALIAN DIVISION.

"I was down there for about two hours. Several other wounded men were dressed, and one poor fellow was brought in who had been shot in the boats. He was quite dead, but still looked a splendid strong fellow. When I again came up on deck, after the doctor had seen to us, it was broad daylight… At about 11.30 a.m. most of us were taken on to a hospital ship, which was well into land, and were immediately attended to and put in comfortable bunks. From then on the doctors and nurses had their hands full. As the wounded were brought in they were attended to with every care, clothes being cut to pieces in cases to save pain to shattered or wounded limbs."[359]

PTE ERIC RATCLIFF, 12TH BATTALION, 3RD BDE, 1ST AUSTRALIAN DIVISION.

"The Second Battalion on our left were having rather a bad time, and Lt Patterson[360] took part of his men to reinforce them. I believe he lost his life on this ridge, as the last I saw of him he was leading and encouraging his men to reinforce the Second Battalion. Captain Lalor afterwards went up on this ridge with part of his company. He sent me down to the beach with Bugler Quantrill[361] for reinforcements and stretcher-bearers. I rolled down into the gully, sniped at all the way, and made my way towards the beach. It was just as much as I could do to get back, as in places the mud was up to our knees, and I was thoroughly exhausted before I left the firing line. I met some stretcher-bearers, and sent them up, and reported that Captain Lalor wanted reinforcements, and then went along to divisional headquarters and reported myself."[362]

LT IVOR MARGETTS, 12TH BATTALION, 3RD BDE, 1ST AUSTRALIAN DIVISION.

"About noon… I was lying in a fold of the ground dodging shrapnel and machine gun bullets, and it seemed as though I had been there for a year; anyway, I had to stop there, as to move a finger was death. All the chaps around me were dead, and we were some little distance ahead of our general firing line, so I was trying to resign myself to

---

358   *Lyttelton Times* (New Zealand), 30th July 1915.
359   *Daily Telegraph* (Launceston, Tasmania), 15th June 1915.
360   Lt Penistan Patterson, 12th Bn, AIF, was killed on 25th April 1915. The 20 year-old son of the Rev. James and Annie Patterson, of Ballarat, Victoria, is commemorated on Lone Pine Memorial.
361   Pte William Quantrill, a hairdresser from Ulverston, Tasmania. Wounded on 7th August 1915, he spent the remainder of the war attached to the Postal Corps.
362   *The Examiner* (Launceston, Tasmania), 4th August 1915.

half a dozen Turkish bayonets, when I thought I heard a voice I knew over a little rise to the right. I decided to make a dash in that direction and see what was doing, as the Turks were on three sides of me, and seemed to be advancing. My first dash was for about 20 yards, and I got drilled through the left hip joint to the groin in front. It was not painful, but bled copiously. Fortunately it missed the femoral artery by half an inch. I was unable to find anyone alive, so returned and found Erle, with his company just beyond the crest of the hill. I believe that a few minutes later Erle[363] was hit, and died heroically."[364]

LT NOEL FETHERS, 12TH BATTALION, 3RD BDE, 1ST AUSTRALIAN DIVISION.

"As nearly as I can estimate about midday we got an order to retire by threes from the right. I passed the word along to retire and re-form in a gully to our rear, but when I followed my men I found they had fallen right back on to Captain Lalor's position. We were under heavy shrapnel fire, and the shells were bursting right over our heads. By this time I felt pretty well exhausted — I had been up all night, and had been going at high pressure all day from 4 a.m. — and my legs simply refused to carry me any further. When I got back to Captain Lalor he was just about as dead-beat as I was, and we decided to spell for a few minutes. I stretched out under some bushes beside him, and in spite of the roar of guns, bursting of shrapnel and rattle of rifles, I dropped off, and was fast asleep almost as soon as my head touched the ground. How long I slept I don't know; it could not have been many minutes, when I was awakened by hearing an order given to support an Indian Mountain Battery on our left.

"On that part of our front there was a spur separated from us by a thickly-wooded gully. Three times we crossed the gully and rushed the hill, and three times orders were passed along to retire; back the men went three times before there was a chance of having the order confirmed or countermanded. These orders to retire were not given by any of our officers. I am certain; I am just as sure that they were passed along by German officers who had got into our lines dressed in our uniforms. At the fourth attempt we were more successful, and managed to form up a firing line."[365]

LT EDWARD BUTLER, 12TH BATTALION, 3RD BDE, 1ST AUSTRALIAN DIVISION.

"Our orders were to prolong the left of the Australians. I found General Bridges on the beach. His leading troops had met with considerable opposition, but had pushed on with great gallantry... some of them had penetrated a considerable distance across the Peninsula. But, in the broken and difficult country, they had got into situations where it was impossible to locate or reinforce them. My battalions had therefore to be sent up as fast as they arrived, without it being possible to indicate exactly where and how they were to connect with, and support, those who had first landed."[366]

MAJOR-GENERAL SIR ALEXANDER GODLEY, COMMANDING NZ & A DIVISION.

363   Noel's brother, Major Erle Fethers, 5th Bn, AIF, was killed on 25th April 1915. Buried in Lone Pine Cemetery, he was the 27 year-old son of James Denton Fethers and Amelia Charlotte Fethers, of "Weyanoke," Kooyong Road, Caulfield, Victoria.
364   *The Examiner* (Launceston, Tasmania), 7th July 1915.
365   *The Sun* (Sydney, New South Wales), 22nd August 1915.
366   Godley, Alexander, *Life of an Irish Soldier*, p. 170, John Murray (London) 1939.

"We reached our landing place about 1 p.m... and waited until the battleships bombarded the place a bit and then landed. Talk about big gun fire! I never heard the like before. We heard it four hours' sail away from Lemnos."[367]

PTE LESLIE LATIMER, OTAGO BATTALION, NEW ZEALAND BDE, NZ & A DIVISION.

"We were put straight into the firing line on the left flank with the Australians. We were only there about an hour when we got word to shift and reinforce the centre. We had to drop our packs and all we owned, bar rifles and ammunition and bully beef and biscuits, so that we would be light to get about, as the country was so awfully rough. We got to the position all right, but we had just landed when they started to burst the shrapnel just over our heads. It was a real hell, I can tell you."[368]

L/CPL JOHN McLEAY, OTAGO BATTALION, NEW ZEALAND BDE, NZ & A DIVISION.

"About 1 o'clock a rifle bullet hit me close to the left thigh, making a clean flesh wound almost six inches in length. I dropped back about 50 yards, had a look at it, and then went back into the firing line. Almost immediately another bullet hit my haversack and went clean through a full tin of bully beef. Ten minutes later a third bullet struck me on the top inside of the left leg and travelled about twelve inches clown, coming out within a couple of inches of the inside of the knee. I thought it was time to get out then. I started to crawl away, but had only got about a dozen yards when another bullet hit me on the right arm above the elbow, making a flesh wound... I saw Rowe getting bandaged by a doctor. I just said "shrapnel," and he nodded his head... I am beginning to think there is a poor chance of ever getting back to Hamilton if the war keeps on at this rate."[369]

L/CPL ARTHUR HAYBITTLE, AUCKLAND BN, NEW ZEALAND BDE, NZ & A DIVISION.[370]

"Three shells burst very close to me, and it nearly made me feel sick to hear the groans which went up after each one. Stretcher bearers were being called up all along the line, but there was not the slightest hope of stretcher bearers ever getting up there to their assistance... There was a major of the 10th battalion here, and he ordered us to go behind the trench and lay low until nightfall. It was now about 1 pm (as far as I could guess. Orders had been sent along the firing line to hold on till nightfall, so we settled ourselves down, and presently I found that I could hardly keep awake, in spite of bursting shells, etc. I was gradually getting to sleep when a major came back from the firing line and asked us could we fight, and I answered that we could, and that there were none of us wounded. I had hardly got the words out of my mouth when whack! I felt a trickle down my leg and a funny sort of pain, so I rolled over and slithered down the gully and made my way along a rough track, and eventually reached the dressing station, and had my leg bandaged up."[371]

PTE VICTOR PINKSTONE, 3RD BATTALION, 1ST BDE, 1ST AUSTRALIAN DIVISION.

367 *Otago Witness* (New Zealand), 7th July 1915.
368 *Southland Times* (New Zealand), 2nd August 1915.
369 *Waikato Times* (New Zealand), 3rd July 1915.
370 Haybittle, a former painter, was killed in action on 13th August 1915. Commemorated on the Chunuk Bair (New Zealand) Memorial, he was the 34 year-old son of Richard Frederick and Anne Elizabeth Haybittle, of 4 Bowen Street, Feilding.
371 *The Young Chronicle* (New South Wales), 11th June 1915.

"We had to get out of our position, but got a message from the rear (this was about 12 o'clock) to hang on, reinforcements were coming. At 1 they began to arrive in dribs and drabs — strung out by their long climb up the hill. They were New Zealanders, and the 1st and 2nd Battalions. In that hour we were driven off the crest of our ridge twice, but went back again. It was the last time we were driven back that I got mine. The men had retired, and I was just getting ready to, when I saw five Turks together about 400 yards away. Three stood up and one had a look round with field glasses. I had a rifle and let 'em have five rounds rapid. I dropped one and attracted quite a lot of fire towards myself. Then I started to sprint back, and got it in the leg — tumbled and sprinted like blazes till I got past our own firing line, where I dropped in a heap. I bandaged myself up and lay there for a while till I heard the scrap coming nearer and nearer and thought it better to get away to a dressing station if I could. So, with the help of a pick handle and a rifle I hobbled back to the edge of the ravine. The rest was to slide down for a quarter of a mile to the beach. This took me three hours."[372]

LT ERIC TULLOCH, 11TH BATTALION, 3RD BDE, 1ST AUSTRALIAN DIVISION.

"I woke up at 1.15 p.m. I had a look round, but was still very dazed. I wiped the blood out of my eyes, and felt my head. It had a gash in it, and was bleeding freely. I also felt a pain in my stomach, and rolled over to have a look if anything was wrong, and I found I had a hole there also. Shrapnel was falling like rain all this time, and no sooner had I turned over than a machine gun got to work. The bullets going over me made a noise like a swarm of bees going by. Eventually I got back to the beach, and Dad[373] put me on a boat, and sent me out to the hospital ship."[374]

LT EDWARD McVEA, 5TH BATTALION, 2ND BDE, 1ST AUSTRALIAN DIVISION.

"About 1.30 p.m. our subsection gunners received orders to stand by, to go ashore, so ten of us with our major, sergeant-major, range taker and observer, embarked with one gun and two waggons, and duly landed without any mishap. The gun was manhandled by ourselves and a number of engineers put it into position on top of the first ridge, a position which one would have thought it impossible to reach. We immediately opened fire in the general direction of the enemy, and had no sooner done so than we were fired on ourselves by what we afterwards found out to be a six-gun battery on our right flank. Our gun was swung round and laid on their flashes, and after about 20 rounds the battery was completely silenced. If the Turks had been good gunners I think our detachment would have gone to glory, as we were in an absolutely open position; but we had the luck. We swung back into our first direction and continued to pump the shrapnel on to them until dark, when we started to dig a gun pit for ourselves. We worked all night, and by daybreak were solidly entrenched."[375]

GNR JACK HEYWOOD, 4TH BATTERY, 2ND BDE, AFA.

---

372 *The Moora Herald and Midland Districts Advocate* (Western Australia), 29th June 1915.
373 His father was Lt-Col. Herbert McVea, Headquarters, 1st Australian Division, who was evacuated from the peninsula on 17th May 1915 due to knee trouble. Lt-Col. McVea returned to Australia the following month.
374 *The Sydney Mail* (New South Wales), 23rd June 1915.
375 *Bairnsdale Advertiser and Tambo and Omeo Chronicle* (Victoria), 16th June 1915.

"About 2 p.m. things had eased off a bit, and we were trying to chew a bit of biscuit and some tinned meat, when, 'Gee-whiz, bang!' came a shell right behind us. Another second, and another one arrived a bit closer. The following one landed right behind me, and the man next to me had his leg nearly blown off. I was not able to render the poor fellow any assistance, because it was instant death to move."[376]

PTE JOHN COLLESS, 3RD BATTALION, 1ST BDE, 1ST AUSTRALIAN DIVISION.

Turkish counter-attacks developed on the ANZAC's left, northern flank. It was now that the consequences of Sinclair-Mclagan's focus on his right, southern flank, of not pushing his men towards Hill 971, as per his orders, began to be revealed.

"We were on the extreme left flank, and at 2.30 the Turks put six battalions on to us. We, on the left, got a goodly share of them, as it was the key of the position, for if they beat us back they could have enfiladed the centre and right, so there was no retiring where we were. Major Robertson was doing his utmost to get reinforcements up to us, but the shrapnel was so thick that a certain regiment was unable to come to our aid. Major Robertson told me to hold a trench with 32 men (it was a Turkish trench on the extreme left second ridge), and went away to get reinforcements for us. He was bowled over, by a burst of shrapnel, and died as a brave gentleman.[377] Lieut. Rigby[378] got a bullet of afterwards, and shrapnel completed his short career as a soldier — a very short one, but he died another good example to all, in the very first line of fire, where he had been all day."[379]

SGT FRED COE, 9TH BATTALION, 3RD BDE, 1ST AUSTRALIAN DIVISION.

"It was [at 3.00 p.m.] while Capt. Leer was directing the New Zealanders that a bullet hit him to the right of the windpipe. Sgt Major Phipps[380] on his left, and myself on his right, supported him, while we tried to put a bandage on with my first field dressing. We put it to his throat, and I pressed it over the wound, and it and my hand went right into the throat. An explosive bullet had hit him, and death was instantaneous.[381] I and all of our company were in an absolutely murderous rage, but we had nothing to vent it upon… From then on, we couldn't raise our heads, the fire was so strong. My chum alongside me was hit on the left shoulder. I cut his tunic open, and dug part of the bullet out with my jack-knife."[382]

PTE SAMUEL MARSHALL, 3RD BATTALION, 1ST BDE, 1ST AUSTRALIAN DIVISION.[383]

376 *Mudgee Guardian and North West Representative* (New South Wales) 28th October 1915.
377 Major Sydney Beresford Robertson, 9th Bn, AIF, was killed on 25th April 1915. Buried in Beach Cemetery, he was the son of Joseph and Catherine Ross Robertson, of 49 Redmyre Rd, Strathfield, NSW.
378 Lt William Rigby, 9th Bn, AIF, was killed on 25th April 1915. Commemorated on Lone Pine Memorial, he was the 23 year-old son of William Alfred and Julia Rigby, of Yeronga, Queensland.
379 *Casino and Kyogle Courier and North Coast Advertiser* (New South Wales), 21st August 1915.
380 CSM Walter Phipps, 3rd Bn, AIF, enlisted, aged 45, on 27th August 1914, having previously served 24 years with the RMLI. Commissioned, Capt. Phipps returned to Australia on 18th June 1918.
381 Capt. Charles Leer, 3rd Bn, AIF, was killed on 25th April 1915. He is commemorated on the Lone Pine Memorial.
382 *Molong Express and Western District Advertiser* (New South Wales), 11th September 1915.
383 Marshall, a railway worker originally from London, was wounded the following day. Rejoining his unit on 19th July 1915, he was hospitalised a week later and evacuated to Malta. Promoted Sgt, the premature explosion of a rifle grenade in France on 25th August 1916 led to his discharge.

"The Auckland machine-gun section was ordered 500 yds to the left, and to go up at once to the aid of the Australians. Off came our packs, and each carrying his portion of the machine-gun and its equipment we started up the hill. The engineers had cut a path up the face, and were working hard cutting more. We scattered round and lay down in the bushes for a spell, but was it a rest? Shrapnel screamed over our heads, and we would seek cover, lying close to the ground. I did this half-a-dozen times, until I realised that by the time I had heard the scream the shell had already passed and exploded over the beach. Bullets whistled past or buried themselves in the ground. Was I frightened? In one way I was, but it was more dazed. For a while I seemed as if I was stunned, but as I watched the shells exploding about me the dazed feeling gradually wore away. Lieutenant (Bob) Frater[384] gave the order for us to advance and collect, as we got the chance, over the ridge in the next gully. We were under a perfect hail of shrapnel and bullets, fired at those on the ridge in front. I would jump up, run about 10 yds, and then dive under a bush, or behind a small ridge. Then the bullets would fly, for some sniper would be busy…

"I got to the side of the flat on top of the ridge, where I found a couple of our machine-gun belts in their boxes. I added one to my load, and started to gallop down the track into the gully below… As I looked I was nearly deafened. Smoke and dust were all around me, for a shell had burst a few feet above. Pieces were all about, but not a particle had touched me.

"I got to the bottom, and waited till the others came over. We went on again, up the bed of a small stream. On the next ridge I watched about fifty Australians and New Zealanders collect behind a bank, fix bayonets, creep through the scrubs, and then charge. I could hear them yell out 'Yallah, emshi,' Egyptian for 'Clear out!' The Turks did clear! They ran for their lives for a way and then dropped into trenches. Then the shots belched out, and our fellows had to come back, followed by a strong body of Turks. As soon as our fellows got out of the way, one of our Auckland machine-guns took a hand and poured 500 rounds a minute into them."[385]

PTE ROBERT STEELE, MG SECTION, AUCKLAND BN, NEW ZEALAND BDE, NZ & A DIVISION.

"I found myself left with a mere handful of six or eight men on the extreme left flank of the firing line, about the spot "Gallipoli 237 Z6." This ridge was so strongly occupied by Turks that I had to shelter my men down the side of the cliff on the opposite bank. Shrapnel was raining torrents of lead, and we saw Colonel Stewart nearly struck. He called to us, "What are you doing?" and learning that I had so few men, said, "Lie down where you are; I am sending for reinforcements." Shortly afterwards he joined us, and about 100 reinforcements, Australians, Aucklanders and Canterbury men, soon came up. Colonel Stewart took charge, and with great coolness led us to successive positions till we were within 150 yards of the crest, where he decided to await the enemy. An Australian said to him, "Sir, we took this hill six times to-day, and

---

384 2/Lt Robert Frater, MG Section Auckland Battalion, died of wounds aboard HMT *Seang Choon* on 30th April 1915. Commemorated on the Lone Pine Memorial, he was the 23 year-old son of Robert and Martha Frater, of Auckland.

385 *New Zealand Herald*, 17th June 1915.

six times we have been driven back." With characteristic coolness, the colonel replied, "Very well, we will take it a seventh time. With the help of God and the battery, we will hold it, but this is a better position than on the top, so we will await them here." The hillside was covered with scrub, and as the Turks came on the colonel moved with great daring from bush to bush, controlling fire and encouraging the men. At one time an Australian captain urged him to retire, but he replied, "No, if we lose this hill we are done; we must hold on." There seemed to be snipers about picking off the officers. Possibly the colonel was too unmindful of his own safety, and about 4 p.m. he was killed by a bullet which passed through both temples. Immediately afterwards the Turks made a bayonet charge. Finding myself the only officer on the ridge, I ordered rapid fire, which the men gave heartily, the Turks returning to their trenches in a few minutes, and again a rapid fire hurled them back, but as they fell back they rushed round our left flank. We swung our thin line round, and a third time drove them back. But by now (4.45 p.m,), the ranks were sorely depleted. I had only a dozen men able to fire, and we took the opportunity to get away down into the valley with the wounded. I am sure that Colonel Stewart's cheery coolness under so very hot a fire enabled our men to do valuable and desperate work against vastly superior numbers."[386]

LT RAYMOND LAWRY, CANTERBURY BN, NZEF, NEW ZEALAND BDE, NZ & A DIVISION.

"I climbed up to the top of Plugge's Plateau — where an Indian mountain battery was in action — and, subsequently, to the head of Walker's Ridge, and lay, with Walker, at the Nek on the top of the ridge, watching the fight for its possession. The scrub was then very thick, and it was impossible to get any clear idea as to how things were going."[387]

MAJOR-GENERAL SIR ALEXANDER GODLEY, COMMANDING NZ & A DIVISION.

"I was with Colonel Stewart... when he was killed. It was about 4 in the afternoon, and we were very hard pressed, trying to hold the top of ridge of 227 Z6. We were on the point of retiring when Colonel Stewart arrived with about 30 men, Australians, and New Zealanders, and said, "Hullo, corporal; how's things?" I told him things were pretty hot, and we were thinking of retiring. "Well, come on boys," he said. "We'll give them a bit more before we leave, and although I told him it was not safe for him to come on the top, as men were falling all round us, he came and knelt down beside me; and a few seconds after a bullet struck him in the temple, passing clean through his head, and killing him instantly. I was hit twice myself then, one grazing my shoulder and another hitting the cartridges in my belt, but doing no damage. There were about a dozen of us left when Col. Stewart[388] lost his life, like the brave man he was."[389]

CPL ALEXANDER McINNES, CANTERBURY BN, NEW ZEALAND BDE, NZ & A DIVISION.[390]

386   *Poverty Bay Herald* (New Zealand), 2nd July 1915.
387   Godley, *Life of an Irish Soldier*, p. 171.
388   Lt-Col. Douglas MacBean Stewart, Canterbury Bn, was killed in action on 25th April 1915. Commemorated on the Lone Pine Memorial, he was the 38 year-old son of the late Dr. F. Macbean Stewart and Annie Macbean Stewart, of Christchurch; husband of Edith Macbean Stewart, of "Zeitoun," Merivale Lane, Christchurch.
389   *Poverty Bay Herald* (New Zealand), 2nd July 1915.
390   Appointed L/Sgt on 7th May, McInnes was killed in action at Helles on 8th May 1915. Commemorated on the Twelve Tree Copse (New Zealand) Memorial, he was the son of Malcolm McInnes, of Glenmore, Scargill, Canterbury.

"After waiting an hour which seemed like two hours, and there being no appearance of the Twelfth and Thirteenth Companies, I went up to the plateau, and there met a wounded corporal, who stated that Colonel Stewart was killed and Major Grant [391]badly wounded, but he did not know where the two companies were.

"I reported Colonel Stewart's casualty, and assumed command, but it was not possible to find the men. It seems that when the companies arrived at the top they were sent in all directions, and the track being so narrow, company commanders lost control. I met Major Rowe, and he told me he had only fifteen men, and did not know what had happened to the others. At this time I got a message from the beach to say that my other companies were arriving, so I went down and found part of them already there. I took these up the hill and met our brigadier, who ordered me to reinforce the centre. I was about to do this, when I received another order to descend, march about a mile, occupy a spur, and watch the beach to the north, as the Turks were working their way around. I was to hold this flank to the last."[392]

MAJOR ALBERT LOACH, CANTERBURY BN, NZEF, NEW ZEALAND BDE, NZ & A DIVISION.

"By 4 p.m… we were in a real good fix. One platoon was cut off from our company, and mixed up with the Australians. We were lying in a thin line on an open plateau, with the Turks on a ridge in front of us, and with Turks on the hills on either side. It was a fair death-trap, and the Turks made the most of it. We hung on, our men dropping everywhere."[393]

PTE GEORGE HAMPSON, CANTERBURY BN, NZEF, NEW ZEALAND BDE, NZ & A DIVISION.

"For a little time I must confess I was a little unnerved, but then a spirit of revenge came into me and I went up to the firing line without any hesitation at all. Then word came to us to reinforce a party of Australians who were in a position about thirty yards in advance of the main line. I was one to go out, and had just settled down to commence operations with my rifle when bullets started to make their appearance from our right flank. The Turks had worked round to our right and with a machine-gun started to enfilade us. I can tell you it was perfect hell, and I really thought it was a case of goodbye for the lot of us. There was nothing to do but retire, so those of us who were left did so. It wasn't a very orderly retirement either, as from the Turks' position they could smack at us all the way back. However, some of us got back, and were soon at work digging trenches for ourselves."[394]

PTE LESLIE HILL, AUCKLAND BATTALION, NEW ZEALAND BDE, NZ & A DIVISION.

"Presently an Australian officer appeared out of somewhere and ordered us back to the ridge, and passing on the order, followed him. He fell with a sigh and lay still. We

---

391 Major David Grant, Canterbury Bn, was killed in action on 25th April 1915. Commemorated by a special memorial at Walker's Ridge Cemetery, where he is believed to be buried, he was the 41 year-old son of the late Archibald and Louisa Grant, of Elizabeth Street, Timaru; husband of Ann Grant, of 23 Le Cren Street, Timaru.
392 *Lyttelton Times* (New Zealand), 29th July 1915.
393 *Nelson Evening Mail* (New Zealand), 23rd July 1915.
394 *Taranaki Herald* (New Zealand), 28th July 1915.

lined the crest, somewhat protected from the hail of lead, and waited grimly with our bayonets. The incredulous amazement of it all was past now, our blood was up, and we had a few scores to settle. The Turks, ignorant of our weak numbers, did not advance, and at last came the night and a slackening of fire. We advanced over the crest, removed the wounded, and poured volleys into the advancing Turks until they reached our unprotected flanks. We then entrenched strongly beneath the crest, while the Turks, several times our number, took up a position 10 to 20 yards the other side.

"Volley after volley we poured into them, and into the bushes where we could hear them trying to get us to charge. Major [Thomas] Dawson's leadership here saved us. With nerves strained to the breaking point, we wanted to get at them with our bayonets and end it one way or the other, but, sizing up the situation — that they had not the courage to charge, and that our charging would mean annihilation — he kept us in hand, and used our enthusiasm in other directions."[395]

CPL FREDERICK HALL-JONES, AUCKLAND BN, NEW ZEALAND BDE, NZ & A DIVISION.

"About 5.30 p.m. I was shooting over the edge of a hill. Two Australians were with me. Something landed in the ground ahead of us, and all I remember was that the earth rose and when I came to I realised that I must have rolled 150 feet down the slope. One of the Australians had an arm blown off from the elbow. I asked him where our mate was, and he pointed to pieces of him."[396]

PTE JAMES METTRICK, AUCKLAND BN, NEW ZEALAND BDE, NZ & A DIVISION.

"We were firing as fast as we could sight our rifles. A man next to me in his anxiety to get a good shot knelt up, and put his rifle to his shoulder to fire, I yelled to him to keep down, but it was too late; he was shot through the neck, and fell back. I rolled over to him, but could do nothing for him except give him a drink, and put his hat under his head, and make his position as easy as was possible under the circumstances. He was unconscious, but paralysed and bleeding from the mouth. I took his ammunition, filled my pouches, and passed some on to the men near me. I went to him again a little later. He was still alive, but I could do nothing for him.

"As nearly as I could judge it was about 5.30 in the afternoon when I was hit. I had seen a Turk rush up to, and drop behind, a bush about 90 yards ahead of me, so I aligned my rifle on him, and waited for him to reappear. Presently he did so, and I instantly aimed at him. In my anxiety to get a good view of him I knelt up; just as I pulled the trigger the rifle was knocked out of my hand, and I found my thumb hanging in ribbons and my rifle-shooting was at an end. I remained in the firing line for some time, but feeling exhausted and faint from loss of blood I decided to try and get back to the beach to report, to ask for reinforcements, and to get my thumb (which had been giving me particular "gip") dressed."[397]

LT EDWARD BUTLER, 12TH BATTALION, 3RD BDE, 1ST AUSTRALIAN DIVISION.

---

395   *New Zealand Herald*, 5th July 1915.
396   *Poverty Bay Herald* (New Zealand), 25th June 1915.
397   *The Sun* (Sydney, New South Wales), 22nd August 1915.

"I slipped up on deck about 4.30 p.m. to watch the bombardment by the large battleships. They were only about twenty yards away from us at first, and the noise of the discharge nearly deafened us on the hospital ship. The dense dirty colored smoke of the lyddite shells discharge could be seen long before the noise of the bursting of the discharge could be heard, and it was some seconds ere they burst on or near the enemy's position, belching up a great mass of earth and rubble as the huge shells struck the ground. They fired shrapnel at the time. These can be easily picked out by their bursting high in the air, and the light puff of smoke remains in the air for a long while after the shell has burst and sent its deadly missiles spread over about a mile square. It was a weird but grand sight."[398]

PTE ERIC RATCLIFF, 12TH BATTALION, 3RD BDE, 1ST AUSTRALIAN DIVISION.[399]

"[B]y the afternoon the fight was going against us. It was accentuated by the fact that our infantry were landed by companies and half-companies at a time, and were immediately sent off to where they were most wanted. The need was so great that they could not give the regiments time to collect on the beach. The result was that our lines were held by scattered and intermingled groups of men, without their own officers, and far from their own regiments."[400]

MAJOR ARTHUR TEMPERLEY, WELLINGTON BN, NEW ZEALAND BDE, NZ & A DIVISION.

"The New Zealanders were now passing us up the Gully so we joined up with them and climbed a hill to the left of Pope's Hill and formed a firing line.

"A New Zealander next to me was wounded in the wrist and I turned to bandage him up, when he got another in the stomach. He was in terrible agony and asked me to finish him off. I told him to lie still while I went and sought a stretcher-bearer. But when I looked around me I could see no sign of our former firing-line, nor could I see anyone — they seemed to have vanished completely.

"I tried to get back to Shrapnel Gully for a stretcher; but I had lost all sense of direction and actually went in another direction altogether, and found myself in another Gully, where I fell in with a wounded man from the 1st Brigade. He was shot through the hip, and I asked him if he required assistance. I helped him down the Gully till he could go no further, while I went on to see if I could find a stretcher.

"I continued down this Gully till it opened out on to an old water bed at its junction with another Gully. I had gone about half way across this open space when I heard a shout, and on looking up, saw about 8 or 10 Turks covering me with their rifles. At the same time bullets were coming from the rear on the right. I immediately threw

398 *Daily Telegraph* (Launceston, Tasmania), 15th June 1915.
399 Ratcliff rejoined his unit on the peninsula on 2nd June 1915. The former clerk, then L/Sgt Eric Bird, still with 12th Bn, was killed at Pozieres on 24th July 1916. Believed to be buried in Pozieres British Cemetery, Ovillers-La Boiseselle, commemorated by a special memorial, he was the 23 year-old son of Henry Edwardes Ratcliff and Ethel Annie Ratcliff, of Launceston, Tasmania.
400 *Manawatu Standard* (New Zealand), 9th September 1915.

up my hands and the Turks immediately came forward and knocked me on the head with their rifle-butts, dazing me."[401]

PTE FREDERICK ASHTON, 11TH BATTALION, 3RD BDE, 1ST AUSTRALIAN DIVISION.

As the firing lines were being slowly consolidated, those caught in front of them tried to make their way back as best they could.

"There were twenty or thirty of us just preparing to rush the top of the hill when a shrapnel burst overhead and laid most of us out. One piece went clean through my wrist, bone and all. At first I thought that the whole of my arm had been blown off. I can't describe the agony and torture I suffered the first hour, but after a while it eased off a bit. Shrapnel was still falling all round us and I could not move. I had a field service bandage in my pocket but could not get it out to stop the bleeding. The fire was too hot to try and get back, so I decided to wait until reinforcements came up. It was five and a half hours before I moved and I did not give myself a chance of getting back. The 6th battalion and what was left of the 5th, had formed a firing line 200 yards behind and shots [were] going over me from both sides. I was lying behind some thick scrub and every now and then I would get a shower of leaves on my back that had been cut off by a machine gun. Needless to say I kept my head well down. Suddenly the Turks ceased fire and one of the other wounded chaps not far from me said [about 4 p.m.] he thought the Turks were going to advance. We were not far from them and ran a fair chance of being captured, so we decided to try and [get] back. We crawled for over a hundred yards amongst the scrub towards our firing line. Then we got up and showed ourselves, as we would be mistaken for Turks by our mates. We no sooner stood up than one chap dropped dead. The rest of us managed to get there safely."[402]

PTE FRED RITCHIE, 6TH BATTALION, 2ND BDE, 1ST AUSTRALIAN DIVISION.

"As soon as I could, I went ashore to see the progress made, and clambered around as much as possible of the front line on the heights. Owing to the thick scrub I could see very little, but from a point later known as Walker's Top I got a fairly good idea of the situation, realizing for the first time that a large valley separated the New Zealanders there from the Australians on a ridge to the east."[403]

LT-GEN. SIR WILLIAM BIRDWOOD, COMMANDING
AUSTRALIAN & NEW ZEALAND ARMY CORPS.

"The afternoon seemed a miserable time without beginning and without end. Haymen[404] was killed. There seemed to be little hope for us. When darkness came the Turks seemed to be all around us. Some men who had been on the hill joined us, and then we started to get back. We formed three parties, a carrying party and two protecting parties. The machine gun formed one protecting party and some riflemen

---

401 Frederick Ashton's service record, NAA: B2455, ASHTON F.
402 *Every Week* (Bairnsdale, Victoria), 24th June 1915.
403 Birdwood, *Khaki and Gown*, p. 257.
404 Lt Frank Granville Haymen, 9th Bn Australian Infantry, was killed in action on 25th April 1915. Commemorated on the Lone Pine Memorial, he was the son of Marmaduke Granville Haymen and Florence Maude Lucretia Haymen, of Brisbane.

the other. These two retired alternately, each one covering the other. The men who had been killed in the pits and on the way back we could do nothing with; we could not bury them. All we could do was to cut off their ammunition and bring their rifles with us. How we got out of the mess I don't know. The only thing that I can think is that we made such a noise with fire orders and bursts of fire that the Turks thought there were more of us than there were and did not press on. When we got back near the main position another machine gun joined us. We had been expecting to find a machine gun on us any moment during the trip back, and to hear one start up, and find that it was helping us and not firing at us was a wonderful relief."[405]

LT CHARLES FORTESCUE, 9TH BATTALION, 3RD BDE, 1ST AUSTRALIAN DIVISION.

"The men were naturally very exhausted after so hard a day — and inclined to be despondent, too. Small groups would tell me that they were all that was left of their respective battalions — 'all the others cut up!' On such occasions I would promptly tell them not to be damned fools; that the rest of the battalion was not far distant, having simply been separated in tows. This always had an encouraging effect, though I must confess that I might not yet have seen 'the rest'."[406]

LT-GEN. SIR WILLIAM BIRDWOOD, COMMANDING
AUSTRALIAN & NEW ZEALAND ARMY CORPS.

"A stretcher-bearer then helped me to the beach. Here was a most pitiful sight, wounded and dying. I can just recollect seeing poor Sergeant Gardner, who used to live in Hillgrove, calling out to me. He was very seriously wounded... We had to wait for about two hours for boats to take us across to a transport which they had turned into a hospital ship."[407]

CAPT. CLIFFORD RICHARDSON, 2ND BATTALION, 1ST BDE, 1ST AUSTRALIAN DIVISION.

"During the day the Brigadier received three congratulatory messages over our wire, congratulating him and his brigade. Major Brand, his right-hand man, bore a charmed life. Time and again when the extreme left was driven back he would go and steady them, and take them back and after they had settled down again, he would quietly stroll back to Brigade headquarters."[408]

SPR HARRY MARSHALL, DIV. SIGNAL COY, AUSTRALIAN ENGINEERS, 1ST AUSTRALIAN DIV.

"At half-past five, out of the 33 we had in the trench, on the left, only two were left, and we were forced to retire expecting death at any moment. Then, down below, we heard the glorious cry ring out, 'Come on, the Otago Regiment!' Shall I ever forget it? Up they came, and we dug in on the first first ridge, and with Major Dawson's men held the enemy!"[409]

SGT FRED COE, 9TH BATTALION, 3RD BDE, 1ST AUSTRALIAN DIVISION.

405 *The Western Champion* (Barcaldine, Queensland), 25th April 1925.
406 Birdwood, *Khaki and Gown*, p. 257.
407 *The Armidale Chronicle* (New South Wales), 26th June 1915.
408 *The Bundberg Mail and Burnett Advertiser* (Queensland), 15th June 1915.
409 *Casino and Kyogle Courier and North Coast Advertiser* (New South Wales), 21st August 1915.

"I then went back to a place where a few slightly wounded and exhausted men of the 12th had collected, and here met Mr. Green,[410] who had received orders to collect stragglers. There I lay down utterly finished; for a while I was too stiff and sore to move. Later on the provost marshal told me to form my men up, and report to Major Glasford;[411] he ordered me to get ready to move out to the right, but afterwards told me to stand by for orders. We were later sent off to the right, but met Captain Ross, who ordered us back for the night."[412]

LT IVOR MARGETTS, 12TH BATTALION, 3RD BDE, 1ST AUSTRALIAN DIVISION.

"During one burst of shell I got a bump in the left arm — just sufficiently hard a smack to make you think some practical joker had hit you with a poker or a ruler on the biceps! That's what the feeling was like; only a wee bit of blood came out of the hole in the jacket sleeves, so that was all right. I didn't have to make tracks for the base at the beach, as I would have had to had the bit of lead come about three inches to the right — the shoulder strap of my web equipment (officers all wear the same equipment as the men) was punctured first, and probably it diverted the pellet into the arm; I was lying down using my glasses when hit — good job it didn't hit the glasses, or I'd have had to get a new pair, and perhaps a new face also! Well, those two ridges were hung on to till about 5 p.m., when we slowly had to make our way back to the ridge we had been on about 10 a.m. I'm not referring to my company only when I say "we" — for by this time "we" included many men of many units; the line towards our left had also found it necessary to get back — the shell fire was too hot, and the positions we were holding were not suitable for digging in to hang on to for the night. Losses had been pretty severe too, my best platoon commander, poor old Monro,[413] was killed, and quite a number of the non-com's and men had been hit. We got back to the ridge and joined the troops holding it and dug in with them. I did not remain just there, but went along the line further to the left (or northwards) trying to find our own battalion — no hope — they were all well mixed up. About 6 p.m. I found a corporal and about six men of my company, and we joined in digging in a line; found Rafferty[414] also about this time, and he and Corporal Austin and myself passed the night in one hole together; it was a wee hole at first, but by daybreak it formed part of a long line of trenches; am afraid I didn't do much of the digging though."[415]

CAPT. JOHN WHITHAM, 12TH BATTALION, 3RD BDE, 1ST AUSTRALIAN DIVISION.

"Towards dusk, my gun jammed owing to the copper round the barrel melting with the heat of firing, and soldered it to the muzzle attachment. We got this cleared away, and whilst getting back into position again, the sergeant was shot through the neck,

410   Lt Aylmer Green, 12th Battalion, AIF, was evacuated on 26th April 1915 suffering from 'concussion of spine'. Rejoining his unit on 1st July, he was again evacuated on 6th August 1915 suffering from asthma.
411   Major Duncan Glasfurd, Argyll & Sutherland Highlanders, GSO2, attd 1st Australian Division. He died of wounds as the Brigadier-General commanding 12th Australian Bde in France on 12th November 1916. He is buried in Heilly Station Cemetery, Mericourt-L-Abe.
412   *The Examiner* (Launceston, Tasmania), 4th August 1915.
413   Lt Gordon Munro, 12th Battalion, AIF, was killed on 25th April 1915. Commemorated on the Lone Pine Memorial, he was the son of James and Mary Munro, of Kilkenny, South Australia.
414   2/Lt, later Lt-Col. Rupert Rafferty, DSO, 12th Battalion, AIF, a teacher from Sprent, Tasmania.
415   *The Critic* (Hobart, Tasmania), 2nd July 1915.

putting him out of action for some days. Shortly afterwards another of the section was wounded through the chest. These were our first casualties. During the night we managed to get the gun forward to a better place which we improved by digging a good gun pit before morning."[416]

PTE GEORGE HENDERSON SMITH, MG SECTION, 11TH BATTALION, 3RD BDE, 1ST DIVISION.

"Just at dusk the Turks tried to advance towards the firing line, but we were too strong for them and they retired. Somehow the Turks got around to our left and began an enfilading fire during the night, and we lost some men. Men were hit alongside of me, but I was lucky."[417]

L/CPL JOHN LAWN, AUCKLAND BATTALION, NEW ZEALAND BDE, NZ & A DIVISION.

"[I]n the evening streams of wounded men and stragglers began to pour down to the beach. The reports coming in and the accounts given by staff officers, and others, who visited the most advanced troops, became very disquieting; especially those coming from the head of Monash Gully, and the ridge east of it, which formed the farthest point of penetration."[418]

MAJOR-GENERAL SIR ALEXANDER GODLEY, COMMANDING NZ & A DIVISION.

"It was awful to see so many of one's friends and comrades being slaughtered. We held on to this position without officers till about 6.30 p.m., when we simply had to retire and, unfortunately, we were unable to take our wounded with us. It was just getting dark then, and we decided to make another stand, and advanced slightly again, but our numbers were tremendously reduced by this time. However, we stuck on the top of the ridge, and secured as much cover as possible and improved it slowly with our entrenching tools. Major Dawson, of Auckland, and Lt Ffitch, came to light here, and I was just alongside the latter all night, and he fought like a Briton.[419] The Turks advanced and entrenched themselves 15 or 20 yards from our trench, but as our numbers were so few, a charge was out of the question, but we were expecting one from them all night long. We heard every order passed down their line, but, of course, couldn't understand them. It was here that Norris[420] was killed. He was a fine chap, and I had made a great friend of him."[421]

PTE KEELEY JAMESON, CANTERBURY BN, NZEF, NEW ZEALAND BDE, NZ & A DIVISION.[422]

---

416   *Barrier Miner* (Broken Hill, New South Wales), 25th July 1915.
417   *Inangahua Times* (New Zealand), 24th July 1915.
418   Godley, *Life of an Irish Soldier*, p. 171.
419   Lt Harry Herbert Ffitch, Canterbury Bn, NZEF, was killed the following morning, Jameson recording in the same letter, "at daybreak poor Ffitch was shot dead. He was taking deliberate aim, and a bullet passed through his wrist and into his face, and he dropped without a word." Commemorated on the Lone Pine Memorial, he was the 25 year-old son of Henry and Florence Ffitch, of 16 Snowden Road, Fendalton, Christchurch; a native of Canterbury.
420   Pte Oswald Mark Norris, Canterbury Bn, NZEF, was killed in action on 25th April 1915. The 22 year-old son of the late Thomas Cheal Norris and Mary Maria Norris, of 85 Page's Road, Christchurch, a former scholar at Christ's College, Christchurch, is commemorated on the Lone Pine Memorial.
421   *The Press* (New Zealand), 12th August 1915.
422   Jameson was killed in action on 8th May 1915. Commemorated on the Twelve Tree Copse (New Zealand) Memorial, he was the 27 year-old son of William and Mary Haswell Jameson, of 54 Garden Road, Fendalton, Christchurch.

"On the way to the new position I picked up the stragglers, and feeling our way among the wounded arrived at the foot of a steep spur. Leaving the men to remove their packs, I went ahead with the scouts, and after climbing some 300 ft, met a large party of Australians coming down. They said they were being driven in, and that the place was too hot to hold. It was now about 7.30 p.m., and very dark. I knew that the moon would rise about 9 p.m., and therefore any action taken must be under cover of darkness. So with the West Coasters in the lead, and leaving the Nelson Company to bring along the stragglers, we started for the 900 ft rise. We soon reached the top, and found that the Turks had not come on after their attack. So I decided to dig my men in on the top of the ridge itself.

"After seeing that all was well, I descended the hill and secured the lower part of the ridge, down to the beach, with the Nelson boys in position to check the enemy at that point. Captain Salmonson[423] reported to me at this stage. Stragglers of the First and Second Companies dribbled in."[424]

MAJOR ALBERT LOACH, CANTERBURY BN, NZEF, NZ BDE, NZ & A DIVISION.

"The pressure that evening became very severe. We were losing heavily from the enemy's shrapnel, and it seemed to me doubtful whether we should be able to hang on. Were these raw troops going to endure? The careless clashing of a bayonet charge is nothing. It is the endurance that is required. We had no more reserves, and it was simply a question of hanging on [but] mere scattered and disorganised groups of men, in some cases, without officers, without water, without help or reinforcements, by bayonet charges and by fire, held their own and the situation was saved."[425]

MAJOR ARTHUR TEMPERLEY, WELLINGTON BN, NEW ZEALAND BDE, NZ & A DIVISION.

The 4th Brigade was the last to arrive during the late afternoon.

"Six p.m. came with the order to stand ready to leave the ship, and in a quarter, of an hour, on both decks on both sides of the ship, were the 1200 gallant men ready… Joking, laughing, singing, and cheering, in their turn, they stood this till at 8 p.m. a large rowing boat hove in sight coming towards us. This was the cause for more delight and cheering, thinking that at last their turn had come to go ashore, but I shall never forget the disappointment of those men when they saw that what they had been cheering was [a boat] full of wounded men."[426]

PTE FRED PENHALL, AAMC, HMAT *SEANG CHOON*.

"Two New Zealand chaps assisted me down to the beach, and I was taken with a large number of others on to a tow, and then were taken aboard the *Seang Choon*. There were 600 wounded aboard, and only three doctors and ten A.M.C. orderlies, so you can imagine what sort of a job the doctors had, and they worked like Trojans, too.

---

423   Captain, later Major Arthur Critchley-Salmonson, DSO, was wounded on 7th August 1915.
424   *Lyttelton Times* (New Zealand), 29th July 1915.
425   *Manawatu Standard* (New Zealand), 9th September 1915.
426   *Kalgoorlie Western Argus* (Western Australia), 24th August 1915.

Some of the cases there were very pitiable to see. One chap had a shrapnel bullet pass behind both eyes, and he was completely blind; others had numerous wounds to the face and body. When I saw how things were I felt that there was absolutely nothing wrong with me, and wanted to go back again."[427]

PTE VICTOR PINKSTONE, 3RD BATTALION, 1ST BDE, 1ST AUSTRALIAN DIVISION.

"The boats began to bring off wounded towards evening. The hospital ship that was attached was filled by about 4 p.m., and left to empty at Alexandria. There were some 2,000 casualties altogether, and the hospital ship only held 300 or 400 men. We had nearly 100 wounded on board, and transports that had been emptied were used for wounded... Some of the men were in an awful state. One we had on board had a shrapnel bullet in his eye, and his head badly cut about. Others had arms and legs shattered. They were very cheerful on the whole, and we did our best to make them comfortable."[428]

MIDSHIPMAN HON CHARLES GIFFORD, RN, HMS *LONDON*.

"We hadn't the hospital ready... We certainly did not expect wounded, at least not till the troops had left the ship, so you can understand a bit of commotion at the start... There were 25 wounded in this boat, and it wasn't long before we had those that could walk, up the gangway on to the deck. Then we had to go further along the side under the derrick, where the stretcher and serious cases had to be hauled up by the winch. This was terrible work, as the rain was pouring down; and it was so cold. This with the boat tossing and heaving, and the poor chaps moaning in their agony, and others pitifully crying for water. It was enough to turn anyone's heart cold. It was slow, yet considering we weren't prepared for this horrible surprise, we were doing wonderfully well. Before we had the last on board, I had to go up after getting relieved. I was too sick to stand the tossing and heaving any longer. By this time there were two other boats at the gangway full of wounded. After going on deck I went straight to the hospital, where the doctors were already hard at work. I began dressing those mostly in need of it. The hospital was already almost full, and two of our cases had died. The wounded were still coming in... They were carried down to the mess room decks and placed wherever there was room, and when we had finished we had over 600 on board — more than twice the number we were supposed to carry... We had every fracture imaginable to attend to, but the majority were shot through the legs and arms, others through the bodies, in some cases piercing the lungs and other internal organs. It was a shambles."[429]

PTE FRED PENHALL, AUSTRALIAN ARMY MEDICAL CORPS, HMAT *SEANG CHOON*.

Some of those arriving on the *Seang Choon* came across old friends.

---

427   *The Young Chronicle* (New South Wales), 11th June 1915.
428   *Grantham Journal*, 29th May 1915.
429   *Kalgoorlie Western Argus* (Western Australia), 24th August 1915.

"I was very weak, having lost a lot of blood and not anything to eat since breakfast, at 2 a.m. The [14th] Battalion of the Second Force were on board, and were going ashore next day. I found Reg Bonwick and had a long talk to him[430]... He was rather surprised, and could not do enough for me."[431]

PTE FRED RITCHIE, 6TH BATTALION, 2ND BDE, 1ST AUSTRALIAN DIVISION.

"We arrived at our destination early in the afternoon... to hear our lads, who had arrived earlier, well in action. About sundown they started bringing the wounded on to our boat, as the hospital ship was then full. All night we were bringing them on board, and it rained practically all night."[432]

L/CPL REGINALD BONWICK, 14TH BATTALION, 4TH BDE, NZ & A DIVISION.[433]

Others made new friends after being taken on board warships.

"The wounded soon began to arrive, and the hospital ship attending here was soon full of wounded, and [in the] evening a portion of the wounded were taken on board some of the battleships which were standing by covering the troops. We had about 100 wounded brought on board our ship at 8.30 [p.m.] We dressed their wounds, gave them soup, and the sailors gave up their beds for all the wounded, so that each one had a bed."[434]

SBA HARRY BRADSHAW, RN, HMS *MAJESTIC*.

The senior officers of 16th Battalion advanced inland to try and find out what was happening for themselves. What followed was almost comedic.

"There was great confusion on the beach, and I was getting as many of every people assembled together as possible, when I was ordered to take them and a lot more to relieve a position taken up by the troops who had landed in the morning. At all events, we marched off, and finally arrived at a very big, steep hill, by which time I had about 200 men left. We got on to this all right and found a perfect mixture up there — a few men from all sorts of battalions, with a Captain Jacobs in command. He told us that about 200 yards farther on there were some Indian troops, so I decided to try and find them, with the object of carrying on the line to connect with them. (I may say we were being fired on more or less the whole time since we left the ship.) Your husband was close by when I was talking to Captain Jacobs, and said, "I have a man in my battalion who knows Hindustani. Shall I go with him to locate the Indians? This was about 10 p.m. — moonlight but obscured by clouds. The man's name was Lushington.[435] It seemed all right, so I let them go. Presently he came back, and said he had found the

---

430   *Every Week* (Bairnsdale, Victoria), 24th June 1915.
431   *Bairnsdale Advertiser and Tambo and Omeo Chronicle* (Victoria), 23rd June 1915.
432   *Every Week* (Bairnsdale, Victoria), 10th June 1915.
433   L/Cpl Reginald Bonwick, 14th Bn, AIF, was wounded on 2nd May. Wounded a second time on 21st August 1915, he returned to Australia, leaving England on 8th May 1916.
434   *Farnworth Chronicle*, 19th June 1915.
435   Pte Reginald Lushington, 16th Bn, AIF, born in India, was taken prisoner on 25th April 1915. He survived his captivity and was repatriated to Australia on 21st January 1919.

Indians all right, but they wanted an officer of higher rank to negotiate with, so I went out. Captain McDonald,[436] the adjutant, was following him up, when he called out through the night 'Are you there, colonel?' They want the colonel of this battalion; no one else will do. I called out, 'All right, McDonald; I'll come along.'

"The ground ran along the edge of a very steep, deep gully. I went along about 150 yards, and met Elston[437] and Lushington with six Oriental soldiers, with rifles and fixed bayonets. They were having a sort of informal conversation and standing about. I said at one 'Look here! These people look like Turks to me.' He replied, "Oh, no; they are Gurkhas." I said, 'Well, what do they want me to do? He replied, "To go with them and see their officer." I answered, "Let their officer come to me. I feel sure they are Turks. Look at them." But McDonald kept repeating "Oh, they're all right; they are Indians — Gurkhas!" While this conversation had been going on the soldiers had been gathering round, and at the finish one of the most ferocious-looking grabbed hold of my wrist. That was enough for me. I had purposely kept pretty close to the edge of the gully. I sang, "Look out, you fellows, those beggars are Turks," and pulled out as quickly as ever I could, pulling the Turk after me and jumping over the edge of the gully. The Turk let go, and I fell down. They fired some shots at me. Forty feet down I touched bottom, and then got back to my former position, and carried on. The last I saw of your husband, McDonald, and Lushington, the Turks were closing in on them, and the last impressions I had is that they were both looking at me in astonishment at jumping down the precipice."[438]

LT-COL. HAROLD POPE, 16TH BATTALION, 4TH BDE, NZ & A DIVISION.

"I landed with Col. H. Pope and reported immediately to Brig-Gen. Monash. He took us to General Godley's Headquarters on the beach, where the Colonel received the information that the 3rd Brigade had been hard pressed and had been repulsed, and that Colonel Maclagan wanted reinforcements. The orders given were, to move the 16th Battalion (less 1 Coy. not landed) plus 1 Coy. of New Zealanders to Col. Maclagan's support...

"We came to a place where the gully forked and the colonel took the route along what he considered to be the main gully. A little distance ahead we found a firing line established on the top of a steep incline on our right. There was no sign however of Brigade Headquarters. The Colonel halted, and moved up to the firing line himself. I established a small temporary Headquarters with my staff-sergeant and some H.Q. Signallers. We discovered a signalling post but they were out of touch with everybody. I then moved up towards the firing line where I met an N.C.O., whom I had known previously known in W.A. (Allen?). He told me that there were no Officers alive in the vicinity and that they had been pushed back with loss from their position and that Indian troops were fighting on our left rear; that this firing line also was on the extreme left of the 3rd Brigade. I asked him where the 11th Battalion (his battalion) Headquarters was. He replied "Near Brigade Headquarters." I then asked him where

---

436   Capt. Ronald McDonald, 16th Bn, AIF, was taken prisoner on 25th April 1915.
437   Capt. William Elston, 16th Bn, AIF, was taken prisoner on 25th April 1915.
438   *Morning Bulletin* (Rockhampton, Queensland), 4th November 1915.

Brigade Headquarters were, and he said "Somewhere up that gully", meantime pointing in the direction indicated.

"I asked him who the Indian troops were, as we had no information regarding the presence of any. He said that he did not know, but that Indians had been there since the retirement had commenced, keeping the same relative position.

"I went and found the colonel, who was right on the left. I reported what information I had and he decided that he would reinforce the firing line, which seemed in great need of support.

"He also decided to try and pick up 'touch' with the Indians. Lieut. Elston "A" Coy, was called up and was instructed to reconnoitre troops on the flanks. Lieut. Elston moved off, as I thought, with some men; but he subsequently said that he was alone.

"Pte Lushington, "A" Coy, reported to the Colonel that he could talk a little Tamil and might be useful as an interpreter if we got in touch with the Indians. He moved off and the Colonel and I stood talking. Almost at once we heard voices in the direction in which the patrol had gone. We could see Elston, in the dusk, talking with somebody. Almost immediately a call came for a senior officer.

"The Colonel started to move forward, but I went in his place calling as I went "I'm the Adjutant and I'll come". I moved forward and a man kneeling covered me with his rifle. He had a scarf wrapped round his head, (resembling, very much, a puggaree) I raised one hand and walked towards him stating, "I'm a British Officer, I want your burrashaid". He rose and put his rifle down. Immediately I was seized by both arms by two men who rose out of the undergrowth, and I was immediately surrounded by quite a considerable party with fixed bayonets, some of whom started running to where the Colonel had been.

"I was thrown down and my revolver was taken from me. I was then taken quickly ahead where I joined Elston and Lushington, who were in a similar plight. We were then taken ahead to where an Officer with some Turks were posted in a small hollow. I was searched and had all my equipment and papers taken away. Our men opened heavy fire on us and we were hurried away behind a hill where we found a considerable force of Turks. Turkish reinforcements were arriving at the time, with machine guns."[439]

CAPTAIN RONALD McDONALD, 16TH BATTALION, 4TH BDE, NZ & A DIVISION.

Meanwhile, more wounded were being taken aboard transports pressed into service as hospital ships. Those still ashore tried to dig themselves in as best they could.

"Shortly after dusk we were ordered back to the landing place to take on wounded, and all hands (there were only a handful of us on board consisting of the transport part of Headquarters and a fatigue party who had helped to unload the ship) had to set to and clear the saloons of tables and chairs, and clean up the ship. How we all worked! No one thought of shirking. Even the ship's engineers off duly went into the stokehold and helped the Greek firemen. In half an hour we were back in the firing

---

439   Ronald McDonald's army service record, NAA: B2455, MCDONALD R T A.

zone, and could hear the shells shrieking and machine guns purring, as was the case in the morning. At 11 p.m. the wounded started to come on board, they being brought from land by men-of-war's men in steam pinnaces towing cutters. I shall never forget the cries of the wounded as they were hoisted on board. One poor chap was dead before he was got on board. It was easily seen that the Australians had suffered the most severely."[440]

L/CPL CLARENCE UMBERS, NZASC, ATTD HQ, NZ & A DIVISION, HMT *LUTZOW*.

"As soon as darkness began to set in, all hands got to work deepening the trench. When the Turks noticed the picks and shovels going and the dirt flying the rifle-fire increased to such a cruel extent that we had to knock off. The night was bitterly cold with a slight rain falling, and as we had left our packs and great-coats on the beach we huddled together for warmth. It will give you some idea of the mixture when I tell you that a New South Welshman and I slept on bags close together. I placed a bag across my knees, and a New Zealand officer slept with his head under it."[441]

L/SGT ALEX MARSHALL, 11TH BATTALION, 3RD BDE, 1ST AUSTRALIAN DIVISION.

"The reports coming in became worse and worse, and eventually Bridges and I agreed that it was our duty to let the Army Corps Commander, General Birdwood, know that although the troops were hanging on well, and had so far repulsed all attacks made upon them, their situation was such that, should a heavy attack be delivered by the Turks during the night or at daybreak, it was questionable whether they would be able to hold their own."[442]

MAJOR-GENERAL SIR ALEXANDER GODLEY, COMMANDING NZ & A DIVISION.

"[T]he situation ashore seemed fairly satisfactory when, in the evening, I returned to my headquarters on the *Queen* after discussing matters with Bridges and Godley. I was therefore horrified, about an hour later, to receive a message from Bridges asking me to return at once, as the position was now critical."[443]

LT-GEN. SIR WILLIAM BIRDWOOD, COMMANDING
AUSTRALIAN & NEW ZEALAND ARMY CORPS.

"About 9 p.m... a message came to General Birdwood, who was on board the *Queen*, to say that his presence was necessary on shore. Before he landed, I arranged with him to keep Brig.-General Carruthers, his D.A. and Q.M.G. and Brig.-General Cunliffe Owen, his artillery officer, on board for the present... until the position was thoroughly established on shore."[444]

REAR-ADMIRAL CECIL THURSBY, RN, HMS *QUEEN*.

"I went ashore again and was met by Bridges and Godley, with several senior officers. They told me that the men were so exhausted after all they had gone through, and so unnerved

---

440   *Evening Star* (New Zealand), 23rd June 1915.
441   *Sunday Times* (Perth, Western Australia), 25th July 1915.
442   Godley, *Life of an Irish Soldier*, p. 171.
443   Birdwood, *Khaki and Gown*, p. 257.
444   Thursby quoted in Wester-Wemyss, *The Navy in the Dardanelles Campaign*, p. 93.

by constant shell-fire after their wonderfully gallant work, that they feared a fiasco if a heavy attack should be launched against us next morning. I was told that numbers had already dribbled back through the scrub, and the two Divisional Commanders urged me most strongly to make immediate arrangements for re-embarkation."[445]

LT-GEN. SIR WILLIAM BIRDWOOD, COMMANDING ANZAC.

"10-30 p.m. It has all been so very terrible that I can hardly write what we have seen to-day. Our first boat loads were ashore at 4 a.m. Will I ever forget it? That awful, anxious time! To think of it all and look at that shore with its beach and cliff like that which leads to the Sorrento Hotel, you can't imagine how one ever got a footing. It wasn't long before boat loads of wounded were being brought back. The sight was dreadful."[446]

LT ARTHUR CHESTER COLMAN, AASC HMAT *KATUNA*.

"At last darkness came, and I can assure you it was never more welcome. We set straight away to "dig-in," or, in other words, to make a bit of a trench. By midnight we had some sort of cover, so we started to remove the dead... we had to dig lying down, pushing the dead away as best we could."[447]

PTE GEORGE HAMPSON, CANTERBURY BN, NZEF, NEW ZEALAND BDE, NZ & A DIVISION.

"During that night, whilst lying amongst the scrub, on the side of a deep gully, about fifty yards or so below the firing line, we had what was to us, at the time at any rate, a realistic, an impressive picture of a modern battle as waged in the darkness. Overhead, the sky was clear and the stars twinkled coldly; the bushes around rustled faintly in a slight breeze; beside us lay a dead man. Stray shots continually passed over our heads or buried themselves in the ground close by, while above rattled and spluttered the rifles of the firing line, which was evidently hard pressed on this Sunday evening. Through all the noise of conflict could be heard the voice of an officer, clear and commanding, which I shall never forget. "Hold your fire, men, keep cool," he would say; "they can't hurt you! Let them come right up!"' Silence. Then suddenly there would be a terrific outburst of firing. Rifles would flash and bullets sing through the air in dozens. After two or three minutes, as the whistle sounded "Cease fire," and the last desultory shots rang out, the officer's voice would again be heard. "Stretcher bearers! Stretcher bearers wanted here!" Then the prostrate form of some wounded man would be borne past us down the track."[448]

PTE CECIL YORKE, CANTERBURY BN, NZEF, NEW ZEALAND BDE, NZ & A DIVISION.

DARDANELLES

Meanwhile, the *AE2*, after passing through the minefields protecting the Narrows, attracting some attention in the process, sent a wireless message to confirm it had broken through.

445   Birdwood, *Khaki and Gown*, p. 257.
446   *Sporting Judge* (Melbourne, Victoria), 11th September 1915.
447   *Nelson Evening Mail* (New Zealand), 23rd July 1915.
448   *Lyttelton Times* (New Zealand), 23rd July 1915.

"At 8.45 [p.m.] *AE2* rose to the surface, having been submerged over sixteen hours.

"Our position was about half a mile from the Asiatic shore, in the sweep of the bay which lies above Nagara Point... Now, too, we could signal to the fleet. A dramatic moment this, while one watched the damp aerial wire throwing purply blue sparks as the longs and shorts of the call sign were flashed. But — myriads of maledictions — the answering call never came."[449]

LT-CMDR HENRY STOKER, RN, HMAS *AE2*.

ANZAC

The growing darkness, added to the fatigue, the unfamiliar, unforgiving ground, the casualties and confusion, led to evacuation being given serious consideration. There was real concern that the foothold they had secured might be overrun. It was decided to refer the matter to Sir Ian Hamilton.

"At first I refused to take any action. I argued that Turkish demoralisation was in all probability considerably greater than ours, and that in any case I would rather die there in the morning than withdraw now. But on thinking things over I felt myself bound to place the position before Sir Ian, if only because every report I had sent him so far (and these reports had been largely based on what Bridges himself had told me) had been entirely optimistic."[450]

LT-GEN. SIR WILLIAM BIRDWOOD, COMMANDING ANZAC.

"Birdwood, who had been on the *Queen*, came ashore, and though of course we were all most reluctant to contemplate the possibility of re-embarkation, we agreed that the situation was such as could only be decided by the Commander-in-Chief. I therefore, at Birdwood's dictation, wrote the following message, which was taken by Admiral Thursby to Sir Ian on board the *Queen Elizabeth*.

"Both my divisional generals and brigadiers have represented to me that they fear their men are thoroughly demoralized by shrapnel fire to which they have been subjected all day after exhaustion and gallant work in the morning. Numbers have dribbled back from firing line and cannot be collected in this difficult country. Even the New Zealand Brigade, which has only recently been engaged, lost heavily, and is to some extent demoralized. If troops are subjected to shell fire again to-morrow morning there is likely to be a fiasco, as I have no fresh troops with which to replace those in the firing line. I know my representation is most serious, but if we are to re-embark, it must be at once."[451]

MAJOR-GENERAL SIR ALEXANDER GODLEY, COMMANDING NZ & A DIVISION.

Understandably, the request to evacuate came as a shock to the Navy.

---

449 Stoker, *Straws in the Wind*, pp. 118–119.
450 Birdwood, *Khaki and Gown*, p. 257.
451 Godley, *Life of an Irish Soldier*, pp. 171–172.

"At about 11 p.m. Captain [Arthur] Vyvyan, RN, the Beach Master, came off to see me. He seemed rather agitated and said he had an important letter to give me from General Birdwood. In the letter he said that his Divisional Generals advised him that they did not think it possible for them to maintain their positions on shore, and that if the troops had to be re-embarked, the sooner it could be done the better. Captain Vyvyan further informed me that he had sent messages to the transports to send in their boats. I was quite taken aback, for I had no idea that things were in such a critical state on shore. A moment's consideration convinced me that to re-embark under such conditions would be disastrous and could not be thought of, especially as we did not yet know what had happened at Helles. The night had turned dark and stormy, our men were tired and disorganized, the confusion in any attempt to re-embark would have been indescribable and our losses must have been appalling. Besides which, I felt quite confident that when daylight came, I could, with the guns of the Fleet, hold back the enemy, while our men dug themselves in and reorganized, either for a further effort to advance or for an orderly re-embarkation under the guns of the Fleet.

"I at once countermanded the order for the boats to be sent on shore and told Captain Vyvyan I would go on shore at once and see General Birdwood and tell him of my decision. I asked Generals Carruthers and Cunliffe Owen to come on shore with me to see what arrangements we could make, naval or otherwise, to enable our men to hold on to the positions they had gained. We had just got into a steam-boat and were proceeding towards the beach when the officer of the watch hailed me and said that the *Queen Elizabeth* was standing towards us. I knew that both Admiral de Robeck and General Sir Ian Hamilton were on board her, so I went to her to report the situation. On arriving on board I had my first news of the southern landing which, so far, had only been partially successful. I sent General Birdwood's letter in to Sir Ian, who was in his cabin."[452]

REAR-ADMIRAL CECIL THURSBY, RN, HMS QUEEN.

"At 12.5 a.m. I was dragged out of a dead sleep by Braithwaite who kept shaking me by the shoulder… "Sir Ian, you've got to come right along — a question of life and death — you must settle it!" Braithwaite is a cool hand, but his tone made me wide awake in a second. I sprang from bed; flung on my "British Warm" and crossed to the Admiral's cabin — not his own cabin but the dining saloon — where I found de Robeck himself, Rear-Admiral Thursby (in charge of the landing of the Australian and New Zealand Army Corps), Roger Keyes, Braithwaite, Brig-General Carruthers (Deputy Adjutant and Quartermaster-General of the Australian and New Zealand Army Corps) and Brig-General Cunliffe Owen (Commanding Royal Artillery of the Australian and New Zealand Army Corps). A cold hand clutched my heart as I scanned their faces. Carruthers gave me a message from Birdwood written in Godley's writing."[453]

GENERAL SIR IAN HAMILTON, C-IN-C, MEF.

"[Hamilton] immediately came out and held a hurried Council of War during which

---

452 Thursby quoted in Wester-Wemyss, *The Navy in the Dardanelles Campaign*, pp. 93–94.
453 Hamilton, *Gallipoli Diary*, vol. 1, pp. 142–143.

he asked me for my opinion, which I gave and further told him that I was on my way on shore to see General Birdwood, and if he would give me a letter to him, I would deliver it personally and would explain to him the necessity of holding on at all costs. I was soon down in my boat again with the letter from Sir Ian and making for the shore, the position of which could only be made out by the flashes from the rifles and explosive shells. It was pitch dark, so we stood in until we could see the white breakers and hear the noise of the surf. It looked as if it would be impossible to land in the steam-boat, when, fortunately, we saw a small merchant-ship's boat rowed by two naval seamen and called them alongside. They said they had just landed two officers and were returning to their ship. They were just as unconcerned as if they had been coming off from a routine trip after dinner in peace time, although bullets were coming over our lines and dropping all around them. We got into their boat and after several attempts and getting very wet, we got on shore. We found ourselves on the right of our position in what was afterwards known as Anzac Cove. Rows of wounded men were laid out in front of us under the shelter of some rising ground. The doctors in charge, who evidently expected that the beach might be rushed at any moment, implored me to have them taken off as it was murder to leave them there. I promised to do my best for them and went off to find Birdwood who, I was told, was on the beach, a little further along. The beach was crowded with men; some, exhausted after the strenuous day, had just thrown themselves down and slept like logs; some were getting food and drink for the first time for many hours, others were being collected by officers and N.C.O.'s and being formed into organized units, being sent either to reinforce the fighting line or to prepare positions to fall back on in case of necessity. I found General Birdwood sitting down with his Divisional Generals and Staff. Birdwood, whom nothing could daunt and who is never so happy as when in the fighting line, and for preference in a tight place, was cheerful but not very hopeful. I gave him Sir Ian's letter."[454]

REAR-ADMIRAL CECIL THURSBY, RN, HMS *QUEEN*.

There was no possibility that an orderly withdrawal could be organised. Hamilton could do little other than tell them to stick it out.

"Your news is indeed serious. But there is nothing for it but to dig yourselves right in and stick it out. It would take at least 2 days to re-embark you as Admiral Thursby will explain to you. Meanwhile the Australian submarine has got up thro' the Narrows and has torpedoed a gun boat at Chanak. Hunter-Weston despite his heavy losses will be advancing tomorrow which should divert pressure from you. Make a personal appeal to your men and Godley's to make a supreme effort and to hold their ground.

"P.S. You have got through the difficult business. Now you have only to dig dig dig until you are safe."[455]

GENERAL SIR IAN HAMILTON, C-IN-C, MEF.

---

454 Thursby quoted in Wester-Wemyss, *The Navy in the Dardanelles Campaign*, pp. 94–96.
455 MEF Headquarters, General Staff War Diary, TNA WO 95/4264.

> "During the night, when the big guns were quiet, and only the miners were busy, we were able to 'dig in.'... The poor beggars wounded, moaning, and calling out was pitiful; but very little could be done for them, and many died for want of attention and exposure to the rain, which fell and wet us all to the skin (our overcoats, etc., were discarded on the beach)."[456]
>
> CAPT. ROSS JACOB, 10TH BATTALION, 3RD BDE, 1ST AUSTRALIAN DIVISION.

The rain added to the misery of men still in shock after experiencing what was, for most of them, their first action. And if their hopes had been high that morning, things were very different by night. What, on paper, was an overwhelming force had been first brought to a halt and then taken to brink of disaster by a stubborn, skilled and determined defence.

SUVLA

One minor mystery on a day when much was unclear surrounded the fate of a single missing Australian soldier. Months after the landings a note in a bottle washed up in Egypt.

> "Am prisoner about two miles from where we landed between dried lake and other one. — Signed, A. [sic] R. C. Adams, 8th, A.I.F."[457]
>
> PTE EDGAR ROBERT COLBECK ADAMS, 8TH BN, 2ND BDE, 1ST AUSTRALIAN DIVISION.[458]

LEMNOS

Those left behind at Mudros were eager for news. They were soon to discover things for themselves.

> "Never since we joined the A.I.F. were our glasses used so much, but in these days of long-range shooting we could only theorise as to the effects of the bombardment, and wonder how our brave fellows were faring. There was no doubt that the 3rd Infantry Brigade had effected a landing, for the rest of the troops were being disembarked. Now and then a troopship that had unloaded its living cargo would come across to our anchorage, and then there would be vigorous flag-wagging until our signallers had learned (or not learned) how things went ashore. The 3rd Brigade did its job as covering party — others landed — enemy pushed back — losses heavy, but the landing scheme was carried out...

---

456 *Barrier Miner* (Broken Hill, New South Wales), 25th July 1915.
457 *The Mildura Telegraph and Darling and Lower Murray Advocate* (Victoria), 10th December 1915.
458 Reported missing on 25th April 1915, Adams' note was found in a bottle by Tpr Sydney Gluth, 4th Australian Light Horse, washed up on the Egyptian shore on 1st November 1915. Efforts to trace his whereabouts proved fruitless, never being recorded as a prisoner of war. His death was presumed to have taken place on or since 25th April 1915 and he is commemorated on the Lone Pine Memorial.

"I had been anxious to go ashore early, but that was not permitted. Here we were, in a safe anchorage by Imbros, out of range of the enemy's guns, fretting to know the fate of all those whom we had watched going to do their duty… But there is small time for soliloquy. A message comes for the *Ionian* to steam in and become a hospital ship."[459]

COLONEL CHAPLAIN JOHN McPHEE, CHAPLAIN'S CORPS, AIF, HMT *IONIAN*.

GERMANY

If news of the Australian *AE2*'s success in entering the Dardanelles was cause for celebration, another submarine's departure from Germany that day would prove rather more portentous for the Allied cause.

"On April 25, 1915, we moved out of the harbour of Wilhelmshaven and set a course north. The English Channel by now, with its entanglements of nets and mines, was exceedingly dangerous for U-boat navigation, and we were not to take any more chances that we had to until we reached the scene of action. So we took the long route around Scotland, the northern tip of the Orkney Islands. We went along minding our own business. Any ships that hove in sight might be good game for some other U-boat but they meant nothing to us."[460]

KAPITAN-LEUTNANT OTTO HERSING, U-21.

TENEDOS

Many of the ships heading for Kum Kale and Helles passed by Tenedos.

"All through the night transports had been arriving, and when morning broke… one looked upon a strange and wonderful sight indeed. Right up from where we lay to the entrance of the Dardanelles were vessels of every description, from tiny minesweeping trawlers to gigantic ships of war. Almost as soon as it became light the air began to vibrate with the booming of heavy guns… and the doors shook with the vibration. Along the coast to our right huge French and English transports, armed with 4 and 6-inch guns, were firing at the shore…

"But nearer in, with the French and a Russian cruiser, our other ships were carrying on the deadly work. Kum Kale, on the Asiatic side, appeared to be continually deluged with shell; the town itself and the hills around were one mass of smoke, great white puffs appearing in the air and upon the hillside, and terrific black spouts of earth and smoke, which told of a good hit."[461]

ASSISTANT-PAYMASTER WILLIAM EDWARD MOLINEUX GUY, RNVR, HMS *BLENHEIM*.

---

459   *Daily Herald* (Adelaide, South Australia), 21st July 1915.
460   Otto Hersing quoted in Thomas, Lowell, *Raiders of the Deep*, p. 57, Heineman (London) 1930.
461   *Taranaki Herald* (New Zealand), 3rd July 1915.

KUM KALE

Over on the Asian shore, the French prepared to play their part, a diversionary operation in support of the British landings across the straits at Helles.

> "[5.00 a.m.] British battleships start operations. We hear a dull thud like heavy cases being moved in an upstairs room. Then the firing becomes more rapid, and one imagines a distant storm with unending echoes.
>
> "Everybody is on deck. We look on in awe and wonder. Smoke is now mixed with the early mists, and the *Pointe d'Europe* steams like a cauldron."[462]
>
> MO JOSEPH VASSAL, 6TH COLONIAL REGT, BDE *COLONIALE*, 1ST DIVISION, CEO.

> "British guns are already firing on our left. On our side, silence reigns. We move slowly towards the Asian coast and drop anchor, quite close to land…
>
> "It's five o'clock, and broad daylight. The smooth sea is barely wrinkled by a light breeze. The signal to "open fire" goes up to the mainmast of the *Charlemagne*."[463]
>
> LT HENRI FEUILLE, 52ND BATTERY, 30TH REGT, 1ST DIVISION, CEO.

> "Our regiment, commanded by Lt-Col Nogues, has received orders to land at Koum Kaleh in order to silence the Asiatic batteries and to protect the British landing-parties. We do not know the number of Turkish troops which will be opposed to us. The artillery works have been partly destroyed by the Allied Fleet. A detachment of British marines landed some weeks ago, and put the guns of the castle out of action."[464]
>
> MO JOSEPH VASSAL, 6TH COLONIAL REGT, BDE *COLONIALE*, 1ST DIVISION, CEO.

> "The squadron is composed of *Jeanne d'Arc, Henri IV, Charlemagne, Jauréguiberry, Latouche Tréville* and a squadron of eleven torpedo boats, preceded by minesweepers, all under the command of Admiral Guépratte. The Russian cruiser *Askold* is attached.
>
> "The squadron deployed in line of battle, rounded the tip of Kum Kale and entered the strait… Kum Kale can now be seen in a light mist, emerging from the calm sea."[465]
>
> LT HENRI FEUILLE, 52ND BATTERY, 30TH REGT, 1ST DIVISION, CEO.

> "We left Lemnos at 5 a.m. on a simply perfect day in April, and could plainly hear the heavy firing. As we came out several other ships were on the move, and when we got outside they were like a swarm of bees, all making for Tenedos… When we got near… several armed French cruisers, and the Russian cruiser, *Askold* — known in the service as the packet of woodbines, owing to her five funnels — [were] blazing away at the forts on Kum Kale, and sweeping the trenches for all they were worth."[466]
>
> SUB-DISPENSER (JAMES) EDWIN SCOTT, RAMC, HMHS *SOUDAN*.

---

462   Vassal, *Uncensored Letters from the Dardanelles*, p. 48.
463   Feuille, *Capitaine* Henri, *Face aux Turcs. Gallipoli 1915*, pp. 28–29, Payot (Paris) 1934.
464   Vassal, *Uncensored Letters from the Dardanelles*, p. 48.
465   Feuille, *Face aux Turcs*, p. 29.
466   *Richmond and Twickenham Times*, 12th June 1915.

"Admiral Guepratte ordered the disembarkation to commence [at 6.00 a.m.], but, owing to various causes, mainly the inability of the French steamboats to tow the heavily laden boats against the strong current, the troops did not reach the beach until four hours later. This delay was probably an advantage, as it enabled the ships to bombard the villages and low land to such an extent that the enemy fled to the opposite side of the Mendere River."[467]

COMMODORE ROGER KEYES, RN, CHIEF OF STAFF, EMS, HMS *QUEEN ELIZABETH*.

"We are on the Transatlantic liner *Savoie*, converted into an auxiliary cruiser. Her guns shell Yenisher... The *Askold* joins us. Now their fire is directed on the castle of Koum Kaleh. Enormous fountains of steam, dust, fragments, and flames spring up out of the old fortress...

"The soldiers on our ship prepare to land; and we approach the Asiatic coast. Lt-Col Nogues says "good-bye" to me. We remain standing together for a few minutes, but without speaking...

"[6.15 a.m.] The soldiers are waiting, armed and fully equipped... Their comrades are filling the boats... The sea is absolutely calm, without the slightest ripple."[468]

MO JOSEPH VASSAL, 6TH COLONIAL REGT, BDE *COLONIALE*, 1ST DIVISION, CEO.

"During the bombardment, preparations were made for the landings... The signal to go was given, and the boats at once headed towards the shore. The troops landing included the 6th Mixed Regiment, under the command of Colonel Ruef, composed of two battalions of Senegalese, a European battalion, an engineer company, a machine-gun section of the 4th Chasseurs Regiment, as well as three 75mm guns of the 33rd Battery of the 8th Artillery Regiment, under the orders of Captain Reich, in all 3,000 men."[469]

LT HENRI FEUILLE, 52ND BATTERY, 30TH REGT, 1ST DIVISION, CEO.

"The Gallipoli Peninsula seems to be on fire. The old castle of Sedd-el-Bahr is in flames. We hear that the English are beginning landing operations over there.

"On our side the... warships fire unceasingly; and the ship I am on trembles. The *Askold*, quite close to us, lets off each time a broadside of four guns. Green lights flash out from the thick yellow smoke. The shelling becomes quicker and fiercer. The noise is prodigious."[470]

MO JOSEPH VASSAL, 6TH COLONIAL REGT, BDE *COLONIALE*, 1ST DIVISION, CEO.

"At 8.30 a.m. the boats landed on the beach near the ruined pier... They came under heavy fire, suffering significant losses...

---

467   Keyes, *Naval Memoirs 1910–1915*, pp. 301–302.
468   Vassal, *Uncensored Letters from the Dardanelles*, pp. 48–49.
469   Feuille, *Face aux Turcs*, p. 30.
470   Vassal, *Uncensored Letters from the Dardanelles*, p. 50.

"[O]ur men rushed into the tangle of the smoking rubble of the village, while the Turkish machine guns continued to inflict heavy casualties.

"The cemetery, a refuge for the Turks, was cleared by the Lejeune and Distel companies, supported by a machine-gun section.

"The houses that remained standing, despite the bombardment, became dangerous centres of resistance… Our 75s were dragged into this maze… shells do their work and clear the way. Our men advanced through the smouldering ruins, amidst the fire of the enemy snipers."[471]

LT HENRI FEUILLE, 52ND BATTERY, 30TH REGT, 1ST DIVISION, CEO.

"The first boats return to the *Savoie*. In one I see, as it comes alongside, a wounded man who is being brought to us. It is a Russian sailor who has been hit by several bullets. He lies unconscious at the bottom of the boat; his clothes are covered with blood. One bullet has gone through his arm, a second right through the abdomen, a third has broken one of his legs. Nothing to be done for him; and he dies at 12 o'clock without regaining consciousness."[472]

MO JOSEPH VASSAL, 6TH COLONIAL REGT, BDE *COLONIALE*, 1ST DIVISION, CEO.

"Turkish machine-guns, stationed in a mill, caused heavy casualties [but] accurate naval shellfire blew it into the air.

"Aboard the *Ceylon* we could see the Turks massing above the village and in a cemetery. They were spotted by the gunners of the *Savoie* and soon completely covered by the smoke from the explosion of the shells.

"At the village crossroads, the Senegalese, having dropped their Lebels [rifles], fought hand-to-hand with the Turks… Everywhere there are screams, sounds of gunfire and… the cries of the wounded."[473]

LT HENRI FEUILLE, 52ND BATTERY, 30TH REGT, 1ST DIVISION, CEO.

"1 p.m. We received confirmation from a French soldier (badly wounded, the lung perforated) that the landing was successful, but that one boat hit by a large shell has sunk with a certain number of Senegalese.

"1-45 [p.m.] The French torpedo-boat *Trident* comes to fetch munitions for our troops ashore. The *Trident* tows to land the boats containing the last contingent of troops. Two huge boxes containing artillery horses are also landed on Turkish soil. I am told that the first guns have already gone an hour before.

"2 [p.m.] A corporal bugler wounded in the leg returns to the *Savoie*. He says that the Turks are driven back beyond the village of Koum Kaleh, but that we have many wounded.

"4 [p.m.] Another convoy of wounded arrives. They are fourteen together, and the spectacle is already distressing. Among them is a lieutenant of the machine-gun

---

471   Feuille, *Face aux Turcs*, p. 31.
472   Vassal, *Uncensored Letters from the Dardanelles*, pp. 51–52.
473   Feuille, *Face aux Turcs*, p. 31.

French artillery being landed.

section whose arm has been broken. All human miseries seem to be united in this little group of warriors lying at the bottom of the boats.

"5 [p.m.] The second section of artillery is sent off to Koum Kaleh. Three 75-mm. guns had been placed at the disposal of Colonel Ruef, the 3rd Battery of the 8th Artillery Regiment; but one gun was not landed."[474]

MO JOSEPH VASSAL, 6TH COLONIAL REGT, BDE COLONIALE, 1ST DIVISION, CEO.

"No sooner had our defensive position been arranged than, at 8.30 p.m., the Turks attacked along the whole line with the bayonet… [It] was broken by a barrage of machine guns and the incessant fire of a 75…

"At this time, we occupied Kum Kale. General d'Amade, aboard the flagship *Charlemagne* with Admiral Guépratte, made an urgent request for reinforcements from… Ian Hamilton, in order to exploit our success and to establish ourselves firmly on this strategic point on the Asian coast."[475]

LT HENRI FEUILLE, 52ND BATTERY, 30TH REGT, 1ST DIVISION, CEO.

"The convoys of wounded follow each other rapidly. From twilight of the 25th till the first rays of dawn the next day we are leaning over wounded in an atmosphere of blood, of groans, and of indescribable horrors. We do not stop for a single minute."[476]

MO JOSEPH VASSAL, 6TH COLONIAL REGT, BDE COLONIALE, 1ST DIVISION, CEO.

---

474   Vassal, *Uncensored Letters from the Dardanelles*, p. 55.
475   Feuille, *Face aux Turcs*, pp. 32–34.
476   Vassal, *Uncensored Letters from the Dardanelles*, p. 56.

HELLES

The men on their way to the five beaches around the tip of Cape Helles were tasked with moving rapidly inland, joining together before advancing to occupy the heights of Achi Baba that evening. By the third day, they were expected to have taken the heights dominating the Narrows, the Kilid Bahr Plateau.

V BEACH

Concerns about the vulnerability of men landing on a defended shore in open boats in broad daylight led to the adoption of the *River Clyde* as a large troop carrier. Could the old collier be run aground, allowing those onboard to move swiftly onto the beach and to overpower its defenders?

> "Some Companies went on battleships and other in torpedo boats. My Company, with the Munster Fusiliers and 2nd [Bn] of the Hants. Regiment, went on board an old boat named the *River Clyde*. We arrived off the Dardanelles about 3 a.m."[477]
>
> SGT CHARLES SMART, 1ST BN ROYAL DUBLIN FUSILIERS, 86TH BDE, 29TH DIVISION.

> "Astir at three, gorgeous morning. We moved up just before dawn, and lay about a mile and a half off the land, full steam up. The sight was then wonderful, and the mighty Fleet surrounded us. At about [5.00 a.m.] the terrible bombardment of our ships and the French started. It was too terrible to describe, and awful in its grandeur. The very heavens shook. Commander Unwin on our bridge navigated us in a wonderful manner, circling about in a most bewildering way, waiting for the signal from the Admiral to go right in. All this time the enemy peppering us with large shells, some of which came dangerously close, but none hit us so far. At last, amid the terrible deafening, roaring hell, I heard the Commander say, "Look out, I am going right in," and so, with full steam ahead, we rushed for the shore. We little suspected what we were in for, as opinion was divided, some saying it was certain death, others said we would not have a shot fired, as nothing could live in the terrible fire of the navy."[478]
>
> SURGEON-LIEUTENANT PETER KELLY, RN, SS *RIVER CLYDE*.

> "Each man had two days' rations and two bottles of water, and 200 rounds of ammunition, with our field kits, packs, etc., all told about 60 lbs., to carry. About 5 a.m. when we were getting into the boats the ships started the bombardment of the Gallipoli peninsular, the crashing of the guns was like one continuous roll of thunder coming from every direction. We were about four miles from the land at this time and in the boats, which were towed to the shore by naval picquet boats. As we got nearer the land we were escorted by destroyers, and as we passed the warships the bands played "Tipperary," etc., and cheered us, our lads responded with "Are we downhearted. No!" and cheered and sang all the latest songs they knew. As we got nearer to the land we could see our shells smashing down walls, houses, and blowing

---

477   *Midland Counties Tribune*, 25th June 1915.
478   *Galway Express*, 28th August 1915.

up the Turks' trenches in all directions; in fact, one would think it impossible for a house to stand in the place. When we got close in to the land the picquet boats threw off, and the sailors in our boats dropped their oars and pulled for the shore."[479]

SGT DAVID McCREARY, 1ST BN ROYAL DUBLIN FUSILIERS, 86TH BDE, 29TH DIVISION.

"We were all in lively spirits, smoking and laughing and saying what we would do when we landed in Gallipoli. Six sailors sat middle of the boat oars in their hands, ready to pull the boat ashore the moment the picket-boat let us go. I myself sat next to the officer.

"Suddenly the order came, "Out oars. Give way together" — the picket boat being unable to take us any further on account of the shallowness of the water."[480]

OS WILLIAM TROTTER, RN, HMS *CORNWALLIS*.

"These boats... were behind their time, and it was 6.40 a.m. when the steam boats slipped their tows and the boats began to pull the short remaining distance to the shore. Until this moment the *Albion* had been bombarding the beach and the surrounding amphitheatre, but, alas with no better success than the *Swiftsure* had obtained at W. The moment therefore she ceased bombarding, the boats were subjected to a furious storm of fire before they even reached the land."[481]

REAR-ADMIRAL ROSSLYN WEMYSS, RN, HMS *EURYALUS*.

"All went well till we were within about 100 yards of landing. Then the Turks commenced. Men were shot in the boats; boats were sunk; men jumping over the side; everyone was trying to get ashore; boats adrift — such a scramble I have never seen before. It was like hell let loose."[482]

SGT WILLIAM MASLIN, 1ST BN ROYAL DUBLIN FUSILIERS, 86TH BDE, 29TH DIVISION.

"Bullets and shells began flying about, shrapnel shells, and those large explosive bullets all burst in the boat at once. The sailors were hit first, being in blue, which made us very conspicuous among the khaki. I remember seeing four of my chums killed. Then the officer next to me was hit.

"The next moment a terrific explosion took place... One large piece shattered my right knee completely. The boat by this time had grounded on the beach, and only five soldiers got out of it."[483]

OS WILLIAM TROTTER, RN, HMS *CORNWALLIS*.

"I was in the same boat as the captain, and about 30 comrades. All I could hear was, 'Oh, I am shot.' I got out of the boat, breast deep in water, and when I got about two yards away I got a bullet through my leg just above the ankle. I fell into the water, and

---

479 *Nuneaton Observer*, 11th June 1915.
480 *Fife Free Press & Kirkcaldy Guardian*, 19th February 1916.
481 Wester-Wemyss, *The Navy in the Dardanelles Campaign*, p. 76.
482 *Burnley Express*, 29th May 1915.
483 *Fife Free Press & Kirkcaldy Guardian*, 19th February 1916.

would have drowned, but my equipment slipped off me, and I managed to get on my feet and reach the beach."[484]

L/CPL PATRICK FLANAGAN, 1ST BN ROYAL DUBLIN FUSILIERS, 86TH BDE, 29TH DIVISION.

"We lost practically all of our officers… Our Colonel[485] was one of the first to go. He, poor chap, like many of the others, was killed in the boat before he got the chance to step ashore. Some of the boats capsized and a lot of chaps were drowned. Others got wounded and fell out of the boats, and, of course, having their heavy equipment on, they never got a chance. The Turks had constructed barbed wire entanglements under the water, and the men, when they jumped out of the boats into the water, got their feet entangled in the wire and were picked off."[486]

SGT CHARLES SMART, 1ST BN ROYAL DUBLIN FUSILIERS, 86TH BDE, 29TH DIVISION.

"Add to this wire entanglements in the sea, men hung and stuck on this, and you will have some idea of our landing. Anyway, through all this I managed to struggle, and just as I was getting on dry land I was hit above the knee. Nothing for it but down on all fours and struggle on and under the bank."[487]

SGT WILLIAM MASLIN, 1ST BN ROYAL DUBLIN FUSILIERS, 86TH BDE, 29TH DIVISION.

"O'Hanlon[488]… had his hand blown off. He and I were trying to save another when he got hit again in his foot and fell back into the boat.

"Two bullets went through my pack and I dived into the sea. Then came the job to swim with the pack and one leg useless. So I managed to pull out the knife and cut straps and swim to shore… All the time bullets were ripping around me."[489]

CPL JAMES COLGAN, 1ST BN ROYAL DUBLIN FUSILIERS, 86TH BDE, 29TH DIVISION.

"I thought my time was up every minute… the worst of it all was, we had a priest[490] who came along with us. He was in the boat; he insisted on coming along with us, as he said he would be wanted for the poor boys. They were all calling for him, but the poor priest could do nothing for them. He got out of the boat afterwards and made great run for the beach; but the Turks got him as soon as he landed, for he was hit four times."[491]

PTE ROBERT MARTIN, 1ST BN ROYAL DUBLIN FUSILIERS, 86TH BDE, 29TH DIVISION.

"Father Finn was killed beside me, also the C.O. and Adjutant. It is simply a marvel

---

484 *Midland Counties Tribune*, 2nd July 1915.
485 Lt-Col. Richard Rooth, 1st Bn Royal Dublin Fusiliers, was killed in action on 25th April 1915. Buried in V Beach Cemetery, he was the 49 year-old son of John Rooth and Elizabeth Rooth; husband of Amy Rooth, of Dawlish, Devon.
486 *Midland Counties Tribune*, 25th June 1915.
487 *Burnley Express*, 29th May 1915.
488 Pte Arthur O'Hanlon, 1st Bn Royal Dublin Fusiliers, was discharged due to wounds on 2nd December 1915.
489 *Midland Counties Tribune*, 21st May 1915.
490 Chaplain 4th Class The Rev. William Joseph Finn, Army Chaplains' Department, attached 1st Bn Royal Dublin Fusiliers, was killed in action on 25th April 1915. He is buried in V Beach Cemetery.
491 *Coventry Herald*, 9th July 1915.

"that I am alive. You can guess what it was like — only six of us got away alive out of a boatload of 32. One fellow's brains were shot into my mouth as I was shouting to them to jump for it."[492]

CPL JAMES COLGAN, 1ST BN ROYAL DUBLIN FUSILIERS, 86TH BDE, 29TH DIVISION.

"It was cruel to see our poor, helpless wounded, and the efforts to save them were of no use, for everyone who went to try and do so were killed. I myself was lying beside six of our company, all wounded, and it was pitiful to hear Harris... crying, "Masterson, dress me quick," What was I to do? If I stirred, the two of us were certain to be killed, so I left him, and he managed to get in a boat. I was fully two hours lying in the water, because I knew death was facing me if I moved."[493]

PTE WILLIAM MASTERSON, 1ST BN ROYAL DUBLIN FUSILIERS, 86TH BDE, 29TH DIVISION.

"I ran and got under cover, but not before I got another one in the chest, which just missed my heart, but passed through my left lung. I fell on my face and lay there flat for twelve hours, snipers potting all the time, but they did not manage to hit me again."[494]

L/CPL PATRICK FLANAGAN, 1ST BN ROYAL DUBLIN FUSILIERS, 86TH BDE, 29TH DIV.[495]

"Our own shells were flying over us and the Turks scattering shrapnel bullets all round us; rifle and machine gun fire from every direction; men getting knocked over by the dozen; some firing away where they could see any sign of a Turk: others building head cover and digging themselves in. It was nothing to see a man with his leg half off trying to dress some other chap with his arm or head smashed, while others were trying to pull the wounded out of the water and get them under cover."[496]

SGT DAVID McCREARY, 1ST BN ROYAL DUBLIN FUSILIERS, 86TH BDE, 29TH DIVISION.

"On came the *River Clyde* through a storm of fire from both European and Asiatic batteries. She was hit several times, but sustained no material damage, though a few men were killed, but her Commander's nerve and steady hand never failed him and the vessel took the ground at 6.30 a.m. in her assigned position, but, unfortunately, further out than had been hoped owing to the beach being more shelving than expected."[497]

REAR-ADMIRAL ROSSLYN WEMYSS, RN, HMS *EURYALUS*.

"I never noticed the grounding, for the horror in the water, on the beach. Five tows of five boats each, loaded with men, were going ashore alongside of us. One moment it had been early morning in a peaceful country, with thoughts of smells of cows and hay and milk; and the next, while the boats were just twenty yards from the shore, the blue sea round each boat was turning red. Is there anything more horrible than to see men wading through water waist-high under a heavy fire? You see where each bullet

---

492   *Galway Express*, 29th May 1915.
493   *Midland Counties Tribune*, 2nd July 1915.
494   *Midland Counties Tribune*, 2nd July 1915.
495   Flanagan enlisted on 10th April 1911; discharged due to wounds on 25th March 1916.
496   *Nuneaton Observer*, 11th June 1915.
497   Wester-Wemyss, *The Navy in the Dardanelles Campaign*, pp. 77-78.

hits the water, which, like a nightmare, holds back the man for the next shot, which will not miss. Of all those brave men two-thirds died, and hardly a dozen reached unwounded the shelter of the five-foot sand dune."[498]

LT-CDR JOSIAH WEDGWOOD, RNVR, NO. 3 SQN ROYAL NAVAL ARMOURED CAR DIVISION.

"The *River Clyde* beached according to plan at 6.30. None of us felt it, there was no jar. As she beached 2 Companies of the Dublins in "tow" came up on the Port side and were met with terrific… fire. They were literally slaughtered like rats in a trap. The steamer hopper towing the lighters from the *Clyde* was either shot away or broke loose, anyway it beached alone."[499]

CAPT. GUY GEDDES, 1ST BN ROYAL MUNSTER FUSILIERS, 86TH BDE, 29TH DIVISION.

"Directly she touched, Commander Unwin rushed to the side and was horrified to see that the hopper, instead of going ahead as intended, was hung up alongside… The wooden lighters, however, shot ahead, and Commander Unwin, realizing the possible results of the failure of the hopper to fulfil her part, with no other thought than the success of the enterprise, jumped into the lighter and with the assistance of Lieutenant Morse, RN, Midshipmen Malleson, RN, and Drewry, RNR, Petty Officers Russell and Rummings [sic] and Able Seamen Williams and Samson got them into position under the ship's bows and, standing up to their waists in water, held them in position whilst the troops began to pour out of the ship."[500]

REAR-ADMIRAL ROSSLYN WEMYSS, RN, HMS *EURYALUS*.

"Here I saw Lieutenant Morse, RN.[501] He called to me to lend him a hand in securing a lighter. So we hauled the lighter astern, giving the stern a kick out so as to meet the other lighters. We both jumped into the lighter; but as she was moving, Morse said: "Have you secured the hawser?" My reply was: "No, sir, I thought you had." So again I jumped out on to the hopper, before the lighter swung out, and secured the hawser round a bollard. Just in time, as I got another bullet through my lung. I spun round and fell down, managing to get more or less under cover."[502]

MIDSHIPMAN MAURICE LLOYD, RN, HMS *CORNWALLIS*.

"The collier s.s. *River Clyde* beached herself proper style in the best possible place. Immediately twelve machine-guns in the *River Clyde* started opening fire, and the steamboats with their tows drew in (four lifeboats per tow)… We towed in four boats, but could only manage to bring out two, and then we proceeded to Fleet-sweeper No. 2 for another load, which consisted of two boats, one of soldiers, and the other of our beach party. We towed these in on the starboard side of the *River Clyde*, as the midshipman of a cutter sang out to us not to go the port side. The last I saw of

---

498   Wedgwood, Josiah Clement, *With machine-guns in Gallipoli*, pp. 4–5, Darling & Son Ltd. (London) 1915.
499   TNA WO95/4310.
500   Wester-Wemyss, *The Navy in the Dardanelles Campaign*, p. 78.
501   Lt John Morse, RN, HMS *Cornwallis*, later Vice-Admiral Sir John Morse, KBE, DSO.
502   Maurice Lloyd quoted in Stewart, Archibald & Pershall, Charles, *The Immortal Gamble*, pp. 107–108, A. C. Black Ltd (London) 1917.

those two boat-loads they were in the water taking cover behind the boats. I could do nothing, however, to help them."⁵⁰³

MIDSHIPMAN PERCY WAIT, RN, HMS *CORNWALLIS*.

"Within five minutes of the "Clyde" beaching "Z" Company got away on the Starboard side. The gangway on the Port-side jammed, and delayed X. Company for a few seconds, and off we went the men cheering wildly, and dashed ashore with Z. Company."⁵⁰⁴

CAPT. GUY GEDDES, 1ST BN ROYAL MUNSTER FUSILIERS, 86TH BDE, 29TH DIVISION.

"Then the order came from our Commanding Officer (Colonel Tizard), 1st R.M.F., 'Now, boys, out after them!' Our lads ran down the gangways, which were not quite one yard wide. As they did so they were met with a terrible fire, which caused us heavy losses, while many of our wounded dropped into the water and were drowned."⁵⁰⁵

PTE CHARLES MAYNARD, 1ST BN ROYAL MUNSTER FUSILIERS, 86TH BDE, 29TH DIVISION.

"There was a hole cut in her port and starboard bow, big enough for one man to pass through, and from which ran a narrow gangway forming an angle. Down this gangway you had to run and jump into a barge alongside, under heavy fire — also pom-poms — and if you got hit in your passage from the ship to the barge, you fell into about 12 feet of water, so that you lost your life by shot and drowning. I could not tell you how I got down there. I remember making a wild dive out of the opening, and the next place I found myself was on [shore] which I can assure you was a very pleasant sensation."⁵⁰⁶

PTE EDWARD BARTON, 1ST BN ROYAL MUNSTER FUSILIERS, 86TH BDE, 29TH DIVISION.

"It was a good job we had the protection of the steel sides of the ship otherwise not a man would have been left. Then the landing operations started. The gangway was put out, and a few pontoon bridges stretched to the beach. All this time the Turks were sending a hail of lead at anything that appeared.

"The captain of my company asked for 200 volunteers, and as I was in his company I volunteered. We got ready inside on the deck, and opened the buckles of our equipment, so that every man might have a chance of saving himself if he fell into the water. He gave the order to fix bayonets when we should get ashore.

"He then led the way, but fell immediately at the foot of the gangway. The next man jumped over him, and kept going until he fell on the pontoon bridge...

"I was about the twenty-seventh man out. I stood counting them as they were going through. It was then I thought of peaceful Macroon and wondered if I should ever see it again. When my turn came I was wiser than some of my comrades. The moment I stood on the gangway I jumped over the rope on the the pontoon. Two more did the same, and I was already flat on the ground. Those two chaps were at each side of me, but not for long, as the shrapnel was bursting all around.

---

503   *East Anglian Times*, 7th June 1915.
504   TNA WO95/4310.
505   *Kent Messenger & Gravesend Telegraph*, 26th June 1915.
506   *Galway Express*, 26th June 1915.

"I was talking to the chap on my left and saw a lump of lead enter his temple, I turned to the chap on my right. His name was Fitzgerald.[507] He was from Cork, but soon he was over the border. The one piece of shrapnel had done the job for the two.

"In the meantime the water was full of dead. Some got slightly wounded and fell off the barge. The water did the remainder. The remainder of the ships stopped there until night, as it would be wholesale slaughter to try and get out. The men would be shot without seeing a Turk."[508]

PTE TIMOTHY BUCKLEY, 1ST BN ROYAL MUNSTER FUSILIERS, 86TH BDE, 29TH DIVISION.

"We lost our first two men while we were waiting to disembark… We had to run along the gangway on the side of the ship, and then we had to jump into the water and get ashore as best we could. There was a man in front of me and a man behind, and they were both killed, and I thought it would be my turn next, but I managed to get through, though the bullets were singing round me like peas, so you can guess what I felt like."[509]

SPR EDWARD SILVESTER, 1/1ST (WEST RIDING) FIELD COY, ROYAL ENGINEERS, 29TH DIV.

"McKnight[510]… was drowned. If a man got hit in the water there wasn't much chance for him. You could never realise what war is like until actually in it."[511]

PTE JOHN PATTERSON, 1ST BN ROYAL MUNSTER FUSILIERS, 86TH BDE, 29TH DIVISION.

"Then came some of the "Turkish delight." The bullets hit against her steel sides with a lively "ping," and a shell came right through her, killing three of the Royal Army Medical Corps who were down in the hold."[512]

PTE MAURICE SINDEN, 2ND BN HAMPSHIRE REGT, 88TH BDE, 29TH DIVISION.

"We had to run… over masses of dead, jump into the water, and try to get ashore, so with full packs it was a rough time…

"There were some of the chaps hit while in two feet of water and could not move, so when the tide started to come in it was awful to see those chaps getting drowned and roaring for someone to save them. The Navy saved a lot of them, as we could not help, being ashore firing."[513]

PTE WILLIAM MORIARTY, 1ST BN ROYAL MUNSTER FUSILIERS, 86TH BDE, 29TH DIVISION.

"I looked ashore [and] I saw five Munsters. They at some moment had got ashore; they had been told off to cut the wire entanglements; they had left the shelter of the bank, charged fifteen yards to the wire, and there they lay in a row at two yards interval.

---

507   Pte James Fitzgerald, 1st Bn Royal Munster Fusiliers, was killed in action on 25th April 1915. He is commemorated by a special memorial in V Beach Cemetery, where he is believed to be buried.
508   *Nottingham and Midland Catholic News*, 24th July 1915.
509   *Penistone, Stocksbridge and Hoyland Express*, 17th July 1915.
510   Pte Edward McKnight, 1st Bn Royal Munster Fusiliers, was killed in action on 25th April 1915. He is commemorated on the Helles Memorial.
511   *Coventry Evening Telegraph*, 24th May 1915.
512   *Western Gazette*, 23rd July 1915.
513   *Coventry Evening Telegraph*, 14th May 1915.

One could hardly believe them dead."[514]

LT-CDR JOSIAH WEDGWOOD, RNVR, NO. 3 SQN ROYAL NAVAL ARMOURED CAR DIVISION.

"I volunteered to go in and try to get a better landing-stage rigged to get the soldiers ashore, but the firing was too heavy: it was impossible for me and my men to work. I had all of my men and a soldier wounded in my boat within a few minutes of my arrival. I saw about 30 soldiers leave the *River Clyde* — only two got safely on the beach.

"I busied myself doing what I could amongst the wounded dressing wounds, getting some out of the water, etc., but I had to give up the work, not that I valued my own life, a thing I never thought of at the time, but the enemy had shot two of the wounded whom I had already got out of the water. One bullet shaved my hair, while another grazed my little finger. After that I had 13 wounded and seven dead in two boats put into my care for transmission to the hospital ship...

"An officer, with two men, brought a boat to me, and we managed to get 11 off and safely back to the *River Clyde*. When rowing back the second time I placed a large box full of biscuits at my back for safety. The box got three bullets into it."[515]

PO FREDERICK GIBSON, RFR, HMS *ALBION*.[516]

"Midshipman Drury[517] of the *Clyde* swam to the hopper, was wounded in the head, got a line of the hopper & got somehow back to the ship with it. Commander Unwin,[518] maddened by the failure of his landing plan, stood up alone on the hopper & hauled surrounded by dead and wounded. He was slightly wounded. He went in again and rowed to the wounded on the rock jetty & loaded them into the boat under fire. One of our men, J. H. Russell,[519] seeing him wounded and unable to lift the men out of the water went overboard, swam to him. He was shot through the stomach and with Unwin lay in the water as though dead for a while, got somehow into the boat in a lull and were pulled back to the ship by a line.

"The wounded were still crying and drowning on that awful spit.

---

514 Wedgwood, *With machine-guns in Gallipoli*, p. 6.
515 *Hartlepool Northern Daily Mail*, 1st November 1915.
516 Gibson was awarded the Conspicuous Gallantry Medal: "Petty Officer, Second Class Frederick Gibson... jumped overboard with a line and got his boat beached to complete bridge from "River Clyde" to shore. He then took wounded to "River Clyde" under heavy fire." (*Edinburgh Gazette*, 17th August 1915.) Gibson later received the French *Médaille Militaire* (*London Gazette*, 28th August 1918).
517 Midshipman George Drewry, RN, was awarded the VC. In the words of his official citation: "Assisted Commander Unwin at the work of securing the lighters under heavy rifle and maxim fire. He was wounded in the head, but continued his work and twice subsequently attempted to swim from lighter to lighter with a line." (*London Gazette*, 13th August 1915.)
518 Capt. Edward Unwin, RN, received the VC: "While in SS *River Clyde*, observing that the lighters which were to form the bridge to the shore had broken adrift, Commander Unwin left the ship, and under a murderous fire attempted to get the lighters into position. He worked on, until suffering from the effects of cold and immersion, he was obliged to return to the ship, where he was wrapped up in blankets. Having in some degree recovered, he returned to his work against the doctor's order and completed it. He was later attended by the doctor for three abrasions caused by bullets, after which he once more left the ship, this time in a lifeboat, to save some wounded men who were lying in shallow water near the beach. He continued at this heroic labour under continuous fire, until forced to stop through physical exhaustion." (*London Gazette*, 13th August 1915.)
519 Petty Officer Mechanic John Hepburn Russell, RNACD, received the Conspicuous Gallantry Medal for his conduct on V Beach on 25th April 1915 (*London Gazette*, 16th August 1915).

"Lieut. Tidsdale[520] (?) RND. took a boat, one of the *Clyde*'s sailors and one of my men, Rumming.[521] They got 4 men on board before Tidsdale and the sailor were shot and wounded. Hiding behind the side of the boat they walked & swam it back. I saw one of the wounded stretch out his hand[,] stroke Rumming as he hung on to the side, the most pathetic thing I have ever seen."[522]

LT-CDR JOSIAH WEDGWOOD, RNVR, NO. 3 SQN ROYAL NAVAL ARMOURED CAR DIVISION.

"The brave sailors belonging to HMS *Swiftsure* did all they could to help our wounded by rowing and swimming, being shot while doing so. The captain of the *Clyde* was one of the bravest there; I should like to know his name, so as to mention it, for he did many things to help the wounded of the regiments."[523]

PTE CHARLES MAYNARD, 1ST BN ROYAL MUNSTER FUSILIERS, 86TH BDE, 29TH DIVISION.

"I saw among other things one poor fellow shot in very shallow [water] and, fearing that he would drown, I decided (spurred on by the doings of Commander Unwin and young Drewry) to go after him, and if I could get a volunteer to assist me, and as soon as others were attended to. Having done what I had to do, I called for another, and whilst talking to a petty officer on the deck about what we could, someone shouted, "Take cover, you fool," and I hesitated. Almost simultaneously a bullet hit my cap and one went either side of my head, shivering my ears. As I dropped flat one chipped me on the right shin and another entered my right foot, but came out again. So of our staff I was the first to go. I was dragged to the Chart room and attended to, and found I could carry on. This was, I think, about 7.30 a.m.

"At 8 o'clock Commander Unwin was carried in suffering from prolonged immersion and shock and hit badly. Two hours I stood by attending him. He, man that he is, got on fresh clothes and returned once more to the beach, though hardly able to walk. Young Drewry was next brought in, exhausted to the last, and I really thought he was done for. His head was a ghastly sight."[524]

SURGEON-LIEUTENANT PETER KELLY, RN, SS *RIVER CLYDE*.

"I went to the hole cut in No. 4 hold on the starboard side aft and climbed up to see if I could do anything the other side. There was a man there trying to swim off to the ship with bullets dropping all round him, and so I went and brought him in. He was

---

520 Sub-Lt Arthur Walderne St Clair Tisdall, RNVR, Anson Bn, was awarded the VC, *London Gazette*, 31st March 1916: "During the landing from the S.S. "River Clyde" at V Beach in the Gallipoli Peninsula on the 25th April, 1915, Sub-Lt Tisdall, hearing wounded men on the beach calling for assistance, jumped into the water and, pushing a boat in front of him, went to their rescue. He was, however, obliged to obtain help, and took with him on two trips Leading Seaman Malia and on other trips Chief Petty Officer Perring and Leading Seamen Curtiss and Parkinson. In all Sub-Lt Tisdall made four or five trips between the ship and the shore, and was thus responsible for rescuing several wounded men under heavy and accurate fire. Owing to the fact that Sub-Lt Tisdall and the platoon under his orders were on detached service at the time, and that this Officer was killed in action on the 6th May, it has only now been possible to obtain complete information as to the individuals who took part in this gallant act. Of these, Leading Seaman Fred Curtiss, O.N. Dev 1899 has been missing since the 4th June, 1915"
521 Petty Officer Mechanic Geoffrey Rumming, RNAS, received the Conspicuous Gallantry Medal (*London Gazette*, 16th August 1915) and the French *Médaille Militaire*.
522 Private letter, Lt-Cdr Josiah Wedgwood, RNVR, author's collection.
523 *Kent Messenger & Gravesend Telegraph*, 26th June 1915.
524 *Galway Express*, 28th August 1915.

badly wounded in both legs and had only managed to keep afloat by means of his swimming collar. He had paddled about 15 yards from the shore.

"While we were on our way three shots dropped within a foot of his head, and he couldn't duck as he had the collar on... After we had brought this man to the *River Clyde* we saw a lifeboat right under the cliff trying to paddle out from the shore. We picked her up and proceeded to the *Albion* with her. She had six dead and eight wounded in her, and one man who wasn't quite certain whether he would like to die or live."[525]

MIDSHIPMAN PERCY WAIT, RN, HMS *CORNWALLIS*.

"I sat on top of the wireless cabin, whence I had a fine view of the battle on both sides of the Dardanelles. The battleships on each side of us were pouring shot and shell into the Turkish trenches. The shrapnel fell everywhere, and sent up the water as though it was raining stones. The big shells as they struck the water sent columns quite 30 feet high. When the *Queen Elizabeth* began firing with her 15-inch guns the shells ploughed their way through the forts, reducing them and a village — which was full of snipers — to the ground. Near Cape Helles a cargo boat was landed to act as a protection to our troops when landing...

"We were at one time very close to the beach on the European side, where our fellows were landing under fire. The Turks were well entrenched, and some of our poor chaps were killed before they could set foot on land. I could see them advancing to the attack quite plainly with the naked eye."[526]

WIRELESS OPERATOR ELIJAH DAUBNEY, HMHS *SOUDAN*.

"It is quite clear now, and I can just see through my glasses... "V" Beach, where the *River Clyde* is lying beached, all seems hell and confusion. Some fool near me says, "Look, they are bathing at "V" Beach." I get my glasses on to it and see about a hundred khaki figures crouching behind a sand dune close to the water's edge. On a hopper which somehow or other has been moored in between the *River Clyde* and the shore I see khaki figures lying, many apparently dead. I also see the horrible sight of some little white boats drifting, with motionless khaki freight, helplessly out to sea on the strong current that is coming down the Straits."[527]

CAPT. JOHN GILLAM, DIV. TRAIN, ARMY SERVICE CORPS, 29TH DIVISION, HMT *DONGOLA*.

Those on the *Minnewaska* got a closer view of one of the boats.

"On shore a ship had been beached, and regiments were scrambling ashore under heavy fire. About an hour afterwards some boats drifted past us, full of soldiers and sailors, who had been wounded when trying to land. We picked up one of the boats. Some of the men were very badly hurt."[528]

RFN HARRY MASON, CEYLON PLANTERS' RIFLE CORPS, ATTD ANZAC HQ, HMT *MINNEWASKA*.

---

525   *East Anglian Times*, 7th June 1915.
526   *Nottingham Daily Express*, 1st June 1915.
527   Gillam, *Gallipoli Diary*, p. 32.
528   *Sheffield Independent*, 20th May 1915.

"Some fifty yards off we noticed an English Bluejacket stand up in the cutter's bow and semaphore to us, "All wounded aboard here — can we come alongside?" As the boat drew alongside a sailor was seen to hold the ship's tiller in the right hand, whilst his left hand had been blown away with shrapnel. Our ship's officer shouted "Port your helm," and the bluejacket answered, "Aye, aye, sir," with a forced smile. It was a pitiful sight that met our eyes. The boat was half full of water, and the wounded were lying half drowned. We soon had them aboard. One had his eye blown out, another shot in the groin, others in the lungs, etc."[529]

DVR CHARLES HEMMING, NO 1 COY, AASC, AIF, HMT *MINNEWASKA*.

There was no let-up on the shore.

"Seeing that Sed-el-Bahr and the beach to our Right was unoccupied, and fearing the Turks might come down I called for volunteers to make a dash for it, and make good the Right of the Beach. The men responded gallantly. Picking Sergt Ryan and 6 men we had a go for it. Three of the men were killed, one other and myself wounded. However we got across, and later picked up 14 stragglers from the Company of the Dublins who had landed at Cambian Bay.

"This little party attempted to get a lodgement inside the Fort but we couldn't do it so we dug ourselves in as well as we could with our intrenching [sic] tools.

"Sergt Ryan made some daring reconnaissances during the day, reporting the further side of the Fort well held.

"I reported to Colonel Tizard by semaphore from the shore that I could do nothing, as I had no men left. He told me to go for my objective Fort No.1. but it could not be done."[530]

CAPT. GUY GEDDES, 1ST BN ROYAL MUNSTER FUSILIERS, 86TH BDE, 29TH DIVISION.

A mixed group of Worcesters and Hampshires got into the remaining boats, boats bearing the evidence of the fate of their previous occupants, and made another attempt to land. Estimates of the time this took place vary but all are agreed that Brig.-General Napier, despite being warned about the impossibility of the attempt, carried on.

"The [Munsters] tried to land in front of us, and got very heavily fired on, losing a lot of men. We tried the same, with a similar result, and had to stop. A little time after we were hung up, about 10 a.m. the General (General Napier[531]) and Costeker[532] came along in a boat and got on to a lighter alongside us to try and land. They were both hit, and the fire was so heavy we could not get to them. At last I managed to get over

---

529   *Nepean Times* (Penrith, New South Wales), 7th August 1915.
530   TNA WO95/4310.
531   Brig. Gen. Henry Napier, commanding 88th Infantry Bde, was killed in action on 25th April 1915. Buried at sea, he is commemorated on the Helles Memorial.
532   Major John Costeker, DSO, Royal Warwickshire Regt, Brigade Major, 88th Infantry Bde, 29th Div., was killed on 25th April 1915. Buried in V Beach Cemetery, he was the 36 year-old son of William and Clara Costeker, of South Kensington, London; husband of Margaret Costeker, of Elm Park Gardens, London.

V Beach & Sedd-el-Bahr photographed from the *River Clyde* on 25 April 1915.

to them with three men, but unfortunately found both dead. The General looked quite peaceful, as if he had not suffered; Costeker had been dead some time, and he, too, looked as if he had not felt it. We were all very grieved about it; they were both very popular."[533]

CAPT. & ADJ. GEORGE REID, 2ND BN HAMPSHIRE REGT, 88TH BDE, 29TH DIVISION.

"[A]bout ten o'clock, the transport with the troops could not get in to land us, owing to submerged rocks, so boats were used to take us ashore, and they had to do so under a heaving firing. I was shot while in the boat and fell in the water. I had the presence of mind to throw off my pack and tunic, or I certainly should have been drowned. I then clung to a rock for nearly an hour, until the firing ceased a little, when I was brought ashore and laid on the beach."[534]

PTE ALFRED BAKER, 2ND BN HAMPSHIRE REGT, 88TH BDE, 29TH DIVISION.

"A dash about nine o'clock was led by General Napier and his Brigade-Major. Would they ever get to the end of the lighter and jumping into the sheltering water? No; side by side they sat down on the engine coaming. For one moment one thought they might be taking cover; then their legs slid out and they rolled over."[535]

LT-CDR JOSIAH WEDGWOOD, RNVR, NO. 3 SQN ROYAL NAVAL ARMOURED CAR DIVISION.

---

533   *Stratford-upon-Avon Herald*, 28th May 1915.
534   *Hampshire Telegraph*, 4th June 1915.
535   Wedgwood, *With machine-guns in Gallipoli*, p. 5.

Napier and Costeker were brought back to *River Clyde*. Napier's body was later taken aboard a hospital ship.

> "When we got within five miles of Cape Helles the G.W.R. steamer *Reindeer* came alongside and brought five wounded, including Brig-General Napier and two dead men. She had been engaged in landing the troops, and they were shot as they were going down the gangway."[536]

SUB-DISPENSER (JAMES) EDWIN SCOTT, RAMC, HMHS *SOUDAN*.

> "I lay where I was till about 11 a.m., when, coming under machine gun fire, I crawled round to the rear of the hopper… The only bit of me that was exposed was my ankle, which caught another bullet.
>
> "I was rescued by a seaman from the *Hussar* — Samson,[537] A.B., RNR. He came out of the engine-room and carried me below."[538]

MIDSHIPMAN MAURICE LLOYD, RN, HMS *CORNWALLIS*.[539]

> "By the time the *Cornwallis* took up her position off V Beach [around noon] there was no attacking and no further landing of large numbers of troops, as the impossibility of taking Sedd-ul-Bahr and the ridge was realized. The remainder of the men allotted to V Beach were deflected to W Beach, afterwards called Lancashire Landing.
>
> "It was comparatively quiet. There was no movement whatever, except the constant plying to and fro of boats removing the wounded from the *River Clyde*. Numberless dead bodies lay along the beach and in the wash of the waves; the only living men were those who crouched together under a small bank which separated the land from the water's edge."[540]

LT-CDR ARCHIBALD STEWART, RN, HMS *CORNWALLIS*.

> "One of my men came to me: "May I go over and help get in those wounded… I can't stand hearing them crying." He went with the second lot, but another of my men had been before him, and he had dived in, without leave… He was shot through the stomach but lives. The Turks could easily have killed all those who went to the wounded. They did not fire on them sometimes for ten minutes, and then a burst of fire would come."[541]

LT-CDR JOSIAH WEDGWOOD, RNVR, NO. 3 SQN ROYAL NAVAL ARMOURED CAR DIVISION.

536 *Richmond and Twickenham Times*, 12th June 1915.
537 OS George Samson, RNR, was awarded the VC. His citation read: "Worked on a lighter all day under fire, attending wounded and getting out lines; he was eventually dangerously wounded by maxim fire." (*London Gazette*, 16th August 1915.)
538 Maurice Lloyd quoted in Stewart, Archibald & Pershall, Charles, *The Immortal Gamble*, p. 108, A.C. Black Ltd (London) 1917.
539 Midshipman Maurice Lloyd, DSC & Bar, RN, HMS *Cornwallis*, was wounded at V Beach on 25th April 1915. Promoted Sub-Lt, HMS *Iphigenia*, he was fatally wounded during the Zeebrugge Raid on 23rd April 1918 and died the following day. Buried in Dover (St James') Cemetery, he was the 20 year-old son of Lt.-Col. Charles Patterson Lloyd (late of The Buffs) and Mrs. Aileen Rosa Lloyd, of 28 Church Square, Rye, Sussex.
540 Stewart & Pershall, *The Immortal Gamble*, p. 91.
541 Wedgwood, *With machine-guns in Gallipoli*, pp. 7–8.

"At 1 o'clock we started in earnest to take the wounded on, and by 8 p.m. we had taken on board 354 wounded and 22 dead, as they came alongside in trawlers, picket boats, whalers, and transport boats. Most of them, it appears, were shot before they got out of the boats, the Turks having entrenched right down on the beach. One boat had four dead soldiers and who had not been out of the boat.

"We are only fitted up with 200 cots, and as we previously had about sixty on board, so we had to find sleeping accommodation for about 220. We were all struck with the cheerfulness of the men — not one was heard to complain — and those who were able were assisting the others; but I never want to see such a sight again. Their one grievance was they had not had a chance to fire a shot at the Turks and Germans."[542]

SUB-DISPENSER (JAMES) EDWIN SCOTT, RAMC, HMHS *SOUDAN*.

"About noon, a Naval Officer (Lieutenant Smith), a fine fellow, came off to get some more small arms ammunition for the machine guns on the *River Clyde*. He said the state of things on and around that ship was "awful," a word which carried twentyfold weight owing to the fact that it was spoken by a youth never very emotional, I am sure, and now on his mettle to make his report with indifference and calm."[543]

GENERAL SIR IAN HAMILTON, C-IN-C, MEF.

"At one o'clock I got 20,000 more rounds from the fleet, and the Lancashires were appearing over the "Lancashire Landing." We saw fifteen men in a window in the castle on the right by the water. They signalled that they were all that remained of the Dublins who had landed at the Camber at Seddel Bahr."[544]

LT-CDR JOSIAH WEDGWOOD, RNVR, NO. 3 SQN ROYAL NAVAL ARMOURED CAR DIVISION.

"At two o'clock a large number of our wounded who had taken refuge under the base of the arches of the old Fort at Sedd-el-Bahr began to signal for help. The *Queen Elizabeth* sent away a picket boat which passed through the bullet storm and most gallantly brought off the best part of them."[545]

GENERAL SIR IAN HAMILTON, C-IN-C, MEF.

At 2.25 p.m. Hunter-Weston was informed of Napier's death and the impossibility of moving off the beach.

"Enemy's position around Beach V very strong and tremendous lot of wire. Every attempt to reconnoitre ground has so far failed. My casualties are very heavy, including General Napier and his Brigade Major both killed. Unless in the meantime the high ground to the north-west of me is taken by other troops, I propose to wait till dark and then attack position 141. To carry this out can you send me a barrel pier, as I am very short of boats and our pontoon pier does not reach the shore. I have a great

---

542 *Richmond and Twickenham Times*, 12th June 1915.
543 Hamilton, *Gallipoli Diary*, vol. 1, pp. 134–135.
544 Wedgwood, *With machine-guns in Gallipoli*, pp. 7–8.
545 Hamilton, *Gallipoli Diary*, vol. 1, pp. 135–136.

number of wounded on the beach. I have 900 infantry still on the ship. Of the half company of Dublins at Camber only 25 remain. O.C. Munsters roughly estimates his casualties at 200. Mark VI ammunition for RNAS machine guns is required."[546]

LT-COL. HERBERT CARINGTON SMITH, 2ND BN HAMPSHIRE REGT, 88TH BDE, 29TH DIV.

Smith was dead thirty minutes later.

"It was now realized that there was no use in trying to land any more men, for to cross the lighters was certain death. Colonel Carrington [sic] Smith, commanding the troops on board, therefore gave orders to stop the disembarkation. At that moment a party of British troops was seen trying to make its way from W Beach with the object of outflanking the enemy defending V Beach. Cheering on these men, a cheer taken up by the troops on board, the Colonel followed Commander Unwin to the bridge, but while the latter reached the upper bridge, the Colonel paused on the lower one, which had no protection, looking towards the shore through his glasses. Commander Unwin, turning round to warn him of his danger, saw Colonel Carrington Smith[547] fall down dead, shot through the mouth."[548]

REAR-ADMIRAL ROSSLYN WEMYSS, RN, HMS *EURYALUS*.

"I was near him when he was killed, on the bridge of a ship which we had been sent to use as a landing-place. She had been run ashore, and we were hung up under a heavy fire, unable to land. The Colonel was standing on the bridge trying to locate a machine-gun which was troubling us very much, when a brute of a sniper shot him through the head, killing him instantaneously."[549]

CAPT. & ADJ. GEORGE REID, 2ND BN HAMPSHIRE REGT, 88TH BDE, 29TH DIVISION.[550]

"[G]reat chunks were flying out of the old castle as the "15" shells from the *Elizabeth* plastered the ten-foot walls. We watched our men working to the right and up into the castle ruins — at each corner the officer crouching with revolver in rest. One watched them through the fire zone, and held one's breath and pressed the button of the maxim."[551]

LT-CDR JOSIAH WEDGWOOD, RNVR, NO. 3 SQN ROYAL NAVAL ARMOURED CAR DIVISION.

"It was about 4 p.m. when most of the wounded began to arrive. We soon took in between 400 and 500. The way in which they bore their injuries is beyond all praise, some of them smoking cigarettes as though war was an everyday occurrence.

---

546 *Hampshire Telegraph and Post*, 6th May 1949.
547 Lt-Col. Herbert Carington Smith, 2nd Bn Hampshire Regt, was killed in action on 25th April 1915. He is commemorated on the Helles Memorial.
548 Wester-Wemyss, *The Navy in the Dardanelles Campaign*, pp. 77–78.
549 *Stratford-upon-Avon Herald*, 28th May 1915.
550 Reid was killed on 1st May 1915. Buried in Redoubt Cemetery, the Boer War veteran was the 35 year-old son of George and Alice Reid, of Villa Nova de Gaya, Portugal.
551 Wedgwood, *With machine-guns in Gallipoli*, pp. 8–9.

The body of General Napier ... was brought on board ... General [Hare] was wounded in the leg and also brought here."⁵⁵²

WIRELESS OPERATOR ELIJAH DAUBNEY, HMHS *SOUDAN*.

A renewed attempt to land men from the *River Clyde* was made at 4.00 p.m. led by Lt Guy Nightingale but this did little more than add to the casualty list. There was nothing to do but wait for night.

"6.30 p.m ... About 60 men from RMF, Dublins & Hants took up outpost position along the Old Castle & cliff under command of Major Jarrett, RMF."⁵⁵³

CAPT. & ADJ. HARRY WILSON, 1ST BN ROYAL MUNSTER FUSILIERS, 86TH BDE, 29TH DIV.

"About 8.30 p.m. when it was quite dark Major Jarrett, Lieut's Russell[,] Lee & Nightingale with the remnants of W. Y. & Z. Companies came over to me without molestation. I suggested to Jarrett that the best thing was to establish oneself in the Fort and try and get the village of Sed-el-Bahr. He was killed alongside me, and shortly after Lieut's Russell and Lee were wounded.

"Suddenly Major Williams and Beckwith appeared out of the dark with two Companies of the Hampshires from the *River Clyde*. Not a shot being fired as they were unobserved."⁵⁵⁴

CAPT. GUY GEDDES, 1ST BN ROYAL MUNSTER FUSILIERS, 86TH BDE, 29TH DIVISION.

"Major Jarrett⁵⁵⁵ was killed outright. We had informed him of the snipers but he stood up on the top of a hill to look for the enemy's position. He had just turned round to call for signallers when he was shot straight through the head, and died without uttering a word. He was a proper gentleman."⁵⁵⁶

PTE HERBERT DALTON, 1ST BN ROYAL MUNSTER FUSILIERS, 86TH BDE, 29TH DIVISION.

"[A]t dusk we had a few more come to our assistance, which we so badly needed. This assistance was composed of two Companies of the Hants. We could do nothing for our wounded to save them from the exposure of the cold through the night, although we gave our great coats to cover them. To hear their cries was painful to us who had been so lucky as to land safe in this heavy rain of shells and bullets. Through the night everything seemed to be quiet with the Turks, although their snipers would come down a little closer, and kept snapping away, doing but very little damage to us."⁵⁵⁷

PTE CHARLES MAYNARD, 1ST BN ROYAL MUNSTER FUSILIERS, 86TH BDE, 29TH DIVISION.

---

552   *Nottingham Daily Express*, 1st June 1915.
553   1st Bn Royal Munster Fusiliers War Diary, TNA WO 95/4310.
554   TNA WO95/4310.
555   Major Charles Jarrett, 1st Bn Royal Munster Fusiliers, was killed in action on 25th April 1915. Buried in Lancashire Landing Cemetery, he was the 40 year-old son of Colonel Henry Sullivan and Agnes Delacour Jarrett, of South Lodge, East Grinstead, Sussex.
556   *Coventry Graphic*, 18th June 1915.
557   *Kent Messenger & Gravesend Telegraph*, 26th June 1915.

"We were the only hospital ship present, and there must have been a great mistake somewhere or on somebody's part, as three military hospital ships were left idle at Lemmos.

"I told off for one, at 9 p.m., to take the numbers, ratings, &c., of the 86 men in the ward I had to do. It was a rare job finding their discs, and strange to find some of the articles they had round their necks — such as lockets, crosses, charms, buttons, &c."[558]

SUB-DISPENSER (JAMES) EDWIN SCOTT, RAMC, HMHS *SOUDAN*.

"Then night came, but a house in Seddul Bahr was burning brightly, and there was a full moon. We disembarked men at once. All around the wounded cried for help and shelter against the bullets, but there was no room on boats or gangway for anything but the men to come to shore. For three hours I stood at the end of the rocks up to my waist in water, my legs jammed between dead men, and helped men from the last boat to the rocks. Every man who landed that night jumped on to the backs of dead men, to the most horrible accompaniment in the world."[559]

LT-CDR JOSIAH WEDGWOOD, RNVR, NO. 3 SQN ROYAL NAVAL ARMOURED CAR DIVISION.

"At 10.45 p.m. the enemy opened fire on the old *River Clyde*, and we lived in a perfect hell… I lay on the after deck flat on my face in great agony surrounded by wounded and dead. That night I shall never as long as I live forget. The mental strain was terrible and it also rained, and I was already saturated from being in the water during the day. l really thought I should go under, as I was very weak and had had no food since Saturday night. I evidently collapsed."[560]

SURGEON-LIEUTENANT PETER KELLY, RN, SS *RIVER CLYDE*.

"As night came we were told to land, and half of No. 1 Section, under Lt Ronksley,[561] landed about 11.30, and moved forward. However, they had not been ashore very long before there was a fresh outburst of firing, which held the remainder of us up for the rest of the night, as there was no getting off under the heavy machine gun fire which was being poured on the boats. We were on the qui vive the whole night, expecting the enemy to rush the ship, but they thought better of it."[562]

SPR HERBERT HEATH, 1/1ST (WEST RIDING) COY, ROYAL ENGINEERS, 29TH DIVISION.

"Our regiment got orders to go ashore [at] night… We got ashore all right and did not lose a man. We advanced in line with a fort which they had been trying to take all day, and we entrenched ourselves. I did not get a wink of sleep as the firing went on all night."[563]

PTE STEPHEN FALLA, 2ND BN HAMPSHIRE REGT, 88TH BDE, 29TH DIVISION.

---

558 *Richmond and Twickenham Times*, 12th June 1915.
559 Wedgwood, *With machine-guns in Gallipoli*, p. 9.
560 *Galway Express*, 28th August 1915.
561 2/Lt Francis Ronksley, Royal Engineers. He transferred to the Royal Air Force, rising to Flight-Lieutenant during the Second World War.
562 *Sheffield Daily Telegraph*, 3rd June 1915.
563 *Guernsey Evening Press and Star*, 26th May 1915.

"The Turkish force which held up the three battalions at "V" Beach consisted of one company, reinforced later in the day by two more companies. There can be no doubt that the dogged defence of this small detachment, despite the ordeal they suffered from the tremendous explosions of heavy shells, was the main cause of the collapse of the British."[564]

COMMODORE ROGER KEYES, RN, CHIEF OF STAFF, EMS.

S BEACH

The landing in Morto Bay, inside the Dardanelles, involved three companies of South Wales Borderers, supported by some Royal Engineers and a naval detachment. Their objective was the commanding position known as De Tott's Battery, on the eastern arm of the bay, securing the right flank at Helles. They were running late.

"The *Cornwallis* was detailed to take about eight hundred men of the South Wales Borderers, Colonel [Hugh] Casson in command… The *Cornwallis* had to go two and a half miles up the Straits, and after landing the Borderers, and giving necessary support, return to an anchorage laid down to the south-west of Sedd-ul-Bahr…

"The trawlers, four in number, each with six transport boats in tow, were to run ashore as best they could; and the battalion had, by their order, to pull ashore, not only in heavy marching gear, but with boats laden with ammunition, water, and provisions…

"I made myself responsible for landing as much water, ammunition, and stores as two cutters could take; a beach party from the ship was told off to unload these on arrival, also kits belonging to the Borderers, as it was considered that by their flying light (except for ammunition and water) casualties would be reduced to a minimum, and from having often examined De Tott's Battery, I was of opinion that it could be scaled. This was the key of the position and the right flank."[565]

CAPT. ALEXANDER DAVIDSON, RN, HMS *CORNWALLIS*.

"Well, we got called at four o'clock, and had breakfast at half-past. The bombardment started at five. About half-past we were transferred to mine-sweepers, and had our small boats round the side."[566]

PTE JAMES SMITH, 2ND BN SOUTH WALES BORDERERS, 87TH BDE, 29TH DIVISION.

"We had about 150 aboard. We landed ours with three more trawlers full inside the Dardanalles in Morton [sic] Bay, the furthest up. I shall never forget it. It was a lovely calm morning."[567]

AB ALBERT SHARPIN, RN, HM TRAWLER LORD WIMBOURNE.

---

564 Keyes, *Naval Memoirs 1910–1915*, pp. 310–311.
565 Davidson quoted in Stewart, Archibald & Pershall, Charles, *The Immortal Gamble*, pp. 98–100, A.C. Black Ltd (London) 1917.
566 *Coventry Evening Telegraph*, 24th May 1915.
567 *Lynn Advertiser*, 18th June 1915.

"Events went as jointly planned, but owing to French mine-sweepers and the *Agamemnon* getting in the way, the *Cornwallis* had to reduce speed inside the Straits, and the trawlers, which had embarked the Borderers at 4.45 a.m…, crept ahead, and grounded as ordered about the same time as the *Cornwallis* anchored, about 7.30 a.m."[568]

CAPT. ALEXANDER DAVIDSON, RN, HMS *CORNWALLIS*.

"We had to run aground, and then put the soldiers into cutters we had in tow. As we got close inshore they turned their fire on us."[569]

AB ALBERT SHARPIN, RN, HM TRAWLER LORD WIMBOURNE.

"There were 35 in my boat, and I was one of the rowers. We pulled in as far as we could… About ten yards from where my boat stopped there was about 20 yards of stone landing. We all jumped out into the water not lower than our chests."[570]

PTE JAMES SMITH, 2ND BN SOUTH WALES BORDERERS, 87TH BDE, 29TH DIVISION.

"When we got to the shore we had to get out in the water. It was up to our necks. I went headlong under and was soaked to the skin. The cliffs towered above us 260 feet high… They could have prevented us from landing, but they were taken by surprise, at our daring feat perhaps. Anyhow, they got up and fled to a ridge 750 yards in front. That left us safe to reach the top… We just had a bit of wall for cover, and the Turks kept up a perfect fusilade… Our Major was the first man killed of our company,[571] and he just reached the top."[572]

PTE JAMES FISHER, 2ND BN SOUTH WALES BORDERERS, 87TH BDE, 29TH DIVISION.

"We had to carry all our equipment, in addition to which I carried a machine gun. All the time the enemy were firing upon us, but once we got ashore we fixed bayonets and charged their trenches.

"My chum got our gun into action, and we succeeded in driving the enemy back. We let them have it for all we were worth, but after a few minutes my chum received a bullet wound through the thigh, so I had to act by myself."[573]

PTE JOSEPH HILL, MG SECTION, 2ND BN SOUTH WALES BORDERERS, 87TH BDE, 29TH DIV.

"What was expected of us before landing was done in a few hours. The three companies of regiment were given 48 hours to take our position, but took it about three hours…

"Of course we had to make all preparations for food and water before landing. We carried a good stock of food with us, water was put in ammunition boxes and big oil cans well cleaned out. We had to bring it in case we could not get water for about four

---

568 Davidson quoted in Stewart & Pershall, *The Immortal Gamble*, p. 101.
569 *Lynn Advertiser*, 18th June 1915.
570 *Coventry Evening Telegraph*, 24th May 1915.
571 Major Edward Margesson, 2nd Bn South Wales Borderers, was killed on 25th April 1915. He is commemorated on the Helles Memorial.
572 *Coventry Evening Telegraph*, 8th June 1915.
573 *County Express for Worcestershire and Staffordshire*, 12th June 1915.

or five days, as the water springs were in the possession of the enemy. We very soon shifted them and found one in the hill that stands on the left entering the narrows."[574]

PTE JAMES SMITH, 2ND BN SOUTH WALES BORDERERS, 87TH BDE, 29TH DIVISION.[575]

"Shortly after the party had landed, the Asiatic batteries became active, and sent a great many shells into the boats that were being cleared by a small party from the Borderers, and into the supports who were waiting for orders to go ahead and taking cover on the wrong side of a ridge, as they naturally thought they were being shelled by guns on Gallipoli."[576]

CAPT. ALEXANDER DAVIDSON, RN, HMS *CORNWALLIS*.

"[T]he Sappers continued to wade out & bring ashore water and ammunition from the boats under shell fire from [the] Asiatic shore. Some 30 cases each had been safely landed when the men were ordered under cover as the enemy had attained [the] exact range of beach. The work of unloading was continued later during the day & only one boat got adrift without being discharged."[577]

LT ANDREW LAIRD, 1/2ND (LONDON) FIELD COY, ROYAL ENGINEERS, 29TH DIVISION.

"It will always be a matter of regret to me, in a limited sense, that I was unable to be present and see for myself the conduct of our little Naval Brigade as it disappeared over the ridge with the Borderers. I had no option, not only because of the heavy shelling of the beach by the Asiatic batteries, which had to be silenced, but it was my duty to keep in touch with the ship in case of any further developments, and also I had arranged with Colonel Casson that I would act as beachmaster. However, it was afterwards reported to me by Major Frankis[578] and Lt Minchin[579] that our men behaved in accordance with the best traditions of the Navy; that they assisted in the capture of two Turkish trenches, showed dash and energy, and sustained a loss of two killed and four wounded."[580]

CAPT. ALEXANDER DAVIDSON, RN, HMS *CORNWALLIS*.

The objective was secured after a brief action against the light Turkish forces covering the area. Some accounts talk of encountering considerably more formidable opposition but, in reality, only one platoon opposed those who landed here.

Though the landings had already been delayed, Captain Davidson now spent time, as he mentions, undertaking the duties of beachmaster, a responsibility far below his pay scale. As a result, *Cornwallis* did not leave to support the men at V Beach until 10.00 a.m. And, believing there were significantly larger Turkish forces close at hand, no move was made

574 *Coventry Evening Telegraph*, 24th May 1915.
575 Pte James Smith, 2nd Bn South Wales Borderers, was killed in action on 28th June 1915. He is buried in Twelve Tree Copse Cemetery, where he is commemorated by a special memorial.
576 Davidson quoted in Stewart & Pershall, *The Immortal Gamble*, pp. 101–102.
577 Commander Royal Engineers, 29th Division War Diary, TNA WO95/4307.
578 Major Walter Frankis, RMLI, who commanded the naval contingent at S Beach, was Mentioned in Despatches for his work that day (*London Gazette*, 16th August 1915).
579 Lt Henry Minchin, RN, HMS *Cornwallis*, was Mentioned in Despatches (*London Gazette*, 13th August 1915).
580 Davidson quoted in Stewart & Pershall, *The Immortal Gamble*, p. 103.

by those now occupying De Tott's Battery to march to the sound of the guns. Unless relief came from W Beach, those pinned down before Sedd-el-Bahr were on their own.

W BEACH

One of the two main Helles landing sites, from here units were to advance north, linking up with the men at X Beach, taking Hill 114 along the way, and from there establish contact with those further along the coast at Y Beach. Others would move east (right) and attack Hill 138, connecting with men from V Beach before joining a general movement towards Achi Baba.

> "We were aroused at 2-30 a.m., had a good breakfast given to us by the crew of the ship we were on, and then were ordered to dress as quickly and as quietly as possible. We were then ordered into the small boats. This was just before daybreak. As soon as the boats were full up we were towed away from the side of the battleship. As we neared the shore there was not a sound. Day was just breaking. It was a beautiful night, the sea was as calm as a pond in a park."[581]
>
> PTE THOMAS HUME, 1ST BN LANCASHIRE FUSILIERS, 86TH BDE, 29TH DIVISION.

> "At 5.15 [a.m.] we went to action stations and started a terrible bombardment of the enemy trenches with our guns. This lasted a good hour, and hundreds of the Turks must have been killed. All this time our troops were arriving in transports, and the boats and trawlers were all ready to land them."[582]
>
> CPO FREDERICK NEAL, RN, HMS *SWIFTSURE*.

> "At 5.30 we had to be all ready, and paraded and started to get into the ship's boats, each containing forty-five, including sailors to row. There were several pinnaces, and each towed four boats. As the sun rose straight in our eyes over the place we had to land, the pinnaces started for the shore with about 100 yards between, and each towing its four boats."[583]
>
> CAPT. HAROLD CLAYTON, 1ST BN LANCASHIRE FUSILIERS, 86TH BDE, 29TH DIVISION.

> "Suddenly she opened fire with her 6-inch guns, sending the shells over our heads. By this time the noise was deafening… It was a sight one will never forget. While all this was going on we were getting nearer to the shore. We could see lumps of rock being hurled into the air by the fire of our ships. The whole place seemed to be one mass of fire and smoke and the noise was awful… We thought that after the number of shells the Navy dropped there would not be a Turk left."[584]
>
> PTE THOMAS HUME, 1ST BN LANCASHIRE FUSILIERS, 86TH BDE, 29TH DIVISION.

> "As we got nearer the naval guns had to cease fire for fear of hitting us, and in a solemn

---

581 *Hartlepool Northern Daily Mail*, 24th June 1915.
582 *Mid-Sussex Times*, 6th July 1915.
583 Harold Robert Clayton quoted in Creighton, *With the Twenty-Ninth Division in Gallipoli*, p. 58.
584 *Hartlepool Northern Daily Mail*, 24th June 1915.

and awe-inspiring silence we approached. As all the boats were beached in three feet of water and some 20 yards from the land, a perfect hail of bullets... encircled us."[585]

MAJOR WALTER PEARSON, 1ST BN LANCASHIRE FUSILIERS, 86TH BDE, 29TH DIVISION.

"About 300 yards from the beach the pinnaces stopped their tows, and the seamen in each boat pulled us in. Almost directly an awful... fire broke out from the cliffs each side of our little bay. The sailors were splendid, pulling on under the awful fire, while they and our men were being hit."[586]

CAPT. RICHARD HAWORTH, 1ST BN LANCASHIRE FUSILIERS, 86TH BDE, 29TH DIVISION.

"The fire was too fierce for the men to try and stick in the boats, and men who were hit were falling across the oars or on the sailors, so that it was stopping them using the oars, and the boats were almost at a standstill."[587]

PTE ALFRED GILLIBRAND, 1ST BN LANCASHIRE FUSILIERS, 86TH BDE, 29TH DIVISION.

"I was in charge of No. 12 boat, and I told the men to lie down in the bottom of the boat, leaving myself and six oarsmen exposed to the enemy's fire. I then ordered them all to jump out, and to get under cover as quickly as they could. As soon as we touched the beach we could see wire entanglements. The fire was terrible, just like a hail-storm. I jumped out of the stern up to my arms in water and pushed the boat in. The sergeant jumped in front of me, and got mortally wounded. The cries of the wounded were terrible."[588]

LS JOHN GILLIGAN, RNR, HMS *EURYALUS*.

"Anyone who witnessed the bombardment of the Peninsula by our fleet previous to our landing would have wondered how anyone could have lived through it. But the Turks were well-prepared, having had plenty of time to entrench themselves... they had even put trip wires out in the sea, and barbed wire just at the edge of the water, and then again on the beach."[589]

PTE HARRY RIGBY, 1ST BN LANCASHIRE FUSILIERS, 86TH BDE, 29TH DIVISION.

"Eventually the boat grounded about 30 yards out, and I shouted to the men to get out. A good many were finished, and I jumped out up to my chest in the water. We were carrying two days' rations, 200 rounds, and full packs, which was the cause of several poor fellows, who were hit while in the water, being drowned. I fell down twice before reaching the beach when I found we were under a terrible cross-fire. I was getting the Company together when Captain Heard[590] was very badly hit in the arm."[591]

CAPT. RICHARD HAWORTH, 1ST BN LANCASHIRE FUSILIERS, 86TH BDE, 29TH DIVISION.

585   *Taunton Courier and Western Advertiser*, 16th June 1915.
586   *Lakes Herald*, 21st May 1915.
587   *Midland Counties Tribune*, 16th July 1915.
588   *Dundee Evening Telegraph*, 16th June 1915.
589   *The Macclesfield Times & East Cheshire Observer*, 25th June 1915.
590   Capt. Robert Heard, 1st Bn Lancashire Fusiliers, died of wounds in Egypt on 4th May 1915. Buried in Alexandria (Chatby) Military and War Memorial Cemetery, he was the 26 year-old son of Prebendary Heard, of Caterham Rectory, Caterham, Surrey.
591   *Lakes Herald*, 21st May 1915.

"We were helpless in the boats; we were knocked down right and left, so we had to get out the best way we could; then (on leaving the boats) we were up to our necks in the water. Those who were not shot were drowned."[592]

PTE AMOS TODD, 1ST BN LANCASHIRE FUSILIERS, 86TH BDE, 29TH DIVISION.

"It was a sight I shall never forget. Dawn had broken; our fire was still going on, though it was not so rapid. Some of the boats had beached; others were nearing the shore; and in the rear again were lines of our boats filled with troops. Our shells were bursting over the Turkish trenches, and the rattle of machine guns and pompoms could be plainly heard. We could see our men jumping out of the boats and charging straight away at the enemy trenches, and saw them falling in all directions. The beach was soon strewn with dead and wounded."[593]

CPO FREDERICK NEAL, RN, HMS *SWIFTSURE*.

"A lot were shot in the boats and some boats were smashed up. I was lucky enough to get out of the boat. I jumped into deep water but managed to get out and get on to the beach. As soon as those who were left of us we found our way stopped by barbed wire. Everyone who went near it to cut it was killed."[594]

PTE THOMAS HUME, 1ST BN LANCASHIRE FUSILIERS, 86TH BDE, 29TH DIVISION.

"There were plenty of barbed wire entanglements all along the beach, but a good many were killed as they were trying to get through the entanglements… it was awful to see them fall across the wire."[595]

PTE JOHN ROURKE, 1ST BN LANCASHIRE FUSILIERS, 86TH BDE, 29TH DIVISION.

"Some poor chaps while lying on the barbed wire shot were fired at again… Despite our heavy losses our lads stuck it, and got on to the beach; then they laid down, as the Turks were still firing at us."[596]

PTE AMOS TODD, 1ST BN LANCASHIRE FUSILIERS, 86TH BDE, 29TH DIVISION.[597]

"Several of my company were with me under the wire, one of my subalterns was killed next to me, and also the wire-cutter who was lying the other side of me. I seized his cutter and cut a small lane myself through which a few of us broke and lined up under the only available cover procurable, a small sand ridge covered with bluffs of grass. I then ordered fire to be opened on the crests, but owing to submersion in the water and dragging rifles through the sand, the breech mechanism was clogged, thereby rendering the rifles ineffective."[598]

MAJOR HAROLD SHAW, 1ST BN LANCASHIRE FUSILIERS, 86TH BDE, 29TH DIVISION.

---

592 *The Evening Star and Daily Herald* (Ipswich), 4th June 1915.
593 *Mid-Sussex Times*, 6th July 1915.
594 *Hartlepool Northern Daily Mail*, 24th June 1915.
595 *Midland Counties Tribune*, 14th May 1915.
596 *The Evening Star and Daily Herald* (Ipswich), 4th June 1915.
597 Amos Todd, who enlisted on 11th June 1906 was discharged due to wounds on 16th June 1916.
598 Shaw quoted in Creighton, *With the Twenty-Ninth Division in Gallipoli*, p. 58.

"The worst of it was, we could only see [the Turks] now and then, for they had trenches running in all directions, so we did not do them much harm."[599]

PTE HARRY RIGBY, 1ST BN LANCASHIRE FUSILIERS, 86TH BDE, 29TH DIVISION.

"After I was wounded I had to lie there, as our fellows had to get a footing before anything could be done. After a while some boats from the battleship came and took us away to where we were dressed and looked after."[600]

PTE THOMAS HUME, 1ST BN LANCASHIRE FUSILIERS, 86TH BDE, 29TH DIVISION.[601]

"Some of the boats came back to the transports with half their number dead or wounded. These men had not put a foot on the soil, having been hit before they could disembark."[602]

CPO FREDERICK NEAL, RN, HMS *SWIFTSURE*.

Those yet to land could see there was little point in adding to the carnage within the small bay. But there was an opportunity to outflank the defenders' trenches by scaling the left, or western, headland framing the beach. A party landed just beneath the bluff and made their way up.

"The only thing to do was to get the trench above us on the edge of the cliff. I got together about 30 men, and we were clambering up the cliff when Porter,[603] who was by my side, was shot through the head and died almost at once."[604]

CAPT. RICHARD HAWORTH, 1ST BN LANCASHIRE FUSILIERS, 86TH BDE, 29TH DIVISION.

"Men appear [6.25 a.m.] on western crest of Cape Helles having apparently got round under the cliff and up that way. They gain a footing on the top of the ridge…

"Second line of tows arrives [6.40 a.m.] off the beach… Boats of the first tow seen struggling to get back, with only one or two men fit to pull an oar."[605]

MAJOR OSCAR STRIEDINGER, ASC, GENERAL HEADQUARTERS, 29TH DIVISION.

The beach, later named Lancashire Landing, was a ghastly sight. More than half the battalion had been killed or wounded. Six Lancashire Fusiliers were later awarded the Victoria Cross.[606] No decorations of any kind were made to those sent to help the survivors, a cause of some resentment later.

599   *The Macclesfield Times & East Cheshire Observer*, 25th June 1915.
600   *Hartlepool Northern Daily Mail*, 24th June 1915.
601   Hume was wounded, losing his right eye, and discharged from the Army on 10th January 1916. He had enlisted on 1st June 1906 after the death of his wife and two children, who both died in infancy.
602   *Mid-Sussex Times*, 6th July 1915.
603   Lt Alwyne Porter, 1st Bn Lancashire Fusiliers, was killed on 25th April 1915. Buried in Lancashire Landing Cemetery, he was the son of Joseph Francis and Edith Porter, of Helmsley, Yorkshire.
604   *Lakes Herald*, 21st May 1915.
605   29th Division Staff, General Staff War Diary, TNA WO 95/4304.
606   The so-called six VCs before breakfast were not all approved at the same time. Those awarded to Capt. Richard Willis, Sgt Alfred Richards and Pte Joseph Keneally were published in the *London Gazette* on 24th August 1915. Those to Capt. Cuthbert Bromley, Sgt Frank Stubbs and Cpl John Grimshaw had to wait until the *London Gazette* of 15th March 1917 (Grimshaw had already received a DCM but this was upgraded).

"B Sec. Tent Subdivn. landed in second tow from the "Whitby Abbey" about 6.30 a.m. Owing to large number of wounded Tent Subdivn was split into four dressing parties, one under Lieut Thomson, one under Lieut Whyte, one under Staff-Sgt Gaskin & one under Sgt Gilbert. First aid was rendered & wounded were removed as speedily as possible to the cover afforded by the cliffs west of the beach. Fires were lit & the cooks prepared beef tea &c. for wounded. During the forenoon several boat loads of wounded were got off to the "Implacable" & a hospital ship."[607]

LT JAMES THOMSON, 89TH (1ST HIGHLAND) FIELD AMBULANCE, RAMC, 29TH DIVISION.

"A tug comes alongside with wounded, and they are carefully hoisted on board by slings. They are the first wounded that I have ever seen in my life, and I look over the side with curiosity and study their faces. They are mostly Lancashire Fusiliers from "W" Beach. Some look pale and stern, some are groaning now and again, while others are smoking and joking with the crew of the tug. I talk to one of the more slightly wounded, and he tells me that it was "fun" when once they got ashore, but they "copped it" from machine guns in getting out of the boats into shallow water, where they found venomous barbed wire was thickly laid. He laid out four John Turks and then "copped it" through the thigh, and three hours afterwards was picked up by sailors."[608]

CAPT. JOHN GILLAM, DIVISIONAL TRAIN, ARMY SERVICE CORPS, 29TH DIVISION.

After the defenders were cleared from the western arm of the bay, at 7.15 a.m. Brig.-General Steuart Hare, commanding 86th Brigade, led an attempt to establish contact with the men who had landed at X Beach. His reversion to temporary subaltern did not last long.

"Fortunately two of the tows, carrying a company, with which was General S. W. Hare, CO. of this 86th Brigade, put to shore a little to the left of the central beach, and found shelter under a ledge of rock at the foot of Cape Tekke cliff. Here they escaped the cross-fire, and were able partly to enfilade the enemy's trenches. The Brig-General was severely wounded, either at this time or a little later, but part of the company succeeded in scrambling up the rocks in front of them to the summit."[609]

HENRY NEVINSON, WAR CORRESPONDENT.

"We soon came in sight of the Turks in the trench, and when about 100 yards away one let fly at me and took the top off my right ear, but I am glad to say, I got him with my revolver the moment after, right in the head. Just as I reached the trench there was a terrible explosion — the trench was mined,[610] and I and those near me were sent hustling down to the bottom of the cliff again. By this time I was fairly dazed, and after a bit decided to work round to the right round the cliffs with what I had left of

---

607   89th Field Ambulance, Royal Army Medical Corps War Diary, TNA WO 95/4309.
608   Gillam, *Gallipoli Diary*, p. 33.
609   Nevinson, Henry, *The Dardanelles Campaign*, p. 102, Nisbet & Co. (London) 1918.
610   It was actually a large calibre naval shell striking the area.

the Company, about 40 or so, I was joined there by Tom Cunliffe,[611] whose guns had been left on the boat, which had now backed out again, and by Beaumont[612] with a few of B Company."[613]

CAPT. RICHARD HAWORTH, 1ST BN LANCASHIRE FUSILIERS, 86TH BDE, 29TH DIVISION.

Others attempted to link with men at V Beach, via Hill 138 to their right, though who ought to be coming to help who might have crossed some minds.

"We reached the lighthouse, and pushed on until hung up by a maze of barbed wire. Fortunately there was almost dead ground against the wire for a strip. The Signal Section were established under cover of the lighthouse and they got communication with the Royal Fusiliers, the *Euryalus* and the *River Clyde*. Frankland[614] left me and went to the ridge on the right to see if there was any way on from there. At about 8.45 a.m. he stood up in order to see, and was shot through the heart, neck and head."[615]

CAPT. HAROLD FARMAR, LANCASHIRE FUSILIERS, STAFF CAPTAIN, 86TH BDE, 29TH DIV.

"The beach was an awful sight after they had landed, absolutely piled up with dead and wounded. It was a wonderful show, and I don't believe any other than British troops would ever have done it... All the time we were... watching the troops from W beach trying to get the Turks out of their trenches just behind Tekeh Burna, but they were repulsed several times as the enemy were in a very strong position. It was not till about 10 o'clock that they managed to dislodge them and even then they did not entirely clear them out."[616]

LT FRANK STEPHENSON, RN, HMS *IMPLACABLE*.

"The Essex are disembarking now, going down the rope ladders slowly and with difficulty. One slips on stepping into a boat and twists his ankle... They are landing in small open boats. A tug takes a string of them in tow, and slowly they steam away for "W" Beach."[617]

CAPT. JOHN GILLAM, DIVISIONAL TRAIN, ARMY SERVICE CORPS, 29TH DIVISION.

The 1st Battalion Essex Regiment and most of 4th Battalion Worcestershire Regiment, diverted from V Beach, began landing on W Beach at 10.00 a.m.

611 Lt, later Capt. Thomas Cunliffe, 1st Bn Lancashire Fusiliers, one of the officers serving with the regiment at the time of the Taba Crisis in 1906, was killed in action on 4th June 1915. The 29 year-old son of Helen Cunliffe, of Croft, Ambleside, Westmorland, and the late Robert Cunliffe, is buried in Lancashire Landing Cemetery.
612 2/Lt, later Capt. George Beaumont, 1st Bn Lancashire Fusiliers.
613 *Lakes Herald*, 21st May 1915.
614 Capt., Brevet Major Thomas Hugh Colville Frankland, Royal Dublin Fusiliers, Brigade Major 86th Infantry Bde, was killed while undertaking a reconnaissance towards V Beach on 25th April 1915. As a 2/Lt during the Boer War, Frankland served with 2nd Bn Royal Dublin Fusiliers. He was taken prisoner on 15th November 1899 at Blaauwkrantz — in the same incident involving the armoured train as Churchill. He was released 6th June 1900 at Waterval.
615 Capt. Harold Farmar, Lancashire Fusiliers, Staff Captain, 86th Bde, quoted in Bush, Eric, *Gallipoli*, pp. 138–139, George Allen & Unwin Ltd. (London) 1975.
616 *The Press* (New Zealand), 22nd April 1938.
617 Gillam, *Gallipoli Diary*, p. 31.

> "[W]e left the *Dongola*, and got on a smaller boat and went nearer the land. The bullets started coming at us while we were on this boat, and a corporal standing close to me was hit in the thigh. We then got into some smaller boats and rowed towards the beach."[618]
>
> PTE ALBERT AMOS, 1ST BN ESSEX REGT, 88TH BDE, 29TH DIVISION.

> "As we reached the beach the ships shelled all the positions on top, and when we landed on the beach we could see the awful struggle that had taken place an hour or so previous, for dead and wounded were lying in batches and many in the water who had been drowned after being wounded."[619]
>
> PTE HERBERT LODGE, 1ST BN ESSEX REGT, 88TH BDE, 29TH DIVISION.

> "We waded to the beach and took cover for a few minutes, and then made for the firing line. We had a bit of a climb to get there, and when I reached the top I was that exhausted that I fell in a trench on top of a dead Turk! I can tell you I felt a bit "queasy" for a minute."[620]
>
> PTE ALBERT AMOS, 1ST BN ESSEX REGT, 88TH BDE, 29TH DIVISION.[621]

> "I shall never forget the sight. The sea was discoloured with blood for fully a hundred yards out, and the beach was covered with dead, some just at the water's edge, others wholly submerged."[622]
>
> L/CPL SAM BYRD, 4TH BN WORCESTERSHIRE REGT, 88TH BDE, 29TH DIVISION.

Officer casualties had seriously weakened command and control. At 11.30 a.m., 29th Division's senior staff officer, Colonel Owen Wolley-Dod, was sent to W Beach to take charge.

> "With all our Brigadiers and two of the three brigade majors hit, someone was badly required ashore to run the show in this corner… therefore, Colonel Wolley Dod, the GSO1, went ashore, but unfortunately for me, he decided to take one of the Administrative Staff with him, to look after water supplies and ammunition arrangements. A lull now ensued, and the General insisted on our going to lie down for an hour at this time."[623]
>
> CAPT. CLEMENT MILWARD, 53RD SIKHS, GSO3, GENERAL STAFF, 29TH DIVISION.

Hunter-Weston wanted to know what was holding things up. The Essex's Adjutant gave a classic reply.

> "11.35 a.m. A message was received from G.H.Q. to report progress and reasons for not pushing on. Reply was sent that we were waiting for our left to come up and would

---

618   *Essex Weekly News*, 28th May 1915.
619   *Banbury Advertiser*, 10th June 1915.
620   *Essex Weekly News*, 28th May 1915.
621   Promoted, Sgt Albert Amos was killed in action in France on 14th April 1917. Commemorated on the Arras Memorial, he was the 27 year-old son of Harriet Amos, of "Dawn-Kloof," Hall Road, Heybridge, Maldon, Essex.
622   *Evesham Standard & West Midland Observer*, 26th June 1915.
623   Capt. Clement Milward, 53rd Sikhs, G.S.O.3 29th Division General Staff, CAB 45/259.

then advance. This was attempted but the advance was held up by very heavy fire and many casualties occurred."[624]

CAPT. & ADJ. ALGERNON WOOD, 1ST BN ESSEX REGT, 88TH BDE, 29TH DIVISION.

"At 12 noon word came that two guns from each Battery were wanted. Major Allan, the Adjutant, and a Subaltern landed from our ship, and Major Mackelvie (Argyll) and one subaltern, with two guns, landed from the other transport, which had just come up — the first artillery to land, and they [did] good work."[625]

LT ANGUS MACDONALD, ROSS BTY, 4TH HIGHLAND MTN BDE, RGA, 29TH DIVISION.

"We... were in action as soon as we landed. I was a bit shaky at first, but it soon passed away, when we saw the shells taking deadly effect... We were the first Artillery ashore and the first in action. No one can realize the life out here in the firing line till they come out; it can't be described — bullets whistling all over you and shells bursting around."[626]

WHEELER THOMAS McPHERSON, ARGYLL BTY, 4TH HIGHLAND MTN BDE, RGA, 29TH DIV.

The 1/5th Royal Scots began to arrive at 12.30 p.m. The casualties from that morning were still there.

"The first sight we had was several long rows of dead lying on the beach — the bodies of the brave men who had forced a landing for us. Some of them were simply riddled with bullets."[627]

PTE JOHN McKILLOP, 1/5TH BN ROYAL SCOTS (LOTHIAN REGT), 88TH BDE, 29TH DIV.

"Landed on W beach [12.45 p.m.]. Position seems to be somewhat as follows... Troops are occupying the high ground... every man in the firing line — no supports — battalions considerably disorganized, most of them having been severely handled earlier in the day; all have been up since 3.30 a.m. carrying heavy equipment, fighting all day and no time for meals."[628]

MAJOR OSCAR STRIEDINGER, ASC, GENERAL HEADQUARTERS, 29TH DIVISION.

"As many of the wounded as possible were brought on board to give relief to the hospital ships and amongst others Brig.-General Hare, who was placed in my cabin. During the afternoon I snatched a minute to go down and see how he was getting on and to my consternation discovered the cabin empty with only a few traces of blood, nobody had found time to remove, to show that he had been there. Nobody knew anything about him and I was much concerned at his disparition and wondered whether he were still alive. It was only much later that I learned, greatly to my relief, that he had been taken off to a hospital ship without the knowledge of any of my people."[629]

REAR-ADMIRAL ROSSLYN WEMYSS, RN, HMS *EURYALUS*.

624 1st Bn Essex Regt War Diary, TNA WO 95/4312.
625 *Oban Times and Argyllshire Advertiser*, 5th June 1915.
626 *Campbeltown Courier*, 29th May 1915.
627 *ibid.*, 5th June 1915.
628 29th Division Staff, General Staff War Diary, TNA WO 95/4304.
629 Wester-Wemyss, *The Navy in the Dardanelles Campaign*, p. 103.

"During the afternoon, the work of identifying the dead was carried out. These had been collected into groups earlier in the day. Thus were three officers & eighty-two men identified. Valuables were secured before burial by Sgt Clark. The rations carried by the dead & their first field dressings were used to supplement our stores for the wounded. We were left with about 180 wounded over night which is testimony to the large number that must have been evacuated."[630]

LT JAMES THOMSON, 89TH (1ST HIGHLAND) FIELD AMBULANCE, RAMC, 29TH DIVISION.

"We were never safe to move, but still went bandaging, unheeding the dangers. My worst time was when, on the boat, a sailor came and asked if any of us would go right over to a flat barge and take up two serious cases; so I got four men, and went. We got there all right, but could not get back. We lay in the barge for an hour, waiting a chance, clear from the snipers. The sailor went up to see if it was safe, but never returned. He got struck in two places — very badly wounded, poor chap. For three hours after that I had to lie on that barge, unable to move, for the bullets were flying over my head like peas. Still, we managed at last. We got the cases on to the stretchers, and had to crawl back on our hands and knees with them."[631]

SGT WILLIAM STEPHEN, 89TH (1ST HIGHLAND) FIELD AMBULANCE, RAMC, 29TH DIV.

Renewed attempts were made to take Hill 138 to link up with the men at V Beach.

"We... charged the hill just above us. Then we laid down and started to fire. A few hours later we made an advance towards another ridge, where the Turks had given us a lot of trouble, our navy firing to cover our advance, and, my word! didn't we have a warm time of it. I think I must have borne a charmed life... There were chaps laying besides me getting wounded, and not one touched me."[632]

PTE ARTHUR WHITE, 1ST BN ESSEX REGT, 88TH BDE, 29TH DIVISION.

"At 2 p.m., after the hill had been subjected to a heavy bombardment, the Worcestershire Regiment advanced and reached the wire entanglements without much loss. Passages were cut through the wire at places where it had been damaged by shell fire, and by 4 p.m. the hill was in our hands... An attempt was now made to dominate the trenches overlooking the Sedd-el-Bahr (V) landing by a movement along the cliffs eastward from the lighthouse, but though the edge of the cliffs overlooking them were reached, it was found that the enemy could still dominate the beach, and the troops there were unable to afford any assistance to the movement."[633]

MAJOR OSCAR STRIEDINGER, ASC, GENERAL HEADQUARTERS, 29TH DIVISION.

"Extended formations were adopted when the first platoon came under effective fire. The advance was slow on account of the opposition. About 2.30 p.m. the Battalion reached the S.E. slope on Hill Point 138 being held up for two hours. It was then

---

630   89th Field Ambulance, Royal Army Medical Corps War Diary, TNA WO 95/4309.
631   *Aberdeen Evening Express*, 27th May 1915.
632   *Banbury Advertiser*, 3rd June 1915.
633   29th Division Staff, General Staff War Diary, TNA WO 95/4304.

decided to advance as one or two openings had been made in the barbed wire which had obstructed the advance.

"The whole of the Hill 138 was well protected with wire entanglements.

"Captain Ray of "Z" Company decided to push through at all costs towards the enemy, holding trenches on the N.E. slope. He was successful in getting through the first lot of wire entanglements, but met his death endeavouring to cut the second lot of wire which was only forty yards from the enemy's trenches. There were also 21 other casualties of his Company."[634]

CAPT. & ADJ. EDWARD KERANS, 4TH BN WORCESTERSHIRE REGT, 88TH BDE, 29TH DIV.

"An attempt was then made to relieve the terrible situation at V Beach by advancing along the top of the headland north-east… The distance to V Beach was not great — barely half a mile — and if it could have been covered, the enemy must have abandoned their V Beach trenches. Wire-cutters fearlessly advanced. From headquarters on the *Queen Elizabeth* they could be watched, clipping the powerful entanglements as though pruning a garden at home. But the rows of wire were too thick, the fire from the ruins of No. 1 Fort too deadly. Exhausted by a sleepless night and the hot day's fighting, these bravest of men abandoned the attempt, and sought rest in the trenches along the summit of the cliffs now deserted by the enemy."[635]

HENRY NEVINSON, WAR CORRESPONDENT.

"It was in the afternoon that we took the very important ridge, and there we took cover till night time; then we started making trenches, and finished about nine p.m."[636]

PTE ARTHUR WHITE, 1ST BN ESSEX REGT, 88TH BDE, 29TH DIVISION.

But progress remained slow. One contributory factor was that sixty three of the eighty naval ratings manning the boats were killed or wounded by day's end.[637]

"Reinforcements seem to take a long time landing. I am not clear when the Worcester Regiment arrived on the right, nor when touch was obtained with a Fusiliers battalion on the left. But the centre could not be reinforced until about 5 p.m. when the 5th Royal Scots landed."[638]

MAJOR OSCAR STRIEDINGER, ASC, GENERAL HEADQUARTERS, 29TH DIVISION.

As the defenders were pushed back then the build up of the supporting services could begin. The medics had cleared the wounded but the dead had to make way for the stores. And the dead had names.

"I have a smoke, and view the scenes on shore. Gradually the beach is becoming filled

---

634   4th Bn Worcestershire Regt War Diary, TNA WO 95/4312.
635   Nevinson, Henry, *The Dardanelles Campaign*, p. 104, Nisbet & Co. (London) 1918.
636   *Banbury Advertiser*, 3rd June 1915.
637   Bush, Eric, *Gallipoli*, p.158.
638   29th Division Staff, General Staff War Diary, TNA WO 95/4304.

with medical stores and supplies. It is gruesome seeing dozens of dead lying about in all attitudes. It becomes eerie as it gets darker....

"I am sitting on the boxes now, and "ping" goes something past my head, and then "ping-ping" with a long ringing sound, follows one after the other. The crackle of musketry begins again, and faster and faster the bullets come. At last I know what bullets are like.

"The feeling at first is weird. We get behind the pile of boxes, and bullets hit bully beef and biscuit boxes or pass harmlessly overhead. At last, boats come alongside and we unload the boxes into them, and I go ashore with the first batch, and there I meet 86th and 87th Supply Officers, who landed two hours earlier. My servant meets me and asks where shall I sleep. What a question! What does he expect me to answer "Room 44, first floor"? I say, "Oh, shove my kit down there," pointing to some lying figures on the sand. Five minutes after he comes up, and with a scared voice says, "Them is all stiff corpses, sir; you can't sleep there." I reply, "Oh, damn it; go and sit down on my kit till I come back." I start to work to get the stores higher up the cliff. Oh! the sand. It is devilish heavy going, walking up and down with my feet sinking in almost ankle-deep. It is quite dark now, and I stumble at frequent intervals over the dead. Parties are removing them, not for burial, but higher up the beach out of the way of the working parties. I run into the Brigade quartermaster-sergeant and ask him, "How's the Brigadier?" He replies, "Killed, sir." I can't speak for a moment. "And the Brigade Major?" "Killed also, sir." That finishes me. It is my first experience of the real horrors of war losing those who had become friends, whom one respected."[639]

CAPT. JOHN GILLAM, DIVISIONAL TRAIN, ARMY SERVICE CORPS, 29TH DIVISION.

"We... proceeded to build a jetty[640] for the purpose of landing artillery and heavy guns, for our infantry badly needed support, I assure you."[641]

SPR JOHN PUGH, 1/2ND (LOWLAND) FIELD COY, ROYAL ENGINEERS, 29TH DIVISION.

"We left our ship at 7 p.m. in small boats, and were drawn by a tug to the beach. Unlike the troops before us, we were saved wading waist deep: we had the luxury of a plank. Here we arrived, and that night there was hard work of all sorts to be done."[642]

PTE ARTHUR SIZER, 88TH (1ST EAST ANGLIAN) FIELD AMBULANCE, RAMC, 29TH DIV.

"Dusk was now coming on and active operations on our part ceased. We entrenched ourselves on a line from X Beach across the corner of the Peninsula some 600 yards inland to Hill 138. On W Beach they were working feverishly getting water, ammunition and supplies ashore and rigging up piers. The enemy were still in strength in Seddul Bahr and in the trenches round V Beach."[643]

CAPT. CLEMENT MILWARD, 53RD SIKHS, GSO3, GENERAL STAFF, 29TH DIVISION.

---

639 Gillam, *Gallipoli Diary*, pp. 36-37.
640 The company war diary records that they built a barrel pier in the afternoon (TNA WO 95/4309).
641 *Motherwell Times*, 4th June 1915.
642 *Cambridge Independent Press*, 18th June 1915.
643 Capt. Clement Milward, 53rd Sikhs, GSO3 29th Division General Staff, CAB 45/259.

"I was wounded... and had my wound dressed under fire by Jagger,[644] who was by me at the time, and who helped me down to the dressing station in the evening."[645]

CAPT. RICHARD HAWORTH, 1ST BN LANCASHIRE FUSILIERS, 86TH BDE, 29TH DIVISION.

"[At] dusk we commenced to disembark. A grim old salt manning the boat in which I went ashore said to me, as he slowly pulled on his oar: 'Mate, the last trip we did ashore in this boat there were 16 men killed.' Cheerful! At last we touched shore and had a brief respite, before being ordered to take up a position on the left of 'W' Beach or 'Lancashire Landing.' It was our duty to reinforce the sadly-thinned ranks of the... Lancashire Fusiliers. We took up our position at midnight. I shall never forget that awful night march up the steep cliffs. My chum and I were told off to take up 1500 rounds of ammunition, two machine-gun spare-part boxes and other things — a terrible load for two men in ordinary circumstances, and seemingly impossible under the conditions. It had to be done, however, and we did it in four heartbreaking hours."[646]

AB ALBERT ROWSON, RNVR, DRAKE BATTALION, ROYAL NAVAL DIVISION.

Dusk had brought some degree of relief to those at V Beach. On W Beach, by contrast, it seems to have conjured up phantom hordes bearing down upon them.

"Heavy fire opened by the Turks who attempted a counter-attack in force. Anson battalion sent up to reinforce. Every man again in the firing line and very heavy firing continued all night.

"Position appeared serious. There is no doubt that our men, whose nerves were naturally shaken by their day's work, were firing away an unnecessary quantity of ammunition, but all managed to hang on to the ground gained. Considerable difficulties in the way of replacing ammunition owing to lack of transport. Most of it had to be carried up by hand, every available man being seized upon for this purpose."[647]

MAJOR OSCAR STRIEDINGER, ASC, GENERAL HEADQUARTERS, 29TH DIVISION.

"We worked hard all day and all night till about 10 p.m., when suddenly the enemy delivered a severe counter-attack with the hope, I suppose, of driving us all back into the sea. Now there was some excitement; all lights were extinguished and the local R.E.'s were asked to join the Ammunition Column and carry ammunition to our heroes in the trenches we had taken during the day. We put in a terrible time now, but... succeeded in delivering 20,000 rounds between us, so we did our bit. We had to crawl up to the trenches dragging the ammunition after us as best we could."[648]

SPR JOHN PUGH, 1/2ND (LOWLAND) FIELD COY, ROYAL ENGINEERS, 29TH DIVISION.

---

644 Pte, later 2/Lt Harry Jagger, 1st Bn Lancashire Fusiliers, was killed in action on 10th June 1918. Buried Cinq Rues British Cemetery, Hazebrouck, France, he was the 24 year-old son of Irwin and Mary Jane Jagger, of Hebden Bridge, Yorkshire.
645 *Farnworth Chronicle*, 17th July 1915.
646 *Penistone, Stocksbridge and Hoyland Express*, 9th December 1916.
647 29th Division Staff, General Staff War Diary, TNA WO 95/4304.
648 *Motherwell Times*, 4th June 1915.

"At first the Battery was without ponies, as they were all required for carrying up ammunition and water for the troops, and the young drivers did their duty nobly, never thinking of danger…

"The first night ashore the Turks made a most determined attack upon us, trying to drive us into the sea, but we held on, although I thought it was all up with us."[649]

MAJOR THOMAS MACKELVIE, ARGYLL BTY, 4TH HIGHLAND MTN BDE, RGA, 29TH DIV.

"We did not go off the ship until the evening… What a sight we witnessed when landing — dead and dying were strewn all over the beach… at about 11 o'clock in the evening, we were very busy looking after other things in general. More was to come, our troops had driven them back, but of course the counter-attack had to come, this happening about 1 or 2 on the morning… They drove our lot back some distance, so they were only about half-a-mile from the beach. We had to seek cover as best we could. The ordeal was awful for us, as we were absolutely defenceless."[650]

PTE GILBERT RESBURY, 88TH (1ST EAST ANGLIAN) FIELD AMBULANCE, RAMC, 29TH DIV.

"It is awfully quiet and uncanny. Suddenly the musketry and rifle fire breaks out with a burst which develops into a steady roar. The beach becomes alive with people once more. All seems confusion. The Naval officer goes on steadily smoking, and we sit still, wondering how things are going to develop. The Fleet is silent. But I can just see the outline of the warships, with a few lights showing.

"Then I hear an officer shouting angrily, "Now then, fall in, you men! Who are you? Well, fall in. Get a rifle. Find one then, and damn quick!" Then another officer shouts, "All but R.A.M.C. fall in. Who are you? Fall in. Into file, right turn, quick march." About a dozen or two march off into the night up the cliff — officers' servants, A.S.C., seamen, RND. — every man who was not either R.A.M.C. or working on the dozen or so lighters that had been beached. I pause a bit. I feel a worm skulking behind these boxes while these events are happening. I express my feelings to Foley, and he says he feels the same. I say, "We must do something," and he replies, "Let's get rifles," and off we go searching for rifles, but can find none in the dark. I lose my temper — why, Heaven only knows. I see some men falling in, and I go up to them and say, "Fall in, you men; why aren't you falling in?" — although I know they are, and I find an officer in charge and feel an ass. They move up to man the third-line trench just running along the edge of the cliff. All the beach parties have moved up to this trench. I have lost Foley, and so I follow up with no rifle and no revolver, and shivering with cold. But I feel much better, although I am still in a temper. Extraordinary this! I am annoyed with everybody I see. Nerves, I suppose. Then a petty officer comes along and shouts, "Now then, you men, where the ——— is the ——— ammunition?" and in the darkness I discern some seamen carrying boxes of S.A.A. I go to the first pair, carrying a box between them, and take one side of the box from one of the seamen, and immediately feel delighted with myself, the sailor, and everybody. I have

---

649 *Campbeltown Courier*, 29th May 1915.
650 *Cambridge Independent Press*, 11th June 1915.

got a definite job. Up we pant; half-way up the cliff, I find Foley[651] on the same job. A voice shouts, "Have you got the ammunition, Foley?" It is O'Hara's[652] voice, our D.A.Q.M.G., and he comes running down to us."[653]

CAPT. JOHN GILLAM, DIVISIONAL TRAIN, ARMY SERVICE CORPS, 29TH DIVISION.

"On the Sunday night we all thought the Turks would drive our soldiers back into the sea, as they were driven right down to the beach, and here I must say the Implacable scored; for we got our guns to bear on the Turks and drove them back, and the same thing occurred several times."[654]

AB HERBERT BOX, RN, HMS *IMPLACABLE*.

"Suddenly the Fleet open fire, and the infernal din begins all over again, the flashes lighting up the beach, silhouetting men on shore and ships lying off, and all the time the song of bullets. Red Hell and a Sunday night! And this is war at last! I never thought I should ever get as near it as this, when I was a civilian. O'Hara says, "Who's that?" to me, and I answer my name, and he says, "Righto! give us a hand with this little lot, lad." He bends down, and he and a sailor lift a box. Foley and I lift another, and six seamen (I find they are off the Implacable) lift the others, and off we pant up the cliff over that third-line trench, lined with men of the beach parties with fixed bayonets. It's a devil of a walk to the second line, and it reminds me of hurrying to the railway station with a heavy portmanteau to catch a train. Foley and I constantly change hands.

"The seamen too find it heavy going. We arrive at the second line and run into the Adjutant of the Lancashire Fusiliers, calmly walking up and down his trench with a stick. We halt, open the boxes, and hang the strings of ammunition around our necks and over our shoulders. I am almost weighed down with the load. We have a rest, taking cover in the trench now and again as bullets come rather thicker than usual. The firing is frightful — now a roar of musketry, and now desultory firing — while the ships' guns boom away in the same spasmodic way. O'Hara then says, "Come along; follow me," and we go, headed by the Adjutant of the Lancashire Fusiliers to show us the way, and on over the grass and gorse into the blackness beyond. We are lucky, for it is a quiet moment and we have only to go three or four hundred yards, but just as we approach the first line, out bursts a spell of machine gun and rifle fire — rapid — and I fall headlong into what I think is space, but which proves to be our front-line trench, I fall clean on top of a Tommy, who is the opposite of polite, for my ammunition slings had tapped his nose painfully. I apologize, and feeling a bit done, he down in the mud like a frog, the coolness of the mud soon reviving me. We pass the ammunition along, each man keeping two or three slings. O'Hara wanders along the trench, having to keep his head low, for it is none too deep and bullets are pretty free overhead, while I remain and chat to the Tommy, another Lancashire Fusilier,

---

651 Capt. Francis Foley, Army Service Corps, 29th Divisional Train.
652 Major Emil O'Hara, Army Service Corps, DAQMG, 29th Division.
653 Gillam, *Gallipoli Diary*, pp. 38-39.
654 *Kent Messenger & Gravesend Telegraph*, 26th June 1915.

who is shivering, with teeth chattering and wet through, for it is raining. A Tommy on the other side of me is fast asleep and snoring loudly…

"I am enjoying myself now, for I am in the front-line trench with a regiment which has just added a few more laurels to its glorious collection. It is curious, but no shells are coming from the Turks, and bullets are such gentlemanly little things that they do not worry me. It is funny, but everybody up here appears very cool and confident, while on the beach they all are inclined to be jumpy."[655]

CAPT. JOHN GILLAM, DIVISIONAL TRAIN, ARMY SERVICE CORPS, 29TH DIVISION.

"During the night a violent counter-attack was made by the enemy against us. But we stood firm and drove them back again. Our casualties were very heavy. In my company alone I had 95 casualties out of 205 men. One of my platoons captured 13 Turks in one trench. The officers and men who were killed were buried together, close to the beach in an enclosure."[656]

MAJOR HAROLD SHAW, 1ST BN LANCASHIRE FUSILIERS, 86TH BDE, 29TH DIVISION.[657]

"Thirty Turkish prisoners were put on to dig a huge grave for the men and a small one for the officers. As I was the only chaplain ashore I took the service. I buried them on Sunday night in the dark by the aid of torches — 95 men in one grave and five officers in the other."[658]

CHAPLAIN HENRY FOSTER, RN, 2ND (NAVAL) BRIGADE, ROYAL NAVAL DIVISION.

"With the troops ¼ mile from the beach the ammunition supply could just be coped with, but had they been 3 or 4 miles away, as was intended, on the first night, and then had to withstand a counter-attack, the situation would have been most precarious and probably disastrous. It is quite impossible that ¼ the quantity of ammunition which they asked for on the first night could have reached them even if they had only been 2 miles away."[659]

MAJOR OSCAR STRIEDINGER, ARMY SERVICE CORPS,
GENERAL HEADQUARTERS, 29TH DIVISION.

Ever since some commentators have discussed what might have been at W Beach. If only more determined leadership been shown, if the men had been driven further forward more aggressively, what opportunities were lost. If Striedinger was correct, the chances missed included more British soldiers to see the inside of Ottoman POW camps.

---

[655] Gillam, *Gallipoli Diary*, pp. 39–41.
[656] Shaw quoted in Creighton, *With the Twenty-Ninth Division in Gallipoli*, p. 58.
[657] Shaw was killed in action on 4th June 1915. Commemorated on the Helles Memorial, he was the 39 year-old son of Henry Shaw, JP, DL, of Whitehall, Buxton, Derbyshire.
[658] *Yorkshire Post*, 22nd May 1915.
[659] 29th Division Staff, General Staff War Diary, TNA WO 95/4304.

X BEACH

After establishing themselves ashore, the first priority was the capture of Hill 114 and to establish contact with W Beach. Sea conditions were calm and as they approached the shore and then the noise of the supporting bombardment at Anzac reached them.

> "At about 3.45 a.m. the ship stopped and the boats came alongside and were given.... hot cocoa, and filled up with troops. All very cold and expectant, at about 4 a.m. we heard heavy firing and saw flashes away to the North."[660]

MIDSHIPMAN HUGH TATE, RN, HMS *IMPLACABLE*.

> "About 4 o'clock we started getting the troops into the boats and at 5 o'clock the bombardment commenced. It was carried out by about six battleships and was one of the finest sights I have ever seen."[661]

LT FRANK STEPHENSON, RN, HMS *IMPLACABLE*.

> "Everybody was about by 3 a.m... and fell in by 3.15, and how real to see them charge their magazines before they embarked. By 4.45 all the ships had come into line, opposite their respective positions, and the bombardment began. We ran close in to our beach, and poured salvo after salvo; 12 in. shrapnel searched the cliffs, while 6 in. did ditto to beach, and you can understand what an 800 lb. shell will do at a 1,000 yards range."[662]

STOKER PETTY OFFICER CHARLES COOK, RN, HMS *IMPLACABLE*.

> "The sun rose about 5.30 and when the bombardment commenced the sky was reddening for the sunrise right behind the peninsula. After they had been at it for about 20 minutes the whole of the southern end of the peninsula was hidden under an enormous cloud of smoke from the bursting shells and dust which, with the sun rising behind it, looked all red, and all through it you could see the flashes of the bursting shell. Not a single shot was fired at the ships...
>
> "We went on in as far as we could go with the boats following us until we were not more than 500 yards or so from the beach, and all the time we were easing off for all we were worth with the fore turret, as many 6 in guns as would bear, and also a lot of 12-pounders. The latter we kept on the top of the cliffs all along the top of the beach. When we could go no farther in we stopped and the boats went past us, but we still kept on firing over their heads until they were actually on the beach."[663]

LT FRANK STEPHENSON, RN, HMS *IMPLACABLE*.

> "[W]hen we began to close the beach... things got warmer; there was a whine overhead and a large shell fell far astern by the "Dartmouth." Then one came down about five yards away and drenched us and caused a good deal of bad language. All this time

---

660 Hugh Tate quoted in Lockyer, *The Tragedy of The Battle of the Beaches*, p. 10.
661 *The Press* (New Zealand), 23rd April 1938.
662 *Cambridge Independent Press*, 4th June 1915.
663 *The Press* (New Zealand), 23rd April 1938.

we were approaching the beach, the ship with her anchor hanging down. When she reached 500 yards she went astern and anchored and we went out of her shelter into the open; the blast of the 12 inch over our heads was most unpleasant, as were the bullets which were now coming down all round like little wasps…

"According to plan the picket boat slipped us and we landed our little load. I was on the extreme left and found a sandy patch and got right in, third boat ashore, nearly all the rest ran on a reef about 20 yards out and the troops had to jump overboard and swim. As we touched land one of our 12 inch shells hit the cliff just overhead and filled the boat with earth."[664]

MIDSHIPMAN HUGH TATE, RN, HMS *IMPLACABLE*.

The twelve Turks watching the beach were persuaded it was an unhealthy place and withdrew, leaving two Nordenfelts behind.[665]

"We got off very lightly while getting ashore; I can only put it down largely to the way our mother-ship plastered the beach for us at close range; however, we had our bad time later on. About 100 yards from the shore the launches cast us off and we rowed in for all we were worth till the boats grounded, then jumped into the water, up to our chests in some places, waded ashore and swarmed up the cliff, very straight but, fortunately, soft enough for a good foothold. We then came under fire from front and both flanks. I sent one company, under Frank Leslie, to left front to hold them back there, and one straight ahead and to the right front. The fire was very hot, rifles, machine guns and shrapnel, and our losses heavy at once. I can never say half enough for the gallantry of the men under these trying circumstances. They lost most of the leaders, but fought on splendidly just the same."[666]

LT-COL. HARRY NEWENHAM, 2ND BN ROYAL FUSILIERS, 86TH BDE, 29TH DIVISION.

"Well, the first to land were the Royal Fusiliers, at about 5 a.m… They got on the land all right, and made a fine advance, but the enemy held a strong position, and had the ranges of all places of value, so were able to put a very heavy rifle fire on us. But… the Royal Fusiliers pushed on until the numbers against them were too great, and they had to retire."[667]

PTE CECIL WOODS, 1ST BN BORDER REGT, 87TH BDE, 29TH DIVISION.

"I had orders to join up with the Lancashire Fusiliers, who were landed on "W" beach. I knew they had had a terrible time in the boats, as they were next to us going ashore. I collected all I could, after holding the left and front and leaving a reserve Company, to bring ammunition, water, etc., up the cliff, and moved to attack Tekke Hill; this we eventually captured on our side with the bayonet, losing heavily, at about noon. The *Implacable* was so close in that we heard her crew cheering us after the attack."[668]

LT-COL. HARRY NEWENHAM, 2ND BN ROYAL FUSILIERS, 86TH BDE, 29TH DIVISION.

664 Hugh Tate quoted in Lockyer, *The Tragedy of The Battle of the Beaches*, pp. 10–11.
665 One of these can be seen today in the Tower of London.
666 Harry Newenham quoted in Creighton, *With the Twenty-Ninth Division in Gallipoli*, p. 56.
667 *Middlesex County Times*, 5th June 1915.
668 Harry Newenham quoted in Creighton, *With the Twenty-Ninth Division in Gallipoli*, pp. 56–57.

"Approaching "X" beach everything seems to be at peace; the beach party was quietly at work and HMS *Implacable*, standing close-in, was no longer using her guns. Brigade Headquarters went ashore in the first tow, and, instead of the expected struggle through the water to gain the shore, I walked down a plank, courteously assisted by a bluejacket, and landed dry-shod.

"On gaining the top of the low cliff I found a company of the Royal Fusiliers, whose commander informed me that they had landed without opposition; that his C.O. (Col. Newenham) had gone to the right with two companies to clear some sand-hills of a few Turks and connect up with the covering force (Lancashire Fusiliers) at "W" beach; that the remaining company was reconnoitring to the front, and that he had been left in reserve.

"I asked him whether any arrangements had been made to form a bridgehead in order to protect the landing-place, to which he replied "No." Whilst still talking to him I noted that the reconnoitring company seemed to be somewhat heavily engaged with the enemy, and an orderly appeared carrying a note from the O.C. of that company, asking for reinforcements, stating that he was heavily engaged and that the Turks were getting round his left flank. I told the O.C. Reserve that he had better push out and reinforce, and that I would take over the defence of the landing.

"I then sent for Colonel Home (O.C. Border Regt.), but, as he was busily engaged, Brevet Lieut.-Colonel Vaughan (2nd in command) came forward instead. I pointed out the position to him and asked him to get two companies of his battalion lined up under cover of the edge of the cliff, with bayonets fixed, ready to move out. This was done; but meanwhile we saw parties of the Royal Fusiliers gradually retiring, followed by parties of Turks, and the fire was increasing in intensity.

"A ragged volley fired by the Turks had fatal effects; the gallant Vaughan,[669] shot through the brain, pitched over the cliff, and several bullets grazed me, thought only one caused a flesh wound.

"Just then the retirement of the Fusiliers developed into a rapid retreat followed closely by the Turks. The latter having got within 120 yards, I ordered the Borders to charge, and over the cliff-edge they went with ringing cheers. The enemy fired heavily, causing some 50 casualties, but, not waiting for the bayonet, they fled in every direction as soon as our men got within 50 yards or so. The Borders then occupied some old Turkish trenches."[670]

BRIG.-GENERAL WILLIAM MARSHALL, COMMANDING 87TH BDE, 29TH DIVISION.

"Our Commanding Officer, seeing what was the matter, at once gave the order for D Company of the 1st Border Regiment to advance, and with a cheer we went off in one mad rush, under a very heavy fire, to support the Royal Fusiliers."[671]

PTE CECIL WOODS, 1ST BN BORDER REGT, 87TH BDE, 29TH DIVISION.

---

669 Major Charles Davies Vaughan, 1st Bn Border Regt, was killed in action on 25th April 1915.
     Buried in Pink Farm Cemetery, commemorated by a special memorial, he was the 47 year-old husband of Dorothy Jean Vaughan, of Crabtree Cottage, Wanborough, Guildford, Surrey.
670 Marshall, Sir William, *Memories of Four Fronts*, pp. 56–57, Ernest Benn Ltd. (London) 1929.
671 *Middlesex County Times*, 5th June 1915.

"About 12 noon our Fleet Surgeon was shot on the quarterdeck and died at 1 p.m.[672] He was preparing to go ashore to attend the wounded. We also lost two sailors, who were in the boats landing the troops."[673]

CPL ALBERT HAYLER, RMLI, HMS *IMPLACABLE*.

"Pending further orders and information from the Division, I thought it best to form a bridgehead to protect "X" beach. The remnants of the two companies of Royal Fusiliers were dispatched to join up with Colonel Newenham, and in addition a company of the Borders was sent, as his request, to reinforce him in his attempt to link up with "W." The Inniskillings and the remainder of the Borders set to work to make a bridgehead, part of which was to consist of the old Turkish trenches, reversed.

"This work was in progress when a man of the Borders came in carrying a dead tortoise, discovered in one of the trenches, and said: "I've found one of these here land-mines, sir.""[674]

BRIG.-GENERAL WILLIAM MARSHALL, COMMANDING 87TH BDE, 29TH DIVISION.

"On our left was Major [Geoffrey] May's platoon of the Border Regiment, and when we got a good way from the beach, he told his men to dig in, which means dig trenches. My platoon at once joined up, and made the front line of trenches, and when we had finished our work and looked round we found that the other part of the regiment had joined our trenches."[675]

PTE CECIL WOODS, 1ST BN BORDER REGT, 87TH BDE, 29TH DIVISION.

"We did not have much excitement till the afternoon, when we went to action stations and shelled the Turks who were pressing the Royals very badly just above X beach. They (the Royals) were good enough to say that we saved the whole situation… Please don't think I am bragging or anything of that sort; I don't think there is much doubt that the ship did very good work indeed, but, of course, all the credit is due to the captain, commander, and gunnery lieutenant."[676]

LT FRANK STEPHENSON, RN, HMS *IMPLACABLE*.

"About 3 p.m. our centre, which was unavoidably rather weak, was being driven back. I got a message to "X" beach, where 87th Brigade were now landing, and eventually we were supported by some of the Border Regiment.

"I had been wounded earlier, and now managed, with the help of Crowther, my servant, to get into cover and get "first aided"."[677]

LT-COL. HARRY NEWENHAM, 2ND BN ROYAL FUSILIERS, 86TH BDE, 29TH DIVISION.

---

672 Fleet Surgeon Adrian Forrester, RN, was killed on the quarter-deck just before he was about to leave for the beach to attend to wounded. He is commemorated on the Chatham Memorial.
673 *Norwood News*, 2nd July 1915.
674 Marshall, Sir William, *Memories of Four Fronts*, pp. 57–58, Ernest Benn Ltd. (London) 1929.
675 *Middlesex County Times*, 5th June 1915.
676 *The Press* (New Zealand), 23rd April 1938.
677 Harry Newenham quoted in Creighton, *With the Twenty-Ninth Division in Gallipoli*, p. 57.

"The signallers meanwhile tried to get into communication with "Y" but without success, though we had received an early message from that beach to say that the landing there had also been unopposed.

"The day dragged on with no information as to the progress of the landings at "S," "V," or "W"; although at the last-named we could see parties of our men making gallant, and eventually successful, attempts to cut the high wire entanglement.

"Heavy firing had started at "Y" and I was very much tempted to move in that direction, but my hands were tied; because, being Divisional Reserve, I felt it incumbent to keep the force intact, pending an order from the Division. Besides, one never knew. "X" was now secure from any attack and, if all the other landings proved to be failures, everything might have to be transferred there."[678]

BRIG.-GENERAL WILLIAM MARSHALL, COMMANDING 87TH BDE, 29TH DIVISION.

"Hundreds were lying on the beach helpless and shells and bullets were flying all round them and into them, trenches having to be dug for their safety. I got to the base hospital on the beach, and was taken to the hospital ship, "London," [*Soudan*?] about 5 p.m. I was fairly done up and had a good sleep."[679]

PTE REGINALD PARRISH, 2ND BN ROYAL FUSILIERS, 86TH BDE, 29TH DIVISION.

"At night the Turks made a counter-attack, but as we had now a good line of trenches, we were able to put a good fire into them, making them retire again."[680]

PTE CECIL WOODS, 1ST BN BORDER REGT, 87TH BDE, 29TH DIVISION.

"I got my lot at night. We were getting entrenched, as an attack was expected. I was sent with three men about a hundred yards in front on an advanced post, and told not to come in until four the next morning. But the enemy opened a deadly fire at eleven, and we four men were only about fifty yards from their trenches. So I said, "Come on, we'll retire." Well, I got about ten paces when I got 'biffed,' and down I went. I was not going to stop there for them to skylark with me, so I crawled. Bullets and shrapnel were flying all roads, and I was crawling between our own fire and the enemy's. I don't know how I got back to our trenches alive."[681]

L/CPL THOMAS FITZPATRICK, 1ST BN ROYAL INNISKILLING FUSILIERS, 87TH BDE, 29TH DIVISION.

"We dug ourselves in about 200 yards from the cliff. The Turks made three attacks on us that night, but were repulsed each time. All the wounded that could use a rifle had to reinforce the firing line, some acting as ammunition carriers."[682]

PTE HERBERT BURRIDGE, 2ND BN ROYAL FUSILIERS, 87TH BDE, 29TH DIVISION.[683]

678 Marshall, Sir William, *Memories of Four Fronts*, p. 58.
679 *Midland Counties Tribune*, 25th June 1915.
680 *Middlesex County Times*, 5th June 1915.
681 *Rugby Advertiser*, 19th June 1915.
682 *Middlesex Chronicle*, 5th February 1916.
683 Burridge transferred to 4th Bn Lancashire Fusiliers and was killed in action on 28th March 1917. He is buried in Faubourg d'Amiens Cemetery, Arras.

> "To give you some idea of the amount of ammunition used, I was told by an officer who ought to know, that the Border Regiment and Inniskilling Fusiliers, who were also landed on X beach after the Royals, together with some of the Royals, used up all the ammunition they took ashore with them, 230 rounds a man, and in addition got through 115,000 rounds during the first 24 hours."[684]
>
> LT FRANK STEPHENSON, RN, HMS *IMPLACABLE*.

> "At midnight the enemy again made a fierce attack on Hill 114 (Teke-Burnu) and we received a signal from X Beach that they were "very hard pressed." The only point of aim we had was where the star shell rose and fell behind the coast ridge and the range of the "Mushroom Tree" which was behind the enemy. We opened fire at the coast ridge sky line and worked the guns along this area setting the sights for the shell to drop in the valley. We fired 60 rounds of six in Common Shell and the attack ceased gradually."[685]
>
> CAPT. HUGHES LOCKYER, RN, HMS *IMPLACABLE*.

> "Of course we couldn't see what damage we were doing — we were simply firing away with the same range and in the same direction as we had been during the day; but we heard afterwards that our fire was very effective and as I said before, more or less saved the situation, but naturally that was mostly luck. Many officers I talked to afterwards said that if the Turks had made a really determined attack that night they could have pushed us into the sea with the greatest ease."[686]
>
> LT FRANK STEPHENSON, RN, HMS *IMPLACABLE*.

Y BEACH

There were no Turks within a mile of the landing site, a very narrow strip of beach at the foot of a cliff looming 150 feet or more above them. It was just as well. There were no clear objectives, more broad suggestions of what this force was supposed to do. It could interfere with the movement of any Turkish reinforcements seen heading towards beaches further to the south; it could eliminate an artillery piece, though its precise location was unknown. The situation was confused before the landing. And when they got ashore no-one was sure who was in command. Was it Koe, commanding the King's Own Scottish Borderers or Matthews of the Marines?[687]

> "From the *Sapphire* we were transferred to a trawler for landing purposes, and the hum of shells with a loud crack as they passed overhead commenced at 4.55."[688]
>
> PTE DANIEL WALKER, 1ST BN KOSB, 87TH BDE, 29TH DIVISION.

---

684   *The Press* (New Zealand), 23rd April 1938.
685   Lockyer, *The Tragedy of The Battle of the Beaches*, p. 17.
686   *The Press* (New Zealand), 22nd April 1938.
687   A footnote in the British Official History explained: "Actually the point had been settled at a divisional conference on 21st April, when it was found that Colonel Matthews was senior to Colonel Koe. But Colonel Koe was in bad health at this time; he was too ill to attend the conference; and for some time after landing he imagined himself in command. It was also thought at G.H.Q. that Colonel Koe was in command." (Aspinall-Oglander, *Military Operations. Gallipoli*, vol. 1, p. 202n.)
688   *Jedburgh Gazette*, 25th June 1915.

"We got into trawlers, and when dawn was just breaking the trawlers, or mine sweepers, went in close to the coast. We got into rowing boats and were pulled ashore. As soon as the boats touched the bottom we jumped overboard, waded ashore, and ran across the bit of beach into the shelter of the cliff, which was very high and steep.

"As soon as the men were ashore we mounted the cliff, scouts going in front, and manned the ridge at the top. Fortunately for us the landing at this particular point was a complete surprise for the Turks."[689]

PTE ALBERT WRIGHT, PLYMOUTH BN, RMLI, 2ND (NAVAL) BDE, ROYAL NAVAL DIVISION.

"We... [were] brought to a stop thirty yards from the shore by big boulders. We jumped into the sea, up to the armpits in some cases, and after a good deal of spluttering, falling and splashing we arrived on the sandy beach and started to climb the cliff in front of us. Our landing party was composed as follows — 1st Battalion K.O.S.B., a battalion RMLI, and a company of S.W. [South Wales] Borderers."[690]

PTE DANIEL WALKER, 1ST BN KOSB, 87TH BDE, 29TH DIVISION.

"As soon as we landed we were faced by a big cliff and we only had about 15 yards of a beach. Our boys swarmed up that cliff like flies, and we took the position at the top of the cliff without opposition."[691]

CSM FRED BROWN, 1ST BN KOSB, 87TH BDE, 29TH DIVISION.

Y Beach had been Hamilton's pet project. He was delighted to see the men established ashore.

"Both Battalions, the Plymouth and the K.O.S.B.s, had climbed the high cliff without loss; so it was signalled; there is no firing; the Turks have made themselves scarce; nothing to show danger or stress; only parties of our men struggling up the sandy precipice by zigzags, carrying munitions and large glittering kerosine tins of water. Through the telescope we can now make out a number of our fellows in groups along the crest of the cliff, quite peacefully reposing — probably smoking. This promises great results to our arms — not the repose or the smoking, for I hope that won't last long — but the enemy's surprise... we have brought off our tactical coup and surprised the enemy Chief. The bulk of the Turks are not at Gaba Tepe; here, at "Y," there are none at all!"[692]

GENERAL SIR IAN HAMILTON, C-IN-C, MEF.

But there they, for the most part, remained. Beyond sending out a few scouts, they did not advance, instead, as per their orders, expected others to join them. If that was understandable, the failure to dig in properly was not. Instead, time was wasted while they watched the spectacle to the south.

"All this time the guns of the Fleet were bombarding Kilid Bahr and Sedd-ul-Bahr

---

689 *Bradford Daily Telegraph*, 8th September 1915.
690 *Jedburgh Gazette*, 25th June 1915.
691 *Retford and Worksop Herald and North Notts Advertiser*, 15th June 1915.
692 Hamilton, *Gallipoli Diary*, vol. 1, p. 129.

Forts and a village below Achi Baba, and the damage done must have been awful. Fires were continually breaking out, and at other times a shell would land among the Turks, who must have been badly knocked about."[693]

PTE DANIEL WALKER, 1ST BN KOSB, 87TH BDE, 29TH DIVISION.

Watching the unfolding disaster at V Beach, Hamilton considered reinforcing the men standing idle at Y.

"When we saw our covering party fairly hung up under the fire from the Castle and its outworks, it became a question of issuing fresh orders to the main body who had not yet been committed to that attack. There was no use throwing them ashore to increase the number of targets on the beach. Roger Keyes started the notion that these troops might well be diverted to "Y" where they could land unopposed and whence they might be able to help their advance guard at "V" more effectively than by direct reinforcement if they threatened to cut the Turkish line of retreat from Sedd-el-Bahr. Braithwaite was rather dubious from the orthodox General Staff point of view as to whether it was sound for G.H.Q. to barge into Hunter-Weston's plans, seeing he was executive Commander of the whole of this southern invasion. But to me the idea seemed simple common sense. If it did not suit Hunter-Weston's book, he had only to say so. Certainly Hunter-Weston was in closer touch with all these landings than we were; it was not for me to force his hands: there was no question of that: so at 9.15 I wirelessed as follows:

"G.O.C. in C. to G.O.C. *Euryalus*. Would you like to get some more men ashore on "Y" beach? If so, trawlers are available."[694]

GENERAL SIR IAN HAMILTON, C-IN-C, MEF, HMS *QUEEN ELIZABETH*.

Hunter-Weston's attention was focused on V and W beaches. He did not reply to Hamilton, so the Commander-in-Chief repeated his 'suggestion.'

"Although on board the *Euryalus* we knew that the party at Y Beach had landed without opposition, we had no information of their subsequent movements, but about 10 a.m. General Hunter Weston received a proposal from Sir Ian Hamilton that the troops diverted from V Beach should be landed to reinforce those at Y. Captain Dent,[695] the principal naval transport officer, pointed out how such a change would involve considerable alteration in the original plans for landing guns, stores, etc., and after consulting with me the General came to the conclusion that the delay inevitably arising from such a change would be greater than he could afford and he adhered to his decision of putting these men on to W Beach whose proximity to their original destination would not create any such dislocation of arrangements."[696]

REAR-ADMIRAL ROSSLYN WEMYSS, RN, HMS *EURYALUS*.

693 *Jedburgh Gazette*, 25th June 1915.
694 Hamilton, *Gallipoli Diary*, vol. 1, pp. 132–133.
695 Capt. Douglas Dent, RN, was Mentioned in Despatches (*London Gazette*, 16th August 1915) for his services as PNTO in April 1915.
696 Wester-Wemyss, *The Navy in the Dardanelles Campaign*, pp. 83–84.

"At 11 a.m. I got this answer: —

"From General Hunter-Weston to G.O.C., *Queen Elizabeth*.

"Admiral Wemyss and Principal Naval Transport Officer state that to interfere with present arrangements and try to land men at "Y" Beach would delay disembarkation."[697]

GENERAL SIR IAN HAMILTON, C-IN-C, MEF, HMS *QUEEN ELIZABETH*.

This was around the time that the first shot was fired in their direction.

"We did not get a round fired on us until 11 a.m… In fact we were watching the other troops landing, also watching the naval bombardment of Sedd-al-Bahr, which they reduced to a heap of scrap iron."[698]

SGT WILLIAM VICKERTON, 1ST BN KOSB, 87TH BDE, 29TH DIVISION.

"We heard… so peaceful was that area… that the commanding officer, attended only by his adjutant, had walked to within 500 yards of Krithia without sighting an enemy… It is only fair, however, to the harassed Commander to record that he was not given any instructions as to his conduct in the event of the advance from the south not materialising; and a request for instructions signalled about noon, remained unanswered."[699]

COMMODORE ROGER KEYES, RN, CHIEF OF STAFF, EMS, HMS *QUEEN ELIZABETH*.

The men who did venture south came across the barrier of Gully Ravine. A detachment of Turks defending Gully Beach, at the mouth of the ravine, was the closest of the defenders to Y Beach.

"We were there for seven hours and did not have a shot fired at us. In this place there is a big valley something like Wallis Vale, but without a river running through it. The Turks had command of this valley, and while we were there it fairly rained bullets and shrapnel, so we got back to the edge of the cliff and dug ourselves in. The Turks, however, managed to keep coming up and charging our trenches, but we always succeeded in driving them back again."[700]

PTE ARTHUR PIKE, PLYMOUTH BN, RMLI, 2ND (NAVAL) BDE, ROYAL NAVAL DIVISION.

"[I]t was not until 3 p.m. that any steps were taken to entrench. At about 4 p.m. the enemy brought a gun to bear on them, and at 5.40 p.m. the first attack was delivered, which was broken up by the fire of the ships."[701]

COMMODORE ROGER KEYES, RN, CHIEF OF STAFF, EMS.

---

697 Hamilton, *Gallipoli Diary*, vol. 1, pp. 133–134.
698 *Dumfries and Galloway Standard*, 4th August 1915.
699 Keyes, *Naval Memoirs 1910–1915*, p. 307.
700 *Somerset Standard*, 8th October 1915.
701 Keyes, *Naval Memoirs 1910–1915*, p. 307.

> "[W]e commenced to dig trenches, [but] were attacked by the enemy before we could get properly started, and the majority had to lie in the open with no shelter."[702]
>
> SGT ANDREW KATER, 1ST BN KOSB, 87TH BDE, 29TH DIVISION.

> "About 4.30 the Turkish and German [sic] troops got on the move and advanced parties opened fire on us. A steady fire was kept up till about 7 p.m., when a large Turkish and German force commenced an attack on us."[703]
>
> CSM FRED BROWN, 1ST BN KOSB, 87TH BDE, 29TH DIVISION.

> "In the afternoon the Turks attacked our left, which was held by A and D Companies… As darkness began to descend the rifle fire on both sides became hotter, and we had to maintain a rapid fire so as to keep the Turks from attacking. All Sunday night it was attack and counter-attack, and how we stuck it fairly beats one. We have to thank the men on HMS Goliath for saving us from being cut up to a man. Some Turks got on the shore behind us, and but for those watchful eyes of the Navy I would not be writing this — with the aid of a steam launch and a quick-firer they wiped those Turks out."[704]
>
> PTE DANIEL WALKER, 1ST BN KOSB, 87TH BDE, 29TH DIVISION.

The attacks would begin in earnest after dark, the Turks then being safe from naval gunfire, and would continue throughout the night. Those on the receiving end were convinced that up to 13,000 Turks were attacking them. In reality, the forces were approximately equal.

BULAIR

The last of Hamilton's attempts to sow confusion in the defenders' mind was the feint at Bulair. It was an area that particularly concerned Liman von Sanders.

> "[W]e silently crept up the Turkish coast to the northernmost point of Gallipoli, to a place named Bulari [sic]; here we learned we were to undertake a feint landing whilst the real attempt was to take place at the southern end. At 4.30 a.m. HMS *Doris* and *Dartmouth* opened fire on the Turkish positions, and all that day was spent in shelling the coast and fortifications, transferring men from the transports to small boats, and cruising up and down, keeping the wily Turk on the 'qui vive.'"[705]
>
> PO (M) LESLIE POWELL, ROYAL NAVAL ARMOURED CAR DIVISION.

> "After alarming the 7th Division in the town of Gallipoli and instructing it to march at once in the direction of Bulair, I rode ahead to the heights of Bulair with my German adjutants.

---

702 *Rugby Advertiser*, 19th June 1915.
703 *Retford and Worksop Herald and North Notts Advertiser*, 15th June 1915.
704 *Jedburgh Gazette*, 25th June 1915.
705 *Western Daily Press*, 22nd April 1916.

"On the narrow ridge of Bulair where neither tree nor bush impedes view or gives cover, we had a full view of the upper Saros (Xeros) Gulf. About twenty large hostile ships, some war vessels, some transports, could be counted in front of us. Some individual vessels were lying close in under the steep slopes of the coast. Others were farther out in the gulf or were still underway. From the broadsides of the war vessels came an uninterrupted stream of fire and smoke and the entire coast including our ridge was covered with shells and shrapnel. It was an unforgettable picture. Nowhere, however, could we see any debarking of troops from the transports."[706]

GENERAL OTTO LIMAN VON SANDERS, COMMANDING OTTOMAN 5TH ARMY.

26 APRIL 1915

BULAIR

As the arrival of a number of troops transport and warships made a limited impression upon those watching from the shore, it was decided to land a platoon of the Hood Battalion to make as much of a row as possible at night. The plan was changed and Bernard Freyberg, a particularly strong swimmer, volunteered to go it alone.

"At 12.40 a.m. this morning, I started swimming to cover the remaining distance towing a waterproof canvas bag containing three oil flares and five calcium lights, a knife, signalling light and a revolver. After an hour and a quarter's hard swimming in bitterly cold water, I reached the shore and lighted my first flare, and again took to the water and swam towards the East, and landed about 300 yards away from my first flare where I lighted my second and hid among some bushes to await developments, nothing happened, so I crawled up a slope to where some trenches were located the morning before. I discovered they were only dummies, consisting of only a pile of earth about 2 feet high, and a hundred yards long and looked to be quite newlt [sic] made. I crawled in about 350 yards and listened for some time, but could discover nothing. I now went to the beach where I lighted my last flare and left on a bearing due South. After swimming for a considerable distance I was picked up by Lt. Nelson in our cutter some time after 3 a.m. Our Cutter in company with the Pinnace and the T.B.D. "Kennet" searched the shore with 12 pdrs and Maxim fire, but could get no answer from the shore.

"It is my opinion that the shore was not occupied, but from the appearance and lights on the tops of the hills during the early hours of the morning, I feel certain that numbers of the Enemy were there, but owing to [the] chance of being captured, and as I had cramp badly, I could get no further."[707]

LT-CDR BERNARD FREYBERG, RNVR, HOOD BATTALION, HMT *GRANTULLY CASTLE*.

---

706 Liman von Sanders, *Five Years in Turkey*, pp. 63–64.
707 Royal Naval Division General Staff War Diary, TNA WO 95/4290.

"Mr. Peppiatt[708] and I along with two or three others were in the same boat with him all night. The Commander had swum ashore and lit some lights so as to draw the enemy's attention. He caught cramp close to land, however, and we had to row after him and bring him back."[709]

LS HORACE ELSOM, RFR, HOOD BATTALION, HMT *GRANTULLY CASTLE*.

"I remained on the heights of Bulair…. the night remained quiet excepting for artillery fire; the ships frequently changed position. As the concentration of the hostile ships in the upper Saros Gulf was fully recognized as a demonstration by morning of April 26, I ordered that forenoon that further units of the 5th and 7th Divisions should be transported to Maidos by boat during the next night and that the field artillery of the two divisions march to Maidos by land. I placed Lt Colonel Kiazim Bei, chief of staff of the Fifth Army, in command on the upper Saros Gulf with instructions to start all the remaining troops of the 5th and 7th Divisions for Maidos, if no landing was made during the next twenty-four hours."[710]

GENERAL OTTO LIMAN VON SANDERS, COMMANDING OTTOMAN 5TH ARMY.

If Freyberg's efforts had not impressed Liman von Sanders, they did his commanding officer.

"I cannot speak too highly of the manner in which this Officer carried out the difficult task assigned to him. He threw his whole energies into his preparations and carried the adventure through to a successful conclusion in an exceptional manner.

"As an example of resource, pluck, physical strength and endurance, his performance is worthy of the traditions of both services. I would like to mention the assistance and support he received from Lt. Nelson, although his part was a secondary one, it as an important one, for a mistake on his part would probably have sacrificed the life of Lt. Comdr. Freyberg.[711] He is an extremely cool and reliable Officer and I selected him for the part for those reasons."[712]

LT-COL. ARTHUR QUILTER, GRENADIER GUARDS, ATTD HOOD BN, HMT *GRANTULLY CASTLE*.

KUM KALE

It was far from quiet over on the Asian shore.

"Furious Turkish attacks were made at 2:30 a.m., at 3:30 a.m. and 4 a.m. Each time the enemy was repulsed with heavy casualties by the 6th Colonial Regiment, who fired 11,000 rounds of ammunition during the night."[713]

LT HENRI FEUILLE, 52ND BATTERY, 30TH REGT, 1ST DIVISION, CEO.

---

708 Stoker 1st Class Eugene Peppiatt, RFR, was wounded with Hood Bn on 11th June 1915.
709 *Boots Comrades in Khaki* (September 1915), p. 153, Boots (Nottingham) 1915.
710 Liman von Sanders, *Five Years in Turkey*, p. 67.
711 Freyberg was awarded the DSO for his actions at Bulair (*London Gazette*, 3rd June 1915).
712 Royal Naval Division Headquarters, General Staff War Diary, TNA WO 95/4290.
713 Feuille, *Face aux Turcs*, p. 32.

"About 6 o'clock the attack on the cemetery was to have been delivered, but for some unknown reason there was a delay. At 7 o'clock a fresh attack seemed to be in preparation from the cemetery. As a matter of fact Turks came out from their trenches with their arms [rifles] in their hands, but they waved at the same time white handkerchiefs and flags. Suspecting a ruse common enough at other fronts, the Colonel ordered the firing to be kept up.

"Notwithstanding this, some 60 to 70 Turks left the trench, threw down their weapons and raised their arms above their heads. The firing was stopped and the prisoners led into our lines. Following their example, other Turks quitted their trenches and came forward with many gestures of amity, without, however, raising their arms."[714]

MO JOSEPH VASSAL, 6TH COLONIAL REGT, BDE *COLONIALE*, 1ST DIVISION, CEO.

"The enemy began to wave flags and showed a wish to give themselves up. Eighty Turkish soldiers approached unarmed and were conducted inside our lines. Immediately afterwards many more Turks (several hundred) arrived in succession but refused to lay down their arms."[715]

COLONEL RUEF, HEADQUARTERS, BRIGADE *COLONIALE*, 1ST DIVISION, CEO.

"To facilitate surrender talks, Captain Raeckel, accompanied by a Turkish-speaker, went forward… But the Turks, noticing our small numbers, and given the general confusion, considered the French to be their prisoners, and not themselves ours. A general brawl ensued…blows with rifle butts and rifle shots were quickly exchanged. Lieutenant Lefort, who was in the midst of this skirmish… was seized and taken away by the Turks."[716]

LT HENRI FEUILLE, 52ND BATTERY, 30TH REGT, 1ST DIVISION, CEO.

"Others, jostling our men, succeeded in seizing and making off with two machine guns. Our men did not dare to open fire for fear of wounding their own comrades."[717]

COLONEL RUEF, HEADQUARTERS, BRIGADE *COLONIALE*, 1ST DIVISION, CEO.

"Profiting by the disorder, some Turks slipped into the village and barricaded themselves in the houses. To begin to fire again on the dissentients seemed a matter of urgency, but that would have been to abandon our two representatives to their fate, and perhaps to deprive us of the last arrangement possible.

"While this was going on General d'Amade arrived at the *Chateau d'Asie*. He gave the order to form our lines at all hazards, and consequently to begin firing again. Captain Roeckel, accompanied by two Turkish officers and two soldiers, coming back, made signs not to fire. He waved his kepi. At the first salvos Captain Roeckel was taken quickly back towards the Turkish lines."[718]

MO JOSEPH VASSAL, 6TH COLONIAL REGT, BDE *COLONIALE*, 1ST DIVISION, CEO.

---

714 Vassal, *Uncensored Letters from the Dardanelles*, p. 68.
715 Colonel Ruef quoted in Aspinall-Oglander, *Military Operations Gallipoli*, vol. 1, p. 261.
716 Feuille, *Face aux Turcs*, p. 33.
717 Colonel Ruef quoted in Aspinall-Oglander, *Military Operations Gallipoli*, vol. 1, p. 261.
718 Vassal, *Uncensored Letters from the Dardanelles*, p. 69.

"The Turks in the houses defended themselves energetically, killing several of our troops, amongst them the captain of engineers and his lieutenant, and it was only in the afternoon that we were able to drive these men out by bombarding the houses they occupied."[719]

COLONEL RUEF, HEADQUARTERS, BDE *COLONIALE*, 1ST DIVISION, CEO.

"Captains de Queral and Distel were mortally wounded. A "75" gun swept away everything that stood upright and at length we remained masters of the field. The Turks ran in every direction. Next we had to dislodge some 80 to 100 men who were concealed in the ruins of the village. One after another the houses were surrounded...

"The Turks sold their lives dearly. There was one house where a small number were entrenched, which could not be silenced. Cannon had to be brought up. A "75" gun shelled the place, but during the storm of shells the besieged men continued to fire. Here Captain Ferrero, commandant of engineers, was killed. His colleague, Lieutenant Lefort, was dragged into a house and disappeared. Elsewhere calm settled over the whole line till 11 o'clock a.m."[720]

MO JOSEPH VASSAL, 6TH COLONIAL REGT, BDE *COLONIALE*, 1ST DIVISION, CEO.

"It was not long before tragic stories... were circulated among the troops. It was rumoured that Lieutenant Lefort had been tortured and hung by his feet from a tree. Who had seen it? No one, of course."[721]

LT HENRI FEUILLE, 52ND BATTERY, 30TH REGT, 1ST DIVISION, CEO.

"During the the recapture of the village a captain and eight men were taken prisoners. This officer, who was a German, spoke French quite well. He was asked what had become of Lieutenant Lefort. It was generally believed that he had been tortured and hung.

"Colonel Nogues questioned the prisoners one after the other. No one could give any information about him. Several said it was no use looking for him... The Colonials, who had suffered much in this new affair of the village, and who believed their adversaries were acting in bad faith, excitedly surrounded the prisoners. Already shining blades were flashing from their scabbards. Some of the Senegalese stamped with their feet and gesticulated furiously. Stray shots were let fly.

"At this critical moment Colonel Nogues used his great authority over the [Senegalese] to calm them. For all that, they demanded immediate execution — a course in full accord with all the laws of war after sentence delivered."[722]

MO JOSEPH VASSAL, 6TH COLONIAL REGT, BDE *COLONIALE*, 1ST DIVISION, CEO.

"About sixty men were captured. The officer in command, and eight of his men, were then shot."[723]

COLONEL RUEF, HEADQUARTERS, BRIGADE *COLONIALE*, 1ST DIVISION, CEO.

719   Colonel Ruef quoted in Aspinall-Oglander, *Military Operations Gallipoli*, vol. 1, p. 261.
720   Vassal, *Uncensored Letters from the Dardanelles*, pp. 69–70.
721   Feuille, *Face aux Turcs*, p. 33.
722   Vassal, *Uncensored Letters from the Dardanelles*, p. 70.
723   Colonel Ruef quoted in Aspinall-Oglander, *Military Operations Gallipoli*, vol. 1, p. 261.

"Since dawn the warships had been engaged in the bombardment of In Tepe. They had also rained a great number of projectiles on the ruins of Yenisher, and directed attacks on the reinforcements coming by different routes to the succour of Koum Kaleh. The *Savoie* and the *Askold* fired principally against the Turkish troops entrenched along the Mendereh and in the cemetery.

"At 2 o'clock, after some very fortunate shots from the *Savoie* and the "75" guns it became impossible for them to hold the trenches in the cemetery and we saw the Turks waving their white flags. We took 600 prisoners."[724]

MO JOSEPH VASSAL, 6TH COLONIAL REGT, BDE *COLONIALE*, 1ST DIVISION, CEO.

"General d'Amade came on board and told us of the gallant capture of Kum Kale, of desperate counter-attacks during the night, repulsed with great loss to the enemy, and of prisoners taken — a brave story.

"In view of Lord Kitchener's very definite orders to Sir Ian, on no account to embark on a campaign in Asia, the occupation of Kum Kale was only intended to be a temporary diversion, to prevent interference from there with the landing at Sedd el Bahr; and d'Amade was very anxious to clear out, as he said he could not remain at Kum Kale without capturing Yeni Shehr, and that would need a whole division. Sir Ian, being anxious to land as many French troops as possible in Gallipoli, concurred in the evacuation."[725]

COMMODORE ROGER KEYES, RN, CHIEF OF STAFF, EMS, HMS *QUEEN ELIZABETH*.

"At 7 Colonel Ruef gave the order for re-embarkation. Very strict arrangements had been made to avoid any sort of surprise, but the enemy were too dispirited, and did not trouble us. Not one rifle was fired; but the batteries of In Tepe dropped some shells on the shore whilst the troops and stores were being got together."[726]

MO JOSEPH VASSAL, 6TH COLONIAL REGT, BDE *COLONIALE*, 1ST DIVISION, CEO.

"The survivors of this daring coup.... re-embarked in the afternoon and night of the 26th... During this arduous operation shells fell on the beach, wounding Colonel Noguès in the arm and decapitating Lieutenant Weinpling, killing and wounding some men."[727]

LT HENRI FEUILLE, 52ND BATTERY, 30TH REGT, 1ST DIVISION, CEO.

"I am half inclined to believe, and so is Colonel Noguès, that the whole thing was a misunderstanding, and that the intention was really to surrender, but that our interpreters did not succeed in explaining things properly. The numerous surrenders during the afternoon (about 500) seem to confirm this opinion."[728]

COLONEL RUEF, HEADQUARTERS, BRIGADE *COLONIALE*, 1ST DIVISION, CEO.

---

724 Vassal, *Uncensored Letters from the Dardanelles*, pp. 70–71.
725 Keyes, *Naval Memoirs 1910–1915*, p. 306.
726 Vassal, *Uncensored Letters from the Dardanelles*, p. 71.
727 Feuille, *Face aux Turcs*, p. 34.
728 Colonel Ruef quoted in Aspinall-Oglander, *Military Operations Gallipoli*, vol. 1, p. 262.

ENGLAND

A paper was presented to the Royal Geographical Society by a noted archaeologist and Near East expert.

> "As for the Dardanelles, all the western end of the Gallipoli peninsula [Cape Helles] is of broken, hilly character, which combines with lack of water and consequent lack of population and roads to render it an unfavourable area for military operations. None would choose to land a considerable force on it at any spot below the Narrows."[729]
>
> DAVID HOGARTH, KEEPER OF THE ASHMOLEAN MUSEUM, OXFORD.

HELLES

Men were, of course, fighting in this "unfavourable area." Overnight, those on S Beach held tight; the men on X and W beaches, by then linked, maintained their tentative toe-holds. For them it would be a day of consolidation and regrouping. But the fate of the men on V and Y beaches, whose experience of landing could scarcely have been more different, would be decided this day.

X BEACH

> "That night was an anxious one for me; not from any trouble at "X," where there was little or no firing, but from the uncertainty of the whole situation. Firing seemed to be general at all other beaches, rising to a perfect crescendo at "Y," and I greatly feared that my K.O.S.B., and S.W.B. Company, were being overwhelmed."[730]
>
> BRIG.-GENERAL WILLIAM MARSHALL, COMMANDING 87TH BDE, 29TH DIVISION.

Y BEACH

One immediate concern for those hanging on here was their inability to respond to one weapon in the Turks' armoury.

> "During the night the enemy used hand grenades against us, and we were driven from our trenches no fewer than three times, but we always re-took them and inflicted severe losses on the enemy. In this engagement we were reinforced. Well, it was 'hell on earth' all through the night."[731]
>
> CSM FRED BROWN, 1ST BN OSB, 87TH BDE, 29TH DIVISION.

> "We kept the Turks at bay, but at night we had a frightful job to hold them back. At one time they got so close as to throw a couple of bombs in one of our trenches. During

---

729 *The Weekly Oxfordshire Weekly News*, 5th May 1915.
730 Marshall, Sir William, *Memories of Four Fronts*, p. 59.
731 *Retford and Worksop Herald and North Notts Advertiser*, 15th June 1915.

the night I was carrying ammunition to the trenches and the bullets were sending the earth up all around me."[732]

PTE FRED HAWORTH, PLYMOUTH BN, RMLI, 2ND (NAVAL) BDE, ROYAL NAVAL DIVISION.

"I don't think there was a man but said his prayers that night… The Borderers fought like lions, and I do not think there could have been any braver deeds done by any other regiment. Gibson,[733] my chum from Mauchline, was killed [that] night…, and I never knew the moment I might fall myself, but we never think of that…

"At first the Turks tried to break through our lines at night, but were cut down in hundreds; it was awful to see their dead bodies in the morning."[734]

PTE ROBERT EDGAR, 1ST BN KOSB, 87TH BDE, 29TH DIVISION.[735]

Hamilton struggled to understand what was happening there.

"About this time, also, i.e., somewhere about 9 a.m., we picked up a wireless from the O.C. "Y" Beach which caused us some uneasiness. "We are holding the ridge," it said, "till the wounded are embarked." Why "till"? So I told the Admiral that as Birdwood seemed fairly comfortable, I thought we ought to lose no time getting back to Sedd-el-Bahr, taking "Y" Beach on our way. At once we steamed South and hove to off "Y" Beach at 9.30 a.m. There the *Sapphire*, *Dublin* and *Goliath* were lying close inshore and we could see a trickle of our men coming down the steep cliff and parties being ferried off to the Goliath: the wounded no doubt, but we did not see a single soul going up the cliff whereas there were many loose groups hanging about on the beach. I disliked and mistrusted the looks of these aimless dawdlers by the sea. There was no fighting; a rifle shot now and then from the crests where we saw our fellows clearly. The little crowd and the boats on the beach were right under them and no one paid any attention or seemed to be in a hurry. Our naval and military signallers were at sixes and sevens. The *Goliath* wouldn't answer; the *Dublin* said the force was coming off, and we could not get into touch with the soldiers at all."[736]

GENERAL SIR IAN HAMILTON, C-IN-C, MEF, HMS *QUEEN ELIZABETH*.

"Our officers and men realised the position we were in, and every one knew he was fighting for his very existence… About 10 a.m. the order was passed along for us to retire to the beach and embark. Everything was carried out calmly."[737]

CSM FRED BROWN, 1ST BN KOSB, 87TH BDE, 29TH DIVISION.

---

732 *Accrington Observer and Times*, 19th June 1915.
733 Pte Samuel Gibson, 1st Bn KOSB, was killed on 26th April 1915. Commemorated on the Helles Memorial, he was the 26 year-old son of John and Agnes Gibson, of Clelland Park, Mauchline, Ayrshire.
734 *Dumfries and Galloway Standard*, 19th June 1915.
735 Robert Edgar was killed on 1st July 1916. Commemorated on the Thiepval Memorial, he was the son of Sarah Edgar, of Bridge of Dee, Castle Douglas.
736 Hamilton, *Gallipoli Diary*, vol. 1, p. 146.
737 *Retford and Worksop Herald and North Notts Advertiser*, 15th June 1915.

"All through the night they kept up the attack, and we could just see them coming up, sometimes within about thirty yards the trench we had dug. Then we 'saw' the beggars off with rifle and maxim fire. If they had broken our line in the darkness goodness knows what would have happened. I expect we should have been fighting everyone we saw, as it was black dark, and we had nowhere to retire to.

"About midday ... the Turks came up in thousands, and it was seen that with the few men we had it was impossible to hang on, so a retirement to the back commenced. I have a dim recollection of the retirement first to the cliff edge and hanging on there while some of the men and the wounded were taken aboard, and then rushing down after and getting aboard myself. The boats had to be kept afloat, and we had to wade out up to our necks in water to get into them. It was a struggle to get into them with full equipment on and a rifle. After we had filled the boats we were taken aboard a warship and afterwards put aboard a transport."[738]

PTE ALBERT WRIGHT, PLYMOUTH BN, RMLI, 2ND (NAVAL) BDE, ROYAL NAVAL DIVISION.

"Passing "Y" Beach [around 12.20 p.m.] the re-embarkation of troops was still going on. All quiet, the Goliath says: the enemy was so roughly handled in an attack they made last night that they do not trouble our withdrawal — too pleased to see us go, it seems! So this part of our plan has gone clean off the rails. Keyes, Braithwaite, Aspinall, Dawnay, Godfrey are sick — but their disappointment is nothing to mine. De Robeck agrees that we don't know enough yet to warrant us in fault-finding or intervention. My orders ought to have been taken before a single unwounded Officer or man was ferried back aboard ship."[739]

GENERAL SIR IAN HAMILTON, C-IN-C, MEF, HMS *QUEEN ELIZABETH*.

"At daylight when, as was thought, the enemy was dispersed by ships' fire, the worn-out party began to carry the wounded down to the beach and asked for ships' boats to embark them. Colonel Matthews, who was in command, sent a message to General Hunter Weston saying he could not hold his position without fresh ammunition and reinforcements. This message never reached its destination. Whether the men mistook the order for evacuating the wounded as one for evacuating the beach is unknown, but they did begin to leave the trenches, when the enemy promptly attacked them once more, but once more was beaten off. The ships, believing that evacuation had been ordered, were doing their utmost to further what they concluded to be the General's order and sent every available boat to the beaches and the Sapphire asked the *Queen Elizabeth*, approaching from Gaba Tepe, to open fire on the ridge to cover the re-embarkation. This the *Queen Elizabeth* did for some little time, and Sir Ian Hamilton, not understanding the reason for this movement, but believing that the order for evacuation had been given by General Hunter Weston, came on to us to have the matter elucidated. Until this moment we on board the *Euryalus* had been

---

738   *Bradford Daily Telegraph*, 8th September 1915.
739   Hamilton, *Gallipoli Diary*, vol. 1, p. 158.

in blissful ignorance of what was occurring at Y, and Sir Ian Hamilton's surprise at learning that no orders for such a move had been given was no greater than was ours at hearing what had taken place."[740]

REAR-ADMIRAL ROSSLYN WEMYSS, RN, HMS *EURYALUS*.

"We were ashore for 36 hours and then retired. We could not hold the number of Turks we had drawn on to us, but thanks to the guns of the "Queen Elizabeth" we managed to retire in safety. We got into boats and were taken to the battleship "Goliath," and then we went aboard a transport."[741]

PTE ARTHUR PIKE, PLYMOUTH BN, RMLI, 2ND (NAVAL) BDE, ROYAL NAVAL DIVISION.

"At 4.50 we were opposite Krithia passing "Y" Beach. The whole of the troops, plus wounded, plus gear, have vanished. Only the petrol tins they took for water right and left of their pathway up the cliff; huge diamonds in the evening sun. The enemy let us slip off without a shot fired. The last boatload got aboard the Goliath at 4 p.m., but they had forgotten some of their kit, so the Bluejackets rowed ashore as they might to Southsea pier and brought it off for them — and again no shot fired!"[742]

GENERAL SIR IAN HAMILTON, C-IN-C, MEF, HMS *QUEEN ELIZABETH*.

"When we passed "Y" we heard from the Goliath that the evacuation of that beach was completed by 4 p.m. Not quite, for a little later a boat was seen to land on the beach and, leaving his crew in the boat, a naval officer was seen to climb the cliff and spend an hour or so on the scene of the action. It was my brother Adrian,[743] miserably unhappy, making sure that no wounded had been left behind."[744]

COMMODORE ROGER KEYES, RN, CHIEF OF STAFF, EMS, HMS *QUEEN ELIZABETH*.

By late afternoon all the men were withdrawn from Y Beach but, unlike the French at Kum Kale, that was not part of the plan. The whole affair spoke of the muddle and lack of clarity behind the whole planning process: who gave the order to evacuate; why was Hunter-Weston so indifferent to it; and why, if Hamilton was so invested in its success, did he fail to take more direct intervention?

V BEACH

The beach was covered with dead and wounded men and it had been almost impossible to recover them before night fell. Darkness allowed the men to finally disembark from the *River Clyde*; for the link to W Beach to be consolidated; and for help to be given to those still able to benefit from the attentions of the Royal Army Medical Corps.

"1 a.m... Advance of "W" Coy RMF & 2 Coys Hants under Major Beckwith, Hants

---

740 Wester-Wemyss, *The Navy in the Dardanelles Campaign*, pp. 109–110.
741 *Somerset Standard*, 8th October 1915.
742 Hamilton, *Gallipoli Diary*, vol. 1, p. 156.
743 Lt-Cdr Adrian Keyes, RN.
744 Keyes, *Naval Memoirs 1910–1915*, p. 307.

Regt, for 200 yards. This force had to fall back 2 hours later; "W" Coy joining up with the Worcestershire Regt. on left flank."[745]

CAPT. & ADJ. HARRY WILSON, 1ST BN ROYAL MUNSTER FUSILIERS, 86TH BDE, 29TH DIV.

"As soon as we were all off we fixed our bayonets and made an advance of about 200 yards, cutting down barbed wire entanglements which were all over the hill."[746]

CPL CHARLES WATERMAN, 2ND BN HAMPSHIRE REGT, 88TH BDE, 29TH DIVISION.

"I volunteered to go ashore to see the wounded on the beach. The dead and dying were here in hundreds. Before I got back to the ship at 4 this morning I had a very hot time of it, and cannot understand why I am not a dead man. We were told yesterday that a counter-attack was to be made and that the Turks intended to blow the ship to pieces with cannon, which they were to bring up in the night. When the attack did come I gave up all hopes of anything but slaughter, as the men we had on land were insufficient in number to meet a large force.

"About fifty men were leaving the ship when this started, and at the sound of the firing all fell flat on their faces, and if any one dared to move he was at once fired at. Someone on a barge next the small boat in which I had taken shelter asked if he could crawl into our boat, but I dared him or anyone else to move as such movement would only draw fire on every one of us. Not a man stirred, but lay on his face from midnight to 4 o'clock. It was not till the end of the attack that I learned these men had an officer with them. As I lay in the boat I shouted to them that an assault on us was likely, and ordered them to load and fix bayonets, and to see that all had plenty of ammunition. Extra bandoliers of cartridges were passed up from the rear, each pushing the sea long with a clatter. All this with the red cross on my arm! And with loaded revolver in hand I was prepared to die game."[747]

LT GEORGE DAVIDSON, 89TH (1ST HIGHLAND) FIELD AMBULANCE, RAMC, SS *RIVER CLYDE*.

"All night long it was nothing but thousands of rifles going off. We could render no aid with our guns, and it would have been unwise to use our searchlights, as they would have shown up the position of our troops. So we waited till daylight, and then we gave them beans with our big guns, and so helped our men to take Sedd-eh-Bahr. We had sustained a large number of casualties by this time, but had succeeded in establishing ourselves on shore, and were landing more troops, stores and horses."[748]

CPO FREDERICK NEAL, RN, HMS *SWIFTSURE*.

"We cannot believe ourselves that it was so bad. It was only after we had overcome the difficulty and put the enemy to flight that we thought we should live. The only man I was really sorry for was Dan Mulhall[749]... He had escaped the worst part of

---

[745] 1st Bn Royal Munster Fusiliers War Diary, TNA WO 95/4310.
[746] *Stratford-upon-Avon Herald*, 11th June 1915.
[747] Davidson, *The Incomparable 29th and the River Clyde*, p. 44.
[748] *Mid-Sussex Times*, 6th July 1915.
[749] Pte Daniel Mulhall, 1st Bn Royal Dublin Fusiliers, was killed in action on 26th April 1915. He is buried in V Beach Cemetery, Cape Helles.

the business and had got into a trench with me. I was beside him when he got killed; he never spoke a word, for he was shot through the brain. He was lying down beside me behind a nice cover, which you could just look over and take a good aim, and was saying 'Just look at that sniper 80 yards in front there. Watch him go down.' But the next moment poor Dan went down, never to rise nor speak again."[750]

PTE CORNELIUS FOX, 1ST BN ROYAL DUBLIN FUSILIERS, 86TH BDE, 29TH DIVISION.

"When dawn came, between 5 and 5.30, I got shot in the calf of the left leg. I tell you I stopped a beauty; it cut my leg for about four inches. That put me out of action, and I retired on the beach and waited for orders to go on the hospital ship."[751]

PTE STEPHEN FALLA, 2ND BN HAMPSHIRE REGT, 88TH BDE, 29TH DIVISION.

"7 a.m… 1st RMF, 2 Coys Hants, 1 Coy Dublins ordered to take & occupy Old Castle … 8 a.m…. Old Castle occupied. Advance through Sedd-el-Bahr village commenced."[752]

CAPT. & ADJ. HARRY WILSON, 1ST BN ROYAL MUNSTER FUSILIERS, 86TH BDE, 29TH DIV.

"[8.00 a.m.] The Dublins and ourselves occupied the ruins of Sed-el-Bahar Fort. Our progress was stopped now by snipers in the village beyond. Here Russell and I had a very narrow escape. I went to speak to him at our opening in the wall, when a bullet whizzed between our heads. I got my cheek cut by a splinter of stone — that was all."[753]

LT FREDERICK WALDEGRAVE, 1ST BN ROYAL MUNSTER FUSILIERS, 86TH BDE, 29TH DIV.

Thus far, it had been difficult for senior commanders to gain an understanding of how things were progressing onshore. Though wireless communications had not been wholly successful, the failure to move any distance inland meant that direct observation could still be carried out.

"Whilst having a hurried meal, Jack Churchill rushed down from the crow's nest to say that he thought we had carried the Fort above Sedd-el-Bahr. He had seen through a powerful naval glass some figures standing erect and silhouetted against the sky on the parapet. Only, he argued, British soldiers would stand against the skyline during a general action."[754]

GENERAL SIR IAN HAMILTON, C-IN-C, MEF, HMS *QUEEN ELIZABETH*.

"[A]ll the ridge of the hill was wire entanglements. I remember five of our regiment volunteered to make for this wire and cut it, but the poor fellows were done over in their dash side by side."[755]

PTE JOHN WALTERS, 1ST BN ROYAL MUNSTER FUSILIERS, 86TH BDE, 29TH DIVISION.[756]

750 *Coventry Herald*, 25th June 1915.
751 *Guernsey Evening Press and Star*, 26th May 1915.
752 1st Bn Royal Munster Fusiliers War Diary, TNA WO 95/4310.
753 *Northern Scot and Moray & Nairn Express*, 10th July 1915.
754 Hamilton, *Gallipoli Diary*, vol. 1, p. 153.
755 *Coventry Evening Telegraph*, 16th June 1915.
756 Walters was wounded, probably on 1st May. He was killed at Suvla on 21st August 1915. Commemorated on the Helles Memorial, he was the son of Timothy Walters, of Watch House, Long Pavement, Limerick, Ireland.

> "The dash was quite 100 yards, and I don't know whether I ran or prayed the faster. We started to cut the wire entanglements with pliers, but you might as well try and snip Cloyne Round Tower with a lady's scissors. 'Pull them up,' I roared; 'put your arms round them and pull them out of the ground.' I dashed at the first one; heaved and strained, and it came away.
>
> "I could not tell you how many I pulled up. The boys that were left with me were every bit as good as myself, and I wish they all got some recognition. When the wire was down the rest of the lads came on, and, under pulverising fire reached the trenches, and won about 200 yards length by 20 yards deep, and 700 yards from the shore."[757]
>
> CPL WILLIAM COSGROVE, VC, 1ST BN ROYAL MUNSTER FUSILIERS, 86TH BDE, 29TH DIV.[758]

> "The Turkish barbed wire was too thick for our wire cutters to cut through in time, and Corpl. Cosgrove immediately went forward and pulled the barbed wire stakes out of the ground, thereby clearing a way for the Battalion to advance. He was severely wounded at the time."[759]
>
> QMS STEPHEN AHERN, 1ST BN ROYAL MUNSTER FUSILIERS, 86TH BDE, 29TH DIVISION.

Breaking out from the fort and beach, the men had to fight through the village, fire coming from the first floors and cellars of ruined buildings, from behind piles of rubble, from all directions.

> "Early [on we] started to clear the fort and a village behind it, both of which had been knocked to bits by the fleet; but they were full of snipers all the same, and we had to clear the place out by getting from one house to another. Addison[760] was killed in the village and three subalterns wounded."[761]
>
> CAPT. & ADJ. GEORGE REID, 2ND BN HAMPSHIRE REGT, 88TH BDE, 29TH DIVISION.

> "The troops that went in that attack had already lost half their strength; the officers that led up those narrow streets, dodging first through gateways, across the openings, and beckoning when safe for their men to come on, were nearly all killed."[762]
>
> LT-CDR JOSIAH WEDGWOOD, RNVR, NO. 3 SQN ROYAL NAVAL ARMOURED CAR DIVISION.

> "That was when we found all the snipers, some of them were under the ground, lifting up trap doors, firing as many rounds as they wished, and then pulling the doors

---

757 *Irish Independent*, 26th August 1915.
758 Cosgrove's official citation for the VC was published in the *London Gazette* on 23rd August 1915: "For most conspicuous bravery leading this section with great dash during our attack from the beach to the east of Cape Helles on the Turkish positions on 26 April 1915. Cpl Cosgrove on this occasion pulled down the posts of the enemy's high wire entanglements single-handed, notwithstanding a terrible fire from both front and flank, thereby greatly contributing to the successful clearing of the heights."
759 *Coventry Herald*, 1st October 1915.
760 Capt. Alfred Addison, 2nd Bn Hampshire Regt, was killed on 26th April 1915. Buried in V Beach Cemetery, he was the 35 year-old son of Annie Kate Addison, of "St. Lawrence," Queen's Crescent, Southsea, Portsmouth, and the late Albert Addison.
761 *Stratford-upon-Avon Herald*, 28th May 1915.
762 Wedgwood, *With machine-guns in Gallipoli*, pp. 10 – 11.

down after them. We found most of these in the village we captured. This is where we lost Captain Addison."⁷⁶³

CPL CHARLES WATERMAN, 2ND BN HAMPSHIRE REGT, 88TH BDE, 29TH DIVISION.

"[W]e advanced for the village of Shed-ut-Bhar [sic] and it was a house-to-house fight there. The castle, fort, and village were all aflame, and were nothing but a heap of ruins, but it was infested with Turkish snipers, and we suffered very heavy losses clearing the Turks out of it. At last they were cleared out, and [we] entrenched on the outside of the village."⁷⁶⁴

PTE MAURICE SINDEN, 2ND BN HAMPSHIRE REGT, 88TH BDE, 29TH DIVISION.

"No one would credit the destruction that can be caused by our big guns. Whole blocks of masonry caved in, guns big as a tramcar were hurled from their foundations right across street. Destruction and ruin were to seen at every turn. One cannot help admiring a nation who would stop and fight amidst the ruin of one of their hopes such as that must have been. The first village was just above the fort, and presently the command came for the Hampshires to move forward, with us in support, and to rush it at all costs. I can tell you we looked to our rifles and fixed our bayonets and moved forward at as near a run as we could manage, although it was like going up a house side."⁷⁶⁵

SPR HERBERT HEATH, 1/1ST (WEST RIDING) COY, ROYAL ENGINEERS, 29TH DIVISION.

"Slowly — house by house — we could see our people taking possession of the village. At one house on top of an eminence we could see a figure throwing bombs or hand grenades. So our troops withdrew, and the *Lord Nelson* 'let rip' three 12-pounders into the house. There was more trouble from this quarter."⁷⁶⁶

TELEGRAPHIST JAMES HUTCHINSON, RN, HMS *LORD NELSON*.

"11 a.m... Forced village which was strongly held by snipers & took up a position at further end; preparations to storming Hill 141."⁷⁶⁷

CAPT. & ADJ. HARRY WILSON, 1ST BN ROYAL MUNSTER FUSILIERS, 86TH BDE, 29TH DIV.

Entering the village before it was cleared completely, George Davidson took in the scene.

"The only living things I saw in the village were two cats and a dog. I was very sorry for a cat that had cuddled close to the face of a dead Turk in the street, one leg embracing the top of his head. I went up to stroke and sympathise with it for the loss of what I took to be its master, when I found that the upper part of the man's head had been blown away, and the cat was enjoying a meal of human brains. The dog followed till I came upon three Dublin Fusiliers, who wished to shoot it straight away when I

---

763   *Stratford-upon-Avon Herald*, 11th June 1915.
764   *Western Gazette*, 23rd July 1915.
765   *Sheffield Daily Telegraph*, 3rd June 1915.
766   *Dundee Courier*, 11th June 1915.
767   1st Bn Royal Munster Fusiliers War Diary, TNA WO 95/4310.

pleaded for it, but one of them had a shot at it when my back was turned and the poor brute went off howling."[768]

LT GEORGE DAVIDSON, 89TH (1ST HIGHLAND) FIELD AMBULANCE, RAMC, 29TH DIV.

"As we swept through the village, gaining the top of the hill, and reaching the enemy's trenches we saw a scene I shall never forget. Presently the firing ceased, and we were able to look round and get an idea of things. The Turks had bolted and left everything — their food, kit, bedding, rifles, communications, clothing, and every conceivable thing."[769]

SPR HERBERT HEATH, 1/1ST (WEST RIDING) COY, ROYAL ENGINEERS, 29TH DIVISION.

"2 p.m... RN commenced bombardment of Hill 141."[770]

CAPT. & ADJ. HARRY WILSON, 1ST BN ROYAL MUNSTER FUSILIERS, 86TH BDE, 29TH DIV.

Supported by fire from 26th Battery, Royal Field Artillery, which had landed at W Beach, as well as the fleet, the assault on Old Fort No. 1 went in led by Charles Doughty-Wylie.

"Dead-beat, at [2.30 p.m.], before the final rush they hesitated. Then our last colonel, a Staff man, Colonel Doughty Wylie,[771] ran ashore with a cane, ran right up the hill, ran through the last handful of men sheltering under the crest, took them with that rush into the trench, and fell with a bullet through his head. But the Turks ran and the ridge was ours.

"I had to take the maxim guns up, skirting the village. If you have never felt afraid, try crawling up a gutter, crawling over dead men, with every wall and corner hiding a marksman trying to kill you."[772]

LT-CDR JOSIAH WEDGWOOD, RNVR, NO. 3 SQN ROYAL NAVAL ARMOURED CAR DIVISION.

Doughty-Wylie failed to take cover. He was, as Jack Churchill would have agreed, a proper British officer and paid the price for it.

"Getting the village entirely cleared of them was hot work, and a lot of life was lost. From the village we could get an enormously strong position of the enemy's on Hill 141 (a redoubt). The Navy smashed at it for about an hour with terrific effect, and the Dublins and ourselves took it by assault. We got right through their barbed wire.

---

768 Davidson, *The Incomparable 29th and the River Clyde*, p. 46.
769 *Sheffield Daily Telegraph*, 3rd June 1915.
770 1st Bn Royal Munster Fusiliers War Diary, TNA WO 95/4310.
771 Lt-Col. Charles Doughty-Wylie, Royal Welsh Fusiliers, attached as GSO2 to HQ, MEF, was killed in action on 26th April 1915. Awarded a posthumous VC for his bravery, his official citation was published in the *London Gazette* on 23rd June 1915: "On 26th April 1915 subsequent to a landing having been effected on the beach at a point on the Gallipoli Peninsula, during which both Brigadier- General and Brigade Major had been killed, Lt-Colonel Doughty-Wylie and Captain Walford organised and led an attack through and on both sides of the village of Sedd el Bahr on the Old Castle at the top of the hill inland. The enemy's position was very strongly held and entrenched, and defended with concealed machine-guns and pom-poms. It was mainly due to the initiative, skill and great gallantry of these two officers that the attack was a complete success. Both were killed in the moment of victory." Buried at Seddel-Bahr Military Grave, he was the 46 year-old son of H. M. Doughty, of Theberton Hall, Suffolk, and Edith, his wife (nee Cameron); husband of Lilian O. Doughty-Wylie.
772 Wedgwood, *With machine-guns in Gallipoli*, p. 11.

They would not face our bayonets, but bolted. It was excellent fun shooting at them as they ran like rabbits down a communication trench. We entrenched ourselves on the position thus gained, and remained there for the night."[773]

LT FREDERICK WALDEGRAVE, 1ST BN ROYAL MUNSTER FUSILIERS, 86TH BDE, 29TH DIV.

"4 p.m… 1st RMF less "W" Coy, 1 Coy Dublins & 2 Coys Hants under Major Grimshaw captured Hill 141 & drove enemy from the redoubt on top. Turks in full retreat for 2 miles."[774]

CAPT. & ADJ. HARRY WILSON, 1ST BN ROYAL MUNSTER FUSILIERS, 86TH BDE, 29TH DIV.

"We have lost the bravest officers the world could produce. Major Grimshaw,[775] D.S.O., my old Company officer, was a hero. He led his men to a ridge, which they eventually captured. He was in the forefront of the fight and advanced amid a hail of shells and rifle fire. The men shouted to him and warned him that a sniper was firing at him. He took no notice, and when appealed to by the men to lie down he refused. The result was that he was fatally shot — killed among the men who adored him and whom he had led to a glorious victory. Major Grimshaw's bravery was thrilling and his death spurred his men to take swift revenge of the Turks. His deed will live in history as one of the bravest ever performed in the ranks of the Royal Dublin Fusiliers."[776]

L/CPL CHARLES ELLOWAY, 1ST BN ROYAL DUBLIN FUSILIERS, 86TH BDE, 29TH DIVISION.

"Remainder of guns & wagons of 92nd Bty landed and went into action late in the day, about 4.0 p.m., and shelled retreating Turkish infantry moving across our front, ranges 3500 [yards] to 4100 [yards]."[777]

CAPT. FREDERICK MORGAN, 17TH BDE, ROYAL FIELD ARTILLERY, 29TH DIVISION.

"I had done my best, when going along the fosse of the "Old Fort," to save a badly wounded Turk from three of another battalion who were standing over him and discussing the advisability of putting an end to him, but I am afraid my interference was in vain."[778]

LT GEORGE DAVIDSON, 89TH (1ST HIGHLAND) FIELD AMBULANCE, RAMC, 29TH DIV.

"5 p.m… Moved into an outpost position extending from Worcesters on the left at Beach [W] to the Dublins on the right at Hill 141. Line held by Division — Hill 114 — Lighthouse on 138 — to Hill 141."[779]

CAPT. & ADJ. HARRY WILSON, 1ST BN ROYAL MUNSTER FUSILIERS, 86TH BDE, 29TH DIV.

---

773 *Northern Scot and Moray & Nairn Express*, 10th July 1915.
774 1st Bn Royal Munster Fusiliers War Diary, TNA WO 95/4310.
775 Major Cecil Grimshaw, DSO, 1st Bn Royal Dublin Fusiliers, was killed on 26th April 1915. Buried in V Beach Cemetery, he was the 40 year-old son of Thomas Wrigley Grimshaw, CB; husband of Agnes Violet Grimshaw, of "Grattons," Dunsfold, Godalming, Surrey.
776 *Nuneaton Observer*, 25th June 1915.
777 17th Brigade Royal Field Artillery War Diary, TNA WO 95/4308.
778 Davidson, *The Incomparable 29th and the River Clyde*, p. 46.
779 1st Bn Royal Munster Fusiliers War Diary, TNA WO 95/4310.

"[W]e attacked the Turks with the help of the naval guns, and gradually we dug them out of their holes with bayonet and shot. We fought all that day until about five that evening when the Turks broke and fled for their lives. Thus we gained our first victory and a good position. That night we dug trenches and made ourselves secure against attack, but John Turk had enough for one day, and left us alone."[780]

PTE EDWARD BARTON, 1ST BN ROYAL MUNSTER FUSILIERS, 86TH BDE, 29TH DIVISION.

With the fighting around the beach, fort and village effectively over, men took in what was left behind.

"The Dublins, whose officers I have associated most with, have only three of these left out of twenty-seven. I came across two of these to-day — Padre Finn, R. C. Chaplain, whom I knew well and greatly respected, I found at the edge of the sea, with his clothes thrown open exhibiting a wound in the chest. And in the village, all huddled up among long weeds and nettles I found a lieutenant who sat at my table on the "Ausonia" — Bernard.[781] In both cases death must have been instantaneous."[782]

LT GEORGE DAVIDSON, 89TH (1ST HIGHLAND) FIELD AMBULANCE, RAMC, 29TH DIV.

Those who had only observed it from the ships offshore were keen to see things for themselves. Some saw more than they bargained for.

"In the afternoon we received a signal that ammunition was needed, and presently a pinnace came along with a lighter in tow. Then there was turmoil. Every officer itched and clamoured to go. But O.C. troops was on board, and he went himself, and took my second in command with him. I silently consigned them both to a place which, after all, was probably cool and comfortable compared with the spot where they landed.

"They both came back safely towards sunset, and we gathered about them like schoolboys round a toffee-box. But they wouldn't talk. I believe they couldn't. Wilson said that it was indescribable, and that was every word that I could get out of him."[783]

CAPT. ALBERT MURE, 1/5TH BN ROYAL SCOTS (LOTHIAN REGT), HMT *DONGOLA*.

Some of the dead were taken out to sea for burial.

"To-day we had a naval funeral of General Napier and Colonel [Carington-Smith]. The former was killed on a barge attached to us, and the other on the bridge. No one is to be present but the Catholic padre. A number of men are to be buried at the same time. The orders I received stated that all bodies had to be got rid of before we advanced. A pinnace from a warship was signalled for and all were taken out to sea."[784]

LT GEORGE DAVIDSON, 89TH FIELD AMBULANCE, RAMC, SS *RIVER CLYDE*.

780 *Galway Express*, 26th June 1915.
781 Lt Robert Bernard, 1st Bn Royal Dublin Fusiliers, was killed in Seddülbahir village on 26th April 1915. Buried in V Beach Cemetery, he was the 23 year-old son of the Most Rev. and Rt. Hon. J. H. Bernard, D.D., and Maud, his wife, of Provost's House, Trinity College, Dublin.
782 Davidson, *The Incomparable 29th and the River Clyde*, p. 47.
783 Mure, Albert Haye, *With the Incomparable 29th*, pp. 32–33, W. & R. Chambers (Edinburgh) 1919.
784 Davidson, *The Incomparable 29th and the River Clyde*, p. 45.

"We were thankful we had the light of a splendid Oriental moon to cheer us up. Of course, we had had no sleep since Friday night, and it was now Monday, and we were all beginning to feel a bit jiggered. However, about 10 p.m., the concert started with a few preliminary rifle shots from the enemy. No one in the trenches minds rifle fire, as they are bound to come straight at the trench, and so long as you keep your head down you are as safe as if you were at home in bed, but as the firing increased in volume and our maxims came into play things began to look bit lively. After a while the Turks retired and left us alone."[785]

SPR HERBERT HEATH, 1/1ST (WEST RIDING) COY, ROYAL ENGINEERS, 29TH DIVISION.

W BEACH

After overcoming the near panic of the previous night, the priority here was for the establishment of an operational base.[786] Others tried to support their comrades over at V Beach.

"Dawn broke, and with it the respite we needed came. The enemy had evidently spent their strength in trying to hold us off, and we gripped the ground we had so hardly won. Going down to the beach in the morning I saw what it had cost to win 'W' Beach alone. The ground was one mass of death."[787]

AB ALBERT ROWSON, RNVR, DRAKE BATTALION, ROYAL NAVAL DIVISION.

"I awake about seven and find myself nestling up close against Foley, who is still asleep. I wake him, and he promptly falls asleep again, murmuring something about "that machine gun."

"The beach quickly becomes alive with men all working for dear life, and we get to our feet, go down to the water's edge and bathe our faces, and start to finish the work of making a small Supply depot which we left last night. My servant comes to tell me that breakfast is ready, and we go up the cliff and join Way and Carver at a repast of biscuits, jam, bacon, and tea. But the tea tastes strong of sea water. All water had been carried with us in tins, and we had struck a bad batch, for most of them leaked. And then our day's work begins in all seriousness…

"At frequent intervals the Fleet bombard, but we are quite used to the roar of the guns now. I am covered and coated with clayey mud and have no time to clean myself properly. We have to take cover continually from snipers — unknown enemies who fire at us from Lord knows where. One open part of the beach is especially dangerous, and I cross that part about six times during a day — not a very wide space, but I feel each time I go across that I am taking a long journey. The dead are still lying about, and as there is no time to bury them, we pass to and fro by their bodies unheedingly. In addition to these snipers who pick off one of our number now and again, we have

---

785 *Sheffield Daily Telegraph*, 3rd June 1915.
786 Orders to that effect were issued at 11.00 a.m. (29th Division General Staff War Diary, TNA WO 95/4304)
787 *Penistone, Stocksbridge and Hoyland Express*, 9th December 1916.

spent bullets flying in all directions, for our firing-line is but a few hundred yards away. The Turk, however, does not appear to have a proper firing-line; he only seems to have advanced posts strongly held, and must have retreated well inshore."[788]

CAPT. JOHN GILLAM, DIVISIONAL TRAIN, ARMY SERVICE CORPS, 29TH DIVISION.

"The Regiment was ordered to attack Hill 141 in conjunction with the… Munster Fusiliers, except "Z" Company who held their original line. Hill 141 was stormed at 2.30 p.m., this relieving the pressure on "V" Beach which had been held up since the morning of the 25th."[789]

CAPT. & ADJ. EDWARD KERANS, 4TH BN WORCESTERSHIRE REGT, 88TH BDE, 29TH DIV.

"Heavy firing sounds, however, from "V" Beach, a rattle of musketry and a roar of the battleships and torpedo-boat destroyers lying at the mouth. Colonel Beadon and Major Striedinger are getting a proper system of supply and transport working.

"We become venturesome in the late afternoon, and many of us, quite two to three hundred, go up on the high land on the right and left of the beach and make a tour of the lately captured trenches. Turkish dead are lying about in grotesque attitudes; the trenches are full of equipment, and I notice particularly bundles of remarkably clean linen, and many loaves of bread, one loaf sticking out of a dead Turk's pocket. Several of the dead are dressed in a navy-blue uniform with brass buttons, but most are in khaki with grey overcoats and cloth hats. Suddenly a whistle blows, and several cry "Get off the skyline!" and we all run helter-skelter for the safety of the beach."[790]

CAPT. JOHN GILLAM, DIVISIONAL TRAIN, ARMY SERVICE CORPS, 29TH DIVISION.

Y Beach had been abandoned; those at S, X and W beaches had been unable to make ground against modest opposition despite their preponderance in numbers; and V Beach had been a near catastrophe. But that was not how Hunter-Weston viewed the situation.

"Today the men on Beach V who had disembarked during the night, very gallantly led by Lt. Colonel Doughty Wylie, worked through Seddel Bahr village and on both sides of it: Hants. on the right and Dublins and Munsters on the left. This attack, excellently carried out and well covered by gunfire of the fleet, was entirely successful, and the fort was taken about 2 pm. The four pom poms on the left were captured en route. The success of the attack was due to Lt. Colonel Doughty Wylie was shot through the head as he approached the fort. Beach V is now safe except for a little occasional distant shell fire and our line is continuous from X to V. The French will land on Beach V on their arrival. The result of day's operations is therefore satisfactory. My line is thin and the men tired, but we shall be ready to push forward as soon as the French are in line and we have weight enough for further advance…

"Fighting all along front of Division last night and today. Attacks during the night were repulsed by rifle fire and gun fire from ships. The troops landed at Y Beach

---

788   Gillam, *Gallipoli Diary*, pp. 43–44.
789   4th Bn Worcestershire Regt War Diary, TNA WO 95/4312.
790   Gillam, *Gallipoli Diary*, p. 44.

did not effect junction with troops at X Beach, and, being hard pressed, were re-embarked at Y and brought round to main anchorage off W this afternoon. The troops at S are maintaining themselves successfully. The troops at X and W are well situated in a continuous position which enemy attacked during the night and again twice by day. They were driven off each time with heavy loss and our line pushed forward. The capture of the redoubt on Northern side of hill 138 by the Worcesters was an excellent piece of work. The losses on V Beach on landing yesterday were very severe, many boat loads being entirely wiped out, both soldiers and sailors by machine gun fire from both flanks and from pom poms in concealed positions near Fort No. 1. The "Wooden Horse" *River Clyde* saved the situation, the men being able to stay in it under cover until nightfall. Commander Unwin and Midshipman Drury of the *River Clyde* did marvels of work and valour."[791]

MAJOR-GENERAL SIR AYLMER HUNTER-WESTON, COMMANDING 29TH DIVISION.

The gallantry of the men was not in question. But much more than that would be needed.

ANZAC

"During the night one brigadier asked if he might tell his men that our guns and howitzers had been landed, as it would cheer them greatly. Another asked urgently that all available guns, both of the ships and on shore, should be brought to bear on the enemy's artillery. Only one of our Australian 18-pounder field guns had been landed so far — that was brought ashore about half-past 2 in the afternoon, and was hauled by a team of artillery-men up a steep slope into position. There the men dug themselves gunpits in the heath, under fire all the time. At about 5 o'clock, the infantry out on the hilltops had been cheered to hear a solitary bang alternating with the six quick bangs of the mountain gun salvoes. The troops were eagerly looking to hear a dozen bangs, instead of that one in the morning. But down on the beach, as the night wore on, one could not help feeling anxious, for, although hour after hour the artillery staff awaited them on the beach, none of the barges that came from the transports contained guns. Fresh troops were landing and marching off — we must have been at least four battalions stronger at daybreak than we were at sun down the night before. But there was not a sign of a gun. Water and ammunition were being sent up from the beach all the time — the scene was one of restless motion under the half-moon."[792]

CHARLES BEAN, AUSTRALIAN WAR CORRESPONDENT.

"At 1.30 a.m. I took my place on picket and at the same time they were calling for volunteers to row four boats ashore to take Red Cross reinforcements owing to heavy casualties following on a night attack by the Turks. I tried to induce several to do my picket for me, so that I could volunteer, but there were too many men offering for the trip ashore."[793]

GUNNER JOHN PARK, 3RD BTY NEW ZEALAND FIELD ARTILLERY, HMT *CALIFORNIAN*.

---

791 29th Division General Staff War Diary, TNA WO 95/4304.
792 *The West Australia* (Perth, Western Australia), 12th July 1915.
793 *Manawatu Standard* (New Zealand), 7th July 1915.

> "At … 5 a.m. we take on more. Three hospital ships have already sailed for [Alexandria] … The weather has been simply perfect, and yet there are piles of wounded lying on the beach. All ships' boats have been ordered ashore in case our boys have to retire. What a sad day this has been to us on board… what pitiful sights they are! Most of them suffered from shrapnel."[794]
>
> L/CPL CLARENCE UMBERS, NZASC, ATTD HQ, NZ & A DIVISION, HMT *LUTZOW*.

Ordered up from Lemnos, the *Ionian* arrived offshore that morning to assist with the evacuation of the wounded.

> "Items of news began to drift in. First one and then another of our comrades was reported wounded or dead or missing as the case might be. Later we found that a number of these reports were not true, but the excitement was intense when the first contingent of wounded came alongside. Among them was Major Beevor, of the 10th, with a comparatively slight wound in the foot. The wounds were of varying degrees of severity, and one poor fellow breathed his last just before the boats got to us….
>
> "A section of the clearing hospital had been left on board, but not sufficient to cope with the immense demand on the surgical and nursing resources of the ship. Then was seen a glorious sight! Our fellows have proved themselves "handy" men at various times, and now they are called upon to prove their versatility in a new direction. While the doctors and their trained corps give immediately attention to the more serious cases transport men, members of the hold party, soon proved themselves adepts in caring for their wounded comrades."[795]
>
> COLONEL CHAPLAIN JOHN McPHEE, CHAPLAIN'S CORPS, AIF, HMT *IONIAN*.

> "I went along to have a dressing put on my arm about 9 o'clock… — Captain Butler,[796] medical officer of the 9th battalion, had a dressing station in a gully not far from our trenches (they were not the elaborate trenches you see pictures of as used in France and Belgium, you know — they take time to dig). Butler sent me on down to the beach, as the bullet was still in the arm — found the several field ambulances hard at work there, also the A.S.C. getting food, ammunition, and water ashore (no water available where we landed), and a nice old time they were having — horses, guns, troops, all being landed as fast as ever the navy launches could tow them in, and every few minutes, bang-bang-bang, and shrapnel pellets would fly all over the foreshore and water, barges and boats full of wounded were also being towed back to the transports anchored off shore; our friend the enemy knew where to drop his shells. A steep little ridge of sand ran up immediately from the beach, and this gave some protection to the wounded in the temporary dressing stations — but not much…

---

794 *Evening Star* (New Zealand), 23rd June 1915.
795 *Daily Herald* (Adelaide, South Australia), 21st July 1915.
796 Capt. Arthur Butler, 9th Bn, AIF, would be awarded the DSO: "During operations in the neighbourhood of Gaba Tepe on 25th April, 1915, and subsequent dates, for conspicuous gallantry and devotion to duty in attending wounded under heavy fire, continuously displaying courage of a high order." (*London Gazette*, 3rd July 1915). Promoted Lt-Col., he eventually commanded 3rd Australian General Hospital in France. Butler would later write Australia's official medical history of the war.

"They wouldn't chop my bullet out on shore, so I had to journey out to one of the ships. Dr. Campbell (of Hobart) got it out about 8 p.m... 'twas quite easy, as it had gone right through and was just beneath the skin on the other side."[797]

CAPT. JOHN WHITHAM, 12TH BATTALION, 3RD BDE, 1ST AUSTRALIAN DIVISION.

"The morning was beginning to break, and the Turkish artillery with all the reinforcements that would probably be got into position during the night might be expected to open at any moment. Artillery officers were waiting on the beach — but still no sign of the guns. At last — about 4.30 — with the light growing fast, some of the New Zealand guns were landed. Ages seemed to pass, and then the punts came in containing animals. Two were full of mules, but the third contained gun horses for our guns. During that day five of our guns came ashore. Twelve men of the artillery headquarters had made a road over the worst of the slopes. The guns were taken along the beach by gun teams and hauled into position across hillsides and up steep pinches by teams of artillery on to a position from which they afterwards did tremendous work sometimes at a range of fifty yards or less."[798]

CHARLES BEAN, AUSTRALIAN WAR CORRESPONDENT.

"The shore in front gradually opened up as the sun rose, although, shining as it did, directly in the eyes of the ships' gunners, they were not in a position to support the attack in the early hours of the morning.

"It was then discovered that the boats had landed rather further north of Gaba Tepe than was originally intended, at a point where the sandstone cliffs rise very sheer from the water's edge. As a matter of fact, this error probably turned out a blessing in disguise, because there was no glacis down which the enemy's infantry could fire, and the numerous bluffs, ridges, and broken ground afford good cover to troops once they have passed the 40 or 50 yards of flat, sandy beach...

"It is indeed a formidable and forbidding land. To the sea it presents a steep front, broken up into innumerable ridges, bluffs, valleys, and sandpits, which rise to a height of several hundred feet. The surface is either a kind of bare yellow sandstone, very soft, which crumbles when you tread on it, or else is covered with very thick shrubbery about 6 ft in height."[799]

ELLIS ASHMEAD-BARTLETT, BRITISH WAR CORRESPONDENT, HMS *LONDON*.

"About two miles down along the beach we could see our dead where we had left them the previous day... At once the stretcher bearers (New Zealanders) made for him with the assistance, of some of our chaps who were not on duty. They succeeded in getting him, but the snipers, who were plentiful in that position, made it hard. We then learnt that there were still another seven lying out wounded; two had died during the night. The New Zealanders at once proceeded to the place with stretchers. One of them was killed and two wounded in the attempt to rescue them. They succeeded in getting all

---

797 *The Critic* (Hobart, Tasmania), 2nd July 1915.
798 *The West Australia* (Perth, Western Australia), 12th July 1915.
799 *The Referee* (Sydney, New South Wales), 23rd June 1915.

the wounded who were suffering from the night out. One to die was Sgt Collins[800] ... and had he been taken away his chance was very bright. He had some bad wounds, but I have seen them since get better with worse."[801]

L/CPL JOHN KIELY, 7TH BATTALION, 2ND BDE, 1ST AUSTRALIAN DIVISION.

"Our battalion landed about 15 hours after the fight began, and as soon as we left the ship, the shrapnel was falling all around us. We got ashore with our boat with the loss of one man killed and one wounded…

"Boats, filled with dead were lying all along the beach. We buried 50 men in one grave; they belonged to the 7th Battalion; poor chaps, they never got ashore."[802]

PTE ALFRED POWER, 14TH BATTALION, 4TH BDE, AIF, NZ & A DIVISION.

"Whilst being towed in we had shrapnel bursting all around us. Arrived at the beach safely and stayed that day. Whilst there we buried a lot of the chaps that had been shot on the beach and in the rowing boats."[803]

PTE JOHN STRAUGHAIR, 14TH BATTALION, 4TH BDE, AIF, NZ & A DIVISION.

"By the morning… the whole of my division was on shore, including a howitzer battery, which came into action. The *Queen Elizabeth* and other battleships were raking the Turkish positions beyond the head of Monash Gully, and the situation looked definitely more hopeful. The landing had been very disappointing. First, there was the loss of direction, which resulted in landing at the wrong place. Then, the irregular arrival on shore of the troops — some from one ship, some from another, just as they could be despatched — and a mistake about the landing of guns which resulted in their being sent back to the ships, had further contributed to the difficulties of consolidating, and properly reinforcing, the positions originally gained."[804]

MAJOR-GEN. SIR ALEXANDER GODLEY, COMMANDING NZ & A DIVISION.

"By dawn… I was on top of the ridge over our left flank. Turkish artillery was soon pounding us, but a few well-placed 15-inch shells from the *Queen Elizabeth* quickly silenced the enemy. During the night we landed two Australian 15-pounder batteries; then our precious New Zealand howitzers arrived; and with the rising sun the spirit of the weary men revived miraculously. Gradually and with great difficulty I wormed my way along to our various posts in the scrub, never knowing for certain where the next post might be, and conscious that it would be all to easy to end up in the Turkish lines instead."[805]

LT-GEN. SIR WILLIAM BIRDWOOD, COMMANDING ANZAC.

800 L/Sgt Alfred Collins, 7th Bn, AIF, was wounded while attempting to reach the beach in the morning of 25th April. He died of those wounds later that day or early the next. He is commemorated by a special memorial in No. 2 Outpost Cemetery.
801 *Echuca and Moama Advertiser and Farmers' Gazette* (Victoria), 21st October 1915.
802 *The Birchip Advertiser and Watchem Sentinel*, 23rd June 1915.
803 *The Bendigo Independent* (Victoria), 23rd November 1915.
804 Godley, *Life of an Irish Soldier*, p. 173. A combination of congestion and concern that the whole position might be evacuated helped to delay the landing of guns on 25th April.
805 Birdwood, *Khaki and Gown*, p. 261.

As the work to remove casualties continued, so more men landed. Some of the most important were the artillery and those who supplied them.

> "[A]t 7 a.m... a trawler came along with some pontoons, which were immediately loaded with one 18-pound gun, and two ammunition carriages, and then got orders to dress for going ashore. I think in all we had about fifteen minutes to get aboard. There were about 15 artillerymen and 25 A.A.S.C. men, and I was one of the latter to go ashore. Everything went well till we got about 50 yards from shore, and then we could feel the bullets flying all round us, and the steam pinnaces cast us of to get ashore. It was impossible for us to move, so the landing officer called out to the steamboat to tow us in closer. They did so, and we were able to throw a line out to shore, and so were hauled in. We landed all right, and there was a team of eight horses waiting to hook on to the gun which we had in the pontoon. Orders came to take our gear off and help to pull the gun ashore. Every man put his weight on the rope, and we got it ashore with the ammunition carriages. Then we proceeded along the beach in search of our Supply Depot, where our Colonel was anxiously waiting our arrival.
>
> "Our first job when we got there was to level the beach so that we could store some cases and start a supply depot. Lieut. McHattie[806] was showing us what we had to do, when a shell burst, and a piece of shrapnel hit him in the leg. He was taken straight to the hospital, and treated, but it was only a slight wound, so he returned when he had been attended to."[807]

CPL ARTHUR WOOLMAN, DIVISIONAL TRAIN, AASC, 1ST AUSTRALIAN DIVISION.

> "Heavy bombardment commenced this morning by the battleships. The *Queen Elizabeth*'s guns silenced the Turkish batteries in an hour and a half. Noise of the guns is deafening. Heavy casualties from all battalions are reported. Those in the trenches are under continuous shrapnel and rifle fire. Some of the artillery has been landed and placed in concealed positions."[808]

PTE APCAR DE VINE, 4TH BATTALION, 1ST BDE, 1ST AUSTRALIAN DIVISION.

Inland, the fighting continued and more men had to be taken to the beach.

> "It was Monday morning when I got my dose of lead, and of course that stopped my share in the fight. At this time we were on a ridge, and had been subjected to a heavy fire for some time, and it was too hot for stretcher bearers to be near, as I had either to stay there or drop over the ridge into a gulley. This I did, and after sliding and rolling about 50 ft. I stopped in a narrow hollow in a water course down the hillside. Here two medical men found me and dressed me up, and went away for a stretcher. It took five of them to get me from there to the beach, so you can have an idea of the rough nature of the country, and of the hard work of the stretcher bearers here. However,

---

806 Lt, later Captain Donald McHattie, AASC, died of wounds while serving with 3rd Divisional Train on 17th July 1917. Buried in Kandahar Farm Cemetery, Belgium, he was the 24 year-old son of James Cameron McHattie and Frances McHattie, of 4 Hunter Street, Newcastle, New South Wales.
807 *Leicester Daily Post*, 30th December 1915.
808 *The Sun* (Sydney, New South Wales), 25th April 1935.

everybody worked like heroes and no doubt deserve all the praise I understand has been given to them."[809]

PTE WALTER WRIGHT, 11TH BATTALION, 3RD BDE, 1ST AUSTRALIAN DIVISION.

"All the time the Turks kept firing at us both with shrapnel and rifles, as hard as they could shoot, but our crowd hung on like Britons. Sunday night was as bad, not a wink of sleep, expecting to be attacked every minute, but, as luck would have it, they would not tackle us with the bayonet. Monday morning was just as bad, and at about 9 a.m. they got me in the right shoulder, which finished me as far as fighting was concerned. Henderson-Smith[810] was with me at the time, and with a good deal of risk to himself from bullets, he got me out from the firing line-down to the ambulance people."[811]

PTE CHRISTOPHER FORREST, MG SECTION, 11TH BN, 3RD BDE, 1ST AUSTRALIAN DIVISION.

"About 10 o'clock on Monday morning I was having a smoke when all of a sudden my eyes went black, and when I woke again I was one mass of blood. I got down somehow, and one of the R.A.M.C. men bandaged me. I was then taken to the beach, and shipped on an hospital ship for Alexandria."[812]

PTE ALFRED ISACKSON, 2ND BATTALION, 1ST BDE, 1ST AUSTRALIAN DIVISION.

"All day long wounded continue to be brought on board. A lot of New Zealanders are now coming on. Two naval doctors have been on all day, and worked like heroes. The Navy men deserve great credit. Although they are in no great danger on the warships that are shelling the land, the officers and men on the destroyers and pinnaces come under fire whenever they approach the shore. Some have already been shot dead. We are not now seeing the smart, clean-shaven naval officer, but an unshaven, clothes-soiled man, who still, however, uses plenty of "Mister," and "Would you mind doing such-and-such a thing, please?" Some of the sailors doing wounded-carrying have not had food for 23 hours. They are not growling. See them handle the dying and wounded with the care of a woman! My word! What a smack for those anti-navy preaching people!... During the day nine wounded died, and were transferred to HMS ——— for burial. We few on board have got all the looking after of these poor souls. We now have over 500 wounded on board. As soon as we can get some doctors from shore we are to steam at full speed for ———. There is not much sleep for us at night, as the moaning of the suffering is so loud."[813]

L/CPL CLARENCE UMBERS, NZASC, ATTD HQ, NZ & A DIVISION, HMT *LUTZOW*.

The Turkish artillery was an enemy it was difficult to fight. Counter-battery fire was practically impossible; no-one could spot where the guns were.

"The country is hilly, but not very rough, and is covered with light scrub and trees.

---

809   *Bradford Weekly Telegraph*, 18th June 1915.
810   Commissioned on 5th May 1915, 2/Lt George Holt Henderson Smith, 11th Bn Australian Infantry, was killed in action on 25th May 1915. He is buried in Lone Pine Cemetery, Anzac.
811   *The Times* (Perth, Western Australia), 13th June 1915.
812   *Evening News* (Sydney, New South Wales), 19th July 1915.
813   *Evening Star* (New Zealand), 23rd June 1915.

Where we are landing hills rise straight up from the beach and are covered with scrub, affording splendid protection for guns and trenches. It abounds with small gullies, and this is where the enemy have their guns hidden — so well hidden that you cannot make out even the flash as they fire. Out to the left is a nice bit of country gently rising from the beach, but the authorities evidently had their reasons for not landing us there...

"The aeroplanes are a very important asset, although it is a very difficult matter even for them to locate the enemy's batteries. It is a fine sight to see them skim along the water and then rise gracefully into the air, circling round over the sea until they have reached a sufficient height and then flying away over the Peninsula."[814]

GNR JOHN PARK, 3RD BATTERY, NEW ZEALAND FIELD ARTILLERY, HMT *CALIFORNIAN*.

"The amount of ground that we had occupied was a mere cheese-bite. The total perimeter was only three miles. The Turkish trenches were only about 1,000 yards from the beach... and our guns were in action and firing over open sights, at that range. Our positions were about as bad as they could be. The men were hanging on by their teeth and eyelids to the top of the scrub-covered hills, and were digging, for all they were worth... Fortunately the fire of the *Queen Elizabeth* and *Bacchante*, both of which... stood close in and raked the head of Monash Gully and Gaba Tepe, gave us something of a respite."[815]

MAJOR-GENERAL SIR ALEXANDER GODLEY, COMMANDING NZ & A DIVISION.

"We learnt that of our 32 officers, only six were left on the field. Heavy shrapnel fire took place all day, especially toward nightfall; but the splendid design of our trenches saved us from much loss. But some of us had to take chances in carrying ammunition or water. Our warships were now able to support us. You should hear the "Lizzie" [HMS *Queen Elizabeth*] let go. The very earth trembles."[816]

PTE GEORGE GIDEON, 6TH BATTALION, 2ND BDE, 1ST AUSTRALIAN DIVISION.

The support of heavy naval gunfire, doubtless, boosted morale but the men in the line needed food, ammunition and water.

"Then the Colonel came along, and said he wanted several men and two N.C.O.'s, and I was one of the N.C.O.'s detailed for the task. The Colonel said he wanted us to take 16 mules each and find out where the headquarters were. There was an Indian soldier in charge of every three mules and every mule had eight gallons of water on it. There were three Australian soldiers and myself, and we had not the faintest idea where we should find them, but we proceeded up Shrapnel Gully, and had not got many yards up before we found we were in the danger zone. I can tell you it was a living hell, what with snipers and shells that were flying about, and the stray bullets that whizzed by us. I really thought I should not see the beach again, but I made a way up the gully. We

---

814   *Manawatu Standard* (New Zealand), 7th July 1915.
815   Godley, *Life of an Irish Soldier*, p. 173.
816   *Flemington Spectator* (Victoria), 23rd September 1915.

passed dead, and wounded men in dozens, the latter coming down from the trenches. Unfortunately we two N.C.O.'s did not agree with the road, so he took half and I took the remainder, and I made my way to the top of the gully.

"I got three-parts of the way up, and saw a road through some bushes, where there were dozens of cables laid. Thinking it would be a good road to try, I went up, and got about 50 yards up. The mules got tangled in the cables, and some of them rolled down the hill, so I had to stop and think which would be the best to do. I got the mules out of the cables, and picked up those that had rolled over, and put their water receptacles on them again, and returned into the gully, going about 500 yards further up. There I met a lot of tired-out Australians, who asked me where I was going with the water. I told them, but they said there was no chance of getting mules up there, and that they could do with a drink. I gave them all some and then came another mob of men, who said they were looking for water for the 1st Brigade. I told them I had some for them, and handed it over to the warrant officer of the 1st Australian Brigade who conveyed it up some steep hills into the trenches. Then I retuned with my mules to the beach, and reported that I had delivered it all right. The officer immediately sent me up again with another load, which I took to the same place. After a good feed of bully beef and and biscuits, and a drink of tea that had been made of condensed [?] water, which was salty, so that you could not take much of it, I got my mules loaded with ammunition and delivered that to H.Q."[817]

CPL ARTHUR WOOLMAN, DIVISIONAL TRAIN, AASC, 1ST AUSTRALIAN DIVISION.

"After a cold and uneventful night, the day broke fine, and found the troops still clinging to the high ledges that had been so hardly won… Every hour more troops and guns, ammunition, donkeys, and mules were landing, and the beach began to reproduce the scene I witnessed at Mudros. Gear and fodder and food were piled up everywhere, turning the outlandish place into a kind of seaside holiday resort…

"Away to the right the troops had advanced about 800 yards and a field-battery of howitzers had been mounted in a good position, from which they were able to blaze away at Turks hidden behind a ridge where the ships' guns could not penetrate…

"About noon… the wounded from the "Majestic" were transferred to a hospital ship which steamed away…, the men giving three cheers for the "Majestic" as they went off."[818]

DECK HAND JOHN COWIE, RNR, HMS *MAJESTIC*.

"About midday… I was wounded. The Turks sent a party on our right with a machine gun and rifles. There we were enfiladed. First of all I got a smack on the heel. I thought it went right through, but it only took a piece clean out of my boot. It bruised my heel a bit. About half an hour later I got an awful smack on the left thigh. It was just as if someone gave me a crack with a big stick. I put my hand down and soon felt the warm blood. I could not move my leg an inch. It was impossible to get up and try to put my field service dressing on it. So far, so good. About a quarter of an hour later

---

817   *Leicester Daily Post*, 30th December 1915.
818   John Cowie quoted in Goodchild, George, *The Last Cruise of the Majestic*, pp. 126–128, Simpkin, Marshall, Hamilton, Kent & Co. Ltd. (London) 1917.

the ping of a rifle bullet went past my right arm 'stern.' The missile went in, but after travelling for about two inches passed out again. It then became impossible for me to resume shooting."[819]

L/CPL JAMES WEDERELL, 3RD BATTALION, 1ST BDE, 1ST AUSTRALIAN DIVISION

"On the Monday we had to take some of our guns ashore. We could not land horses, so every available man had to land to draw the guns up the steep hills into position. When we got ashore — awful sights met our eyes — everywhere dead and wounded lay in thousands. We soon got to work hauling the guns into position. Our work was not yet done — we had to carry ammunition to the guns. You cannot imagine what action is like, as you are fighting and carrying ammunition. All at once someone would drop or fall forward, shot or wounded, and cry out "I am done!" or "My God, I am hit!" Some would drop without a word, dead. The worst of all is that if you see your best comrade fall the rules are you cannot stay to help him — pass him by, for it is the A.M.C. work to look after the wounded. It seems hard to go by a comrade without trying to help him."[820]

DVR CHARLES HEMMING, NO. 1 COY, AASC AIF.

"We entered the firing line under a hail of shrapnel and machine gun bullets, and many of our fellows were shot down. I threw myself flat into a dug-out, and the machine guns were hitting up the dust all around me. When I received my first hit in the leg it was like a big electric shock, and I cried out, "I'm hit." The pain suddenly went away and I thought it had been just a graze, and so I looked round for some Turks to have a 'whang' at. I couldn't see anything but scrub until l noticed something moving just ahead. I let drive but didn't know the result. I happened to glance down at my leg, and found the ground covered in blood. Of course I thought I was seriously hurt, and so I got off my equipment and piled up my ammunition on the ground and made to crawl back. Just as I left my dug-out the machine gun which had our whole line enfiladed opened again and I received the second shot. It caught me in the muscle of the foot, and my leg, drew up under me and quivered like a leaf. The pain was like a prolonged high voltage shock. I lay for about 10 minutes, with the blood pouring out of my boot, and then essayed a move again. I crawled painfully, elbowing my way forward, and dragging my legs like two wet socks after me. I must have lost my direction, for I landed on the edge of a gully. There I stopped, and applied my first field dressing to the foot wound and cut away the clothes from the knee wound. I rested, and then plunged into the gully head first, brushing aside the brambles as I went. I never thought I'd get there but after about two miles — it seemed to me two miles — I came across the 3rd Field Ambulance. There they dressed the wound and laid me in the shade. Wounded men were coming down like flies, and presently the Turks began sweeping our gully with shrapnel. They burst all around us, but fortunately none was hit. It was getting too warm so the doctor sent all who could walk on. A stretcher-bearer offered to carry me, as no stretchers were available. For two miles

---

819   *Dubbo Dispatch and Wellington Independent* (New South Wales), 11th June 1915.
820   *Nepean Times* (Penrith, New South Wales), 7th August 1915.

he carried me down the bed of a muddy creek falling heavily with me twice, and I was glad when on reaching the more open parts to meet a stretcher party. From here I went to a rest hut and there saw some of our unlucky fellows."[821]

SGT ERIC EVANS, 13TH BATTALION, 4TH BDE, AIF, NZ & A DIVISION.

At 4.00 p.m. 4th Battalion was ordered to attack. How it came about, what it was supposed to achieve was unclear. No objective was set and things quickly became confused.

"I do not know that the exact facts of this charge can even now be fully ascertained, because many of those who alone know them have been killed. But this much is fairly definite.

"General Bridges visited the line in this part early during the second afternoon. He approved of it, but suggested that two platoons should be moved up in one part of it in order to strengthen and straighten a short stretch of it. After he had left this order was about to be carried out. What happened next is not clear. But somehow or another the order that came along the line was one for a general advance. We learnt during those first two days that the enemy was very clever at getting such orders passed verbally along our line, and it may have been some such occurrence, or it may have been some mistake in the passing of our own order. Anyhow, what happened was that the 4th Battalion dashed forward most gallantly, under the impression that the whole of our line was moving forward. It charged right through two valleys — through a small Turkish camp — charged with the bayonet through such part of a line of Turkish infantry as waited for it; and only brought up when it came out on an open hillside, with machine guns somewhere in front of it — a battery of artillery not a thousand yards away — and rifle fire coming at it from all sides from an enemy which it could not see. It hung on there for a few minutes, when an order was given by someone to retire. There was the natural confusion, but the battalion came back without losing its head."[822]

CHARLES BEAN, AUSTRALIAN WAR CORRESPONDENT.

"[A] general advance of the whole line was ordered. Everyone moved forward, and soon we had met with a withering fire of machine-guns, shrapnel, and rifle fire. It was terrific. I was lying with my signalling officer, Lieut. Smith, and a machine gun played all over us. He was wounded in the knee. When there was a lull I got him back to cover. We advanced with the colonel, and everyone was ordered into the firing line. I collected a platoon of stragglers and moved through the hail of lead. I felt no fear, and was only anxious to get forward. Soon the order to retire was given, and I was one who got back to the trench without a scratch. Our colonel was killed retiring,"[823]

---

821 *Sydney Morning Herald* (New South Wales), 16th July 1915.
822 *The West Australia* (Perth, Western Australia), 12th July 1915.
823 Lt-Col. Astley Thompson, VD, 4th Bn, AIF, was killed on 26th April 1915. Buried in 4th Battalion Parade Ground Cemetery, he was the 50 year-old son of Astley and Udea Thompson, of Glamorgan, Wales.

and the second in command[824] was badly wounded. My officer bound up his knee, collected a few men, and charged forward to his death.[825] He was brave to a degree. We lost the pick of our non-commissioned officers."[826]

SSM JOHN SLOAN, 4TH BATTALION, 1ST BDE, 1ST AUSTRALIAN DIVISION.

"Our adjutant died with fifteen bullets in him. The colonel (Colonel Onslow Thompson) had the top of his head blown off by a shell.

"Eight officers, including ours, went down in half an hour. We were firing as fast as we could load. My rifle was red-hot, and I had to wait for it to cool, as it had jammed. I could not see my foresight for the heat waves rising from it.

"A mate alongside me had his head blown off, whilst another lost his arm… The machine guns were chipping the twigs from the bushes a foot above our heads, and we dared not lift our heads an inch. We have no other officer or non-com. in charge, as they have been shot.

"We have found that instead of being fired on from 650 yards they are only 100 yards in front… The scrub is so dense that we can't see what we are firing at.

"Just at this minute a shell landed in amongst twenty of us and put ten out of action. A lump of shrapnel went into my back and a bullet through my leg, which was burning and helpless. I had to crawl a mile and a half from the firing line to the A.M. Corps…

"The wounded were being taken away in dozens, and I am afraid the stretcher-bearers and the A.M.C. will be unable to stand the strain much longer. From here I was taken by stretcher to the hospital pontoons alongside the piers that have been built by our engineers."[827]

PTE ALAN EDWARDS, 4TH BATTALION, 1ST BDE, 1ST AUSTRALIAN DIVISION.[828]

"We were ordered to advance, and did so, but, after suffering very heavy casualties, had to retire to the original line. We lost many officers and men — Colonel Thompson killed, and Major McNaughton wounded.

"It is reported that the English and French troops are hourly expected, and we are told that we must hang on at all costs until they arrive…

---

824 Major Charles MacNaghten, 4th Bn, AIF, was wounded on 26th April 1915. He returned to the unit on 9th June and assumed command on 14th July. Evacuated on 10th August due to wounds and dysentery, MacNaghten resumed command on 6th December but was admitted to hospital again on 20th December 1915. He returned to Australia, leaving Suez on 10th June 1916.

825 Lt Muir Paul Smith, 4th Bn, AIF, was killed in action on 26th April 1915. Commemorated on the Lone Pine Memorial, he was the 24 year-old son of Walter Alexander Smith and Grace Anne Smith, of "Selhurst," 368 Alfred Street, North Sydney.

826 *Casino and Kyogle Courier and North Coast Advertiser* (New South Wales), 3rd July 1915.

827 *The Star* (Christchurch, New Zealand), 17th June 1915.

828 Edwards, a former clerk from Sydney, was evacuated to Egypt. He returned to his unit on 19th July but was evacuated sick on 15th August, returned to Egypt and left Suez to return home on 17th September 1915.

"Many reports are about that the Turks have been abusing the white flag and have shot a lot of our men, but did not observe anything myself. As a matter of fact, we all agree that the Turks play the game fairly.

"We have been holding our own all day along the battlefront. Everything is in a state of mix-up, and nobody can find his company: so we just link up with anybody we happen to be near."[829]

PTE APCAR DE VINE, 4TH BATTALION, 1ST BDE, 1ST AUSTRALIAN DIVISION.

"[At] 5 p.m. on Monday, when my platoon, which, besides being in the thick of the fighting, had been carrying water and ammunition up to the firing line since we landed, was ordered to go forward. Here's where my game ended. Shrapnel burst about ten yards to my left, and all at once I felt a stinging pain on the ankle, then the blood began to flow. How I got back to the ambulance is a miracle. You see some awful sights on a battlefield, and they make you turn sick at your first experience. Men are lying dead with their heads blown to pieces, their legs blown off, and their bodies mutilated by shells."[830]

PTE FREDERICK WRIGHT, 4TH BATTALION, 1ST BDE, 1ST AUSTRALIAN DIVISION.

"Still collecting wounded under fire. Sniping very bad. Mountain batteries and artillery ashore. Cruisers very active. About 250 casualties. Had a good rest at night; much quieter."[831]

PTE SYDNEY PENHALIGON, 3RD FIELD AMBULANCE, AAMC.

"We had all Monday for a spell in the same place until four o'clock in, the afternoon, when word came to our officer that the 16th Battalion wanted two gunners I was chosen as one. My first experience in the firing line made me feel funny especially when my mate got shot through the body and they dragged him out dead. But that day made me an old soldier, and l was not pleased unless the Turks were having a shot at us. I had a fair run at some Turks, but it was not enough to please me."[832]

PTE JACK ADAMS, MG SECTION, 15TH BATTALION, 4TH BDE, AIF, NZ & A DIVISION.

"Great work done by Third Brigade, who were terribly cut up. First gun mounted and doing good work — New Zealand 4.5 howitzers 9 a.m.; Australian 18-pounders 7 p.m."[833]

CSM ALBERT TURNER, DIV. SIGNAL COY, AUSTRALIAN ENGINEERS, 1ST AUSTRALIAN DIV.

"I could not twist one way or the other, and had to wait patiently for darkness to come on to enable the stretcher-bearers to come to my aid, which they did about 8 o'clock. The wound on my left though was the worst. The bullet went in just about level with

---

829 *The Sun* (Sydney, New South Wales), 25th April 1935.
830 *The Press* (New Zealand), 18th June 1915.
831 *The Queenslander* (Brisbane, Queensland), 21st April 1917.
832 *The Maitland Daily Mercury* (New South Wales), 24th July 1915.
833 *The Brisbane Courier* (Queensland), 19th August 1915.

my knee. It travelled upwards, but luckily it came out about half way up the thigh, leaving a hole large enough to put an egg in."[834]

L/CPL JAMES WEDERELL, 3RD BATTALION, 1ST BDE, 1ST AUSTRALIAN DIVISION.[835]

"There were constant attacks against most parts of our front [that] night, and this night the enemy's shrapnel did not altogether cease with nightfall. Several times during the early hours of the night he burst shells over the sea, apparently under the impression that this would hinder any disembarkation that might be going on under cover of nightfall. The next day, for the first time, he began really to rain shells upon the beach and the sea just outside it — as many as a dozen smoke clouds of shrapnel shell would be flying in the air at once. But never for one moment, as far as I saw, was the constant work of bringing troops and stores in from the transports in lighters and ships' boats interrupted for one minute by this fire, either by day or night."[836]

CHARLES BEAN, AUSTRALIAN WAR CORRESPONDENT.

"We left our ship, an old coolie boat called the *Surada*, on the Monday night, [26th April 1915] in a flat-bottomed boat with two of our guns, gunners, and about 300 rounds of ammunition, another punt with a similar load following us. When we got near the beach we found that the congestion of mules, men, stores, and so on, was so great that we stopped a few hundred yards from the shore, tied up to a small trawler, and all night long the bullets were whistling over us, sometimes missing us by an inch. First thing in the morning we landed, and no sooner got our guns ashore than the Turks started a terrible bombardment of shrapnel all over the beach. We had to take cover at once, and how half of us were not wiped out there and then I cannot make out yet. Not one of us was hit. When at last the fire slackened, with the help of some infantry men we dragged A and B guns up the hill at Anzac cove, and took up our first position. Our battery, No. 2, was the first New Zealand 18-pounder gun ashore, and my gun, A, was the first to fire at the Turks, and by all reports that shot found its mark and accounted for a few of the enemy."[837]

GNR WALTER LESTER, NEW ZEALAND FIELD ARTILLERY, NZ & A DIVISION.

"I laid down to rest that night, wondering how it was possible to be hit like I had been and to be still alive. I thought myself very fortunate. All the time the enemy's shells were popping over in great style, and after tossing about for some time I at last fell asleep. I had not been in the land of slumber very long when a shell landed just over my head and buried one in about three feet of earth. I took a long breath, and said to myself, "Gordon, old man, you can thank your stars you are still in the land of the living." I set about to free myself, for everyone was too intent on looking after their own skins to think of noticing anything so paltry as a man being buried. It was some

---

834 *Dubbo Dispatch and Wellington Independent* (New South Wales), 11th June 1915.
835 Served as Prior, the former labourer was killed between 7th and 12th August 1915. Commemorated on the Lone Pine Memorial, he was the son of Charles and Maria Wederell; husband of Margaret Wederell, of "Shanballa," Rose Street, Timaru, New Zealand.
836 *The West Australia* (Perth, Western Australia), 12th July 1915.
837 *The Press* (New Zealand), 16th November 1915.

minutes before I could free myself, for both arms were penned down, but at last I emerged into the open, stood up, took a deep breath, but suddenly ducked, for I heard the same old crack of the shells. It was then that I found that I was shaking like an aspen leaf, though in no other way hurt. At last from sheer exhaustion I fell asleep, huddled in a corner like a bundle of rags."[838]

SGT GORDON BAIN, NEW ZEALAND FIELD ARTILLERY, NZ & A DIVISION.

EGYPT

"We are all feeling forlorn over all the boys going to Turkey… The night the men were going away we all felt sad. Those left behind cheered them as they passed… I think they were all glad to get to the front. The flies in Egypt are most appalling, and simply crawl all over one. The Light Horsemen are the only ones left in Egypt, and they have gone trekking for eight days. All the infantry have gone to Turkey… I don't know what they intend doing with me when I get better, but I'd like to go on to the Dardanelles on one of the hospital ships. I'll know in a few weeks."[839]

STAFF NURSE MYRA WYSE, AANS, 1ST AUSTRALIAN GENERAL HOSPITAL.

27 APRIL 1915

DARDANELLES

"We had just got our sweep out and gone forward to get a little rest when we were called up again to man the action stations.

"We were under a very heavy fire from big guns on both sides, and we were firing as fast as we could load, but it is not much use firing at what you can't see. They talk about "coal boxes" at the front, and I wonder if they are half as big as those we get here. You can't hear them coming until they explode. Then they deafen one and throw out hundreds of bullets. You can pick them up all over the deck.

"A big shell went through our forecastle into the mess deck, where it burst. We were at our guns at the time, and there were only two men on the mess deck. It cleared the deck, and a piece of shell went right through my new boots and through my locker of clothes. I had only been gone from the locker about five minutes.

"But we had done some good. We brought up a few mines and exploded them, and we were highly commended."[840]

AB FREDERICK WEST, RN, HMS *SCORPION*.

---

838   *Southland Times* (New Zealand), 15th September 1915.
839   *National Advocate* (Bathurst, New South Wales), 9th June 1915.
840   *Newcastle Journal*, 21st May 1915.

LEMNOS

Amidst the self-congratulation, no senior commander at Helles acknowledged the failure to secure the objectives they had set for themselves; the modest nature of the forces they had faced on landing; or the real challenges that still had to be overcome.

"Sunday was indeed a marvellous day. We attacked at 6 a.m., and by 9 a.m. actually had a fairly secure footing on the Gallipoli Peninsula. Now, even two days afterwards, I can hardly believe it, and I know now that I really never believed that we should succeed. So far as I was concerned I was determined that everything should be done to avert disaster, and, thank God, everybody in that was alike and the result is that the apparently impossible has been attained. Look at the map and then try and imagine that we actually landed 6,000 men on two small beaches on the end of the Peninsula in a few minutes — on beaches which were one mass of barbed-wire entanglements and covered from every quarter by maxims and well-concealed rifle fire from trenches which were invisible. The Lancashire Fusiliers covered themselves with glory and so did the bluejackets that pulled them in the boats. On one of the beaches the first lot landing were practically annihilated, and sinking boats with nothing but dead in them were the result. Luckily, thank God, I had fitted up an old collier to hold 2,000 men and she was run on shore on this beach, but the fire brought to bear was so absolutely annihilating that the men couldn't get out. All the first ones that tried were killed. Eventually they remained in her till dark, when they were able to land. Had it not been for this I really doubt if we should have captured the end of the Peninsula and been where we are now."[841]

REAR-ADMIRAL ROSSLYN WEMYSS, RN, HMS *EURYALUS*.

CAPE HELLES

"[T]hey had another go at us about 2 a.m., and then again just before dawn, but we beat them back handsomely each time, and when daylight eventually came there was not a live Turk to be seen. There were a number of dead ones. These they had not had time to take back, as is their usual custom.

"It is surprising how at daybreak everyone seems to get up out of the trenches and have a walk round, despite the fact that half an hour before you dare not show your head above the top of the parapet. After breakfast we were taken back to the seashore for a general sort out and rest, and we were all ready for it, none of us had had our boots off or had wash for four days."[842]

SPR HERBERT HEATH, 1/1ST (WEST RIDING) COY, ROYAL ENGINEERS, 29TH DIVISION.

---

841 Wemyss quoted in Wemyss, Lady Wester, *The Life and Letters of Lord Wester Wemyss*, pp. 216–217, Eyre and Spottiswoode (London) 1935.
842 *Sheffield Daily Telegraph*, 3rd June 1915.

Those seeing the battlefield for the first time marvelled at what they saw; of what had been done to overcome the obstacles the men faced two days previously. Those who had not had the luxury of viewing things as detached observers aboard ships offshore had different reactions. Albert Mure was on V Beach.

> "I now went, as cautiously as possible, leaving my pinnace beside the *River Clyde*, and scrambling as best I could from boat to boat. The moon had risen by this time, and the beastly evidences of the relentless conflict were thick about; you could not fail to see them clearly, and they looked all the ghastlier in the theatrical limelight of the Orient moon. The heroism of the troops who built that bridge of boats, in daylight, under tremendous, hellish fire, must have been superlative. It beggars all words, and I will attempt none. But we thought of them, and our thoughts were eloquent. For we found it no small thing to pick our way, at our own pace, the Turks temporarily inactive, over those swaying, bobbing craft. To go over them in full marching order must have been a difficult feat in itself, let alone building the way as they went, doing it under shot and shell raining down at the rate of ten thousand shots a minute.
>
> "On reaching the beach, I clambered over the lighters to see where the ammunition was to be dumped first, and began to slip and slide all over the place. I bent down to examine the wood on which I was skidding, and I saw — well, it wasn't water that was making me slide about! It was something thicker than water.
>
> "On the shore I found a very tired-looking assistant-beachmaster. He seemed 'all in,' but he directed me alertly enough where to go and what to do… I remarked… 'You seem to have had a pretty thick time.' He answered not a word. He only looked at me. It was enough. I shall remember that look while I live. There were words, and more than words, in his eyes. They seemed to say, 'I'd far rather suffer the tortures of the damned than go through that again.' I turned and went away quietly, rather sheepishly, I suspect, back over the lighters to my pinnace to give the necessary orders, thinking hard the while…
>
> "I admit being a trifle excited at having at last put my foot on the enemy's soil, and any number of things, no immediate part of 'my job,' escaped me. After finishing giving instructions for unloading, I noticed for the first time a continual spattering in the water beside me, not many feet away, and it dawned on me that it was bullets, a rain of bullets from the machine-guns and the rifles of the enemy on the cliff above. I was safe enough in the pinnace at the moment, for we were under the lee of the *River Clyde*, and the bullets were going over us. But they made an uncanny sound, and again I did a little thinking. It was all right enough on the pinnace, but our work there would be over presently, and it was all very wrong indeed going across the lighters to the beach. However, I was favoured with beginner's luck, and had no one hit."[843]
>
> CAPT. ALBERT MURE, 1/5TH BN ROYAL SCOTS (LOTHIAN REGT), 88TH BDE, 29TH DIV.

Those arriving on W Beach saw things for themselves and heard from those who had survived the experience.

---

[843] Mure, Albert Haye, *With the Incomparable 29th*, pp. 36–39, W. & R. Chambers (Edinburgh) 1919.

"The pathos of war. Lancashire Fusiliers tell their story; a great victory but heavy toll. Some terrible sights in the trenches. A good night's rest in the open."[844]

OS WILLIAM BURNETT, RNVR, DRAKE BATTALION, ROYAL NAVAL DIVISION.

The bodies laid out for burial were husbands, brothers, sons and friends.

"Then we dip down to "V" Beach, a much deeper and wider beach than "W," and walk towards the sea. Then I see a sight which I shall never forget all my life. About two hundred bodies are laid out for burial, consisting of soldiers and sailors. I repeat, never have the Army and Navy been so dovetailed together. They lie in all postures, their faces blackened, swollen, and distorted by the sun. The bodies of seven officers lie in a row in front by themselves. I cannot but think what a fine company they would make if by a miracle an Unseen Hand could restore them to life by a touch. The rank of major and the red tabs on one of the bodies arrests my eye, and the form of the officer seems familiar. Colonel Gostling, of the 88th Field Ambulance, is standing near me, and he goes over to the form, bends down, and gently removes a khaki handkerchief covering the face. I then see that it is Major Costaker,[845] our late Brigade Major. In his breast-pocket is a cigarette-case and a few letters; one is in his wife's handwriting. I had worked in his office for two months in England, and was looking forward to working with him in Gallipoli."[846]

CAPT. JOHN GILLAM, DIVISIONAL TRAIN, ARMY SERVICE CORPS, 29TH DIVISION.

Meanwhile, the advance inland continued, supported by what artillery had landed so far; opposed by, in some instances, lighter forces.

"26th Bty fired at short range on (1) Infantry in nullah W of Old Castle (2) Trench 40 left of No.1 [Fort] (3) Trench NE of Old Castle.

"92nd Bty fired at (1) Scattered infantry advancing E to W, range 3500 [yards] to 3000 [yards] (2) supporting expected advance of our infantry from X beach to NE, range 3500 [yards] to 3600 [yards] (3) enemy battery, range 6000 [yards] (4) support of infantry advancing NE from X beach.

"Our infantry made a general advance about 4.0 p.m.; orders received to support advance as strongly as possible."[847]

CAPT. FREDERICK MORGAN, 17TH BDE, ROYAL FIELD ARTILLERY, 29TH DIVISION.

"The Sedd-ed-Bahr — Krithia road was the one and only road to the front; it was reputed to be mined, so it had to be avoided. It was a bright starlight night, and the going across country was not too difficult close to the track, although there was plenty of scrub and banks and ditches. The scattered trees were mostly dwarf oaks. As there

---

844 *Dundee Courier*, 4th June 1915.
845 Major John Costeker, DSO, Royal Warwickshire Regt, Brigade Major, 88th Infantry Brigade, 29th Division, was killed in action on 25th April 1915. Buried in 'V' Beach Cemetery, he was the 36 year-old son of William and Clara Costeker, of 46 Evelyn Gardens, South Kensington, London; husband of Margaret Picton Grant Poole (formerly Costeker), of 6 Mallord Street, Church Street, Chelsea, London.
846 Gillam, *Gallipoli Diary*, pp. 46–47.
847 17th Brigade Royal Field Artillery War Diary, TNA WO 95/4308.

had been no 'mopping up' parties behind the infantry advance that afternoon, and as rumour had been busy about enemy snipers still concealed in various crannies, the above party advances warily. There was no shell fire, but rifle bullets from the enemy far ahead came along at intervals.

"Suddenly the party were pulled up by an undoubted movement in some bushes on a bank just ahead. A few tense moments followed for the small party of rather tired men uncertain what to do. But the spell was quickly broken by Major Gibbon, who had the inspiration to shout "charge" and the courage to launch himself forward. Like one man the rest followed suit, those carrying rifles luckily refraining from using them. The enemy — a stray dog — fled in terror.

"From the sublime to the ridiculous, the reader may ejaculate. To those possessed of a sense of humour such incidents in such a setting are a godsend, and the above adventure enabled the reconnaissance to be resumed in even higher spirits than at the start."[848]

MAJOR DOUGLAS FORMAN, "B" BATTERY, 15TH BRIGADE RHA, 29TH DIVISION.

"[W]ith "X" as the pivot, a line was established across the toe of the Peninsula. The French force had successfully withdrawn from Kum Kale and landed at "V." The S.W.B. from "S" and the K.O.S.B. from "Y" rejoined the 87th Brigade. I was the only surviving Brigadier of the Division; Napier and his Brigade Major had been killed; Hare had been severely wounded and his Brigade Major killed.

"Pending the arrival of General d'Amade, General Hunter-Weston was left in command of all the troops (French and British) in the Helles area. It devolved therefore on me to assume command (temporarily) of the 29th Division, and for this purpose I handed over command of the Brigade to Casson and proceeded to "W" beach, where Divisional Headquarters were established.

"Orders for an advance on the following day were received from General Headquarters. The French were to advance with their right on the Dardanelles and the 29th, with their left on the Aegean, were to keep touch with them."[849]

BRIG.-GENERAL WILLIAM MARSHALL, COMMANDING 29TH DIVISION.

Opposition had been relatively light that day but the men were exhausted; dogged by doubts about what was in front of them. The transport of all stores, that had not been man-handled, had been dependent upon the ponies of the Highland Mountain batteries and other artillery horses, which were themselves played out by this time.

That evening Hunter-Weston met Hamilton and Keyes aboard the *Queen Elizabeth*.

"General Hunter-Weston came on board that evening, and it was decided to attack on the morrow, with the object of capturing Krithia and Achi Baba before Turkish reinforcements arrived. Although the attack on Kum Kale and the feints at Bulair and in Besika Bay, had undoubtedly had the effect of holding a considerable

---

848 *The Sun* (Sydney, New South Wales), 14th July 1915.
849 Marshall, Sir William, *Memories of Four Fronts*, p. 65, Ernest Benn Ltd. (London) 1929.

force of the enemy in those areas, the arrival of strong reinforcements could not be long delayed."[850]

COMMODORE ROGER KEYES, RN, CHIEF OF STAFF, EMS, HMS *QUEEN ELIZABETH*.

"At 7 Hunter-Weston came on board and dined. He is full of confidence and good cheer. He never gave any order to evacuate "Y"; he never was consulted; he does not know who gave the order. He does well to be proud of his men and of the way they played up to-day when he called upon them to press back the enemy. He has had no losses to speak of and we are now on a fairly broad three-mile front right across the toe of the Peninsula; about two miles from the tip at Helles. Had our men not been so deadly weary, there was no reason we should not have taken Achi Baba from the Turks, who put up hardly any fight at all. But we have not got our mules or horses ashore yet in any numbers, and the digging, and carriage of stores, water and munitions to the firing line had to go on all night, so the men are still as tired as they were on the 26th, or more so…

"Tired or not tired, we attack again to-morrow. We must make more — much more — elbow room before the Turks get help from Asia or Constantinople.

"Are we to strike before or after daylight? Hunter-Weston is clear for day and we have made it so. The hour is to be 8 a.m."[851]

GENERAL SIR IAN HAMILTON, C-IN-C, MEF, HMS *QUEEN ELIZABETH*.

So, the meeting ended, all was now clear. Yet at 10.30 p.m., 29th Division Headquarters received this message, presumably from Braithwaite, Hamilton's Chief of Staff:

"General Hamilton is anxious to hear from General Hunter-Weston as to his proposals for tomorrow, especially as regards timing of commencing, which he thinks should be deferred as late as practicable in order to give the men more rest and also the supporting ships the best conditions of light."[852]

For Hunter-Weston, things were going well. He might have kept some things to himself but he was satisfied with their progress.

"We have managed it, we have achieved the impossible! We are established at the South end of the Gallipoli Peninsula. Wonderful gallantry on the part of regimental officers & men has done it. On the 25th, the Lancashire Fusiliers "W" Beach did under my eyes the most marvellous feat in landing in boats on an open beach which was covered with a broad belt of barbed wire & fired on by machine guns from both flanks [sic], & by riflemen from deep trenches dominating it all round.

"By the evening of the 25th, we had made good the S.W. end, but had not yet managed the S.E. end at Seddel Bahir. On the 26th, owing to the action of Colonel Doughty Wylie & Captain Walford, we seized the S.E end… & were safely established. A

---

850   Keyes, *Naval Memoirs 1910–1915*, pp. 313–314.
851   Hamilton, *Gallipoli Diary*, vol. 1, pp. 163–164.
852   29th Division General Staff War Diary, TNA WO 95/4304.

glorious & wonderful performance. Alas that Walford (and Doughty Wylie) should have been killed. I am going to try & get him a V.C. but it is not easy & I may not succeed…

"We have a very tough job before us; but ½ our difficulties are over now we have landed. When we get the strong covering position on the Achi Baba range of hills 6 miles from Cape Helles, ¾ of our difficulties will be over & the remaining quarter, though still being very difficult, will have very strong chances in its favour. Achi Baba will, I fear, take us some time to gain; but after what we have done, I am sure we shall do it."[853]

MAJOR-GENERAL SIR AYLMER HUNTER-WESTON, COMMANDING 29TH DIVISION.

Meanwhile, the men at Helles spent another uncomfortable night.

"On Tuesday we were relieved by the French, and we then prepared for a general advance towards that pretty little hill known as "Achi Baba." We got to within half a mile of the village of Krithia, when we entrenched for the night. The Turks let us have a fairly quiet night that night, but I for one did not feel very comfortable. We could hear them in the distance talking, and they kept calling out "Allah, Allah.""[854]

PTE MAURICE SINDEN, 2ND BN HAMPSHIRE REGT, 88TH BDE, 29TH DIVISION.

ANZAC

"At about 12 o'clock at night a message came from H.Q. that they required ammunition, so I was relieved from guard, and in company with my sergeant and our officer set out to find our way in the dark. I can tell you I never had such an experience in my life. We got lost, and wandered about till we came to the right place. Eventually we delivered it, and returned to the beach. It was then about 3.30 a.m… and I went and lay down, thinking I would get an hour or so's sleep, but it was impossible owing to the roar of the guns and the cracking of rifles.

"I was just about frozen with cold, and then it started to rain, and at 6 a.m. we were called out to repeat our previous day's work of carrying water and ammunition. Every man should have had three days' rations, but very few of them had one day's, because when they landed they threw all their gear off, and made up the hills for all they were worth. It was not the time to stop and look for your rations, I can tell you…

"I must remark what a marvellous lot of men the Indian Transport Corps and the Indian Mountain Batteries are. They worked day and night, and you could always rely on them to do anything you asked them to do. I don't know what we should have done but for the Indian Mule Corps. Every Australian worships them, and they get on well together, and if you treat them all right you can't get anybody to do the work better. I had not seen any mule transport before I landed on the peninsula, and I owe a lot to the Indian Mule Corps in the work we had to do. They never grumbled at

---

853 Extract from letter, 27th April 1915, Maj. Gen. Hunter-Weston, British Library, MS 48364.
854 *Western Gazette*, 23rd July 1915.

turning out in the middle of the night, as they knew they were only doing their duty, and were only too pleased to accomplish anything they were told to do."[855]

CPL ARTHUR WOOLMAN, DIVISIONAL TRAIN, AASC, 1ST AUSTRALIAN DIVISION.

"The battle again commenced at 4.30 am and no response was made by us as we wished to draw the enemy but they would not engage in a bayonet charge. They, early in the day, commenced a vigorous attack on us with shrapnel & throughout the day we had to defend ourselves against this. At times (especially Major Lamb's Coy) was threatened by 2 Battalions of Turks, but the Navy shelled these men & during the day Major Lamb held his own.[856] The trenches on both right & left were threatened & severely oppressed at several points but up to 5.15 pm we kept the enemy at bay despite the fact that we had been so insidiously shelled. About 4 pm Col. Owen was called upon to take temporary command of the Brigade, the Brigadier[857] and Brigade Major[858] both being shot dead. Major Bennett D.S.O, 2/c [second-in-command] took command of the Battalion."[859]

LT OWEN HOWELL-PRICE, 3RD BATTALION, 1ST BDE, 1ST AUSTRALIAN DIVISION.

"I was told to have a sleep but I was not asleep long when the sergeant-major touched me and said, "Your gun is b———. I will send it to you, and you strip it." I took it to pieces, and found the packing gland was too tight, making a pull of 6 lbs, instead of 3½ lbs. I sent it back, and no sooner was it fixed in its place than bang went a bullet through the barrel-casing. That settled it. We sent for a new gun, but could not get one."[860]

PTE JACK ADAMS, MG SECTION, 15TH BATTALION, 4TH BDE, AIF, NZ & A DIVISION.

"We were all very busy preparing breakfast, boiling our dixies, and it is possible that the smoke from our fires attracted the enemy's attention to us. Anyhow, they put a machine gun on to us; but, as we had had shrapnel bursting over us the day before, we took no notice of it, and no one had been hit up to then. Arthur[861] was standing on the trench talking to one of the boys.[862] I heard him let out a few words quickly, and then he stepped down beside me, and started undoing his shirt. He had been shot in the back, the bullet coming right through the centre of his breast-bone and stopping, without breaking through the skin. He was very game. He hardly changed color until he started to weaken.

855   *Leicester Daily Post*, 30th December 1915.
856   Major Malcolm Lamb, 3rd Bn, AIF, was wounded on 3rd May 1915. He received a "Special mention for acts of conspicuous gallantry or valuable service, 25/4/15 to 5/5/15" (*London Gazette*, 3rd August 1915). Promoted Lt-Col., Lamb later commanded the 3rd Battalion in France.
857   Colonel, A/Brig.-General, Henry Norman MacLaurin, Commanding 1st Australian Bde, was killed in action on 27th April 1915. He is buried in 4th Battalion Parade Ground Cemetery, Anzac.
858   Major Francis Irvine, Royal Engineers, an Exchange Officer, Brigade Major, 1st Australian Bde, was killed in action on 27th April 1915. He is commemorated on the Lone Pine Memorial.
859   3rd Bn Australian Infantry War Diary, TNA WO 95/4342.
860   *The Maitland Daily Mercury* (New South Wales), 24th July 1915.
861   Sgt Arthur O'Neale, Wellington Bn, NZEF, died of his wounds on 27th April 1915. Buried in Ari Burnu Cemetery, the former farmer was the 24 year-old son of Anthony and Louisa O'Neale, of Featherston.
862   Pte Oswald Meenken, Wellington Bn, stated that he was talking to O'Neale at the time he was wounded. The former shepherd was wounded on 8th August 1915 and again on the Western Front in 1916.

"Another chap and I carried him to the dressing station. He did not have anything to say on the way; he was too sick. When we put him down he just asked whether it was shrapnel or machine gun that got him, and then he must, have died right away when we left him, because I saw later that the doctor had touched him, and they knew in the trench almost as soon as we got back that he was dead."[863]

PTE HARRY WHISHAW, WELLINGTON BN, NZEF, NEW ZEALAND BDE, NZ & A DIVISION.[864]

"One of the saddest days I have experienced. The enemy attacked with shrapnel fire since day-break. Our signal sections were scattered a little, five of us, Sergeant Thorpe, Captain Wallis, Sapper Denney, Sapper King, and myself were around a fire, making some tea, when a shrapnel fell and burst in the middle of us. Captain Wallis[865] was wounded in the right thigh, Sergeant Thorpe[866] had half his back ripped off, Sapper Denney's[867] head was blown off, and King[868] and myself escaped without a scratch. It was an awful sight to see your friends of so many months blown to pieces, but we will make them pay for it. It is hard for our boys to have to face all the shrapnel and not see a sign of where it is coming from."[869]

CPL ROBERT HUNTER, DIV. SIGNAL COY, AUSTRALIAN ENGINEERS, 1ST AUSTRALIAN DIV.

"Griffith[870] (section cook) killed by shell. Bailey wounded severely, and several slightly wounded. Engaged putting wire entanglements up round guns on right flank."[871]

CSM ALBERT TURNER, DIV. SIGNAL COY, AUSTRALIAN ENGINEERS, 1ST AUSTRALIAN DIV.

Snipers loomed large in the imagination, understandably, and the imagination embellished the image of the bogeyman.

"On Tuesday morning about twenty of us were out after snipers on the ridge at the rear of us, but did not get any. The nature of the ground and the thick scrub gave excellent cover for this branch of the Turkish army, and they accounted for plenty of our chaps. One sniper who was caught had two boxes of ammunition and a week's rations. It was

---

863   *Evening Star* (New Zealand), 25th June 1915.
864   Whishaw was wounded later on 27th April 1915. Promoted L/Sgt, he was killed in action in France with 2nd Bn Wellington Regt, NZEF, on 3rd July 1916. Buried in Cite Bonjean Military Cemetery, Armentieres, he was the 32 year-old son of John Henly Whishaw and Katherine Elizabeth Whishaw, of Featherston, New Zealand.
865   Capt. Herbert Wallis, 2nd Infantry Bde HQ, was wounded on 27th April 1915. He returned to the peninsula on 7th June but was evacuated sick on 8th August 1915. He was granted a commission in the Imperial Army in November 1915.
866   Sgt Frederick Thorpe, formerly a cavalryman in the British Army, was severely wounded on 27th April 1915. His wounds led to his discharge in Perth on 30th September 1916.
867   Spr John Denney, 1st Divisional Signal Company, Australian Engineers, was killed in action on 27th April 1915. Buried in Lone Pine Cemetery. His brother, Pte James Arthur Denney, 11th Bn, was killed on 28th June 1915 and is buried in Shell Green Cemetery. They were the sons of John and Annie Denney, of 133 7th Avenue, Maylands, Western Australia, originally of Middlesbrough, England.
868   Spr Francis Patrick King, a former telephone operator from Bundaberg, Queensland.
869   *The Evening Star* (Boulder, Western Australia), 26th June 1915.
870   Spr Lewis Griffiths, 3rd Field Coy, Australian Engineers, was killed on 26th April 1915 according to official sources. Commemorated on the Lone Pine Memorial, he was the husband of Catherine Griffiths, of Queens Buildings, Birkenhead, Cheshire.
871   *The Brisbane Courier* (Queensland), 19th August 1915.

stated that another one caught had his face painted green, also his rifle, and pieces of scrub tied to him. I don't know if the above is true or not, but it is quite possible."[872]

SGT JOHN DALGLEISH, OTAGO BATTALION, NEW ZEALAND BDE, NZ & A DIVISION.

"At 11 a.m. we moved away to relieve the Canterbury men on the left flank. The day was getting hot, so we were allowed to shed our packs at the foot of the hill, and then we had to get up at the double as the Turks were pressing and were only 300 yards away from our trenches. At the top we were given a spell and then the order: "Fix bayonets!" — Oh Hell! I wondered what ——— would have thought of it. Well, we had to charge through the scrub and this we did, and cleared 300 yards. The country is covered with thick scrub, and it is impossible to see more than 15 yards in front of oneself. Satisfying ourselves that the Turks had retired a bit, we got the order to dig ourselves in and prepare for the night. First of all we had a biscuit as lunch. Then we set to and though the day was extremely hot, none complained, and we got on well with the trench. However, at 3.30 we got the order to advance and after all that useless digging, once more with bayonets fixed, we advanced as fast as our legs could cary [sic] us, but once again the Turks were faster than we — not once have they stood to the bayonet. This time we advanced 600 yards, and getting into some kind of a line we got down in the scrub. Our platoon was on the left of the company, and Hawke's Bay being on the flank, we were at the end of the line — a place I never did like. From our position we could not see any of the enemy, but we fired all the same, and rapidly too.

"When trying to get in touch with the right we were politely told that, they had been blown out, and then we realised that we were not the only side in this scrap. There were only forty of us left at our end, and the two officers decided to hang on. The Turks were pouring a hot fire into us, the bullets fairly whizzing through the scrub, while our men were also being sniped from behind. Soon the Turks turned the shrapnel on, and it was "Finis Hawke's Bay," while to add to our sorrows a few shells from the warships fell short."[873]

PTE CHARLES CRISPIN, WELLINGTON BN, NZEF, NEW ZEALAND BDE, NZ & A DIVISION.

"About eleven o'clock they sorted us out, and got busy with shrapnel, and the next three hours were the worst I have experienced in my life. My dug-out wasn't very good as I was seeing that the chaps in the section were well fixed up before I got busy with my own, with the result I was not finished when the fun began... The shrapnel was a continual rain for three hours and one dare not shift. I did nothing else but smoke cigarette after cigarette. A kerosene tin alongside my dug-out was riddled, and another shell struck the earth above me and covered me with earth; but a miss is as good as a mile. About 2.30 we were told to reinforce the firing line, and about three o'clock I took about a dozen chaps into the trench.

"We had good fun getting on to the Turks at about 400 to 500 yards. We had things all our own way till about five o'clock, when they got busy on us with rifle, machine gun

---

872   *Lyttelton Times* (New Zealand), 22nd June 1915.
873   *Dominion* (New Zealand), 23rd June 1915.

and shrapnel. Two of my chaps were killed, and five left the trench wounded before… I got hit about eight o'clock."[874]

SGT JOHN DALGLEISH, OTAGO BATTALION, NEW ZEALAND BDE, NZ & A DIVISION.

"We had advanced to take up a position on the left front, when we came under heavy fire from the enemy, and as the result two bullets put me out of action. Notwithstanding the heavy fire, your son[875] most gallantly attempted to move me out of the firing line, and had almost got me to a place of comparative safety, when he himself was struck by a bullet. Even after he was wounded I am informed that he insisted upon my being moved first.

"I have been trying for some time to write you upon this subject, but really it is so hard to adequately express one's feelings. I know, however, that your son will have the ever-lasting gratitude of my wife, my family, and myself for his noble action. In conclusion please let me say that ever since I have met your son he has proved a fine comrade and a gallant soldier. You will be pleased to know that your son is progressing favourably, and will, I trust, soon again be fit and well."[876]

LT KARL FOURDRINIER, 2ND BATTALION, 1ST BDE, 1ST AUSTRALIAN DIVISION.[877]

"Our section was making its way up Shrapnel Gully towards the firing line. It was raining shrapnel, and every yard we went almost cost us a man. We were just nearing the firing line when poor old Capt. McWhae[878] was hit just above the eye with one of the shrapnel balls, and although he could hardly see he wanted to keep going, but Bill Lindsay[879] took him back. His parting words to us were: "Keep going, boys; your mates are up there, and they need you." We eventually got to the firing line, although there were not many of us felt fit to do much, but when we got to the infantry line we found plenty of work, and then we forgot that we had been so long without sleep, and set to."[880]

PTE EDWARD LANGOULANT, 3RD FIELD AMBULANCE, AAMC.

"Things got serious again to-day. A new Turkish division came up and attacked several times; but they were repulsed each time. Our artillery had landed, and with some difficulty had got into the firing line. About a hundred of us to each gun, with long ropes, dragged them up from the beach. Night is the favourite time for the Turkish attacks. They came at us several times, but our waiting game proved very effective, and the morning showed them lying dead in hundreds."[881]

PTE GEORGE GIDEON, 6TH BATTALION, 2ND BDE, 1ST AUSTRALIAN DIVISION.

874 *Lyttelton Times* (New Zealand), 22nd June 1915.
875 Pte Neil Murray, 2nd Bn, AIF, was wounded on 25th April 1915. The former grocer from Cowra, New South Wales, returned to Australia as a result of his wounds, leaving Suez on 15th August 1915.
876 *The Sun* (Sydney, New South Wales), 14th July 1915.
877 Fourdrinier was evacuated, first to England, then sent home to Australia. Returning to his civilian employment as a solicitor, he was found dead in his office on 27th January 1917, having taken his own life.
878 Capt. Douglas McWhae, "C" Section, 3rd Field Ambulance, AAMC, was sent out to "evacuate the collecting posts of the 9th and 10th Infantry Battalions" (unit war diary, AWM4 26/46/4) on the night of 27th April. Hit by shrapnel, he was evacuated the next morning. His severely damaged right eye was later removed.
879 Pte William Lindsay, "C" Section, 3rd Field Ambulance, AAMC. He was evacuated from the peninsula on 19th August 1915 suffering from chrystitis.
880 *Sunday Times* (Perth, Western Australia), 23rd January 1916.
881 *Flemington Spectator* (Victoria), 23rd September 1915.

"To-day is the third day of the battle, and we have gradually pushed them back. The first day (Sunday) was awful. I cannot describe it to you. Our major is wounded, our captain shot through the lungs. Lts Allen (shot dead),[882] Badley (missing, supposed to be wounded in the bush somewhere); my platoon Lt Westmacott[883] (an old Canterbury boy) seriously wounded in arm and back; and Lt Peake, arm torn nearly off — so you see all our officers are out. Nearly the whole of the Waikatos have been killed or wounded. My word, the men were game against great odds.

"A lot of us could not see the Turks, but on came the bullets, downing first one and then the other. The shrapnel was the worst, and nearly drove many to distraction. There will be no doubt about the final issue of the great battle, as we must win, for Lord Kitchener's last message to General Sir Ian Hamilton was: "Once you set foot on Gallipoli Peninsula, you must fight to a finish!' I believe further up the French, on one side, and the Tommies on the other, are giving the Turks hell."[884]

PTE WILLIAM RHODES, AUCKLAND BATTALION, NEW ZEALAND BDE, NZ & A DIVISION.

"The gun was firing continually, simply eating up ammunition and water as fast as it could be brought up. When water was slow in coming up, clouds of steam from the boiling water gave our position away, and we then came in for a hot fire. Our No. 2 gun was put out of action on [Monday], and another section took its tripod, as it was useless to us, and they badly need one, theirs having been shot to pieces by a shell. It was not till [the] morning that the gun was repaired, just in time to replace mine which had had the feed block shot away. This would not stop us in the ordinary course of events, but we had lost our spare parts box first day, and it meant a trip to the beach to pick up what we could find, and spares we could borrow before we could get going again. Each time something went wrong, it meant I had to sprint over the crest of the ridge with the gun to the corporal, and take back the other one."[885]

PTE GEORGE HENDERSON SMITH, MG SECTION, 11TH BATTALION, 3RD BDE, 1ST DIVISION.

The position that came to be known as Quinn's Post developed its reputation very quickly.

"Under my orders they took up their positions, but were subjected to a tremendous fire from machine-guns and rifles. Your brother[886] fell with many comrades at this post. Early in the action at 2 p.m. he was shot through the forehead with a bullet from a machine-gun. Poor lad, he did not suffer, death being instantaneous.

---

882 2/Lt Harold Allen, Auckland Bn, NZEF, was killed on 25th April 1915. Buried in Baby 700 Cemetery, he was the 21 year-old son of George Allen, of 10 Warrington Road, Remuera Road, Auckland, New Zealand, and the late Lucy Allen of Liverpool, England. Passed examination 1st Lt. before outbreak of war, and had been two years at Royal Military College, Duntroon, New South Wales.
883 Lt Herbert Westmacott, Auckland Bn, NZEF, was wounded on 25th April 1915, leading to the amputation of his right arm.
884 *Auckland Star* (New Zealand), 21st June 1915.
885 *Barrier Miner* (Broken Hill, New South Wales), 25th July 1915.
886 Cpl James Butterworth, 14th Bn, AIF, was killed in action on 27th April 1915. Buried in Quinn's Post Cemetery, he was the 22 year-old and youngest son of William and Margery Butterworth, of Further Heights, Rochdale, Lancashire.

His comrade, Corporal Thompson,[887] was shot at the same time, and they were both buried together."[888]

MAJOR JOHN ADAMS, 14TH BATTALION, 4TH BDE, NZ & AUSTRALIAN DIVISION.

"About 2 o'clock… we got orders to mount one gun as quickly as possible, and I ran up with the stand of the gun. When I got there we had no place to put it but on the top of the trench, a very exposed position. We mounted it and started firing at mobs of Turks about 100 yards away, and they fell like flies. Then the snipers got range of us, and the bullets were flopping into the ground around us. I had hold of the handles, and was firing at the time, when all of a sudden I got an awful smack in my left hand. I helped to dismount the gun, and then got to the rear and got my wound dressed. I felt no pain, only a burning sensation."[889]

PTE ALFRED POWER, MG SECTION, 14TH BN, AIF, 4TH BDE, NZ & AUSTRALIAN DIVISION.

"My first trench experience was in one which pointed straight at the enemy, [Quinn's Post] and consequently came in for a great deal of enfilade fire. Here I came across the first dead man of my experience lying in a trench. Later, when reinforcements were sent to us, two of these happened to jump into the spot filled by my dead companion. They landed fair on him, which seemed horrible, but as he was past all feeling I suppose it mattered little. Then the second man from me was shot in head and badly wounded. I used his bandage and puttee to bind his wound and when we vacated the trench some time later we took him with us. Already two of our fellows were killed — one, a tent mate of mine, a Corporal McLaren,[890] a man about 30 years of age, and with a big and varied experience, the most practical man in our company, and possessing the full confidence of our officers. He was absolutely without fear, and need not have been hit, but persisted in refusing to take proper cover, and was shot while standing on the top of the trench. What hurt me was the fact that, no matter how popular a fellow was, the moment he was killed he was forgotten; but one soon gets accustomed to every phase of the ghastly business, and hardened to such things."[891]

PTE JAMES MORIESON, 14TH BATTALION, AIF, 4TH BDE, NZ & AUSTRALIAN DIVISION.[892]

"It was our first day of really strenuous fighting, and Captain Hoggart[893] had part of his company up with ours when we rushed and occupied an important forward position.

---

887   L/Cpl John Thomson, 14th Bn, AIF, was killed in action on 27th April 1915. Buried in Quinn's Post Cemetery, he was the 27 year-old son of John Dickson Thomson and Margaret Helen Thomson, of Melbourne, Victoria, Australia.
888   *Rochdale Observer*, 31st July 1915.
889   *Birchip Advertiser and Watchem Sentinel* (Victoria), 23rd June 1915.
890   Cpl Hector McLaren, 14th Bn Australian Infantry, was killed in action on 27th April 1915. Buried in Quinn's Post Cemetery, he was the 31 year-old son of Peter and Fanny McLaren, of Willow Grove, Victoria.
891   *Warrnambool Standard* (Victoria), 29th July 1915.
892   Morieson was admitted to hospital in Malta on 2nd June 1915 suffering from enteric fever; a condition which led to his eventual return to Australia and discharge from the Army on 14th May 1918.
893   Capt. William Hoggart, 14th Bn, AIF, was killed on 27th April 1915. The former teacher was the 27 year-old son of Alexander and Elizabeth Hoggart; husband of Mrs. R. J. Hoggart, of 35 Domain Road East, South Yarra, Victoria; commemorated by a special memorial in Quinn's Post Cemetery, Anzac, where he is believed to be buried.

The enemy was pressing us very strongly, and our casualties were pretty numerous, but we were holding on well when, suddenly, a machine gun opened fire on our flank, grazing the fingers of my left hand and smashing my rifle. Our senior subaltern[894] was hit in the chest, and it is doubtful whether he will recover, and several men went down. It was impossible to locate the gun, and another burst of fire played havoc with our hastily constructed trenches. [Hoggart] then most bravely ran round to see if he could find out where the fire was coming from, and was immediately fired on and killed instantly."[895]

CAPT. FERDINAND WRIGHT, 14TH BATTALION, 4TH BDE, AIF, NZ & A DIVISION.[896]

"Suddenly a machine gun began to play upon our flank (at Quinn's Post), so we lay down. No one could tell from what direction the shots were coming. Lance-Corporal Skilton[897] was about the first to pick the gun out about 300 yards from our position, but he was hit himself almost immediately. Capt. Hoggart, of Brighton, stood up with his field glasses in his hand to try and locate, it more clearly, but he was killed instantly. Eight bullet wounds were discovered in his side, and he was hit by the machine-gun fire most probably."[898]

PTE THOMAS ROSS, 14TH BATTALION, 4TH BDE, AIF, NZ & A DIVISION.

"[A] sergeant called me to the top of the trench to help him carry a dead captain's body[899] to the rear. This was when our trenches were too narrow to admit of passing anyone therein. Bullets were very plentiful in the vicinity, but our luck was in, and though exposed for more than five minutes, we and our burden got safely to the rear."[900]

PTE JAMES MORIESON, 14TH BATTALION, 4TH BDE, AIF, NZ & A DIVISION.

"I hid behind a box of ammunition. The bullets were flying, all round me, and I knew I could not last long where I was. All of a sudden I lost count of where I was, and when I came to was lying in a pool of blood. I then had an idea that I had been hit, but did not know where. However, I found that a bullet had passed through the box of ammunition and hit me just above the left ear, pieces of it passing along my forehead and left eye. The main piece of bullet was embedded in my head just above the left ear, and about fifty small pieces were scattered over the side of my face. I

---

894   Probably Capt. John Hanby (he had been promoted the previous day), who was severely wounded on 27th April. Admitted to No. 17 General Hospital, Alexandria, on 1st May, he received further treatment in England and left Portsmouth to return to Australia on 7th November 1915.
895   *Geelong Advertiser* (Victoria), 17th June 1915.
896   Capt. Wright was admitted to the Deaconess Hospital, Alexandria, on 8th May suffering from concussion. His mental health was such that he was invalided to Australia on 9th June 1915.
897   Promoted Cpl, William Skilton, 14th Bn, AIF, was killed on 21st August 1915. Commemorated on the Lone Pine Memorial, the former fitter was the 28 year-old son of William Skilton, Hannan Street, Williamstown, Victoria, originally of Manchester.
898   *Geelong Advertiser* (Victoria), 19th July 1915.
899   Capt. William Hoggart, 14th Bn, AIF, was killed in action on 27th April 1915. The former teacher was the 27 year-old son of Alexander and Elizabeth Hoggart; husband of Mrs. R. J. Hoggart, of 35 Domain Road East, South Yarra, Victoria; commemorated by a special memorial in Quinn's Post Cemetery, Anzac, where he is believed to be buried.
900   *Warrnambool Standard* (Victoria), 29th July 1915.

managed to get away and reached our doctor, who bandaged me up and sent me down to the beach."[901]

PTE ROBERT HUGHES, 16TH BATTALION, 4TH BDE, AIF, NZ & A DIVISION.

"McLaurin's[902] death happened on April 27. McLagan, McLaurin, and I had been together only an hour before, discussing a plan for a rearrangement of our respective lines and limits of our sections. It appears that he and Irwin, his brigade major, went to a place which was known to be dangerous, and Lieut.-Col. Owen called out to come down, and he called down that he would when he had finished seeing what he wanted to. Almost immediately after he got a bullet through the head and was killed instantly, and a few seconds after Irwin was killed in the same way. While regrettable, it was undoubtedly avoidable, and such unnecessary exposure not only does no possible good, but seriously impairs morale. While it is true that, like everybody else, I have had many narrow escapes such as, for example, passing a spot where a few minutes after, a shrapnel burst, yet I have always insisted on all my people exercising reasonable caution in not remaining stationary in spots which are obviously dangerous and in doing their observations and reconnaissances from covered places."[903]

COLONEL JOHN MONASH, COMMANDING 4TH BDE, AIF, NZ & A DIVISION.

"By the middle of the afternoon we had a full complement of wounded, and were able to get up anchor and steam away to Alexandria. But first we must solemnly send over the side on to the deck of a trawler seven bodies wrapped each in a Union Jack — seven deaths even before we left our anchorage."[904]

COLONEL CHAPLAIN JOHN McPHEE, CHAPLAIN'S CORPS, AIF, HMT *IONIAN*.

"At 5 o'clock I stopped a bit of shrapnel, and when I looked down at the rip in my trousers I thought my leg must have been blown off, but I was very much relieved to find that it was still on, and that only a piece of flesh was missing — lucky boy. There were only a few men left, so our officers decided to retire. I was told to throw all my gear away, as it would catch in the scrub, and then I crawled back to the trenches. It doesn't sound much in writing, but I little expected I would get there. Didn't the bullets ping? Not half!

"After being bandaged up, I got down the hill, and at the foot picked out my valise, as I had some shaving tackle, etc., I wanted. The packs lay just where we had left them five hours previously. We were then a proud company; how different now. At the dressing station I was told to get into a boat, and then we were conveyed to a transport out in the bay, where we arrived some time about 10 p.m. All the time shrapnel was bursting until we got well away from the shore."[905]

PTE CHARLES CRISPIN, WELLINGTON BN, NZEF, NEW ZEALAND BDE, NZ & A DIVISION.

901 *Port Pirie Recorder and North Western Mail* (South Australia), 24th June 1915.
902 Colonel, A/Brig.-General, Henry Norman MacLaurin, 1st Australian Brigade, was killed in action on 27th April 1915. He is buried in 4th Battalion Parade Ground Cemetery, Anzac.
903 *The Courier-Mail* (Brisbane, Queensland), 17th November 1934.
904 *Daily Herald* (Adelaide, South Australia), 21st July 1915.
905 *Dominion* (New Zealand), 23rd June 1915.

"Rest most of day till 6 p.m. Went collecting wounded under heavy shrapnel fire. Captain M'What [McWhae] wounded. Returned late at night, very cold. Three thousand two hundred casualties. Had a swim under fire."[906]

PTE SYDNEY PENHALIGON, 3RD FIELD AMBULANCE, AAMC.

"I was sent away in a boat to fetch wounded from shore and take them to the "Derflinger." She was a steamer that had been captured during the war. We left at 9 p.m., and… were under pretty heavy shrapnel fire all the time. We were near the shore and loading up with wounded from the field hospital…. I went ashore with some gear for the three midshipmen who had landed with the beach parties. I was standing in front of their dug-out, when one of them said: "You had better come inside as there is a lot of shrapnel about." We had hardly got inside when a shell struck just where we had been standing. We filled the boat with wounded and were towed to the ———, where, after a good deal of delay, the wounded on stretchers were hoisted inboard on a cradle. There were only two doctors and ten Red Cross men to look after between 500 and 600 men. These did not count about 18 untrained orderlies and stewards who assisted when possible. There were Greeks working the winches which were hoisting the wounded out of the boats, and they could not be made to understand the word of command, so that the wounded came in for a lot of unnecessary bumping. I went on board and had a look round. The wounded were still on their stretchers on the promenade deck. The medical staff did what they could, making a quick round of the patients and giving attention in very urgent cases. After ourselves and another boat had been emptied, the two were towed back to the shore by a picket boat, and we filled up again. This time we had a seriously wounded lot. We did what we could for them by giving them water and covering them with our overcoats, &c. (Although it is warm in the day-time, it gets quite cold at night.) We had to wait alongside about two hours before we could get cleared. At last the other boats were empty, and we were able to get alongside, and even then it was slow work, as each man had to be hoisted separately. "[907]

MIDSHIPMAN HON CHARLES GIFFORD, RN, HMS *LONDON*.

"The doctors were very chary about me at first, as they thought my bullet was lodged in the stomach, but I was in good spirits and after a number of hours lying on a cot I complained of a pain on my left side some few inches from where the bullet went in. On examination the doctors found that it had not penetrated the stomach, but on striking me it had turned its sharp nose and lodged between the muscles and intestines of the stomach. They operated on me and got it out, and here right now I am getting on A1 and want to get back and give the Turks and their German officers some change. I am sending you the bullet to keep for me till I get back, when I will get him bored. The devils here have been firing explosive bullets."[908]

PTE WILLIAM RHODES, AUCKLAND BATTALION, HMAT *SEANG CHOON*.

---

906 *The Queenslander* (Brisbane, Queensland), 21st April 1917.
907 *Grantham Journal*, 29th May 1915.
908 *Dominion* (New Zealand), 19th June 1915.

"A pal and I were carrying a case of ammunition down the side of the gully and a sniper shot him dead, and a few seconds later he had a pink at me. The bullet entered through the front of my coat and shirt but never touched my skin. I had a tin of jam and a tin of beef under my coat, and it went through both and even out the top out of my identification disc, which was on string round my neck. Two officers have taken photos of me with the shattered disc. A bullet also punctured my water bottle, and I got one through my puttee which did no more harm than put a pink tint on the calf of my leg. Their shrapnel fire, though, was particular hell. I got three bullets from a shell, but as they all got me in the left hand I came off light. I shall probably lose half of the forefinger, but to come out of that lot with a head on one's body one is lucky."[909]

PTE CHARLES WALTON, 3RD BATTALION, 1ST BDE, 1ST AUSTRALIAN DIVISION.[910]

The isolated positions were deepened, extended and linked together. Aircraft flew over the lines as quickly as stories of Turkish tricks and atrocities.

"Seaplane reconnaissance reported no movements of troops to be seen to N [North] and NE [North East] of our position for a distance of from 4 miles.

"Enemy full of ruses. A message was sent to Battn. asking for an officer to go and see the Commander of the Indian Troops on North flank. One was sent. The Adjutant was asked for and sent. The C.O. was sent for and went but saw that the first two were prisoners, and got back.

"Our dead are stripped and clothing worn by enemy. Party advanced against part of 1st Bde line with white flag. But were fired on."[911]

MAJOR CHARLES VILLIERS-STUART, 56TH PUNJABI RIFLES (FRONTIER FORCE), ATTD HQ, ANZAC.

28 APRIL 1915

ENGLAND

The British Prime Minister believed that men landed on V Beach from the *River Clyde* without loss. Who told him that?

"To-day's news of the Dardanelles is quite good & illustrates the truth that in this extraordinary war all the oldest, as well as the most modern, devices come in. The Trojan Horse, for instance. 2000 soldiers who were unable to land were shut up in a collier, the decks covered with coal, & the sides made openable & flappable. And then when night came on, she was run ashore, and they emerged from their cupboard, & were all safely landed. I dare say they told you this at lunch. It is quite one of the

---

909 *Leader* (Orange, New South Wales), 26th June 1915.
910 Walton was returned to Australia, discharged as no longer physically fit for service on 14th August 1915.
911 Intelligence, Headquarters Australian and New Zealand Army Corps, AWM4 1/27/2.

romantic by-episodes of the War. I am afraid there have been very heavy losses, but so far we hear nothing of the Naval Division."[912]

HERBERT ASQUITH, MP, BRITISH PRIME MINISTER.

HELLES

"About 12.30 a.m… I was hove out of my little bunk, consisting of a blanket stretched out on the sand, and sent about 200 yards along the bottom of the cliffs with half a dozen British blues, in consequence of a rumour that reached us to the effect that some Turkish snipers were trying to make their way round under the cliffs to the beach. We stayed there till daylight and saw nothing, so we came to the conclusion that there was something wrong about the message which had been delivered in rather a roundabout way, so I had my night out for nothing."[913]

LT FRANK STEPHENSON, RN, HMS *IMPLACABLE*, ASSISTANT BEACHMASTER, X BEACH.

"I awake feeling very fit and refreshed, and find a beautiful morning awaiting me. Opposite our tent is a little "bivvy" made of oil-sheets and supported by rope to one of the walls of the house and a lilac-tree. A head pokes out from under this "bivvy" with a not very tidy beard growing on its chin, and the owner loudly calls for his servant… I find later that he is Josiah Wedgwood, M.P., and being interested, get into conversation with him. He is a most entertaining man, and tells me that he is O.C. Armoured Cars, but that as it is not possible for his cars to come on shore, he had been instructed to use his intelligence and make himself useful."[914]

CAPT. JOHN GILLAM, DIVISIONAL TRAIN, ARMY SERVICE CORPS, 29TH DIVISION.

With Turkish resistance finally weakening, as they withdrew towards Krithia, it was time for the British to move inland towards their first day objective of Achi Baba. There were no interconnected trench systems crossing the peninsula at this time. Precisely where the Turks were was far from obvious. It would be, to use the military term for having little idea where their enemy was, an advance to contact. It became known as the First Battle of Krithia.

"At 8 a.m. on the morning of April 28th [the] weary but indomitable troops were called upon for a further effort. On the extreme left of the line the 87th Brigade, minus the King's Own Scottish Borderers, but reinforced by the Drake Battalion of the Naval Division, pushed on rapidly, and at 10 a.m. had advanced some two miles."[915]

ELLIS ASHMEAD-BARTLETT, WAR CORRESPONDENT.

"26th Bty opened fire [8.15 a.m.] on hostile infantry supporting 88th Inf Bde. Hostile battery replied without effect. This battery could not be located.

---

912 Herbert Asquith quoted in Brock, Eleanor & Brock, Michael (Eds.), *Letters to Venetia Stanley*, p. 574, Oxford University Press (Oxford) 1982.
913 *The Press* (New Zealand), 23rd April 1938.
914 Gillam, *Gallipoli Diary*, p. 50.
915 Ashmead-Bartlett, *Uncensored Dardanelles*, p. 76.

"92nd on left flank supported 87th Inf Bde."[916]

CAPT. FREDERICK MORGAN, 17TH BDE, ROYAL FIELD ARTILLERY, 29TH DIVISION.

"We moved forward in the usual "snake" formation for about two miles, being heavily shelled by the enemy, but getting few casualties. The order came along to extend before we came under heavy rifle fire. The Turks were very hard to locate amongst the thick shrubs. My Company was in position behind a low bank, and about 50 yards in front was a crest which was alive with snipers. Anyone exposing themselves the least bit got hit. A great number of us got our packs, which showed up over the small bushes and gave our positions away, riddled."[917]

SGT BERTIE CASTLETON, 2ND BN HAMPSHIRE REGT, 88TH BDE, 29TH DIVISION.

"Our battleships and the artillery behind gave them a taste of what British guns can do, and so we advanced. When the infantry came into contact, then commenced what will perhaps be the most awful day in our lives. The enemy's machine guns were terrible and so also their rifle fire, and it mowed down our men, who had no cover except bushes, etc...

"For three hours I lay behind a bush watching the progress; and all the while bullets whistled around. Dozens struck the ground within inches of me, and I've a few for keepsakes."[918]

PTE JOHN McKILLOP, 1/5TH BN ROYAL SCOTS (LOTHIAN REGT), 88TH BDE, 29TH DIV.

"While we were advancing the enemy's artillery... were shelling us... but the losses were inconsiderable until we came under rifle fire. How I got through without a bullet I cannot quite yet realise."[919]

L/CPL HERBERT MIDDLETON, 1ST BN ROYAL INNISKILLING FUSILIERS, 87TH BDE, 29TH DIV.

"I was hit in the left knee, and after lying where I fell for over two hours, I was taken away by the stretcher-bearers. After a rough and tumble journey, I arrived at the dressing station, where I lay in the rain without any covering."[920]

PTE DANIEL WALKER, 1ST BN KOSB, 87TH BDE, 29TH DIVISION.

"The 88th Brigade, on the right of the 87th, advanced steadily until 11.30 a.m., when the enemy's stubborn resistance and a shortage of ammunition brought the movement to a stop. Thereupon the 86th Brigade, under Lieut.-Colonel Casson, which had been held in reserve, was ordered to push forward through the 88th Brigade and to occupy Krithia. This movement commenced at 1 p.m. but was unsuccessful, for the broken nature of the ground and the enemy's machine-gun posts held up the advance."[921]

ELLIS ASHMEAD-BARTLETT, WAR CORRESPONDENT.

916   17th Brigade Royal Field Artillery War Diary, TNA WO 95/4308.
917   *Hampshire Telegraph and Post*, 2nd July 1915.
918   *Campbeltown Courier*, 5th June 1915.
919   *Nottingham Evening Post*, 28th May 1915.
920   *Jedburgh Gazette*, 25th June 1915.
921   Ashmead-Bartlett, *Uncensored Dardanelles*, p. 77.

"11.45 a.m. 86th Brigade ordered to advance through 88th Bde & seize Krithia but this [was] not accomplished as units became much mixed and were exhausted."[922]

MAJOR-GENERAL SIR AYLMER HUNTER-WESTON, COMMANDING 29TH DIVISION.

"The Turks were retiring from all places, and we were close on the village of Krithia, which is not so far from Achi Baba. But, alas! we had to retire on account of having no ammunition and no fresh troops to make a counter-attack... There are few of my Battalion left now."[923]

PTE HERBERT BURRIDGE, 2ND BN ROYAL FUSILIERS, 86TH BDE, 29TH DIVISION.[924]

"Frett[925] got wounded in the foot when we were out in the advanced fighting line and trenches were being being dug about half a mile behind. When we retired he was left there with hundreds of others... A lot of others were also shot down when we were coming back. Lawrence,[926] of my company, was shot down by my side. He only said two words, 'Oh, God,' and did not move again."[927]

PTE JOHN PATTERSON, 1ST BN ROYAL MUNSTER FUSILIERS, 86TH BDE, 29TH DIVISION.

"Rushing forward with a line of men, we lay down and were immediately fired on by a sniper from behind, at a distance of not more than twenty yards. I was wounded in the wrist by his first shot, which splintered on the rock. The second went through my arm. This must have been about 11 a.m. A fellow victim was our excellent Mess Sergeant — Sgt Allsopp[928] — who lay mortally hit a few paces away."[929]

LT-COL. JAMES WILSON, 1/5TH BN ROYAL SCOTS (LOTHIAN REGT), 88TH BDE, 29TH DIV.

"We pressed forward about two miles, and all got into the firing line. During the day Colonel Wilson and Major M'Donald[930] were wounded, and Captain Hepburn,[931] the Adjutant, was killed. We had other heavy casualties, several junior officers being killed and wounded, and practically the whole of the headquarters of the battalion being wiped out."[932]

MAJOR DOUGLAS McLAGAN, 1/5TH BN ROYAL SCOTS (LOTHIAN REGT), 88TH BDE, 29TH DIV.

922 Diary entry, 28th April 1915, Maj. Gen. Hunter-Weston, British Library, MS 48364.
923 *Middlesex Chronicle*, 5th February 1916.
924 Burridge transferred to 4th Bn Lancashire Fusiliers, and was killed in action on 28th March 1917. He is buried in Faubourg d'Amiens Cemetery, Arras.
925 Pte Robert Frett, 1st Bn Royal Munster Fusiliers, was killed on 28th April 1915. From Bow, Middlesex, the 24 year-old is commemorated on the Helles Memorial.
926 Pte Charles Lawrence, 1st Bn Royal Munster Fusiliers, was killed on 28th April 1915. Commemorated on the Helles Memorial, he was the husband of Emily Lawrence of West Street, Somerton, Somerset.
927 *Coventry Evening Telegraph*, 24th May 1915.
928 Sgt George Allsopp, 1/5th Bn Royal Scots (Lothian Regt), was killed in action on 28th April 1915. Commemorated on the Helles Memorial, he was the 44 year-old son of Mrs. W. Allsopp, of Ford Street, Derby; husband of Mary Allsopp (nee Campbell), of 34 North Castle Street, Edinburgh.
929 Lt-Col. James Wilson, quoted in Weaver, Lawrence, *The Royal Scots*, p. 230, Country Life (London) 1915.
930 Major Alexander McDonald, 1/5th Bn Royal Scots (Lothian Regt), was wounded in action on 28th April 1915.
931 Capt. William Hepburn, Seaforth Highlanders, attached 1/5th Bn Royal Scots (Lothian Regt), was killed in action on 28th April 1915. He is commemorated on the Helles Memorial.
932 Major Douglas McLagan, quoted in Weaver, *The Royal Scots*, p. 233.

"For another hour or two I was running from the headquarters to the firing line with ammunition, and all the time those deadly bits of lead played round me. I'll never be nearer death than I was that day. The Turkish position seemed impregnable, but with the true British instinct we kept at it, and I'm positive we would have taken it if reinforcements had arrived in time. As it was we were forced to retire, but we didn't lose ground; on the contrary we gained most ground that day. Their shrapnel fairly chased us up, but we re-formed and held them back. Our Adjutant was killed when in the thick of it, and our Colonel and Major wounded."[933]

PTE JOHN McKILLOP, 1/5TH BN ROYAL SCOTS (LOTHIAN REGT), 88TH BDE, 29TH DIV.

"Then began the murder. Men who went through Mons say that Mons was a picnic to that awful Wednesday. We continued to advance under the Turks' heavy rifle and shrapnel fire, and then I happened to stop a bit of Turkish delight through the thigh. Soon after our men were forced to retire owing to the heavy fire from the Turks, who were strongly entrenched. After that I saw no more of the battle, but our troops entrenched themselves for the night, and remained there the next day. I crawled back from the firing line for some distance, and was picked up by some of the R.A.M.C. stretcher bearers."[934]

PTE MAURICE SINDEN, 2ND BN HAMPSHIRE REGT, 88TH BDE, 29TH DIVISION.

"The Turks were waiting for us about 500 yards from our position, and, of course, we had to drive them out of it. Well, this is where I got knocked over, and the bullets, they were coming in thousands, but anyway, I got out of it all right after a bit of a struggle. I got to our trench, and had it dressed up, and there I stopped till our chaps had got further on, because it was not safe to bob your head up before they got you."[935]

PTE SYDNEY STREET, 4TH BN WORCESTERSHIRE REGT, 88TH BDE, 29TH DIVISION.[936]

"I was shot through my right shoulder... [and not] having had any sleep for four nights in succession, [I was] frightfully thirsty, though I had plenty of water in my bottle, which I had carried two days. But this was too precious to drink as there was a possibility of getting no more for some time. The springs and wells we did come across were probably poisoned, so we dare not touch them. However, I suddenly discovered that I was nearly on top of the Turks' lines, which meant certain death if I got captured. So I got up, exhausted as I was, through want of sleep, food and water and loss of blood, and made a run for it through a deadly hail of bullets, but I did not get any further wounds, though every fraction of a second I expected to go under. I might here mention that we had no trenches to protect ourselves... All the

---

933 *Campbeltown Courier*, 5th June 1915.
934 *Western Gazette*, 23rd July 1915.
935 *Evesham Standard & West Midland Observer*, 29th May 1915.
936 Street was killed in action during the diversionary attack at Helles on 6th August 1915. Commemorated on the Helles Memorial, he was the son of Elizabeth Street, Bewdley Street, Worcester.

way along… as I stumbled and fell along I was sniped at, but managed to fall into the hands of the Red Cross."[937]

SGT PERCIVAL KNELLER, 2ND BN HAMPSHIRE REGT, 88TH BDE, 29TH DIVISION.[938]

"Poor Deane[939] was killed, shot through the temple: two other officers killed[940] and five wounded, including Beckwith slightly in the foot. The men have behaved splendidly. and we are very proud of them; they have kept up the credit of the regiment well."[941]

CAPT. & ADJ. GEORGE REID, 2ND BN HAMPSHIRE REGT, 88TH BDE, 29TH DIVISION.

"Gripping our rifles, we got up and rushed forward through a storm of bullets. On the crest we found two small trenches containing snipers. We soon made short work of them, but we paid very dearly for it. On getting on the high ground we must have appeared in full view of the enemy, and the fire which was directed at was awful.

"We seemed to draw their fire, the shots coming from all directions. It was impossible to live in it, and yet the Company went forward as if on parade. Some of the men were actually smoking. Half of the Company was knocked over before we had gone another 50 yards, and all our officers were down. Our Major [John Deane], who was just in front of me, got shot. He spun round and sank to the ground without a word. Suddenly an order came down the line to get back and rejoin the line. It was certain death to go on any further. I had just turned round when I received a bullet which broke my left leg, and down I went."[942]

SGT BERTIE CASTLETON, 2ND BN HAMPSHIRE REGT, 88TH BDE, 29TH DIVISION.

"Some small parties did succeed in pushing through and arriving within a few hundred yards of the village, but the bulk of the 86th never passed beyond the line held by the 88th, and the units of the one became absorbed in those of the other. The French, on the right flank of the 29th Division, in the face of strong opposition, pushed forward along the spurs of the western bank of the Kereves Dere and arrived within a mile of Krithia with their right wing thrown back and their left in touch with the 88th Brigade. Here they were unable to make any further progress in face of the enemy's ever-increasing opposition.

"By 2 p.m. the whole of the troops available had been drawn into the struggle with the exception of the Drake Battalion of the Naval Division. It must be remembered that the advance had been made almost without artillery support, for only a few guns

---

937 *Stratford-upon-Avon Herald*, 28th May June 1915.
938 Kneller was killed on 12th July 1915; commemorated by a special memorial in Twelve Tree Copse Cemetery, where he is believed to be buried.
939 Major John Deane, 2nd Bn Hampshire Regt, was killed in action on 28th April 1915. Commemorated on the Helles Memorial, he was the 41 year-old son of the late William and Matilda Deane; husband of Iris Deane.
940 The other officers killed were Lt Charles Pakenham and 2/Lt Arthur Howard. Both are commemorated on the Helles Memorial; their deaths recorded, in error, by the Commonwealth War Graves Commission, as taking place on 30th April 1915.
941 *Stratford-upon-Avon Herald*, 28th May 1915.
942 *Hampshire Telegraph and Post*, 2nd July 1915.

had been landed and these had very little ammunition. The line was too far inland for the warships to be able to render adequate support in a country broken into nullahs, ravines, dry river-beds, and covered with scrub. The men were completely exhausted by four days of continuous fighting almost without sleep, and the available transport hardly sufficed to keep the infantry supplied with small-arm ammunition. Thus all hope of reaching Krithia and of occupying Achi Baba had perforce to be abandoned."[943]

ELLIS ASHMEAD-BARTLETT, WAR CORRESPONDENT.

Back on V Beach, the work to land supplies, to meet all the requirements of an army in the field, continued.

"I was on the beach all day long, hard at it. Fighting, actual personal encounter or contribution to battle, is but one part of soldiering. The tangible brief 'fight' is the concentration of months of indescribably arduous and intricate preparation and transport, which is quite another part of soldiering. Things are thought out at home, munitions are made, stores gathered and packed, men trained and equipped. The simply enormous transport work is accomplished, no matter at what cost, over what distance. The awful goal of the imminent carnage reached, literally ten thousand indispensable, nerve-racking, back-breaking tasks confront and fatigue the soldier, who must work his hard way through them to his hour of supreme trial...

"We had by this time made considerable advance both inroad on the peninsula and in preparation of all sorts. What we had gained, how far we had penetrated in this deadly, warded place, I knew as yet but scantily and in disjointed scraps. News filtered through, of course, but I had little leisure to listen. But of what the men in my immediate charge were doing, and the splendid spirit in which they sweated on at a job as uninteresting as it was gigantic, and as perilous as any actual battle could be, I saw and knew all. Back and forth they waded all day long, from the beach to the small boats, from the boats to the shore, unloading, carrying, stacking up, sorting munitions, food, water, stores of every sort; shells and bullets falling thick, fast, constantly. It was one rain of death. Not a box reached the sand without being a target to the Turk. All day long the men worked and carried and waded, walking over the dying and the dead, when they had to. Have you ever walked over dead men, still warm and quivering?"[944]

CAPT. ALBERT MURE, 1/5TH BN ROYAL SCOTS (LOTHIAN REGT), 88TH BDE, 29TH DIV.

Once brought ashore, those stores had to be transported to the firing line and casualties brought back.

"An officer was wanted to go to the firing line with ammunition. I was sent with 14 ponies and their loads about four miles away from our base, right up to the trenches.

---

943   Ashmead-Bartlett, *Uncensored Dardanelles*, p. 77.
944   Mure, Albert Haye, *With the Incomparable 29th*, pp. 41–43, W. & R. Chambers (Edinburgh) 1919.

We had an exciting time, shells bursting all around and bullets whistling past us. but I got there and back without losing man or beast. On the way up I passed one of our sections (Ross) getting on well, and wonderful to relate, we have had only six men wounded so far, none seriously. They were glad to see me when I got up with my little lot to the firing line...

"On the way back I picked up eight wounded, and put them on the ponies and brought them in. Our guns and ponies have been invaluable — the guns were easy to land, and the ponies were the only means of getting water and ammunition to the firing lines in our neighbourhood. All our spare ponies, about 160, are doing transport work at present. We have only kept enough to move the guns from place to place."[945]

LT ANGUS MACDONALD, ROSS BTY, HIGHLAND MOUNTAIN BDE, RGA, 29TH DIVISION.

"Bearers were pushed out following the advance along the Krithia road & during the evening they were carrying in wounded on stretchers... This work though very hard on account of the distance (about 2 miles) was continued without a break until 5 a.m. Owing to congestion at the Casualty Clearing Station, 36 cases had to be retained at the lighthouse & shelters were built for these. The night was bitterly cold & very wet & the bearers gave up their overcoats & ground sheets & even tunics to cover the wounded. Hot tea was given to wounded during the evening & night."[946]

LT-COL. THOMAS FRASER, 89TH (1ST HIGHLAND) FIELD AMBULANCE, RAMC, 29TH DIV.

"My first field case was one shot in both legs, and while I was treating him four shots from a sniper showed that they were not far off. All evening when it was late they tried to take us, but they missed. I had to take shelter in the trench from another sniper, who fired about ten shots at us, but we lay low, then slipped away quietly."[947]

SGT WILLIAM GIBB, 89TH (1ST HIGHLAND) FIELD AMBULANCE, RAMC, 29TH DIVISION.

"The object of the Commander-in-Chief was now no longer to seize these positions, but to hold the ground he had won in face of the determined Turkish counter-attacks which were launched against the centre and right of his line at 3 p.m. The 88th Brigade, stubbornly resisting, was nevertheless forced to yield some ground, and thus our line was bent, but not broken. More serious in its results was the enforced retirement of the French, who were driven back along the western slopes of the Kereves Dere and thus uncovered the right flank of the 88th Brigade. This unfortunately caused heavy casualties to the Worcestershire Regiment."[948]

ELLIS ASHMEAD-BARTLETT, WAR CORRESPONDENT.

"About four in the afternoon I had a spare hour, and felt entitled to use it as I chose. The wounded were beginning to come down, and I thought I'd see if I could find any of our fellows...

---

945 *Oban Times and Argyllshire Advertiser*, 5th June 1915.
946 89th Field Ambulance, Royal Army Medical Corps War Diary, TNA WO 95/4309.
947 *Aberdeen Evening Express*, 26th May 1915.
948 Ashmead-Bartlett, *Uncensored Dardanelles*, pp. 77–78.

"I started off on a voyage of discovery along the cliff that rose abruptly behind our narrow sand-strip of beach. Before I had gone far I saw coming towards me a figure that I seemed to recognise. It turned out to be Captain Lindsay, and I was as glad to see him as if we'd been foster-brothers parted for years. It is extraordinary how your heart leaps at the sight of a familiar face at the front, when things are a bit thick, and your own people scattered... I bombarded Lindsay with questions, and he told me his news, some of it none too good. The CO., Lt-Col Wilson, and the senior major, Major McDonald, were wounded. The adjutant, Captain W. D. Hepburn, had been killed, and so had the regimental sergeant-major, R.S.M. F. Bailey; and several of our subalterns were knocked out, some for the time being, others for all time. Lindsay was pretty thoroughly done up, so I took him down to a pal of mine on one of the trawlers that by luck was just at the beach then, and the skipper produced tinned salmon, biscuits galore, and rum. I had to hurry back now and get on with my job, but I left Lindsay in good hands, tucking in vigorously. Heaven knows what or when he had eaten before! He was feeding as if it had been at some remote period, and little enough at that."[949]

CAPT. ALBERT MURE, 1/5TH BN ROYAL SCOTS (LOTHIAN REGT), 88TH BDE, 29TH DIV.

"Owing to the centre of our line wavering, [at 4.30 p.m.] O.C. 26th battery brought one section up into action in the open. The infantry were steadied to a great extent, but the section suffered some casualties, including the Major[950] killed."[951]

CAPT. FREDERICK MORGAN, 17TH BDE, ROYAL FIELD ARTILERY, 29TH DIVISION.

"5.30 p.m. Obvious that further advance impossible, so troops ordered to connect & establish themselves on existing line for the night.

"Most of the night spent in linking up."[952]

MAJOR-GENERAL SIR AYLMER HUNTER-WESTON, COMMANDING 29TH DIVISION.

"At 6 p.m. orders were issued for the whole line to entrench where it stood and to hold on at all costs. The net gains and losses of the day can be shortly summarised. The attack had absolutely failed in its objective, heavy losses had been suffered, but some ground had been gained. On the evening of the 28th, the line held by the British extended from a point on the coast three miles north-west of Tekke Burnu to a point one mile north of Eski Hissarlik, whence it was continued by the French south-east to the coast."[953]

ELLIS ASHMEAD-BARTLETT, WAR CORRESPONDENT.

---

949 Mure, Albert Haye, *With the Incomparable 29th*, pp. 43–47, W. & R. Chambers (Edinburgh) 1919.
950 Major John Pattison, 26th Battery, 17th Brigade, Royal Field Artillery, was killed in action on 28th April 1915. Commemorated on the Helles Memorial, he was the 40 year-old son of Admiral John Pattisson and Emma Pattisson, of Rodbourne, Tonbridge, Kent.
951 17th Brigade Royal Field Artillery War Diary, TNA WO 95/4308.
952 Diary entry, 28th April 1915, Maj. Gen. Hunter-Weston, British Library, MS 48364.
953 Ashmead-Bartlett, *Uncensored Dardanelles*, p. 78.

"We were firing all day, as our infantry were trying to hold against terrible odds, but ran out of ammunition, and as fast as men left the firing-line to try and fetch some up they were popped off by snipers, and so eventually they had to retire to the trenches they held the day before. We were firing hard all day to support them. The French retired on our right, and the Turks were within a few hundred yards of our guns. We fully expected that we should have to have a go at close quarters to save the guns, but, luckily, our chaps held on, and things quietened down a little towards nightfall."[954]

GUNNER PETER McGUIRE, "Y" BATTERY, ROYAL HORSE ARTILLERY, 29TH DIVISION.

"We were told off… at 7 p.m. to go and provide a water supply for the brigade holding the advanced positions. Well, we worked hard all day and were rewarded at last with success, and sunk a couple of wells.

"We were not to be finished work yet, for just as we were gathering our gear together an Infantry Major asked our officer if we would volunteer to support the 3rd line of fire. This we did, but got more than we bargained for. The Turks got through the French lines, causing our men also to retire and this leaving us in the front firing line, but we stuck to our guns and emerged safely next morning, with a few casualties, including an officer."[955]

SPR JOHN PUGH, 1/2ND (LOWLAND) FIELD COY, ROYAL ENGINEERS, 29TH DIVISION.

"Shells are passing over my head from the Fleet, but the rifle fire appears to have died down. Wounded are straggling back in twos and threes, and bearers carrying the more serious cases, with great fatigue to themselves. To carry a man two and a half miles over rough ground on a stretcher is hard work. Nearing the line, I pass police forming battle posts, and these, together with the badges of the wounded men, which are sewn on their tunics, returning to the beaches, helps me to steer my course. Now and again I am warned not to go near where snipers are said to be, and perpetually I trip over thin black wires, which serve for the nonce for signallers' cables. Passing a cluster of farm buildings, I arrive at last at a scene of great activity and feel relieved that I am once more amongst men. A trench is being dug with forced energy, orderlies are passing to and fro, signallers at work laying cables, doctors dressing wounded, and bearers carrying them to the rear. I discover that we have had a set-back."[956]

CAPT. JOHN GILLAM, DIVISIONAL TRAIN, ARMY SERVICE CORPS, 29TH DIVISION.

"We had to go out three miles to collect the men, and having only two men to each stretcher, it was terrible work. We had to go over trenches, ditches and fields. The dead were lying all about us… [and] it rained heavily, handicapping us severely."[957]

PTE GILBERT RESBURY, 88TH (1ST EAST ANGLIAN) FIELD AMBULANCE, RAMC, 29TH DIV.

---

954 *Boston Guardian*, 4th September 1915.
955 *Motherwell Times*, 4th June 1915.
956 Gillam, *Gallipoli Diary*, p. 52.
957 *Cambridge Independent Press*, 11th June 1915.

Those beyond the reach of stretcher-bearers had to make their way to safety as best they could.

> "I lay on my back till dusk fell, hearing the sounds of battle wavering to and fro, but all the while believing that the British had reached Krithia. All was quiet, and after a struggle of almost two hours, I got rid of my equipment, and leaving a few water-bottles with my fellow sufferers, I went to seek the ambulance we had hoped for so long. Reaching the road, I turned towards Krithia, and passed badly wounded Turks who had crawled there... I turned and ran along with what speed my condition allowed. A cry from the sentry brought at least thirty men out, who ran parallel with the road, and cut off my way to the right, which I judged to be the direction of our camp. Desperately I dived into the low scrub on [the] left of road, though the bright moonlight gave little hope of cover, but hardly had I gone a hundred yards when I saw a small hole and dropped in exhausted, pulling the earth and vegetation round me. Almost miraculously, it seemed, their search failed, and after much discharging of rifles, silence reigned."[958]

LT-COL. JAMES WILSON, 1/5TH BN ROYAL SCOTS (LOTHIAN REGT), 88TH BDE, 29TH DIV.

> "The cries and groans of the men, badly wounded, were awful to hear. As I soon as it got dark, I decided to try and get back to our lines. My leg was broken just above the ankle. Having cut my boot off, I put on my field bandage, improvised my note books as splints, and, winding my putties round the lot, made quite a respectable job of it. I had just finished, when I saw someone crawling towards me. Covering him with my rifle, I said, "Who's that?" and a voice replied "Me, Sergeant Batchelor."[959] There was another man with him, both being wounded. While we were talking a shot skimped over our heads; so we moved off to try and get back. I could not keep up with the other two, who could crawl on their hands and knees, whilst I could only drag myself along; so I told them to get on and I would take my time, as the lines could not be far away... It now began to rain in torrents, and I was soon wet through."[960]

SGT BERTIE CASTLETON, 2ND BN HAMPSHIRE REGT, 88TH BDE, 29TH DIVISION.

> "Ordered to the firing line. Our first taste of shrapnel. Sudden orders to retire. A Company straight ahead. Guarding the hill. Rain comes on. A bitter cold and miserable night."[961]

OS WILLIAM BURNETT, RNVR, 1ST (NAVAL) BDE, DRAKE BN, ROYAL NAVAL DIVISION.

> "The 2-mile advance had the effect of somewhat easing the congestion on the beaches that up till now had ever been growing greater as more guns, ammunition, mules and all the paraphernalia of a modern army were landed. The beach parties of naval

---

958 Lt-Col. James Wilson, quoted in Weaver, Lawrence, *The Royal Scots*, pp. 230–231, Country Life (London) 1915.
959 Boer War veteran, L/Sgt Walter Batchelor, 2nd Bn Hampshire Regt, had enlisted on 21st June 1898. Landing at Helles on 25th April 1915, he was discharged from the Army due to wounds on 27th October 1916. His army service record gives two dates for his wounding – 4th May and 28th June 1915.
960 *Hampshire Telegraph and Post*, 2nd July 1915.
961 *Dundee Courier*, 4th June 1915.

ratings under Captain Phillimore, of the wounded *Inflexible*, had been hard at work night and day, clearing lighters, building piers, carrying ammunition to the fighting line and helping with the wounded. Theirs was a task of hard manual labour carried out with the proverbial cheerfulness of the British bluejacket, who is apt to regard any work performed outside his own ship, whatever its nature, in the light of a picnic. On this occasion there was an extra spur to exertion, if indeed that were needed, in their admiration for the gallantry of their comrades of the Sister Service who for so many hours had been fighting, fighting, fighting. Without arms they could not follow their hearts into the trenches, but they could, and did, assist by supplying ammunition and helping to evacuate the wounded."[962]

REAR-ADMIRAL ROSSLYN WEMYSS, RN, HMS *EURYALUS*.

The extension to the beachhead was hardly the objective of the attack or compensation for the losses incurred.

ANZAC

The lack of sleep, stress of battle, the death and wounding of friends and constant threat the men faced had taken its toll. The men were in desperate need of relief.

"The whole scene on Anzac Beach reminded one irresistibly of a gigantic shipwreck. It looked as if the whole force and all the guns and material had not been landed, but had been washed ashore."[963]

ELLIS ASHMEAD-BARTLETT, WAR CORRESPONDENT.

"[The day] dawned with a fierce attack from the enemy, immediately followed with terrible artillery fire from both sides. Time after time the Turks were driven back with heavy losses. The warships were firing all day. It seems marvellous that the battleships fire over hills miles and miles away, and never see what they are hitting. Yet they land with wonderful accuracy. It is all done by communication with aeroplanes. For some time we were under the fire of six pieces of field artillery and our guns could not silence them. At last the *Queen Elizabeth* got to work. A shell from the *Queen Elizabeth* weighs about half a ton, and contains thousands of bullets. Where its shrapnel explodes, it rakes the ground to the extent of two acres. I mean to say, it would be almost certain to kill or wound anything within an area of two acres. I am glad the enemy had no such guns. The *Queen* started firing at the six guns that were tormenting us. The first two shells put four of the guns out of action and the next two got the remaining two guns, and the fifth shot set fire to the ammunition. So, you see, when a 15-inch gun of the *Queen Elizabeth* speaks, it doesn't speak for nothing.

---

962   Wester-Wemyss, *The Navy in the Dardanelles Campaign*, p. 111.
963   Ashmead-Bartlett, *Uncensored Dardanelles*, p. 57.

"As I said previously, fierce fighting continued throughout the whole of the day. Talk about the horrors of war. It was cruel. Our poor chaps lay groaning and moaning everywhere. Others were dead. It is too fearful to write about, so I will pass it over, because... my nerves are all unstrung."[964]

PTE JOHN COLLESS, 3RD BATTALION, 1ST BDE, 1ST AUSTRALIAN DIVISION.

"Still holding our ground; the men in the best of spirits, and entrenched well... We (the Signalling Section) buried poor Jack Denney last night, and Sergeant Tuckett[965] and I collected all his personal effects and addressed them to his mother. We are also sending her a letter, with full details. More, I am sorry to say, we cannot do."[966]

CPL ROBERT HUNTER, DIV. SIGNAL COY, AUSTRALIAN ENGINEERS, 1ST AUSTRALIAN DIV.

"[W]e went along the beach in front of the Turkish snipers to bury some dead Australians, who had been lying on the beach for some days, and of course the Turks had a good fly at us again. We could not see one of them to have a shot at back. In the finish, we got three Australians buried and had two of our men killed and five wounded, and we had to leave our dead without burying them until later. General Walker, in charge of us, came and congratulated the 13th on the way they had behaved, but there is not much satisfaction in losing lives trying to bury the dead, especially when you cannot hit back."[967]

PTE WILLIAM O'BRIEN, CANTERBURY BN, NZEF, NEW ZEALAND BDE, NZ & A DIVISION.

"We... were told off to bury some of our dead along the beach. We were fired at by snipers in the thicket on the terrace. It was here Bob Currie,[968] of Ross, was shot dead, and two others slightly wounded. We had only buried three of our dead when we were ordered to retire, having the same number of dead as when we started."[969]

PTE ARTHUR BROWN, CANTERBURY BN, NZEF, NZ BDE, NZ & A DIVISION.

"I can only describe it as an awful battle. Imagine the worst thunderstorm you ever heard... You would have have thought after the last two days' battle no one could have emerged alive from the inferno. The Turks sank one of our trawlers yesterday, but it was near the beach, and did not go out of sight."[970]

DVR GORDON McSKIMMING, 9TH BTY, 3RD FA BDE, AFA, 1ST AUSTRALIAN DIVISION.

"We got up more artillery; also some Indian mountain guns. They are only little guns, but they can do some damage. We also got a transport section of Russian and French Jews

---

964 *Mudgee Guardian and North West Representative* (New South Wales) 28th October 1915.
965 Sgt, later Capt. Lewis Tuckett, MC, MM, Australian Engineers, a former Civil Servant from Victoria Park, Western Australia.
966 *The Evening Star* (Boulder, Western Australia), 26th June 1915.
967 *Lyttelton Times* (New Zealand), 1st July 1915.
968 Pte Robert Hunt Currie, Canterbury Bn, NZEF, was killed in action between 25th April and 1st May 1915, according to official records. Commemorated on the Lone Pine Memorial, the Boer War veteran was the 37 year-old son of the late John Hunt Currie and Helen Currie, of Ross, Westland, Greymouth.
969 *Wanganui Chronicle* (New Zealand), 17th June 1915.
970 *The Examiner* (Hobart, Tasmania), 30th June 1915.

— refugees from Palestine. They were very handy with their pack mules. Also some Greeks from the islands, with their handy little donkey teams. It was moonlight early, and there was a little green crop in front of us. A scout came in and told us the enemy was forming up there. Our skipper passed the word that the signal to fire would be our machine-gun. He waited for a cloud to come off the moon, and there they were. The rifles spoke as one in answer to the signal and in the morning the crop would hardly be called a green one."[971]

PTE GEORGE GIDEON, 6TH BATTALION, 2ND BDE, 1ST AUSTRALIAN DIVISION.

In addition to the extra guns and mules, Royal Marines arrived to relieve some of those so much in need of a chance to rest and reorganise themselves.

"And this day there steamed up about midday to the anchorage off the shore several transports, which brought a certain number of troops of the naval and marine brigades as a temporary relief. These troops had been through one of the most trying ordeals in the war — or rather, a portion of them had — the siege of Antwerp. They enabled the brigade of the first Australian division to be taken out of the trenches one at a time — mustered, reorganised, and in some cases rested. It was then for the first time that the battalion commanders obtained real particulars as to what the losses of their battalions were. Those losses showed what those who had been in France told us to be the fact — that the fighting we had been through was just as hard as the hardest fighting in the war. No battalion had been wiped out, but every battalion had suffered heavily — the losses running extraordinarily evenly through the whole of the first Australian division. So ended the second episode in the campaign of the Australian and New Zealand force in Gallipoli. The battle which had begun in the scrub had gradually crystallised into a battle of trenches, which it has remained."[972]

CHARLES BEAN, AUSTRALIAN WAR CORRESPONDENT.

"We were on the mess deck of the "Gloucester Castle," getting tea, when the order came to don our marching order. We were immediately packed on torpedo boats. All went well till we got about seven hundred yards off shore, then a bullet from shrapnel would hit the funnel, and a regular "storm" began. As we got about a hundred yards from the beach, someone fell by my side. I looked round to see what was the matter, and, to my horror, I found one of my platoon mates dead, shot through the head. As we rushed off the boats, the Australians gave us a hearty cheer. Never shall I forget that night. It was raining "cats and dogs." As we marched up the first hill, to the trenches, (Portsmouth Battalion going first), the Australians were continually saying, "Take no prisoners, lads, bayonet the lot of them." And bet your life on it, we did as they asked us to. Eventually, after going up hill and own dale for about five miles [sic], we landed in the trenches, which were knee deep in mud and water. The

---

971   *Flemington Spectator* (Victoria), 23rd September 1915.
972   *The West Australia* (Perth, Western Australia), 12th July 1915.

Turks must have spotted us, for they gave us shrapnel and hand grenades, in addition to infantry fire, which was terrible. It was just like a "living hell."[973]

PTE FRANK TIMMINS, PORTSMOUTH BN, RMLI, 3RD (ROYAL MARINE) BDE, RND.[974]

"We had no difficulty in landing, as the Australians had paved the way for us. We had not been on shore long before it commenced to rain very hard, this making matters worse, and our progress to the trenches was very difficult, for at places we could only proceed by Indian file, at times we were up to our boot tops in mud and at others we were crawling on all fours up a hill and making way for the wounded to pass by to the base hospital. Reaching the trenches, we lay in them… All the time the firing of rifles was going on… Our greatest danger was with the sniper, as the country afforded good cover for him."[975]

PTE ISAAC EAGLE, CHATHAM BATTALION, RMLI, 3RD (ROYAL MARINE) BDE, RND.

"I went ashore… three days after the first landing. The night was very miserable. We did not land until about 7 p.m., when we took up a position on the beach to await the reassembling of the battalion the next day. It rained for the greater part of the night, so we crawled into some dug-outs and tried to sleep. I could not, as my feet were as cold as ice."[976]

LT ERIC STUTCHBURY, 3RD BATTALION, 1ST BDE, 1ST AUSTRALIAN DIVISION.

"I left the trenches to see another officer. A sniper had a shot at me, but hit a stone instead. The bullet ricochetted and struck me on the left leg, but only caused a bruise. At 8 p.m. I was relieved to join my battalion, and an hour later met my company commander,[977] now Lt-Col, and in charge of the battalion, and was informed of the death of four of our officers, including the C.O. (Colonel Clarke[978]), and the second in command (Colonel Hawley[979]) had been seriously wounded and had never got ashore."[980]

LT HERBERT ORBELL, 12TH BATTALION, 3RD BDE, 1ST AUSTRALIAN DIVISION.

"Rest all day. Collecting at night. Captain Buchanan in command of our section. *Goeben* [sic] shelled ships in bay; no damage. A few men shot by snipers. Reinforcements arrived. Third Brigade 48 hours' spell from trenches. Things much better. Charge repulsed."[981]

PTE SYDNEY PENHALIGON, 3RD FIELD AMBULANCE, AAMC.

973 *Mansfield Reporter & Sutton Times*, 2nd July 1915.
974 Served as Grosvenor.
975 *Lynn News & County Press*, 12th June 1915.
976 *The Daily Telegraph* (Sydney, New South Wales), 9th July 1915.
977 Major, later Lt-Col. Ernest Hilmer Smith, was appointed to the command of 12th Bn, AIF, on 26th April 1915.
978 Lt-Col. Lancelot Clarke, 12th Bn, AIF, was killed in action on 25th April 1915. Buried in Beach Cemetery, Anzac, he was the 57 year-old son of the late Joseph Johnston Clarke and Charlotte Elizabeth Clarke; husband of Beatrice Fox Clarke, of Holebrook, Tasmania, originally of Melbourne.
979 Lt-Col. Sydney Hawley, 12th Bn, AIF, was wounded, a gunshot wound to his spine, on 25th April 1915. He returned to Australia on 8th May 1915.
980 *Daily Telegraph* (Launceston, Tasmania), 8th September 1915.
981 *The Queenslander* (Brisbane, Queensland), 21st April 1917.

"Tonight has been nothing but work for our section. It seems that the Turks have been sounding the alarm, and afterwards the retreat, on bugles, with our own calls, thereby causing men to leave their trenches. In consequence we had to cut out bugle calls altogether. Later on verbal and written orders were passed down the lines, with staff office signatures on, telling unit commanders to leave their lines at a certain time and also saying that such and such a battalion would relieve them."[982]

CPL ROBERT HUNTER, DIV. SIGNAL COY, AUSTRALIAN ENGINEERS, 1ST AUSTRALIAN DIV.

Some Australians were obliged to get used to very different surroundings.

"I and Lt Elston, of my battalion, with two privates, Lushington and Bugler Ashton, both of Western Australia, have been taken prisoners. We have received splendid treatment, and all the Turkish officers have done everything that could be expected of them to make us feel at home. We get awfully good food and plenty of cigarettes, have comfortable beds, and a good room. As it is possible that we have been posted as missing by our people it would be as well to notify the Department that we are not killed, but are receiving excellent treatment. I am well and all the party are. Our soldiers are being treated just as well as Elston and myself."[983]

CAPT. RONALD McDONALD, 16TH BATTALION, 4TH BDE, AIF, NZ & A DIVISION.

AEGEAN

"[That] afternoon five more were gone, including one of our brave Indian soldiers — we had three on board altogether. It was the first time that I had conducted a burial service at sea, and it affected me very deeply. The ceremony was carried out as quickly as possible under an awning on the after well deck, the ship's engines being stopped for a couple of minutes at the time of the committal of the bodies to the deep."[984]

COLONEL CHAPLAIN JOHN McPHEE, CHAPLAIN'S CORPS, AIF, HMT *IONIAN*.

"I have been helping to nurse the poor officers and men on our ship. We work day and night attending to them. Some of them have died and others are dying in front of our eyes. And, mother, the best bit of work that ever happened was that we were left on the boat... otherwise these poor creatures, most of them at any rate, would not have anyone to attend to them on the way back to the hospitals in Cairo and Alexandria.

"Some have faces blown away, legs off, arms shattered, others have great holes in the lungs, head, or stomach, feet missing, and skulls smashed in. Oh, it has quite upset me having to help to bandage up these poor comrades. You have no idea what the terrible sights are like. Dear mother, you can say that Bill[985] and I and the chaps left on our ship have done our duty towards those dying and wounded; in fact we could not have done this had we been ashore with the first lot. Wait until you see the large

---

982 *The Evening Star* (Boulder, Western Australia), 26th June 1915.
983 *Newcastle Evening Chronicle*, 10th September 1915.
984 *Daily Herald* (Adelaide, South Australia), 21st July 1915.
985 His brother, Pte William Edward Darby. Later promoted to Cpl, he was invalided from the army after being gassed on 19th February 1918 while serving with the Wellington Regt on the Western Front.

list of dead and wounded. Three hours before the boys left the ship they were all eager; six hours after, most of them were dead or wounded."[986]

PTE JOHN DARBY, NZ PERMANENT STAFF, HQ, NZ & A DIVISION, HMT *LUTZOW*.

29 APRIL 1915

ENGLAND

In conversation with Sir George Riddell, Churchill made the astonishing claim that the Turks had not fired upon the *River Clyde*; the men aboard it disembarked unmolested. Was Churchill Asquith's source when he recorded the same tale the previous day?

> "I think the Dardanelles Expedition will be successful. The most heroic deeds have been accomplished during the past few days. Our men have been landed in the face of terrible difficulties and dangers. Miles of barbed wire; some of it sunk beneath the sea. We performed remarkable feats. The whole thing was most carefully arranged. For example, an old collier ran on to the shore. The Turks took no notice of the supposed wreck; they did not even fire at it. At the right moment 2,000 men marched out of the collier on to the shore.

> "This is one of the great campaigns of history. Think what Constantinople is to the East. It is more than London, Paris, and Berlin all rolled into one are to the West. Think how it has dominated the East. Think what its fall will mean. Think how it will affect Bulgaria, Greece, Rumania, and Italy, who have already been affected by what has taken place. You cannot win this war by sitting still. We are merely using our surplus ships in the Dardanelles. Most of them are old vessels. The ammunition, even the rifle ammunition, is different from that which we are using in France — an older type — so there is no loss of power there. I am not responsible for the Expedition; the whole details were approved by the Cabinet and Admiralty Board. I do not shirk responsibility, but it is untrue to say that I have done this off my own bat. I have followed every detail. Fisher and I have a perfect understanding."[987]

WINSTON CHURCHILL, MP, FIRST LORD OF THE ADMIRALTY.

SEA OF MARMORA

Allied submarines had made an impact by their simple presence in these narrow waters. On his way to take command of the Ottoman 9th Division, Hans Kannengiesser saw evidence of their activity.

> "About 2.30 I suddenly heard rifle shots, which I could not understand. I hurried to the bow of the ship and saw close ahead, and squarely crossing our bows, a bubbling

---

986 *Wanganui Herald* (New Zealand), 21st June 1915.
987 Winston Churchill quoted in Riddell, Sir George, *Lord Riddell's War Diary 1914–1918*, pp. 51–52, Ivor Nicholson & Watson (London) 1933.

line in the water — the track of a torpedo, the first I had seen in its own element. Ah! An English submarine!

"Our torpedo-boats drove round us like wild dogs, but there was nothing to see. The ships were now ordered to sail close to the shore so that the torpedo-boats had only one side of the transports to protect. Three-quarters of an hour later I drew the attention of my interpreter, Major Zia Bey, to a four-cornered black flag which was apparent on the Asiatic shore, a fairly long way off in the direction of Kara Bigha. He had hardly the words out of his mouth, "the sails here look like that," when several shells exploded in the neighbourhood of the apparent sail, which was actually an English submarine and which quickly submerged. It was a splendid sight for us to watch how the torpedo-boats steamed at full speed towards the place where the submarine had been. Without success! We had, however, a comfortable and thankful feeling of safety."[988]

MAJOR HANS KANNENGIESSER, COMMANDING 9TH OTTOMAN DIVISION.

The British submarine HMS *E14* had passed through the Narrows earlier that day. Its commander, Lt-Cdr Edward Boyle, recorded a successful attack upon a Turkish transport that afternoon. He seems a more likely candidate for Kannengiesser's close encounter than the only other submarine in those waters, HMAS *AE2*.

HELLES

That night men isolated since the previous day's fighting, some of them wounded, attempted to make their way back to British lines. At that time the peninsula was not covered by more than isolated sections of trench, making it difficult to tell who was who.

"I... made up my mind to crawl away. It were just as well to be caught getting away as to wait and be captured. I kept going until daybreak, when I saw coming towards me a sniper... I could not get away, so I simply laid flat out as if I was dead. I heard cries to my right, which I reckon saved my life. Evidently the sniper heard the cries also, because he made towards that direction.

"I heard talking going on; it was a wounded Turk. I then heard an awful yell. I guess that the sniper was trying to move him, but gave it up as a bad job. After waiting a few minutes I slowly turned my head to have a look round, and I saw the muzzle of a rifle resting on a small tripod poking through a bush not 25 yards away and pointing straight in my direction.

"It was certain death for me to move, so I laid there all day expecting to get shot any minute. I can't explain to you what I went through that day. I was wet through to the skin and very cold. I suffered agonies from my leg, which was not improved by being dragged over the rough ground. I was feeling pretty well exhausted from the want of food and drink. I also got the cramp through lying in one position, and to add to my misery small ants began to crawl up my left arm, and over my face, which was lying

---

988   Kannengiesser, Hans, *The Campaign in Gallipoli*, pp. 123–124, Hutchinson & Co. (London) 1928.

on the ground, and during the day the sun was unbearable. I did not dare move as much as a finger. I thought I should go mad. Several times I had a hard struggle to prevent myself from jumping up and ending it all."[989]

SGT BERTIE CASTLETON, 2ND BN HAMPSHIRE REGT, 88TH BDE, 29TH DIVISION.

"The long night wore on, an icy rain fell for two hours, my wounds stiffened, and my hunger was appeased by lozenges. At last the dawn, an hour after which I judged would be the safest hour to escape. I found myself unable to move, owing to the earth having caked with rain. Digging with my clasp-knife at length released me, and I crawled, now unable to walk, to the bed of a little stream, and with many pauses and much care, wriggled thence to near the road, where I could look round the country. I was spotted several times, but the Turks were too busy looting bodies to come after me. The sun by this time had revived my strength, and when at least one and a half miles away, I saw British troops moving in regular lines (it turned out they were systematically hunting snipers), I determined to risk all, and got to my feet. All went well until I was within half a mile of our troops, when two bullets in succession whizzed by. Experience had taught me that a sniper will not fire on a dead or badly wounded man, and when the third bullet came, I fell, simulating disablement, in such a way as to be able to watch our men. Four hours later two approached within a hundred yards, and I shouted and waved. They were about to shoot me, as my very dishevelled condition suggested a Turk, but curiosity prevailed. They were men of the 1st Essex — I was saved..."[990]

LT-COL. JAMES WILSON, 1/5TH BN ROYAL SCOTS (LOTHIAN REGT), 88TH BDE, 29TH DIV.

Everyone was tired. But some could make more comfortable arrangements to take their rest.

"I wake at eight, but am given permission to sleep all the morning. I have breakfast. Getting fed-up with biscuit. My servant rigs me up a "bivvy" and I roll up and go fast asleep. Lord, what a gorgeous sleep it was! I slept till one, and then had lunch, and after, a shave and a wash. I did little all day but watch the Fleet firing and the transports unloading everything imaginable necessary for an army."[991]

CAPT. JOHN GILLAM, DIVISIONAL TRAIN, ARMY SERVICE CORPS, 29TH DIVISION.

Roger Keyes accompanied Sir Ian Hamilton ashore, seeing for themselves what they had only seen from aboard ship up to then.

"Sir Ian and some of his Staff, Godfrey, Ramsay and I embarked in the destroyer *Kennet* attached to the *Queen Elizabeth*, and landed at "W" Beach.

"The naval and military beach parties were working with feverish energy and landing animals, guns, ammunition and stores; a considerable swell, which had risen during the night, did not lighten their labours. Crowds of wounded were limping down to the beach, or being carried in stretchers to await transhipment to the transports for

---

989   *Hampshire Telegraph and Post*, 2nd July 1915.
990   Lt-Col. James Wilson, quoted in Weaver, Lawrence, *The Royal Scots*, p. 231, Country Life (London) 1915.
991   Gillam, *Gallipoli Diary*, p. 59.

passage to Egypt or Malta, or to fleet sweepers, to be carried to the field hospitals at Mudros. The hospital ships had already sailed, crowded far beyond their normal capacity. The patience, fortitude and endurance of the wounded, under conditions of incredible hardship and suffering, is a lasting memory...

"We were amazed at the strength of the defences which our troops had stormed. In addition to the deep concealed trenches, two caves had been dug into the cliffs, and it was from these that their maxims [sic] had enfiladed the main wire entanglement. We were told that the first rush of four officers and 75 men fell against it in a straight line.

"Our boats' crews had had a good many casualties when the boats were beached, but without exception they had succeeded in getting off and pulling back for reinforcements, in some cases with only two or three survivors, in a boat full of killed and wounded. Directly the latter had been removed, and the casualties made good, they started back again. When the flank was turned, and the main defences were captured, the situation was, of course, much improved, but casualties were suffered all day from the fire of snipers, until the advance on the 26th freed the beach from rifle fire.

"From "W" Beach we walked over the hill to Sedd el Bahr, only a mile away, but a bloody mile if ever there was intersected by trenches and wired redoubts, the latter like the maze at Hampton Court. We went to the place where we had watched the men of the Worcestershire Regiment cutting wire, and found that they had been working within a few yards of the Turkish trenches. Then down to "V" Beach, it was here the Dublins lost most of their people. We were told that after the action of the previous day they were reduced to one officer and about 300 men, and had joined up with the remnant of the Munster Fusiliers and called themselves the 'Dubsters.' We came across one grave with 350 in it, and many others. Of the naval party employed here 63 were killed or wounded out of the 80 engaged. The position is aptly described in our official telegram as an amphitheatre with the beach as a stage. The whole circumference of the beach had been honeycombed like a rabbit warren with trenches, tunnels and caves, converging on the landing place, and protected by the most appalling barbed wire, far more formidable than ours at that date, and practically uncutable by our wire cutters. "V" Beach was impregnable to frontal attack, and there is nothing else to be said about it."

"After three hours on shore, we returned to *Queen Elizabeth* with much to reflect on. The appalling nature of military operations under modern conditions, and the magnitude of the Army's task was now only too apparent, and I felt that the sooner we took a hand the better, an opinion most fully shared by Godfrey. We felt that a naval attack could not fail to turn the position for the Army, and signals were prepared for dispatch to the commanding officers of units, directing them to complete with fuel, stores, and ammunition to the full capacity of their ships."[992]

COMMODORE ROGER KEYES, RN, CHIEF OF STAFF, EMS.

All the time, the landing of stores and equipment continued.

---

[992] Keyes, *Naval Memoirs 1910–1915*, pp. 317–318.

"Our people are advancing slowly. They are, I know, very tired, but every moment we are putting on shore more guns, more men and more ammunition, and I sincerely hope many more men will come from England. I am sure we shall want them. Our losses must be very, very severe, but one hasn't had time to count. Everything has to be pushed forward so rapidly… I shall have the pleasure of getting two of our officers the V.C. for acts of gallantry and self-abnegation seldom if ever equalled. There must be many such, but only these two so far come under my notice. I do hope we may break the Turkish resistance soon. If we do, and can get at Constantinople, it should, I am sure, help very greatly to bring this beastly war to an end."[993]

REAR-ADMIRAL ROSSLYN WEMYSS, RN, HMS *EURYALUS*.

"I was again running stores to the beach, and having a little time to spare in the afternoon, I went up to the hospital tents. They always seem to call one. I came across one of my own men, who told me that the CO. was in a small bell-tent close by, and I immediately found it and him. He was lying on a stretcher, looking rather limp and very much bandaged, but wonderfully cheery — as cheerful as I had ever seen him. Yet he had just been through hell, and had come out of it not unscathed. It seemed to ease him to talk, and to do him no harm, and I was keenly glad to listen.

"First of all, he gave me half-a-dozen urgent instructions, and when I had assured him that they should be carried out scrupulously, he plunged into his story… While our commander was still dictating Captain Lindsay came in, and we three had a good old regimental pow-wow. But when we thought the CO. really ought to talk no more, Lindsay and I said, 'Cheer-o, sir,' and 'See you later, sir.' I walked a little way back with Lindsay along the cliff, and then, as I had to get back to my own 'odd jobs,' I said how much I hoped to be beside him in a day or two, and we exchanged a careless, friendly good-night. He went off, whistling 'Annie Laurie.' I never saw him again. I walked back to my job. He[994] walked on to his death."[995]

CAPT. ALBERT MURE, 1/5TH BN ROYAL SCOTS (LOTHIAN REGT), 88TH BDE, 29TH DIV.

"We fell in at 5.30 p.m. Shells were screaming round occasionally, but we had no casualties. HMS "Cornwallis" (battleship) had sent her boats for us to land in. These boats were manned by some of her crew.

"We were towed by a steam pinnace in strings of six boats. On nearing the shore we landed on a hastily constructed pontoon landing stage which the blue jackets were completing under a heavy shell fire.

"What a scene met my eyes! The beach, which was very sandy, was strewn with men's gear, picks, rifles, clothes and dead bodies. Under the shelter of the cliff was a first aid station where a crowd of wounded men were waiting for attention. Some were hurt slightly, some badly. Two long lines of stretcher bearers, with their human burdens, were also there. The men bore their wounds as cheerfully as they could.

---

993   Wemyss quoted in Wemyss, *The Life and Letters of Lord Wester Wemyss*, p. 217.
994   Capt. Douglas Lindsay, 1/5th Bn Royal Scots, was killed in action on 2nd May 1915. He is commemorated by a special memorial in Redoubt Cemetery, where he is believed to be buried.
995   Mure, Albert Haye, *With the Incomparable 29th*, pp. 51–55, W. & R. Chambers (Edinburgh) 1919.

"When all had landed we fell in under the friendly shelter of the cliffs, while an occasional shell whistled overhead and fell with a gurgling report in the sea, throwing up a column of water. Darkness was now coming on and we "fell in." All were told to keep silent. The platoon were ordered to march off one at a time and assemble with the battalion at the top off the cliffs. The Turkish guns were not firing, the reason being (I suppose) that it would have given away their positions. But there was intermittent rifle firing down in the valley. We were halted.

"My officer informed me that he was at a loss to know the whereabouts of the rest of the battalion. He asked me to get away and have a look around for them while he waited with the rest of my mates. I found them in a very few minutes and directed my officer to them.

"When we had all assembled on the top of the cliff it was pitch dark. Picks and shovels were served out, and we were told to make holes — in which we were told to sleep — as protection from shells. Cyril Ford sprained his ankle and was taken away.

"We dug in for quite an hour, and then I was told off for outpost work. Oh, the irony of it! I was so tired, too. The outposts were placed in pairs at each corner of the dug-out camp, and our orders were to report anything unusual. The headquarters were distinguished by two red lights just under a bank. G. Hurst and I were together, our post being at a corner of a Turkish trench which had been abandoned on the 25th, the day of the landing; and I may add that the odour was none too good all night."[996]

AB FREDERICK HURD, RNVR, HOWE BN, 2ND (NAVAL BRIGADE), ROYAL NAVAL DIVISION.

Amongst those landing were the French, given V Beach as their operational base.

"We have been waiting to go ashore on the European side since 11 o'clock last night. After lunch (the Colonel stood us champagne) Colonel Ruef and his staff go ashore. Colonel Noguès and staff of our regiment and a few men landed at 3 p.m.

"We passed the *Cornwallis*, which was bombarding over our heads. Krithia was in flames. Columns of smoke kilometres long hid the sky on one side.

"We approached a small boat which is up against what the Tommies call "the horse of Troy." This is the *River Clyde*, of Glasgow, which was run ashore to make a landing stage. I cross "the horse of Troy," and I step on the sacred Turkish soil. Digging has begun on the left where the English are sleeping.

"A great deal of movement, many troops, tents, horses, artillery. Human detritus of every kind. The Château of Europe is in ruins. An enemy seaplane has just dropped four or five bombs on the camp.

"We return to the *Savoie*. Krithia is burning. Shells from In Tepe are falling near the camp of the 4th Colonial."[997]

MO JOSEPH VASSAL, 6TH COLONIAL REG, BDE *COLONIALE*, 1ST DIVISION, CEO.

---

996 *Eastbourne Gazette*, 29th December 1915.
997 Vassal, *Uncensored Letters from the Dardanelles*, p. 73.

After spending days onshore, one sailor, Adrian Keyes, came back on ship. He described the experience to his brother on *Queen Elizabeth*.

> "Adrian... attached himself to the troops which were to assault Achi Baba, where he was to establish a naval observation station directly it was captured. He came on board to report himself on the 29th. I think his feelings were mixed; he said he could hardly bear to tear himself away from the Army. We could get very little out of him, except his intense admiration for the 29th Division and his sorrow at seeing most of the officers of the Scottish Borderers, with whom he had made great friends, killed alongside him. We gathered from him that Brig-General Marshall, who was wounded on the 25th but remained in action, like the other two Brigadiers of the Division, was always in the thick of every action. I think my brother's condition was typical of that of the 29th Division — dead dog-tired. He had been fighting incessantly since the 25th, and had hardly slept since the night of the 23rd. His new ship was undergoing repairs, half her bridge having been shot away, when her captain was killed,[998] so I made him lie on my bed, where he lay like a log for several hours."[999]
>
> COMMODORE ROGER KEYES, RN, CHIEF OF STAFF, EMS.

Bertie Castleton was spending his second night, wounded, in the open.

> "How I longed for the night to come! At last it came. I began to drag myself away. At first I could hardly move. I felt properly done up. My mouth and tongue were so dry I could hardly swallow. In the distance I could see the lights of the ships in the bay, but my chance of reaching them was a very poor one. At any rate, I made for their direction. After going some distance I could see dozens of bodies lying about, and I heard someone crying out "Oh, my God! [t]he pain is awful, someone get me a drink!" I made towards him, and I found he was a bugler of my company and belonged to Portsmouth. He had a water bottle, but worst luck it was empty. He was badly hit. A shot entered his back, and passed out of his stomach. He told me that a sergeant was about 50 yards further on, but he could not move, so we decided to join him. On our way we searched the water bottles of the dead that were lying about, but found them all empty...

> "I... found myself practically on the same ground, only on the bank where my company had suffered so much. I didn't expect to find much water, owing to it being so scarce. I took a haversack from one poor chap — who had died for his country — a bandolier and a rifle, not wishing to be without one after my experience of the previous night. The bugler carried the haversack, which I have now, and I took the gun.

> "We could only get along now a few yards at a time. Both of us were panting like a broken-winded horse. Calling out quietly to the sergeant, we managed to get to him. He was lying on his back in some thick shrubs. He seemed to be in a bad way, having been shot through both knees, and as he fell he got another shot, which broke the upper

---

998 The bridge of HMS *Wolverine* had been struck by shellfire on 28th April while operating inside the Dardanelles. Three men were killed: Cdr Osmond Prentice, RN; A/Sub-Lt Ivor Jones-Parry, RNR; and CPO William Endean, RN. All are commemorated on the Portsmouth Naval Memorial.

999 Keyes, *Naval Memoirs 1910–1915*, pp. 320–321.

part of his right arm. On seeing me he said, "Hullo, Cas, got any water?" I am sure we would have drunk any water, poisoned or not, if we could have found some."[1000]

SGT BERTIE CASTLETON, 2ND BN HAMPSHIRE REGT, 88TH BDE, 29TH DIVISION.

Not everyone was short of something to drink that night.

"On hearing my name, he asked me why I had come to the *Implacable*. I explained that I had been sent on board by Commodore Keyes, who had promised to make a signal. He replied, "Well, sit down and make yourself at home. Have you dined?" I replied, "No." He then ordered dinner, a whisky and soda, and when the repast was finished made me drink three or four glasses of port with him. This was my first introduction to the famous Captain "Tubby" [Hughes Campbell] Lockyer, one of the best-known characters in the Navy, of whom I had heard so much. He is about the only captain who dines in the wardroom with his officers every night, and who absolutely declines to be shut up like a "Blooming hermit," as he puts it, in his own quarters. He is a terrible man for sitting up late, and your only chance of ever getting to bed is to sneak out quietly without saying "Good night," otherwise you are certain to be collared for "just one more." This being my first evening, there was no escape, and it was 3 a.m. before I got to bed, having had far more drink than was good for me in the heat of the Mediterranean."[1001]

ELLIS ASHMEAD-BARTLETT, WAR CORRESPONDENT.

ANZAC

"We landed at Gaba Tepe the previous evening, and at daybreak we entered the trenches to relieve the Australian troops. Within half-an-hour Ellis[1002] received one through the head as he was filling his pipe, probably fired from the flank, as the enemy's snipers were concealed everywhere. He fell without a murmur, death being instantaneous. He was the first man of the Chatham Battalion to fall, although we lost terribly during the day. I was wounded during the afternoon."[1003]

SGT THOMAS BALL, CHATHAM BN, RMLI, 1ST (NAVAL) BDE, ROYAL NAVAL DIVISION.

"Found that we had passed 1005 messages through our section in four days showing how important a factor is the Divisional Signal Service in warfare… Things are fairly quiet this morning… Our position is now very secure. We are going to have a spell tomorrow, and the Marine Brigade is taking our place. Another one of our section was wounded last night, but only slightly. Just heard that the other British troops are having a touch-and-go landing about 16 miles down the Peninsula, at Sedd-el-Bahr but they have now established themselves firmly."[1004]

CPL ROBERT HUNTER, DIV. SIGNAL COY, AUSTRALIAN ENGINEERS, 1ST AUSTRALIAN DIV.

1000 *Hampshire Telegraph and Post*, 2nd July 1915.
1001 Ashmead-Bartlett, *Uncensored Dardanelles*, p. 59.
1002 Pte John Sheppard Ellis, Chatham Bn, RMLI, RND, was killed in action at Gallipoli on 29th April 1915. He is commemorated on the Helles Memorial.
1003 *Notts Free Press*, 16th July 1915.
1004 *The Evening Star* (Boulder, Western Australia), 26th June 1915.

For those leaving the line for the first time since the landing, it was time to regroup and reorganise.

> "The RMLI commenced our relief at 4 am, which owing to the difficulty of access to the trenches was not completed till late that date. As relieved the men proceeded to a rendezvous on the beach. Here we were enabled, at last, to get some idea of our losses. We were unable to do this previously as in such an operation as we had to carry out no definite allotments of frontage could be made & arriving troops must be sent into the firing line where required, irrespective of unit or brigade.
>
> "It was ascertained that the Bn stood as follows: —
>
> "Killed Officers 3 O.R. 36. Wounded 40 O.R. Missing 69 O.R. The total of officers wounded was thirteen.
>
> "The Bn bivouacked on beach for the balance of the day."[1005]

LT OWEN GLENDOWER HOWELL-PRICE, 3RD BN, 1ST BDE, 1ST AUSTRALIAN DIVISION.

> "[W]e took up a position on the beach to await the reassembling of the battalion… the survivors of the first four days fighting began to arrive at the beach, having been relieved by the marines. All day long men came in, and how tired and worn everyone looked, too. When the rolls were called, the strength of the battalion, it was found, had been reduced by 14 officers and between 300 and 400 men."[1006]

LT ERIC STUTCHBURY, 3RD BATTALION, 1ST BDE, 1ST AUSTRALIAN DIVISION.

Not everyone got to the beach without incident.

> "I was carrying despatches, and an unsympathetic mob of the enemy let go at me. I took cover behind a disgustingly dead Turk for a while, and then had another go. This time they got me, and I fell into a shell-hole where I stopped to get my wind. Then I sneaked along about another 60 yards like a worm and got into a trench abandoned by the enemy. I went along this for a while, and round the first bend there was a Turk, clad only in a shirt, looking straight at me. I should think I was frightened. I had a good look at him, and he was dead — very much so. But his eyes were wide open, and he seemed to be laughing at me. The rest of the road was easy, as I never had a shot fired at me."[1007]

PTE COULTON 'DICK' GREY, 3RD BATTALION, 1ST BDE, 1ST AUSTRALIAN DIVISION.[1008]

> "I learnt that the First Battalion had all been sent to the beach, so I went down and joined them. Members of the Battalion came rolling in one by one, and in the evening a monster parade and roll call was held. It was a very sad group that assembled; there were only some six hundred… left out of our strength of a thousand odd, the

---

1005 3rd Bn Australian Infantry War Diary, TNA WO 95/4342.
1006 *The Young Chronicle* (New South Wales), 13th July 1915.
1007 *The Uralla Times and District Advocate* (New South Wales), 17th July 1915.
1008 Returned to Australia, the wound to his left hand was serious enough for Pte Grey to discharged as unfit for further service on 23rd November 1916, aged 21.

remainder being killed, wounded or missing. In vain we looked for many familiar faces — congratulating those still alive, and mutually sorrowing for many who we would see no more. The survivors, too, made a very forlorn spectacle: unwashed and unshaven, with clothes stained with dirt, and splashed with blood; uniform and equipment cut and and torn with shrapnel."[1009]

PTE FREDERIC MUIR, 1ST BATTALION, 1ST BDE, 1ST AUSTRALIAN DIVISION.

But the beach was not a safe place either.

"We had a good wash and a shave after breakfast. After a bit of a spell we received orders to go back to the firing line. Just about the time these orders were received, poor old Ossie... who was about the last man to go and have a wash, arrived at the water's edge. He had just stooped. I believe I was about 100 yards away at the time down he went. One of the soldiers who were handy went to his assistance. Ossie said to him, "I'm shot in the back." As far as I know these were the last words he spoke. He only lived a few minutes. I heard the call for stretcher bearers, and I took no notice for a minute or two, as I was getting so used to the call, as we used to rescue all the wounded we could at night time. Any way when I got up I got a great shock to see that it was Ossie[1010] who had been hit, as he was bleeding from the nose and mouth, and he was apparently unconscious. I could see by the pallor of his face and the look in his eyes that he was settled. He only lived about a minute after I arrived on the scene."[1011]

PTE HENRY EDELSTEN, DCM, 15TH BATTALION, 4TH BDE, NZ & A DIVISION.[1012]

After his morning excursion to Helles, Roger Keyes was keen to see the positions held by the Australians and New Zealanders.

"The *Queen Elizabeth* then steamed up the coast to Anzac. The firing was incessant, but no one seemed to be paying much attention to it, and work was proceeding on the beaches without check. I was very anxious to go on shore, and see if anything was wanted, and Sir Ian whispered to me, 'Don't you think we might land and look at it.' A good deal of cold water was thrown on the suggestion. It seems that in modern war, the Staff consider that the Commander-in-Chief should not be subjected to any risk; they may be right, though I think otherwise myself. In any case the idea can be overdone, as there must be occasions when the personal touch and leadership are nine-tenths of the battle.

"We embarked in the *Colne* and steamed in towards Anzac Cove, transferring to a picket boat when about a mile from the beach. The enemy burst shrapnel over Anzac

---

1009 *South Coast Times and Wollongong Argus* (New South Wales), 20th August 1915.
1010 Pte Oswald Stewart Wemyss, 15th Bn, AIF, was killed in action on 29th April 1915. Commemorated on the Lone Pine Memorial, the former shearer was the 32 year-old son of Arthur Charles and Mary Jane Wemyss, of "Fair View," Carapooee, Victoria.
1011 *Darling Downs Gazette* (Queensland), 28th July 1915.
1012 Edelsten was awarded the DCM for his actions on 25th April 1915: "For conspicuous gallantry on the 25th April, 1915, near Gaba Tepe (Dardanelles). After the landing he passed frequently from the supports to the firing line under a very heavy fire, to keep the communications open. Later on he showed great bravery on three occasions in carrying wounded men to a place of safety." (*London Gazette*, 3rd September 1915.)

all day, but the position we hold is now wonderfully protected by tunnels, trenches, dug-outs, etc. The only direct fire one meets is in the front line trenches and at sea, otherwise the fire is all indirect and not very dangerous. But for the firing, we might have been landing on Margate beach on a warm sunny day; 2,000 to 5,000 men were about, many bathing, others all over the hillsides and in the gullies in small parties making tea. They have made wonderful roads and galleries, have got guns up into apparently inaccessible positions, and seem very happy and contented. The bathing and picnic parties occasionally have casualties, but no one seems to mind. General Birdwood took us up to the front line trenches; most of the way we were under cover, every now and then we were told to keep our heads down, and the bullets zipped overhead; we seemed to draw a lot of fire, and they attributed this to our white cap covers, so after that I wore khaki when I landed."[1013]

COMMODORE ROGER KEYES, RN, CHIEF OF STAFF, EMS.

"A full General landing to inspect overseas is entitled to a salute of 17 guns — well, I got my dues. But there is no crisis; things are quieter than they have been since the landing, Birdie says, and the Turks for the time being have been beat. He tells me several men have already been shot whilst bathing but there is no use trying to stop it; they take the off chance. So together we made our way up a steep spur, and in two hours had traversed the first line trenches and taken in the lie of the land. Half way we met Generals Bridges and Godley, and had a talk with them, my first, with Bridges, since Duntroon days in Australia. From the heights we could look down on to the strip of sand running Northwards from Ari Burnu towards Suvla Bay."[1014]

GENERAL SIR IAN HAMILTON, C-IN-C, MEF.

"About three o'clock... after a very fierce set-to, I was getting back into the trenches, when 'whack,' and I doubled up and fell into the trench, shot through the right side. After I got my senses back, I thought I would try and walk back to the base. But it was no good. As soon as I would get half up on end I would fall over. So I waited till a couple of my mates came along, and they just picked me up and ran like blazes to the main trench with me... I saw a lot of my old chums as I was carried through the trench, and nearly all of them were wounded. When we reached the further end of the trench I was just about done, and I began to be a bit frightened, although I still tried to bear up, and tried to keep my mates from seeing that I was. I was lucky, though, as just as we started to descend the hill a couple of stretcher-bearers came along. So they put me on the stretcher and carried me to the beach, where the doctor dressed my wound and gave me some bovril. After that I felt a lot better."[1015]

PTE JOHN COLLESS, 3RD BATTALION, 1ST BDE, 1ST AUSTRALIAN DIVISION.

"In the afternoon parties were sent out to collect rifles, etc., that had been dropped by wounded men and came across letters and photos. from friends. These were

---

1013 Keyes, *Naval Memoirs 1910–1915*, p. 319.
1014 Hamilton, *Gallipoli Diary*, vol. 1, p 179.
1015 *Mudgee Guardian and North West Representative* (New South Wales) 28th October 1915.

collected and burnt so that no traces of the units could be gathered by spies. I might mention that during the first four days I had been under fire 88 hours, and, had not got more that four hours' sleep, as the strain on our nerves was too great to think of trying to rest."[1016]

LT HERBERT ORBELL, 12TH BATTALION, 3RD BDE, 1ST AUSTRALIAN DIVISION.

"You cannot possibly imagine how horrible it is to see your chums and comrades, just before laughing and joking, dropping round you with all kinds of horrible wounds. We lost Colonel Stewart, shot through the head, and our own Major (Grant), shot in legs and stomach, before we had been under fire an hour, and Mr Hill, my platoon commander, was shot through the face. Our roll-call to-day was only eighty-six out of 256; but, of course, there may be a lot still to come. It is a hard job to find your own particular company.

"I have had to pull my stripes off my tunic, as they seem to be sniping all our officers and non-coms. We have already caught two or three I can tell you they get short shrift. Yesterday the Turks were reinforced, and we had a great fight to hold them, but the Wellington Company came up in time, and we repelled them with heavy loss after two or three bayonet charges. They don't like the steel, and we give it them every opportunity we get. Our reinforcements arrived today, Marines and Indians, so no doubt we will be advancing."[1017]

CPL MOSTYN JONES, CANTERBURY BN, NZEF, NEW ZEALAND BDE, NZ & A DIVISION.

"[W]e got word that twenty two drivers were to go ashore, and I was one of the lucky twenty two. A torpedo destroyer came alongside, and we boarded her and left for the much-talked of beach. After we had visited other ships for troops and had got a load, we started, and about half a mile from the beach we transferred to barges towed by pinnaces. We travelled half the way safely, and then shrapnel started to drop. The sensation was, curious to say, not funk, but you wished to goodness the barge would shake it up and get alongside. Anyway we all got on the beach without a casualty. We were ordered to go to our headquarters and get in dug-outs. I (with my chum) scored a decent dug-out, and half an hour after turning in, we were both fast asleep."[1018]

DVR WILLIAM SULLIVAN, 1ST FIELD COY, NEW ZEALAND ENGINEERS, NZ & A DIVISION.

"We left before nightfall, having enjoyed our outing very much, the whole atmosphere was exhilarating, one inhaled confidence and optimism, and I am sure it did the Anzacs a lot of good to see Sir Ian in their trenches. Admiral Thursby joined us for our walk. The Anzacs owe a good deal to him, for his confidence in them on the night of the 25th. It would have been an awful tragedy if Sir Ian had acted on the message Birdwood sent on board that night."[1019]

COMMODORE ROGER KEYES, RN, CHIEF OF STAFF, EMS.

1016 *Daily Telegraph* (Launceston, Tasmania), 8th September 1915.
1017 *Lyttelton Times* (New Zealand), 6th July 1915.
1018 *Grey River Argus* (New Zealand), 9th September 1915.
1019 Keyes, *Naval Memoirs 1910–1915*, p. 320.

"After some time we were put on punts, and no sooner did we get a little way from the beach than the artillery of the Turks began to fire on us. I think the shells landed on every side of us. They kept it up until we got out of range. Finally we reached the s.s. *Galeka*, and once safely on board we were pretty right. But load after load of wounded came on board all day, and by noon the next day the ship was full. On account of so many wounded coming on board it was impossible for the doctors to give us all the proper attention, and it was Saturday before the medical man could find time to attend to my wound. By Jove, I was crook. All I could do was to lie on my back. I could hardly move at all. After the wound was dressed I felt much better."[1020]

PTE JOHN COLLESS, 3RD BATTALION, 1ST BDE, 1ST AUSTRALIAN DIVISION.

"Another noisy awakening. Arrived at our destination during the night and they are still going at it; not so heavy as before, though. Shells were bursting often in the sea, but we were at a safe distance this time. Got orders to prepare to make this an hospital ship to take on wounded temporarily, so at last we had something do.

"After dinner-time three boat loads came alongside and our real work began. What a terrible sight some of the poor fellows were. They had been lying for three days in the firing line with their wounds. It was impossible to get at them. I got a good job and saw something that I won't forget in a hurry. I was helping in the operating theatre, and some of the wounds the fellows had were frightful. The Turks were using dum-dum bullets[1021] and they made a terrible mess. Well, we worked up till two in the morning, going as hard as we could, one taken off the table and another taken on as quick as they could be done. Some had no chance of surviving, but did our best for them. I won't describe any of the sights to you, as they be too gruesome. Shrapnel and snipers seemed the cause of most of the wounds. I helped at thirty operations that night and at two o'clock I was told to go to bed, and I was glad as I was dead beat."[1022]

PTE ALEX EWEN, 2ND FIELD AMBULANCE, AAMC, HMAT *MASHOBRA*.

"We have received orders to at once prepare our ship to receive the wounded. It is now midnight. What an awful day it has been! Hundreds have been put aboard suffering from all sorts of wounds. It has unnerved me a bit, but I did things that I thought I could never do. It is wonderful the way the poor fellows bore it. Our doctors, Col. Sturdy and Chas. Ryan, Major Shaw and Capt. Quick did wonderful work."[1023]

LT ARTHUR COLMAN, AASC, HMAT *KATUNA*.

"A big fire, reported to be in a village called Maidos, with a population of 25,000, could be seen on the peninsula… the burning of Maidos created a wonderful glare in the sky, and quietness, on the whole, prevailed."[1024]

CPL JOHN BUTLER, 3RD AUST. FIELD BDE AMMUNITION COLUMN, HMAT *WILTSHIRE*.

1020 *Mudgee Guardian and North West Representative* (New South Wales) 28th October 1915.
1021 It was a common misconception amongst inexperienced troops that the wounds inflicted by modern rifles were due to explosive ammunition.
1022 *Perthshire Advertiser*, 12th June 1915.
1023 *Sporting Judge* (Melbourne, Victoria), 11th September 1915.
1024 *The Advertiser* (Adelaide, South Australia), 26th June 1915.

GALLIPOLI

HMS *Triumph* shelled Maidos from the Gulf of Saros with unintended consequences.

"The few villages of the peninsula situated in sectors of the two battle fronts or in their rear, suffered severely from the fire of British ships. The prosperous port of Maidos went up in flames on April 29 under the fire of the British ships. The first building to become the victim of British naval shells was the large local hospital which was crowded with wounded. In spite of every effort many Turks and some twenty-five wounded British became victims of the conflagration, which spread with irresistible force. Many peaceful inhabitants likewise perished. Men, women and children, who were trying to save their most indispensable possessions, had to be removed from their houses under a rain of projectiles. From Kilid Bay they were transferred to the Asiatic side to save their lives. In Maidos, which was not fortified or occupied by staffs or troops, not a single house or wall remained intact."[1025]

GENERAL OTTO LIMAN VON SANDERS, COMMANDING 5TH OTTOMAN ARMY.

"After being taken prisoner, I was carried to hospital at Maidos, but only stopped there a few days, owing to a shell from our own guns bringing the hospital down on top of us. Very few got out, and I happened to be one of the lucky ones, also Kelly[1026] (RDF) and Corporal Brooke,[1027] 11th [sic, 12th] A.I.F."[1028]

PTE ALFRED RAWLINGS, 2ND BATTALION, 1ST BDE, 1ST AUSTRALIAN DIVISION.

"[A]t half-past four we reached the burning town of Maidos, whose inhabitants we could see crawling along the coast, in miserable plight, with what few articles they had been able to save from the fire. A perfect picture of misery.

"Our anchor had scarcely rattled into the harbour of Kilia when shells began to burst all round us. At first we thought these were the bombs of the aeroplanes circling overhead. It was soon clear to us, however, that these were only controlling the indirect fire of warships lying on the far side of the peninsula in the Aegean, who were thus greeting us in such a friendly fashion."[1029]

MAJOR HANS KANNENGIESSER, COMMANDING 9TH OTTOMAN DIVISION.

MEDITERRANEAN

Transports, pressed into service as hospital ships, made their way to Egypt, the lack of proper facilities meant many did not receive the treatment they required.

1025 Liman von Sanders, *Five Years in Turkey*, p. 72.
1026 Pte John Kelly, 1st Bn Royal Dublin Fusiliers. He survived his captivity.
1027 L/Cpl Vivian Brooke, 12th Bn, AIF, died of wounds at Biga on 4th May 1915. Buried at Ari Burnu Cemetery, he was the 27 year-old son of the late Robert Parkinson Brooke and Amy Beaumont Brooke, of Newtown, Tasmania.
1028 *The Argus* (Melbourne, Victoria), 14th June 1918.
1029 Kannengiesser, Hans, *The Campaign in Gallipoli*, pp. 124–125, Hutchinson & Co. (London) 1928.

"Our ship started to unload about 4 o'clock on Sunday afternoon [25th April 1915]. About 6, four barges came alongside with 80 wounded. We were supposed to unload stores and join the troops. Hospital ships had gone away full up with wounded men, so we had to start to get the poor beggars on board. We signalled the warships for doctors, and obtained three and two assistants. These were soon worked to a standstill. The barges continued to come alongside with wounded men. We had only four ship officers (the crew were black) to help, and they worked like Trojans. Had no sleep and little to eat from Sunday to Tuesday morning, and others were the same. The only thing lacking was ability to attend wounds, as we possessed neither medicine or equipment. A lot of poor beggars lay there and nothing could be done to relieve them. We were obliged to send for more doctors, and three more came from the warships. Later on three army doctors and 16 Red Cross men arrived, and soon got things going better, and the wounded made more comfortable. We now have 500 wounded men aboard. Left Gallipoli for Alexandria on Tuesday evening with all possible speed. A lot of men died on board simply because we could not treat them. We expect to arrive at Alexandria on Thursday evening. So far we have had 20 deaths on board, and these were buried at sea. Three were New Zealand boys, but did not know them. I helped to carry several dead men to the room and search them for personal belongings. It was an awful time for me to have to attend to the poor chaps. My heart used to nearly stop at some of the sights when we took the dressings off and commenced to dress them again. The men that died were nearly all wounded in the stomach — few in the head. There seems little hope for a man hit in the stomach. Our men are doing grand work on the boat here, as well as the medical men, only we don't know enough. We were trained for fighting — not Red Cross work."[1030]

SGT HAROLD LEE, WELLINGTON BN, NZEF, HMAT *ITONUS*.[1031]

"The burial of the one… at 9 [p.m.] provided a pathetic scene. A group light (electric) was hung where I stood, the body lay in a place on a stretcher convenient for the plunge over the side. The awning hid the scene from the wounded men lying on the deck above. The ship's chief officer was in charge of arrangements, and standing in solemn and reverent ranks were members of the ship's company and of the transport and hold party of the 10th Battalion. But here and there in the dim light might be seen the white bandage on head, hand, arm, or leg, which indicated that some wounded but more fortunate comrade had come to offer his last respects to one who had died fighting gallantly for his Empire. At a sign from myself the stretcher was raised at one end, and at the words of committal the body, sewn up in its blanket, slid over the ship's side and entered the water with a splash which somehow struck a sense of chill in one. And so the service went on to the pronouncement of the benediction, and, its amen was spoken not only by the chaplain but in reverent, deep-throated fashion by the mourners who stood by."[1032]

COLONEL CHAPLAIN JOHN McPHEE, CHAPLAIN'S CORPS, AIF, HMT *IONIAN*.

1030 *Taranaki Herald* (New Zealand), 13th July 1915.
1031 Sgt Harold Lee, Wellington Bn, NZEF, was killed in action on 17th June 1915. Buried in Shrapnel Valley Cemetery, the former bootmaker was the 23 year-old son of John James and Rosina Christable Lee, of Grafton, New South Wales.
1032 *Daily Herald* (Adelaide, South Australia), 21st July 1915.

EGYPT

The first wounded began to arrive for treatment in Egyptian hospitals.

"We came back post haste here with the wounded. I never want such another experience. Among other things I had to obtain full particulars of each wounded man on board and so had to deal with every one of them. They are a game lot. New Zealand need not have any fear as to the grit of the boys. Some poor chaps were in a terrible condition, but not a grouse could be heard from them."[1033]

SGT CHARLES GAMBLE, HQ STAFF, NZ & A DIVISION, HMT *LUTZOW*, ALEXANDRIA.

"I'm not dead yet, not by a bit! but have been wounded, shot through the side.[1034] The bullet came clean round the shoulder, making a deep flesh wound, and so preventing me using the rifle for a day or so.

"I am at present at the above address, and was to be sent to hospital, but owing to a wireless [message] requesting all men who could manage to get around at all to reinforce, myself and about thirty others on board are going straight back again voluntarily. It will take us two days to get there, and as we are two days on board, we will be most of us pretty fit again by that…

"I believe there is only 15% fighting strength of our Brigade left, and the other Brigade suffered heavily too, at least those that landed on that eventful Sunday… No time to write more. The ship is sailing for the Dardanelles."[1035]

PTE JOHN O'DONNELL, 10TH BATTALION, AIF, HMAT *CLAN MAGILLVRAY*.

"It took all day unloading our patients. Most of them went to Cairo. Have worked all the morning getting the ward ready again. Washed about 40 pyjama suits… we had to wash them in sea water… And we hung them out on lines improvised from rope."[1036]

SISTER ELLA JANE TUCKER, AUSTRALIAN ARMY NURSING SERVICE, HMHS *GASCON*.

"News has reached us of heavy casualties. The hospitals are all prepared for the wounded who are expected to-morrow. The chaplains' work is increasing. We had a conference of chaplains to-day. Three more chaplains are wanted for British regiments on loan for hospital work. It seems I am required as Base Senior Chaplain to administer chaplains' work here for a time."[1037]

CHAPLAIN CAPTAIN HENRY LAWRENCE BLAMIRES, BASE SENIOR CHAPLAIN, NZEF, CAIRO.

As casualties arrived in the country, their replacements began to leave.

"When we got to Zeitoun on April 28 I knew some of the Kapunda lads were there in

---

1033 *The Press* (New Zealand), 18th June 1915.
1034 O'Donnell was wounded on 25th April 1915.
1035 *Carlow Sentinel*, 22nd May 1915.
1036 *The News* (Adelaide, South Australia), 10th November 1936.
1037 *Mataura Ensign* (New Zealand), 17th June 1915.

the 4th [Reinforcements] of the 10th, and I found Fred. Martin,[1038] Stan. Burgess,[1039] and Wally Trinne.[1040] They moved off on the 29th and I slipped down to see them away. It was half past nine at night and I called out for Fred Martin. He was standing right beside me and I was able to say good-by [sic] to him and wish him luck."[1041]

TPR OSCAR ROSSER, 3RD AUSTRALIAN LIGHT HORSE REGT, AIF.

30 APRIL 1915

DARDANELLES

"We have been firing all day at invisible guns that have been firing at us… Our troops are getting on as well as can be expected, and the Narrows are being shelled from 'somewhere.'

"Chanak was well alight… from one end of the town to the other. Navy firing overland and balloon spotting the firing — a pretty sight from our way of looking at things. I expect the Kaiser won't feel too comfortable when he hears of this lot, nor will the people of Constantinople."[1042]

LEADING SIGNALMAN BERNARD MOORE, RN, HMS *AGAMEMNON*.

Many of the men were exhausted. Denied sleep, sometimes food, understrength and facing Turkish counter-attacks, Hamilton circulated a message to all those under his command.

"I rely on all officers and men to stand firm and steadfast to resist the attempts of the enemy to drive us back from our present position, which has been so gallantly won. The enemy is evidently trying to obtain a local success before reinforcements can reach us. The first portion of them will arrive to-morrow, and will be followed shortly by a division from Egypt. It behoves us all, French and British, to stand fast, hold what we have gained, wear down the enemy, and thus prepare for a decisive victory. Our comrades in Flanders have had the same experience of fatigue after hard-won fights. We shall, I know, emulate them steadfastly, and so achieve a result which will confer added laurels to French and British arms."[1043]

GENERAL SIR IAN HAMILTON, C-IN-C, MEF.

1038   Pte Frederick Martin, 10th Bn, AIF, landed on the peninsula on 7th May 1915; was wounded on 12th May, a gunshot wound; and more seriously, a bayonet wound to his forehead, on 9th August 1915. He transferred to the AASC on 1st November 1915.
1039   Pte Stanley Burgess, 10th Bn, AIF, joined his unit on 7th May 1915. Dental problems saw him evacuated to Mudros on 24th July 1915.
1040   Pte Benjamin Trinne, 10th Bn, AIF, served throughout the Gallipoli campaign after landing there on 7th May 1915. He was killed in action on 22nd July 1916 and is commemorated on the Villers Bretonneux Memorial.
1041   *Kapunda Herald* (South Australia), 24th September 1915.
1042   *Nottingham Evening Post*, 5th June 1915.
1043   *Northern Advocate* (New Zealand), 24th July 1915.

SEA OF MARMORA

"The submarine was diving easily and comfortably at a depth of 50 feet; not a suspicion of impending disaster lay in our minds.

"Suddenly, and for no accountable reason, the boat started rising rapidly in the water. We were in the presence of enemy ships, and so it was essential we should not come to the surface. But all efforts at regaining control proved futile, the diving rudders had not the slightest effect towards bringing the boat back to the horizontal position or stopping her rising in the water.

"We increased the motors to full speed in order to give the rudders their maximum power and shifted the water ballast forward as quickly as possible, but still she continued to rise, and at last broke surface.

"Through the periscope I saw an enemy torpedo boat a bare 100 yards off, firing hard. At all costs we must get under again at once. I ordered one of the forward tanks to be flooded, and a few minutes later the submarine took an inclination down by the bows and slipped under water.

"Closing off the forward tank and stopping the movement of water ballast from aft to forward, we endeavoured to catch her at 50 ft., but now again the diving rudders seemed powerless to right her, and with an ever-increasing inclination down by the bows she went to 60 and then to 70 feet, and was obviously quite out of control.

"Water ballast was expelled as quickly as possible, yet down and down she went — 80 ft., 90 ft., and 100 ft. Here was the limit of our depth gauges; when that depth was passed she was still sinking rapidly — blindly, for us, as we could not tell to what depths she was reaching.

"As a last desperate chance I ordered full speed astern on the motors... In a few moments — moments in which death seemed close to every man — there came a cry from the coxswain: — "She's coming up, sir!" and the needle of the depth gauge seemed to jerk itself reluctantly away from the 100 ft. limit mark and then rise rapidly.

"The amount of water expelled from the ballast tanks had now made the boat light, so with increasing speed she jumped to the surface, and remained there an appreciable time. While I attended to the reflooding of the tanks another officer looked through the periscope. He reported the enemy pressing us hard.

"Under we must get again — and away we went with the same terrible inclination, down by the bows, this time expelling water ballast immediately she began to dive in desperate attempts to regain control. But down and down she went, faster even that before, 60 ft., 80 ft., and 100 ft.

"The inclination down by the bows became more and more pronounced — she seemed to be trying to stand on her nose. Eggs, bread, food of all sorts, knives, forks, plates, came tumbling forward from the petty officers' mess. Everything that could fall over fell, men, slipping and struggling, grasped hold of valves, gauges, rods, anything to hold them up in position to their posts.

"Full speed astern again... A thousand years passed — well, this time we were gone for ever.... In heaven's name what depth were we at?... Why did not the sides of the boat cave in under the pressure and [crush] it?

"And then, once again, that fateful needle jumped back from its limit mark, and *AE2* rushed stern first to the surface. This time the enemy made no mistake. Shortly afterwards *AE2* slid away on her last and longest dive."[1044]

LT-CDR HENRY STOKER, RN, HMAS *AE2*.

"Then the torpedo boat lowered a boat to take us off in which there was a German officer, but she could only take five hands so we had to swim for it. When we got aboard we saw that her torpedo tubes were empty and a German sailor who could speak English told us they had both been fired at us but missed. Aboard the torpedo boat the officers were kept in the dark cabin while we were in the forward mess deck. While our clothes were being dried on deck the torpedo boat proceeded to Gallipoli and made fast alongside a hospital ship, where we were interviewed by General Liman von Sanders who was in command of the Peninsula. At 8.0 pm the torpedo boat proceeded to Constantinople."[1045]

AB ALBERT KNAGGS, RAN, HMAS *AE2*.[1046]

CAPE HELLES

Bertie Castleton was still trying to make his way back to British lines.

"During the night we heard a lot of firing going on, as if there was another night attack taking place on our right, and bullets kept whizzing past us, some of them too near to be healthy. Just to the right of the village of Kritting [sic], which appeared to be quite close to us, was in flames. To add to our danger, the poor bugler got delirious, and kept on talking. He thought he was on guard, and kept on saying, "Sergeant, when are we going to be relieved?" Next morning he was very quiet. I could see he was not long for this world. In the haversack I found some biscuits, but it was impossible to eat them, owing to thirst; a tin of jam, and half a pound of Cadbury's milk chocolate, which was a God-send, and a tin of Oxo cakes, which would have been all right if we had had some water. We kept our strength up by eating small quantities of the chocolate."[1047]

SGT BERTIE CASTLETON, 2ND BN HAMPSHIRE REGT, 88TH BDE, 29TH DIVISION.

The approach of some unidentified men towards British lines caused a stir before dawn.

"At 12 midnight we were relieved but we had to sleep beside our post. The party of sentries at the other corner fired and raised the camp. It turned out to be only a batch of Turks who were being brought in as prisoners. We were finally relieved at 4 a.m.

1044 *Saturday Journal* (Adelaide, South Australia), 26th December 1925.
1045 Albert Knaggs quoted in Wilson, Michael, *Destination Dardanelles*, p. 134, Leo Cooper (London) 1988.
1046 Knaggs died during his captivity on 22nd October 1916. Buried in Baghdad (North Gate) War Cemetery, he was the 34 year-old son of Henry Stephen Knaggs and Louisa Knaggs; husband of Sarah Annie Knaggs, of 11 Canton Street, St. Paul's, Bristol, England.
1047 *Hampshire Telegraph and Post*, 2nd July 1915.

this morning. The night was awfully cold, and dawn broke with a cold shroud of mist holding everything beyond a radius of five yards. It was just getting light. We got it. Yes, we got it in the form of a little bombardment.

"The Turkish gunners must be very erratic, as they were firing shells all over the place and did not seem to be aiming at any object in particular, most of the missiles passing with a scream overhead into the sea."[1048]

AB FREDERICK HURD, RNVR, HOWE BN, 2ND (NAVAL) BDE, ROYAL NAVAL DIVISION.

The heavy losses sustained led to the creation a 'new' temporary regiment, the 'Dubsters.'

"Our casualties and the Dublins' have been so heavy that the remainder of the two battalions have been formed into one, consisting of only three companies. However, everyone is very cheerful and confident.

"Tomlinson and I have a ripping little hut dug under the ground. He is sleeping by my side while I am writing this just now, and the artillery on each side popping away at each other over our heads. Although we have only been at it for five days, we are all quite accustomed to it. We are getting plenty of food and tobacco, but the luxuries like chocolate, oxo, and cigarettes are very scarce."[1049]

LT FREDERICK WALDEGRAVE, 1ST BN ROYAL MUNSTER FUSILIERS, 86TH BDE, 29TH DIV.

The lack of luxuries was not the main concern for some.

"During the day shells were whistling over our heads. We could see them burst on the side of the hill and in [Krithia] village. On our extreme left we saw shells pitching in some trenches, but could not tell whether they were our troops or the Turks. Whoever they were they were getting a warm time of it.

"It was about mid-day and very hot, when I thought about digging for water. When digging trenches we often found water about two feet down. Taking the bugler's knife I began to dig. After working for some time I managed to get down about two feet, but found no signs of water although the soil was very moist."[1050]

SGT BERTIE CASTLETON, 2ND BN HAMPSHIRE REGT, 88TH BDE, 29TH DIVISION.

Others had their breakfasts before beginning one of the less attractive but most necessary tasks on a battlefield.

"For breakfast this morning we had bully beef and hard biscuits. We managed (C. Fox and I) to "rake" out enough water from our bottles to make a billy can of "cafe au lait," which I had in my pack. After "brekker" the battalions fell in just under the brow of the hill so as not to be conspicuous on the sky line. They were to bury the dead. C. Fox and I stayed behind, as we were out last night as outposts."[1051]

AB FREDERICK HURD, RNVR, HOWE BN, 2ND (NAVAL) BDE, ROYAL NAVAL DIVISION.

1048 *Eastbourne Gazette*, 29th December 1915.
1049 *Northern Scot and Moray & Nairn Express*, 10th July 1915.
1050 *Hampshire Telegraph and Post*, 2nd July 1915.
1051 *Eastbourne Gazette*, 29th December 1915.

The detritus of battle was visible everywhere, while men worked to bring some semblance of order to the confusion.

"I landed at X Beach and worked my way along the coast to W Beach, Lancashire Landing, V Beach, and saw the *River Clyde*. Everywhere are evidences of the bitter struggle. The forlorn nature of the undertaking is clearly shown by the strength of the enemy's works, and only the magnificent courage of the 29th Division enabled them to get ashore. Nevertheless, the fact remains that such a landing should never have been undertaken, as it does not come within the category of legitimate operations of war, those which a general is entitled to ask his men to undertake. The first landing parties, knowing there was no retreat behind, stormed these entrenchments and redoubts in sheer despair, and thus saved the reserves from annihilation. Our line is now about two miles inland, stretching from the Gully Ravine in front of Krithia to Totts Battery on the Dardanelles. The right of the line is held by the French, while we occupy the centre and left. I went on board the *River Clyde* and met Captain Unwin, who gave me a first-hand account of what had happened. In fact, by getting eyewitnesses on each beach to describe exactly what they saw I was able to visualise the whole attack, and to write an account of these events."[1052]

ELLIS ASHMEAD-BARTLETT, WAR CORRESPONDENT.

"[It] was again a quiet day and the unloading of guns, horses, and material was continued, The Navy did a certain amount of shelling, but the Turks remained almost quiescent.

"We found that "The Pink Farm" was being used as a guiding post for all and sundry, viz. men, horses, mules, and vehicles on their way up to the front line, with tools, stores, ammunition, etc., and knowing what the effect would probably be, we tried in vain to keep them away from our Headquarters."[1053]

BRIG.-GEN. WILLIAM MARSHALL, COMMANDING 87TH BRIGADE, 29TH DIVISION.

"Things went on much the same till Friday, when the R.A.M.C. decided to go up to a beach, known as Gully Beach, about a mile above X beach, as there was a big gully leading out on to the beach which was a natural way for the wounded to come down. They went up that evening, and next day the sub. of the *Implacable*, myself, and two midshipmen were told off to go up there with about 12 seamen to act as beach party. I was sent up as beachmaster. Major Jones was in charge of both beaches, but as he lived down at X, I was pretty well on my own, which was very nice."[1054]

LT FRANK STEPHENSON, RN, HMS *IMPLACABLE*, BEACHMASTER, GULLY BEACH

"The scene presented by the beaches was unparalleled, presenting as it did activity in its most intense form. At one and the same time piers were under construction, roads being made, dug-outs excavated, stores and ammunition landed and stowed away in such small space as might be considered clear of shell-fire, whilst horses and mules were

---

1052 Ashmead-Bartlett, *Uncensored Dardanelles*, pp. 59–60.
1053 Marshall, Sir William, *Memories of Four Fronts*, p. 65, Ernest Benn Ltd. (London) 1929.
1054 *The Press* (New Zealand), 23rd April 1938.

Foredeck of *River Clyde* at V Beach showing the improvised machine-gun casements on her bow.

being coaxed or shoved from the water's edge to the shelter of the cliffs. Looking down upon this teeming hive from the cliffs above produced the impression of perpetual motion, for nothing in the crowded space was ever still for one moment, except that line of stretchers bearing wounded men awaiting their turn of embarkation."[1055]

REAR-ADMIRAL ROSSLYN WEMYSS, RN, HMS *EURYALUS*.

"I am writing this beside a large grave. Seventy-two non-commissioned officers and men were buried here yesterday — all of the Essex Regiment. There is a grave alongside here with twenty-one men of the Lancashire Fusiliers. A rough wooden cross marked with an indelible pencil serves as a memorial, and there is a rough circle of stones around the base. All around at intervals are little wooden crosses marking the graves of others who have fallen. There are several unknown graves too. Lots of gear is laying about here too. On landing the British threw off their packs in order to fight more easily.

"Just in front of the trench by where we had made our coffee this morning there was a broken rifle (English) and a pack with some poor chap's clothes in it. His water bottle was lying close by and all around were empty cartridge cases. We drew our own conclusions. Some brave man had made a last stand there. Just in front of the trench were little holes rudely dug in the earth — about a dozen or two in a straight line at intervals. This, I suppose, was where our men started to dig in their bodies as there were empty cartridge cases all around. It certainly seems as though we are for it. The chaps are back from their burying job. They have been burying dead Turks, and I may add that their opinions of it are not fit to hear."[1056]

AB FREDERICK HURD, RNVR, HOWE BN, 2ND (NAVAL) BDE, ROYAL NAVAL DIVISION.

1055 Wester-Wemyss, *The Navy in the Dardanelles Campaign*, p. 116.
1056 *Eastbourne Gazette*, 29th December 1915.

Hamilton and Hunter-Weston toured the area, while Commander Samson of the Royal Naval Air Service scouted for a location for the planned airfield.

> "The audacious Commander Samson cheers us up. He came aboard at 9.15 a.m. and stakes his repute as an airman that his fellows will duly spot these guns and that once they do so the ships will knock them out. I was so pleased to hear him say so that I took him ashore with me to "W" Beach, where he was going to fix up a flight over the Asiatic shore, as well as select a flat piece of ground near the tip of the Peninsula's toe to alight upon.
>
> "Saw Hunter-Weston: he is quite happy. Touched on "Y" Beach; concluded least said soonest mended… Our men are now busy digging themselves into the ground they gained on the 28th. The Turks have done a good lot of gunnery but no real counter-attack. Hunter-Weston's states show that during the past twenty-four hours well over half of his total strength are getting their artillery ashore, building piers, making roads, or bringing up food, water and ammunition into the trenches. This does not take into account men locally struck off fighting duty as cooks, orderlies, sentries over water, etc., etc. Altogether, it seems that not more than one-third of our fast diminishing total are available for actual fighting purposes. Had we even a Brigade of those backward Territorial reserve Battalions with whom the South of England is congested, they would be worth I don't know what, for they would release their equivalent of first-class fighting men to attend to their own business — the fighting."[1057]

GENERAL SIR IAN HAMILTON, C-IN-C, MEF.

> "Breakfast 5.30. Started 6 am with Commodore Backhouse, his Brigade Major & two battalion Commanders of the 2nd Bde Royal Naval Division to choose a Torres Vedras lines. I chose as lines for trenches the ground forming the mouth of the river N. of Seddul Bahr. The West to X Beach. Saw General d'Amade at Seddul Bahr old castle on hill. Sir Ian Hamilton & Braithwaite came to see me. Admiral Wemyss came & we rode together to see a beach at the mouth of the river [presumably Gully Ravine] on west coast…"[1058]

MAJOR-GENERAL SIR AYLMER HUNTER-WESTON, COMMANDING 29TH DIVISION.

> "At 2.30 we received our orders to go ashore. At 3 we left the *Savoie*. We said good-bye to Captain Tourrette. Half an hour later we were in the much-bombarded Sedd-el-Bahr Bay. Work is going on steadily. There are a great many troops, much material, and a field hospital. I photograph one of the holes made by a "marmite." I visit the director of the *Service de Sante*. We go straight on, mounting the semi-circular hills which form an arena where so much heroism was displayed that the place will be for ever immortal. We climb up the paths which cross the town of Sedd-el-Bahr. The sappers are repairing them. Artillery horses and men are engaged on the work. Houses are burst open, many are in atoms. There is not a corner that has escaped shells and bullets. We find, however, a little orchard intact. There are already five or six tents there. A cemetery is destroyed, also a mosque close to it.

1057 Hamilton, *Gallipoli Diary*, vol. 1, pp. 181–183.
1058 Extract from diary, 30th April 1915, Maj. Gen. Hunter-Weston, British Library, MS 48364.

"Nature remains beautiful and serene. The weather is delicious. Flowers everywhere where man has not destroyed them. We arrive on the crest of the plateau which dominates the bay where the landings took place in spite of a determined resistance. From here one can see everything. In this clear Eastern atmosphere it is easy to distinguish each man, each horse. The ships make an ideal target, and the *River Clyde* the best of all. The great ruined castle, whiter than ever from the breaches, fills one whole side of the horizon. At the foot, under its walls, the staff of General d'Amade is camping. We are led along roads full of flowers and bordered with trees, through very pretty narrow lanes, to our post. It is in a grove of young olive trees. Digging begins immediately, and a shelter is made for me. A few "marmites" fall. Fusillade and cannonade in front. A hole one yard deep is dug and my camp-bed placed in it. There is a stone wall on one side, and over the hole tree trunks and branches. We dine in absolute darkness, with shells flying over our heads. A ferocious "75" behind us fires and fires. Sharp and disagreeable noise. After having established the dressing station I go to bed. It is very cold. An enormous red moon rises above Morto Bay. For a time all is quiet, and a night bird begins to sing."[1059]

MO JOSEPH VASSAL, 6TH COLONIAL REGT, BDE *COLONIALE*, 1ST DIVISION, CEO.

The presence of French troops added another language to those to be heard at Helles. But some industrial English helped bring Bertie Castleton home.

"When having a rest I saw some distance away two men beckoning to someone. I could not tell whether they were enemies or not. I then saw another two on my right. Telling the sergeant to watch these two, I kept my eyes on the two in front. They began to get closer, and I could see that they had no equipment. We thought at the time they must be enemy, for they were in front of us. We were looking towards the enemy's position. We could not see our own lines.

"They were still getting closer when I heard one of them in front shout out, using very strong language to some one on our left. We both heard it, and feeling very excited I said "That's English." Raising myself up I shouted and waved my arm, when a shot came from the left throwing the sand over us. The shot was by a private, who, seeing me get up suddenly and wearing a stocking cap, took me for a Turk. These men belong to my company, and were the lucky ones that got back on the Wednesday. These brave fellows, knowing that there were snipers everywhere, had volunteered with a stretcher party to come out and search for wounded. The first thing we asked them for was water, but they would only give us a mouthful, and we both felt as if we could have drunk a bucket full. This was Friday evening.

"They also found two more men, a corporal and private. The former was shot through the right breast, and died the following night. We were carried to the first dressing station, which was being shelled. I saw Corporal Bartlett,[1060] who brought me a canteen of tea, and didn't that taste good! He said that there were very few of the

---

1059 Vassal, *Uncensored Letters from the Dardanelles*, pp. 74–75.
1060 Possibly, Cpl, later Sgt Edward Bartlett, 2nd Bn Hampshire Regt, who was killed in action on 1st June 1915. He is commemorated on the Helles Memorial.

Company left… I was carried about half-way by our own stretcher bearers — who had been working hard ever since they landed, and looked fit to drop — and then handed over to the R.A.M.C."[1061]

SGT BERTIE CASTLETON, 2ND BN HAMPSHIRE REGT, 88TH BDE, 29TH DIVISION.[1062]

"You should see me now; the wild man from Borneo is not in it. I have not had a shave for six days. Glad to let you know that I have been recommended for the Distinguished Conduct Medal, and am being made sergeant, but I could not put that in my address until the commanding officer tells me."[1063]

CPL JAMES WYCH, 4TH BN WORCESTERSHIRE REGT, 88TH BDE, 29TH DIVISION.[1064]

"There have been one or two casualties through chaps being wounded with splinters of shell. We have got to shift from our dug-outs as there is too much shelling going on. We have to get nearer to the cliffs, just under the brow. There is a fine sunset. I must say that, so far, all the lads are cheerful and aching to know when they are going up to the firing line. Some of them have brought back some curios from the dead Turks — all sorts of things, Turkish Korans and beads. Well, it is almost dark now so I must pack up. Besides I am going to have some tea which Charlie is brewing for two of us. All fires are to be out in half-an-hour's time. If it is as cold as it was last night it will be cold. It is so damp in the morning that everything is wet through. We have gathered some branches to-night as it will be so wet in the morning that we will not be able to light our fire. We are laying in for a warm time of it now, I think. There have been some casualties in the "Anson" Battalion to-day through shells dropping in their dug-outs. Two chaps were blown to pieces. The ships are letting drive at the Turks now. They are partly raising a dust. There is a lot of rifle firing. I expect it will last all night again. Probably the Turks will make a night attack to-night. Then we shall have to go up. Well, let it come if it has got to."[1065]

AB FREDERICK HURD, RNVR, HOWE BN, 2ND (NAVAL) BDE, ROYAL NAVAL DIVISION.

1061 *Hampshire Telegraph and Post*, 2nd July 1915.
1062 Castleton's wounded left leg led to his eventual discharge from the Army on 29th January 1919.
1063 *Cotton Factory Times*, 28th May 1915.
1064 Wych did not receive the DCM but was promoted Sgt. He was killed in action on 6th August 1915. Commemorated on the Helles Memorial, he was the 29 year-old son of John William Wych, of 25 Marine Street, Oldham, and the late Alice Wych.
1065 *Eastbourne Gazette*, 29th December 1915.

ANZAC

Things were relatively quiet, providing an opportunity for reforming units and resting men after nearly a week of intense fighting. Not everyone felt relaxed, however.

"We were up at four [in the] morning and at five we started for the firing line [Quinn's Post]. Along the beach for a quarter of a mile, and then up a gully (called Shrapnel Gully) for about one and a half miles; shrapnel and rifle fire on us all the time. The sight we saw as we trudged up the gully fairly sickened us. Dead men and mules were lying at intervals all the way up. Of course, at this time there were no roads, just the rough sloppy gully. Every now and again we would meet stretcher-bearers with wounded, bound for the beach. We arrived at the foot of the position we were to work in, and were ordered to make dug-outs for ourselves. The rattle of rifles, and the whistle of bullets and shells was going on all the time. That evening I had my first experience of the trenches (which were then nothing more than dug-outs connected together). We drove a sap out towards the Turks, who were only thirty yards away, while bombs were landing from the Turks where we were working. As soon as they would see the dirt being thrown up, they would make that their objective. In a sap on our left a bomb landed in the trenches, and one of the drivers, who landed with us, picked it up and threw it out, but the thing went off and killed him[1066] on the spot, his mate being also hit. So out of the eight of us who went on shift six came off."[1067]

DVR WILLIAM SULLIVAN, 1ST FIELD COY, NEW ZEALAND ENGINEERS, NZ & A DIVISION.

"By jove, we have had our baptism of fire, shrapnel, shells, and ordinary rifle fire, and it is hot. My company have behaved splendidly. We landed Monday morning, and just pushed straight into the firing line, and have by now appreciated the value of a pick and shovel, which is next in importance to the rifle; also the value of a sandbag.

"The first Queenslanders have made a name for themselves, and we intend to keep up their name. Australians and New Zealanders are great fighters. We have been at it continually day and night since Monday, and it is hot. One cannot describe the siege. At any rate, the enemy can fight, and are very brave. However, they have gradually been driven back. Everywhere at present my company is in a very tight corner, and are fighting well, although the enemy have the advantage of sniping us off. We will have a cross-fire on to them to-day. Jack Walsh (captain)[1068] was shot dead. He and myself were very close pals; also Sam Harry.[1069]

1066 Spr Piers Acton Eliot Warburton, British Section, New Zealand Engineers, was killed in action on 30th April 1915. Commemorated on Lone Pine Memorial, he was the 30 year-old son of George Hartop Eliot Warburton and Isabel Warburton, of 97 Alexandra Street, Palmerston North, Wellington.
1067 *Grey River Argus* (New Zealand), 9th September 1915.
1068 Major John Walsh, 15th Bn, AIF, was killed in action on 28th April 1915. He is commemorated on the Lone Pine Memorial.
1069 Capt. Samuel Harry, 15th Bn, AIF, was killed, according to 2/Lt William Twynam Mundell, on 9th May 1915; according to the Commonwealth War Graves Commission, on 10th May 1915. He is commemorated on the Lone Pine Memorial.

"I am very well; only a little knocked out after strenuous and continued fighting. One of my Lts (Robertson)[1070] is missing. I think he is in all probability wounded, and in the hospital. It is wonderful how one gets used to fire.

"Climate here very hot during the day, but extremely cold at night. Ere long I think I will be grey-headed from worry. It is a big responsibility to be in charge of a position, and to hold same under disadvantages."[1071]

MAJOR HUGH QUINN, 15TH BATTALION, 4TH BDE, AIF, NZ & A DIVISION.

"I am in a deep trench, with my head in a hole in one side and my feet in a hole in the other. Bullets have been pouring in all the morning mostly from snipers, and dirt pours down on us continually. Unshaven, and grubby in the extreme, we are perfectly happy, enjoying a joke at the expense of a man who is going off because his equipment has been shot to bits. We are lying low and have done practically no shooting. It is slightly cold at nights. Our overcoats and extras were left on the beach when we arrived, so we have remained since then in what we stand up in. We doze during the day when the shooting is on, especially shrapnel. This little lot should be over soon. This is the sixth day of the battle and we must be pushing on."[1072]

PTE ERIC RYBURN, OTAGO BATTALION, NEW ZEALAND BDE, NZ & A DIVISION.

"Parade in morning. Casualty list read. Two dead, four missing, many wounded. Went to aid post (Captain Goldsmith) at night. Two casualties for us. Still holding our own trenches."[1073]

PTE SYDNEY PENHALIGON, 3RD FIELD AMBULANCE, AAMC.

"A holiday for our brigade. Alas! They were 4,000 strong six days ago, but now they are about 1800. We went down to have a swim. There were about 5,000 in, all enjoying themselves, although now and then we had to take cover while the Turks wasted a bit of ammunition. Have not heard how George Sharp is, but Dr. Matheson told me he was not in a dangerous condition. Many heroic deeds have been performed, and are still being done daily. As the casualties of officers have been so severe men have been led by corporals in charging trenches, and in holding them against big odds. Some battalions are in charge of captains instead of Lt-Cols."[1074]

CPL ROBERT HUNTER, DIV. SIGNAL COY, AUSTRALIAN ENGINEERS, 1ST AUSTRALIAN DIV.

Casualties, the inter-mixing of units created not only confusion but the perfect breeding ground for rumour. One of the most prevalent was that Germans, in particular, were wearing uniforms taken from the dead giving false orders to men at Anzac.

1070 Lt Thomas Robertson, 15th Bn, AIF, was killed in action on 27th April 1915. Commemorated on the Lone Pine Memorial, he was the 21 year-old son of Robert Cochran Robertson and Sophia Robertson, originally of Brisbane.
1071 *The Western Australian Record* (Perth, Western Australia), 25th September 1915.
1072 *Southland Times* (New Zealand), 3rd August 1915.
1073 *The Queenslander* (Brisbane, Queensland), 21st April 1917.
1074 *The Evening Star* (Boulder, Western Australia), 26th June 1915.

"Heavy rifle fire all day, and warm work for A.M.C. Turks and Germans, dressed in our uniforms, giving orders to cease fire in our trenches."[1075]

CSM ALBERT TURNER, DIV. SIGNAL COY, AUSTRALIAN ENGINEERS, 1ST AUSTRALIAN DIV.

"[W]e had orders to fall in to go to the firing line, as an attack was expected that night. You can imagine our feelings as we proceeded up the gully in darkness, not knowing in the least where we were going, except that there was the chance of having our first dinkum go with the Turks. However, we lay behind the trenches all night, with the bullets flying over our heads. In the morning we were taken further up the gully to rejoin our battalion. On the way up, Bert Redpath and I were detailed for duty with the telephone section. I have been with that section every since. The position we had in this gully was very bad, for it was commanded by two hills occupied by Turkish snipers, and they levied a heavy toll upon us. Being linesman in my section my work necessitated my constant knocking about, keeping the lines in repair, and... I feel that God in His great goodness, took especial care of me."[1076]

PTE EDWIN PENNELL, 13TH BATTALION, 4TH BDE, AIF, NZ & A DIVISION.

As the reorganisation continued, it afforded some the time to reflect on their recent experiences and how little of the wider context they understood at the time.

"As to what is feels like to be under fire, that depends, I believe, as much upon one's knowledge as upon one's nerve. Judging from general conversation on the subject, every man for the first time coming under fire, especially if he cannot reply to it, asks himself why he joined the force, and wishes that the ground would open up and receive him; but after he has been under fire two or three times he gets to know the meaning of each constituent noise making up the grand tumult, and is able to single out the comparatively few noises that are really menacing. When he thus understands fire, the strain on his nerves is much less. Indeed, he almost enjoys the thundering of his own guns, while those of the enemy trouble him only when they shoot anywhere near, which, ordinarily, is not often. It is surprising how a man will jump or wince when a shell lands close by and then, when it is all over, laugh at himself and the whole business and think no more about it.

"With respect to the progress being made by the Allied troops in these operations it is, of course, impossible, even were it at this stage allowable, for any mere individual unattached to headquarters to give any account of the slightest value."[1077]

PTE CECIL YORKE, CANTERBURY BN, NZEF, NEW ZEALAND BDE, NZ & A DIVISION.[1078]

1075 *The Brisbane Courier* (Queensland), 19th August 1915.
1076 *Cootamundra Herald* (New South Wales), 7th September 1915.
1077 *Lyttelton Times* (New Zealand), 23rd July 1915.
1078 Yorke, a former journalist, was killed in action on 10th August 1915. Believed to be buried in Embarkation Pier Cemetery, where he is commemorated by a special memorial, he was the 24 year-old son of Joseph Courtenay Yorke and Ella Yorke, of Gonville, Wanganui, New Zealand, originally of Manchester, England.

"To-night our chief officer came to me and asked me to see a poor fellow who was asking for a priest. There is none aboard, and as I am the only Catholic officer on the ship they thought I might do something. As I neared the bench the dear chap turned his eyes and in a gurgling voice said: "Are you a Catholic?" I said, yes, and speaking as comforting as I could gave him my beads. He kissed the crucifix and hung them round his neck, at the same time thanking me most sincerely. His name is E. J. Orr,[1079] and he lived at Coburg."[1080]

LT ARTHUR COLMAN, AASC, HMAT *KATUNA*.

EGYPT

More ships bringing wounded back from Gallipoli arrived.

"By 7 o'clock [in the] morning we were alongside the quay at Alexandria, at the same spot as we had put in when the *Benalla* arrived last December. Friday was still a busy day for the wounded could not be taken ashore till Saturday, but I found time to go along to the *Clan MacGillivray* to see how they were faring there. Chaplain Dexter had come down on that ship and had done fine service in helping to look after the wounded."[1081]

COLONEL CHAPLAIN JOHN McPHEE, CHAPLAIN'S CORPS, AIF, HMT *IONIAN*.

"We arrived here last night after two days' fighting on the Gallipoli peninsula. I am writing this in the Wireless room of the German steamer *Lutzow*, which is at present acting as a troopship and hospital ship.

"At present I cannot tell you how many casualties there are, but by the time this reaches you, you will know all about it. The fighting was awful for the two days I was there. We had to land on an open beach and then take a position with the bayonet which was supposed to be impregnable. The Turks were waiting for us, with machine-guns, artillery, and rifle-fire, and we did get it. It was terrible, wholesale slaughter. We were all mixed up with the Australians, and it was just a case of every man fight for all he knew how.

"I think the New Zealanders and Australians proved what they were worth. I am returning to the firing line in a few days now, and will be at it again."[1082]

PTE GEORGE VANCE, CANTERBURY BN, NZEF, HMT *LUTZOW*, ALEXANDRIA.

"At 4.30 wounded commenced to arrive. I shall never forget the shock when we saw the men arrive covered in blood[,] most of them with half their uniform shot or torn away.

1079 L/Sgt Ernest James Orr, 7th Bn, AIF, was wounded on 25th April 1915. Evacuated to Alexandria and then transferred to England, the former baker was admitted to the 3rd County of London Hospital, Wandsworth, on 15th August. He died there on 14th November 1915 and is buried in Wandsworth (Earlsfield) Cemetery; the 21 year-old son of James and Elizabeth Orr, of 14 Park Street, Coburg, Victoria. Listed in his personal effects was a rosary.
1080 *Sporting Judge* (Melbourne, Victoria), 11th September 1915.
1081 *Daily Herald* (Adelaide, South Australia), 21st July 1915.
1082 *Marlborough Express* (New Zealand), 17th June 1915.

We found then that 700 badly wounded had arrived. All the cases who could walk were sent down to us. By night we had over 500 patients & only 3 nurses & two orderlies to cope with the work. The meals had to be got & the wounded were clamouring to be dressed. They had been wounded on Sunday & had not been dressed since. We had no guard to keep order... When tea time came the men pushed the barriers & those who were strongest got the most tea. It was a wild beast show. Men went to bed without any nourishment at all."[1083]

SISTER ALICE ROSS-KING, AANS, 1ST AUSTRALIAN GENERAL HOSPITAL.

"This evening I visited the Heliopolis Hospital. All visitors, including staff officers, were excluded. As chaplains we went through. Most of the men here have shrapnel and shot wounds in the arms, legs, and feet, and the wounds are not very serious, At 10.00 p.m. I met a train load of fresh arrivals. Our men are very cheery, The train stops at the hospital gates. We do all we can to care for the sick and wounded, and arrange social enjoyment and sport for the spare hours of the more fit. Letters may be brief now."[1084]

CHAPLAIN CAPT. HENRY BLAMIRES, BASE SENIOR CHAPLAIN, NZEF, CAIRO.

The returned wounded found a ready audience for their tales, tall or otherwise.

"Things are moving here now. Nearly every place has been commandeered as a hospital, and yesterday 500 of our boys lobbed back here wounded — a lot of them were from the 10th Battalion. The poor old 9th Battalion (Queensland) was cut to pieces; only 27 answered the roll call afterwards. But a lot, of course, were only wounded. Two thousand wounded are expected to arrive to-day. There is an everlasting stream of Red Cross motors to the hospitals. The trains they use look very nice. They are white with a red cross on the top of each carriage and a big red moon and star on the sides. Our boys got a ticklish landing. HMS *Triumph* went right under a battery and blew it to pieces, and then went to Kum Kale fort. While that was being bombarded the 9th Battalion landed under heavy fire and broke down the Turkish defence on top of the cliffs. Many of our boys were bayoneted in the legs through running over the trenches, C Section of the 1st Australian Field Ambulance had just landed and run up their flag when they were annihilated. I only wish I were with them, Take my tip, they are making a name for themselves and for Australia. All the 3rd Reinforcements leave here to-day, bound for the firing line."[1085]

TPR ARTHUR HOLT, 3RD AUSTRALIAN LIGHT HORSE REGT, AIF, HELIOPOLIS.

Others were still taking in the sights and stories of ancient Egypt.

"This country is the most marvellous place in the world. It is far more wonderful than one could imagine. The people dress just the same as the illustrations I have seen in Bibles at the old school, and nearly every building has quite a history, such an the

---

1083 Diary relating to the First World War service of Sister Alice Ross-King, AANS. AWM PR02082.
1084 *North Otago Times*, 18th June 1915.
1085 *Barrier Miner* (Broken Hill, New South Wales), 24th October 1915.

well where Jesus and the Virgin Mary sat. The Pyramids are only a short distance away from us. Many interesting hours may be spent looking at these things. I hope all the old scholars are well. I often think about the old school, especially when I hear one of my old favourite hymns being sung. I am also glad that the old boys have — like myself — responded to the call for soldiers. At least we are doing our little bit for our country."[1086]

SPR ANDREW GERAGHTY, DIVISIONAL SIGNAL COMPANY,
ROYAL ENGINEERS, EAST LANCASHIRE DIVISION.

ENGLAND

Those less than sympathetic to Winston Churchill proved to be an eager audience for tales of the miscarriage of operations at Gallipoli. While, for some, it was mere confirmation of long-established views, the First Lord of the Admiralty had decidedly not helped his own cause. And those around him had failed to rein him in.

> "His Antwerp adventure, which resulted in an irreparable loss of naval personnel for the period of this war, and did not delay the progress of the German troops by a single hour, would alone suffice to give him a place in the historical pillory... If the War Office had approved of the adventure it would have been carried out by properly trained soldiers — not by naval recruits, some of whom had only had a week's training, and did not even understand the mechanism of their rifles...

> "The costly failure of the hasty attempt to rush the Dardanelles must also be put to the discredit of Mr. Churchill... The lost ships would not have been sacrificed if the amateur High Admiral had allowed them to be equipped with "dress improvers" before going into a region of floating mines coming with the current. But that precaution and the provision of adequate support on land would have taken time. And when Mr. Churchill has an idea, it must become an act within 24 hours. He cannot wait even for second thoughts. He is one of your high-explosive politicians.

> "Despite all his colossal blunders, Mr. Churchill still has a great chance of serving his country. If he could forget himself and cease to be a showman of self for a few immortal moments, he might earn the nation's gratitude. Let him make way for a seaman who... would know what to do in any and every naval emergency. Such an act of common sense would cause us all to forgive and forgo all his deadly cleverness and colossal egoism. Let him hasten thus to make his peace with history. Otherwise it will not be long before the whole nation is crying out — Churchill must go."[1087]

EDWARD OSBORN, BRITISH JOURNALIST.

1086 *Todmorden Advertiser and Hebden Bridge Newsletter*, 30th April 1915.
1087 *The West Australian* (Perth, Western Australia), 18th June 1915.

AFTERWORD

At the end of April 1915 it was clear that the landings on the Gallipoli peninsula had failed to secure any of their objectives. Instead of sweeping across from Kabatepe and up from Seddülbahir to the Kilitbahir plateau to dominate the Narrows, the allied forces held slivers of the coastline, narrow toeholds vulnerable to Ottoman artillery and counter-attack. The French had achieved some success at Kum Kale but that diversion could not disguise how far short of expectations the entire operation had fallen.

Ever since, the reasons for the defeat have been debated with the focus placed overwhelmingly upon a combination of allied errors and Turkish topography. One of the authors of the 1906 feasibility study, Major-General Charles Callwell, knew better.

> "The plan chosen failed for all practical purposes, was not so much the consequence of topographical conditions nor of the disposition of the enemy forces nor of bad luck, as it was the upshot of a factor that had not been sufficiently taken into account. This factor was the rare fighting qualities that the Osmanli soldier was to display in the campaign. The troops who had come so badly out of the struggle with the Bulgars and Serbs and Greeks two years before, turned out to be an extremely tough proposition. But if Sir I. Hamilton and his staff at the outset possibly underrated Ottoman valour and grit, if they assumed too readily that the opposition that would be offered by this soldiery would not be of the most whole-hearted type, they were only following the lead of Governments which, in a happy-go-lucky mood and confident that the enemy would crumble up before a show of bluff, had despatched the expedition on a mission of which they had failed to realise the danger, and for which suitable preparations had not been made by them in advance."[1]

After the American Civil War, General George Pickett was asked why his famous charge at Gettysburg had been defeated. He pondered the question before replying, "I've always thought the Yankees had something to do with it."

Substitute Turks for Yankees and we get closer to the truth of what happened in 1915.

* * *

---

[1] Callwell, Major-General Sir Charles, *The Dardanelles*, pp. 130–131, Constable & Co. (London) 1924.

11th (Northern) Division leaving Grantham on 4 April 1915, the beginning of its journey to Gallipoli.

ANCHOR INN
E. RICHARDSON
ALES, BOTTLED
BEER AND STOUT

HAM
4.4.15   N°3

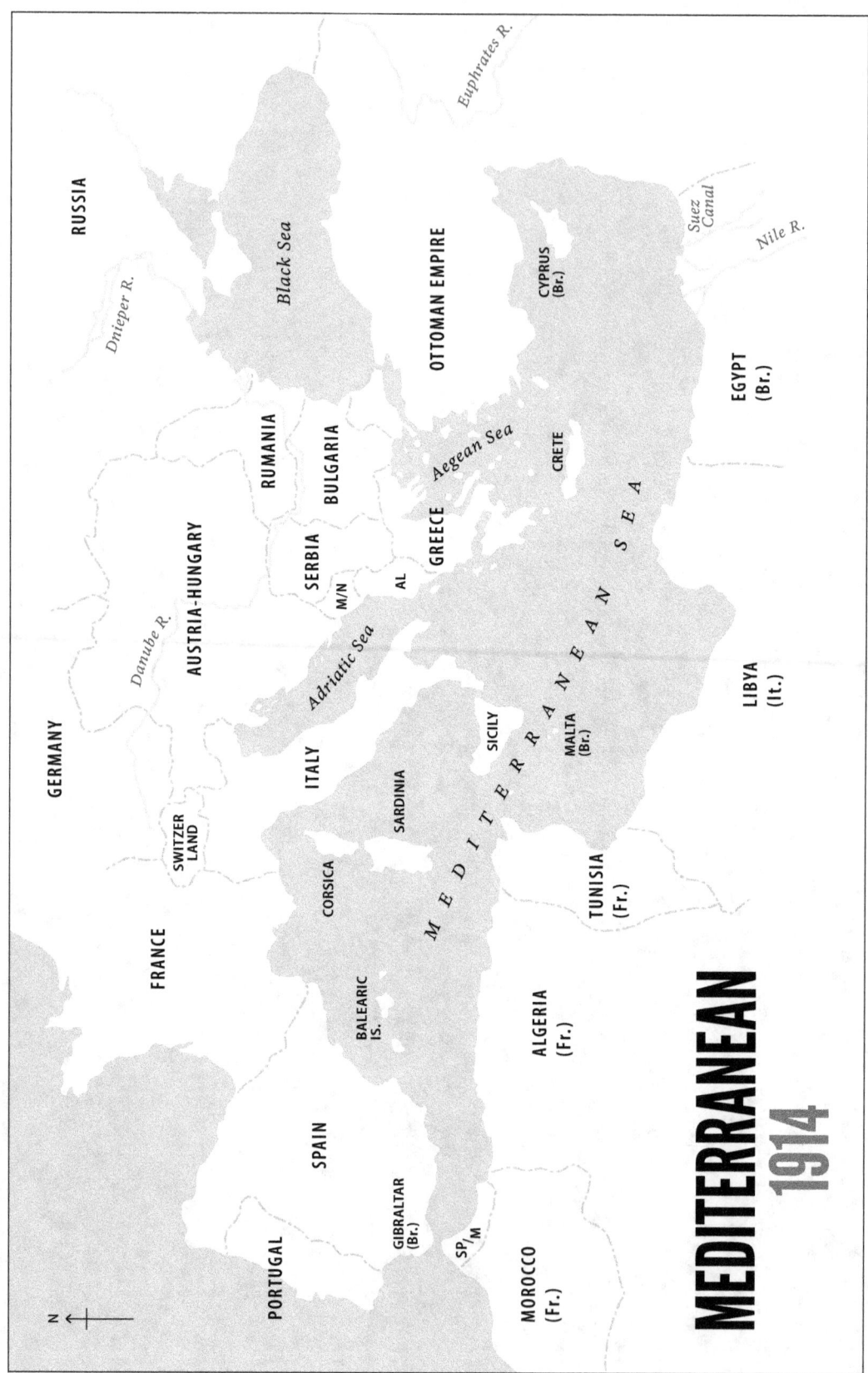

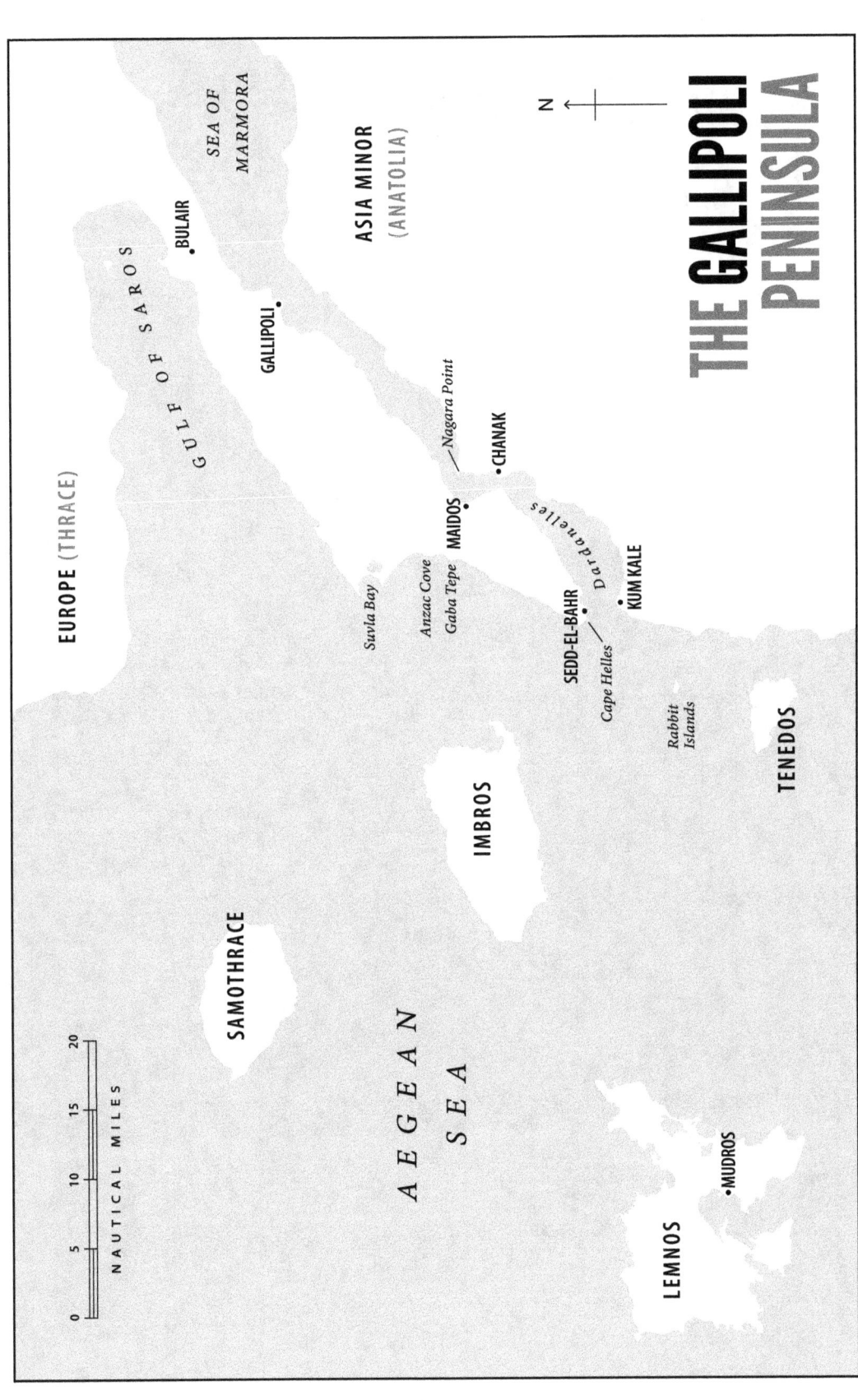

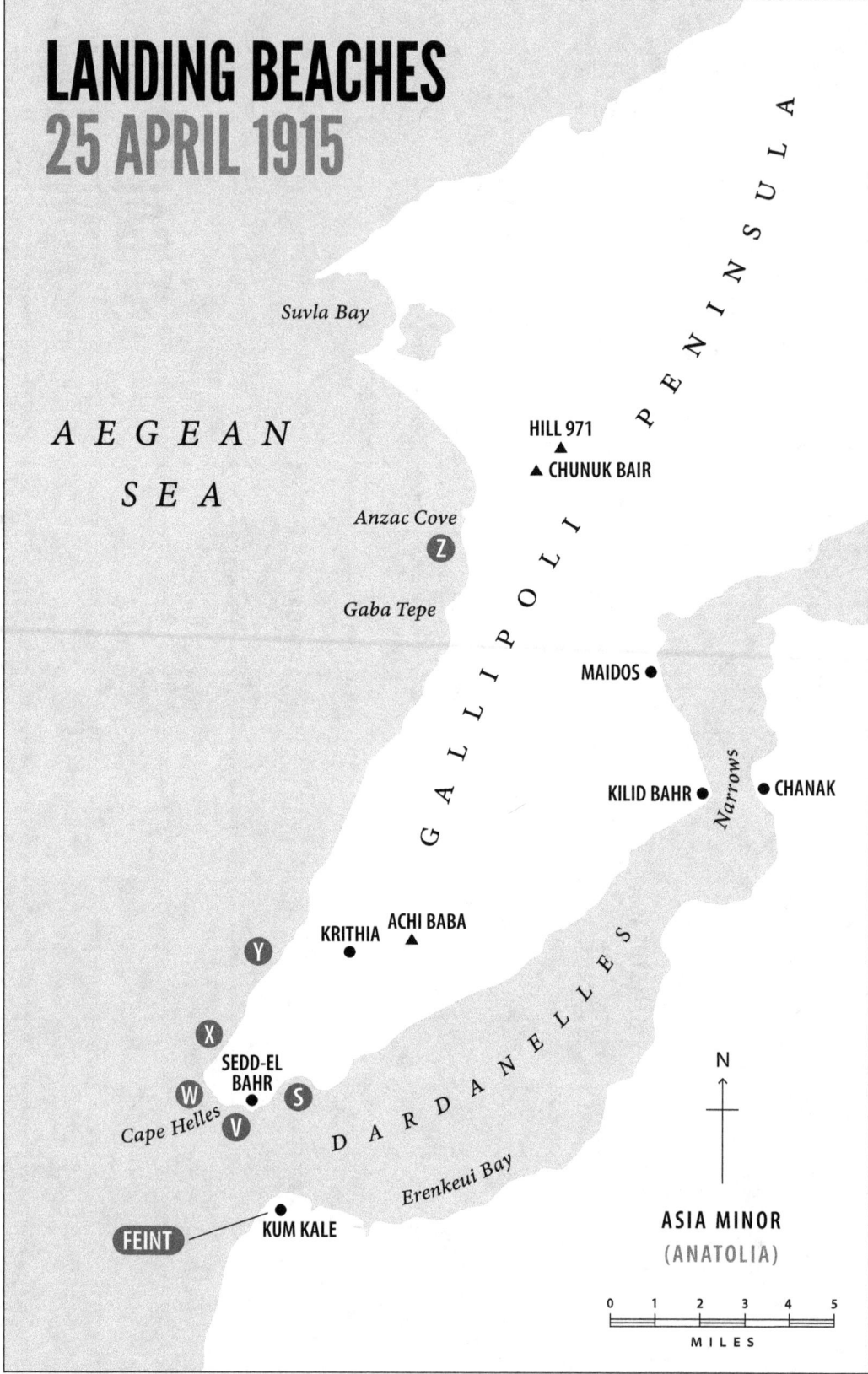

ABBREVIATIONS & ACRONYMS

2/Lt	Second Lieutenant
A/	Acting
AA & QMG	Assistant Adjutant and Quartermaster General
AAMC	Australian Army Medical Corps
AANS	Australian Army Nursing Service
AASC	Australian Army Service Corps
AB	Able Seaman
ADC	Aide-de-camp
Adj.	Adjutant
AFA	Australian Field Artillery
AIF	Australian Imperial Force
ANZAC	Australian & New Zealand Army Corps
attd	attached
AWM	Australian War Memorial
Bde	Brigade
BEF	British Expeditionary Force
Bn	Battalion
Brig.	Brigadier
Bty	Battery
C-in-C	Commander-in-Chief
CAB	Records of the Cabinet Office
Capt.	Captain
CERA	Chief Engine Room Artificer
Cpl	Corporal
CEO	*Corps Expéditionnaire d'Orient*
CB	Companion of the Order of the Bath
CGS	Chief of the General Staff
C-in-C	Commander in Chief
CID	Committee of Imperial Defence
CO	Commanding Officer
Col.	Colonel
Coy	Company
CPO	Chief Petty Officer
CSM	Company Sergeant Major

CWGC	Commonwealth War Graves Commission
DAC	Divisional Ammunition Column
DAQMG	Deputy Adjutant and Quartermaster General
DCM	Distinguished Conduct Medal
Div.	Division, Divisional
DSC	Distinguished Service Cross
Dvr	Driver
EMS	Eastern Mediterranean Squadron
ERA	Engine Room Artificer
FO	Foreign Office
GCB	Knight Grand Cross
Gen.	General
GOC	General Officer Commanding
GSO	General Staff Officer
HMAS	His Majesty's Australian Ship
HMAT	His Majesty's Australian Transport
HMHS	His Majesty's Hospital Ship
HMNZT	His Majesty's New Zealand Transport
HMS	His Majesty's Ship
HMSO	His Majesty's Stationary Office
HMT	His Majesty's Transport, Hired Military Transport
KOSB	King's Own Scottish Borderers
L/Cpl	Lance Corporal
L/Sto.	Leading Stoker
Lt	Lieutenant
Lt-Cdr	Lieutenant Commander
Lt-Col.	Lieutenant Colonel
MC	Military Cross
MEF	Mediterranean Expeditionary Force
MM	Military Medal
MO	Medical Officer
MP	Member of Parliament
NAA	National Archives of Australia
NZAC	New Zealand Army Corps
NZASC	New Zealand Army Service Corps
NZEF	New Zealand Expeditionary Force
OR	Other Ranks
OS	Ordinary Seaman

PO	Petty Officer
Pte	Private
PNTO	Principal Naval Transport Officer
QMG	Quartermaster General
QMS	Quartermaster Sergeant
RAMC	Royal Australian Medical Corps
RAN	Royal Australian Navy
RE	Royal Engineers
Regt	Regiment
RFA	Royal Field Artillery
RFR	Royal Fleet Reserve
RGA	Royal Garrison Artillery
RHA	Royal Horse Artillery
RMA	Royal Marine Artillery
RMF	Royal Munster Fusiliers
RMLI	Royal Marine Light Infantry
RMS	Royal Mail Ship
RN	Royal Navy
RNAS	Royal Naval Air Service
RND	Royal Naval Division
RNR	Royal Naval Reserve
RNVR	Royal Naval Volunteer Reserve
RSM	Regimental Sergeant Major
SBA	Sick Berth Attendant
Sgt	Sergeant
SMS	*Seiner Majestät Schiff* (His Majesty's Ship)
Spr	Sapper
SQMS	Squadron Quartermaster Sergeant
Sqn	Squadron
SS	Steam Ship
SSM	Squadron Sergeant Major
Sto.	Stoker
TNA	The National Archives
Tpr	Trooper
USMC	United States Marine Corps
USN	United States Navy
USS	United States Ship
VC	Victoria Cross

SELECT BIBLIOGRAPHY

Ashmead-Bartlett, Ellis, *Uncensored Dardanelles*, Hutchinson & Co. (London) 1928.

Aspinall-Oglander, C. F., *Military Operations. Gallipoli, Vol. 1*, William Heinemann (London) 1929.

Asquith, Herbert, *Memories and Reflections, 1852–1927. The Earl Oxford and Asquith, K.G., Vol. 2*, Cassell and Co. (London) 1928.

Bean, C.E.W., *The Story of ANZAC: From The Outbreak of War to the End of the First Phase of the Gallipoli Campaign May 4, 1915*, Angus & Robertson (Sydney) 1941.

Bacon, Admiral Sir R.H., *The Life of Lord Fisher of Kilverstone, Admiral of the Fleet, Vol. 2*, Hodder & Stoughton (London) 1929.

Berkeley Milne, Admiral Sir Archibald, *The Flight of the 'Goeben' and the 'Breslau.' An Episode in Naval History*, Eveleigh Nash Company (London) 1921.

Birdwood, Lord, *Khaki and Gown. An Autobiography*, Ward, Lock & Co. (London) 1942.

Broadbent, Harvey, *Defending Gallipoli. The Turkish Story*, Melbourne University Press (Melbourne) 2015.

Bush, Eric, *Gallipoli*, George Allen & Unwin (London) 1975.

Callwell, Major-General Sir Charles, *The Dardanelles*, Constable & Co. (London) 1924.

Chatterton, E. Keble, *Dardanelles Dilemma*, Rich & Cowan (London) 1935.

Churchill, *The World Crisis 1911–1914*, Thornton Butterworth (London) 1923.

—, *The World Crisis 1915*, Thornton Butterworth (London) 1923.

Corbett, Sir Julian S., *Naval Operations Vol. 1*, Longman, Green and Co. (London) 1920.

—, *Naval Operations Vol. 2*, Longman, Green and Co. (London) 1921.

Creighton, Oswin, *With the Twenty-Ninth Division in Gallipoli. A Chaplain's Experiences*, Longmans, Green & Co. (London) 1916.

Davidson, George, *The Incomparable 29th and the River Clyde*, James Gordon Bisset (Aberdeen) 1920.

Denham, H.M., *Dardanelles. A Midshipman's Diary*, John Murray (London) 1981

Djemal Pasha, *Memories of a Turkish Statesman 1913–1919*, George H. Doran Company (New York) 1922.

Einstein, Lewis, *Inside Constantinople. A Diplomatist's Diary During the Dardanelles Expedition April – September, 1915*, John Murray (London) 1917.

Feuille, Henri, *Face aux Turcs. Gallipoli 1915*, Payot (Paris) 1934.

Gibbon, Frederick P., *The 42nd (East Lancashire) Division: 1914–1918*, Country Life (London) 1920.

Gillam, John Graham, *Gallipoli Diary*, George Allen & Unwin (London) 1918.

Godley, Alexander, *Life of an Irish Soldier*, John Murray (London) 1939.

Goodchild, George, *The Last Cruise of the "Majestic"*, Simpkin, Marshall, Hamilton, Kent & Co (London) 1917.

Hadaway, Stuart, *Pyramids and Fleshpots. The Egyptian, Senussi and Eastern Mediterranean Campaigns, 1914–16*, The History Press (Stroud) 2014.

Hamilton, Sir Ian, *Gallipoli Diary Vol. I*, Edward Arnold (London) 1920.

Hart, Peter, *Gallipoli*, Profile Books (London) 2011.

Kopp, Georg, *Two Lone Ships. 'Goeben' & 'Breslau'*, Hutchinson & Co. (London) 1931.

Liman von Sanders, Otto, *Five Years in Turkey*, United States Naval Institute (Annapolis) 1927.

Lockyer, Hughes C., *The Tragedy of The Battle of the Beaches*, Privately Published (London) 1936.

Lumby, E.W.R. (Ed.), *Policy and Operations in the Mediterranean 1912–14*, Navy Records Society (London) 1970.

McFarland, E., *A Slashing Man of Action. The Life of Lieutenant-General Sir Aylmer Hunter-Weston MP*, Peter Lang (Oxford) 2014.

Marder, Arthur Jacob (Ed.), *Portrait of an Admiral: The Life and Papers of Sir Herbert Richmond*, Harvard University Press (Cambridge, Massachusetts) 1952.

Montgomery, Ina, *John Hugh Allen of the Gallant Company. A Memoir*, Edward Arnold (London) 1919.

Morton-Jack, George, *The Indian Empire at War. From Jihad to Victory The Untold Story of the Indian Army in the First World War*, Little Brown (London) 2018.

Nevinson, H.W., *The Dardanelles Campaign*, Nisbet (London) 1918.

Olson, Wes, *The Last Cruise of a German Raider. The Destruction of SMS Emden*, Seaforth Publishing (Barnsley) 2018.

Oral, Haluk, *Gallipoli Through Turkish Eyes*, Türkiye Bankasi (Istanbul) 2007.

Plowman, Peter, *Voyage to Gallipoli*, Transpress (Wellington) 2013.

Price, William Harold, *With the Fleet in the Dardanelles. Some impressions of naval men and incidents during the campaign in the spring of 1915*, Andrew Melrose (London) 1915.

Pugsley, Christopher, *Gallipoli. The New Zealand Story*, Sceptre (Auckland) 1984.

Roberts, Chris, *The Landing at Anzac 1915*, Army History Unit (Canberra) 2011.

Roskill, Stephen, *Hankey. Man of Secrets, Volume I, 1877–1918*, Collins (London) 1970.

Schreiner, George Abel, *From Berlin to Baghdad*, Harper & Brothers (New York) 1918.

Scott, James Brown, (Ed.) *Diplomatic Documents Relating to the Outbreak of the European War*, Oxford University Press, American Branch (New York) 1916.

Snelling, Stephen, *Voices From The Past. The Wooden Horse of Gallipoli. The Heroic Saga of SS River Clyde, a WW1 Icon, Told Through the Accounts of Those Who Were There*, Frontline Books (Barnsley) 2017.

Stanley, Peter, *Die in Battle, Do Not Despair. The Indians on Gallipoli, 1915*, Helion & Co. (Solihull) 2015.

Stoker, Commander H.G., *Straws in the Wind*, Herbert Jenkins (London) 1925.

Thomas, Lowell, *Raiders of the Deep*, Heineman (London) 1930.

Vassal, Joseph, *Uncensored Letters from the Dardanelles*, Heineman (London) 1916.

Waite, Fred, *The New Zealanders at Gallipoli*, Whitcombe & Coombs (Auckland) 1919.

Weymss, Lady Wester, *The Life and Letters of Wester Wemyss*, Eyre and Spottiswoode (London) 1935.

Wester-Wemyss, Lord, *The Navy in the Dardanelles Campaign*, Hodder & Stoughton (London) 1924.

Wilson, Keith (Ed.), *The Rasp of War. The Letters of H.A. Gwynne to The Countess Bathurst 1914–1918*, Sidgwick & Jackson (London) 1988.

Wolf, Karl (Trans. Iredale, Thomas P.), *Victory at Gallipoli. The German-Ottoman Alliance in the First World War*, Pen & Sword (Barnsley) 2020.

Woods, H.C., *Washed by Four Seas. An English Officer's Travels in the Near East*, T. Fisher Unwin (London) 1908.

IN 3 VOLUMES

HELL & CONFUSION

GALLIPOLI DAY *by* DAY

VOL. 2

'Always Nearly Winning'

MAY — AUGUST 1915

VOL. 3

'Between the Devil and the W.C.'

SEPTEMBER 1915 — JANUARY 1916

www.ingramcontent.com/pod-product-compliance
Lightning Source LLC
Chambersburg PA
CBHW080321080526
44585CB00021B/2422